Hopf Algebras

Series on Knots and Everything — Vol. 49

Hopf Algebras

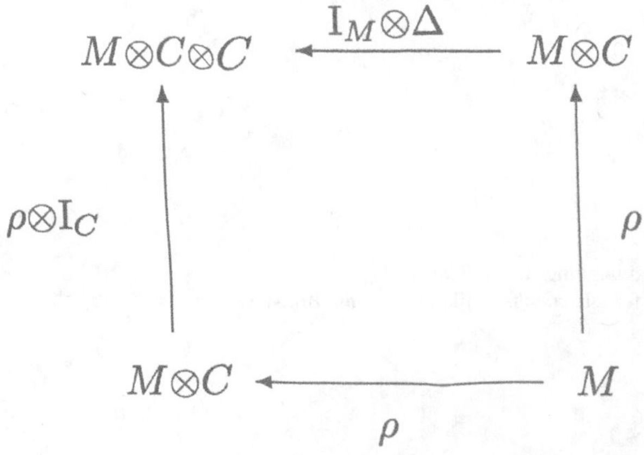

David E Radford
University of Illinois at Chicago, USA

World Scientific

NEW JERSEY • LONDON • SINGAPORE • BEIJING • SHANGHAI • HONG KONG • TAIPEI • CHENNAI

Published by

World Scientific Publishing Co. Pte. Ltd.

5 Toh Tuck Link, Singapore 596224

USA office: 27 Warren Street, Suite 401-402, Hackensack, NJ 07601

UK office: 57 Shelton Street, Covent Garden, London WC2H 9HE

Library of Congress Cataloging-in-Publication Data

Radford, David E.

Hopf algebras / by David E. Radford.

p. cm. -- (Series on knots and everything ; v. 49)

Includes bibliographical references and index.

ISBN-13: 978-981-4335-99-7 (hardcover : alk. paper)

ISBN-10: 981-4335-99-1 (hardcover : alk. paper)

1. Hopf algebras. I. Title.

QA613.8 .R33 2012

512'.55 2011039959

British Library Cataloguing-in-Publication Data

A catalogue record for this book is available from the British Library.

Dedicated to the memory of my parents, Albert and Laurie Radford. Their love of the nature was an inspiration to me.

Preface

The subject of Hopf algebras originated in an algebraic topology paper by Heinz Hopf [75] in 1941. Not long after a notion of Hopf algebra was formulated for the category of vector spaces over a field k. Hopf algebras of this type, which we refer to as Hopf algebras over a field, were seen to arise in a number of settings: as underlying algebras of affine groups in the theory of algebraic groups, as formal groups in number theory, and as universal enveloping algebras in Lie theory. This book is about Hopf algebras over a field k.

A general theory for them was beginning to emerge by the time of Sweelder's book [201] in 1969. Against the background of its development, applications to other areas were being discovered, for example to combinatorics and computer science [79, 185]. Other areas of mathematics were being treated in the context of Hopf algebras, for example Galois theory [30]. Ideas from ring theory were being applied to aspects of Hopf algebra theory [217]. Hopf algebra techniques were being used to provide purely algebraic proofs of results on algebraic groups [199, 202, 212, 213]. Affine group structures were being studied as algebraic objects related to Hopf algebras [215, 216].

The paper by Drinfel'd [44] in 1986 on quantum groups opened the floodgates for applications of Hopf algebras to physics, invariant theory for knots and links, and representations closely connected to Lie theory. The Hopf algebras involved are referred to as quantum groups, a term still lacking formal definition. Quantum group theory includes much more than Hopf algebras. A great number of mathematicians have contributed to the subject. Some contributions and references are given in chapter notes.

The subject of Hopf algebras continues to grow in many directions. More recent considerations have led to generalizations of the notion of Hopf algebra. We do not pursue the threads of evolution here; a quick foray into the literature will begin to reveal them.

This book is intended to be a graduate text and to be used by researchers in Hopf algebras and related areas. It does not replace standard texts, such as the book by Abe [1], by Dăscălescu, Năstăsescu, and Raianu [35], by Montgomery [133], or by Sweedler [201]. Each has some material not included in this one. This book reflects the deep influence of quantum groups on the subject and recent developments in the theory of pointed Hopf algebras.

Prerequisites are few. The reader is expected to be familiar with linear algebra, elementary abstract algebra including the tensor product and basic representation theory of algebras, and also with rudimentary knowledge of the language of category theory. What follows is an overview of the book chapter by chapter.

In Chapter 1 we set basic notation conventions. We cover linear algebra needed for the book in two brief sections, in discussion and through accompanying exercises.

Two notions from linear algebra play a basic role in this text, the rank of tensors and the concept of closed subspaces of a dual vector space. Most structures in Hopf algebras have an underlying vector space and give rise to important structures on the dual space. For any vector space V over k there is an inclusion reversing bijective correspondence between the subspaces of V and the closed subspaces of its linear dual V^*. The connection between structures on V and those on V^* via the bijection will be examined over and over again.

Most sections in this text come with a generous supply of exercises. Some exercises develop new ideas, new directions, or further results. Exercises for completing proofs tend to be accompanied by generous hints. Categorical aspects of material are usually developed in the exercises.

Chapter 2 is the first of a three-chapter treatment of coalgebras and their representations. Hopf algebras have an algebra and a coalgebra structure. Why the coalgebra structure is important. It affords the Hopf algebra with locally finite properties which algebras generally do not possess. It provides a way to form the tensor product of representations of the Hopf algebra's underlying algebra structure.

Algebras and coalgebras are dual structures in a categorical sense. We think of an algebra A over k formally as a triple (A, m, η), where

$$m : A \otimes A \longrightarrow A \qquad \text{and} \qquad \eta : k \longrightarrow A$$

are linear maps, m describes multiplication, η describes the unity of A via

$\eta(1_k) = 1_A$, such that certain commutative diagrams are satisfied which express that multiplication is associative and 1_A is the multiplicative neutral element of A. Thus a coalgebra C over k is formally a triple (C, Δ, ϵ), where

$$\Delta : C \longrightarrow C \otimes C \qquad \text{and} \qquad \epsilon : C \longrightarrow k$$

are linear maps, such that the commutative diagrams for A hold for C with arrows reversed, where C replaces A and ϵ replaces η. Δ is referred as the comultiplication map and ϵ to as the counit map.

Suppose A is an algebra over the field k. Then the dual vector space A^* contains a unique maximal coalgebra A^o derived from the algebra structure on A, called the dual coalgebra. When A is finite-dimensional $A^o = A^*$. Let C be a coalgebra over k. Then C^* has an algebra structure derived from the coalgebra structure of C, called the dual algebra.

A finite-dimensional subspace of C generates a finite-dimensional subcoalgebra of C. Therefore the theory of finite-dimensional algebras over k can be used to study coalgebras over k. And simple coalgebras are finite-dimensional.

The bijective correspondence between the subspaces of C and the closed subspaces of C^* is an important means for relating the coalgebra structures of C and the algebra structures of C^*. A significant operation in C is the wedge product which corresponds to multiplication of subspaces in C^*.

There is a very important finite-dimensional coalgebra which makes an appearance over and over again in the theory of Hopf algebras. This is $\mathrm{C}_n(k) = \mathrm{M}_n(k)^*$, where $\mathrm{M}_n(k)$ is the algebra of $n \times n$ matrices over the field k. Just as every finite-dimensional algebra over k is a subalgebra of $\mathrm{M}_n(k)$ for some $n \geq 1$, every finite-dimensional coalgebra over k is a quotient of $\mathrm{C}_n(k)$ for some $n \geq 1$.

Chapter 2 ends with the first of many dual constructions, here with the construction of the cofree coalgebra on a vector space. This coalgebra is the categorical dual of the free algebra on a vector space. The cofree coalgebra and free algebra play important roles in the structure theory of Hopf algebras.

Representation theory of coalgebras is a very extensive subject as is the representation theory of algebras. In Chapter 3 we focus on aspects of it needed for the sequel. Objects of study are comodules of C. These can be thought of as the rational C^*-modules, that is locally finite C^*-modules whose elements are annihilated by a *closed* cofinite ideal of C^*. Here the notion of closed subspace enters the theory.

Injective rational modules form an important class which we treat in detail. Results on them are applied to the study of indecomposable coalgebras.

Simple rational modules, or equivalently simple comodules, are finite-dimensional and are closely related to simple subcoalgebras. We begin to study the coradical of C, the sum of all simple subcoalgebras of C, in this chapter.

One of the most important structures associated with the coradical is the coradical filtration of C. Chapter 4 is devoted to its study. The terms of the coradical filtration of C correspond to the closures of powers of the Jacobson radical of C^*. Its terms can be expressed by the wedge product. Certain families of idempotents of C^* provide useful decompositions of the terms of the coradical filtration.

Several coalgebras are associated with C which have a simpler coradical and whose coradical filtrations are related to that of C in a good way. We show how they can be used to study C.

Chapter 5 introduces bialgebras and Chapter 7 Hopf algebras. Perhaps the relationship between bialgebras and Hopf algebras can be thought of as the relationship between monoids and groups. Group theory is very rich. For monoids assumptions are made at times to compensate for the lack of inverses.

Bialgebras are vector spaces with compatible algebra and coalgebra structures. A bialgebra H over k is formally a tuple $(H, m, \eta, \Delta, \epsilon)$, where (H, m, η) is an algebra over k and (H, Δ, ϵ) is a coalgebra over k, such that Δ and ϵ are algebra maps. Since algebra and coalgebra are dual concepts, the notion of bialgebra is self-dual. Bialgebras have the important property that the tensor product of representations of their underlying algebra structures can be formed. Hopf algebras are bialgebras with an antipode, an endomorphism analogous to the map of groups which takes elements to their inverses. For bialgebras assumptions are made at times to compensate for the lack of an antipode.

Two examples of Hopf algebras. The first is the group algebra $k[G]$ of a group G over k. Here

$$\Delta(g) = g \otimes g \qquad \text{and} \qquad \epsilon(g) = 1$$

for all $g \in G$. An element g of any bialgebra H over k which satisfies the two preceding equations is called grouplike and the set of grouplike elements of H is denoted $\mathrm{G}(H)$. The antipode S of $k[G]$ is determined by $S(g) = g^{-1}$ for all $g \in G$.

The second is the universal enveloping algebra $U(L)$ of a Lie algebra L over k. Here

$$\Delta(\ell) = 1 \otimes \ell + \ell \otimes 1 \qquad \text{and} \qquad \epsilon(\ell) = 0$$

for all $\ell \in L$ which determines the coalgebra structure of $U(L)$ since L generates $U(L)$ as an algebra and Δ, ϵ are algebra maps. An element ℓ of any bialgebra H over k which satisfies the two preceding equations is called primitive and the set of primitive elements of H is denoted $\mathrm{P}(H)$. The antipode of $U(L)$ is an algebra anti-endomorphism of $U(L)$. The antipode S is determined by $S(\ell) = -\ell$ for all $\ell \in L$.

Note that $1 \in \mathrm{G}(H)$ for any bialgebra H over k since Δ and ϵ are algebra maps. In particular $k1$ is a simple subcoalgebra of B. In terms of the coradical $k[G]$ and $U(L)$ are at the opposite ends of the spectrum. The coradical of $k[G]$ is all of $k[G]$. The coradical of $U(L)$ is the one-dimensional simple subcoalgebra $k1$ and is therefore as small as possible.

Chapter 5 emphasizes certain universal constructions for bialgebras. These have applications to Hopf algebras. They are based on the free algebra and the cofree coalgebra described earlier. The free algebra on a vector space over k is used to construct the free bialgebra on a coalgebra over k. The cofree coalgebra on a vector space over k is used to construct the dual counterpart to the free bialgebra, the cofree bialgebra on an algebra over k.

As algebras are quotients of free algebras on vector spaces, bialgebras are quotients of free bialgebras on coalgebras. We discuss a quotient construction in great detail since it is a model for construction of Hopf algebras from a typical presentation, a coalgebra which generates the Hopf algebra and relations among its elements.

Enter $\mathrm{C}_n(k)$ into the theory of bialgebras. Every finite-dimensional coalgebra over k is a quotient of $\mathrm{C}_n(k)$ for some $n \geq 1$. A finitely generated bialgebra over k is generated by a finite-dimensional subcoalgebra and therefore is the quotient of the free bialgebra on $\mathrm{C}_n(k)$ for some $n \geq 1$.

We have observed the notion of bialgebra is self dual. As there is a dual coalgebra and a dual algebra, there is dual bialgebra. Suppose A is a bialgebra over k. We have noted the algebra structure of A gives rise to a coalgebra A^o and the coalgebra structure of A accounts for an algebra structure on A^*. The subspace A^o of A^* is a subalgebra. The coalgebra structure of A^o together with this subalgebra structure makes A^o a bialgebra, called the dual bialgebra.

Chapter 6 is a short technical chapter which expands a bit on results used to analyze the antipode. Chapter 7 introduces Hopf algebras with an extensive discussion of the antipode and proceeds right away to the construction of two families of finite-dimensional Hopf algebras. Involved is the relation $xa = qax$, where q is a non-zero scalar. This relation arises in many well-known families of Hopf algebras. As a result Gaussian integers and q-binomial symbols play a role in the theory of Hopf algebras. We realize Hopf algebras in our two families as quotients of free bialgebras. A construction detail we discuss in depth is how to find a basis for the quotient. Generally there are various methods which might work. We demonstrate application of the Diamond Lemma for the first of many times.

The free and cofree bialgebra constructions of Chapter 5 are modified for Hopf algebras and variants of the cofree construction are introduced, notably the shuffle algebra. The shuffle algebra is used to study Hopf algebras whose coradical is one-dimensional.

In some cases an algebra multiplication can be "twisted" to form another algebra. There are notions of twisting for Hopf algebras, alterations of the algebra or the coalgebra structure, which give rise to other Hopf algebras. Twistings are important in the classification of certain families of Hopf algebras. Hopf algebras also give rise to others via the dual. The dual bialgebra H^o of a Hopf algebra H over k is a Hopf algebra. In particular H^* is a Hopf algebra over k when H is a finite-dimensional.

Chapter 8 describes the complete theory of Hopf modules; there is but one main result. Its power lies in numerous important applications. Let H be a Hopf algebra over k. A left H-Hopf module is a vector space with a left H-module and a left H-comodule structure which are compatible in a certain way. There are other compatibility conditions. These are discussed in Chapter 11.

An example of a left H-Hopf module is H with multiplication and comultiplication as module and comodule structures. Direct sums of Hopf modules are Hopf modules. Every non-zero left H-Hopf module M is isomorphic to the direct sum of copies of H. In particular M is a free left module over the algebra H. There is a subspace of $M_{co\,inv}$ of M described in terms of the comodule structure and any linear basis for M is a module basis.

When M is finite-dimensional the dual vector space M^* has the structure of a co-Hopf module. The Hopf module and co-Hopf module structures go a long way in accounting for the deeper properties of finite-dimensional Hopf algebras.

In Chapter 9 we consider when H is a free module over its Hopf subalgebras. Various conditions are given, some involving the coradical of H. The most important result is that finite-dimensional Hopf algebras are free over their Hopf subalgebras. This is established by using a modified notion of Hopf module. This result has very important implications for the structure of finite-dimensional Hopf algebras. There are examples of commutative Hopf algebras over the field of complex numbers which are not free over one of their Hopf subalgebras.

Chapter 10 is the most important chapter in the book on the structure of a finite-dimensional Hopf algebra H over k. The story is told in terms of non-zero integrals. A non-zero integral for H is a generator or a certain one-dimensional ideal of H. Existence and uniqueness (up to scalar multiple) of non-zero integrals follow as an application of the structure theorem for Hopf modules. This theorem also implies that only finite-dimensional Hopf algebras can possess a non-zero finite-dimensional ideal.

Semisimplicity of H or H^* as an algebra is determined by integrals. The trace function for endomorphisms of H can be computed in terms of an integral for H and an integral for H^*. Important formulas which express the trace of the square of the antipode in terms of integrals for H and H^* follow as a result.

There is a distinction between left and right integral for H. Suppose H is finite-dimensional. The relationship between the two accounts for a certain Hopf algebra automorphism of H. The composition of this automorphism with the transpose of its counterpart for H^* is the fourth power of the antipode of H. As a consequence the antipode of H has finite order as an endomorphism.

Generally one-dimensional ideals of H and H^* are important. Their generators are referred to as non-zero generalized integrals. Suppose H is not finite-dimensional. Then left (and right) integral can be defined for the algebra H^*. A non-zero left or right integral for H^* may not exist, but if so it is unique.

Chapter 11 is devoted to actions by a bialgebra or Hopf algebra H over k; especially actions on algebras or coalgebras. We discuss these actions since they account for useful constructions on tensor products where H as a factor.

Let ${}_H\mathcal{M}$ be the category whose objects are left H-modules and morphisms are module maps under function composition. Since H is a bialgebra the tensor product of two objects of ${}_H\mathcal{M}$ is defined and k has the structure of an object of ${}_H\mathcal{M}$. A left H-module algebra A is an algebra in ${}_H\mathcal{M}$; that is a k-algebra which is also a left H-module such that the algebra structure maps of A are module maps. For such an algebra A the smash product

$A\sharp H$ can be formed. The smash product is a k-algebra and is $A\otimes H$ as a vector space. A left H-module coalgebra is a k-coalgebra which is also a coalgebra in ${}_H\mathcal{M}$.

Now let ${}^H\mathcal{M}$ be the category whose objects are left H-comodules and whose morphisms are comodule maps under function composition. Since H is a bialgebra over k the tensor product of objects of ${}^H\mathcal{M}$ is defined and k has the structure of an object of ${}^H\mathcal{M}$. The notions of left H-comodule algebra and left H-comodule coalgebra are defined as above. If C is a left H-comodule coalgebra then the smash coproduct $C\natural H$ is defined. It is a k-coalgebra and as a vector space is $C\otimes H$. Smash products and smash coproducts are dual structures.

A natural question to ask at this point is when a smash product and smash coproduct structure form a bialgebra. The question brings us to a rather complicated setting which turns out to be important for several reasons.

Let ${}^H_H\mathcal{M}$ be the category whose objects are left H-modules and left H-comodules and whose morphisms are module and comodule maps. Suppose A is an object of ${}^H_H\mathcal{M}$ which is both a left H-module algebra and a left H-comodule coalgebra. Then $A\otimes H$ is a bialgebra with algebra structure $A\sharp H$ and coalgebra structure $A\natural H$ if and only if A is a left H-module coalgebra, a left H-comodule algebra, and there is a certain compatibility between the module and comodule structures on A. We let ${}^H_H\mathcal{YD}$ be the full subcategory category of ${}^H_H\mathcal{M}$ whose objects satisfy this compatibility condition. ${}^H_H\mathcal{YD}$ is called a Yetter-Drinfel'd category and its objects are called left Yetter-Drinfel'd modules. Our conclusion: $A\otimes H$ is a k-bialgebra with the smash product and the smash coproduct structures if and only if A is a bialgebra in ${}^H_H\mathcal{YD}$. In this case we denote $A\otimes H$ with its bialgebra structure by $A\times H$ and call this bialgebra a biproduct.

We are glossing over the subtle points of how to form the tensor product of algebras and coalgebras in ${}^H_H\mathcal{YD}$. These important details are thoroughly examined in Chapter 11.

Now suppose that H is a Hopf algebra. Then ${}^H_H\mathcal{YD}$ is a braided monoidal category. These categories are a natural starting point for the construction of invariants of knots and links.

There is a way of characterizing biproducts of the form $B\times H$ when H is a Hopf algebra over k. Suppose A is a bialgebra over k. Then A is a biproduct of the form $A = B\times H$ if and only if there are bialgebra maps $A \underset{\pi}{\overset{\jmath}{\leftrightarrows}} H$ which satisfy $\pi\circ\jmath = \mathrm{I}_H$; loosely speaking if and only if there is a bialgebra projection from A onto H.

Chapter 12 is an in depth analysis of quasitriangular Hopf algebras, structures which produce invariants of knots and links, and sometimes of 3-manifolds. A quasitriangular Hopf algebra over k is a pair (H, R), where H is a Hopf algebra over k and $R \in H \otimes H$, which satisfies axioms guaranteeing that R is a solution to the quantum Yang–Baxter equation. Solutions to this equation provide a means of constructing invariants of knots and links. For invariants of 3-manifolds an extra bit of structure is needed, an element $v \in H$ called a ribbon element. The antipode of a quasitriangular Hopf algebra over k is bijective.

For the purposes of computing knot or link invariants we may assume (H, R) is minimal quasitriangular, meaning H is the smallest Hopf subalgebra K of H such that $R \in K \otimes K$. In this case H is finite-dimensional.

Finite-dimensional quasitriangular Hopf algebras abound. Suppose H is a finite-dimensional Hopf algebra over k. A quasitriangular Hopf algebra $(D(H), \mathcal{R})$ can be constructed, where $\mathrm{Dim}(D(H)) = \mathrm{Dim}(H)^2$, and H is a Hopf subalgebra of $D(H)$. The pair $(D(H), \mathcal{R})$ is called the Drinfel'd, or quantum, double.

Perhaps $D(H)$ is one of the most important examples of a finite-dimensional Hopf algebra. Chapter 13 is devoted to the study of $(D(H), \mathcal{R})$. It is minimal quasitriangular and factorizable. $D(H)$ is unimodular and the square of its antipode is an inner automorphism. Whether or not $(D(H), \mathcal{R})$ has a ribbon element is related to the formula for the fourth power of the antipode of H mentioned above.

Chapter 14 deals with the dual notions of quasitriangular algebra, bialgebra, and Hopf algebra; these are coquasitriangular coalgebra, bialgebra, and Hopf algebra respectively. Here R is replaced by a bilinear form and the quasitriangular axioms are replaced by axioms for the form. A proof that the antipode of a quasitriangular Hopf algebra is bijective can be modified to give a proof that the antipode of a coquasitriangular Hopf algebra is bijective. The free bialgebra on a coalgebra is the basis for the construction of the free coquasitriangular bialgebra on a coquasitriangular coalgebra over k.

Two classes of Hopf algebras have been studied extensively; pointed and semisimple. A Hopf algebra is pointed if its simple subcoalgebras are one-dimensional. This is equivalent to saying its coradical is a Hopf subalgebra isomorphic to a group algebra $k[G]$. Pointed Hopf algebras figure prominently in the theory of quantum groups.

Classification of pointed Hopf algebras is complete in the finite-dimensional case when k is algebraically closed of characteristic zero, G is abelian, and minor constraints are placed on orders of elements of G. For many important classes of pointed Hopf algebras G is abelian.

The purpose of Chapter 15 is to set the stage for the study of pointed Hopf algebras and their classification. We provide a detailed construction of the quantized enveloping algebras and their generalizations. One of the earliest results for Hopf algebras was the structure theorem for cocommutative pointed Hopf algebras when k has characteristic zero. Cocommutative means the dual algebra is commutative. Such a Hopf algebra is a smash product $U\sharp k[G]$, where U is the universal enveloping algebra of a Lie algebra. Generally a pointed Hopf algebra over any field k is a crossed product $\mathcal{U}\sharp_\sigma k[\mathcal{G}]$, where $\mathcal{U}$ is an indecomposable pointed Hopf algebra and $\mathcal{G}$ is a quotient of G.

Now we consider a recipe for classification. Since the coradical $k[G]$ of H is a Hopf subalgebra of H, there is an associated graded Hopf algebra $\mathrm{gr}(H)$ with coradical $k[G]$ and $\mathrm{gr}(H) = B\times k[G]$ is a biproduct, where B is a pointed irreducible Hopf algebra in the Yetter-Drinfel'd category ${}^{k[G]}_{k[G]}\mathcal{YD}$. By definition B is a Nichols algebra when it is generated as an algebra by its space of primitive elements $B(1)$.

The recipe for classification is to pass from H to $\mathrm{gr}(H)$, show that B is a Nichols algebra, find defining relations for B, and use them to determine defining relations for H. The analysis of Nichols algebras is well beyond the scope of this book. The space $B(1)$ is an object of ${}^H_H\mathcal{YD}$. We do show for every object V of ${}^H_H\mathcal{YD}$ there is a Nichols algebra $B(V)$ in ${}^H_H\mathcal{YD}$ with $B(V)(1) = V$ and show that Nichols algebras can be defined by a universal mapping property.

Suppose that k is algebraically closed of characteristic zero. Chapter 16 treats a few topics in the theory of finite-dimensional Hopf algebras over k. In the first section semisimple Hopf algebras are characterized in various ways and the fact that their antipodes are involutions is established.

For a given dimension there are only finitely many isomorphism classes of semisimple Hopf algebras. This is not the case for pointed Hopf algebras as shown in the second section. The third and last section of the chapter states some very useful results for classification of semisimple Hopf algebras over k. Just a few proofs are included.

The chapter notes contain a lengthy discussion of the current status of classification of finite-dimensional Hopf algebras, in particular of semisimple ones. In contrast to the pointed case, classification of semisimple Hopf algebras over k is still in beginning stages.

It has been my great pleasure to get to know many who have contributed to Hopf algebras and related areas. Their collective contributions are far to numerous to completely list here. I would like to recognize my coauthors: Nicholás Andruskiewitsch, Robert Grossman, Robert Heyneman, Louis Kauffman, Leonid Krop, Larry Lambe, Richard Larson, Kenneth Newman, Stephen Sawin, Hans-Jürgen Schneider, Earl Taft, Jacob Towber, Sara Westreich, and Robert Wilson. Work with them was always interesting and fruitful.

A word of thanks to Margaret Beattie, Susan Montgomery, and Hans-Jürgen Schneider who helped me with certain references. Finally, I express deep gratitude to my wife Jean for her unwavering patience and support, especially during the last months of work on the manuscript.

David E. Radford

Contents

Chapter 1

Preliminaries

In this chapter we establish basic conventions and discuss some results from linear algebra which will be very useful to us, especially in Chapter 3.

1.1 Notation and terminology conventions

Throughout this text k is a field and most objects of study are vector spaces over k. Definitions of many of the structures we consider can be made over a commutative ring. The resulting theory, however, is not nearly as rich as when the commutative ring is a field. The category of vector spaces over k and their linear transformations under function composition is denoted by k-Vec. We refer to linear transformations informally as linear maps, maps of vector spaces, or vector space maps. There are many categories whose morphisms are functions. We refer to these morphisms as maps of "objects" or "object" maps and use these terminologies interchangeably.

Let U and V be vector spaces over the field k. We drop the subscript k from $U \otimes_k V$, $\mathrm{Hom}_k(U, V)$, and $\mathrm{End}_k(U)$ and write $U \otimes V$, $\mathrm{Hom}(U, V)$, and $\mathrm{End}(U)$ instead. If $u^* \in U^*$ and $u \in U$ the result of evaluation of u^* on u is denoted by either $u^*(u)$ or $<u^*, u>$. The notation I_U stands for the identity map of U. When U is understood, we at times write I for I_U. When U is an algebra with unity over k, we denote the multiplicative neutral element of U by 1 or 1_U.

Suppose that I is a non-empty set and $i, j \in I$. We use the notations

$$\delta_{i,j} = \delta^i_j = \begin{cases} 1 : i = j \\ 0 : i \neq j \end{cases}$$

for the Kronecker delta function. Ordinarily 1 and 0 are the multiplicative and additive neutral elements respectively of an algebra over k.

The Einstein summation convention is a shorthand for expression sums and we frequently use it. In this convention $s = \sum_{i=1}^{m}\sum_{j=1}^{n} a^i_j b_i c^{(j)}$ is expressed $s = a^i_j b_i c^{(j)}$. Expressions such as the preceding are understood to represent sums whose summation indices are the common upper and lower indices and are summed over the full range of their values.

Let $f : X \longrightarrow Y$ be a function and $Z \subseteq X$. We denote the restriction of f to Z by $f|Z : Z \longrightarrow Y$. Sometimes the domain of a restriction map results from operations. Viewing restriction as an operation it is performed last. For example $f|X\wedge Y$ is read $f|(X\wedge Y)$.

Suppose A is an algebra over k and M is a left (respectively right) A-module. Then the linear dual M^* is a right (respectively left) A-module under the transpose action which we denote by $\prec$ (respectively $\succ$). Let $m^* \in M^*$, $a \in A$, and $m \in M$. Thus if M is a left A-module $<m^*\prec a, m> = <m^*, a\cdot m>$ and if M is right A-module $<a\succ m^*, m> = <m^*, m\cdot a>$.

Throughout this book U and V are vector spaces over the field k. Usually each chapter or section has notation conventions which are normally repeated in formal statements so the latter can stand alone as much as possible.

1.2 Rank of a tensor

Ranks of tensors are integers which are useful in the theory of coalgebras and comodules.

Definition 1.2.1. Let U, V be vector spaces over the field k and $\nu \in U\otimes V$. If $\nu = 0$ then $\mathrm{Rank}(\nu) = 0$. If $\nu \neq 0$ then $\mathrm{Rank}(\nu)$ is equal to the smallest positive integer r arising from representations of $\nu = \sum_{i=1}^{r} u_i\otimes v_i \in U\otimes V$ as a sum.

Given a representation $\nu = \sum_{i=1}^{r} u_i\otimes v_i \in U\otimes V$ of a non-zero $\nu \in U\otimes V$ there is a very simple test to determine whether or not $r = \mathrm{Rank}(\nu)$.

Lemma 1.2.2. *Let U, V be vector spaces over the field k, $\nu \in U\otimes V$ be non-zero, and write $\nu = \sum_{i=1}^{r} u_i\otimes v_i \in U\otimes V$. Then the following are equivalent:*

(a) $r = \mathrm{Rank}(\nu)$.
(b) *The sets $\{u_1, \ldots, u_r\}$ and $\{v_1, \ldots, v_r\}$ are linearly independent.*

Proof. Part (a) implies part (b). For suppose $r = \text{Rank}(\nu)$. If $\{u_1, \ldots, u_r\}$ is linearly dependent then $r > 1$, since $\nu \neq 0$, and one of the u_i's can be written as a linear combination of the others. By rearranging summands we may assume that $i = r$. Thus $u_r = \alpha_1 u_1 + \cdots + \alpha_{r-1} u_{r-1}$ for some $\alpha_1, \ldots, \alpha_{r-1} \in k$. As a result $\nu = \sum_{i=1}^{r-1} u_i \otimes (v_i + \alpha_i v_r)$. But this is not possible in light of the minimality of r. Therefore $\{u_1, \ldots, u_r\}$ is linearly independent. By a similar argument $\{v_1, \ldots, v_r\}$ is linearly independent also. We have shown part (a) implies part (b).

To show part (b) implies part (a) assume that $\{u_1, \ldots, u_r\}$ and $\{v_1, \ldots, v_r\}$ are linearly independent. Write $\nu = \sum_{j=1}^{r'} u'_j \otimes v'_j \in U \otimes V$ where $r' = \text{Rank}(\nu)$. Fix $1 \leq \ell \leq r$. Since $\{u_1, \ldots, u_r\}$ is linearly independent, there is a functional $u^* \in U^*$ which satisfies $u^*(u_\ell) = \delta_{\ell,i}$ for all $1 \leq i \leq r$. Applying $u^* \otimes \mathrm{I}_V$ to both sides of the equation $\sum_{i=1}^{r} u_i \otimes v_i = \sum_{j=1}^{r'} u'_j \otimes v'_j$ gives $v_\ell = \sum_{j=1}^{r'} u^*(u'_j) v'_j$. Therefore v_ℓ is in the span of $\{v'_1, \ldots, v'_{r'}\}$. As $\{v_1, \ldots, v_r\}$ is linearly independent $r \leq r'$. Since $r' \leq r$ by definition of $\text{Rank}(\nu)$ it follows that $r = \text{Rank}(\nu)$. Thus part (b) implies part (a). □

As elementary as it is, the lemma has some interesting applications. Let $\pi : U^* \otimes V \longrightarrow \text{Hom}(U, V)$ be the linear map defined by $\pi(u^* \otimes v)(u) = u^*(u)v$ for all $u^* \in U^*$, $u \in U$, and $v \in V$. Suppose that $\nu \in U^* \otimes V$ is not zero and write $\nu = \sum_{i=1}^{r} u_i^* \otimes v_i \in U^* \otimes V$, where $r = \text{Rank}(\nu)$. The set $\{v_1, \ldots, v_r\}$ is linearly independent by Lemma 1.2.2. Since $\nu \neq 0$ it follows that $u_i^* \neq 0$ for some $1 \leq i \leq r$. Therefore $\pi(\nu) \neq 0$. We have shown that π is one-one.

Let $\mathcal{V}$ be the span of $\{v_1, \ldots, v_r\}$. Then $\text{Im}(\pi(\nu)) \subseteq \mathcal{V}$. Choose a basis $\{w_1, \ldots, w_s\}$ for $\text{Im}(\pi(\nu))$. Then there are functionals $w_1^*, \ldots, w_s^*$ of U^* such that $\pi(\nu)(u) = w_1^*(u)w_1 + \cdots + w_s^*(u)w_s$ for all $u \in U$. Since π is one-one $\nu = w_1^* \otimes w_1 + \cdots + w_s^* \otimes w_s$. Now $r \leq s$ since $r = \text{Rank}(\nu)$. On the other hand, $s \leq r$ since $\text{Im}(\pi(\nu)) \subseteq \mathcal{V}$. Therefore $r = s$ and $\text{Im}(\pi(\nu)) = \mathcal{V}$. For a map of vector spaces $T : U \longrightarrow V$ recall that $\text{rank}(T) = \text{Dim}(\text{Im}(T))$. We have shown:

Proposition 1.2.3. *Let U, V be vector spaces over the field k and suppose that $\pi : U^* \otimes V \longrightarrow \text{Hom}(U, V)$ is the linear map defined by $\pi(u^* \otimes v)(u) = u^*(u)v$ for all $u^* \in U^*$, $u \in U$ and $v \in U$. Then:*

(a) *π is one-one.*
(b) *Suppose $\nu \in U^* \otimes V$ is not zero and $\nu = \sum_{i=1}^{r} u_i^* \otimes v_i \in U^* \otimes V$ where $r = \text{Rank}(\nu)$. Then $\text{rank}(\pi(\nu)) = \text{Rank}(\nu)$ and $\{v_1, \ldots, v_r\}$ is a basis*

for $\mathrm{Im}(\pi(\nu))$.

□

Apropos of the proposition we note that $\mathrm{rank}(\pi(\nu)) = \mathrm{Rank}(\nu)$ for all $\nu \in U \otimes V$.

Let $\imath_U : U \longrightarrow U^{**}$ be the linear map defined by $\imath_U(u)(u^*) = u^*(u)$ for all $u \in U$ and $u^* \in U^*$. Suppose that $\{u_1^*, \ldots, u_r^*\}$ is a linearly independent set of functionals in U^* and $\alpha_1, \ldots, \alpha_r \in k$. Then there is a functional $f \in U^{**}$ such that $f(u_i^*) = \alpha_i$ for all $1 \leq i \leq r$. By virtue of the following corollary there is such an $f \in \mathrm{Im}(\imath_U)$.

Corollary 1.2.4. *Suppose U is a vector space over the field k. Let $\{u_1^*, \ldots, u_r^*\}$ be a linearly independent set of functionals of U^* and $\alpha_1, \ldots, \alpha_r \in k$. Then there is an element $u \in U$ such that $u_i^*(u) = \alpha_i$ for all $1 \leq i \leq r$.*

Proof. Let $\nu = \sum_{i=1}^{r} u_i^* \otimes u_i^* \in U^* \otimes V$, where $V = U^*$, and consider the map $\pi : U^* \otimes V \longrightarrow \mathrm{Hom}(U, V)$ of Proposition 1.2.3. Now $r = \mathrm{Rank}(\nu)$ by Lemma 1.2.2. Therefore $\{u_1^*, \ldots, u_r^*\}$ is a basis for $\mathrm{Im}(\pi(\nu))$ by part (b) of Proposition 1.2.3. Consequently $\alpha_1 u_1^* + \cdots + \alpha_r u_r^* = \pi(\nu)(u)$ for some $u \in U$. As $\pi(\nu)(u) = u_1^*(u)u_1^* + \cdots + u_r^*(u)u_r^*$ we conclude that $u_i^*(u) = \alpha_i$ for all $1 \leq i \leq r$. □

When $U = V$ and is finite-dimensional we usually identify $U^* \otimes U$ and $\mathrm{End}(U)$ by the mapping π of Proposition 1.2.3. Let $\mathrm{Tr} : \mathrm{End}(U) \longrightarrow k$ be the trace function. Then

$$\mathrm{Tr}(u^* \otimes u) = u^*(u)$$

for all $u^* \in U^*$ and $u \in U$.

Exercises

In the following exercises U, V are vector spaces over the field k.

Exercise 1.2.1. Suppose that U is finite-dimensional. Show that $\mathrm{Tr}(u^* \otimes u) = u^*(u)$ for all $u^* \in U^*$ and $u \in U$.

Exercise 1.2.2. Consider the isomorphism $\imath : (U \otimes U) \otimes U \longrightarrow U \otimes (U \otimes U)$ defined by $\imath((u \otimes v) \otimes w) = u \otimes (v \otimes w)$ for all $u, v, w \in U$. Show that the Rank function, defined on the tensor product of two vector spaces as above, is *not* invariant under the isomorphism $\imath$ when $\mathrm{Dim}(U) > 1$.

Exercise 1.2.3. Reconsider the argument for part (a) implies part (b) in the proof of Lemma 1.2.2. Show that $\{u_1, \ldots, u_r\}$ linearly independent *implies* that $\{v_1, \ldots, v_r\}$ is linearly independent. [Hint: Observe that Rank is invariant under the "twist map", the isomorphism $\tau_{U,V} : U \otimes V \longrightarrow V \otimes U$ defined by $\tau_{U,V}(u \otimes v) = v \otimes u$ for all $u \in U$ and $v \in V$.]

Exercise 1.2.4. Let $\pi : U^* \otimes V \longrightarrow \mathrm{Hom}(U, V)$ be the linear map defined by

$$\pi(u^* \otimes v)(u) = u^*(u)v$$

for all $u^* \in U^*$, $u \in U$, and $v \in V$. Show that:

(a) π induces a linear isomorphism $U^* \otimes V \simeq \mathrm{Hom}_{\mathrm{f}}(U, V)$, where the latter is the subspace of $\mathrm{Hom}(U, V)$ consisting of all linear maps $f : U \longrightarrow V$ of finite rank.

(b) π is an isomorphism if and only if U or V is finite-dimensional.

Exercise 1.2.5. Let $V_1, \ldots, V_n$ be vector spaces over the field k and let

$$\imath : V_1^* \otimes \cdots \otimes V_n^* \longrightarrow (V_1 \otimes \cdots \otimes V_n)^*$$

be the linear map defined by

$$\imath(v_1^* \otimes \cdots \otimes v_n^*)(v_1 \otimes \cdots \otimes v_n) = v_1^*(v_1) \cdots v_n^*(v_n)$$

for all $v_i^* \in V_i^*$ and $v_i \in V_i$, $1 \leq i \leq n$. Show that:

(a) $\imath$ is one-one.

(b) $\imath$ is an isomorphism if $V_1, \ldots, V_n$ are finite-dimensional.

[Hint: To show part (a), first assume that $n > 2$. Let

$$\imath_2 : V_1^* \otimes (V_2 \otimes \cdots \otimes V_n)^* \longrightarrow (V_1 \otimes \cdots \otimes V_n)^*$$

and

$$\imath_{n-1} : V_2^* \otimes \cdots \otimes V_n^* \longrightarrow (V_2 \otimes \cdots \otimes V_n)^*$$

be defined in the same manner as $\imath$. Show that $\imath = \imath_2 \circ (\mathrm{I}_{V_1^*} \otimes \imath_{n-1})$ which reduces part (a) to the case $n = 2$ by induction. For the case $n = 2$ use Corollary 1.2.4 or note that $\imath$ is the composite

$$V_1^* \otimes V_2^* \xrightarrow{\pi} \mathrm{Hom}(V_1, V_2^*) \simeq \mathrm{Hom}(V_1 \otimes V_2, k),$$

where π is the one-one map of Proposition 1.2.3.]

For $\nu \in U \otimes V$ let $L_\nu = (\mathrm{I}_U \otimes V^*)(\nu)$ and $R_\nu = (U^* \otimes \mathrm{I}_V)(\nu)$. Observe that L_ν and R_ν are subspaces of U and V respectively.

Exercise 1.2.6. Suppose $\nu \in U \otimes V$ is not zero and write $\nu = \sum_{i=1}^{r} u_i \otimes v_i \in U \otimes V$, where $r = \mathrm{Rank}(\nu)$. Show that:

(a) $\{u_1, \ldots, u_r\}$ is a basis for L_ν and that $\{v_1, \ldots, v_r\}$ is a basis for R_ν.

(b) $\nu \in L_\nu \otimes R_\nu$ and $\mathrm{Dim}(L_\nu) = \mathrm{Rank}(\nu) = \mathrm{Dim}(R_\nu)$.

(c) If $\mathcal{U}, \mathcal{V}$ are subspaces of U, V respectively and $\nu \in \mathcal{U} \otimes \mathcal{V}$ then $L_\nu \subseteq \mathcal{U}$ and $R_\nu \subseteq \mathcal{V}$.

(d) If $\{u'_1, \ldots, u'_r\}$ is a basis for L_ν then there are unique $x_1, \ldots, x_r \in V$ such that $\nu = \sum_{i=1}^{r} u'_i \otimes x_i$ and furthermore $\{x_1, \ldots, x_r\}$ is a basis for R_ν.

Exercise 1.2.7. Suppose that $\mathcal{U}$ and $\mathcal{V}$ are subspaces of U and V respectively and $\mathrm{Dim}(\mathcal{U}) = \mathrm{Dim}(\mathcal{V}) > 0$. Show that there is some $\nu \in U \otimes V$ such that $\mathcal{U} = L_\nu$ and $\mathcal{V} = R_\nu$.

Exercise 1.2.8. Suppose that $\{\mathcal{U}_i\}_{i \in I}$ and $\{\mathcal{V}_i\}_{i \in I}$ are families of subspaces of U and V respectively indexed by a set I. Use Exercise 1.2.6 to show that $\bigcap_{i \in I} (\mathcal{U}_i \otimes \mathcal{V}_i) = (\bigcap_{i \in I} \mathcal{U}_i) \otimes (\bigcap_{i \in I} \mathcal{V}_i)$.

Definition 1.2.5. Let U be a finite-dimensional vector space over the field k with basis $\{u_1, \ldots, u_n\}$. The *dual basis for* U^* is $\{u^1, \ldots, u^n\}$ where $u^i(u_j) = \delta^i_j$ for $1 \leq i, j \leq n$.

Exercise 1.2.9. Suppose U is finite-dimensional. Let $\{u_1, \ldots, u_n\}$ be a basis for U. Show that:

(a) $\sum_{i=1}^{n} <u^i, v> u_i = v$ for all $v \in U$.

(b) $\sum_{i=1}^{n} u^i <p, u_i> = p$ for all $p \in U^*$.

(c) $\sum_{i=1}^{n} T^*(u^i) \otimes u_i = \sum_{i=1}^{n} u^i \otimes T(u_i)$ for all $T \in \mathrm{End}(U)$.

(d) If $\{v_1, \ldots, v_n\}$ is also a basis for U then $\sum_{i=1}^{n} u^i \otimes u_i = \sum_{i=1}^{n} v^i \otimes v_i$.

[Hint: For parts (c) and (d), apply the isomorphism $\pi : U^* \otimes U \longrightarrow \mathrm{End}(U)$ of Exercise 1.2.4 to both sides of the equations.]

The sum $\sum_{i=1}^{n} u^i \otimes u_i$ of part (d) is called the *canonical element of* $U^* \otimes U$.

Exercise 1.2.10. Suppose that U is finite-dimensional and $\{u_1, \ldots, u_n\}$ is a basis for U. For $T \in \mathrm{End}(U)$ show that

$$\mathrm{Det}(T) = \sum_{\sigma \in S_n} \mathrm{sgn}(\sigma) \imath (u^{\sigma(1)} \otimes \cdots \otimes u^{\sigma(n)})(T(u_1) \otimes \cdots \otimes T(u_n)),$$

where $\imath$ is the linear map of Exercise 1.2.5.

Recall that a *diagonalizable endomorphism of* V is a linear endomorphism $T \in \mathrm{End}(V)$ such that V is the sum of the eigenspaces of T.

Exercise 1.2.11. Suppose $T, T' \in \mathrm{End}(V)$ are diagonalizable. Show that:

(a) If U is a T-invariant subspace of V then the restriction $T|U \in \mathrm{End}(U)$ is diagonalizable.

(b) If T is an automorphism of V then T^{-1} is diagonalizable.

(c) If T and T' commute then the composite $T \circ T'$ is diagonalizable.

(d) If $\mathcal{T} \subseteq \mathrm{End}(V)$ is a family of commuting diagonalizable automorphisms of V then the subgroup $\mathcal{G}$ of the linear automorphisms of V which $\mathcal{T}$ generates is a family of commuting diagonalizable automorphisms of V.

(e) If $\mathcal{T} \subseteq \mathrm{End}(V)$ is a family of commuting diagonalizable endomorphisms of V, $V \neq (0)$, and V is finite-dimensional, then V has a basis $\{v_1, \ldots, v_n\}$ such that each v_i is an eigenvector for all $T' \in \mathcal{T}$. [Hint: We may assume there is a $T \in \mathcal{T}$ which has at least two eigenvalues. Show that $V = V' \oplus V''$ is the direct sum of proper subspaces which are invariant under all $T' \in \mathcal{T}$.]

Exercise 1.2.12. Let $\imath_V : V \longrightarrow V^{**}$ be given by $\imath_V(v^*)(v) = v^*(v)$ for all $v^* \in V^*$ and $v \in V$. Show that $(\imath_V)^* \circ \imath_{V^*} = \mathrm{I}_{V^*}$.

1.3 Topological aspects of vector space duals

As we shall see in Chapter 2, basic structures on a coalgebra C over k are in one-one correspondence with the subalgebras, ideals, left ideals, and right ideals of the dual algebra C^* which are *closed* subspaces of C^*. If $f : C \longrightarrow D$ is a coalgebra map then the linear transpose $f^* : D^* \longrightarrow C^*$ is continuous, meaning that pre-images of closed subspaces of C^* are closed subspaces of D^*. In this section we study the closed subspaces and continuous linear maps of vector space duals.

The closed subspaces of U^* we discuss are closed subspaces in the weak-* topology on U^*. We develop topological ideas in this text only to the extent that they are useful for us.

Suppose that V is a subset of U and X is a subset of U^*. Then

$$V^{\perp} = \{u^* \in U^* \mid u^*(V) = (0)\}$$

and

$$X^{\perp} = \{u \in U \mid X(u) = (0)\}$$

are subspaces of U^* and U respectively.

Notice that $X^{\perp}$ has a different meaning when our discussion is based on $\mathcal{U} = U^*$ instead of U. Let $\perp'$ be defined for $\mathcal{U}$ as $\perp$ is defined for U and let $\imath_U : U \longrightarrow U^{**}$ be the one-one linear map defined by $\imath_U(u)(u^*) = u^*(u)$ for all $u \in U$ and $u^* \in U^*$. Identifying U with its image under $\imath_U$ in $U^{**} = \mathcal{U}^*$ observe that

$$X^{\perp} = X^{\perp'} \cap U \tag{1.1}$$

for all subsets X of $\mathcal{U} = U^*$.

Basic properties of the correspondence $X \mapsto X^{\perp}$ follow from:

Lemma 1.3.1. *Let U be a vector space over the field k. Suppose X, Y are both subspaces of U or are both subspaces of U^*. Then:*

(a) *If $X \subseteq Y$ then $X^{\perp} \supseteq Y^{\perp}$.*
(b) *$X \subseteq X^{\perp\perp}$.*
(c) *$X^{\perp} = X^{\perp\perp\perp}$.*

Proof. Parts (a) and (b) follow immediately by definition. Part (c) is a consequence of parts (a) and (b). For $X^{\perp} \subseteq X^{\perp\perp\perp}$ and $X \subseteq X^{\perp\perp}$ by part (b). Part (a) applied to the last inclusion gives $X^{\perp} \supseteq X^{\perp\perp\perp}$ and therefore $X^{\perp} = X^{\perp\perp\perp}$. □

Definition 1.3.2. Let U be a vector space over the field k. A *closed subspace of* U is a subspace X of U^* such that $X = X^{\perp\perp}$.

Observe that $U^* = (0)^{\perp}$ and $(0) = U^{\perp}$ are closed subspaces of U^*. Notice that *all* subspaces V of U satisfy $V = V^{\perp\perp}$. For given $u \in U \backslash V$ there is a functional $u^* \in U^*$ which satisfies $u^*(V) = (0)$ and $u^*(u) = 1$. By part (c) of Lemma 1.3.1 a subspace X of U^* is closed if and only if $X = V^{\perp}$ for some subspace V of U. Thus:

Corollary 1.3.3. *Let U be a vector space over the field k. Then $V \mapsto V^{\perp}$ describes an inclusion reversing bijective correspondence between the set of subspaces of U and the set of closed subspaces of U^* whose inverse is given by $X \mapsto X^{\perp}$ for all closed subspaces X of U^*.* □

Suppose that X, Y are subspaces of U^* where Y is closed and $X \subseteq Y$. Then $X^{\perp\perp} \subseteq Y^{\perp\perp} = Y$ by part (a) of Lemma 1.3.1. Our conclusion: among all of the closed subsets of U^* which contain X there is a unique minimal one which is $X^{\perp\perp}$.

Definition 1.3.4. Let U be a vector space over the field k and let X be a subspace of U^*. The subspace $X^{\perp\perp}$ of U^* is the *closure of* X and is denoted $\overline{X}$.

By parts (a) and (c) of Lemma 1.3.1 the closure operation is an inclusion preserving idempotent operation; that is if X, Y are subspaces of U^* then $X \subseteq Y$ implies $\overline{X} \subseteq \overline{Y}$ and $\overline{(\overline{X})} = \overline{X}$.

If U is not finite-dimensional then some subspace of U^* is not closed, for:

Proposition 1.3.5. *Let U be a vector space over the field k. Then the following are equivalent:*

(a) *All subspaces of U^* are closed.*
(b) *U is finite-dimensional.*

Proof. Choose a basis B for U and let X be the subset of U^* consisting of all functionals which vanish on all but finitely many of the elements of B. Then X is a subspace of U^* and $X^\perp = (0)$, or equivalently $\overline{X} = U^*$. If X is closed then B must be finite. We have shown that part (a) implies part (b).

Suppose that U is finite-dimensional. Let $\mathcal{U} = U^*$ and identify U and $\mathcal{U}^* = U^{**}$ via the linear isomorphism $\imath_U : U \longrightarrow U^{**}$ defined by $\imath_U(u)(u^*) = u^*(u)$ for all $u \in U$ and $u^* \in U^*$. Let X be a subspace of $U^* = \mathcal{U}$. We use (1.1) to calculate $X^{\perp'\perp'} = X^{\perp\perp}$. Thus $X = X^{\perp'\perp'} = X^{\perp\perp}$ by Corollary 1.3.3 as required. We have shown part (b) implies part (a). □

Suppose that X and Y are subspaces of U^*. There is a test on the finite-dimensional subspaces of U which determines whether or not X and Y have the same closure. For a subspace $V \subseteq U$ let $X|V$ denote the set of restrictions $f|V : V \longrightarrow k$ where f runs over X. Observe that $X|V$ is a subspace of V^*.

Proposition 1.3.6. *Let U be a vector space over the field k and let X, Y be subspaces of U^*. Then the following are equivalent:*

(a) $\overline{X} = \overline{Y}$.
(b) *$X|V = Y|V$ for all finite-dimensional subspaces V of U.*
(c) *$X|V = Y|V$ for all one-dimensional subspaces V of U.*

Proof. Suppose first of all that $\overline{X} = \overline{Y}$. Then $X^\perp = Y^\perp$ by part (c) of Lemma 1.3.1. Let V be a finite-dimensional subspace of U. Since $X^\perp = Y^\perp$ we conclude that $(X|V)^\perp = (Y|V)^\perp$, where the latter calculation takes place in V^*. Now V finite-dimensional means that all subspaces of V^* are closed by Proposition 1.3.5. Therefore $X|V = Y|V$. We have shown that part (a) implies part (b). That part (b) implies part (c) is trivial.

Assume that $X|V = Y|V$ for all one-dimensional subspaces of U. To show part (c) implies part (a) it is sufficient to show that $X \subseteq \overline{Y}$. For then $\overline{X} \subseteq \overline{Y}$. As $Y|V = X|V$ the inclusion $\overline{Y} \subseteq \overline{X}$ follows as well; thus $\overline{X} = \overline{Y}$.

Let $f \in X$, $v \in Y^\perp$ and $V = kv$. We may as well assume that $v \neq 0$. By assumption $f|V = g|V$ for some $g \in Y$. Since $g(v) = 0$ necessarily $f(v) = 0$. Thus $f \in Y^{\perp\perp} = \overline{Y}$ by definition. We have shown that $X \subseteq \overline{Y}$. Thus part (c) implies part (a) and the proof of the proposition is complete. □

Definition 1.3.7. Let U be a vector space over the field k and suppose Y is a subspace of U^*. A *dense subspace of* X is a subspace Y of U such that $Y \subseteq X \subseteq \overline{Y}$.

Let X, Y be subspaces of U^* and suppose that $Y \subseteq X$. Using Lemma 1.3.1 we see that Y is a dense subspace of X if and only if $\overline{Y} = \overline{X}$, or equivalently $Y^\perp = X^\perp$. In particular, X is a dense subspace of U^* if and only if $X^\perp = (0)$. By Proposition 1.3.6 it follows that X is a dense subspace of U^* if and only if for any finite-dimensional subspace V of U and any $f \in U^*$ there exists a $g \in X$ such that $g|V = f|V$.

We next consider intersections and sums of closed subspaces of U^*.

Proposition 1.3.8. *Let U be a vector space over the field k and suppose that $\{V_i\}_{i\in I}$ is a family of subspaces of U. Then:*

(a) $\bigcap_{i\in I} V_i^\perp = (\sum_{i\in I} V_i)^\perp$. *Thus the intersection of any family of closed subspaces of U^* is closed.*
(b) *If I is finite then* $\sum_{i\in I} V_i^\perp = (\bigcap_{i\in I} V_i)^\perp$. *Thus the sum of a finite number of closed subspaces of U^* is closed.*

Proof. We first show part (a). It is clear that $\bigcap_{i\in I} V_i^\perp \subseteq (\sum_{i\in I} V_i)^\perp$. For $j \in I$ the inclusion $V_j \subseteq \sum_{i\in I} V_i$ implies that $V_j^\perp \supseteq (\sum_{i\in I} V_i)^\perp$ by part (a) of Lemma 1.3.1. Therefore $\bigcap_{i\in I} V_i^\perp \supseteq (\sum_{i\in I} V_i)^\perp$. Part (a) is established.

To show part (b) we may as well assume that $I = \{1, \ldots, n\}$ for some $n \geq 1$. Suppose that part (b) holds for $n = 2$. Then for $n > 2$ the calculation

$$\sum_{i=1}^{n} V_i^\perp = \sum_{i=1}^{n-1} V_i^\perp + V_n^\perp = (\bigcap_{i=1}^{n-1} V_i)^\perp + V_n^\perp = ((\bigcap_{i=1}^{n-1} V_i) \cap V_n)^\perp = (\bigcap_{i=1}^{n} V_i)^\perp$$

shows that part (b) follows by induction on n.

Suppose that $n = 2$. Since $V_1 \cap V_2 \subseteq V_1, V_2$ we have $(V_1 \cap V_2)^\perp \supseteq V_1^\perp + V_2^\perp$ by part (a) of Lemma 1.3.1. It remains to show that $(V_1 \cap V_2)^\perp \subseteq V_1^\perp + V_2^\perp$.

Let $W_0, \ldots, W_3$ be subspaces of U which satisfy

$$W_0 = V_1 \cap V_2, \quad W_0 \oplus W_1 = V_1, \quad W_0 \oplus W_2 = V_2, \quad \text{and} \quad (V_1 + V_2) \oplus W_3 = U.$$

It is easy to see that $W_0 \oplus \cdots \oplus W_3 = U$. Let $u^* \in (V_1 \cap V_2)^\perp$ and let $v^* \in U^*$ be defined by $v^*|W_i = \delta_{2,i} u^*|W_2$ for $0 \leq i \leq 3$. Since $u^*(W_0) = (0)$ it follows that $u^* = v^* + (u^* - v^*) \in V_1^\perp + V_2^\perp$. Thus $(V_1 \cap V_2)^\perp \subseteq V_1^\perp + V_2^\perp$ and the proof of part (b) is complete. □

Let $X = ku^*$ where $u^* \in U^*$. Then X is closed since $X = (\mathrm{Ker}(u^*))^\perp$. Thus one-dimensional subspaces of U^* are closed and therefore by part (b) of Proposition 1.3.8:

Corollary 1.3.9. *Let U be a vector space over the field k. Then the finite-dimensional subspaces of U^* are closed.* □

We now turn our attention to continuous linear maps of vector space duals.

Theorem 1.3.10. *Let U and V be vector spaces over the field k and suppose that $F : V^* \longrightarrow U^*$ is the transpose of a linear map $f : U \longrightarrow V$. Let $\mathcal{U}$ and $\mathcal{V}$ be subspaces of U and V respectively. Then:*

(a) $F^{-1}(\mathcal{U}^\perp) = f(\mathcal{U})^\perp$. *Thus the pre-image of a closed subspace of U^* under F is a closed subspace of V^*.*
(b) $F(\mathcal{V}^\perp) = (f^{-1}(\mathcal{V}))^\perp$. *Thus the image of a closed subspace of V^* under F is a closed subspace of U^*.*

Let J and I be subspaces of V^* and U^* respectively. Then:

(c) $F(J) \subseteq I$ *implies* $f(I^\perp) \subseteq J^\perp$.
(d) $F(\overline{J}) = \overline{F(J)}$.
(e) $F(J)^\perp = f^{-1}(J^\perp)$.

Proof. Let $v^* \in V^*$. From the relation $F(v^*)(\mathcal{U}) = v^*(f(\mathcal{U}))$ we see that $F(v^*) \in \mathcal{U}^\perp$ if and only if $v^* \in f(\mathcal{U})^\perp$. Part (a) is established. To show part (b) we first observe that

$$F(\mathcal{V}^\perp)(f^{-1}(\mathcal{V})) = \mathcal{V}^\perp(f(f^{-1}(\mathcal{V}))) \subseteq \mathcal{V}^\perp(\mathcal{V}) = (0)$$

which implies $F(\mathcal{V}^\perp) \subseteq f^{-1}(\mathcal{V})^\perp$. Let $u^* \in f^{-1}(\mathcal{V})^\perp$. Part (b) will be established once we show that $u^* \in F(\mathcal{V}^\perp)$.

Since $\mathrm{Im}(F) = (\mathrm{Ker}(f))^\perp \supseteq f^{-1}(\mathcal{V})^\perp$ there is a $v^* \in V^*$ such that $u^* = F(v^*)$. Choose subspaces $W_0, \ldots, W_3$ of V which satisfy

$$W_0 = f(f^{-1}(\mathcal{V})), \quad W_0 \oplus W_1 = \mathcal{V}, \quad W_0 \oplus W_2 = f(U),$$

$$\text{and } (\mathcal{V} + f(U)) \oplus W_3 = V.$$

Since $\mathcal{V} \cap f(U) = f(f^{-1}(\mathcal{V}))$ it is easy to see that $V = W_0 \oplus \cdots \oplus W_3$. Let $w^* \in V^*$ be defined by $w^*|W_i = \delta_{1,i}v^*|W_1$ for all $0 \leq i \leq 3$. Then $v^* - w^* \in \mathcal{V}^\perp$, since $v^*(W_0) = w^*(W_0) = (0)$, and $F(w^*) = 0$, since $\mathrm{Ker}(F) = f(U)^\perp$. Thus $u^* = F(v^* - w^*) \in F(\mathcal{V}^\perp)$ and part (b) follows.

Part (c) is a consequence of the calculation

$$J(f(I^\perp)) = F(J)(I^\perp) \subseteq I(I^\perp) = (0).$$

To show part (d) we use the fact that $F^{-1}(\overline{F(J)})$ is a closed subspace of V^*, which follows by part (a), to deduce that $\overline{J} \subseteq F^{-1}(\overline{F(J)})$. Thus $F(\overline{J}) \subseteq \overline{F(J)}$. We use the fact that $F(\overline{J})$ is a closed subspace of U^*, which follows by part (b), to deduce that $\overline{F(J)} \subseteq F(\overline{J})$. Therefore $F(\overline{J}) = \overline{F(J)}$ and part (d) is established.

Part (e) follows from the calculation

$$F(J)^\perp = (\overline{F(J)})^\perp = F(\overline{J})^\perp = (f^{-1}(J^\perp))^{\perp\perp} = f^{-1}(J^\perp),$$

the steps of which are left for the reader to justify. □

When C is a coalgebra over the field k, certain closed subspaces I of C^* such that C^*/I is finite-dimensional correspond to basic finite-dimensional structures of C.

Definition 1.3.11. Let U be a vector space over the field k. A *cofinite subspace of* U is a subspace V of U such that U/V is finite-dimensional.

The reader is encouraged to work out Exercises 1.3.10–1.3.15 which detail basic properties of cofinite subspaces.

For a subspace V of U let $\mathrm{res}^U_V : U^* \longrightarrow V^*$ be the restriction map which is thus defined by $\mathrm{res}^U_V(u^*) = u^*|V$ for all $u^* \in U^*$. Notice that $\mathrm{Ker}(\mathrm{res}^U_V) = V^\perp$. Hence $U^*/V^\perp \simeq V^*$ as vector spaces. Therefore we have the formula

$$\mathrm{Dim}(U^*/V^\perp) = \mathrm{Dim}(V^*). \tag{1.2}$$

In particular $V^{\perp}$ is a cofinite subspace of U^* if and only if V is a finite-dimensional subspace of U. Also notice that $\mathrm{res}^U_V = \imath^*$, where $\imath : V \longrightarrow U$ is the inclusion map.

Proposition 1.3.12. *Let U be a vector space over the field k and let I, J be subspaces of U^*, where I is closed and is a cofinite subspace of U^*. If $J \supseteq I$ then J is closed and is a cofinite subspace of U^*.*

Proof. Let $V = I^{\perp}$. Then $V^{\perp} = I$ since I is closed. We have noted that V is finite-dimensional since I is a cofinite subspace of U^*.

Let $F = \mathrm{res}^U_V : U^* \longrightarrow V^*$. Since V is finite-dimensional $F(J)$ is a closed subspace of V^* by Proposition 1.3.5. Since $J \supseteq I = V^{\perp} = \mathrm{Ker}(F)$ we conclude that $J = F^{-1}(F(J))$ is a closed subspace of U^* by part (a) of Theorem 1.3.10. Since I is a cofinite subspace of U^* and $J \supseteq I$, we conclude that J is a cofinite subspace of U^* by Exercise 1.3.13. □

Corollary 1.3.13. *Let U be a vector space over the field k and suppose that $\imath_U : U \longrightarrow U^{**}$ is the one-one linear map defined by $\imath_U(u)(u^*) = u^*(u)$ for all $u \in U$ and $u^* \in U^*$. For $f \in U^{**}$ the following are equivalent:*

(a) $f \in \mathrm{Im}(\imath_U)$.
(b) $f(I) = (0)$ *for some closed cofinite subspace I of U^*.*

Proof. If $f = \imath_U(u)$ for some $u \in U$ then $f((ku)^{\perp}) = (0)$. Thus part (a) implies part (b). On the other hand, if $f(I) = (0)$ for some closed cofinite subspace of U^* then $\mathrm{Ker}(f)$ is a closed subspace of U^* by Proposition 1.3.12. This means $\mathrm{Ker}(f) = (ku)^{\perp}$ for some $u \in U$, and consequently $f = \alpha\imath_U(u) = \imath_U(\alpha u)$ for some $\alpha \in k$. Therefore part (b) implies part (a). □

Transpose maps of vector space duals are characterized in terms of their relationship to closed subspaces.

Theorem 1.3.14. *Let U, V be vector spaces over the field k and suppose that $F : V^* \longrightarrow U^*$ is linear. Then the following are equivalent:*

(a) $F = f^*$ *for some linear map* $f : U \longrightarrow V$.
(b) *If I is a closed subspace of U^* then $F^{-1}(I)$ is a closed subspace of V^*.*
(c) *If I is a cofinite closed subspace of U^* then $F^{-1}(I)$ is a closed subspace of V^*.*
(d) *If $I = (ku)^{\perp}$ for some $u \in U$ then $F^{-1}(I)$ is a closed subspace of V^*.*

Proof. Part (a) implies part (b) by virtue of part (a) of Theorem 1.3.10. To complete the proof of the theorem we need only show that part (d) implies part (a).

Assume the hypothesis of part (d), let $u \in U$ and $I = (ku)^{\perp}$. Then $F^{-1}(I)$ is a closed subspace of V^* by assumption, and $F^{-1}(I)$ is a cofinite subspace of U^* by part (b) of Exercise 1.3.14. Let $\mathrm{ev}_u : U^* \longrightarrow k$ be defined by $\mathrm{ev}_u(u^*) = u^*(u)$ for all $u^* \in U^*$ and consider the composite $F_u : V^* \longrightarrow k$ defined by $F_u = \mathrm{ev}_u \circ F$. Since $F^{-1}(I) \subseteq \mathrm{Ker}(F_u)$, by Corollary 1.3.13 there is a $v \in V$ which satisfies

$$F(v^*)(u) = F_u(v^*) = v^*(v)$$

for all $v^* \in V^*$. Set $f(u) = v$. Then $f : U \longrightarrow V$ is a function which satisfies the equation $F(v^*)(u) = v^*(f(u))$ for all $v^* \in V^*$ and $u \in U$. Since $F(v^*)$ is linear for all $v^* \in V^*$ necessarily f is linear. Since $F = f^*$ our proof that part (d) implies part (a) is complete. □

Definition 1.3.15. Let U, V be vector spaces over the field k. Then a *continuous linear map* $F : V^* \longrightarrow U^*$ is a linear map F which satisfies any of the equivalent conditions of Theorem 1.3.14.

We close with the observation that for vector spaces U, V over k the map $\mathrm{Hom}(U, V) \longrightarrow \mathrm{Hom}(V^*, U^*)$ given by $f \mapsto f^*$ determines a linear isomorphism between $\mathrm{Hom}(U, V)$ and the subspace of linear maps in $\mathrm{Hom}(V^*, U^*)$ which are continuous.

Exercises

In the following exercises U, V, and W are vector spaces over the field k.

Exercise 1.3.1. Complete the proof of part (e) of Theorem 1.3.10.

Exercise 1.3.2. Show that the following are equivalent:

(a) Arbitrary sums of closed subspaces of U^* are closed.

(b) U is finite-dimensional.

(Thus some restriction on the index set I in part (b) of Proposition 1.3.8 is necessary.)

Exercise 1.3.3. Let $F, G : V^* \longrightarrow U^*$ be continuous linear maps. Show that $F = G$ if and only if F and G agree on a dense subspace of V^*.

Exercise 1.3.4. Let $F : V^* \longrightarrow U^*$ be a continuous linear map. Show that:

(a) If X is a dense subspace of V^* then $F(X)$ is a dense subspace of $F(V^*)$.

(b) If Y is a dense subspace of $F(V^*)$ then $F^{-1}(Y)$ is a dense subspace of V^*.

Exercise 1.3.5. Let X be a dense subspace of U^* and $u, v \in U$. Show that $u = v$ if and only if $u^*(u) = u^*(v)$ for all $u^* \in X$.

In the next four exercises $V_1, \ldots, V_n$ are non-zero vector spaces over the field k and $\imath : V_1^* \otimes \cdots \otimes V_n^* \longrightarrow (V_1 \otimes \cdots \otimes V_n)^*$ is the one-one linear map of Exercise 1.2.5.

Exercise 1.3.6. Show that $\mathrm{Im}(\imath)$ is a dense subspace of $(V_1 \otimes \cdots \otimes V_n)^*$. [Hint: Suppose that $\nu \in V_1 \otimes \cdots \otimes V_n$ is not zero and write $\nu = \sum_{i=1}^{r} u_i \otimes v_i \in V_1 \otimes (V_2 \otimes \cdots \otimes V_n)$ where $r = \mathrm{Rank}(\nu)$. Show that there is a $v_1^* \in V_1^*$ which satisfies $v_1^*(u_i) = \delta_{i,1}$ for all $1 \leq i \leq r$ and then find $v_j^* \in V_j^*$ for $2 \leq j \leq n$ such that $(v_1^* \otimes \cdots \otimes v_n^*)(\nu) = 1$.]

Exercise 1.3.7. There are important ways in practice of distinguishing tensors. Let $\nu, \nu' \in V_1 \otimes \cdots \otimes V_n$. Show that:

(a) $\nu = \nu'$ if and only if for all $1 \leq i \leq n$ the equation

$$(v_1^* \otimes \cdots \otimes v_n^*)(\nu) = (v_1^* \otimes \cdots \otimes v_n^*)(\nu')$$

holds for all $v_i^* \in V_i^*$.

(b) $\nu = \nu'$ if and only if for some $1 \leq i \leq n$ the equation

$$(\mathrm{I}_{V_1} \otimes \cdots \otimes v_i^* \otimes \cdots \otimes \mathrm{I}_{V_n})(\nu) = (\mathrm{I}_{V_1} \otimes \cdots \otimes v_i^* \otimes \cdots \otimes \mathrm{I}_{V_n})(\nu')$$

holds for all $v_i^* \in V_i^*$.

Exercise 1.3.8. Show that $\imath$ is an isomorphism if and only if at most one of $V_1, \ldots, V_n$ is not finite-dimensional. [Hint: See Exercise 1.2.5.]

Exercise 1.3.9. Suppose X_i is a dense subspace of V_i^* for all $1 \leq i \leq n$. Show that the conclusions of Exercise 1.3.7 hold when V_i^* is replaced by X_i for all $1 \leq i \leq n$. In particular $X_1 \otimes \cdots \otimes X_n$ is a dense subspace of $(V_1 \otimes \cdots \otimes V_n)^*$.

The next set of exercises deals with the algebra of cofinite subspaces of a vector space over k.

Definition 1.3.16. Let V be a subspace of a vector space U over the field k. Then $[U{:}V] = \mathrm{Dim}(U/V)$.

Exercise 1.3.10. Suppose V, W are subspaces of U. Show that:

(a) If $V \subseteq W \subseteq U$ then $[U{:}V] = [U{:}W] + [W{:}V]$.

(b) $[U{:}V \cap W] = [U{:}V] + [U{:}W] - [U{:}(V+W)]$.

(c) If U is finite-dimensional then $[U{:}V \cap W] = [U{:}V] + [U{:}W]$ if and only if $U = V + W$.

(d) $[W{:}(W \cap V)] \leq [U{:}V]$.

[Hint: For part (a) write $W_0 = V$, $W_0 \oplus W_1 = W$ and $W \oplus W_2 = U$ for some choice of subspaces W_0, W_1, W_2 of U. Show that $W_0 \oplus W_1 \oplus W_2 = U$. Note, for example, that $[W{:}V] = \mathrm{Dim}(W_1)$. For part (b), first choose subspaces $W_0, \ldots, W_3$ of U which satisfy

$$W_0 = V \cap W, \quad W_0 \oplus W_1 = V, \quad W_0 \oplus W_2 = W \text{ and } (V+W) \oplus W_3 = U.$$

Show that $W_0 \oplus \cdots \oplus W_3 = U$.]

Exercise 1.3.11. Let $f : U \longrightarrow W$ be a linear map. Show that:

(a) If V is a subspace of U then $[f(U){:}f(V)] \leq [U{:}V]$.

(b) If V is a subspace of W then $[U{:}f^{-1}(V)] \leq [W{:}V]$.

Exercise 1.3.12. Suppose that X, Y are subspaces of U^* which satisfy $X \subseteq Y$ and $[Y{:}X]$ is finite. Show that $[\overline{Y}{:}\overline{X}] \leq [Y{:}X]$. [Hint: $X \oplus Z = Y$ for some finite-dimensional subspace Z of U^*. See Proposition 1.3.8 and Corollary 1.3.9.]

Exercise 1.3.13. Suppose that V, W are subspaces of U and $V \subseteq W$. Show that if V is a cofinite subspace of U then W is a cofinite subspace of U.

Exercise 1.3.14. Let $f : U \longrightarrow W$ be linear. Show that:

(a) If V is a cofinite subspace of U then $f(V)$ is a cofinite subspace of $f(U)$.

(b) If V is a cofinite subspace of W then $f^{-1}(V)$ is a cofinite subspace of U.

Exercise 1.3.15. Show that:

(a) The sum of cofinite subspaces of U is a cofinite subspace of U.

(b) The intersection of a finite number of cofinite subspaces of U is a cofinite subspace of U.

(c) If V is a cofinite subspace of U and W is any subspace of U then $W \cap V$ is a cofinite subspace of W.

Exercise 1.3.16. Let $\beta : U \times V \longrightarrow k$ be a bilinear form. Define $\beta_\ell : U \longrightarrow V^*$ and $\beta_r : V \longrightarrow U^*$ by

$$\beta_\ell(u)(v) = \beta(u, v) = \beta_r(v)(u)$$

for all $u \in U$ and $v \in V$. Show that:

(a) β_ℓ is one-one, that is β is left non-singular, if and only if $\beta_r(V)$ is a dense subspace of U^*.

(b) β_r is one-one, that is β is right non-singular, if and only if $\beta_\ell(U)$ is a dense subspace of V^*.

(c) If $U = V$ and U is finite-dimensional then β_ℓ and β_r are both isomorphisms, that is β is non-singular, or neither is.

Exercise 1.3.17. Let $\beta : U \times V \longrightarrow k$ be a bilinear form. Show that parts (a) and (b) of Exercise 1.3.16 are equivalent. [Hint: Consider the bilinear form $\beta^{op} : V \times U \longrightarrow k$ defined by $\beta^{op}(v, u) = \beta(u, v)$ for all $v \in V$ and $u \in U$.]

Exercise 1.3.18. Suppose that $\beta : U \times V \longrightarrow k$ is a bilinear form. Then we define $\beta_{lin} : U \otimes V \longrightarrow k$ by

$$\beta_{lin}(u \otimes v) = \beta(u, v)$$

for all $u \in U$ and $v \in V$. Show that the following are equivalent:

(a) $\beta_{lin} = \sum_{i=1}^{n} u_i^* \otimes v_i^*$, where $u_i^* \in U^*$ and $v_i^* \in V^*$ for all $1 \leq i \leq n$.

(b) $\mathrm{Dim}(\mathrm{Im}(\beta_\ell))$ is finite.

(c) $\mathrm{Dim}(\mathrm{Im}(\beta_r))$ is finite.

If any of (a) – (c) hold then β *is of finite type*.

The more ambitious reader may want to consider Exercises 2.3.2, 2.4.1, and 4.1.1 at this point.

Chapter notes

The material of this chapter is an expansion of Appendix A.4 of [99]. For topological setting of Section 1.3 the reader is referred to Gillman and Jerison [60]. A calculus of cofinite subspaces in the context of algebras is found in a paper by Heyneman and the author [71].

Chapter 2

Coalgebras

Hopf algebras have both an algebra and a coalgebra structure. We will assume that the reader is familiar with the basic concepts for algebras and their representations. In this chapter we study coalgebras and in the next their representations.

Algebras and coalgebras are very closely related. Let C be a coalgebra over the field k. The linear dual C^* is an algebra over k. Coalgebra structures of C correspond to algebra structures of C^* which are *closed* subspaces of C^*. Information about C is encoded in the algebra C^*. Conversely, information about the algebra C^* is encoded in the coalgebra C. If A is an algebra over k there is a coalgebra A^o which is a subspace of the linear dual A^*. When A is finite dimensional $A^o = A^*$. The theories of finite-dimensional algebras over k and finite-dimensional coalgebras over k are one in the same.

2.1 Algebras and coalgebras, basic definitions

Conceptually, coalgebras are dual to algebras. Coalgebras are important in their own right and they have locally finite properties which algebras ordinarily do not possess. The algebras of primary interest to us in this text have a coalgebra structure also which plays a significant role in their structure theory.

The purpose of this section is to motivate the concept of coalgebra, to define coalgebra and its substructures, and to discuss a few elementary examples. One of these examples is a major player in the theory of coalgebras: the counterpart $\mathrm{C}_n(k)$ of the algebra $\mathrm{M}_n(k)$ of $n\times n$ matrices over k.

Formulation of the axioms for an associative algebra over k in terms of commutative diagrams, which express the equality of certain function compositions, suggests how to define a coalgebra. An associative algebra over k is a triple (A, m, η), where A is a vector space over the field k and $m : A\otimes A \longrightarrow A$, $\eta : k \longrightarrow A$ are linear maps, such that the compositions

$$A\otimes A\otimes A \simeq (A\otimes A)\otimes A \stackrel{m\otimes \mathrm{I}_A}{\longrightarrow} A\otimes A \stackrel{m}{\longrightarrow} A$$

and

$$A\otimes A\otimes A \simeq A\otimes (A\otimes A) \stackrel{\mathrm{I}_A\otimes m}{\longrightarrow} A\otimes A \stackrel{m}{\longrightarrow} A$$

are equal, and the compositions

$$A \simeq k\otimes A \stackrel{\eta\otimes \mathrm{I}_A}{\longrightarrow} A\otimes A \stackrel{m}{\longrightarrow} A$$

and

$$A \simeq A\otimes k \stackrel{\mathrm{I}_A\otimes \eta}{\longrightarrow} A\otimes A \stackrel{m}{\longrightarrow} A$$

are both the identity map I_A. Taking the above isomorphisms for granted the axioms can be expressed as

$$m\circ(m\otimes \mathrm{I}_A) = m\circ(\mathrm{I}_A\otimes m) \tag{2.1}$$

and

$$m\circ(\eta\otimes \mathrm{I}_A) = \mathrm{I}_A = m\circ(\mathrm{I}_A\otimes \eta). \tag{2.2}$$

It is customary to refer to (2.1) as the associative axiom and to (2.2) as the unit axiom.

The axioms (2.1) and (2.2) translate to very familiar equations when expressed in terms of elements. Write $m(a\otimes b) = ab$ for all $a, b \in A$ and set $\eta(1_k) = 1$. Then (2.1) is equivalent to $(ab)c = a(bc)$ for all $a, b, c \in A$ and (2.2) is equivalent to $1a = a = a1$ for all $a \in A$. These axioms are equivalent to the diagrams

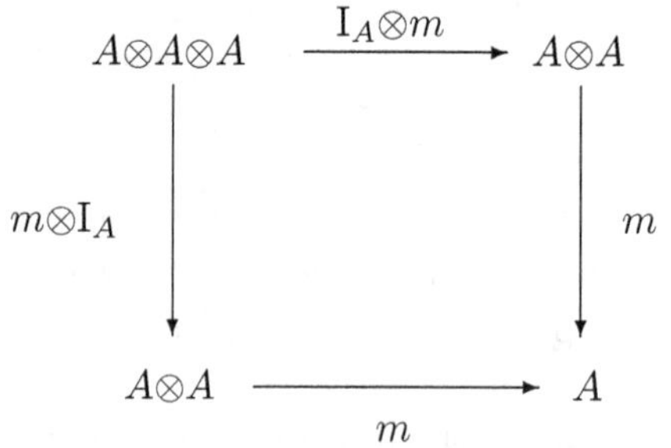

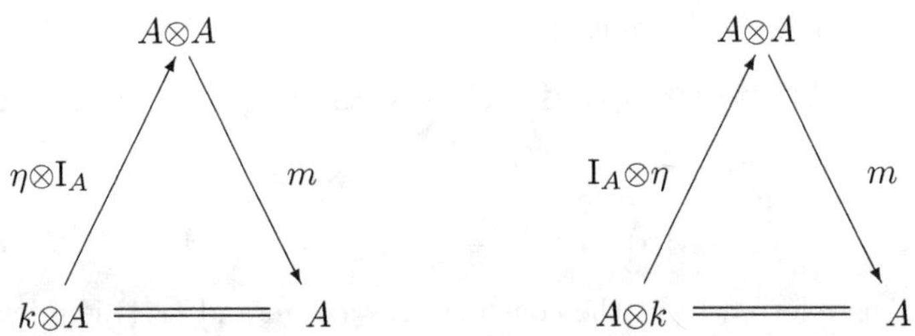

commute, where "equals" is identification by scalar multiplication. Throughout this text all algebras over k, which we refer to as k-algebras as well, are associative unless otherwise stated.

Let (A, m, η) be an algebra over k and write $m^{op}(a\otimes b) = a\bullet b$ for all $a, b \in A$, where $m^{op} = m\circ\tau_{A,A}$. Thus $a\bullet b = ba$ for all $a, b \in A$. It is easy to see that (A, m^{op}, η) is an algebra over k.

Definition 2.1.1. Let (A, m, η) be an algebra over k. The algebra (A, m^{op}, η) is the *opposite algebra.*

By definition $ab = ba$ for all $a, b \in A$ if and only if $ab = a\bullet b$ for all $a, b \in A$.

Definition 2.1.2. A *commutative algebra over* k is an algebra (A, m, η) over k such that $m = m^{op}$.

We most often denote an algebra (A, m, η) over k by A and denote (A, m^{op}, η) by A^{op}. Thus A is commutative if and only if $A = A^{op}$.

The notion of coalgebra over k is derived by replacing A with a vector space C over k, replacing the linear maps $m : A\otimes A \longrightarrow A$ and $\eta : k \longrightarrow A$ with linear maps $\Delta : C \longrightarrow C\otimes C$ and $\epsilon : C \longrightarrow k$ respectively, requiring the compositions

$$C \xrightarrow{\Delta} C\otimes C \xrightarrow{\Delta\otimes \mathrm{I}_C} (C\otimes C)\otimes C \simeq C\otimes C\otimes C$$

and

$$C \xrightarrow{\Delta} C\otimes C \xrightarrow{\mathrm{I}_C\otimes\Delta} C\otimes(C\otimes C) \simeq C\otimes C\otimes C$$

to be the same, and requiring

$$C \xrightarrow{\Delta} C\otimes C \xrightarrow{\epsilon\otimes \mathrm{I}_C} k\otimes C \simeq C$$

and

$$C \xrightarrow{\Delta} C\otimes C \xrightarrow{\mathrm{I}_C\otimes\epsilon} C\otimes k \simeq C$$

to be the identity map I_C. Taking the above isomorphisms for granted, in terms of equations we are requiring

$$(\Delta \otimes I_C) \circ \Delta = (I_C \otimes \Delta) \circ \Delta \tag{2.3}$$

and

$$(\epsilon \otimes I_C) \circ \Delta = I_C = (I_C \otimes \epsilon) \circ \Delta \tag{2.4}$$

Usually (2.3) is referred to as the coassociative axiom and (2.4) is referred to as the counit axiom. In terms of diagrams, the coalgebra axioms are obtained by "dualizing" the diagrams describing an associative algebra, that is formally reversing the arrows and renaming the maps. Thus (C, Δ, ϵ) is a coalgebra over k if and only if the diagrams

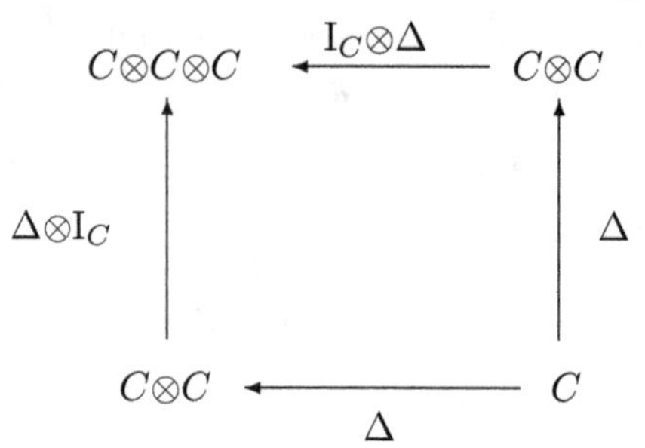

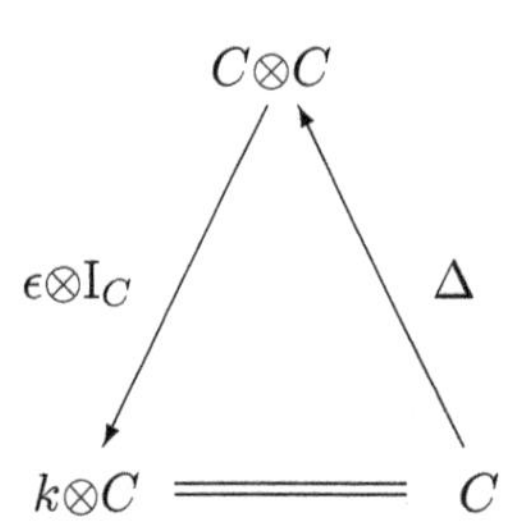

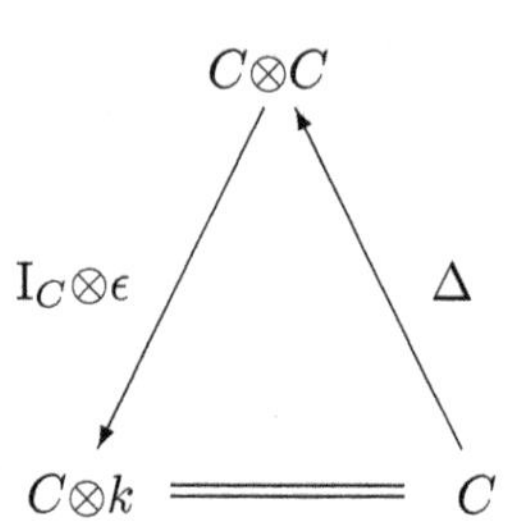

commute, where again "equals" is identification by scalar multiplication.

Definition 2.1.3. A *coalgebra over the field* k is a triple (C, Δ, ϵ), where C is a vector space over k and $\Delta : C \longrightarrow C \otimes C$, $\epsilon : C \longrightarrow k$ are linear maps which satisfy (2.3) and (2.4).

Thus coalgebras are objects which are formally dual to algebras, or vice versa. From time to time we refer to a coalgebra over k as a k-coalgebra.

Let (C, Δ, ϵ) be a coalgebra over k and set $\Delta^{cop} = \tau_{C,C} \circ \Delta$. The reader can check directly from definitions that $(C, \Delta^{cop}, \epsilon)$ is a coalgebra over k.

Definition 2.1.4. Let (C, Δ, ϵ) be a coalgebra over k. The coalgebra $(C, \Delta^{cop}, \epsilon)$ is the *opposite coalgebra.*

The notion dual to commutative is cocommutative.

Definition 2.1.5. A *cocommutative coalgebra over* k is a coalgebra (C, Δ, ϵ) over k such that $\Delta = \Delta^{cop}$.

We often denote a coalgebra (C, Δ, ϵ) over k by C and denote $(C, \Delta^{cop}, \epsilon)$ by C^{cop}. In particular C is cocommutative if and only if $C = C^{cop}$. The map Δ is referred to as comultiplication, or the coproduct, and the map ϵ is referred to as the counit.

Let $c \in C$. In practice it is sufficient to denote $\Delta(c) \in C \otimes C$ by a *symbolic* sum. A widely used notation is the variation of the Heyenman-Sweedler notation $\Delta(c) = \sum c_{(1)} \otimes c_{(2)}$. We drop the summation symbol and write

$$\Delta(c) = c_{(1)} \otimes c_{(2)}.$$

In this notation (2.3) is expressed

$$c_{(1)(1)} \otimes c_{(1)(2)} \otimes c_{(2)} = c_{(1)} \otimes c_{(2)(1)} \otimes c_{(2)(2)} \tag{2.5}$$

and (2.4) is expressed

$$\epsilon(c_{(1)})c_{(2)} = c = c_{(1)}\epsilon(c_{(2)}) \tag{2.6}$$

for all $c \in C$.

The coassociative axiom (2.3) has implications for iterated applications of the coproduct. For positive integers i and j which satisfy $j \leq i$ let $\Delta^{(i,j)} = \mathrm{I}_C \otimes \cdots \otimes \Delta \otimes \cdots \otimes \mathrm{I}_C$ be the tensor product of i linear operators, all of which are the identity map of C except for the one in the jth position which is Δ. Thus

$$\Delta^{(1,1)} = \Delta, \quad \Delta^{(2,1)} = \Delta \otimes \mathrm{I}_C, \quad \Delta^{(2,2)} = \mathrm{I}_C \otimes \Delta, \quad \text{and } \Delta^{(3,2)} = \mathrm{I}_C \otimes \Delta \otimes \mathrm{I}_C$$

for example. It is a very good exercise to show the coassociative axiom implies all of the composites $\Delta^{(n,j_n)} \circ \Delta^{(n-1,j_{n-1})} \circ \cdots \circ \Delta^{(1,1)}$ are the same for fixed $n > 1$ and all $1 \leq j_i \leq i \leq n$. Let $\Delta^{(n)}$ denote their common value. We extend our notation for $\Delta(c)$ and write

$$\Delta^{(n-1)}(c) = c_{(1)} \otimes c_{(2)} \otimes \cdots \otimes c_{(n)}$$

for all $c \in C$.

Subalgebras, left ideals, and right ideals of an algebra have dual counterparts in coalgebras. Let (A, m, η) be an algebra over k and suppose that V is a left ideal of A. Then $m(A \otimes V) \subseteq V$. Thus the restriction of m to $A \otimes V$ determines a map $A \otimes V \longrightarrow V$. A reasonable definition of left coideal of a coalgebra C, therefore, is a subspace V of C such that the coproduct Δ restricts to a map $V \longrightarrow C \otimes V$.

Definition 2.1.6. Let V be a subspace of a coalgebra C over k. Then V is a *subcoalgebra* (respectively *left coideal, right coideal*) *of* C if $\Delta(V) \subseteq V \otimes V$ (respectively $\Delta(V) \subseteq C \otimes V$, $\Delta(V) \subseteq V \otimes C$).

Let C be a coalgebra over k. By definition C and (0) are subcoalgebras, left coideals, and right coideals of C. By Exercise 2.1.6 the intersection of a family of subcoalgebras (respectively left coideals, right coideals) of C is again a subcoalgebra (respectively left coideal, right coideal) of C. Thus any subspace V of C is contained in a unique minimal subcoalgebra, left coideal, and right coideal of C.

Definition 2.1.7. Let V be a subspace of a coalgebra C over k. The unique minimal subcoalgebra (respectively left coideal, right coideal) of C which contains V is the *subcoalgebra* (respectively *left coideal, right coideal*) *of* C *generated by* V.

One more definition before we present examples.

Definition 2.1.8. A *simple coalgebra* (respectively *left coideal, right coideal*) is a coalgebra (respectively left coideal, right coideal) which has exactly two subcoalgebras (respectively left coideals, right coideals).

In particular C is a simple coalgebra over k if and only if $C \neq (0)$ and if D is a subcoalgebra of C then $D = (0)$ or $D = C$.

The ground field k has a unique k-coalgebra structure which is given by $\Delta(1) = 1 \otimes 1$ and $\epsilon(1) = 1$. This example has an immediate and important generalization.

Example 2.1.9. Let S be a set. Then the free k-module $C = k[S]$ on S has a coalgebra structure determined by

$$\Delta(s) = s \otimes s \quad \text{and} \quad \epsilon(s) = 1 \tag{2.7}$$

for all $s \in S$. If $S = \emptyset$ we set $C = k[\emptyset] = (0)$.

Definition 2.1.10. Let C be a coalgebra over k. A *grouplike element of* C is a $c \in C$ which satisfies (2.7). The set of grouplike elements of C is denoted $\mathrm{G}(C)$.

Definition 2.1.11. Let S be a set. The coalgebra $k[S]$ of Example 2.1.9 is the *grouplike coalgebra of* S *over* k.

Let C be a coalgebra over k. By virtue of the following lemma the grouplike coalgebra $k[\mathrm{G}(C)]$ is always a subcoalgebra of C.

Lemma 2.1.12. *The set of grouplike elements* $\mathrm{G}(C)$ *of a coalgebra* C *over the field* k *is linearly independent.*

Proof. Suppose that $\mathrm{G}(C)$ is not linearly independent. Then there is a dependency relation $\alpha_1 s_1 + \cdots + \alpha_r s_r = 0$ among some distinct r elements $s_1, \ldots, s_r \in \mathrm{G}(C)$, where $\alpha_i \in k$ for $1 \leq i \leq r$. We may assume that r is as small as possible. Thus $\alpha_i \neq 0$ for all $1 \leq i \leq r$. Since $0 \notin \mathrm{G}(C)$, it follows that $r > 1$ and $\{s_1, \ldots, s_{r-1}\}$ is linearly independent.

We may write $s_r = \beta_1 s_1 + \cdots + \beta_{r-1} s_{r-1}$, where $\beta_i \in k \backslash 0$ for all $1 \leq i < r$. Now $\mathrm{Rank}(\Delta(s_r)) = 1$ since $\Delta(s_r) = s_r \otimes s_r$. On the other hand $\Delta(s_r) = \beta_1 s_1 \otimes s_1 + \cdots + \beta_{r-1} s_{r-1} \otimes s_{r-1}$ which means that $\mathrm{Rank}(\Delta(s_r)) = r - 1$ by Lemma 1.2.2. Therefore $r - 1 = 1$ and hence $s_2 = \beta_1 s_1$. Since $\epsilon(s_1) = 1 = \epsilon(s_2)$ we conclude that $\beta_1 = 1$ and consequently $s_2 = s_1$. Since $s_1, \ldots, s_r$ are distinct we have a contradiction. Therefore $\mathrm{G}(C)$ is linearly independent after all. □

Let $C = k[S]$ be the grouplike coalgebra on a set S over k. Then

$$\mathrm{G}(k[S]) = S \tag{2.8}$$

by Lemma 2.1.12. Thus the set S is recovered from the coalgebra structure of $k[S]$. Let C be any coalgebra over k. If $\mathrm{G}(C)$ spans C then $C = k[\mathrm{G}(C)]$ is the grouplike coalgebra on $\mathrm{G}(C)$ over k by virtue of the lemma also.

There may be other ways of putting a coalgebra structure on the free k-module $k[S]$ when S has algebraic structure.

Example 2.1.13. Let S be a (multiplicative) monoid with neutral element e. Suppose that all $s \in S$ have only finitely many factorizations $s = ab$ where $a, b \in S$. Then the free k-module $C = k[S]$ on S has a coalgebra structure determined by

$$\Delta(s) = \sum_{ab=s} a \otimes b \quad \text{and} \quad \epsilon(s) = \delta_{s,e}$$

for all $s \in S$.

The factorization requirement of Example 2.1.13 is met when S is a finite monoid or $S = \boldsymbol{N}$ is the set of non-negative integers under addition. The latter is a basic example and is usually described in another notation. The basis $\boldsymbol{N}$ is denoted $\{c_0, c_1, c_2, \dots\}$. Thus

$$\Delta(c_n) = \sum_{\ell=1}^{n} c_{n-\ell} \otimes c_\ell \quad \text{and} \quad \epsilon(c_n) = \delta_{n,0} \tag{2.9}$$

for all $n \geq 0$. For each $n \geq 0$ observe that the span of $c_0, \dots, c_n$ is a subcoalgebra of $C = k[S]$.

Definition 2.1.14. The coalgebra C over k with basis $\{c_0, c_1, c_2, \dots\}$ whose coproduct and counit satisfy (2.9) is denoted $P_\infty(k)$. The subcoalgebra which is the span of $c_0, \dots, c_n$ is denoted $P_n(k)$.

Example 2.1.15. Let S be a finite non-empty set. Then the free k-module $C = k[S{\times}S]$ on the Cartesian product $S{\times}S$ has a coalgebra structure determined by

$$\Delta((i,j)) = \sum_{\ell \in S} (i,\ell) \otimes (\ell, j) \quad \text{and} \quad \epsilon((i,j)) = \delta_{i,j}$$

for all $(i,j) \in S{\times}S$.

Let $\mathrm{C}_S(k)$ denote the coalgebra of Example 2.1.15 and write $e_{i,j}$ for $(i,j) \in S{\times}S$. Then

$$\Delta(e_{i,j}) = \sum_{\ell \in S} e_{i,\ell} \otimes e_{\ell,j} \tag{2.10}$$

and

$$\epsilon(e_{i,j}) = \delta_{i,j} \tag{2.11}$$

for all $i,j \in S$. Set $C_\emptyset(k) = (0)$.

Definition 2.1.16. A *comatrix coalgebra over* k is a coalgebra over k isomorphic to $\mathrm{C}_S(k)$ for some finite set S. The *comatrix identities* are equations (2.10) and (2.11).

Definition 2.1.17. Let S be a non-empty finite set. A *standard basis for* $\mathrm{C}_S(k)$ is a basis $\{c_{i,j}\}_{i,j \in S}$ for $\mathrm{C}_S(k)$ which satisfies the comatrix identities.

Thus the set $\{e_{i,j}\}_{i,j \in S}$ described above is a standard basis for $\mathrm{C}_S(k)$.

When $S = \{1, \dots, n\}$ we write $\mathrm{C}_n(k)$ for $\mathrm{C}_S(k)$. In this case the comatrix identities are expressed as

$$\Delta(e_{i,j}) = \sum_{\ell=1}^{n} e_{i,\ell} \otimes e_{\ell,j} \quad \text{and} \quad \epsilon(e_{i,j}) = \delta_{i,j}$$

for all $1 \le i, j \le n$.

We next consider the notion of coalgebra map, taking our cue from the description of algebra maps in terms of compositions. Let (A, m_A, η_A) and (B, m_B, η_B) be algebras over the field k. An algebra map $f : A \longrightarrow B$ is a linear map of underlying vector spaces such that the compositions

$$A \otimes A \xrightarrow{m_A} A \xrightarrow{f} B$$

and

$$A \otimes A \xrightarrow{f \otimes f} B \otimes B \xrightarrow{m_B} B$$

are equal and the composite

$$k \xrightarrow{\eta_A} A \xrightarrow{f} B$$

is

$$k \xrightarrow{\eta_B} B.$$

These conditions translate to $f : A \longrightarrow B$ is an algebra map if and only if $f(ab) = f(a)f(b)$ for all $a, b \in A$ and $f(1) = 1$. These conditions are equivalent to the diagrams

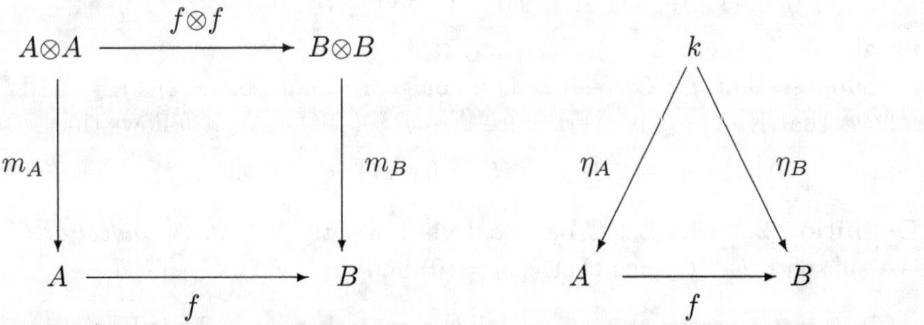

commute.

The dual notion of algebra map is a linear map $f : C \longrightarrow D$ of coalgebras such that the compositions

$$C \xrightarrow{f} D \xrightarrow{\Delta_D} D \otimes D$$

and

$$C \xrightarrow{\Delta_C} C \otimes C \xrightarrow{f \otimes f} D \otimes D$$

are equal and the composite

$$C \xrightarrow{f} D \xrightarrow{\epsilon_D} k$$

is

$$C \xrightarrow{\epsilon_C} k\,;$$

that is the diagrams

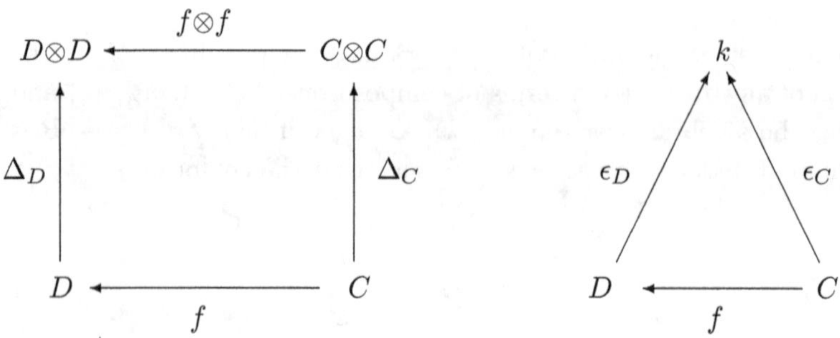

commute.

Definition 2.1.18. Let $(C, \Delta_C, \epsilon_C)$ and $(D, \Delta_D, \epsilon_D)$ be coalgebras over the field k. A *coalgebra map* $f : C \longrightarrow D$ is a linear map of underlying vector spaces such that $\Delta_D \circ f = (f \otimes f) \circ \Delta_C$ and $\epsilon_D \circ f = \epsilon_C$. An *isomorphism of coalgebras* is a coalgebra map which is a linear isomorphism.

In terms of our notation for the coproduct a linear map $f : C \longrightarrow D$ is a coalgebra map if and only if

$$\Delta_D(f(c)) = f(c_{(1)}) \otimes f(c_{(2)}) \quad \text{and} \quad \epsilon_D(f(c)) = \epsilon_C(c)$$

for all $c \in C$, where $\Delta_C(c) = c_{(1)} \otimes c_{(2)}$.

Suppose that $f : C \longrightarrow D$ is a coalgebra map. Since $\epsilon_D \circ f = \epsilon_C$ it follows that $\mathrm{Ker}(f) \subseteq \mathrm{Ker}(\epsilon)$. Since $\Delta_D \circ f = (f \otimes f) \circ \Delta_C$ it follows that

$$\Delta_C(\mathrm{Ker}(f)) \subseteq \mathrm{Ker}(f \otimes f) = \mathrm{Ker}(f) \otimes C + C \otimes \mathrm{Ker}(f).$$

Definition 2.1.19. Let C be a coalgebra over the field k. A *coideal of C* is a subspace I of C such that $\epsilon(I) = (0)$ and $\Delta(I) \subseteq I \otimes C + C \otimes I$.

It is easy to see that $\epsilon : C \longrightarrow k$ is a coalgebra map. Thus $\mathrm{Ker}(\epsilon)$ is a coideal of C.

Definition 2.1.20. The coideal $\mathrm{Ker}(\epsilon)$ of a coalgebra C over k is denoted C^+.

By definition kernels of coalgebra maps are coideals. The converse is true as well by the fundamental homomorphism theorem for coalgebras, the last result of this section. The reader is left with the exercise of proving it.

Theorem 2.1.21. *Suppose that C is a coalgebra over the field k.*

(a) *Let I be a coideal of C. Then the quotient space C/I has a unique coalgebra structure so that the linear projection $\pi : C \longrightarrow C/I$ is a coalgebra map.*

Suppose D is also a coalgebra over k and that $f : C \longrightarrow D$ is a coalgebra map. Then:

(b) $\mathrm{Ker}(f)$ *is a coideal of C, and if I is a coideal of C such that $I \subseteq \mathrm{Ker}(f)$ then there is a coalgebra map $F : C/I \longrightarrow D$ determined by $F \circ \pi = f$.*

(c) *Suppose that f is onto and let $I = \mathrm{Ker}(f)$. Then there is an isomorphism $F : C/I \longrightarrow D$ of coalgebras determined by $F \circ \pi = f$.*

□

Definition 2.1.22. Let I be a coideal of a coalgebra C over k. The unique coalgebra structure on C/I such that the projection $\pi : C \longrightarrow C/I$ is a coalgebra map is the *quotient coalgebra structure on C/I.*

Exercises

In the following exercises C, D, and E are coalgebras over the field k.

Exercise 2.1.1. Let $C = k[S]$ be the grouplike coalgebra on a non-empty set S. Show that a subcoalgebra D of C has the form $D = k[T]$, where T is a subset of S. [Hint: Suppose $d = \sum_{i=1}^{r} \alpha_i s_i \in D$, where $s_1, \ldots, s_r \in S$ are distinct and $\alpha_1, \ldots, \alpha_r \in k$. Show that $\alpha_1 s_1, \ldots, \alpha_r s_r$ are among the elements $c^* \rightharpoonup d = \sum_{i=1}^{r} \alpha_i c^*(s_i) s_i$, where $c^* \in C^*$.]

Exercise 2.1.2. Let $c \in C$. Show that $c \in \mathrm{G}(C)$ if and only if $\Delta(c) = c \otimes c$ and $c \neq 0$.

Exercise 2.1.3. Suppose U, V and W are subspaces of C. Show that $\Delta(U) \subseteq V \otimes W$ implies $U \subseteq V \cap W$.

Exercise 2.1.4. Let V be a subspace of C. Show that V is a subcoalgebra of C if and only if V is a left coideal and a right coideal of C. [Hint: See Exercise 1.2.8.]

Exercise 2.1.5. Suppose that C is cocommutative. Show that the left coideals and the right coideals of C are subcoalgebras of C. [Hint: See Exercise 2.1.4.]

Exercise 2.1.6. Show the following:

(a) The sum of any family of subcoalgebras (respectively left coideals, right coideals) of C is a subcoalgebra (respectively left coideal, right coideal) of C.

(b) The intersection of any family of subcoalgebras (respectively left coideals, right coideals) of C is a subcoalgebra (respectively left coideal, right coideal) of C. [Hint: See Exercise 1.2.8.]

Exercise 2.1.7. Interpreting results in C^{cop} derived for all coalgebras can simplify proofs or yield new results for C.

(a) Show that the left (respectively right) coideals of C^{cop} are the right (respectively left) coideals of C.

(b) In light of part (a) and Exercise 2.1.4, show parts (a) and (b) of Exercise 2.1.6 are established once they are established for left (or right) coideals for C.

Exercise 2.1.8. Let D be a subcoalgebra of C and suppose that U is a subspace of C. If U is a coideal (respectively a subcoalgebra, left coideal, right coideal) of C, show that $D\cap U$ is a coideal (respectively a subcoalgebra, left coideal, right coideal) of D. [Hint: Suppose that U is a coideal of C. Then the linear map $\pi : D \longrightarrow C/U$ defined by $\pi(d) = d + U$ for all $d \in D$ is a coalgebra map and $\mathrm{Ker}(\pi) = D\cap U$. Also, see Exercise 1.2.8.]

Exercise 2.1.9. Prove Theorem 2.1.21.

Exercise 2.1.10. This exercise and the next are designed to help the reader understand compositions represented by the coproduct notation and to give the reader practice in using this notation. For parts (a) – (c) below do the following: (1) write out the composition represented by each symbolic sum and use axioms (2.3) and (2.4) to verify the equation, and (2) use formulas (2.5) and (2.6) to verify for $c \in C$ the following equations:

(a) $\epsilon(c_{(1)})\epsilon(c_{(2)}) = \epsilon(c)$.

(b) $c_{(1)}\otimes c_{(2)(1)}\epsilon(c_{(2)(2)(1)})\otimes c_{(2)(2)(2)} = c_{(1)(1)}\otimes c_{(1)(2)}\otimes c_{(2)}$.

(c) $c_{(1)(1)}\otimes c_{(2)(1)}\otimes c_{(1)(2)}\otimes c_{(2)(2)} = c_{(1)}\otimes c_{(2)(2)(1)}\otimes c_{(2)(1)}\otimes c_{(2)(2)(2)}$.

[Hint: For the left-hand side of part (b) we make the calculation

$$c \xrightarrow{\Delta} c_{(1)}\otimes c_{(2)} \xrightarrow{\mathrm{I}_C\otimes\Delta} c_{(1)}\otimes c_{(2)(1)}\otimes c_{(2)(2)} \xrightarrow{\mathrm{I}_C\otimes\mathrm{I}_C\otimes\Delta} c_{(1)}\otimes c_{(2)(1)}\otimes c_{(2)(2)(1)}\otimes c_{(2)(2)(2)}$$

and conclude $((\mathrm{I}_C\otimes\mathrm{I}_C\otimes\Delta)\circ(\mathrm{I}_C\otimes\Delta)\circ\Delta)(c) = c_{(1)}\otimes c_{(2)(1)}\otimes c_{(2)(2)(1)}\otimes c_{(2)(2)(2)}$.]

Exercise 2.1.11. Show that

$$\begin{aligned} c_{(1)}\otimes c_{(4)}\otimes c_{(2)}\otimes c_{(3)} &= c_{(1)(1)}\otimes c_{(2)(2)}\otimes c_{(1)(2)}\otimes c_{(2)(1)} \\ &= c_{(1)(1)}\otimes c_{(3)}\otimes c_{(1)(2)}\otimes c_{(2)} \\ &= c_{(1)}\otimes c_{(3)(2)}\otimes c_{(2)}\otimes c_{(3)(1)} \\ &= c_{(1)}\otimes c_{(2)(2)(2)}\otimes c_{(2)(1)}\otimes c_{(2)(2)(1)} \end{aligned}$$

for all $c \in C$. The subscripts have a lexicographical ordering when treated as formal words, reading left to right. Is there a connection between this ordering and the equations?

Exercise 2.1.12. Show that all ways of computing $\Delta^{(n)}$ produce the same result.

Exercise 2.1.13. Suppose that C is cocommutative and let $n \geq 2$. Show that

$$c_{(1)}\otimes\cdots\otimes c_{(n)} = c_{(\sigma(1))}\otimes\cdots\otimes c_{(\sigma(n))}$$

for all $c \in C$ and permutations $\sigma \in S_n$.

Exercise 2.1.14. Suppose that V is a left coideal, right coideal, or a subcoalgebra of C. Show that $\epsilon(V) = (0)$ if and only if $V = (0)$.

Exercise 2.1.15. Show that $C^+ = C$ if and only if $C = (0)$.

Exercise 2.1.16. Suppose that $f : C \longrightarrow D$ is a map of coalgebras over k. Show that:

(a) $f(C^+) = f(C)^+$.

(b) If E is a subcoalgebra of C then $f(E) = (0)$ if and only if $E = (0)$.

Exercise 2.1.17. Show that the sum of coideals of C is a coideal of C.

Exercise 2.1.18. Let $\{(C_i, \Delta_i, \epsilon_i)\}_{i\in I}$ be an indexed family of coalgebras over k. Show that:

(a) The vector space direct sum $\mathcal{C} = \bigoplus_{i\in I} C_i$ has a coalgebra structure $(\mathcal{C}, \Delta, \epsilon)$ determined by $\Delta|C_i = \Delta_i$ and $\epsilon|C_i = \epsilon_i$ for all $i \in I$.

(b) If $C = k[S]$ is the grouplike coalgebra on a non-empty set S then $C = \bigoplus_{s\in S} k_s$, where $k_s \simeq k$ for all $s \in S$.

Exercise 2.1.19. The tensor product of coalgebras has a natural coalgebra structure. Show that:

(a) The tensor product of vector spaces $C\otimes D$ is a coalgebra over k where

$$\Delta(c\otimes d) = (c_{(1)}\otimes d_{(1)})\otimes(c_{(2)}\otimes d_{(2)}) \quad \text{and} \quad \epsilon(c\otimes d) = \epsilon(c)\epsilon(d)$$

for all $c \in C$ and $d \in D$.

(b) $(C\otimes D)^{cop} = C^{cop}\otimes D^{cop}$.

(c) $\pi_C : C\otimes D \longrightarrow C$ and $\pi_D : C\otimes D \longrightarrow D$ defined by $\pi_C(c\otimes d) = c\epsilon(d)$ and $\pi_D(c\otimes d) = \epsilon(c)d$ for all $c \in C$ and $d \in D$ respectively are coalgebra maps.

(d) If $f : C \longrightarrow C'$ and $g : D \longrightarrow D'$ are coalgebra maps then the tensor product of maps $f\otimes g : C\otimes D \longrightarrow C'\otimes D'$ is a coalgebra map.

Definition 2.1.23. The coalgebra structure of part (a) of Exercise 2.1.19 is the *tensor product coalgebra structure on* $C\otimes D$.

Exercise 2.1.20. Show that the linear isomorphisms $(C\otimes D)\otimes E \simeq C\otimes(D\otimes E)$ and $C\otimes D \simeq D\otimes C$ described by $(c\otimes d)\otimes e \mapsto c\otimes(d\otimes e)$, and $c\otimes d \mapsto d\otimes c$ for all $c \in C$, $d \in D$, and $e \in E$ respectively are in fact coalgebra isomorphisms.

Exercise 2.1.21. Show that $\mathrm{G}(C\otimes D) = \{c\otimes d \,|\, c \in \mathrm{G}(C), d \in \mathrm{G}(D)\}$. [Hint: Let $g \in \mathrm{G}(C\otimes D)$. Then $g \neq 0$. Write $g = \sum_{i=1}^{r} c_i\otimes d_i \in C\otimes D$ where $r = \mathrm{Rank}(g)$. Show that $\sum_{i=1}^{r}(\Delta(c_i))\otimes(\Delta(d_i)) = \sum_{i,j=1}^{r}(c_i\otimes c_j)\otimes(d_i\otimes d_j)$. The left-hand sum has rank at most r whereas the right-hand sum has rank r^2 by Lemma 1.2.2.]

Exercise 2.1.22. Algebraic structures on finite-dimensional vector spaces over k are sometimes described in terms of coordinates with respect to certain bases.

Assume that C is finite-dimensional with basis $\{c_i\}_{i\in I}$. Write

$$\Delta(c_i) = \sum_{j,\ell\in I} \Delta_i^{j,\ell} c_j\otimes c_\ell \quad \text{and} \quad \epsilon(c_i) = \epsilon_i$$

for all $i \in I$, where $\Delta_i^{j,\ell}, \epsilon_i \in k$. The scalars $\Delta_i^{j,\ell}$ and ϵ_i are examples of *structure constants.*

We will write

$$\Delta(c_i) = \Delta_i^{j,\ell} c_j\otimes c_\ell$$

following the Einstein summation convention. Show that:

(a) $(\Delta\otimes \mathrm{I}_C)\circ\Delta = (\mathrm{I}_C\otimes\Delta)\circ\Delta$ is equivalent to

$$\Delta_i^{u,t}\Delta_u^{r,s} = \Delta_i^{r,u}\Delta_u^{s,t}$$

for all $i, r, s, t \in I$.

(b) $(\epsilon\otimes \mathrm{I}_C)\circ\Delta = \mathrm{I}_C = (\mathrm{I}_C\otimes\epsilon)\circ\Delta$ is equivalent to

$$\epsilon_u\Delta_i^{j,u} = \delta_i^j = \Delta_i^{u,j}\epsilon_u$$

for all $i, j \in I$.

(c) C is cocommutative is equivalent to $\Delta_i^{j,\ell} = \Delta_i^{\ell,j}$ for all $i, j, \ell \in I$.

Definition 2.1.24. Let C be a coalgebra over k. A *skew primitive* (more precisely *g:h-skew primitive*) *element of* C is a $c \in C$ which satisfies $\Delta(c) = g\otimes c + c\otimes h$, where $g, h \in \mathrm{G}(C)$. The set of g:h-skew primitive elements of C is denoted by $\mathrm{P}_{g,h}(C)$.

Observe that $\mathrm{P}_{g,h}(C)$ is a subspace of C.

Exercise 2.1.23. Let $g, h \in \mathrm{G}(C)$. Show that:

(a) $\epsilon(c) = 0$ for all $c \in \mathrm{P}_{g,h}(C)$.

(b) If U is a subspace of $\mathrm{P}_{g,h}(C)$ then $D = kg + kh + U$ is a subcoalgebra of C.

Exercise 2.1.24. Let $f : C \longrightarrow D$ be a coalgebra map. Show that:

(a) $f(\mathrm{G}(C)) \subseteq \mathrm{G}(D)$.

(b) $f(\mathrm{P}_{g,h}(C)) \subseteq \mathrm{P}_{f(g),f(h)}(D)$ for all $g, h \in \mathrm{G}(C)$.

Exercise 2.1.25. Show that subcoalgebras, quotients, and tensor products of grouplike coalgebras over k are grouplike coalgebras over k.

Exercise 2.1.26. Let $C = k[S]$ be the grouplike coalgebra of a non-empty set S over k. Show that:

(a) The difference $s - s'$, where $s, s' \in S$, spans a coideal of C.

(b) Every coideal of C is spanned by differences $s - s'$ for certain $s, s' \in S$.

Exercise 2.1.27. Suppose that S and T are sets. Show that the linear isomorphism $k[S{\times}T] \simeq k[S]{\otimes}k[T]$ determined by $(s,t) \mapsto s{\otimes}t$ for all $s \in S$ and $t \in T$ is a coalgebra isomorphism, where $k[S]$, $k[T]$, and $k[S{\times}T]$ are the grouplike coalgebras of Example 2.1.13 and the tensor product has the tensor product coalgebra structure.

Exercise 2.1.28. Let $f : C \longrightarrow D$ be a coalgebra map. Show that:

(a) If U is a subcoalgebra (respectively left coideal, right coideal, coideal) of C then $f(U)$ is a subcoalgebra (respectively left coideal, right coideal, coideal) of D.

(b) If V is a subcoalgebra (respectively left coideal, right coideal) of D then $f^{-1}(V)$ is not necessarily a subcoalgebra (respectively left coideal, right coideal) of C.

(c) If V is a coideal of D then $f^{-1}(V)$ is a coideal of C.

[Hint: For part (b) the reader is referred to Exercise 2.1.14. For part (c) consider the projection $\pi_V : D \longrightarrow D/V$. Show that

$$(\pi_V{\otimes}\pi_V){\circ}(f{\otimes}f){\circ}\Delta(f^{-1}(V)) = (0)$$

and thus

$$\begin{aligned}\Delta(f^{-1}(V)) &\subseteq \mathrm{Ker}((\pi_V{\circ}f){\otimes}(\pi_V{\circ}f))\\ &= \mathrm{Ker}(\pi_V{\circ}f){\otimes}C + C{\otimes}\mathrm{Ker}(\pi_V{\circ}f)\\ &= f^{-1}(V){\otimes}C + C{\otimes}f^{-1}(V).]\end{aligned}$$

Exercise 2.1.29. Let $f, g : C \longrightarrow D$ be coalgebra maps. Show that $\mathrm{Im}(f-g)$ is a coideal of D. [Hint: Derive the formula $\Delta{\circ}(f-g) = ((f-g){\otimes}g + f{\otimes}(f-g)){\circ}\Delta$.]

Definition 2.1.25. Let C be a coalgebra over the field k. A *cocommutative element of* C is a $c \in C$ such that $\Delta(c) = \Delta^{cop}(c)$. The set of cocommutative elements of C is denoted by $\mathrm{Cc}(C)$.

Notice that $\mathrm{Cc}(C)$ is a subspace of C.

Exercise 2.1.30. Show that $\mathrm{Dim}(\mathrm{Cc}(\mathrm{C}_n(k))) = 1$ for all $n \geq 1$.

Exercise 2.1.31. Let $f : C \longrightarrow D$ be a coalgebra map. Show that:

(a) $f(\mathrm{Cc}(C)) \subseteq \mathrm{Cc}(D)$.

(b) It can be the case that f is onto and $f(\mathrm{Cc}(C)) \neq \mathrm{Cc}(D)$.

Exercise 2.1.32. Show that:

(a) The sum of cocommutative subcoalgebras of C is a cocommutative subcoalgebra of C.

(b) C contains a unique maximal cocommutative subcoalgebra D.

(c) $D \subseteq \mathrm{Cc}(C)$.

(d) It can be the case that $D \neq \mathrm{Cc}(C)$.

Exercise 2.1.33. Show that:

(a) $\epsilon : C \longrightarrow k$ is a coalgebra map.

(b) $\Delta : C \longrightarrow C \otimes C$ is a coalgebra map if and only if C is cocommutative, where $C \otimes C$ is given the tensor product coalgebra structure.

Exercise 2.1.34. Show that:

(a) The composition of coalgebra maps is a coalgebra map.

(b) The inverse of a coalgebra isomorphism is a coalgebra isomorphism.

(c) The identity map of a coalgebra is a coalgebra map.

(In particular coalgebras over k and coalgebra maps form a category.)

Definition 2.1.26. The category whose objects are coalgebras over k and whose morphisms are coalgebra maps under function composition is denoted by k-Coalg.

The preceding exercise holds with "algebra" replacing "coalgebra".

Definition 2.1.27. The category whose objects are algebras over k and whose morphisms are algebra maps under function composition is denoted by k-Alg.

Exercise 2.1.35. We look at sums and tensor products in a categorical context.

(a) Show that any family of objects of k-Coalg has a coproduct in k-Coalg. [Hint: See Exercise 2.1.18.]

(b) Let k-CcCoalg denote the full subcategory of k-Coalg whose objects are *cocommutative* coalgebras over k. Show that any finite set of objects of k-CcCoalg has a product in k-CcCoalg. [Hint: See Exercise 2.1.19.]

Exercise 2.1.36. Let K be a field extension of k. Show that:

(a) $K\otimes C$ is a K-coalgebra, where

$$\Delta_{K\otimes C}(\alpha\otimes c) = (\alpha\otimes c_{(1)})\otimes_K(1\otimes c_{(2)}) = (1\otimes c_{(1)})\otimes_K(\alpha\otimes c_{(2)})$$

and

$$\epsilon_{K\otimes C}(\alpha\otimes c) = \alpha\epsilon(c)$$

for all $\alpha \in K$ and $c \in C$.

(b) If $f : C \longrightarrow D$ is a map of k-coalgebras then $\mathrm{I}_K\otimes f : K\otimes C \longrightarrow K\otimes D$ is a map of K-coalgebras.

(c) $F : k\text{-Coalg} \longrightarrow K\text{-Coalg}$ given by $F(C) = K\otimes C$ and $F(f) = \mathrm{I}_K\otimes f$ is a functor.

2.2 Comatrix identities, the fundamental theorem of coalgebras

In this section we prove that a finite-dimensional subspace of a coalgebra C over k generates a finite-dimensional subcoalgebra and explore some consequences of this very important fact. Our proof is based on the comatrix identities of Section 2.1. One advantage of this approach is that connections between C and the coalgebras $\mathrm{C}_n(k)$ are readily apparent. The relationship between $\mathrm{C}_n(k)$ and the various structures studied in this book is a recurring theme. The comatrix identities are satisfied in a very simple context.

Lemma 2.2.1. *Let C be a coalgebra over the field k and suppose that $\{c_1, \ldots, c_n\}$ is a linearly independent subset of C. Suppose further that there are $c_{i,j} \in C$, where $1 \leq i, j \leq n$, which satisfy*

$$\Delta(c_j) = \sum_{i=1}^{n} c_i\otimes c_{i,j}$$

for all $1 \leq j \leq n$. Then:

(a) *The set* $\{c_{i,j}\}_{1\leq i,j\leq n}$ *satisfies the comatrix identities, (2.10) and (2.11).*
(b) *The span* D *of the* $c_{i,j}$*'s is the subcoalgebra of* C *generated by* $c_1, \ldots, c_n$.

Proof. Let $1 \leq j \leq n$. Applying both sides of the equation $(\mathrm{I}_C \otimes \epsilon) \circ \Delta = \mathrm{I}_C$ to c_j we obtain $c_j = \sum_{i=1}^n c_i \epsilon(c_{i,j})$. Since $\{c_1, \ldots, c_n\}$ is linearly independent $\epsilon(c_{i,j}) = \delta_{i,j}$ for all $1 \leq i \leq n$. Applying both sides of the equation $(\mathrm{I}_C \otimes \Delta) \circ \Delta = (\Delta \otimes \mathrm{I}_C) \circ \Delta$ to c_j we obtain

$$\sum_{i=1}^{n} c_i \otimes \Delta(c_{i,j}) = \sum_{\ell=1}^{n} \Delta(c_\ell) \otimes c_{\ell,j} = \sum_{i,\ell=1}^{n} c_i \otimes c_{i,\ell} \otimes c_{\ell,j}.$$

Since $\{c_1, \ldots, c_n\}$ is linearly independent $\Delta(c_{i,j}) = \sum_{\ell=1}^n c_{i,\ell} \otimes c_{\ell,j}$ for all $1 \leq i \leq n$. We have established part (a).

Let D be the span of the $c_{i,j}$'s. It is clear that D is subcoalgebra of C. Applying both sides of the equation $(\epsilon \otimes \mathrm{I}_C) \circ \Delta = \mathrm{I}_C$ to c_j we see that $c_j = \sum_{i=1}^n \epsilon(c_i) c_{i,j} \in D$. Thus $c_1, \ldots, c_n \in D$. Now suppose that E is a subcoalgebra of C which contains the c_i's. Since $\{c_1, \ldots, c_n\}$ is linearly independent there are $c_1^*, \ldots, c_n^* \in C^*$ which satisfy $c_i^*(c_\ell) = \delta_{i,\ell}$ for all $1 \leq i, \ell \leq n$. Since $\Delta(c_j) \in E \otimes E$, the calculation

$$c_{i,j} = \sum_{\ell=1}^{n} c_i^*(c_\ell) c_{\ell,j} = (c_i^* \otimes \mathrm{I}_C)(\Delta(c_j)) \in E$$

for all $1 \leq i \leq n$ completes the proof of part (b). □

One important consequence of the lemma is that every finite-dimensional coalgebra over k is a quotient of $\mathrm{C}_n(k)$ for some $n \geq 0$. More precisely:

Corollary 2.2.2. *Let* C *be an* n*-dimensional coalgebra over the field* k*. Then there is an onto coalgebra map* $\pi : \mathrm{C}_n(k) \longrightarrow C$.

Proof. We may assume that $n > 0$. Let $\{c_1, \ldots, c_n\}$ be a basis for C. For $1 \leq j \leq n$ there are $c_{1,j}, \ldots, c_{n,j} \in C$ such that $\Delta(c_j) = \sum_{i=1}^n c_i \otimes c_{i,j}$. By part (a) of Lemma 2.2.1 the linear map $\pi : \mathrm{C}_n(k) \longrightarrow C$ defined by $\pi(e_{i,j}) = c_{i,j}$ for all $1 \leq i, j \leq n$ is the desired coalgebra map. □

Every coalgebra is the union of its finite-dimensional subcoalgebras by the following very important result, which we refer to as the fundamental theorem of coalgebras.

Theorem 2.2.3. *Let* C *be a coalgebra over the field* k*. Then every finite-dimensional subspace* V *of* C *is contained in a finite-dimensional subcoalgebra of* C.

Proof. Since the sum of subcoalgebras of C is a subcoalgebra of C we may assume that V is spanned by a single non-zero element $c \in C$. Since c is not zero, $\Delta(c)$ is not zero. Write $\Delta(c) = \sum_{i=1}^{r} c_i \otimes d_i \in C \otimes C$ where $r = \mathrm{Rank}(\Delta(c))$. Then $\{d_1, \ldots, d_r\}$ is linearly independent by Lemma 1.2.2.

We note that the equation $(\Delta \otimes \mathrm{I}_C) \circ \Delta(c) = (\mathrm{I}_C \otimes \Delta) \circ \Delta(c)$ translates to

$$\sum_{\ell=1}^{r} \Delta(c_\ell) \otimes d_\ell = \sum_{\ell=1}^{r} c_\ell \otimes \Delta(d_\ell). \tag{2.12}$$

Since $\{d_1, \ldots, d_r\}$ is linearly independent there are $d_1^*, \ldots, d_r^* \in C^*$ which satisfy $d_i^*(d_j) = \delta_{i,j}$ for all $1 \leq i, j \leq r$. Let $c_{i,j} = (\mathrm{I}_C \otimes d_j^*)(\Delta(d_i))$ for all $1 \leq i, j \leq r$. Applying $\mathrm{I}_C \otimes \mathrm{I}_C \otimes d_j^*$ to both sides of (2.12) we deduce $\Delta(c_j) = \sum_{i=1}^{r} c_i \otimes c_{i,j}$ for all $1 \leq j \leq n$.

The span of the $c_{i,j}$'s is the subcoalgebra D of C generated by $c_1, \ldots, c_r$ by part (b) of Lemma 2.2.1. The calculation $c = (\mathrm{I}_C \otimes \epsilon)(\Delta(c)) = \sum_{i=1}^{r} c_i \epsilon(d_i) \in D$ completes the proof of the theorem. □

Combining Corollary 2.2.2 and Theorem 2.2.3 we see that any coalgebra C over k is the union of homomorphic images of the coalgebras $\mathrm{C}_n(k)$. By virtue of Theorem 2.2.3, many questions about coalgebras over k become questions about finite-dimensional coalgebras over k. Another consequence of the theorem is that non-zero coalgebras contain simple subcoalgebras. Every non-zero subcoalgebra of C contains a non-zero finite-dimensional subcoalgebra. Thus:

Corollary 2.2.4. *Let C be a coalgebra over the field k.*

(a) *All simple subcoalgebras of C are finite-dimensional.*
(b) *Every non-zero subcoalgebra of C contains a simple subcoalgebra of C.*

□

Exercises

In the following exercises C is a coalgebra over the field k.

Exercise 2.2.1. Let S and T be finite non-empty sets. Show that a set bijection $S \simeq T$ determines a coalgebra isomorphism $\mathrm{C}_S(k) \simeq \mathrm{C}_T(k)$. (The more categorically inclined reader should show that there is a natural association $S \mapsto \mathrm{C}_S(k)$ which extends to a functor from the category of finite sets and set bijections to k-Coalg.)

Exercise 2.2.2. Show that $\mathrm{C}_n(k)$ is a simple subcoalgebra for all $n \geq 1$. [Hint: Let D be a subcoalgebra of $\mathrm{C}_n(k)$ and $c = \sum_{i,j=1}^{n} \alpha_{i,j} e_{i,j} \in D$, where $\alpha_{i,j} \in k$. Show that $\sum_{j=1}^{n} \alpha_{i,j} e_{\ell,j} \in D$ for all $1 \leq i, \ell \leq n$ and then show $\alpha_{i,j} e_{\ell,m} \in D$ for all $1 \leq i, j, \ell, m \leq n$.]

Exercise 2.2.3. Show that $\mathrm{C}_{mn}(k) \simeq \mathrm{C}_m(k) \otimes \mathrm{C}_n(k)$ for all $m, n \geq 0$, where the tensor product has the tensor product coalgebra structure.

Exercise 2.2.4. Suppose that $c \in C$ is not zero and write $\Delta(c) = \sum_{i=1}^{r} c_i \otimes d_i \in C \otimes C$ where $r = \mathrm{Rank}(\Delta(c))$. Let V be a subspace of C and suppose that $c \in V$. Show that:

(a) If V is a left coideal of C then $d_1, \ldots, d_r \in V$.

(b) If V is a right coideal of C then $c_1, \ldots, c_r \in V$.

(c) If V is a subcoalgebra of C then $c_1, \ldots, c_r \in V$ and $d_1, \ldots, d_r \in V$.

[Hint: In light of Exercise 2.1.7 only part (a) needs to be established.]

Exercise 2.2.5. Show that the following subspaces of $C = \mathrm{C}_n(k)$ are coideals of C:

(a) the span of the $e_{i,j}$'s, where $1 \leq i < j \leq n$;

(b) the span of the $e_{i,j}$'s, where $1 \leq i, j \leq n$ and $i \neq j$; and

(c) the span of the $e_{i,j}$'s, where $1 \leq j < i \leq n$.

Exercise 2.2.6. Let $C = \mathrm{C}_2(k)$ and I be the coideal of C described in part (a) or part (c) of Exercise 2.2.5. Show that $C/I \simeq D$, where g, h and v form a basis for D and

$$\Delta(g) = g \otimes g, \quad \Delta(h) = h \otimes h, \quad \text{and} \quad \Delta(v) = g \otimes v + v \otimes h.$$

The coalgebras of Examples 2.1.9 and 2.1.13, when $|S| = n$, and $\mathrm{P}_{n-1}(k)$ are quotients of $\mathrm{C}_n(k)$ by Corollary 2.2.2. Exercises 2.2.7, 2.2.8, and 2.2.10 below provide a detailed description of the coideal involved in the quotient. It is convenient to make the identification $\mathrm{C}_n(k) \simeq \mathrm{C}_S(k)$ for the first two exercises.

Exercise 2.2.7. Let S be a finite non-empty set and let $C = \mathrm{C}_S(k)$. Show that:

(a) The span of the elements $e_{a,b}$, where $a, b \in S$ and are distinct, is a coideal I of C.

(b) $C/I \simeq k[S]$, where $k[S]$ is the grouplike coalgebra on S over k (the coalgebra of Example 2.1.9).

[Hint: For $s \in S$ write $\Delta(s) = s \otimes s = \sum_{a \in S} a \otimes c_{a,s}$, where $c_{a,s} \in C$. See Lemma 2.2.1 and Corollary 2.2.2.]

Exercise 2.2.8. Let S be a finite (multiplicative) monoid with neutral element 1 and let $C = \mathrm{C}_S(k)$. Show that:

(a) The span of the differences $e_{a,s} - \sum_{b,ab=s} e_{1,b}$, where $a, s \in S$, is a coideal I of C.

(b) $C/I \simeq k[S]$, where $k[S]$ is the coalgebra of Example 2.1.13.

Exercise 2.2.9. Let $S = \boldsymbol{Z}_n$ be a finite (additive) cyclic group of order n and let $C = \mathrm{C}_S(k)$. Show that:

(a) The span of the differences $e_{i,j} - e_{i+\ell,j+\ell}$, where $0 \leq i, j, \ell < n$, is a coideal I of C.

(b) $C/I \simeq k[\boldsymbol{Z}_n]$, where $k[\boldsymbol{Z}_n]$ is the coalgebra of Example 2.1.13.

Exercise 2.2.10. Let $C = \mathrm{C}_n(k)$ where $n \geq 1$. Show that:

(a) The span I of the elements $e_{i,j}$, where $1 \leq i < j \leq n$, and of the differences $e_{i,j} - e_{1,i-j+1}$, where $1 \leq j \leq i \leq n$, is a coideal I of C.

(b) $C/I \simeq \mathrm{P}_{n-1}(k)$.

Exercise 2.2.11. Suppose that $C \neq (0)$ and the characteristic of k is 0. Show that $\mathrm{Cc}(C) \neq (0)$. [Hint: This is the case if $C = \mathrm{C}_n(k)$. See Exercise 2.1.30.]

The remaining exercises concern the coalgebra $\mathrm{P}_n(k)$.

Exercise 2.2.12. Let $C = \mathrm{P}_n(k)$, where $n \geq 0$. Show that every subcoalgebra of C is the span of $\{c_0, \ldots, c_m\}$ for some $0 \leq m \leq n$.

Exercise 2.2.13. Let $C = \mathrm{P}_n(k)$, where $n \geq 0$. Show that:

(a) $\mathrm{G}(C) = \{c_0\}$.

(b) The only solutions $x \in C$ to $\Delta(x) = c_0 \otimes x + x \otimes c_0$ are $x = \alpha c_1$, where $\alpha \in k$.

(c) $I = kc_1$ is a coideal of C and $C/I \simeq \mathrm{P}_m(k)$ for some $m \geq 0$ if and only if $0 \leq n \leq 2$.

The reader may find the next exercise interesting for the combinatorics involved. We attempt to alter the coproduct of the coalgebra of $\mathrm{P}_n(k)$.

Exercise 2.2.14. Suppose $n \geq 1$ and that C has basis $\{c_0, \ldots, c_n\}$. Suppose $\alpha_{i,j} \in k \backslash 0$ for $0 \leq i, j \leq n$ and that

$$\Delta(c_m) = \sum_{\ell=0}^{m} \alpha_{m,\ell} c_{m-\ell} \otimes c_\ell \quad \text{and} \quad \epsilon(c_m) = \delta_{m,0}$$

for all $0 \leq m \leq n$. ($C = \mathrm{P}_n(k)$ when $\alpha_{i,j} = 1$ for all $0 \leq i, j \leq n$.)

(a) Show that $\alpha_{m,0} = 1 = \alpha_{m,m}$ and $\alpha_{m,i+j}\alpha_{i+j,j} = \alpha_{m,j}\alpha_{m-j,i}$ for all $0 \leq i, j \leq m$ which satisfy $i + j \leq m$.

(b) Set $(0)! = 1$ and $(m)! = \binom{m}{1}\binom{m-1}{1}\cdots\binom{1}{1}$ for all $1 \leq m \leq n$, where $\binom{\ell}{1} = \alpha_{\ell,1}$ for all $1 \leq \ell \leq n$. Show that

$$\Delta(\frac{c_m}{(m)!}) = \sum_{\ell=0}^{m} \frac{c_{m-\ell}}{(m-\ell)!} \otimes \frac{c_\ell}{(\ell)!}$$

for all $0 \leq m \leq n$. (Thus $C = \mathrm{P}_n(k)$ after all.)

[Hint: Set $\binom{m}{\ell} = \alpha_{m,\ell}$ for all $0 \leq \ell \leq m$ and note that these "binomial symbols" can be expressed in terms of the "factorials" defined above just as ordinary binomial symbols are expressed in terms of factorials.]

2.3 The dual algebra

The vector space of linear functionals C^* on a coalgebra C over k has an algebra structure derived for the coalgebra structure on C. In this section we describe the algebra structure on C^* and relate it to the coalgebra structures on C defined in Section 2.1. One natural vehicle for relating C and C^* is the correspondence $U \mapsto U^\perp$ of subspaces of C with the closed subspaces of C^* discussed in Section 1.3.

Let $\imath : C^*\otimes C^* \longrightarrow (C\otimes C)^*$ be the one-one linear map of Exercise 1.2.5 which is determined by $\imath(c^*\otimes d^*)(c\otimes d) = c^*(c)d^*(d)$ for all $c^*, d^* \in C^*$ and $c, d \in C$. We identify $C^*\otimes C^*$ with $\mathrm{Im}(\imath)$ and regard $C^*\otimes C^*$ as a subspace of $(C\otimes C)^*$.

Define a linear map $m : C^*\otimes C^* \longrightarrow C^*$ by $m = \Delta^*|C^*\otimes C^*$ and write $m(c^*\otimes d^*) = c^*d^*$ for $c^*, d^* \in C^*$. Then

$$(c^*d^*)(c) = c^*(c_{(1)})d^*(c_{(2)}) \tag{2.13}$$

for all $c \in C$.

Proposition 2.3.1. *Let C be a coalgebra over the field k. Then (C^*, m, η) is a algebra over k, where $m = \Delta^*|C^*\otimes C^*$ and $\eta(1) = \epsilon$.* □

The proof is left to the reader who is referred to Exercise 2.3.1 for a closer examination of the relationship between (C, Δ, ϵ) and (C^*, m, η).

Definition 2.3.2. Let (C, Δ, ϵ) be a coalgebra over k. The algebra (C^*, m, η) of Proposition 2.3.1 is the *dual algebra of* (C, Δ, ϵ).

We will almost always write C^* for (C^*, m, η).

Proposition 2.3.3. *Let $f : C \longrightarrow D$ be a map of coalgebras over the field k. Then the linear transpose $f^* : D^* \longrightarrow C^*$ is a map of dual algebras.* □

The proof is left to the reader who is referred to Exercise 2.3.8 for a closer examination of the relationship between f and f^*.

The dual algebra C^* acts on C both from the left and from the right according to the formulas

$$c^* \rightharpoonup c = c_{(1)} <c^*, c_{(2)}> \quad \text{and} \quad c \leftharpoonup c^* = <c^*, c_{(1)}> c_{(2)} \tag{2.14}$$

respectively for all $c^* \in C^*$ and $c \in C$.

Definition 2.3.4. Let A be an algebra over the field k. A *locally finite A-module* is an A-module M whose finitely generated submodules are finite-dimensional (or equivalently the cyclic submodules of M are finite-dimensional).

Among other things the left and right C^*-module actions on C are locally finite.

Proposition 2.3.5. *Let C be a coalgebra over the field k. Then:*

(a) *C is a C^*-bimodule under the actions of (2.14). Furthermore C is a locally finite left and right C^*-module.*
(b) *Let V be a subspace of C. Then V is a subcoalgebra (respectively left coideal, right coideal) of C if and only if V is C^*-sub-bimodule (respectively right C^*-submodule, left C^*-submodule) of C.*
(c) *Suppose $c \in C$ and let L and R be the left and right coideals respectively generated by c. Then* $\mathrm{Dim}(L) = \mathrm{Rank}(\Delta(c)) = \mathrm{Dim}(R)$.

Proof. The proof of part (a) is left to the reader. We first establish part (b) for left coideals of C.

Suppose that V is a left coideal of C. Then $\Delta(V) \subseteq C \otimes V$ which means $V \leftharpoonup C^* \subseteq V$. Therefore V is a right C^*-submodule of C. Now let V be a right C^*-submodule of C and suppose that $c \in V$ and is not zero. Write $\Delta(c) = \sum_{i=1}^{r} c_i \otimes d_i \in C \otimes C$ where $r = \mathrm{Rank}(\Delta(c))$. Since $c \leftharpoonup c^* = \sum_{i=1}^{r} <c^*, c_i> d_i$ for all $c^* \in C^*$ and $\{c_1, \ldots, c_r\}$ is linearly independent by Lemma 1.2.2, it follows that $d_1, \ldots, d_r \in V$. Therefore $\Delta(c) \subseteq C \otimes V$ and we conclude that V is a left coideal of C.

The fact that V is a right coideal of C if and only if V is a left C^*-submodule of C follows by a similar argument. It also follows by what we have just established applied to the coalgebra C^{cop}. See Exercise 2.1.7. Since V is a subcoalgebra of C if and only if V is both a left coideal and a right coideal of C, the remainder of part (b) follows. See Exercise 2.1.4.

To prove part (c) we may as well assume that c is not zero. Write $\Delta(c) = \sum_{i=1}^{r} c_i \otimes d_i \in C \otimes C$ where $r = \mathrm{Rank}(\Delta(c))$. Now $\{c_1, \ldots, c_r\}$ and $\{d_1, \ldots, d_r\}$ are linearly independent by Lemma 1.2.2. Since $c^* \rightharpoonup c = \sum_{i=1}^{r} c_i \langle c^*, d_i \rangle$ and $c \leftharpoonup c^* = \sum_{i=1}^{r} \langle c^*, c_i \rangle d_i$ for all $c^* \in C^*$ we conclude that $\{c_1, \ldots, c_r\}$ is a basis for $C^* \rightharpoonup c$ and that $\{d_1, \ldots, d_r\}$ is a basis for $c \leftharpoonup C^*$. As $C^* \rightharpoonup c = R$ and $c \leftharpoonup C^* = L$ by part (b), part (c) follows. □

Let $c \in C$. Then $C^* \rightharpoonup c \leftharpoonup C^*$ (respectively $c \leftharpoonup C^*$, $C^* \rightharpoonup c$) is the subcoalgebra (respectively left coideal, right coideal) of C generated by c, and these are all finite-dimensional subspaces of C by Proposition 2.3.5. Thus the proposition implies simple left or right coideals of C are finite-dimensional and also gives a very different proof of Theorem 2.2.3.

Left and right multiplication by $c^* \in C^*$ in C and C^* are important operations. Define endomorphisms $\mathrm{L}(c^*)$ and $\mathrm{R}(c^*)$ of C by

$$\mathrm{L}(c^*)(c) = c^* \rightharpoonup c \quad \text{and} \quad \mathrm{R}(c^*)(c) = c \leftharpoonup c^* \tag{2.15}$$

for all $c \in C$. For an algebra A over k and $a \in A$ define endomorphisms $\ell(a)$ and $\mathrm{r}(a)$ of A by

$$\ell(a)(b) = ab \quad \text{and} \quad \mathrm{r}(a)(b) = ba \tag{2.16}$$

for all $b \in A$. Observe that

$$\ell(c^*) = \mathrm{R}(c^*)^* \quad \text{and} \quad \mathrm{r}(c^*) = \mathrm{L}(c^*)^* \tag{2.17}$$

for all $c^* \in C^*$. In particular $\ell(c^*)$ and $\mathrm{r}(c^*)$ are continuous endomorphisms of the vector space C^*.

Since

$$\begin{aligned}
\Delta(c^* \rightharpoonup c) &= \Delta(c_{(1)} \langle c^*, c_{(2)} \rangle) \\
&= c_{(1)(1)} \otimes c_{(1)(2)} \langle c^*, c_{(2)} \rangle \\
&= c_{(1)} \otimes c_{(2)(1)} \langle c^*, c_{(2)(2)} \rangle \\
&= c_{(1)} \otimes (c^* \rightharpoonup c_{(2)})
\end{aligned}$$

we have

$$\Delta \circ \mathrm{L}(c^*) = (\mathrm{I}_C \otimes \mathrm{L}(c^*)) \circ \Delta \tag{2.18}$$

and likewise

$$\Delta\circ\mathrm{R}(c^*) = (\mathrm{R}(c^*)\otimes\mathrm{I}_C)\circ\Delta \tag{2.19}$$

for all $c^* \in C^*$.

The fact that $\ell(c^*)$ and $\mathrm{r}(c^*)$ are continuous has implications for C^*.

Proposition 2.3.6. *Let C^* be the dual algebra of a coalgebra C over the field k. Then:*

(a) *Suppose I is a closed subspace of C^*. Then c^*I and Ic^* are closed subspaces of C^* for all $c^* \in C^*$.*
(b) *Finitely generated left of right ideals of C^* are closed subspaces of C^*.*
(c) *Suppose that I, J and K are subspaces of C^* and $IJ \subseteq K$. Then $\overline{I}\,\overline{J} \subseteq \overline{K}$ and $\Delta(K^\perp) \subseteq I^\perp\otimes C + C\otimes J^\perp$.*

Proof. Suppose that I is a closed subspace of C^* and $c^* \in C^*$. Since $\ell(c^*)$ and $\mathrm{r}(c^*)$ are continuous, $c^*I = \ell(c^*)(I)$ and $Ic^* = \mathrm{r}(c^*)(I)$ are closed subspaces of C^* by part (b) of Theorem 1.3.10. In particular principal left of right ideals of C^* are closed since C^* is a closed subspace of C^*. Since a finite sum of closed subspaces of C^* is closed by part (b) of Proposition 1.3.8, parts (a) and (b) follow.

To show part (c) let $c^* \in I$. Since $\ell(c^*)$ is continuous we have $c^*\overline{J} = \ell(c^*)(\overline{J}) \subseteq \overline{\ell(c^*)(J)} \subseteq \overline{K}$ by part (d) of Theorem 1.3.10. Therefore $I\overline{J} \subseteq \overline{K}$. By a similar argument, or interpreting thus inclusion for $(C^{cop})^* = C^{*\,op}$, we have $\overline{I}\,\overline{J} \subseteq \overline{K}$. See Exercises 2.1.7 and 2.3.4.

To complete the proof we use part (e) of Theorem 1.3.10 to calculate

$$K^\perp \subseteq (\Delta^*(I\otimes J))^\perp = \Delta^{-1}((I\otimes J)^\perp)$$

and thus $\Delta(K^\perp) \subseteq (I\otimes J)^\perp = I^\perp\otimes C + C\otimes J^\perp$ by part (b) of Exercise 2.3.2. □

Suppose that U, V and W are subspaces of C. Then by definition of the dual algebra product

$$\Delta(U) \subseteq V\otimes C + C\otimes W \quad \text{implies} \quad V^\perp W^\perp \subseteq U^\perp. \tag{2.20}$$

Now we are in a position to discuss very efficiently how the correspondence $U \mapsto U^\perp$ of the subspaces of C and the closed subspaces of C^* connects coalgebra structures of C and algebra structures of C^*.

Proposition 2.3.7. *Let C be a coalgebra over the field k and C^* be the dual algebra. Suppose that A is a subalgebra of C^*. Then:*

(a) *If A is a dense subalgebra of C^* and I is a subalgebra (respectively ideal, left ideal, right ideal) of A then $I^\perp$ is a coideal (respectively subcoalgebra, left coideal, right coideal) of C.*
(b) *If U is a coideal (respectively subcoalgebra, left coideal, right coideal) of C then $U^\perp \cap A$ is a subalgebra (respectively ideal, left ideal, right ideal) of A.*
(c) *If I is a subalgebra (respectively ideal, left ideal, right ideal) of C^* then $\overline{I}$ is a subalgebra (respectively ideal, left ideal, right ideal) of C^*.*
(d) *There is a one-one inclusion reversing correspondence between the set of coideals (respectively subcoalgebras, left coideals, right coideals) of C and the set of closed subalgebras (respectively closed ideals, closed left ideals, closed right ideals) of C^* given by $U \mapsto U^\perp$.*

Proof. Part (a) follows by part (c) of Proposition 2.3.6. Part (b) follows by (2.20) and part (c) follows by part (c) of Proposition 2.3.6 or by parts (a) and (b) with $A = C^*$. As part (d) follows from parts (a) and (b) our proof is complete. □

All subspaces of U^* are closed when U is a finite-dimensional vector space over k by Proposition 1.3.5. Thus as a corollary of part (d) of Proposition 2.3.7:

Corollary 2.3.8. *Let C be a finite-dimensional coalgebra over the field k. Then $U \mapsto U^\perp$ is a one-one inclusion reversing correspondence between the set of coideals (respectively subcoalgebras, left coideals, right coideals) of C and the set of subalgebras (respectively ideals, left ideals, right ideals) of the dual algebra C^*.* □

Corollary 2.3.9. *Let C be a coalgebra over the field k. Then $S \mapsto S^\perp$ is a one-one inclusion reversing correspondence between the set of simple subcoalgebras of C and the set of closed cofinite maximal ideals of C^*.*

Proof. Simple subcoalgebras of C are finite-dimensional by Corollary 2.2.4. Let D be a finite-dimensional subcoalgebra of C. Then $D^\perp$ is a cofinite ideal of C^* by part (c) of the preceding proposition and (1.2). All subspaces of C^* which contain $D^\perp$ are closed and cofinite by Proposition 1.3.12. At this point the corollary follows by part (d) of Proposition 2.3.7 again. □

We have just noted that simple coalgebras over k are finite-dimensional. Let C be a finite-dimensional coalgebra over k. Then C is simple if and only if C^* is a simple k-algebra by Corollary 2.3.9. When k is algebraically closed the finite-dimensional simple k-algebras are those k-algebras isomorphic to $\mathrm{M}_n(k)$ for some $n \geq 1$. See Propositions 8.51 and 8.59 of [186] and the Artin-Wedderburn Theorem [186, Theorem 8.56] for example. As $\mathrm{M}_n(k) \simeq \mathrm{C}_n(k)^*$ by Exercise 2.3.24:

Corollary 2.3.10. *$\mathrm{C}_n(k)$ is a simple coalgebra over the field k for all $n \geq 1$. If k is algebraically closed then any simple coalgebra over k is isomorphic to $\mathrm{C}_n(k)$ for some $n \geq 1$.* □

The operation in C which corresponds to multiplication of subspaces of C^* is important. It is the subject of Section 2.4.

Exercises

Throughout these exercises C, D are coalgebras over the field k and C^*, D^* are their dual algebras.

Exercise 2.3.1. Suppose V is a vector space over the field k, let $\imath : V^*\otimes V^* \longrightarrow (V\otimes V)^*$ be the one-one linear map of Exercise 1.2.5, and let $\jmath : k \longrightarrow k^*$ be the linear map defined by $\jmath(1)(1) = 1$.

Let $\Delta : V \longrightarrow V\otimes V$, $\epsilon : V \longrightarrow k$ be linear and $m : V^*\otimes V^* \longrightarrow V^*$, $\eta : k \longrightarrow V^*$ be defined by $m = \Delta^*\circ\imath$ and $\eta(1) = \epsilon^*\circ\jmath$. Show that (V, Δ, ϵ) is a coalgebra over k if and only if (V^*, m, η) is an algebra over k. [Hint: Write $m(a\otimes b) = ab$ for $a, b \in V^*$ and $\Delta(v) = v_{(1)}\otimes v_{(2)}$ for $v \in V$. Show that $ab(v) = a(v_{(1)})b(v_{(2)})$ and thus

$$((ab)c)(v) = a(v_{(1)(1)})b(v_{(1)(2)})c(v_{(2)})$$

and

$$(a(bc))(v) = a(v_{(1)})b(v_{(2)(1)})c(v_{(2)(2)})$$

for all $a, b, c \in V^*$ and $v \in V$. See Exercise 1.3.7.]

Exercise 2.3.2. Suppose that U and V are vector spaces over the field k, I is a subspace of U^*, and J is a subspace of V^*. Show that:

(a) $I^\perp\otimes J^\perp = (I\otimes V^* + U^*\otimes J)^\perp$.

(b) $(I\otimes J)^\perp = I^\perp\otimes V + U\otimes J^\perp$. [Hint: Write $U = I^\perp\bigoplus U'$ and $V = J^\perp\bigoplus V'$ and note that $U\otimes V = \mathcal{U}\bigoplus(U'\otimes V')$ where $\mathcal{U} = I^\perp\otimes V + U\otimes J^\perp \subseteq (I\otimes J)^\perp$. Show that $(U'\otimes V')\cap(I\otimes J)^\perp = (0)$.]

Exercise 2.3.3. Let A be a finite-dimensional algebra over k. Show that $A \simeq C^*$ for some finite-dimensional coalgebra C over k. [Hint: Let $m : A\otimes A \longrightarrow A$ and $\eta : k \longrightarrow A$ be the product and unit maps. Define $\Delta : A^* \longrightarrow A^*\otimes A^*$ and $\epsilon : A^* \longrightarrow k$ by $\Delta = \imath^{-1}\circ m^*$ and $\epsilon = \jmath^{-1}\circ\eta^*$, where $\imath, \jmath$ are as in Exercise

2.3.1. Use Exercise 2.3.1 to show that (A^*, m, η) is a coalgebra over k and that $\imath_A : A \longrightarrow A^{**}$ defined by $\imath_A(a)(a^*) = a^*(a)$ for all $a \in A$ and $a^* \in A^*$ is an algebra isomorphism.]

Exercise 2.3.4. Show that:

(a) $(C^{cop})^* = C^{*\,op}$.

(b) C is cocommutative if and only if C^* is commutative.

[Hint: C^* is commutative if and only if $c^*d^*(c) = c^*(c_{(1)})d^*(c_{(2)}) = c^*(c_{(2)})d^*(c_{(1)}) = d^*c^*(c)$ for all $c^*, d^* \in C^*$ and $c \in C$. See Exercise 1.3.7.]

Exercise 2.3.5. Let $c \in C$. Show that c is a cocommutative element of C if and only if $c^* \rightharpoonup c = c \leftharpoonup c^*$ for all $c^* \in C^*$. [Hint: See part (b) of Exercise 1.3.7.]

Exercise 2.3.6. Let $k = \boldsymbol{R}$ be the field of real numbers and C be the 2-dimensional coalgebra over $\boldsymbol{R}$ with basis $\{c, s\}$ whose coproduct is determined by

$$\Delta(c) = c\otimes c - s\otimes s \quad \text{and} \quad \Delta(s) = s\otimes c + c\otimes s.$$

Show that:

(a) C is indeed a coalgebra over $\boldsymbol{R}$ and that $\epsilon(c) = 1, \epsilon(s) = 0$.

(b) C^* is the field of complex numbers $\boldsymbol{C}$.

Exercise 2.3.7. Find a finite-dimensional coalgebra C over some field k such that $\mathrm{G}(C) = \emptyset$ and $\mathrm{G}(K\otimes C) \neq \emptyset$ for some field extension K of k.

Exercise 2.3.8. Let $F : D^* \longrightarrow C^*$ be a continuous linear map and $f : C \longrightarrow D$ be the linear map determined by $F = f^*$. Show that F is an algebra map if and only if f is a coalgebra map. [Hint: For $a, b \in D^*$ and $c \in C$ note that $F(ab)(c) = a(f(c)_{(1)})b(f(c)_{(2)})$ and $(F(a)F(b))(c) = f(c_{(1)})f(c_{(2)})$. See Exercise 1.3.7.]

Definition 2.3.11. Let A be an algebra over the field k. A *derivation of* A is a linear endomorphism F of A such that $F(ab) = F(a)b + aF(b)$ for all $a, b \in A$.

For fixed $b \in A$ note that $F : A \longrightarrow A$ defined by $F(a) = [a, b] = ab - ba$ for all $a \in A$ is a derivation of A.

Definition 2.3.12. Let C be a coalgebra over the field k. A *coderivation of* C is a linear endomorphism f of C such that $\Delta\circ f = (f\otimes \mathrm{I}_C + \mathrm{I}_C\otimes f)\circ\Delta$.

Exercise 2.3.9. Let $F : C^* \longrightarrow C^*$ be a continuous linear endomorphism and $f : C \longrightarrow D$ be the linear endomorphism determined by $F = f^*$. Show that F is a derivation of C^* if and only if f is a coderivation of C.

Exercise 2.3.10. Show that $(\mathrm{L}(c^*)\otimes \mathrm{I}_C)\circ\Delta = (\mathrm{I}_C\otimes \mathrm{R}(c^*))\circ\Delta$ for all $c^* \in C^*$.

Exercise 2.3.11. Working from the definition of coderivation show that:

(a) If f is a coderivation of C then $\mathrm{Im}(f)$ is a coideal of C.

(b) If f, g are coderivations of C then $[f, g] = f\circ g - g\circ f$ is a coderivation of C.

(c) For all $c^* \in C^*$ the difference $\delta(c^*) = \mathrm{L}(c^*) - \mathrm{R}(c^*)$ is a coderivation of C.

Exercise 2.3.12. Show that the assertions of parts (a)–(c) of Exercise 2.3.11 translate to assertions about continuous derivations of C^*, verify these assertions (which hold for derivations generally), and then deduce the assertions of parts (a)–(c) of Exercise 2.3.11 from the corresponding assertions about continuous derivations.

Exercise 2.3.13. Let $C = \mathrm{P}_\infty(k)$ be as described in Definition 2.1.14. Show that the linear endomorphism $f : C \longrightarrow C$ defined by $f(c_n) = (n+1)c_{n+1}$ for all $n \geq 0$ is a coderivation of C.

Exercise 2.3.14. Let $C = kg\bigoplus V$, where $\Delta(g) = g\otimes g$ and $\Delta(v) = g\otimes v + v\otimes g$ for all $v \in V$. Show that:

(a) C^* is a local algebra whose unique maximal ideal $\mathcal{M} = \mathrm{Ker}(\epsilon)$ satisfies $\mathcal{M}^2 = (0)$.

(b) If V is *not* finite-dimensional then C^* has ideals which are not closed.

[Hint: Show that the transpose of the projection $\pi : C \longrightarrow V$ is a one-one map $\pi^* : V^* \longrightarrow C^*$ with image $\mathcal{M}$. See Theorem 1.3.10 and Proposition 1.3.5.]

Definition 2.3.13. Let A and B be algebras over the field k. The *tensor product algebra structure on* $A\otimes B$ is determined by $(a\otimes b)(a'\otimes b') = aa'\otimes bb'$ for all $a, a' \in A$ and $b, b' \in B$.

Exercise 2.3.15. Let $\imath : C^*\otimes D^* \longrightarrow (C\otimes D)^*$ be the one-one linear map of Exercise 1.2.5. Show that $\imath$ is an algebra map, where $C^*\otimes D^*$ has the tensor product algebra structure and $C\otimes D$ has the tensor product coalgebra structure.

Exercise 2.3.16. Suppose that $C \simeq \bigoplus_{i\in I} C_i$ is isomorphic to the direct sum of coalgebras. Show that $C^* \simeq \prod_{i\in I} C_i^*$ is isomorphic to the direct product of algebras.

Exercise 2.3.17. Suppose that $C^* = I_1 \bigoplus \cdots \bigoplus I_r$ is the direct sum of subspaces which are left ideals or which are right ideals of C^*. Show that $I_1, \ldots, I_r$ are closed subspaces of C^*.

Exercise 2.3.18. Show that:

(a) If S is a non-empty subset of C^* then $I = \{a \in C^* \,|\, aS = (0)\}$ is a closed left ideal of C^*.

(b) If L is a closed left ideal of C^* then the largest ideal of C^* which is contained in L is a closed subspace of C^*.

[Hint: The left ideal of part (a) is $\bigcap_{s \in S} \mathrm{Ker}(\mathrm{r}(s))$ and the largest ideal contained in the left ideal of part (b) is $\bigcap_{s \in S} \mathrm{r}(s)^{-1}(L)$. See (2.16) and Theorem 1.3.10 in particular.]

Exercise 2.3.19. Let $\mathcal{M}$ be a maximal left, right, or two-sided ideal of C^*. Show that $\mathcal{M}$ is a dense subspace of C^* or $\mathcal{M}$ is a closed cofinite subspace of C^*.

Exercise 2.3.20. Suppose that C is finite-dimensional and C_0 is the sum of the simple subcoalgebras of C. Show that $\mathrm{Rad}(C^*) = C_0^\perp$, where $\mathrm{Rad}(A)$ is the Jacobson radical of an algebra A over k.

Definition 2.3.14. Let X, Y be non-empty subsets of an algebra A over the field k. The *centralizer of* Y *in* X is

$$\mathrm{Z}_X(Y) = \{x \in X \,|\, yx = xy \text{ for all } y \in Y\}.$$

For $y \in A$ the *centralizer of* y *in* X is $\mathrm{Z}_X(y) = \mathrm{Z}_X(\{y\})$.

Definition 2.3.15. The *center of an algebra* A over the field k is $\mathrm{Z}(A) = \mathrm{Z}_A(A)$.

Observe that when B is a subalgebra of A then $\mathrm{Z}_B(Y)$ a subalgebra of B for all subsets Y of B. When $A = C^*$ centralizers are closed subalgebras of C^*.

Exercise 2.3.21. Let S be a non-empty subset of C^*. For $c^* \in C^*$ let $\delta(c^*) = \mathrm{L}(c^*) - \mathrm{R}(c^*)$ be the coderivation of C described in Exercise 2.3.11. Show that:

(a) $\mathrm{Z}_{C^*}(s) = \mathrm{Im}(\delta(s))^\perp$ for all $s \in S$.

(b) $\mathrm{Z}_{C^*}(S) = (\sum_{s \in S} \mathrm{Im}(\delta(s)))^\perp$.

(c) $\mathrm{Z}(C^*) = (\sum_{s \in C^*} \mathrm{Im}(\delta(s)))^\perp$

[Hint: $\mathrm{Z}_{C^*}(S) = \bigcap_{s \in S} \mathrm{Im}(\delta(s))^\perp$. See part (a) of Proposition 1.3.8.]

Exercise 2.3.22. Let $\eta : C^* \longrightarrow k$ be linear. Show that η is a continuous algebra map if and only if there exists a $g \in \mathrm{G}(C)$ such that $<\eta, c^*> = <c^*, g>$ for all $c^* \in C^*$.

Exercise 2.3.23. Let S be a non-empty set and $C = k[S]$ be the grouplike coalgebra on S over k (the coalgebra of Example 2.1.9). Show that C^* is isomorphic to the algebra of functions $f : S \longrightarrow k$ over k under the pointwise operations.

The dual algebra of the coalgebra of Example 2.1.15 is a matrix algebra.

Exercise 2.3.24. Let $C = \mathrm{C}_n(k)$, where $n \geq 1$, and let $\{e_{i,j}\}_{1 \leq i,j \leq n}$ be a standard basis for C. Denote the dual basis for C^* by $\{E_{i,j}\}_{1 \leq i,j \leq n}$. (See Definition 1.2.5.) Show that:

(a) $E_{i,j}E_{\ell,k} = \delta_{j,\ell}E_{i,k}$ for all $1 \leq i, j, k, \ell \leq n$.

(b) $C^* \simeq \mathrm{M}_n(k)$.

Exercise 2.3.25. Let A be an n-dimensional algebra over the field k. Give a proof of the fact that A is isomorphic to a subalgebra of $\mathrm{M}_n(k)$ based on Exercise 2.3.24 and Corollary 2.2.2.

Exercise 2.3.26. We continue with the notation of Exercise 2.3.24. Let $C = \mathrm{C}_n(k)$, where $n \geq 1$, let I be a coideal of C, and let $f : (C/I)^* \longrightarrow \mathrm{M}_n(k)$ be the composite $(C/I)^* \xrightarrow{\pi^*} \mathrm{C}_n(k) \simeq \mathrm{M}_n(k)$, where $\pi : C \longrightarrow C/I$ is the projection and the isomorphism arises from the identification of $E_{i,j}$ with the $n \times n$ matrix $(a_{r,s})$ defined by $a_{r,s} = \delta_{(i,j),(r,s)}$ for all $1 \leq r, s \leq n$. Show that:

(a) If I is the coideal of part (a) of Exercise 2.2.5 then $\mathrm{Im}(f)$ is the subalgebra of all $n \times n$ lower triangular matrices over k.

(b) If I is the coideal of part (b) of Exercise 2.2.5 then $\mathrm{Im}(f)$ is the subalgebra of all $n \times n$ diagonal matrices over k.

(c) If I is the coideal of part (c) of Exercise 2.2.5 then $\mathrm{Im}(f)$ is the subalgebra of all $n \times n$ upper triangular matrices over k.

Let $(S, \leq)$ be a partially ordered set which is *locally finite*, meaning that for all $i, j \in S$ which satisfy $i \leq j$ the interval $[i, j] = \{\ell \in S \mid i \leq \ell \leq j\}$ is a finite set. Let $\mathcal{S} = \{[i, j] \mid i, j \in S,\ i \leq j\}$ and let A be the algebra which is the vector space of functions $f : \mathcal{S} \longrightarrow k$ under pointwise operations whose product is given by

$$(f \star g)([i, j]) = \sum_{i \leq \ell \leq j} f([i, \ell])g([\ell, j])$$

for all $f, g \in A$ and $[i, j] \in \mathcal{S}$ and whose unit is given by $1([i, j]) = \delta_{i,j}$ for all $[i, j] \in \mathcal{S}$.

Definition 2.3.16. The algebra A over k described above is the *incidence algebra* of the locally finite partially ordered set $(S, \leq)$.

Exercise 2.3.27. Let $(S, \leq)$ be a locally finite partially ordered set and let C be the free k-module on a set of formal symbols $\{e_{i,j} \,|\, i, j \in S,\ i \leq j\}$. Show that:

(a) C has a coalgebra structure determined by

$$\Delta(e_{i,j}) = \sum_{i \leq \ell \leq j} e_{i,\ell} \otimes e_{\ell,j} \quad \text{and} \quad \epsilon(e_{i,j}) = \delta_{i,j}$$

for all $i, j \in S$, where $i \leq j$.

(b) C^* is isomorphic to the incidence algebra of $(S, \leq)$.

Definition 2.3.17. The coalgebra of Example 2.3.27 is the *incidence coalgebra* of the locally finite partially ordered set $(S, \leq)$.

Exercise 2.3.28. Let $n \geq 1$ and let $C = \mathrm{P}_n(k)$ be the coalgebra over k described in Definition 2.1.14. Show that $C^* \simeq k[X]/(X^n)$.

Exercise 2.3.29. Suppose that $C \simeq \mathrm{P}_{n_1}(k) \bigoplus \cdots \bigoplus \mathrm{P}_{n_r}(k)$ where $n_1, \ldots, n_r > 0$ and k has at least r elements. Show that C^* is generated by a single element as a k-algebra. [Hint: Let $\lambda_1, \ldots, \lambda_r \in k$ be distinct. Fix $1 \leq i \leq r$. Use Exercise 2.3.28 to see that $\mathrm{P}_{n_i}(k)^* \simeq k[X]/(X^{n_i}) \simeq k[X]/((X - \lambda_i)^{n_i})$. Let $f(X) = (X - \lambda_1)^{n_1} \cdots (X - \lambda_r)^{n_r}$ and note that

$$k[X]/(f(X)) \simeq k[X]/((X - \lambda_1)^{n_1}) \times \cdots \times k[X]/((X - \lambda_r)^{n_r})$$

as algebras over k by the Chinese Remainder Theorem.]

Exercise 2.3.30. Suppose that k is algebraically closed, C^* is generated by a single element as a k-algebra, and that C^* is finite-dimensional. Show that

$$C \simeq \mathrm{P}_{n_1}(k) \oplus \cdots \oplus \mathrm{P}_{n_r}(k)$$

for some $r \geq 1$ and $n_1, \ldots, n_r \geq 0$.

Exercise 2.3.31. Let $C = \mathrm{P}_\infty(k)$ be the coalgebra over k described in Definition 2.1.14 and let $x \in C^*$ be defined by $x(c_m) = \delta_{m,1}$ for all $m \geq 0$. Show that:

(a) $x^n(c_m) = \delta^n_m$ for all $n, m \geq 0$.

(b) For $c \in C$, $x^n(c) = 0$ for all but a finite number of $n \geq 0$. (Thus for any choice of $\alpha_n \in k$ the sum $\sum_{n=0}^\infty x^n(c)$ interpreted as the sum of the non-zero summands is meaningful.)

(c) $\sum_{n=0}^\infty \alpha_n x^n : C \longrightarrow k$ defined by $(\sum_{n=0}^\infty \alpha_n x^n)(c) = \sum_{n=0}^\infty \alpha_n x^n(c)$ for all $c \in C$ is a linear functional for all $\{\alpha_n\}_{n\geq 0} \subseteq k$.

(d) $f : k[[X]] \longrightarrow C^*$ defined by $f(\sum_{n=0}^{\infty} \alpha_n X^n) = \sum_{n=0}^{\infty} \alpha_n x^n$ is an isomorphism of the k-algebra of formal power series $k[[X]]$ in variable X over k and the dual algebra C^*.

(e) $c^* \in C^*$ has an inverse in the dual algebra C^* if and only if $c^*(c_0) \neq 0$. [Hint: Recall that a formal power series $\sum_{n=0}^{\infty} \alpha_n X^n$ has a multiplicative inverse if and only if $\alpha_0 \neq 0$.]

Exercise 2.3.32. Generalize Exercise 2.3.31. Let $r > 1$ and let $S = \boldsymbol{N} \times \cdots \times \boldsymbol{N}$ be the monoid product of r copies of the (additive) monoid $\boldsymbol{N}$. Let $C = k[S]$ be coalgebra of Example 2.1.13 with basis described as $\{c_{\mathbf{n}}\}_{\mathbf{n} \in S}$. Thus

$$\Delta(c_{\mathbf{n}}) = \sum_{\mathbf{r}+\mathbf{s}=\mathbf{n}} c_{\mathbf{r}} \otimes c_{\mathbf{s}} \quad \text{and} \quad \epsilon(c_{\mathbf{n}}) = \delta_{\mathbf{0},\mathbf{n}}$$

for all $\mathbf{n} \in S$.

For $1 \leq i \leq r$ let $\epsilon_i = (0, \ldots, 1, \ldots, 0) \in S$ be the r-tuple whose components are all zero except for the one in the ith position whose value is 1 and define $x_i \in C^*$ by $x_i(c_{\mathbf{m}}) = \delta_{\epsilon_i, \mathbf{m}}$ for all $\mathbf{m} \in S$. Set $x^{\mathbf{n}} = x_1^{n_1} \cdots x_r^{n_r}$ for $\mathbf{n} = (n_1, \ldots, n_r) \in S$. Show that:

(a) $x^{\mathbf{n}}(c_{\mathbf{m}}) = \delta_{\mathbf{n},\mathbf{m}}$ for all $\mathbf{n}, \mathbf{m} \in S$ and for $c \in C$ the sum $\sum_{\mathbf{n} \in S} \alpha_{\mathbf{n}} x^{\mathbf{n}}(c)$ is meaningful since only finitely many summands are not zero.

(b) $f : k[[X_1, \ldots, X_r]] \longrightarrow C^*$ given by $f(\sum_{\mathbf{n} \in S} \alpha_{\mathbf{n}} X^{(\mathbf{n})}) = \sum_{\mathbf{n} \in S} \alpha_{\mathbf{n}} x^{\mathbf{n}}$ is an isomorphism of the k-algebra of formal power series $k[[X_1, \cdots, X_r]]$ in variables $X_1, \ldots, X_r$ over k and the dual algebra C^*, where $X^{(\mathbf{n})} = X_1^{n_1} \cdots X_r^{n_r}$ for $\mathbf{n} = (n_1, \ldots, n_r) \in S$.

(c) $k[x_1, \ldots, x_r]$ is the polynomial algebra over k in $x_1, \ldots, x_r$ and is a dense subspace of C^*.

Exercise 2.3.33. Show that Exercise 2.1.12 can be formulated as an exercise for the dual algebra C^* and do the exercise for C^*.

Exercise 2.3.34. Show that Exercise 2.1.13 can be formulated as an exercise for the dual algebra C^* and do the exercise for C^*.

Exercise 2.3.35. Suppose k is algebraically closed and S is a simple subcoalgebra of C. Show that $S = kg$ for some $g \in \mathrm{G}(C)$ if and only if S is cocommutative. [Hint: Let k be any field and suppose S is simple. Then S is finite-dimensional and S^* is a finite-dimensional simple algebra over k. See Corollary 2.3.9.]

Exercise 2.3.36. Show that $F : k\text{-Coalg} \longrightarrow k\text{-Alg}$ is a contravariant functor, where $F(C) = C^*$ for coalgebras C over k and $F(f) = f^*$ for maps f of coalgebras over k. See Definitions 2.1.26 and 2.1.27.

Just as there is a notion dual to associative algebra there is a notion dual to Lie algebra. Recall that a Lie algebra is a pair (L, m), where $m : L\otimes L \longrightarrow L$ is a linear map traditionally expressed by $m(a\otimes b) = [a, b]$ for all $a, b \in L$, such that the axioms

$$[a, a] = 0 \quad \text{and} \quad [a, [b, c]] + [c, [a, b]] + [b, [a, c]] = 0$$

for all $a, b, c \in L$.

Definition 2.3.18. A *Lie coalgebra over* k is a pair (C, δ), where C is a vector space over k and $\delta : C \longrightarrow C\otimes C$ is a linear map, which satisfies

$$\tau\circ\delta = 0$$

and

$$(\mathrm{I} + (\tau\otimes\mathrm{I})\circ(\mathrm{I}\otimes\tau) + (\mathrm{I}\otimes\tau)\circ(\tau\otimes\mathrm{I}))\circ(\mathrm{I}\otimes\delta)\circ\delta = 0,$$

where $\tau = \tau_{C,C}$, and I is the appropriate identity map.

Exercise 2.3.37. Suppose that (C, δ) is a Lie coalgebra over the field k. Show that:

(a) $(\mathrm{I} + (\tau\otimes\mathrm{I})\circ(\mathrm{I}\otimes\tau) + (\mathrm{I}\otimes\tau)\circ(\tau\otimes\mathrm{I}))\circ(\delta\otimes\mathrm{I})\circ\delta = 0$.

(b) (C^*, m) is a Lie algebra over k, where $m = \delta^*|C^*\otimes C^*$.

Exercise 2.3.38. Suppose that the characteristic of k is not 2 and (L, m) is a finite-dimensional Lie algebra over k. Show that:

(a) (L^*, m^*) is a Lie coalgebra over k.

(b) There is a Lie coalgebra (C, δ) over k such that (L, m) is isomorphic as Lie algebras to (C^*, δ^*). [Hint: See Exercise 2.3.3.]

2.4 The wedge product

Let C be a coalgebra over the field k. The wedge product of subspaces of C corresponds to multiplication of subspaces in the dual algebra C^*. In this section we describe the wedge product intrinsically in terms of C and study some of its elementary properties. The wedge product is our basic tool for the study of filtrations of coalgebras in Chapter 3.7.

Definition 2.4.1. Suppose that C is a coalgebra over the field k. The *wedge product of subspaces* U *and* V *of* C is $U\wedge V = \Delta^{-1}(U\otimes C + C\otimes V)$.

Let U, V be subspaces of C and let $\pi_U : C \longrightarrow C/U, \pi_V : C \longrightarrow C/V$ be the projections. An equivalent description of the wedge product of U and V is $U \wedge V = \mathrm{Ker}((\pi_U \otimes \pi_V) \circ \Delta)$.

Set $\wedge^0 U = (0)$, $\wedge^1 U = U$, and $\wedge^n U = (\wedge^{n-1} U) \wedge U$ for $n > 1$. Suppose that D is a subcoalgebra of C. Then $D \subseteq D \wedge U$ and $D \subseteq U \wedge D$. In particular $D = D \wedge (0) = (0) \wedge D$. If $U, V \subseteq D$ we define $U \wedge_D V = \Delta^{-1}(U \otimes D + D \otimes V)$. Since $D = \Delta^{-1}(D \otimes D)$ it follows that $U \wedge_D V$ is the wedge product of U and V in the coalgebra D. The unadorned symbol $\wedge$ will always mean $\wedge_C$. If U' and V' are subspace of C such that $U \subseteq U'$ and $V \subseteq V'$ notice that $U \wedge V \subseteq U' \wedge V'$. Since Δ is one-one $(0) \wedge (0) = (0)$.

Now let $U_1, \ldots, U_n$ be subspaces of C, where $n > 1$, and set $U_1 \wedge \cdots \wedge U_n = (U_1 \wedge \cdots \wedge U_{n-1}) \wedge U_n$. By part (c) of the following proposition the wedge product is an associative operation. Thus all full parenthesized renditions of the formal expression $U_1 \wedge \cdots \wedge U_n$ compute to the same value.

Proposition 2.4.2. *Suppose that C is a coalgebra over the field k. Then:*

(a) $(KL)^\perp = K^\perp \wedge L^\perp$ *for subspaces K and L of C^*.*
(b) $U \wedge V = (U^\perp V^\perp)^\perp$ *for all subspaces U and V of C.*
(c) $(U \wedge V) \wedge W = U \wedge (V \wedge W)$ *for all subspaces U, V, and W of C.*
(d) *Let $U_1, \ldots, U_n$ be subspaces of C and $I_1, \ldots, I_n$ be subspaces of C^* such that $U_i = I_i^\perp$ for $1 \leq i \leq n$. Then $U_1 \wedge \cdots \wedge U_n = (I_1 \cdots I_n)^\perp$.*
(e) *Suppose that D and E are subcoalgebras of C. Then $D \wedge E$ is a subcoalgebra of C which contains both D and E.*

Proof. We first show part (a). Let K and L be subspaces of C^*. Then $K^\perp \wedge L^\perp \subseteq (KL)^\perp$ by definition of multiplication in C^*. Conversely, suppose that $c \in (KL)^\perp$. Then $\Delta(c) \in (K \otimes L)^\perp = K^\perp \otimes C + C \otimes L^\perp$, where the equation follows by Exercise 2.3.2. Thus $(KL)^\perp \subseteq K^\perp \wedge L^\perp$ and part (a) is established. Since $U = U^{\perp\perp}$ for all subspaces U of C part (b) follows from part (a).

Observe that $\overline{KL} = \overline{\overline{K}L} = \overline{K\overline{L}} = \overline{\overline{K}\,\overline{L}}$ for all subspaces K and L of C^* by part (c) of Proposition 2.3.6. Thus part (c) follows by part (b) and the algebra of closed subspaces of C^*; see part (c) of Proposition 2.3.6. Note that part (c) follows more directly from the equations $(U \wedge V) \wedge W = \mathrm{Ker}((\pi_U \otimes \pi_V \otimes \pi_W) \circ (\Delta \otimes \mathrm{I}_C) \circ \Delta)$ and $U \wedge (V \wedge W) = \mathrm{Ker}((\pi_U \otimes \pi_V \otimes \pi_W) \circ (\mathrm{I}_C \otimes \Delta) \circ \Delta)$.

Assume the hypothesis of part (d) holds. We may suppose that $n > 1$. Using part (b) we have

$$\begin{aligned} U_1\wedge\cdots\wedge U_n &= (U_1\wedge\cdots\wedge U_{n-1})\wedge U_n \\ &= (I_1\cdots I_{n-1})^\perp\wedge U_n \\ &= (\overline{I_1\cdots I_{n-1}}\,\overline{I_n})^\perp \\ &= (I_1\cdots I_{n-1}I_n)^\perp \end{aligned}$$

by induction on n, and thus part (d) follows by induction on n.

Part (e) follows by part (b) and part (b) of Proposition 2.3.5 since $D \subseteq D\wedge U$ and $D \subseteq U\wedge D$ for all subcoalgebras D of C and subspaces U of C. □

We next consider how the wedge product relates to subcoalgebras and coalgebra maps.

Proposition 2.4.3. *Let C be a coalgebra over the field k and let U, V be subspaces of C. Then:*

(a) $(U\wedge V)\cap D = (U\cap D)\wedge_D(V\cap D)$ *for all subcoalgebras D of C.*
(b) *Suppose that S, D, and E are subcoalgebras of C and S is simple. If $S \subseteq D\wedge E$ then $S \subseteq D$ or $S \subseteq E$.*

Let $f : C \longrightarrow D$ be a map of coalgebras over k. Then:

(c) $f(U\wedge V) \subseteq f(U)\wedge f(V)$.
(d) *Suppose that f is onto and* $\mathrm{Ker}(f) \subseteq U\cap V$. *Then* $f(U\wedge V) = f(U)\wedge f(V)$.

Proof. We first show part (a). Suppose that D is a subcoalgebra of C. Since $D = \Delta^{-1}(D\otimes D)$, we can use Exercise 2.4.1 to compute

$$\begin{aligned} (U\wedge V)\cap D &= (\Delta^{-1}(U\otimes C + C\otimes V))\cap(\Delta^{-1}(D\otimes D)) \\ &= \Delta^{-1}((U\otimes C + C\otimes V)\cap(D\otimes D)) \\ &= \Delta^{-1}((U\cap D)\otimes D + D\otimes(V\cap D)) \\ &= (U\cap D)\wedge_D(V\cap D). \end{aligned}$$

Part (a) is established.

Suppose that S, D and E are subcoalgebras of C, where S is simple, and $S \subseteq D\wedge E$. Then $S = (S\cap D)\wedge_S(S\cap E)$ by part (a). Since $S \neq (0)$ and $(0)\wedge(0) = (0)$ either $S\cap D \neq (0)$ or $S\cap E \neq (0)$. Thus S simple implies either $S \subseteq D$ or $S \subseteq E$. We have shown part (b). Part (c) is an exercise in definitions.

Now suppose that $f : C \longrightarrow D$ is an onto coalgebra map and U, V are subspaces of C. To show part (d) we need only show that $f(U) \wedge f(V) \subseteq f(U \wedge V)$ by part (c). Let $d \in f(U) \wedge f(V)$. Since f is onto $\Delta(d) = (f \otimes f)(\nu)$ for some $\nu \in U \otimes C + C \otimes V$ and $d = f(c)$ for some $c \in C$. Since f is a coalgebra map $(f \otimes f)(\Delta(c)) = \Delta(f(c)) = \Delta(d) = (f \otimes f)(\nu)$ which means that $\Delta(c) - \nu \in \mathrm{Ker}(f \otimes f)$. By assumption $\mathrm{Ker}(f) \subseteq U \cap V$. Thus $\mathrm{Ker}(f \otimes f) = \mathrm{Ker}(f) \otimes C + C \otimes \mathrm{Ker}(f) \subseteq U \otimes C + C \otimes V$ from which we deduce $\Delta(c) \in U \otimes C + C \otimes V$. We have shown $c \in U \wedge V$ and thus $f(U) \wedge f(V) \subseteq f(U \wedge V)$. □

The wedge product $\wedge = \wedge_C$ is a binary operation on the set of all subspaces of C. It is natural to ask for which subcoalgebras D of C is it the case that $\wedge$ induces a binary operation on the set of subspaces of D.

Definition 2.4.4. Let C be a coalgebra over the field k. A *saturated subcoalgebra of* C is a subcoalgebra D of C such that $U \wedge V \subseteq D$ for all subspaces U, V of D.

Observe that a subcoalgebra D of C is saturated if and only if $D \wedge D \subseteq D$, or equivalently $D \wedge D = D$. If D is saturated and U, V are subspaces of D then $U \wedge V = U \wedge_D V$ by part (a) of Proposition 2.4.3. In Exercise 2.4.11 we continue our discussion of saturated subcoalgebras.

The last result of this section relates the wedge product to coideals of C. Coideals have a simple description in terms of the wedge product. The sum of the coideals of C contained in a subspace U of C has a simple description in terms of wedge powers.

Suppose that U is a subspace of C and let I be the sum of all the coideals of C contained in U. Since (0) is a coideal of C and any sum of coideals of C is a coideal of C, it follows that I is a coideal of C. Observe that I is the unique coideal of C which is maximal among all of the coideals of C contained in U.

Theorem 2.4.5. *Suppose that C is a coalgebra over the field k.*

(a) *Let I be a subspace of C. Then I is a coideal of C if and only if $\epsilon(I) = (0)$ and $I \subseteq I \wedge I$.*
(b) *Let V be a subspace of C such that $V \subseteq \mathrm{Ker}(\epsilon)$. Then $I = \bigcap_{n=1}^{\infty}(\wedge^n V)$ is a coideal of C.*
(c) *Let U be a subspace of C and set $V = U^+$. Then the (unique) coideal I of C which is maximal among all of the coideals of C contained in U is described by $I = \bigcap_{n=1}^{\infty}(\wedge^n V)$.*

Proof. Part (a) is a matter of definitions. To show part (b) let $J = V^{\perp}$. Since $\epsilon(V) = (0)$ we have $\epsilon \in J$. By part (d) of Proposition 2.4.2 we have $\wedge^n V = (J^n)^{\perp}$ for all $n \geq 1$. Thus

$$I = \bigcap_{n=1}^{\infty} \wedge^n V = \bigcap_{n=1}^{\infty} (J^n)^{\perp} = (\sum_{n=1}^{\infty} J^n)^{\perp} = B^{\perp},$$

where $B = \sum_{n=1}^{\infty} J^n$. Since B is a subalgebra of C^* it follows that I is a coideal of C by part (a) of Proposition 2.3.7.

It remains to show part (c). Let J be any coideal of C which is contained in U. Then $J = J^+$ and $J \subseteq J \wedge J$ by part (a). In particular $J \subseteq V$ and $J \subseteq \wedge^n J$ for all $n \geq 1$. Since $J \subseteq V$ it follows that $J \subseteq \wedge^n V$ for all $n \geq 1$. Thus $J \subseteq \bigcap_{n=1}^{\infty} \wedge^n V = I$. Now I is a coideal of C by part (b). Since $J \subseteq I \subseteq V$ we have established part (c). □

Exercises

In the following exercises C and D are coalgebras over the field k.

Exercise 2.4.1. Let W and $\mathcal{W}$ be vector spaces over k. Suppose that U, V are subspaces of W and $\mathcal{U}, \mathcal{V}$ are subspaces of $\mathcal{W}$. Show that

$$(U \otimes \mathcal{W} + W \otimes \mathcal{U}) \cap (V \otimes \mathcal{V}) = (U \cap V) \otimes \mathcal{V} + V \otimes (\mathcal{U} \cap \mathcal{V}).$$

[Hint: Consider the projections $\pi_W : W \longrightarrow W/U$ and $\pi_{\mathcal{W}} : \mathcal{W} \longrightarrow \mathcal{W}/\mathcal{U}$ and the projections $\pi_V : V \longrightarrow V/(U \cap V)$ and $\pi_{\mathcal{V}} : \mathcal{V} \longrightarrow \mathcal{V}/(\mathcal{U} \cap \mathcal{V})$. Regarding $V \otimes \mathcal{V}$ and $(V/(U \cap V)) \otimes (\mathcal{V}/(\mathcal{U} \cap \mathcal{V}))$ as subspaces of $W \otimes \mathcal{W}$ and $(W/U) \otimes (\mathcal{W}/\mathcal{U})$ respectively observe that $\mathrm{Ker}(\pi_W \otimes \pi_{\mathcal{W}}) \cap (V \otimes \mathcal{V}) = \mathrm{Ker}(\pi_V \otimes \pi_{\mathcal{V}})$.]

Exercise 2.4.2. Suppose that V is a subspace of C. Show that $\wedge^n V = \wedge^{n+1} V$ for some $n \geq 0$ implies that $\wedge^n V = \wedge^{n+1} V = \wedge^{n+2} V = \cdots$.

Exercise 2.4.3. Suppose that $C = \bigoplus_{i \in I} D_i$ is the direct sum of subcoalgebras and $E = \bigoplus_{i \in I} E_i$, $F = \bigoplus_{i \in I} F_i$ are subcoalgebras of C where E_i, F_i are subcoalgebras of D_i for all $i \in I$. Show that $E \wedge F = \bigoplus_{i \in I} E_i \wedge_{D_i} F_i$.

Exercise 2.4.4. Suppose that U and V are subspaces of C and D is a subcoalgebra of C. Show that $U \cap D = (0) = V \cap D$ implies $(U \wedge V) \cap D = (0)$.

Exercise 2.4.5. Let $g, h \in \mathrm{G}(C)$. Show that

$$kg \wedge kh = \mathrm{P}_{g,h}(C) \bigoplus kg = \mathrm{P}_{g,h}(C) \bigoplus kh.$$

Exercise 2.4.6. Let U and V be subspaces of C and suppose that $U \subseteq \mathrm{Ker}(\epsilon)$. Show that $U \wedge V \subseteq V$, $V \wedge U \subseteq V$ and that $(\mathrm{Ker}\,\epsilon) \wedge V = V = V \wedge (\mathrm{Ker}(\epsilon))$.

Exercise 2.4.7. Let $\wedge^{cop}$ denote the wedging operation in C^{cop}. Show that:

(a) $U \wedge^{cop} V = V \wedge U$ for all subspaces U and V of C.

(b) If C is cocommutative then $U \wedge V = V \wedge U$ for all subspaces U and V of C.

Exercise 2.4.8. Suppose that $U \wedge V = V \wedge U$ for all subspaces U and V of C. Show that C is cocommutative. [Hint: For $a, b \in C^*$ let $I = ka$ and $J = kb$. Show that $kab = IJ = JI = kba$ and then show that C^* is commutative.]

Exercise 2.4.9. Suppose that L is a left coideal of C and that R is a right coideal of C. Show that:

(a) $L \wedge U$ is a left coideal of C for all subspaces U of C.

(b) $U \wedge R$ is a right coideal of C for all subspaces U of C.

(c) $L \wedge R$ is a subcoalgebra of C.

Definition 2.4.6. Let C be a coalgebra over k and (N, ρ) be a left C-comodule. Then $U \wedge X = \rho^{-1}(U \otimes N + C \otimes X)$ is the *wedge product of subspaces U of C and X of N.*

Exercise 2.4.10. Let (N, ρ) is a left C-comodule, U, V be subspaces of C and X be a subspace of N. Show that:

(a) $U \wedge X = (U^{\perp} \cdot X^{\perp})^{\perp}$, where we regard N^* as a left C^*-module under the transpose action induced by the rational right C^*-module action on N.

(b) $(U \wedge V) \wedge X = U \wedge (V \wedge X)$.

(c) If L is a left coideal of C then $L \wedge X$ is a left subcomodule of N.

Exercise 2.4.11. In this exercise we continue our discussion of saturated subcoalgebras of C. Show that:

(a) A subcoalgebra D of C is saturated if and only if $I = \overline{I^2}$, where $I = D^{\perp}$.

(b) C is saturated and that the intersection of saturated subcoalgebras of C is a saturated subcoalgebra of C.

(c) Every subspace of C is contained in a unique minimal saturated subcoalgebra of C.

(d) If the direct sum $D = \bigoplus_{i \in I} D_i$ of subcoalgebras of C is saturated then each D_i is saturated but the converse is not true in general.

(e) There is an onto coalgebra map $f : C \longrightarrow C'$ and a saturated subcoalgebra D of C such that $f(D)$ is not a saturated subcoalgebra of C'.

(f) If D is a saturated subcoalgebra of C and E is a saturated subcoalgebra of a coalgebra F over k then $D \otimes E$ is a saturated subcoalgebra of $C \otimes F$.

Definition 2.4.7. Let C be a coalgebra over k and U be a subspace of C. The unique minimal saturated subcoalgebra of C containing U is the *saturated closure of U in C.*

There is a notion for coalgebras corresponding to the notion of prime ideal for algebras. Recall that an ideal P of an algebra A over k is prime if whenever I, J are ideals of A and $IJ \subseteq P$ then either $I \subseteq P$ or $J \subseteq P$.

Definition 2.4.8. Let C be a coalgebra over the field k. A *prime subcoalgebra of C* is a subcoalgebra D of C such that whenever E, F are subcoalgebras of C and $D \subseteq E \wedge F$ then either $D \subseteq E$ or $D \subseteq F$.

Exercise 2.4.12. Let D be a subcoalgebra of C. Show that D is prime if and only if $D^{\perp}$ is a prime ideal of C^*.

2.5 The dual coalgebra

Let (A, m, η) be an algebra over the field k. We regard $A^* \otimes A^*$ as a subspace of $(A \otimes A)^*$ in the same manner that we regarded $C^* \otimes C^*$ as a subspace of $(C \otimes C)^*$ in Section 2.3. We wish to use the product m to define a coproduct on A^* according to

$$m^*(a^*) = \sum_{i=1}^{r} a_i^* \otimes b_i^* \in A^* \otimes A^*. \tag{2.21}$$

Generally (2.21) can not be realized for all $a^* \in A^*$.

Suppose that $a^* \in A^*$ satisfies (2.21). Then

$$a^*(ab) = \sum_{i=1}^{r} a_i^*(a) b_i^*(b) \tag{2.22}$$

for all $a, b \in A$. Observe there is at most one tensor in $A^* \otimes A^*$ which can satisfy (2.22). Set $\Delta_{a^*} = \sum_{i=1}^{r} a_i^* \otimes b_i^*$. Then (2.22) can be rewritten

$$a^*(ab) = \Delta_{a^*}(a \otimes b) \tag{2.23}$$

for all $a, b \in A$.

Let A^o be the subset of all $a^* \in A^*$ such that there exists a tensor $\Delta_{a^*} \in A^* \otimes A^*$ which satisfies (2.23). It is easy to see that A^o is a subspace of A^*. We show that $m^*(A^o) \subseteq A^o \otimes A^o$.

Suppose $a^* \in A^o$ and write $m^*(a^*) = \sum_{i=1}^{r} a_i^* \otimes b_i^* \in A^* \otimes A^*$ where $r = \text{Rank}(m^*(a^*))$. By Lemma 1.2.2 the sets $\{a_1^*, \ldots, a_r^*\}$ and $\{b_1^*, \ldots, b_r^*\}$ are linearly independent. Fix $1 \leq \ell \leq r$. By Corollary 1.2.4 there are $c, d \in A$ which satisfy $a_i^*(d) = \delta_{i,\ell} = b_i^*(c)$ for all $1 \leq i \leq r$. The calculation

$$a_\ell^*(ab) = \sum_{i=1}^{r} a_i^*(ab)b_i^*(c) = a^*(abc) = \sum_{i=1}^{r} a_i^*(a)b_i^*(bc)$$

for all $a, b \in A$ shows that $\Delta_{a_\ell^*}$ exists and $\Delta_{a_\ell^*} = \sum_{i=1}^r a_i^* \otimes (b_i^* \circ r(c))$. Likewise $\Delta_{b_\ell^*}$ exists and $\Delta_{b_\ell^*} = \sum_{i=1}^r (a_i^* \circ \ell(d)) \otimes b_i^*$. Thus $m^*(a^*) \in A^o \otimes A^o$ which means $m^*(A^o) \subseteq A^o \otimes A^o$. Identify k and k^* in the customary way.

Proposition 2.5.1. *Let (A, m, η) be an algebra over the field k. Then:*

(a) $A^o = m^{*-1}(A^* \otimes A^*)$.
(b) (A^o, Δ, ϵ) *is a coalgebra over k, where $\Delta = m^*|A^o$ and $\epsilon = \eta^*$.*

Proof. Part (a) is clear. Regarding part (b), we have shown that $\Delta(A^o) \subseteq A^o \otimes A^o$. By virtue of part (a) of Exercise 1.2.5 we may regard $A^* \otimes A^* \otimes A^*$ as a subspace of $(A \otimes A \otimes A)^*$ by $(a^* \otimes b^* \otimes c^*)(a \otimes b \otimes c) = a^*(a)b^*(b)c^*(c)$ for all $a^*, b^*, c^* \in A^*$ and $a, b, c \in A$. Since m satisfies the associative axiom (2.1) it follows that Δ satisfies the coassociative axiom (2.3). Since η satisfies the unit axiom (2.2) it follows that ϵ satisfies the counit axiom (2.4). □

Definition 2.5.2. The coalgebra (A^o, Δ, ϵ) of Proposition 2.5.1 is the *dual coalgebra of* (A, m, η).

We will ordinarily write A^o for (A^o, Δ, ϵ), denote an element of A^o by a^o, and write $\Delta(a^o) = a^o_{(1)} \otimes a^o_{(2)}$ for all $a^o \in A^o$. By part (b) of Exercise 1.2.5:

Corollary 2.5.3. *If A is a finite-dimensional algebra over the field k then $A^o = A^*$.* □

There are infinite-dimensional algebras A over k such that $A^o \neq A^*$. See Exercise 2.5.1.

Proposition 2.5.4. *Let $f : A \longrightarrow B$ be a map of algebras over the field k. Then $f^*(B^o) \subseteq A^o$ and $f^o : B^o \longrightarrow A^o$ is a map of dual coalgebras, where $f^o = f^*|B^o$.*

Proof. Let $b^o \in B^o$. The calculation

$$\begin{aligned} f^*(b^o)(ab) &= b^o(f(ab)) \\ &= b^o(f(a)f(b)) \\ &= b^o_{(1)}(f(a))b^o_{(2)}(f(b)) \\ &= (f^*(b^o_{(1)})(a))(f^*(b^o_{(2)})(b)) \end{aligned}$$

for all $a, b \in A$ shows that $\Delta_{f^*(b^o)}$ exists and $\Delta_{f^*(b^o)} = f^*(b^o_{(1)}) \otimes f^*(b^o_{(2)})$. Therefore $f^*(b^o) \in A^o$ and $\Delta(f^*(b)) = f^*(b^o_{(1)}) \otimes f^*(b^o_{(2)})$. As

$$\epsilon(f^*(b^o)) = f^*(b^o)(1) = b^o(f(1)) = b^o(1) = \epsilon(b^o)$$

the proposition is established. □

Regard A as a right A-module and as a left A-module under multiplication. Then A is an A-bimodule under these actions. The transpose actions, which are described by

$$a {\succ} a^*(b) = a^*(ba) \quad \text{and} \quad a^* {\prec} a(b) = a^*(ab)$$

respectively for $a, b \in A$ and $a^* \in A^*$, give A^* an A-bimodule structure.

By part (b) of the proposition below A^o consists of those functionals $a^* \in A^*$ whose kernels contain a cofinite ideal of A.

Proposition 2.5.5. *Let A be an algebra over the field k. Then:*

(a) *A^o is a locally finite left (respectively right) A-module. Moreover A^o is the unique maximal locally finite left (respectively right) A-submodule of A^*.*
(b) *Let $a^* \in A^*$. Then $a^* \in A^o$ if and only if $a^*(I) = (0)$ for some cofinite ideal I of A.*
(c) *Let V be a subspace of A^o. Then V is a subcoalgebra (respectively left coideal, right coideal) of A^o if and only if V is a A-sub-bimodule (respectively right A-submodule, left A-submodule) of A^o.*
(d) *Let $a^* \in A^o$. The subcoalgebra (respectively left coideal, right coideal) of A^o generated by a^* is $A {\succ} a^* {\prec} A$ (respectively $a^* {\prec} A$, $A {\succ} a^*$).*
(e) *The one-one inclusion reversing correspondence $U \mapsto U^\perp$ of subspaces of A and closed subspaces of A^* induces a one-one correspondence of the set of cofinite ideals (respectively cofinite left ideals, cofinite right ideals) of A and the set of finite-dimensional subcoalgebras (respectively finite-dimensional left coideals, finite-dimensional right coideals) of A^o.*

Proof. Suppose $a^* \in A^*$ and is not zero. We first show part (a). Let $a^* \in A^o$ and write $m^*(a^*) = \sum_{i=1}^r a_i^* \otimes b_i^* \in A^* \otimes A^*$. Since $a \succ a^*(b) = a^*(ba) = \sum_{i=1}^r a_i^*(b) b_i^*(a)$ for all $a, b \in A$ it follows that $a \succ a^* = \sum_{i=1}^r b_i^*(a) a_i^*$ for all $a \in A$. We have shown $a^* \in A^o$ implies that $A \succ a^*$ is finite-dimensional.

Conversely, suppose that $A \succ a^*$ is finite-dimensional and let $\{a_1^*, \ldots, a_r^*\}$ be a basis for $A \succ a^*$. Then there are $b_1^*, \ldots, b_r^* \in A^*$ such that $a \succ a^* = \sum_{i=1}^r b_i^*(a) a_i^*$ for all $a \in A$ which means Δ_{a^*} exists and $\Delta_{a^*} = \sum_{i=1}^r a_i^* \otimes b_i^*$. Thus $A \succ a^*$ finite-dimensional implies $a^* \in A^o$.

We have shown part (a) holds for the left A-module structure on A^*. This fact interpreted for A^{op} gives part (a) for the right A-module structure on A^*.

To show part (b), suppose that $a^* \in A^o$. Then $A \succ a^* \prec A$ is finite-dimensional by part (a) which means $I = (A \succ a^* \prec A)^\perp = \bigcap_{i=1}^r \mathrm{Ker}(a_i^*)$ is a cofinite subspace of A by Exercise 1.3.15, where $a_1^*, \ldots, a_r^*$ span $A \succ a^* \prec A$. Since $(A \succ a^* \prec A)(AIA) \subseteq (A \succ a^* \prec A)(I) = (0)$ it follows that $AIA \subseteq I$. Therefore I is an ideal of A. We have shown that $a^* \in A^o$ implies $a^*(I) = (0)$ for some cofinite ideal I of A.

Suppose that $a^*(I) = (0)$ for some cofinite ideal I of A. Let $\pi_I : A \longrightarrow A/I$ be the projection. Since A/I is finite-dimensional $(A/I)^o = (A/I)^*$ by Corollary 2.5.3. Now $\pi_I^* : (A/I)^o \longrightarrow A^o$ is a coalgebra map by Proposition 2.5.4. Since $a^* \in I^\perp = \mathrm{Im}(\pi_I^*) \subseteq A^o$ part (b) is established.

To show part (c) we first observe that $\jmath : A \longrightarrow (A^o)^*$ defined by $\jmath(a)(a^o) = a^o(a)$ for all $a \in A$ and $a^o \in A^o$ is an algebra map and that $\mathrm{Im}(\jmath)$ is a dense subspace of $(A^o)^*$. The reader can check that

$$\jmath(a) \rightharpoonup a^o = a \succ a^o \quad \text{and} \quad a^o \prec a = a^o \leftharpoonup \jmath(a)$$

hold for all $a \in A$ and $a^o \in A^o$. Let U be a finite-dimensional subspace of A^*. Since $\mathrm{Im}(\jmath)$ is a dense subspace of $(A^o)^*$, given any functional $f \in (A^o)^*$ there is an $a \in A$ such that $\jmath(a)|U = f|U$ by Proposition 1.3.6. Now it follows by part (b) of Proposition 2.3.5 that V is a subcoalgebra (respectively left coideal, right coideal) of A^o if and only if V is an A-sub-bimodule (respectively right A-submodule, left A-submodule) of A^o. We have shown part (c). Part (d) follows immediately from part (c).

It remains to show part (e). We first recall that all finite-dimensional subspaces of A^* are closed by Corollary 1.3.9. Let U be any subspace of A. Since $(A/U)^* \simeq U^\perp$, under the correspondence $U \mapsto U^\perp$ described in part (e) the cofinite subspaces of A correspond to the finite-dimensional subspaces of A^*. Let I be a subspace of A. It is easy to see that I is an ideal (respectively a left ideal, right ideal) of A if and only if $I^\perp$ is a right

A-sub-bimodule (respectively right A-submodule, left A-submodule) of A^*. We use parts (a) and (c) at this point to conclude the proof of part (e). □

As a consequence of part (b) of Proposition 2.5.5:

Corollary 2.5.6. *Let A be an algebra over the field k. Then*

$$A^o = \sum_I I^{\perp} = \sum_I \mathrm{Im}(\pi_I^*),$$

where I runs over the cofinite ideals of A and $\pi_I : A \longrightarrow A/I$ is the projection. □

We explore the relationship between the algebra structure on A and the coalgebra structure on A^o more fully in Section 2.6.

Exercises

In the following exercises A and B are algebras over the field k.

Exercise 2.5.1. Let A be an infinite-dimensional field extension of k. Show that $A^o = (0)$.

Exercise 2.5.2. Let $A = k1 \bigoplus \mathcal{M}$ where $\mathcal{M}^2 = (0)$. Show that $A^o = A^*$. Compare with Exercise 2.3.14. [Hint: See part (c) of Exercise 1.3.15.]

Exercise 2.5.3. Show that:

(a) $\mathrm{G}(A^o) = \mathrm{Alg}(A, k)$ is the set of k-algebra homomorphisms $\eta : A \longrightarrow k$.

(b) $\mathrm{Alg}(A, k)$ is a linearly independent subset of A^*.

Definition 2.5.7. Let A be an algebra over k. An *η:ξ-derivation of A* is a linear map $f : A \longrightarrow k$ which satisfies $f(ab) = \eta(a)f(b) + f(a)\xi(b)$ for all $a, b \in A$, where $\eta, \xi \in \mathrm{Alg}(A, k)$.

Notice that the set of η:ξ-derivations of A is a subspace of A^*.

Exercise 2.5.4. Let $\eta, \xi \in \mathrm{Alg}(A, k) = \mathrm{G}(A^o)$. Show that $\mathrm{P}_{\eta,\xi}(A^o)$ is the set of η:ξ-derivations of A.

Exercise 2.5.5. Let $\mathrm{tr} \in A^o$. Show that tr is a cocommutative element of A^o if and only if $\mathrm{tr}(ab) = \mathrm{tr}(ba)$ for all $a, b \in A$.

Exercise 2.5.6. Show that A commutative implies that A^o is cocommutative.

Exercise 2.5.7. Let f be a derivation of A. Show that $f^*(A^o) \subseteq A^o$ and that f^o is a coderivation of A^o, where $f^o = f^*|A^o$. Compare with Exercise 2.3.9.

Exercise 2.5.8. Let $k = \boldsymbol{R}$ be the field of real numbers and $A = \boldsymbol{C}$ be the field of complex numbers. Let $\{1, \imath\}$ be the usual basis for A over $\boldsymbol{R}$, where $\imath^2 = -1$, and let $\{c, s\}$ be the dual basis for A^*. Show that $A^o = A^*$ is the coalgebra of Exercise 2.3.6 by using (2.22) to determine the coproduct of A^o.

Exercise 2.5.9. Suppose $A = k[\alpha]$ is a 2-dimensional field extension of k and $\alpha^2 = a \in k$. Show that $A^o = A^*$ has a basis $\{s, c\}$ such that

$$\begin{array}{ll} \Delta(s) = s \otimes c + c \otimes s, & \epsilon(s) = 0, \\ \Delta(c) = c \otimes c + a(s \otimes s), & \epsilon(c) = 1. \end{array}$$

Exercise 2.5.10. Let C be a finite-dimensional coalgebra over k. Show that $C \simeq A^o$ for some finite-dimensional algebra over k. [Hint: See Exercise 2.3.3.]

Exercise 2.5.11. Suppose $n \geq 1$ and $A = \mathrm{M}_n(k)$ is the algebra of $n \times n$ matrices over k. Let $\{E^i_j\}_{1 \leq i,j \leq n}$ be the standard basis for A; thus $E^i_j = (a_{r,s})$ is defined by $a_{r,s} = \delta_{(r,s),(i,j)}$ for all $1 \leq i, j, r, s \leq n$. See Exercise 2.3.24

(a) Let $\{e_{i,j}\}_{1 \leq i,j \leq n}$ be the dual basis for $A^* = A^o$. Show that $A^0 \simeq \mathrm{C}_n(k)$ by using (2.22) to determine the coproduct of A^o on the $e_{i,j}$'s.

(b) Let $\mathrm{Tr} : A \longrightarrow k$ be the trace function. Show that Tr is a cocommutative element of A^o and that $\{\mathrm{Tr}\}$ is a basis for the subspace of cocommutative elements of A^o. Compare with Exercise 2.1.30. [Hint: See Exercise 2.5.5.]

Exercise 2.5.12. Let S be a finite monoid and $A = k[S]$ be the monoid algebra of S over k. Show that A^o is the coalgebra of Example 2.1.13.

Exercise 2.5.13. Regarding $A^* \otimes B^*$ as a subspace of $(A \otimes B)^*$, show that $A^o \otimes B^o = (A \otimes B)^o$, where the tensor products in the equation have the tensor product coalgebra structure and the tensor product algebra structure respectively. [Hint: Let I, J be cofinite ideals of A, B respectively. Show that $I \otimes B + A \otimes J$ is a cofinite ideal of $A \otimes B$. If K is a cofinite ideal of $A \otimes B$ show that $I = \{a \in A \,|\, a \otimes 1 \in K\}$ and $J = \{b \in B \,|\, 1 \otimes b \in K\}$ are cofinite ideals of A and B respectively. See part (b) of Exercise 1.3.14.]

Exercise 2.5.14. Suppose that $A \simeq A_1 \times \cdots \times A_r$ as algebras over k. Show that $A^o \simeq A_1^o \bigoplus \cdots \bigoplus A_r^o$ as coalgebras over k. [Hint: See Exercises 2.1.18 and 2.3.16.]

Exercise 2.5.15. Let $A = k[X]/(X^n)$, where $n \geq 1$ and $k[X]$ is the polynomial algebra in indeterminate X over k. Let $\{c_0, \ldots, c_{n-1}\}$ the basis for A^o dual to the basis $\{1, x, \ldots, x^{n-1}\}$ for A (x is the coset with representative X). Use (2.22) to show that $A^o \simeq \mathrm{P}_n(k)$.

Exercise 2.5.16. Suppose that k is algebraically closed and $A = k[X]$ is the polynomial algebra in indeterminate X over k. Show that:

(a) If C is a non-zero finite-dimensional subcoalgebra of A^o then

$$C \simeq \mathrm{P}_{n_1}(k) \bigoplus \cdots \bigoplus \mathrm{P}_{n_r}(k)$$

for some $r \geq 1$ and $n_1, \ldots, n_r \geq 1$.

(b) $\mathrm{P}_\infty(k)$ is isomorphic to a subcoalgebra of A^o which is a dense subspace of A^*.

[Hint: For part (a) note that $C = I^\perp$ for some proper ideal $I = (f(X))$ of $k[X]$ by part (e) of Proposition 2.5.5. Thus $C \simeq (k[X]/(f(X)))^o$ as a coalgebra. See Exercises 2.5.14, 2.5.15, and 2.3.29.]

Exercise 2.5.17. Let $A = k[X]$ be the polynomial algebra in X over k and let $\mathcal{V}$ be the set of all infinite tuples $(a_0, a_1, a_2, \ldots)$, where $a_i \in k$ for all $i \geq 0$, with vector space structure given by component operations as in the case of k^n. Show that as a vector space $k[X]^o$ can be identified with the subset of $\mathcal{V}$ consisting those tuples $(a_0, a_1, a_2, \ldots)$ which satisfy the following condition: for some $n \geq 1$ there are $\alpha_0, \ldots, \alpha_{n-1} \in k$ such that

$$\alpha_0 a_\ell + \alpha_1 a_{1+\ell} + \cdots + \alpha_{n-1} a_{n-1+\ell} + a_{n+\ell} = 0$$

for all $\ell \geq 0$. [Hint: The association $a^* \mapsto (a^*(1), a^*(X), a^*(X^2), \ldots)$ describes a vector space isomorphism $k[X]^* \simeq \mathcal{V}$. Note that $a^* \in k[X]^o$ if and only if $a^*(I) = (0)$ where $I = (f(X))$ for some monic polynomial $f(X) = \alpha_0 + \alpha_1 X + \cdots + X^n \in k[X]$.]

Exercise 2.5.18. Let A be a finite-dimensional with basis $\{a_i\}_{i \in I}$. Describe the structure maps m and η of A in terms of structure constants by

$$a_i a_j = m_{i,j}^{\ell} a_\ell \quad \text{and} \quad 1 = \eta(1_k) = \eta^\ell a_\ell$$

for all $i, j \in I$. We are following the Einstein summation convention. Show that:

(a) $m \circ (m \otimes \mathrm{I}_A) = m \circ (\mathrm{I}_A \otimes m)$ is equivalent to $m_{r,s}^u m_{u,t}^i = m_{s,t}^u m_{r,u}^i$ for all $r, s, t, i \in I$.

(b) $m \circ (\eta \otimes \mathrm{I}_A) = \mathrm{I}_A = m \circ (\mathrm{I}_A \otimes \eta)$ is equivalent to $\eta^u m_{i,u}^j = \delta_i^j = m_{i,u}^j \eta^u$ for all $i, j \in I$.

(c) A is commutative if and only if $m_{r,s}^i = m_{s,r}^i$ for all $i, r, s \in I$.

Exercise 2.5.19. In this exercise we explore duality in the finite-dimensional case using a structure constant analysis. The reader should work out Exercises 2.1.22 and 2.5.18 before starting this one. We follow the notation conventions of these two exercises.

Let $n \geq 1$ and $\mathcal{A}, \mathcal{C}$ be vector spaces over k with bases $\{a_1, \ldots, a_n\}$, $\{c_1, \ldots, c_n\}$ respectively. Let $\{m_{i,j}^{\ell}\}_{1 \leq i,j,\ell \leq n}$ and $\{\eta^i\}_{1 \leq i \leq n}$ be indexed sets of scalars in k. Set

$$\Delta_{\ell}^{i,j} = m_{i,j}^{\ell} \quad \text{and} \quad \epsilon_i = \eta^i$$

for all $1 \leq i, j, \ell \leq n$.

Define linear maps $m : \mathcal{A} \otimes \mathcal{A} \longrightarrow \mathcal{A}$ and $\eta : k \longrightarrow \mathcal{A}$ by

$$m(a_i \otimes a_j) = m_{i,j}^{\ell} a_{\ell} \quad \text{and} \quad \eta(1_k) = 1 = \eta^{\ell} a_{\ell}$$

for all $1 \leq i, j \leq n$.

Define linear maps $\Delta : \mathcal{C} \longrightarrow \mathcal{C} \otimes \mathcal{C}$ and $\epsilon : \mathcal{C} \longrightarrow k$ by

$$\Delta(c_i) = \Delta_i^{j,\ell} c_j \otimes c_{\ell} \quad \text{and} \quad \epsilon(c_i) = \epsilon_i$$

for all $1 \leq i \leq n$.

(a) Show that $(\mathcal{A}, m, \eta)$ is an algebra over k if and only if $(\mathcal{C}, \Delta, \epsilon)$ is a coalgebra over k.

(b) Suppose that $\mathcal{A} = \mathcal{C}^*$ as vector spaces and that $\{a_1, \ldots, a_n\}$ is the dual basis of $\{c_1, \ldots, c_n\}$. If $(\mathcal{C}, \Delta, \epsilon)$ is a coalgebra over k, show that $(\mathcal{A}, m, \eta)$ is the dual algebra of $(\mathcal{C}, \Delta, \epsilon)$.

(c) Suppose that $\mathcal{C} = \mathcal{A}^*$ as vector spaces and that $\{c_1, \ldots, c_n\}$ is the dual basis of $\{a_1, \ldots, a_n\}$. If $(\mathcal{A}, m, \eta)$ is an algebra over k, show that $(\mathcal{C}, \Delta, \epsilon)$ is the dual coalgebra of $(\mathcal{A}, m, \eta)$.

(d) Use parts (a) and (b) to show that every finite-dimensional algebra over k is the dual algebra of a finite-dimensional coalgebra over k. Compare with Exercise 2.3.3.

(e) Use parts (a) and (c) to show that every finite-dimensional coalgebra over k is the dual coalgebra of a finite-dimensional algebra over k. Compare with Exercise 2.5.10.

Exercise 2.5.20. Show that $G : k\text{-Alg} \longrightarrow k\text{-Coalg}$ is a contravariant functor, where $G(A) = A^o$ for k-algebras A and $G(f) = f^o$ for maps f of k-algebras. See Definitions 2.1.26 and 2.1.27.

Exercise 2.5.21. Let (A, m, η) be an algebra over k. We consider the wedge product in A^o. Show that $(XY)^{\perp} \cap A^o = (X^{\perp} \cap A^o) \wedge (Y^{\perp} \cap A^o)$ for subspaces X, Y of A. [Hint: Let $f \in A^o$. Then $m^*(f) \in V \otimes V$ for some finite-dimensional subspace $V \subseteq A^o$. Consider $\pi(V)$, where $\pi : A^* \longrightarrow V^*$ is the restriction, and rewrite the proof of the Rank-Nullity Theorem applied to $\pi|V : V \longrightarrow \pi(V)$ to find an appropriate basis for V^*.]

2.6 Double duals

Let C be a coalgebra over k and A be an algebra over k. In this section, mainly in the exercises, we explore the relationship between the associations $C \mapsto C^*$ and $A \mapsto A^o$ and related concepts.

Proposition 2.6.1. *Suppose that A is an algebra over the field k and let $\jmath_A : A \longrightarrow (A^o)^*$ be the linear map defined by $\jmath_A(a)(a^o) = a^o(a)$ for all $a \in A$ and $a^o \in A^o$. Then:*

(a) *$\jmath_A : A \longrightarrow (A^o)^*$ is an algebra map.*
(b) *$\mathrm{Ker}(\jmath_A)$ is the intersection of the cofinite ideals of A.*
(c) *$\mathrm{Im}(\jmath_A)$ is a dense subspace of $(A^o)^*$.*

□

The proof follows directly from definitions and is left as an exercise for the reader. The counterpart of Proposition 2.6.1 for coalgebras is:

Proposition 2.6.2. *Suppose that C is a coalgebra over the field k and let $\jmath_C : C \longrightarrow (C^*)^*$ be the linear map defined by $\jmath_C(c)(c^*) = c^*(c)$ for all $c^* \in C^*$ and $c \in C$. Then:*

(a) *$\mathrm{Im}(\jmath_C) \subseteq (C^*)^o$ and $\jmath_C : C \longrightarrow (C^*)^o$ is a coalgebra map.*
(b) *$\jmath_C$ is one-one.*
(c) *$\mathrm{Im}(\jmath_C)$ is the set of all $\alpha \in (C^*)^*$ which vanish on a closed cofinite ideal of C^*.*

Proof. We first show part (a). Let $c \in C$. Then the calculation

$$\jmath_C(c)(ab) = ab(c) = a(c_{(1)})b(c_{(2)}) = (\jmath_C(c_{(1)})(a))(\jmath_C(c_{(2)})(b))$$

for all $a, b \in C^*$ shows that $\Delta_{\jmath_C(c)}$ exists and $\Delta_{\jmath_C(c)} = \jmath_C(c_{(1)}) \otimes \jmath_C(c_{(2)})$. Since

$$\epsilon(\jmath_C(c)) = \jmath_C(c)(1_{C^*}) = \jmath_C(c)(\epsilon) = \epsilon(c)$$

part (a) now follows. Part (b) is easy to see.

To show part (c) we first note that if $\alpha \in (C^*)^*$ vanishes on a cofinite closed ideal of C^* then $\alpha \in \mathrm{Im}(\jmath_C)$ by Corollary 1.3.13. Conversely, suppose that $c \in C$. Then the subcoalgebra D of C which c generates is finite-dimensional by Theorem 2.2.3. Let $\{c_1, \ldots, c_r\}$ span D. Then $I = \bigcap_{i=1}^r \mathrm{Ker}(\jmath_C(c_i))$ is a cofinite subspace of C^* by part (c) of Exercise 1.3.15. By definition I is a closed subspace of C^*. Now I is an ideal of C^* by part (e) of Proposition 2.5.5. Since $\jmath_C(c)(I) = (0)$ the proof of part (c) is complete. □

Since $A^* = A^o$ when A is a finite-dimensional algebra over k we have as a corollary of Propositions 2.6.1 and 2.6.2:

Corollary 2.6.3. *Suppose that A (respectively C) is a finite-dimensional algebra (respectively coalgebra) over k. Then $\jmath_A : A \longrightarrow (A^o)^*$ (respectively $\jmath_C : C \longrightarrow (C^*)^o$) is an isomorphism of algebras (respectively coalgebras).* □

In part (e) of Proposition 2.6.1 we broached the connection between the algebra structure of A and the coalgebra structure of A^o. We are now in a position to pursue this connection much more fully.

Proposition 2.6.4. *Let A be an algebra over the field k. Then:*

(a) *If a subspace I of A is a subalgebra (respectively ideal, left ideal, right ideal) of A then $I^\perp \cap A^o$ is a coideal (respectively subcoalgebra, left coideal, right coideal) of A^o.*
(b) *If a subspace U of A^o is a coideal (respectively subcoalgebra, left coideal, right coideal) of A^o then $U^\perp$ is a subalgebra (respectively ideal, left ideal, right ideal) of A.*

Proof. We let $(\ )^{\perp}$ denote the operation $(\ )^{\perp}$ for the vector space A^o and its dual $(A^o)^*$. First part (a). Let $\jmath : A \longrightarrow (A^o)^*$ be the algebra map of Proposition 2.6.1. By part (c) of the same $\mathrm{Im}(\jmath)$ is a dense subspace of $(A^o)^*$. For a subspace I of A observe that $I^\perp \cap A^o = \jmath(I)^{\perp}$. Thus part (a) follows by part (a) of Proposition 2.3.7. As for part (b), observe that $U^\perp = \jmath^{-1}(U^{\perp})$ for subspaces U of A^o. Thus part (b) follows by part (b) of Proposition 2.3.7. □

Exercises

In the following exercises C is a coalgebra over k and $\jmath_C : C \longrightarrow (C^*)^o$ is the coalgebra map of Proposition 2.6.2. Likewise A is an algebra over k and $\jmath_A : A \longrightarrow (A^o)^*$ is the algebra map of Proposition 2.6.1.

Exercise 2.6.1. Use part (a) of Proposition 2.6.2 and parts (a) and (d) of Proposition 2.5.5 to give yet another proof of Theorem 2.2.3, the Fundamental Theorem of Coalgebras.

Exercise 2.6.2. Show that $\mathrm{P}_\infty(k) \simeq k[[X]]^o$ as coalgebras by constructing an explicit isomorphism.

Exercise 2.6.3. Suppose that $f : A \longrightarrow B$ is a map of algebras over k, let $f^o : B^o \longrightarrow A^o$ be the associated map of dual coalgebras, and let $\mathcal{J}$ denote the set of cofinite ideals of B.

(a) Show that $\mathrm{Ker}(f^o) = \sum_{J\in\mathcal{J}}(J + \mathrm{Im}(f))^{\perp}$.

(b) Show that $\mathrm{Im}(f^o) = \sum_{J\in\mathcal{J}}(f^{-1}(J))^{\perp}$.

Definition 2.6.5. The full subcategory of k-Alg (respectively of k-Coalg) whose objects are finite-dimensional algebras (respectively coalgebras) over k is denoted $k\text{-Alg}_{fd}$ (respectively $k\text{-Coalg}_{fd}$).

Exercise 2.6.4. Let $F : k\text{-Coalg} \longrightarrow k\text{-Alg}$ and $G : k\text{-Alg} \longrightarrow k\text{-Coalg}$ be the functors of Exercises 2.3.36 and 2.5.20 respectively. Show that:

(a) The adjoint relation

$$\mathrm{Hom}_{\mathrm{Coalg}}(C, G(A)) \simeq \mathrm{Hom}_{\mathrm{Alg}}(A, F(C))$$

given by $f \mapsto f^* \circ \jmath_A$ (with inverse given by $f \mapsto f^0 \circ \jmath_C$) is satisfied.

(b) F and G restrict to categorical anti-equivalences of $k\text{-Coalg}_{fd}$ and $k\text{-Alg}_{fd}$.

Definition 2.6.6. A *proper algebra over* k is an algebra A over k such that the intersection of the cofinite ideals of A is (0), or equivalently the algebra map $\jmath_A : A \longrightarrow (A^o)^*$ of Proposition 2.6.1 is one-one.

Exercise 2.6.5. Show that A is a proper algebra if and only if A is a subalgebra of a direct product of finite-dimensional algebras over k.

Exercise 2.6.6. Show that C^* is a proper algebra. [Hint: See Theorem 2.2.3, (1.2), part (b) of Proposition 1.3.8 and part (d) of Proposition 2.3.7.]

Exercise 2.6.7. Suppose that C is the dual coalgebra of some algebra over k. Show that $C \simeq A^o$, where A is a proper algebra over k. [Hint: See Exercise 2.6.3.]

Definition 2.6.7. Let A (respectively C) be an algebra (respectively coalgebra) over k. Then A (respectively C) *is reflexive* if $\jmath_A : A \longrightarrow (A^o)^*$ defined in Proposition 2.6.1 (respectively $\jmath_C : C \longrightarrow (C^*)^o$ defined in Proposition 2.6.2) is an isomorphism.

Notice that reflexive algebras over k are proper.

Exercise 2.6.8. Show that C is reflexive if and only if all cofinite ideals of C^* are closed subspaces of C^*. [Hint: See part (e) of Proposition 2.5.5.]

Exercise 2.6.9. Let C be the coalgebra of Exercise 2.3.14. Show that C is reflexive if and only if C is finite-dimensional.

Exercise 2.6.10. Show that:

(a) $(\jmath_C)^* \circ \jmath_{C^*} = \mathrm{I}_{C^*}$.

(b) C^* is a proper algebra.

(c) C reflexive implies that C^* is reflexive.

Exercise 2.6.11. Show that:

(a) $(\jmath_A)^o \circ \jmath_{A^o} = \mathrm{I}_{A^o}$.

(b) If I is a cofinite ideal of A then there is a cofinite ideal J of $(A^o)^*$ such that $\jmath_A^{-1}(J) \subseteq I$.

(c) A reflexive implies that A^o is reflexive.

Exercise 2.6.12. Suppose that k is algebraically closed. Find (up to isomorphism) all algebras and coalgebras over k of dimension 1, 2 or 3. In each case find a basis and express the structure maps in terms of the basis. See Exercise 2.5.19. [Hint: Let A be a finite-dimensional algebra over k. Then $\mathrm{Rad}(A)$ is a nilpotent ideal of A. If $\mathcal{M}_1, \ldots, \mathcal{M}_r$ are pairwise coprime ideals of A then $A/(\bigcap_{i=1}^r \mathcal{M}_i) \simeq (A/\mathcal{M}_1) \times \cdots \times (A/\mathcal{M}_r)$ by the Chinese Remainder Theorem. Since k is algebraically closed maximal ideals of A have codimension 1.]

The next series of exercises concern sufficient conditions for a coalgebra to be reflexive.

Definition 2.6.8. An *almost left noetherian algebra over* k is an algebra A over k whose cofinite left ideals are finitely generated.

Exercise 2.6.13. Prove the following theorem:

Theorem 2.6.9. *Let A be a finitely generated algebra over the field k. Then A is almost left noetherian.* □

[Hint: Let L be a cofinite left ideal of A and write $A = L \bigoplus U$ for some subspace U of A. Since U is finite-dimensional there is a finite-dimensional subspace V of A which contains U and generates A as an algebra. Let $W = L \cap (k1 + V + V^2)$. Show that $V^n \subseteq AW + U$ for all $n \geq 0$.]

Exercise 2.6.14. Suppose that C^* contains a finitely generated dense subalgebra. Show that:

(a) C^* is almost left noetherian.

(b) C is reflexive.

[Hint: See part (b) of Proposition 2.3.6, Exercises 1.3.10 and 1.3.12.]

Exercise 2.6.15. Suppose that A is a left almost noetherian algebra. Show that the coalgebra A^o is reflexive. [Hint: See part (b) of Proposition 2.3.6 and part (a) of Exercise 2.6.14.]

Exercise 2.6.16. Use part (a) of Exercise 2.6.14 to show that $\mathrm{P}_\infty(k)$, and more generally the coalgebra of Exercise 2.3.32, is reflexive.

Definition 2.6.10. Let $f : U \longrightarrow V$ be a map of vector spaces over k. Then f is an *almost one-one linear map* if $\mathrm{Ker}(f)$ is finite-dimensional, f is an *almost onto linear map* if $\mathrm{Im}(f)$ is a cofinite subspace of V and f is an *almost isomorphism* if f is an almost one-one and an almost onto linear map.

Exercise 2.6.17. Suppose that $f : C \longrightarrow D$ is an almost onto map of coalgebras. Show that D reflexive implies that C is reflexive.

There is a natural way to pair algebras and coalgebras.

Exercise 2.6.18. Let $\beta : A{\times}C \longrightarrow k$ be a bilinear form, $\beta_\ell : A \longrightarrow C^*$ be defined by $\beta_\ell(a)(c) = \beta(a, c)$ and $\beta_r : C \longrightarrow A^*$ be defined by $\beta_r(c)(a) = \beta(a, c)$ for all $a \in A$ and $c \in C$. Show that the following are equivalent:

(a) $\beta_\ell : A \longrightarrow C^*$ is an algebra map.

(b) $\beta(ab, c) = \beta(a, c_{(1)})\beta(b, c_{(2)})$ and $\beta(1, c) = \epsilon(c)$ for all $a, b \in A$ and $c \in C$.

(c) $\mathrm{Im}(\beta_r) \subseteq A^o$ and $\beta_r : C \longrightarrow A^o$ is a coalgebra map.

Definition 2.6.11. Let A be an algebra over k and C be a coalgebra over k. A *pairing of A and C* is a bilinear map $\beta : A{\times}C \longrightarrow k$ which satisfies condition (b) of Exercise 2.6.18.

Exercise 2.6.19. Show that:

(a) $\beta : C^*{\times}C \longrightarrow k$ defined by $\beta(c^*, c) = c^*(c)$ for all $c^* \in C^*$ and $c \in C$ is a pairing of C^* and C.

(b) $\beta : A{\times}A^o \longrightarrow k$ defined by $\beta(a, a^o) = a^o(a)$ for all $a \in A$ and $a^o \in A^o$ is a pairing of A and A^o.

2.7 The cofree coalgebra on a vector space

The construction in this section realizes the dual notion of the free algebra on a vector space.

Definition 2.7.1. Let V be a vector space over k. A *cofree coalgebra on* V is a pair $(\pi, T^{co}(V))$ such that:

(a) $T^{co}(V)$ is a coalgebra over k and $\pi : T^{co}(V) \longrightarrow V$ is a linear map;
(b) If C is a coalgebra over k and $f : C \longrightarrow V$ is a linear map there exists a coalgebra map $F : C \longrightarrow T^{co}(V)$ determined by $\pi \circ F = f$.

If $(\pi, T^{co}(V))$ and $(\pi', T^{co}(V)')$ are cofree coalgebras on V then there exists a coalgebra isomorphism $F : T^{co}(V) \longrightarrow T^{co}(V)'$ determined by $\pi \circ F = \pi'$. See Exercise 2.7.2. Thus we will refer to a cofree coalgebra as the cofree coalgebra.

The construction of the cofree coalgebra is a rather interesting exercise whose details will require some careful explanation. We first suppose that V is a finite-dimensional vector space over k. In this case the linear map $\imath_V : V \longrightarrow V^{**}$, defined by $\imath_V(v)(v^*) = v^*(v)$ for all $v \in V$ and $v^* \in V^*$, is an isomorphism. See Corollary 1.2.4.

Lemma 2.7.2. *Let V be a finite-dimensional vector space over the field k and let $(\iota, T(V^*))$ be a free algebra on the dual vector space V^* of V. Then $(\pi, T(V^*)^o)$ is the cofree coalgebra on V, where $\pi = (\imath_V)^{-1} \circ (\iota^* | T(V^*)^o)$.*

Proof. Let C be a coalgebra over k and suppose that $f : C \longrightarrow V$ is a linear map. Then the transpose $f^* : V^* \longrightarrow C^*$ is a linear map from V^* to the dual algebra C^* of C. By the universal mapping property of the free algebra on V^* there is an algebra map $\mathsf{f} : T(V^*) \longrightarrow C^*$ determined by $\mathsf{f} \circ \iota = f^*$. Let $\mathsf{f}^o : (C^*)^o \longrightarrow T(V^*)^o$ be the coalgebra map of Proposition 2.5.4 and let $\jmath_C : C \longrightarrow (C^*)^o$ be the coalgebra map of Proposition 2.6.2, which is defined by $\jmath_C(c)(c^*) = c^*(c)$ for all $c \in C$ and $c^* \in C^*$. Then the composite $F = \mathsf{f}^o \circ \jmath_C$ is a coalgebra map $F : C \longrightarrow T(V^*)^o$. It is easy to see that $(\iota^* | T(V^*)^o) \circ \mathsf{f}^o \circ \jmath_C = \imath_V \circ f$; thus $\pi \circ F = f$.

Suppose that $F' : C \longrightarrow T(V^*)^o$ is also a coalgebra map which satisfies $\pi \circ F' = f$. To complete the proof of the lemma we need only show that $F' = F$. Let $\jmath_{T(V^*)} : T(V^*) \longrightarrow (T(V^*)^o)^*$ be the algebra map of Proposition 2.6.1, which is defined by $\jmath_{T(V^*)}(a)(a^o) = a^o(a)$ for all $a \in T(V^*)$ and $a^o \in T(V^*)^o$. We first note that $\pi^* = \jmath_{T(V^*)} \circ \iota$. To see this we observe that $\pi^* = ((\imath_V)^{-1} \circ (\iota^* | T(V^*)^o))^* = (\iota^* | T(V^*)^o)^* \circ ((\imath_V)^*)^{-1} = (\iota^* | T(V^*)^o)^* \circ \imath_{V^*}$; the last equation follows by Exercise 1.2.12. Then we show by direct calculation that $(\iota^* | T(V^*)^o)^* \circ \imath_{V^*} = \jmath_{T(V^*)} \circ \iota$.

Now $\pi \circ F = f = \pi \circ F'$ implies that $F^* \circ \pi^* = F'^* \circ \pi^*$, or $F^* \circ \jmath_{T(V^*)} \circ \iota = F'^* \circ \jmath_{T(V^*)} \circ \iota$. Since F^*, F'^* and $\jmath_{T(V^*)}$ are algebra maps, and $\mathrm{Im}(\iota)$ generates $T(V^*)$ as an algebra, it follows that $F^* \circ \jmath_{T(V^*)} = F'^* \circ \jmath_{T(V^*)}$. Now

$\mathrm{Im}(\jmath_{T(V^*)})$ is a dense subspace of $(T(V^*)^o)^*$ by part (c) of Proposition 2.6.1. Since $\mathrm{Ker}(F'^* - F^*) = \mathrm{Ker}((F' - F)^*)$ is a closed subspace of $(T(V^*)^o)^*$ by part (a) of Theorem 1.3.10, it follows that $F'^* = F^*$. Thus $F' = F$. $\square$

We will use Lemma 2.7.2 to construct the free coalgebra on any vector space over k. To keep the flow of the main ideas of the construction clear we dispense with some technical details at the outset.

For a vector space V over k let $(\iota_V, T(V))$ be the free algebra on V. Now let V be a fixed vector space over k and suppose that U is a subspace of V. Let $\mathrm{res}_{V,U} : V^* \longrightarrow U^*$ be the restriction map, which is defined by $\mathrm{res}_{V,U}(v^*) = v^*|U$ for all $v^* \in V^*$. Observe that the restriction $\mathrm{res}_{V,U}$ is the transpose of the inclusion map $\mathrm{inc}_{U,V} : U \longrightarrow V$ of U into V. Let $\mathrm{Res}_{V,U} : T(V^*) \longrightarrow T(U^*)$ be the algebra map determined by

$$\mathrm{Res}_{V,U} \circ \iota_{V^*} = \iota_{U^*} \circ \mathrm{res}_{V,U} \tag{2.24}$$

and let $\mathrm{Res}^o_{V,U} : T(U^*)^o \longrightarrow T(V^*)^o$ be the map of dual coalgebras. Since $\mathrm{inc}_{V,U}$ is one-one, its transpose $\mathrm{res}_{U,V}$ is onto. Thus the restriction $\mathrm{Res}^o_{V,U}$ of $\mathrm{Res}^*_{V,U}$ is one-one.

Let U, U' and U'' be subspaces of V which satisfy $U \subseteq U' \subseteq U''$. Then $\mathrm{Res}_{U',U} \circ \mathrm{Res}_{U'',U'} = \mathrm{Res}_{U'',U}$ which implies that

$$\mathrm{Res}^o_{U'',U'} \circ \mathrm{Res}^o_{U',U} = \mathrm{Res}^o_{U'',U}. \tag{2.25}$$

Now let U be a finite-dimensional subspace of V and set $T^{co}(U) = \mathrm{Im}(\mathrm{Res}^o_{V,U})$. Since $\mathrm{Res}^o_{V,U}$ is a coalgebra map $T^{co}(U)$ is a subcoalgebra of $T(V^*)^o$. Let $(\Pi_U, T(U^*)^o)$ be the cofree coalgebra on U described in Lemma 2.7.2, and let $\pi_U : T^{co}(U) \longrightarrow U$ be the composite $T^{co}(U) \simeq T(U^*)^o \xrightarrow{\Pi_U} U$, where the isomorphism is induced by $\mathrm{Res}^o_{V,U}$. Then $(\pi_U, T^{co}(U))$ is a cofree coalgebra on U as well.

Now suppose that U' is also a finite-dimensional subspace of V and $U \subseteq U'$. Then

$$T^{co}(U) \subseteq T^{co}(U') \qquad \text{and} \qquad \pi_U = \pi_{U'}|T^{co}(U). \tag{2.26}$$

The subset assertion is a consequence of (2.25). The restriction statement is derived from the equation $\iota_{U'^*} \circ \mathrm{Res}^*_{U',U} = (\mathrm{res}_{U',U})^* \circ (\iota_{U^*})^*$, which is a consequence of (2.24).

Proposition 2.7.3. *Let V be a vector space over the field k. Then there exists a cofree coalgebra $(\pi, T^{co}(V))$ on the vector space V.*

Proof. We continue the discussion following the proof of Lemma 2.7.2. Let $T^{co}(V)$ be the sum of the subcoalgebras of $T(V^*)^o$ of the form $T^{co}(U)$, where U runs over the finite-dimensional subspaces of V. Then $T^{co}(V)$ is a subcoalgebra of $T(V^*)^o$. Let $U_1, \ldots, U_r$ be finite-dimensional subspaces of V. Then $T^{co}(U_1) + \cdots + T^{co}(U_r) \subseteq T^{co}(U_1 + \cdots + U_r)$ by (2.26). Therefore every $c \in T^{co}(V)$ is contained in $T^{co}(U)$ for some finite-dimensional subspace U of V.

We next define a linear map $\pi : T^{co}(V) \longrightarrow V$ as follows. Let $c \in T^{co}(V)$. Then $c \in T^{co}(U)$ for some finite-dimensional subspace U of V. Set $\pi(c) = \pi_U(c)$. Using (2.26) we see that π is a well-defined linear map.

Now suppose that C is a coalgebra over k and $f : C \longrightarrow V$ is a linear map. We construct a coalgebra map $F : C \longrightarrow T^{co}(V)$ as follows. Let D be a finite-dimensional subcoalgebra of C and consider the restriction $f|D : D \longrightarrow f(D)$. Since $f(D)$ is a finite-dimensional subspace of V, by the universal mapping property of $(\pi_{f(D)}, T^{co}(f(D)))$ there is a coalgebra map $F_D : D \longrightarrow T^{co}(f(D))$ determined by $\pi_{f(D)} \circ F_D = f|D$.

Suppose D' is also a finite-dimensional subcoalgebra of C which satisfies $D \subseteq D'$. Then $T^{co}(f(D)) \subseteq T^{co}(f(D'))$ and $\pi_{f(D')}|T^{co}(f(D)) = \pi_{f(D)}$ by (2.26). Thus $F_{D'}|D, F_D : D \longrightarrow T^{co}(f(D'))$ are coalgebra maps which satisfy $\pi_{f(D')} \circ (F_{D'}|D) = f|D = \pi_{f(D')} \circ F_D$. Therefore $F_{D'}|D = F_D$. By Theorem 2.2.3 there exists a coalgebra map $F : C \longrightarrow T^{co}(V)$ determined by $F|D = F_D$ for all finite-dimensional subcoalgebras D of C. Thus $\pi \circ F = f$ by this theorem again. We emphasize that $F_D(D) \subseteq T^{co}(f(D))$ for all finite-dimensional subcoalgebras D of C.

As far as uniqueness is concerned, suppose that $F' : C \longrightarrow T^{co}(V)$ is also a coalgebra map which satisfies $\pi \circ F' = f$. Let D be a finite-dimensional subcoalgebra of C. Since $F(D)$ is a finite-dimensional subspace of $T^{co}(V)$ it follows that $F(D) \subseteq T^{co}(U)$ for some finite-dimensional subspace U of V. Since $f(D)$ is also finite-dimensional, by (2.26) we may replace U by $U + f(D)$ and assume that $f(D) \subseteq U$ and that $F'(D), F_D(D) \subseteq T^{co}(U)$. Now $\pi_{f(D)} \circ F_D = f|D$ means that $\pi_U \circ F_D = f|D$. Since $\pi \circ F' = f$ it follows that $\pi_U \circ (F'|D) = f|D$. Therefore $F'|D = F_D = F|D$ which means that $F' = F$ by Theorem 2.2.3 again. □

Just as there is a commutative analog of the free algebra, there is a cocommutative analog of the cofree coalgebra. Let V be a vector space over k and let $(\pi_V, T^{co}(V))$ be the cofree coalgebra on V. Then the sum $C(V)$ of all cocommutative subcoalgebras of $T^{co}(V)$ is a subcoalgebra of $T^{co}(V)$. Let $\pi : C(V) \longrightarrow V$ be the restriction $\pi_V|C(V)$. Since coalgebra maps take cocommutative subcoalgebras to cocommutative subcoalgebras:

Theorem 2.7.4. *Suppose that V is a vector space over the field k. Then pair $(\pi, C(V))$ defined above satisfies the following:*

(a) *$C(V)$ is a cocommutative coalgebra over k and $\pi : C(V) \longrightarrow V$ is a linear map.*
(b) *If C is a cocommutative coalgebra over k and $f : C \longrightarrow V$ is a linear map there exists a coalgebra map $F : C \longrightarrow C(V)$ determined by $\pi \circ F = f$.*

□

Definition 2.7.5. Let V be a vector space over the field k. A *cofree cocommutative coalgebra on V* is any pair $(\pi, C(V))$ which satisfies the conclusion of Theorem 2.7.4.

Exercises

In the following exercises U, V are vector spaces over k and $(\pi_V, T^{co}(V))$ is the cofree coalgebra on V.

Exercise 2.7.1. Show that:

(a) The restriction $\pi_V|\mathrm{G}(T^{co}(V)) : \mathrm{G}(T^{co}(V)) \longrightarrow V$ is a bijection.

(b) For $g, h \in \mathrm{G}(T^{co}(V))$ the restriction $\pi_V|\mathrm{P}_{g,h}(T^{co}(V)) : \mathrm{P}_{g,h}(T^{co}(V)) \longrightarrow V$ is a linear isomorphism.

Exercise 2.7.2. Show that:

(a) $\pi_V : T^{co}(V) \longrightarrow V$ is onto. [Hint: Suppose first of all that V is finite-dimensional. Show that π_V is onto for some cofree coalgebra on V in this case.]

(b) If $f : U \longrightarrow V$ is a linear map then there is a coalgebra map $F : T^{co}(U) \longrightarrow T^{co}(V)$ determined by $\pi_V \circ F = f \circ \pi_U$.

(c) The assignments $V \mapsto T^{co}(V)$ and $f \mapsto F$ determine a functor from k-Vec to k-Coalg.

Exercise 2.7.3. Let I be the sum of all coideals of $T^{co}(V)$ contained in $\mathrm{Ker}(\pi_V)$.

(a) Show that I is a coideal of $T^{co}(V)$.

Give $T^{co}(V)/I$ the quotient coalgebra structure, let $\mathrm{pr} : T^{co}(V) \longrightarrow T^{co}(V)/I$ be the projection, let $\pi' : T^{co}(V)/I \longrightarrow V$ be the linear map determined by $\pi' \circ \mathrm{pr} = \pi_V$, and suppose that C is a coalgebra over k.

(b) Let $f : C \longrightarrow V$ be a linear map. Show that there is a coalgebra map $F : C \longrightarrow T^{co}(V)/I$ which satisfies $\pi' \circ F = f$.

(c) Suppose that $F, G : C \longrightarrow T^{co}(V)/I$ are coalgebra maps which satisfy $\pi' \circ F = f = \pi' \circ G$. Show that $F = G$. [Hint: Note that the only coideal of $T^{co}(V)/I$ which lies in $\mathrm{Ker}(\pi')$ is (0). See part (c) of Exercise 2.1.28. Observe that $\mathrm{Im}(F - G) \subseteq \mathrm{Ker}(\pi')$ and is a coideal of $T^{co}(V)/I$. See Exercise 2.1.29.]

(d) By part (c) the pair $(\pi', T^{co}(V)/I)$ is also a cofree coalgebra over k. Show that $I = (0)$.

By virtue of the last exercise:

Proposition 2.7.6. *Let $(\pi, T^{co}(V))$ be a cofree coalgebra on a vector space V over k. Then the only coideal of $T^{co}(V)$ which is contained in* $\mathrm{Ker}(\pi)$ *is* (0). □

Chapter notes

The material of this chapter is quite standard and is found for the most part in Sweedler's book [201]. Here we emphasize the topological aspects of the dual algebra C^*. The wedge product of coalgebras is developed in great detail in this chapter as was done by Heyneman and the author in [71].

There are many directions the subject of coalgebras has taken which are indicated by exercises. The incidence algebra and coalgebra of Exercise 2.3.27 are two structures which arise in a treatment of combinatorial ideas from the point of view of algebras, coalgebras, or Hopf algebras [43,140,184]. Exercise 2.5.17 is an introduction to the work of Taft and others on linearly recursive sequences [106, 157, 204]. The series of exercises starting with Exercise 2.6.8 deals with reflexive algebras and coalgebras treated by Taft [206,207] and found in [71,158]. In [158] the notions of almost isomorphism (see Exercise 2.6.17) and pairing of an algebra and coalgebra (see Exercise 2.6.18) are introduced.

Cofinite ideals of algebras have been seen to play an important role in the theory of coalgebras and their dual algebras. The notion of almost noetherian algebra, introduced in [71], is a natural concept for the dual algebra. See Exercise 2.6.13. Theorem 2.6.9 is Proposition 1.1.9 of [71].

Chapter 3

Representations of coalgebras

Let C be a coalgebra over the field k. Representations of C are certain C^*-modules which are called rational. Rational C^*-modules are those C^*-modules whose elements are annihilated by a *closed* cofinite ideal of C^*. Thus topological aspects of C^* enter into the discussion of representations of C as well as into the discussion of the relationship between C and C^*. Note also that rational modules are locally finite.

Many module concepts can be adapted to rational modules. By far one of the most important is the notion of injective module. Injective rational C^*-modules play an important role in the study of C. Section 3.5 is devoted to a detailed discussion of them.

Sections 3.1–3.5 should be read carefully. Sections 3.6–3.7 can be considered somewhat technical and skipped on a first pass through the chapter. In this chapter C is a coalgebra and A is an algebra over the field k.

3.1 Rational modules of the dual algebra

Let M be a left C^*-module. We are interested in the set of all $m \in M$ such that the C^* action on m is described by

$$c^*{\cdot}m = (\mathrm{I}_M \otimes c^*)(\rho_m) \tag{3.1}$$

for all $c^* \in C^*$, where $\rho_m \in M \otimes C$. Let $m \in M$. Notice that there is at most one $\rho_m \in M \otimes C$ which satisfies (3.1) for all $c^* \in C^*$ by part (b) of Exercise 1.3.7.

Now suppose that M is a right C^*-module. The counterpart of (3.1) is

$$m{\cdot}c^* = (c^* \otimes \mathrm{I}_M)(\rho_m) \tag{3.2}$$

for all $c^* \in C^*$, where $\rho_m \in C \otimes M$.

Definition 3.1.1. Let C be a coalgebra over the field k and let M be a left (respectively right) C^*-module. The set of all $m \in M$ such that there exists an element $\rho_m \in M \otimes C$ (respectively $\rho_m \in C \otimes M$) which satisfies (3.1) (respectively (3.2)) for all $c^* \in C^*$ is denoted by M_r.

It is easy to see that M_r is a subspace of M. There is a realization of (3.1) which is derived from C itself.

Regard C as a left C^*-module according to (2.14). As $c^* \rightharpoonup c = c_{(1)} < c^*, c_{(2)} >$ it follows that ρ_c exists and $\rho_c = \Delta(c)$ for all $c \in C$.

Proposition 3.1.2. *Let C be a coalgebra over the field k and suppose that M is a left C^*-module. Then:*

(a) *M_r is a locally finite C^*-submodule of M.*
(b) *Let $m \in M$. Then $\mathrm{Dim}(C^* \cdot m) = \mathrm{Rank}(\rho_m)$. Moreover, if m is not zero and $\rho_m = \sum_{i=1}^{r} m_i \otimes c_i$, where $r = \mathrm{Rank}\,\rho_m$, then $\{m_1, \ldots, m_r\}$ is a basis for $C^* \cdot m$.*
(c) *$N_r = M_r \cap N$ for all submodules N of M.*
(d) *Let $f : M \longrightarrow N$ be a map of left C^*-modules. Then $f(M_r) \subseteq N_r$ and the restriction $f_r = f|M_r$ is a module map $f_r : M_r \longrightarrow N_r$.*

Proof. Suppose $m \in M_r$ is not zero and write $\rho_m = \sum_{i=1}^{r} m_i \otimes c_i \in M \otimes C$, where $r = \mathrm{Rank}(\rho_m)$. By Lemma 1.2.2 the sets $\{m_1, \ldots, m_r\}$ and$\{c_1, \ldots, c_r\}$ are linearly independent. Since $c^* \cdot m = \sum_{i=1}^{r} c^*(c_i) m_i$ for all $c^* \in C^*$ it follows that $C^* \cdot m$ is contained in the span of $\{m_1, \ldots, m_r\}$. Fix $1 \leq \ell \leq r$. Since $\{c_1, \ldots, c_r\}$ is linearly independent there is a $c^* \in C^*$ which satisfies $c^*(c_i) = \delta_{\ell,i}$ for all $1 \leq i \leq r$. For such an element c^* note $c^* \cdot m = m_\ell$. Therefore $C^* \cdot m$ is the span of $\{m_1, \ldots, m_r\}$ and part (b) follows.

Let $d^* \in C^*$. To complete the proof of part (a) we need to show that $d^* \cdot m \in M_r$. Since

$$c^* \cdot (d^* \cdot m) = (c^* d^*) \cdot m = \sum_{i=1}^{r} (c^* d^*)(c_i) m_i = \sum_{i=1}^{r} c^*(c_{i(1)}) d^*(c_{i(2)}) m_i$$

for all $c^* \in C^*$, it follows that $\rho_{d^* \cdot m}$ exists and $\rho_{d^* \cdot m} = \sum_{i=1}^{r} m_i \otimes (d^* \rightharpoonup c_i)$. Thus $d^* \cdot m \in M_r$.

Part (c) follows directly from part (b). Part (d) follows as $\rho_{f(m)}$ exists for $m \in M$ and $\rho_{f(m)} = (f \otimes \mathrm{I}_C)(\rho_m)$. □

We note that application of results of this section to C^{cop} and the interpretation of them in terms of C yields results with "left" and "right" interchanged.

Definition 3.1.3. Let C be a coalgebra over k. A *rational C^*-module* is a C-module M such that $M = M_r$.

The left C^*-module $(C, \rightharpoonup)$ mentioned above is rational. Likewise the right C^*-module $(C, \leftharpoonup)$ described in (2.14) is rational.

Let A be an algebra over k, let M be a left A-module, and let $X \subseteq M$. Then $\mathrm{ann}_A(X) = \{a \in A \,|\, a{\cdot}X = (0)\}$ is the annihilator of X. For $m \in M$ we set $\mathrm{ann}_A(m) = \mathrm{ann}_A(\{m\})$. When M is a right A-module we define $\mathrm{ann}_A(X) = \{a \in A \,|\, X{\cdot}a = (0)\}$.

Proposition 3.1.4. *Let C be a coalgebra over the field k and suppose that M is a left C^*-module. For $m \in M$ the following are equivalent:*

(a) $\mathrm{ann}_{C^*}(C^*{\cdot}m)$ *contains a cofinite closed subspace of* C^*.
(b) $\mathrm{ann}_{C^*}(C^*{\cdot}m)$ *is a cofinite closed ideal of* C^*.
(c) $\mathrm{ann}_{C^*}(m)$ *contains a cofinite closed subspace of* C^*.
(d) $\mathrm{ann}_{C^*}(m)$ *is a cofinite closed left ideal of* C^*.
(e) $m \in M_r$.

Proof. Let I, J be subspaces of C^* with $I \subset J$. If I is a cofinite closed subspace of C^* then J is a cofinite closed subspace of C^* by Proposition 1.3.12. Thus part (a) implies part (b) which implies part (c) which in turn implies part (d). We show that part (d) implies part (e).

Assume that $L = \mathrm{ann}_{C^*}(m)$ is a cofinite closed left ideal of C^*. Then $C^*{\cdot}m$ is finite-dimensional. We may as well suppose that m is not zero. Choose a basis $\{m_1, \ldots, m_r\}$ for $C^*{\cdot}m$. There are $c_1^{**}, \ldots, c_r^{**} \in (C^*)^*$ which satisfy $c^*{\cdot}m = c_1^{**}(c^*)m_1 + \cdots + c_r^{**}(c^*)m_r$ for all $c^* \in C^*$. Since $L{\cdot}m = (0)$ necessarily $c_1^{**}(L) = \cdots = c_r^{**}(L) = (0)$. Since L is a closed cofinite subspace of C^* there are $c_1, \ldots, c_r \in C$ such that $c_i^{**}(c^*) = c^*(c_i)$ for all $1 \leq i \leq r$ and $c^* \in C^*$ by Corollary 1.3.13. Therefore ρ_m exists and $\rho_m = \sum_{i=1}^r m_i \otimes c_i$. We have shown part (d) implies part (e).

To complete the proof of the proposition we show that part (e) implies part (a). Let $m \in M_r$, suppose that m is not zero, write $\rho_m = \sum_{i=1}^r m_i \otimes c_i$ and let U be the span of $\{c_1, \ldots, c_r\}$. Then $U^\perp$ is a closed subspace of C^* by definition, is a cofinite subspace of C^* by (1.2), and $U^\perp \subseteq \mathrm{ann}_{C^*}(m)$. Thus $\mathrm{ann}_{C^*}(m)$ contains a cofinite closed subspace of C^*. Now $C^*{\cdot}m \subseteq M_r$ and is finite-dimensional since M_r is a locally finite left C^*-module by part (a) of Proposition 3.1.2. Let $\{n_1, \ldots, n_s\}$ span $C^*{\cdot}m$. Then $\mathrm{ann}_{C^*}(C^*{\cdot}m) = \bigcap_{i=1}^s \mathrm{ann}_{C^*}(n_i)$ and thus $\mathrm{ann}_{C^*}(C^*{\cdot}m)$ contains a finite intersection I of closed cofinite subspaces of C^*. Now I is a closed subspace of C^* by part

(a) of Proposition 1.3.8, and I is a cofinite subspace of C^* by part (b) of Exercise 1.3.15. Hence part (e) implies part (a). □

As noted in the last part of the proof of Proposition 3.1.4, finite intersections of closed cofinite subspaces of C^* are closed cofinite subspaces of C^*. As a corollary to Proposition 3.1.4:

Corollary 3.1.5. *Let C be a coalgebra over the field k. Then submodules, quotients, and direct sums of rational left C^*-modules are rational.* □

We examine when the products of rational C^*-modules are always rational in Exercise 3.1.9. Since (0) is a cofinite subspace of U^* when U is a finite-dimensional vector space over k:

Corollary 3.1.6. *Let C be a finite-dimensional coalgebra over the field k. Then all C^*-modules are rational.* □

The next two results concern the dual vector space M^* of a rational left C^*-module M. Suppose that M is a left A-module. Recall that M^* is a right A-module under the transpose action which is given by

$$(m^* \prec a)(m) = m^*(a \cdot m)$$

for all $m^* \in M^*$, $a \in A$ and $m \in M$. Likewise when M is a right A-module then M^* is a left A-module under the transpose action which is given by

$$(a \succ m^*)(m) = m^*(m \cdot a)$$

for all $a \in A$, $m^* \in M^*$ and $m \in M$.

Definition 3.1.7. Let A be an algebra over k and let M be an A-module. The sum of all finite-dimensional submodules of M is denoted by M_{f} and $M^{\mathrm{f}} = (M^*)_{\mathrm{f}}$.

Observe that M_{f} is a locally finite submodule of M and is the largest locally finite submodule of M.

Definition 3.1.8. Let C be a coalgebra over k and let M be a C^*-module. Then $M^r = (M^*)_r$.

Let M be a C^*-module. Then $M_r \subseteq M_{\mathrm{f}}$. Usually it is the case that $M_r \neq M_{\mathrm{f}}$. See Exercise 3.1.6. If C^* contains a finitely generated dense subalgebra, or more generally if C is reflexive, then $M_r = M_{\mathrm{f}}$. See Exercises 3.1.3 and 3.1.15. When M is rational, $M^r = M^{\mathrm{f}}$ as we now show.

Proposition 3.1.9. *Let C be a coalgebra over the field k and suppose that M is a rational left C^*-module. Then:*

(a) *Finitely generated right C^*-submodules of M^* are closed subspaces of M^*.*

(b) $M^r = M^{\mathrm{f}}$. *In particular finite-dimensional submodules of M^* are rational.*

(c) *The one-one correspondence $U \mapsto U^\perp$ of the set of subspaces of M and the set of closed subspaces of M^* induces a one-one correspondence of the set of cofinite left C^*-submodules of M and the finite-dimensional right C^*-submodules of M^*.*

(d) *Let $m^* \in M^*$. Then $m^* \in M^r$ if and only if $m^*(N) = (0)$ for some cofinite left C^*-submodule of M.*

Proof. Let $m^* \in M^*$ and consider the linear map $\ell_{m^*} : C^* \longrightarrow M^*$ defined by $\ell_{m^*}(c^*) = m^* \prec c^*$ for all $c^* \in C^*$. We first show that ℓ_{m^*} is continuous. To this end let $m \in M$ and $c = (m^* \otimes \mathrm{I}_C)(\rho_m)$. Since

$$\ell_{m^*}(c^*)(m) = m^* \prec c^*(m) = m^*(c^* \cdot m) = m^*((\mathrm{I}_M \otimes c^*)(\rho_m))$$

and

$$m^*((\mathrm{I}_M \otimes c^*)(\rho_m)) = c^*(m^* \otimes \mathrm{I}_C)(\rho_m) = c^*(c)$$

for all $c^* \in C^*$ we have $\ell_{m^*}^{-1}((km)^\perp) = (kc)^\perp$. By Theorem 1.3.14 it follows that ℓ_{m^*} is continuous.

By part (b) of Theorem 1.3.10 we conclude that $m^* \prec C^* = \mathrm{Im}(\ell_{m^*})$ is a closed subspace of M^*. Therefore finitely generated submodules of M^* are closed subspaces of M^* by part (b) of Proposition 1.3.8. We have established part (a).

To show part (b) we first observe that $M^r \subseteq M^{\mathrm{f}}$ by part (a) of Proposition 3.1.2. Let $m^* \in M^{\mathrm{f}}$. Then $m^* \prec C^*$ is finite-dimensional. Since $C^*/\mathrm{ann}_{C^*}(m^*) \simeq m^* \prec C^*$ it follows that $\mathrm{ann}_{C^*}(m^*)$ is a cofinite subspace of C^*. Since ℓ_{m^*} is continuous and (0) is a closed subspace of M^* we conclude that $\mathrm{ann}_{C^*}(m^*) = \ell_{m^*}^{-1}((0))$ is a closed subspace of C^* by part (a) of Theorem 1.3.10. Thus $m^* \in M^r$ by Proposition 3.1.4 applied to the coalgebra C^{cop} and part (b) follows.

To show part (c) we use part (a), or Corollary 1.3.9, to see that the finite-dimensional right C^*-submodules of M^* are closed subspaces of M^*. Let N be a subspace of M. The reader is left to show that N is a left C^*-submodule of M if and only if $N^\perp$ is a right C^*-submodule of M^*. The projection $M \longrightarrow M/N$ induces an isomorphism $(M/N)^* \simeq N^\perp$. Therefore N is a cofinite subspace of M if and only if $N^\perp$ is a finite-dimensional subspace of M^*. We have established part (c). Part (d) is an immediate consequence of part (c). $\square$

Let M be a left A-module. Then M^* is a right A-module under the transpose action arising from the left A-module structure on M and thus M^{**} is a left A-module under the transpose action arising from the right A-module structure on M^*. Note the one-one linear map $\imath_M : M \longrightarrow (M^*)^*$ defined by $\imath_M(m)(m^*) = m^*(m)$ for all $m \in M$ and $m^* \in M^*$ is a module map.

Our next result is a corollary of Proposition 3.1.4. It applies in particular to the left C^*-module $(C, \rightharpoonup)$ described above.

Corollary 3.1.10. *Let C be a coalgebra over the field k, let M be a rational left C^*-module, and let $\imath_M : M \longrightarrow (M^*)^*$ be the map described above. Then:*

(a) $\mathrm{Im}(\imath_M) \subseteq (M^*)^r$ *and* $\imath_M : M \longrightarrow (M^*)^r$ *is a map of (rational) left C^*-modules.*
(b) *If M^* is a finitely generated right C^*-module then $\imath_M : M \longrightarrow (M^*)^r$ is an isomorphism.*

Proof. As for part (a), we need only note at this point that $\rho_{\imath_M(m)}$ exists and that $\rho_{\imath_M(m)} = (\imath_M \otimes \mathrm{I}_C)(\rho_m)$ for all $m \in M$. To show part (b) we let $m^{**} \in (M^*)^r$. Then $I \succ m^{**} = (0)$ for some closed cofinite subspace I of C^* by Proposition 3.1.4. In particular $m^{**}(M^* \prec I) = (0)$. Since I is a closed subspace of C^*, for $m^* \in M^*$ the subspace $m^* \prec I = \ell_{m^*}(I)$ of M^* is closed by part (b) of Theorem 1.3.10, where ℓ_{m^*} is the continuous linear map described in the proof of part (a) of Proposition 3.1.9. By assumption there are $m_1^*, \ldots, m_r^* \in M^*$ which generate M^* as a right C^*-module. Let $N = m_1^* \prec I + \cdots + m_r^* \prec I$. Then N is a closed subspace of M^* by part (b) of Proposition 1.3.8 and N is a cofinite subspace of M^* by Lemma 3.1.13; this result is the content of Exercise 3.1.11. Since $m^{**}(N) = (0)$, by Corollary 1.3.13 we conclude $m^{**} \in \mathrm{Im}(\imath_M)$. □

Exercises

In the following exercises C is a coalgebra over the field k.

Exercise 3.1.1. Find a coalgebra C over k such that there are left C^*-modules M and N which satisfy

(a) $M_r \neq (0)$ and $M^r = (0)$,
(b) $N_r = (0)$ and $N^r \neq (0)$.

Exercise 3.1.2. For an algebra A over k let ${}_A\mathcal{M}$ (respectively $\mathcal{M}_A$) denote the category of left (respectively right) A-modules and module maps under function composition. Let ${}_{C^*}\mathcal{M}_r$ (respectively $\mathcal{M}_{r\,C^*}$) denote the category of rational left (respectively right) C^*-modules and module maps. Show that:

(a) $F : {}_{C^*}\mathcal{M} \longrightarrow {}_{C^*}\mathcal{M}_r$ given by $F(M) = M_r$ for left C^*-modules M and $F(f) = f|M_r$ for maps of left C^*-modules $f : M \longrightarrow N$ is a well-defined functor.

(b) $G : {}_{C^*}\mathcal{M} \longrightarrow \mathcal{M}_{r\,C^*}$ given by $G(M) = M^r$ for left C^*-modules M and $G(f) = f|N^r$ for maps of left C^*-modules $f : M \longrightarrow N$ is a well-defined contravariant functor.

Exercise 3.1.3. Show that the following are equivalent:

(a) All finite-dimensional left C^*-modules are rational.

(b) C is reflexive.

(c) All finite-dimensional right C^*-modules are rational.

[Hint: See Exercise 2.6.8 in particular.]

Exercise 3.1.4. Let D be a subcoalgebra of C. Show that:

(a) If all finite-dimensional left C^*-modules are rational then all finite-dimensional left D^*-modules are rational.

(b) If C is reflexive then D is reflexive.

[Hint: For part (a) note that all D^*-modules are C^*-modules via pullback along the restriction $C^* \longrightarrow D^*$; for part (b) see Exercise 3.1.3.]

Exercise 3.1.5. Let C be the coalgebra of Exercise 2.3.32. Show that all finite-dimensional left $k[[X_1, \ldots, X_r]] = C^*$-modules are rational. [Hint: See Exercise 2.6.16.]

Exercise 3.1.6. Find a coalgebra C over k and a left C^*-module M such that $M_r \neq M_{\mathrm{f}}$. [Hint: See Exercise 2.3.14.]

Exercise 3.1.7. Suppose that all left C^*-modules are rational. Show that $\mathrm{P}_{g,h}(C)$ is finite-dimensional for all $g, h \in \mathrm{G}(C)$. [Hint: See Exercises 1.3.10, 3.1.4 and Theorem 1.3.10.]

Exercise 3.1.8. Let $M = \bigoplus_{i \in I} M_i$ be the direct sum of rational left C^*-modules. Show that $M_r = \bigoplus_{i \in I} (M_i)_r$.

Exercise 3.1.9. Show that direct products of rational left C^*-modules are always rational if and only if C is finite-dimensional. [Hint: Let D be a subcoalgebra of C and regard D^* as a left C^*-module by pullback along the restriction map $C^* \longrightarrow D^*$. Consider $M = \prod_D D$ where D runs over the finite-dimensional subcoalgebras of C. See Theorem 2.2.3.]

Exercise 3.1.10. Let $\mathcal{C} = C \otimes C^{cop}$ have the tensor product coalgebra structure.

(a) Show that C is a rational left $\mathcal{C}^*$-module, where $\alpha \cdot c = c_{(1)}\alpha(c_{(3)} \otimes c_{(2)})$ for all $\alpha \in \mathcal{C}^*$ and $c \in C$.

(b) Show that the $\mathcal{C}^*$-submodules of C are the subcoalgebras of C. [Hint: See part (b) of Proposition 2.3.5.]

The next series of exercises concerns a certain subclass of rational modules.

Definition 3.1.11. Let C be a coalgebra over the field k and let M be a left (respectively right) C^*-module. Then $M_{(r)}$ is the set of elements of M which are annihilated by an ideal of C^* which is a cofinite subspace of C^* and which is also finitely generated as a left (respectively right) ideal of C^*.

Definition 3.1.12. Let C be a coalgebra over the field k. A *strongly rational C^*-module* is a C^*-module such that $M = M_{(r)}$.

We use the following lemma to proceed.

Exercise 3.1.11. Prove the following lemma:

Lemma 3.1.13. *Let A be an algebra over the field k and M be a finitely generated left A-module. If V is a cofinite subspace of A then $V \cdot M$ is a cofinite subspace of M.* □

[Hint: Show that $V \bigoplus U = A$ for some finite-dimensional subspace U of A. If $\{m_1, \ldots, m_r\}$ generates M show that $V \cdot M + U \cdot m_1 + \cdots + U \cdot m_r = M$.]

Exercise 3.1.12. Prove Lemma 3.1.13 in the following manner. Show that for some finite direct sum of copies of A there is an onto map of left A-modules $A \bigoplus \cdots \bigoplus A \longrightarrow M$ such that the composite

$$A \bigoplus \cdots \bigoplus A \longrightarrow M \longrightarrow M/V \cdot M$$

contains $V \bigoplus \cdots \bigoplus V$ in its kernel and thus factors through $(A/V) \bigoplus \cdots \bigoplus (A/V)$.

Exercise 3.1.13. Let M be a left C^*-module. Show that $M_{(r)}$ is a submodule of M and $M_{(r)} \subseteq M_r$. [Hint: Let I, J be cofinite ideals of C^* which are finitely generated as left ideals of C^*. Show that IJ is finitely generated as a left ideal of C^*. Show that IJ is a cofinite subspace of C^*. See Lemma 3.1.13 and part (b) of Proposition 2.3.6.]

Exercise 3.1.14. Show that submodules, quotients and direct sums of strongly rational C^*-modules are strongly rational and, when C is infinite-dimensional, there are direct products of strongly rational C^*-modules which are not strongly rational.

Exercise 3.1.15. If C^* contains a finitely-generated dense subalgebra show that $M_{(r)} = M_r = M_{\mathrm{f}}$ for all C^*-modules M. (In particular locally finite C^*-modules are strongly rational and hence rational.) [Hint: See Exercise 2.6.14.]

Exercise 3.1.16. Let M, N be strongly rational left C^*-modules and suppose that N is a submodule of M. Show that M is strongly rational if and only if N and M/N are strongly rational.

Exercise 3.1.17. Find a coalgebra C over k and left C^*-modules M, N, where N is a submodule of M, such that N and M/N are rational but M is not rational. [Hint: See Exercise 3.1.6.]

3.2 Comodules

The concepts of module and comodule are dual to each other. Comodules of a coalgebra C over k are understood in terms of rational modules of the dual algebra C^*. Just as describing the axioms for an algebra in terms of linear maps leads to the concept of coalgebra, describing the axioms for a module leads to the definition of comodule. This section parallels Section 2.1 in many ways.

Let (A, m, η) be an algebra over the field k and let (M, μ) be a right A-module. We write $m{\cdot}a = \mu(m{\otimes}a)$ for all $m \in M$ and $a \in A$ to describe the module action determined by $\mu : M{\otimes}A \longrightarrow M$. The associative axiom $(m{\cdot}a){\cdot}b = m{\cdot}(ab)$ for all $m \in M$ and $a, b \in A$ can be expressed as

$$\mu\circ(\mu{\otimes}\mathrm{I}_A) = \mu\circ(\mathrm{I}_M{\otimes}m), \tag{3.3}$$

which is equivalent to the composites

$$M{\otimes}A{\otimes}A \xrightarrow{\mu\otimes\mathrm{I}_A} M{\otimes}A \xrightarrow{\mu} M$$

and

$$M{\otimes}A{\otimes}A \xrightarrow{\mathrm{I}_M\otimes m} M{\otimes}A \xrightarrow{\mu} M$$

are the same. The unitary axiom $m{\cdot}1 = m$ for all $m \in M$ is equivalent to

$$\mu\circ(\mathrm{I}_M{\otimes}\eta) = \mathrm{I}_M, \tag{3.4}$$

which is to say that the composite

$$M \simeq M \otimes k \xrightarrow{\mathrm{I}_M \otimes \eta} M \otimes A \xrightarrow{\mu} M$$

is the identity map I_M. These axioms are equivalent to the diagrams

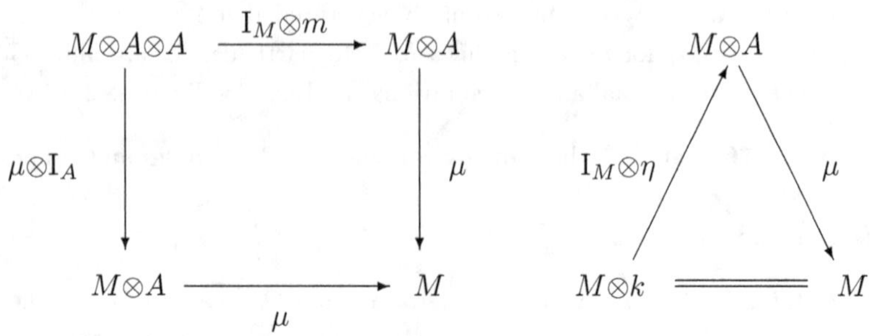

commute, where "equals" is identification by scalar multiplication.

Now let (C, Δ, ϵ) be a coalgebra over k, let M be a vector space over k, and suppose that $\rho : M \longrightarrow M \otimes C$ is linear. The dual formulations of (3.3) and (3.4) are

$$(\rho \otimes \mathrm{I}_C) \circ \rho = (\mathrm{I}_M \otimes \Delta) \circ \rho \tag{3.5}$$

and

$$(\mathrm{I}_M \otimes \epsilon) \circ \rho = \mathrm{I}_M. \tag{3.6}$$

Usually (3.5) is referred to as the coassociative axiom and (3.6) is referred to as the counit axiom.

Definition 3.2.1. Suppose that C is a coalgebra over the field k. A *right C-comodule* is a pair (M, ρ), where M is a vector space over k and $\rho : M \longrightarrow M \otimes C$ is a linear map, which satisfies (3.5) and (3.6).

Thus (M, ρ) is a right C-comodule if and only if the diagrams

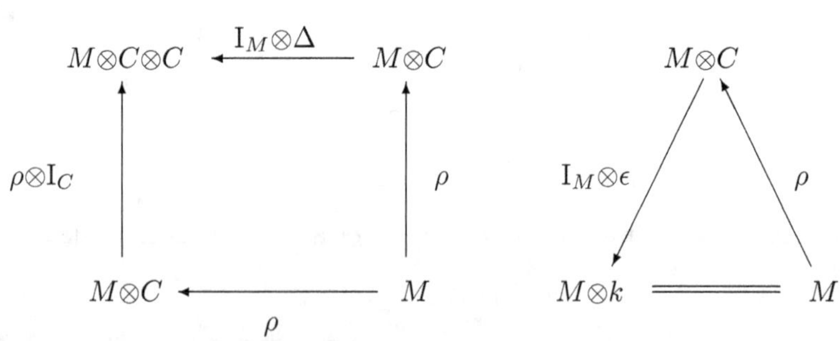

commute, where again "equals" is identification by scalar multiplication.

Observe that the structure map ρ of a right C-comodule (M, ρ) is one-one by (3.6). For $m \in M$ we use the symbolic notation

$$\rho(m) = m_{(0)} \otimes m_{(1)} \tag{3.7}$$

to represent $\rho(m) \in M \otimes C$. In terms of this notation (3.5) is expressed as

$$m_{(0)(0)} \otimes m_{(0)(1)} \otimes m_{(1)} = m_{(0)} \otimes m_{(1)(1)} \otimes m_{(1)(2)}$$

and (3.6) is expressed

$$m_{(0)} \epsilon(m_{(1)}) = m$$

for all $m \in M$. We usually write M for (M, ρ).

Suppose that M is a vector space over k and $\rho : M \longrightarrow M \otimes C$ is linear. Let $\mu_\rho : C^* \otimes M \longrightarrow M$ denote the linear map determined by

$$\mu_\rho(c^* \otimes m) = (\mathrm{I}_M \otimes c^*)(\rho(m)) \tag{3.8}$$

for all $c^* \in C^*$ and $m \in M$. Right C-comodules and left rational C^* modules are very closely related.

Proposition 3.2.2. *Let C be a coalgebra over the field k. Then:*

(a) *If (M, ρ) is a right C-comodule then (M, μ_ρ) is a rational left C^*-module with $\rho_m = \rho(m)$ for all $m \in M$.*
(b) *If (M, μ) is a rational left C^*-module then (M, ρ) is a right C-comodule, where $\rho(m) = \rho_m$ for all $m \in M$.*

□

The reader is referred to Exercise 3.2.1 which outlines a proof of Proposition 3.2.2.

Definition 3.2.3. Let C be a coalgebra over the field k and suppose that (M, ρ) is a right C-comodule. Then (M, μ_ρ) is the *rational left C^*-module associated with* (M, ρ).

Definition 3.2.4. Let C be a coalgebra over the field k and suppose that (M, μ) is a rational left C^*-module. Then (M, ρ) is the *underlying right C-comodule associated with* (M, μ), where $\rho(m) = \rho_m$ for all $m \in M$.

For a given rational left C^*-module (M, μ) there is at most one right C-comodule structure (M, ρ) on M such that $c^* {\cdot} m = (\mathrm{I}_M \otimes c^*)(\rho(m))$ for all $c^* \in C^*$ and $m \in M$. See the discussion at the beginning of Section 3.1.

We have shown that $(M, \rho) \mapsto (M, \mu_\rho)$ is a one-one correspondence between the set of right C-comodules and the set of rational left C^*-modules. We denote the action of μ_ρ by

$$c^* \rightharpoonup m = m_{(0)} < c^*, m_{(1)} > \tag{3.9}$$

for all $c^* \in C^*$ and $m \in M$.

Definition 3.2.5. Let C be a coalgebra over the field k and let (M, ρ) be a right C-comodule. A *subcomodule of* M is a subspace N of M such that $\rho(N) \subseteq N \otimes C$.

One can view subcomodules of a right C-comodule M as rational left C^*-submodules of M or vice versa. Using part (b) of Proposition 3.1.2 we easily see that:

Corollary 3.2.6. *Let C be a coalgebra over the field k, let (M, ρ) be a right C-comodule and let (M, μ_ρ) be the associated rational left C^*-module. Then a subspace N of M is a right C-subcomodule if and only if N is a rational left C^*-submodule of M.* □

Subcomodules are comodules in their own right. If M is a right C-comodule then M and (0) are subcomodules of M.

Definition 3.2.7. Let C be a coalgebra over the field k. A *simple C-comodule* is a C-comodule which has exactly two subcomodules.

Since rational left C^*-modules are locally finite by part (a) of Proposition 3.1.2, by Corollary 3.2.6:

Theorem 3.2.8. *Let C be a coalgebra over the field k and suppose that M is a right C-comodule. Then:*

(a) *Every finite-dimensional subspace V of C is contained in a finite-dimensional subcomodule of M.*
(b) *Simple subcomodules of M are finite-dimensional.*

□

Suppose that M is a right C-comodule. We have noted that M and (0) are subcomodules of M. Using Exercise 1.2.8 we can deduce that the intersection of a family of right C-comodules of M is a right C-comodule of M; this also follows by Corollary 3.2.6. Let V be a subspace of M. Then V is contained in at least one subcomodule of M, and among the family of subcomodules of M which contain V there is a unique minimal one.

Definition 3.2.9. Let C be a coalgebra over the field k, let M be a right C-comodule, and let V be a subspace of M. The unique minimal member of the family of all subcomodules of M which contain V is the *subcomodule of M generated by V.*

By part (a) of Theorem 3.2.8 finite-dimensional subspaces of right C-comodules generate finite-dimensional subcomodules.

Let (M, ρ) be a right C-comodule. By Exercise 1.2.8, among the subspaces V of M which satisfy $\rho(M) \subseteq M \otimes V$ there is a unique minimal one.

Definition 3.2.10. Let C be a coalgebra over the field k and suppose that (M, ρ) is a right C-comodule. The unique minimal member of the family of subspaces V of M which satisfy $\rho(M) \subseteq M \otimes V$ is denoted by $C(\rho)$.

The comatrix identities of Section 2.1 enter our discussion of comodules at this point in a very natural way. The reader should compare the proof of the next result to the proof of Lemma 2.2.1.

Theorem 3.2.11. *Let C be a coalgebra over the field k and let (M, ρ) be a right C-comodule. Then:*

(a) *$C(\rho)$ is a subcoalgebra of C.*
(b) $\mathrm{ann}_{C^*}(M) = C(\rho)^{\perp}$.
(c) *$C(\rho)$ is finite-dimensional if M is finite-dimensional.*
(d) *$C(\rho)$ is a simple subcoalgebra of C if M is a simple right C-comodule.*

Proof. First of all assume that M is finite-dimensional and $\{m_1, \ldots, m_r\}$ is a basis for M. Fix $1 \leq j \leq r$. For $1 \leq i \leq r$ there are $c_{i,j} \in C$ such that $\rho(m_j) = \sum_{i=1}^{r} m_i \otimes c_{i,j}$. Since $\{m_1, \ldots, m_r\}$ is linearly independent $C(\rho)$ is the span of $\{c_{i,j}\}_{1 \leq i,j \leq r}$. Using the equation $(\mathrm{I}_M \otimes \Delta) \circ \rho = (\rho \otimes \mathrm{I}_C) \circ \rho$ we compute

$$\sum_{i=1}^{r} m_i \otimes \Delta(c_{i,j}) = \sum_{i=1}^{r} \rho(m_i) \otimes c_{i,j} = \sum_{\ell=1}^{r} \rho(m_\ell) \otimes c_{\ell,j} = \sum_{i=1}^{r} m_i \otimes (\sum_{\ell=1}^{r} c_{i,\ell} \otimes c_{\ell,j})$$

and using the equation $(\mathrm{I}_M \otimes \epsilon) \circ \rho = \mathrm{I}_M$ we compute

$$\sum_{i=1}^{r} m_i \epsilon(c_{i,j}) = m_j$$

for all $1 \leq j \leq r$. Since $\{m_1, \ldots, m_r\}$ is linearly independent $\{c_{i,j}\}_{1 \leq i,j \leq r}$ satisfies the comatrix identities (2.10) and (2.11). Thus the span of

$\{c_{i,j}\}_{1\leq i,j\leq r}$ is a subcoalgebra of C. Let $c^* \in C^*$. Since $c^*{\cdot}m_j = \sum_{i=1}^{r} m_i c^*(c_{i,j})$ for all $1 \leq j \leq r$ and $\{m_1, \ldots, m_r\}$ is linearly independent, $\mathrm{ann}_{C^*}(M) = C(\rho)^\perp$. We have established parts (a)–(c) when M is finite-dimensional.

Now suppose that M is any right C-comodule and let $\mathcal{F}$ be the set of finite-dimensional subcomodules of M. Then $M = \sum_{N\in\mathcal{F}} N$ by part (a) of Theorem 3.2.8. Since $C(\rho) = \sum_{N\in\mathcal{F}} C(\rho|N)$ part (a) holds for M. Using the fact that $\mathrm{ann}_{C^*}(N) = C(\rho|N)^\perp$ for $N \in \mathcal{F}$ and part (a) of Proposition 1.3.8 we calculate

$$\mathrm{ann}_{C^*}(M) = \bigcap_{N\in\mathcal{F}} \mathrm{ann}_{C^*}(N) = \bigcap_{N\in\mathcal{F}} C(\rho|N)^\perp = (\sum_{N\in\mathcal{F}} C(\rho|N))^\perp = C(\rho)^\perp$$

which establishes part (b) for M. It remains to show part (d).

Suppose that M is a simple right C-comodule. Then M is finite-dimensional by part (b) of Theorem 3.2.8 and thus $C(\rho)$ is finite-dimensional by part (c). For our proof we may assume that C is finite-dimensional. To show part (d) it suffices to show that $\mathcal{M}{\rightharpoonup}M = (0)$ for some maximal ideal $\mathcal{M}$ of C^*. Assume this is the case. Then $\mathcal{M} \subseteq C(\rho)^\perp$ by part (b). Since $M \neq (0)$ and ρ is one-one $C(\rho) \neq (0)$. Therefore the ideal $C(\rho)^\perp \neq C^*$ which means that $C(\rho)^\perp = \mathcal{M}$ since $\mathcal{M}$ is a maximal ideal of C^*. At this point we use Corollary 2.3.9 to conclude that $C(\rho)$ is a simple subcoalgebra of C.

Let $\mathcal{M}_1, \ldots, \mathcal{M}_s$ be the maximal ideals of C^* and let $J = \mathrm{Rad}(C^*)$. Since J is nilpotent and M is a simple left C^*-module $J{\rightharpoonup}M = (0)$. Thus since $\mathcal{M}_1 \cdots \mathcal{M}_s \subseteq J$ either $\mathcal{M}_s{\rightharpoonup}M = (0)$ or for some $1 \leq i < s$ it is the case that

$$(0) = (\mathcal{M}_i \cdots \mathcal{M}_s){\rightharpoonup}M = \mathcal{M}_i{\rightharpoonup}((\mathcal{M}_{i+1} \cdots \mathcal{M}_s){\rightharpoonup}M)$$

and $(\mathcal{M}_{i+1} \cdots \mathcal{M}_s){\rightharpoonup}M \neq (0)$. In the latter case $(\mathcal{M}_{i+1} \cdots \mathcal{M}_s){\rightharpoonup}M = M$. We have shown that $\mathcal{M}{\rightharpoonup}M = (0)$ for some maximal ideal $\mathcal{M}$ of C^*. Part (d) is established and the theorem is proved. □

Comodule maps are understood in terms of the axioms for module maps expressed as equality of certain compositions. Let (A, m, μ) be an algebra over the field k and suppose that $(M, \mu_M), (N, \mu_N)$ are right A-modules. A module map $f : M \longrightarrow N$ is a linear map of underlying vector spaces such that $f(m{\cdot}a) = f(m){\cdot}a$ for all $m \in M$ and $a \in A$, which is to say that the compositions

$$M{\otimes}A \xrightarrow{\mu_M} M \xrightarrow{f} N$$

and

$$M{\otimes}A \xrightarrow{f\otimes \mathrm{I}_A} N{\otimes}A \xrightarrow{\mu_N} N$$

are the same. These axioms are the same as the diagram

$$\begin{array}{ccc} M{\otimes}A & \xrightarrow{\quad f\otimes \mathrm{I}_A \quad} & N{\otimes}A \\ \Big\downarrow{\scriptstyle \mu_M} & & \Big\downarrow{\scriptstyle \mu_N} \\ M & \xrightarrow[\quad f \quad]{} & N \end{array}$$

commutes.

The dual notion of module map is a linear map $f : M \longrightarrow N$ of right comodules such that the composites

$$M \xrightarrow{f} N \xrightarrow{\rho_N} N{\otimes}C$$

and

$$M \xrightarrow{\rho_M} M{\otimes}C \xrightarrow{f\otimes \mathrm{I}_C} N{\otimes}C$$

are the same.

Definition 3.2.12. Suppose C is a coalgebra over the field k and let $(M, \rho_M), (N, \rho_N)$ be right C-comodules. A *map of right C-comodules* $f : M \longrightarrow N$ is a map of underlying vector spaces such that $\rho_N \circ f = (f{\otimes}\mathrm{I}_C) \circ \rho_M$.

Thus $f : M \longrightarrow N$ is a map of right C-comodules if and only if the diagram

$$\begin{array}{ccc} N{\otimes}C & \xleftarrow{\quad f\otimes \mathrm{I}_C \quad} & M{\otimes}C \\ \Big\uparrow{\scriptstyle \rho_N} & & \Big\uparrow{\scriptstyle \rho_M} \\ N & \xleftarrow[\quad f \quad]{} & M \end{array}$$

commutes. One may view comodule maps as maps of associated rational modules and vice versa.

Proposition 3.2.13. *Let C be a coalgebra over the field k, let M and N be right C-comodules, and suppose that $f : M \longrightarrow N$ is a linear map. Then f is a map of right C-comodules if and only if f is a map of associated left rational C^*-modules.* □

We relegate the details of the proof of the proposition to Exercise 3.2.2 which the reader is encouraged to work out.

Let $f : M \longrightarrow N$ be a map right C-comodules. By Proposition 3.2.13 it follows that $\mathrm{Ker}(f)$ is a rational left C^*-submodule of M. Therefore $\mathrm{Ker}(f)$ is a subcomodule of M by Corollary 3.2.6. There is a more elementary proof that $\mathrm{Ker}(f)$ is a subcomodule of M. See Exercise 3.2.3.

Suppose that M is a right C-comodule and N is a subcomodule of M. Then there is a unique right C-comodule structure on the vector space quotient M/N such that the projection $\pi : M \longrightarrow M/N$ is a comodule map. At this point we suggest that the reader formulate and prove a fundamental homomorphism theorem for comodules using Theorem 2.1.21 as a model.

The notion of left C-comodule, map of left C-comodules, and analogs of the other definitions in this section for right C-comodules are made in the expected way. We formally define left C-comodule and map of left C-comodules.

Let M be a vector space over k and $\rho : M \longrightarrow C \otimes M$ be a linear map. Axioms (3.5) and (3.6) are replaced with

$$(\mathrm{I}_C \otimes \rho) \circ \rho = (\Delta \otimes \mathrm{I}_M) \circ \rho \tag{3.10}$$

and

$$(\epsilon \otimes \mathrm{I}_C) \circ \rho = \mathrm{I}_M \tag{3.11}$$

respectively.

Definition 3.2.14. Suppose that C is a coalgebra over the field k. A *left C-comodule* is a pair (M, ρ), where M be a vector space over k and $\rho : M \longrightarrow C \otimes M$ is linear, such that (3.10) and (3.11) hold.

Definition 3.2.15. Suppose that C is a coalgebra over the field k and let $(M, \rho_M), (N, \rho_N)$ be left C-comodules. Then a *map of left C-comodules* $f : M \longrightarrow N$ is a map of underlying vector spaces such that $(\mathrm{I}_C \otimes f) \circ \rho_M = \rho_N \circ f$.

For a left C-comodule (M, ρ) we use the notation

$$\rho(m) = m_{(-1)} \otimes m_{(0)}$$

to represent $\rho(m) \in C\otimes M$. Interpreting the results of this section for right C^{cop}-comodules gives analogs for left C-comodules.

Exercises

In the following exercises C is a coalgebra over the field k.

Exercise 3.2.1. Suppose that M is a vector space over the field k and suppose $\rho : M \longrightarrow M\otimes C$ is a linear map. Write $\rho(m) = m_{(0)}\otimes m_{(1)} \in M\otimes C$ for all $m \in M$ and write $a\rightharpoonup m = m_{(0)}a(m_{(1)})$ for all $m \in M$ and $a \in C^*$. Let $m \in M$. Show that:

(a) If A is a dense subspace of C^* then

$$m_{(0)(0)}\otimes m_{(0)(1)}\otimes m_{(1)} = m_{(0)}\otimes m_{(1)(1)}\otimes m_{(1)(2)}$$

if and only if $a\rightharpoonup(b\rightharpoonup m) = (ab)\rightharpoonup m$ for all $a, b \in A$. [Hint: See Proposition 1.3.6 and Exercise 1.3.7.]

(b) $m_{(0)}\epsilon(m_{(1)}) = m$ if and only if $\epsilon\rightharpoonup m = m$.

Exercise 3.2.2. Let M, N be right C-comodules and $f : M \longrightarrow N$ be a linear map. Suppose that A is a dense subspace of C^*. Show that f is a map of right C-comodules if and only if $f(a\rightharpoonup m) = a\rightharpoonup f(m)$ for all $a \in A$ and $m \in M$. [Hint: See Proposition 1.3.6 and Exercise 1.3.7.]

Exercise 3.2.3. Let $f : M \longrightarrow N$ be a map of right C-comodules. Show that:

(a) If K is a subcomodule of M then $f(K)$ is a subcomodule of N. Work directly from definitions.

(b) $\mathrm{Ker}(f)$ is a subcomodule of M. Establish for $m \in M$ that $f(m) = 0$ if and only if $\rho(m) \in \mathrm{Ker}(f\otimes \mathrm{I}_C) = \mathrm{Ker}(f)\otimes C$.

(c) If L is a subcomodule of N then $f^{-1}(L)$ is a subcomodule of M. Apply part (b) to the composition of comodule maps $M \xrightarrow{f} N \xrightarrow{\pi} N/L$, where π is the projection. Compare with the suggested proof for part (c) of Exercise 2.1.28.

Exercise 3.2.4. Use Proposition 3.2.2 and Corollary 3.2.6 to establish parts (a)–(c) of Exercise 3.2.3 as an exercise in maps of rational modules.

Exercise 3.2.5. Show that:

(a) The composition of right C-comodule maps is a right C-comodule map.

(b) The linear inverse of an isomorphism of right C-comodules is a comodule isomorphism.

(c) The identity map of a right C-comodule is a map of right C-comodules.

(In particular right C-comodules and maps of right C-comodules under function composition form a category which we denote $\mathcal{M}^C$.)

Exercise 3.2.6. Show the conclusions of Exercise 3.2.5 are valid when "left" replaces "right". (In particular left C-comodules and maps of left C-comodules under function composition form a category which we denote ${}^C\mathcal{M}$.)

Exercise 3.2.7. General results about right C-comodules translate to general results about left C-comodules and vice versa. Show that:

(a) If (M, ρ) is a right C-comodule then (M, ρ^{cop}) is a left C^{cop}-comodule according to $\rho^{cop}(m) = m_{(1)}\otimes m_{(0)}$ for all $m \in M$, where $\rho(m) = m_{(0)}\otimes m_{(1)}$.

(b) If (M, ρ) is a left C-comodule then (M, ρ^{cop}) is a right C^{cop}-comodule according to $\rho^{cop}(m) = m_{(0)}\otimes m_{(-1)}$ for all $m \in M$, where $\rho(m) = m_{(-1)}\otimes m_{(0)}$.

(c) ${}^C\mathcal{M} \simeq \mathcal{M}^{C^{cop}}$ and $\mathcal{M}^C \simeq {}^{C^{cop}}\mathcal{M}$ via $(M, \rho) \mapsto (M, \rho^{cop})$ and $f \mapsto f$.

Exercise 3.2.8. On a categorical level there is no difference between rational modules and comodules. Show that $F : \mathcal{M}_C \longrightarrow {}_{C^*}\mathcal{M}_r$ given by $F((M, \rho)) = (M, \mu_\rho)$ for right C-comodules (M, ρ) and $F(f) = f$ for maps $f : M \longrightarrow N$ of right C-comodules is a well-defined isomorphism of categories.

Exercise 3.2.9. Formulate and prove a fundamental homomorphism theorem for comodules which is analogous to Theorem 2.1.21.

Exercise 3.2.10. Let (M, ρ) be a right C-comodule and let D be a subcoalgebra of C.

(a) Show that $\mathcal{M} = \rho^{-1}(M\otimes D)$ satisfies $\rho(\mathcal{M}) \subseteq \mathcal{M}\otimes D$; thus $(\mathcal{M}, \rho|\mathcal{M})$ is a right D-comodule. [Hint: See the proof of Theorem 2.2.3.]

(b) Suppose that $\rho(M) \subseteq M\otimes D$ and let $I = D^\perp$. Show that M is a left C^*/I-module where $(c^* + I)\cdot m = c^*\cdot m$ for all $c^* \in C^*$ and $m \in M$.

(c) Let $f : C^*/I \longrightarrow D^*$ be the isomorphism of algebras induced by the restriction map $C^* \longrightarrow D^*$ given by $c^* \mapsto c^*|D$. Show that the left C^*/I-module action on M of part (b) is the pullback of the rational D^*-module action on M via f.

Exercise 3.2.11. Let (M, ρ) be a right C-comodule. Suppose that $C = \bigoplus_{i\in I} D_i$ is the direct sum of subcoalgebras. Show that:

(a) $M = \bigoplus_{i\in I} M_i$, where $\rho(M_i) \subseteq M_i\otimes D_i$.

(b) For such a decomposition of M necessarily $M_i = \rho^{-1}(M\otimes D_i)$.

[Hint: Since $\sum_{i\in I} M\otimes D_i$ is direct and ρ is one-one, $\sum_{i\in I} M_i$ is direct, where $M_i = \rho^{-1}(M\otimes D_i)$. To show that $\sum_{i\in I} M_i = M$ we may assume that I is finite and without loss of generality let $I = \{1, \ldots, r\}$. For each $i \in I$ define $e_i \in C^*$

by $e_i|D_j = \delta_{i,j}\epsilon|D_i$. For $m \in M$ show that $m = \epsilon \rightharpoonup m = e_1 \rightharpoonup m + \cdots + e_r \rightharpoonup m \in M_1 + \cdots + M_r$.]

Exercise 3.2.12. Let S be a non-empty set and $C = k[S]$ be the coalgebra over k of Example 2.1.9. Let (M, ρ) be a left C-comodule. Show that $M = \bigoplus_{s \in S} M_s$, where $M_s = \{m \in M \mid \rho(m) = s \otimes m\}$ for all $s \in S$.

Exercise 3.2.13. Let $\{M_i\}_{i \in I}$ be an indexed family of right C-comodules. Show that the vector space direct sum $M = \bigoplus_{i \in I} M_i$ has a right C-comodule structure (M, ρ) determined by $\rho|M_i = \rho_i$ for all $i \in I$.

Exercise 3.2.14. Let A be an algebra over the field k and suppose that (M, μ) is a finite-dimensional left A-module. Show that:

(a) If A is finite-dimensional there is a right A^o-comodule structure (M, ρ_μ) on M determined by

$$a{\cdot}m = m_{(0)} <m_{(1)}, a> \tag{3.12}$$

for all $a \in A$ and $m \in M$. [Hint: The algebra map $\jmath_A : A \longrightarrow (A^o)^*$ of Proposition 2.6.1 is an isomorphism by Corollary 2.6.3. Thus M is a left $(A^o)^*$-module by pullback along $\jmath_A^{-1}$. See Corollary 3.1.6 and Proposition 3.2.2.]

(b) There is a right A^o-comodule structure (M, ρ_μ) on M determined by (3.12). [Hint: As for uniqueness, see Propositions 1.3.6, 2.6.1 and Exercise 1.3.7. As for existence, $I = \mathrm{ann}_A(M)$ is a cofinite ideal of A and M is a left A/I-module where $(a+I){\cdot}m = a{\cdot}m$ for all $a \in A$ and $m \in M$. Apply part (a) to A/I and M and note that $(A/I)^o$ can be thought of as a subcoalgebra of A^o.]

(c) $A^o(\rho_\mu) = (\mathrm{ann}_A(M))^\perp$. [Hint: See Corollary 1.3.9.]

Exercise 3.2.15. Show that the comodule structure map $\rho_\mu : M \longrightarrow M \otimes A^o$ of Exercise 3.2.14 is described as follows. Let $\mu^T : M^* \otimes A \longrightarrow M^*$ denote the right transpose action on M^* induced by the left A-module structure $\mu : A \otimes M \longrightarrow M$. Then ρ_μ is the composite

$$M \xrightarrow{\imath_M} M^{**} \xrightarrow{(\mu^T)^*} (M^* \otimes A)^* \xrightarrow{\imath^{-1}} M^{**} \otimes A^* \xrightarrow{\imath_M^{-1} \otimes \mathrm{I}_{A^*}} M \otimes A^*,$$

where $\imath_M : M \longrightarrow M^{**}$ is the map of Corollary 1.3.13 and $\imath : M^{**} \otimes A^* \longrightarrow (M^* \otimes A)^*$ is the map of Exercise 1.3.8. Since M is finite-dimensional both $\imath_M$ and $\imath$ are isomorphisms.

Exercise 3.2.16. Suppose C is finite-dimensional. Show that all left C^*-modules are rational by modifying the argument used in Exercise 3.2.15. Compare with the argument used for Corollary 3.1.6.

Exercise 3.2.17. Let A be an algebra over the field k and $\jmath_A : A \longrightarrow (A^o)^*$ be the algebra map of Proposition 2.6.1. Show that:

(a) Every right A^o-comodule (M, ρ) induces a locally finite left A-module structure $(M, \mu(\rho))$ on M by pullback along $\jmath_A$ given by $a{\cdot}m = \jmath_A(a){\rightharpoonup}m = m_{(0)}{<}m_{(1)}, a{>}$ for all $a \in A$ and $m \in M$.

(b) $(M, \rho) \mapsto (M, \mu(\rho))$ and $f \mapsto f$ describes an isomorphism of the category $\mathcal{M}^{A^o}$ of right A^o-comodules and comodule maps and of the category ${}_A\mathcal{M}_{lf}$ of locally finite left A-modules and maps of locally finite left A-modules. [Hint: See Exercise 3.2.14.]

Exercise 3.2.18. Suppose that (M, ρ) is a right C-comodule and suppose that $f : C \longrightarrow D$ is a coalgebra map. Show that:

(a) (M, ρ_f) is a right C-comodule, where $\rho_f(m) = m_{(0)} \otimes f(m_{(1)})$ for all $m \in M$.

(b) If (M, μ) is the left D^*-module action on M which is the pullback via f^* of the rational left C^*-module structure on M associated with (M, ρ) then $\mu = \mu_{\rho_f}$ is the rational left D^*-module action associated with (M, ρ_f).

Exercise 3.2.19. Suppose that the simple subcoalgebras of C are one-dimensional. Find all 1, 2 and 3-dimensional right C-comodules. [Hint: The factors of a composition series of a finite-dimensional right C-comodule are one-dimensional.]

Definition 3.2.16. A *finitely cogenerated right* C-comodule is a right C-comodule M such that there exists a one-one map of right C-comodules $M \longrightarrow C \oplus \cdots \oplus C$ from M into a finite direct sum of copies of C.

Exercise 3.2.20. Let (M, ρ) be a right C-comodule and regard M^* as a right C^*-module under the right transpose action arising from the left rational C^*-module structure on M. Show that M is finitely cogenerated if and only if M^* is a finitely generated right C^*-module. [Hint: See part (d) of Proposition 3.1.2 and part (b) of Corollary 3.1.10.]

Exercise 3.2.21. Let S be a non-empty set and $C = k[S]$ be the grouplike coalgebra of S over k described in Example 2.1.9. Show that right C-comodules have the form $M = \bigoplus_{s \in S} M_s$, where $M_s = \{m \in M \mid \rho(m) = m \otimes s\}$. [Hint: See Exercise 3.2.11.]

Exercise 3.2.22. Let $n \geq 1$ and $C = \mathrm{C}_n(k)$. Show that all non-zero right C-comodules are direct sums of copies of the n-dimensional right coideal N of C with basis $\{e_{1,i}\}_{1 \leq i \leq n}$. [Hint: $C^* \simeq \mathrm{M}_n(k)$ is a simple algebra.]

Exercise 3.2.23. Let $C = \mathrm{P}_\infty(k)$ be the coalgebra of Definition 2.1.14 and suppose that M is an n-dimensional vector space over k. Show that:

(a) If $T : M \longrightarrow M$ is a nilpotent linear endomorphism of M then (M, ρ_T) is a right C-comodule, where

$$\rho_T(m) = \sum_{s=0}^{n-1} T^s(m) \otimes c_s$$

for all $m \in M$.

(b) $T \mapsto (M, \rho_T)$ determines a one-one correspondence between the set of all nilpotent linear endomorphisms of M and the set of all right C-comodule structures on M. [Hint: Recall that $x \in C^*$ defined by $x(c_m) = \delta_{m,1}$ for all $m \geq 0$ generates a dense subalgebra A of C^*. See part (c) of Exercise 2.3.32. Let (M, μ) be a rational left C^*-module structure on M. Then μ restricts a left A-module action on M which is determined by the action of x on M. Show that $T \in \mathrm{End}\,(M)$ defined by $T(m) = x{\cdot}m$ for all $m \in M$ is nilpotent.]

Exercise 3.2.24. Generalize Exercise 3.2.23. Let C be the coalgebra of Exercise 2.3.32 and suppose that M is an n-dimensional vector space over k.

(a) Suppose that $\mathcal{T} = \{T_1, \ldots, T_r\}$ is an indexed family of commuting nilpotent linear endomorphisms of M. Show that $(M, \rho_{\mathcal{T}})$ is a right C-comodule where

$$\rho_{\mathcal{T}}(m) = \sum_{\mathbf{s} \in S, s_i \leq n} \mathcal{T}^{\mathbf{s}}(m) \otimes c_{\mathbf{s}}$$

for all $m \in M$ and $\mathcal{T}^{\mathbf{s}} = T_1^{s_1} \circ \cdots \circ T_r^{s_r}$ for $\mathbf{s} = (s_1, \ldots, s_r) \in S$, the set of all r-tuples with non-negative integer entries.

(b) Show that $\mathcal{T} \mapsto (M, \rho_{\mathcal{T}})$ determines a one-one correspondence between the collection of sets of commuting nilpotent linear endomorphisms of M indexed by $1, \ldots, r$ and the set of all right C-comodule structures on M.

Exercise 3.2.25. We continue our discussion of structure constants. See Exercises 2.1.22, 2.5.18, and 2.5.19. Let M be a vector space over k with basis $\{m_1, \ldots, m_r\}$.

(a) Let A be a finite-dimensional algebra over k with basis $\{a_1, \ldots, a_n\}$. Write $a_i a_j = m_{i,j}^{\ell} a_\ell$ and $1 = \eta^\ell a_\ell$. Let $\mu : A \otimes M \longrightarrow M$ be a linear map described by $\mu(a_i \otimes m_j) = \mu_{i,j}^{\ell} m_\ell$. Show that (M, μ) is a left A-module if and only if

$$\mu_{j\ell}^{u} \mu_{i,u}^{v} = m_{i,j}^{u} \mu_{u,\ell}^{v} \quad \text{and} \quad \eta^u \mu_{u,j}^{i} = \delta_j^i$$

for all i, j, ℓ, v.

(b) Let C be a coalgebra over k with basis $\{c_1, \ldots, c_n\}$. Write $\Delta(c_\ell) = \Delta_\ell^{i,j} c_i \otimes c_j$ and $\epsilon(c_\ell) = \epsilon_\ell$. Let $\rho : M \longrightarrow M \otimes C$ be a linear map described by $\rho(m_\ell) = \rho_\ell^{i,j} m_i \otimes c_j$. Show that (M, ρ) is a right C-comodule if and only if

$$\rho_i^{u,v} \rho_u^{j,\ell} = \rho_i^{j,u} \Delta_u^{\ell,v} \quad \text{and} \quad \rho_j^{i,u} \epsilon_u = \delta_j^i$$

for all i, j, ℓ, v.

(c) Suppose that (M, μ) is a left A-module. Suppose $\Delta_\ell^{i,j} = m_{i,j}^\ell$, $\epsilon_\ell = \eta^\ell$ and $\rho_\ell^{u,i} = m_{i,\ell}^u$. Show that (M, ρ) is a right C-comodule.

(d) Use part (a) of Exercise 2.5.19 and part (c) to give a proof of Corollary 3.1.6.

3.3 M_r and M^r

Let M be a left C^*-module and regard M^* as a right C^*-module structure under the transpose action. Since $\alpha(c^*{\cdot}m) = (\alpha{\prec}c^*)(m)$ for all $\alpha \in M^*$, $c^* \in C^*$ and $m \in M$ it follows that

$$\alpha(m_{(0)})c^*(m_{(1)}) = c^*(\alpha_{(-1)})\alpha_{(0)}(m)$$

for all $\alpha \in M^r$, $m \in M_r$, and $c^* \in C^*$. Therefore the comodule structures on M_r and M^r are related by the equation

$$<\alpha, m_{(0)}>m_{(1)} = \alpha_{(-1)}<\alpha_{(0)}, m> \tag{3.13}$$

for all $\alpha \in M^r$ and $m \in M_r$.

3.4 The coradical of a coalgebra

Suppose that (M, ρ) is a simple right C-comodule. Then $\rho(M) \subseteq M{\otimes}D$ for some simple subcoalgebra D of C by part (d) of Theorem 3.2.11. Thus the simple C-comodules can be understood in terms of the sum of the simple subcoalgebras of C.

Definition 3.4.1. Let C be a coalgebra over the field k. The *coradical of* C is the sum of the simple subcoalgebras of C and is denoted C_0.

By Corollary 2.2.4 every simple subcoalgebra of C is finite-dimensional and every non-zero subcoalgebra of C contains a simple subcoalgebra of C. Thus $C = (0)$ if and only if $C_0 = (0)$. The coradical of C is characterized in terms of simple right and simple left coideals of C.

Theorem 3.4.2. *Let C be a coalgebra over the field k. Then C_0 is*

(a) *the sum of the simple right coideals of C and is also*

(b) *the sum of the simple left coideals of C.*

Proof. Since the coradicals of C and C^{cop} are the same, we need only establish part (a). Let N be a simple right coideal of C. As noted above $\Delta(N) \subseteq N \otimes D$, where D is a simple subcoalgebra of C. Since $\mathrm{I}_C = (\epsilon \otimes \mathrm{I}_C) \circ \Delta$ it follows that $N \subseteq \epsilon(N)D \subseteq C_0$.

Let D be a simple subcoalgebra of C. To complete the proof of the theorem we need only show that D is the sum of simple right coideals of C. Since D is a non-zero finite-dimensional right coideal of C, it follows that D contains a minimal right coideal N of C. Let $c^* \in C^*$. By (2.19) the linear endomorphism $\mathrm{R}(c^*)$ of C defined by $\mathrm{R}(c^*)(c) = c \leftharpoonup c^*$ for all $c \in C$ is a map of right C-comodules. Thus $N \leftharpoonup c^*$ is a homomorphic image of N. Consequently $N \leftharpoonup c^* = (0)$ or $N \leftharpoonup c^* \simeq N$ since N is simple. Let $E = N \leftharpoonup C^*$. Then $E \subseteq D$ and is the sum of simple right coideals of C. Since

$$C^* \rightharpoonup E \leftharpoonup C^* = C^* \rightharpoonup N \leftharpoonup C^* = N \leftharpoonup C^* = E$$

if follows by part (b) of Proposition 2.3.5 that E is a subcoalgebra of C. Since D is a simple subcoalgebra of C we conclude $D = E$ and thus is the sum of simple right coideals of C. □

Parts (a) and (b) of the following proposition have significant implications for coalgebras in general. They have important implications for the coradical in particular.

Proposition 3.4.3. *Let C be coalgebra over the field k, let $\{C_i\}_{i \in I}$ be a family of subcoalgebras of C and suppose that D is a subcoalgebra of C. Then:*

(a) *If D is simple and $D \subseteq \sum_{i \in I} C_i$ then $D \subseteq C_i$ for some $i \in I$.*
(b) *If $\sum_{i \in I} C_i$ is direct then $D \cap (\bigoplus_{i \in I} C_i) = \bigoplus_{i \in I}(D \cap C_i)$.*
(c) *C_0 is the direct sum of the simple subcoalgebras of C.*
(d) *$D_0 = C_0 \cap D$.*
(e) *D_0 is the direct sum of all simple subcoalgebras of C which are contained in D.*

Proof. Assume the hypothesis of part (a). Since D is simple D is finite-dimensional by part (a) of Corollary 2.2.4. Thus $D \subseteq \sum_{i \in J} D_i$ where J is a finite subset of I. We may assume that J is as small as possible.

Suppose that $|J| > 1$ and let $i_0 \in J$. Then $D \not\subseteq D_{i_0}$. Therefore $D \cap D_{i_0} = (0)$ since D is simple. Choose $c^* \in C^*$ which satisfies $c^*|D = \epsilon|D$ and $c^*|D_{i_0} = 0$. Then

$$D = \epsilon \rightharpoonup D = c^* \rightharpoonup D \subseteq \sum_{i \in J} c^* \rightharpoonup D_i = \sum_{i \in J \setminus i_0} D_i,$$

a contradiction. Therefore $|J| = 1$ and $D \subseteq D_i$, where $J = \{i\}$. We have established part (a).

To show part (b) note first of all that $D\bigcap(\bigoplus_{i\in I} C_i) \supseteq \bigoplus_{i\in I}(D\bigcap C_i)$. Let $d \in D\bigcap(\bigoplus_{i\in I} C_i)$. Then $d \in D\bigcap(\bigoplus_{i\in J} C_i)$ where J is a finite subset of I. For $i \in J$ choose $e_i \in C^*$ which satisfies $e_i|C_j = \delta_{i,j}\epsilon|C_i$ for all $j \in J$. Let $e = \sum_{i\in J} e_i$. Then e and ϵ agree on $\bigoplus_{i\in J} C_i$ and

$$d = \epsilon \rightharpoonup d = e \rightharpoonup d = \sum_{i\in J} e_i \rightharpoonup d \in \bigoplus_{i\in J}(D \cap C_i).$$

Therefore $D\bigcap(\bigoplus_{i\in J} C_i) \subseteq \bigoplus_{i\in I}(D\bigcap C_i)$ and part (b) is established.

By Zorn's Lemma C_0 is the direct sum of simple subcoalgebras of C. By part (b) all simple subcoalgebras of C must be summands. We have shown part (c). To show parts (d) and (e) we first note that the simple subcoalgebras of D are simple subcoalgebras of C which are contained in D. Let $\{D_i\}_{i\in I}$ be the set of simple subcoalgebras of C. Using part (b) we calculate

$$D_0 = \bigoplus_{i\in I}(D \cap D_i) = D \cap (\bigoplus_{i\in I} D_i) = D \cap C_0$$

from which parts (d) and (e) follow. □

At this point we make several definitions based on the coradical.

Definition 3.4.4. A *pointed coalgebra over* k is a coalgebra C over k whose simple subcoalgebras are one-dimensional.

Note that C is pointed if and only if $C_0 = k[\mathrm{G}(C)]$ is the grouplike coalgebra of the set of grouplike elements of C.

Definition 3.4.5. An *irreducible coalgebra over* k is a coalgebra C over k such that C_0 is a simple subcoalgebra of C. A *pointed irreducible coalgebra over* k is a coalgebra C over k such that C_0 is one-dimensional.

Let D be a subcoalgebra of the coradical C_0 of C. Consider the family $\mathcal{F}$ of all subcoalgebras of E of C which satisfy $E_0 = D$. Then $D \in \mathcal{F}$, $\mathcal{F}$ is closed under intersections by part (e) of Proposition 3.4.3 and $\mathcal{F}$ is closed under sums by parts (a) and (e) of the same. In particular $F = \sum_{E\in\mathcal{F}} E$ belongs to $\mathcal{F}$. Thus $\mathcal{F}$ has a unique maximal element. When D is simple F is of particular interest.

Definition 3.4.6. Let C be a coalgebra over k. An *irreducible component of* C is a maximal irreducible subcoalgebra of C.

Corollary 3.4.7. *Let C be a coalgebra over the field k. Then:*

(a) *Every irreducible subcoalgebra of C is contained in a unique irreducible component of C.*
(b) *The sum of the irreducible components of C is direct.*

Proof. Suppose that E is an irreducible subcoalgebra of C and let $D = E_0$. Then D is a simple subcoalgebra of C by definition. Let $\mathcal{F}$ be the family described above for D. Then $E \in \mathcal{F}$ and all of the elements of $\mathcal{F}$ are irreducible subcoalgebras of C. The unique maximal element F of $\mathcal{F}$ is an irreducible component of C and $E \subseteq F$.

Suppose that $E \subseteq G$, where G is an irreducible subcoalgebra of C. Since $E_0 \subseteq G_0$ and G_0 is simple, it follows that $D = E_0 = G_0$. Therefore $G \in \mathcal{F}$ whence $G \subseteq F$. We have shown part (a).

To establish part (b) we first show that if F and G are irreducible components of C then $F \bigcap G = (0)$ or $F = G$. For $F \bigcap G \neq (0)$ means $F \bigcap G$ is irreducible and thus $F = G$ by part (a). To show that the sum of all irreducible components of C is direct we need only show that the sum of a finite number of distinct irreducible components $F_1, \ldots, F_n$ of C is direct. This we do by induction on n. We may assume $n > 1$.

Suppose that $n > 1$ and $F_1 + \cdots + F_{n-1}$ is direct. By part (b) of Proposition 3.4.3 we have $(F_1 + \cdots + F_{n-1}) \bigcap F_n = (F_1 \bigcap F_n) \bigoplus \cdots \bigoplus (F_{n-1} \bigcap F_n) = (0)$. Thus $F_1 + \cdots + F_n$ is direct. □

Generally a coalgebra is not the sum of its irreducible components. See part (b) of Exercise 3.4.4. If C is cocommutative then C is the direct sum of its irreducible components. See Theorem 4.8.8 which is discussed in the context of Exercise 4.8.5.

Irreducible coalgebras have the smallest possible coradical in the sense that their coradicals are simple. At the other end of the theoretical spectrum are the coalgebras whose coradicals constitute the whole coalgebra.

Definition 3.4.8. A *cosemisimple coalgebra over k* is a coalgebra C over k such that $C = C_0$.

Let M be a rational left C^*-module. Recall that M is completely reducible if M is the sum of simple submodules, which must be rational by Corollary 3.1.5. By Proposition 3.2.2 and Corollary 3.2.6 an equivalent formulation of completely reducible for rational C^*-modules is the following:

Definition 3.4.9. Let C be a coalgebra over the field k. A *completely reducible C-comodule* is a C-comodule M which is the sum of its simple subcomodules.

Cosemisimple coalgebras over k are characterized much in the same way as are semisimple artinean algebras over k.

Theorem 3.4.10. *Suppose that C is a coalgebra over the field k. Then the following are equivalent:*

(a) *All right C-comodules are completely reducible.*
(b) $C = C_0$.
(c) *All left C-comodules are completely reducible.*

Proof. We need only show the equivalence of parts (a) and (b). For the equivalence of parts (c) and (b) is the equivalence of parts (a) and (b) for C^{cop}.

Suppose that all right C-comodules are completely reducible. Then C itself is the sum of simple right coideals of C. Therefore $C = C_0$ by part (a) of Theorem 3.4.2.

On the other hand, suppose that $C = C_0$ and let $\{D_i\}_{i \in I}$ be the set of simple subcoalgebras of C. Then any right C-comodule (M, ρ) can be written $M = \bigoplus_{i \in I} M_i$, where $\rho(M_i) \subseteq M_i \otimes D_i$ for all $i \in I$, by Exercise 3.2.11. To complete the proof we may assume that C is simple. In this case C^* is a finite-dimensional simple algebra over k by Corollary 2.3.8 and thus all C^*-modules are completely reducible. Therefore all right C-comodules are completely reducible and the theorem is proved. □

There is more to be said about the coradical of C. First we need to study an associated filtration of C which is the major topic of discussion of Sections 4.1–4.3.

Exercises

Throughout these exercises C is a coalgebra over the field k.

Exercise 3.4.1. Let D be a subcoalgebra of C. Show that $D = (0)$ if and only if $D \bigcap C_0 = (0)$.

Exercise 3.4.2. Suppose that D, E are subcoalgebras of C and $D \subseteq E$. Show that $D_0 \subseteq E_0$.

Exercise 3.4.3. Show that non-zero subcoalgebras of irreducible subcoalgebras of C are irreducible.

Exercise 3.4.4. Let V be a vector space over k and $C = kg \bigoplus kh \bigoplus V$, where $g, h \in \mathrm{G}(C)$ and $\Delta(v) = g \otimes v + v \otimes h$ for all $v \in V$. Show that:

(a) C is pointed and that $\mathrm{G}(C) = \{g, h\}$.

(b) C is the direct sum of its irreducible components if and only if $V = (0)$.

See Exercise 2.1.23.

Exercise 3.4.5. Show that the coalgebra of Exercise 2.3.14 is pointed irreducible.

Exercise 3.4.6. Show that the coalgebra of Exercise 2.3.32 is pointed irreducible.

Exercise 3.4.7. Let $\{D_i\}_{i \in I}$ be a family of subcoalgebras of C. Show that

$$(\sum_{i \in I} D_i)_0 = \sum_{i \in I} (D_i)_0.$$

Exercise 3.4.8. Suppose that C is simple and that N is a simple right coideal of C.

(a) Using the fact that C is the sum of simple right coideals which are isomorphic as right C-comodules, show that all simple right coideals of C are isomorphic to N as right C-comodules.

Suppose that (M, ρ) is a simple right C-comodule.

(b) Let $\alpha \in M^* \backslash 0$ and define $f : M \longrightarrow C$ by $f(m) = <\alpha, m_{(0)}> m_{(1)}$ for all $m \in M$. Show that f is a one-one map of right C-comodules.

(c) Show that $M \simeq N$ as right C-comodules.

Exercise 3.4.9. Suppose that C is finite-dimensional. Show that $C_0^{\perp} = \mathrm{Rad}(C^*)$. [Hint: See Corollary 2.3.8 and part (a) of Proposition 1.3.8.]

Exercise 3.4.10. Show that the irreducible components of C are saturated.

3.5 Injective comodules

Injectives in the category $\mathcal{M}^C$ of right C-comodules have a simple basic theory and they are useful in any number of contexts in the analysis of C. Our treatment of injectives in $\mathcal{M}^C$, which we call injective right C-comodules, parallels in many ways a standard treatment of injective modules over an algebra. Since comodules have a locally finite character which modules generally do not possess there are some interesting differences. There is an important connection between direct sum decompositions of C into right (injective) coideals of C and certain families of idempotents of C^*.

Some definitions and results in this section are stated for right comodules over C. The word "right" can be replaced by "left" since these results applied to the coalgebra C^{cop} give their "left" counterparts for C.

Definition 3.5.1. Let C be a coalgebra over k. An *injective right C-comodule* is a right C-comodule I such that for maps of right C-comodules $\imath : M \longrightarrow N$ and $f : M \longrightarrow I$, where $\imath$ is one-one, there exists a map of right C-comodules $F : N \longrightarrow I$ such that $F \circ \imath = f$.

Apropos of the definition, without loss of generality we may assume that $\imath$ is the inclusion which we will do from time to time as is convenient. The concept of injectivity in $\mathcal{M}^C$ is really about extension of comodule maps.

To show that I is injective, the comodules M and N of Definition 3.5.1 may be assumed to be finite-dimensional. For this reason the direct sum of injective comodules is injective.

Lemma 3.5.2. *Suppose that C is a coalgebra over the field k and I is a right C-comodule. Then the following are equivalent:*

(a) *I is injective.*
(b) *For all finite-dimensional right C-comodules M and N and maps of right C-comodules $\imath : M \longrightarrow N$ and $f : M \longrightarrow I$, where $\imath$ is one-one, there exist a map of right C-comodules $F : N \longrightarrow I$ such that $F \circ \imath = f$.*

Proof. Part (a) implies part (b) by definition. Suppose the condition of part (b) holds. Let M and N be right C-comodules, $f : M \longrightarrow I$ be a comodule map, and suppose that M is a subcomodule of N. We need only show that there is a comodule map $F : N \longrightarrow I$ which extends f. This we do by Zorn's Lemma.

Let $\mathcal{F}$ be the set of all pairs (N', F'), where N' is a subcomodule of N which contains M and $F' : N' \longrightarrow I$ is a comodule map which extends f. Then $(\mathcal{F}, \leq)$ is a partially ordered set, where $(N', F') \leq (N'', F'')$ if $N'' \supseteq N'$ and F'' extends F'. Now $\mathcal{F} \neq \emptyset$ since $(M, f) \in \mathcal{F}$. It is easy to see that every chain of $(\mathcal{F}, \leq)$ has an upper bound. Therefore $\mathcal{F}$ has a maximal element (N', F) by Zorn's Lemma.

Suppose that $n \in N$ and let N'' be the subcomodule of N which n generates. Then N'' is finite-dimensional by Proposition 3.2.8. Thus the comodule map $F|N' \bigcap N'' : N' \bigcap N'' \longrightarrow I$ extends to a comodule map $F' : N'' \longrightarrow I$ by assumption. Since $F|N' \bigcap N'' = F'|N' \bigcap N''$ it follows that $F'' : N' + N'' \longrightarrow I$ given by $F''(n' + n'') = F(n') + F'(n'')$ for all $n' \in N'$ and $n'' \in N''$ is a well-defined comodule map which extends F and F'. In particular $(N', F) \leq (N' + N'', F'')$ by definition. Since (N', F) is maximal it follows that $N' = N' + N''$. Consequently $n \in N'$. We have shown that $N \subseteq N'$. Therefore $N = N'$ and F is the desired extension of f. □

An important corollary of Lemma 3.5.2 is:

Corollary 3.5.3. *Suppose that C is a coalgebra over the field k and $\{I_i\}_{i\in\mathcal{I}}$ is an indexed family of right C-comodules. Then $I = \bigoplus_{i\in\mathcal{I}} \mathrm{I}_i$ is injective if and only if each I_i is injective.*

Proof. For $j \in \mathcal{I}$ let $\imath_j : I_j \longrightarrow I$ be the canonical inclusion and let $\pi_j : I \longrightarrow I_j$ be the canonical projection onto the jth summand. Then $\pi_j \circ \imath_j = \mathrm{I}_{I_j}$.

Suppose that I is injective and fix $j \in \mathcal{I}$. Let M and N right C-comodules, $f : M \longrightarrow I_j$ be a comodule map and M be a subcomodule of N. Then the comodule map $\imath_j \circ f : M \longrightarrow I$ extends to a comodule map $F : N \longrightarrow I$ since I is injective. Thus the comodule map $\pi_j \circ F : M \longrightarrow I_j$ is an extension of $\pi_j \circ \imath_j \circ f = \mathrm{I}_{I_j} \circ f = f$. We have shown that I_j is injective.

Now suppose that I_j is injective for all $j \in \mathcal{I}$. Let M and N be finite-dimensional right C-comodules, $f : M \longrightarrow I$ be a comodule map and suppose that M is subcomodule of N. Since $\mathrm{Im}(f)$ is finite-dimensional $\mathrm{Im}(f) \subseteq I' = \bigoplus_{i\in\mathcal{I}'} I_i$ for some finite subset $\mathcal{I}'$ of $\mathcal{I}$. We may assume that $\mathcal{I}' = \{1, \ldots, r\}$ for some $r \geq 1$. For each $1 \leq j \leq r$ let $F_j : N \longrightarrow I_j$ be a comodule map which extends $\pi_j \circ f : M \longrightarrow I_j$ and let $F : N \longrightarrow I$ be the comodule map defined by $F(n) = F_1(n) \bigoplus \cdots \bigoplus F_r(n)$ for all $n \in N$. Then for $m \in M$ we compute

$$F(m) = F_1(m) \oplus \cdots \oplus F_r(m) = \pi_1(f(m)) \oplus \cdots \oplus \pi_r(f(m)) = f(m).$$

Thus I is injective by Lemma 3.5.2. □

Projective C^*-modules give rise to injective C-comodules. The proof we give uses some elementary facts about duals of modules and associated module maps.

Suppose that A is an algebra over k and M is a left (respectively right) A-module. Then M^* is a right (respectively left) A-module under the transpose action. Observe that $\imath_M : M \longrightarrow M^{**}$ defined by $<\imath_M(m), m^*> = <m^*, m>$ for all $m \in M$ and $m^* \in M^*$ is a module map. Note that $(\imath_M)^* \circ \imath_{M^*} = \mathrm{I}_{M^*}$. See Exercise 1.2.12.

Let $f : M \longrightarrow N$ be a map of left (respectively right) A-modules. Then the transpose $f^* : N^* \longrightarrow M^*$ is a map of right (respectively left) A-modules. Thus $f^{**} : M^{**} \longrightarrow N^{**}$ is a map of left (respectively right) A-modules. Observe that $f^{**} \circ \imath_M = \imath_N \circ f$.

Proposition 3.5.4. *Suppose that C is a coalgebra over the field k.*

(a) *If P is a projective right C^*-module then P^r is an injective right C-comodule.*
(b) *(C, Δ) is an injective right C-comodule.*
(c) *Every right C-comodule is the subcomodule of an injective right C-comodule.*

Proof. We will think of right C-comodules as rational left C^*-modules and of maps of right C-comodules as maps of left rational C^*-modules. Regard C as a left C^*-module under the action $c^* \rightharpoonup c = c_{(1)} < c^*, c_{(2)} >$ for all $c^* \in C^*$ and $c \in C$. Then C^* is a right C^*-module under the transpose action which is right multiplication in C^*. Hence $\imath_C : C \longrightarrow (C^*)^r$ is an isomorphism of rational left C^*-modules by part (b) of Corollary 3.1.10. Thus part (b) follows from part (a). Let M be a rational left C^*-module and let $f : P \longrightarrow M^*$ be an onto map of right C^*-modules, where P is free. Then the composite $f^* \circ \imath_M : M \longrightarrow P^*$ is a one-one map of left C^*-modules. Since M is rational $\mathrm{Im}(f^* \circ \imath_M) \subseteq P^r$. Thus M is a submodule of P^r and part (c) follows from part (a) as well.

It remains to show part (a). Suppose that P is a projective right C^*-module. Let $f : M \longrightarrow P^r$ and $\imath : M \longrightarrow N$ be maps of rational left C^*-modules, where $\imath$ is one-one, and let $\jmath : P^r \longrightarrow P^*$ be the inclusion. Then $(\jmath \circ f)^* \circ \imath_P : P \longrightarrow M^*$ and $\imath^* : N^* \longrightarrow M^*$ are maps of right C^*-modules and $\imath^*$ is onto. Since P is a projective right C^*-module there is a map of right C^*-modules $F_0 : P \longrightarrow N^*$ such that $\imath^* \circ F_0 = (\jmath \circ f)^* \circ \imath_P$. Thus $F_0^* \circ \imath^{**} = (\imath_P)^* \circ (\jmath \circ f)^{**}$. Since $F_0^* \circ \imath^{**} \circ \imath_M = F_0^* \circ \imath_N \circ \imath$ and

$$(\imath_P)^* \circ (\jmath \circ f)^{**} \circ \imath_M = (\imath_P)^* \circ \imath_{P^*} (\jmath \circ f) = \mathrm{I}_{P^*} \circ \jmath \circ f = f,$$

the map of left C^*-modules, $F : N \longrightarrow P^*$ defined by $F = F_0^* \circ \imath_N$ satisfies $F \circ \imath = f$. Since N is rational $\mathrm{Im}(F) \subseteq P^r$ by part (a) of Corollary 3.1.10. Thus $F : N \longrightarrow P^r$ is a map of rational left C^*-modules which extends f. We have established part (a) and have thus completed the proof of the proposition. □

Definition 3.5.5. Suppose that C is a coalgebra over the field k and M is a right C-comodule.

(a) An *essential extension of* M is a right C-comodule N which contains M as a subcomodule and which satisfies the following: whenever L is a subcomodule of N and $L\bigcap M = (0)$ then $L = (0)$.
(b) An *injective hull of* M is an injective essential extension of M.

We note that any comodule is an essential extension of itself. We observe that an injective hull is a maximal essential extension.

Proposition 3.5.6. *Suppose that C is a coalgebra over the field k, N is a right C-comodule and M is a subcomodule of N. Then:*

(a) *N is an essential extension of M if and only if $N_0 = M_0$.*
(b) *Any essential extension of M in N is contained in a maximal essential extension of M in N. In particular there exists a maximal essential extension of M in N.*
(c) *Suppose that N is injective. Then any maximal essential extension of M in N is an injective hull of M. Thus N contains an injective hull of M.*

Proof. We first show part (a). Suppose that N is an essential extension of M and L is a simple subcomodule of N. Since $L \neq (0)$ it follows that $L\bigcap M \neq (0)$. Therefore $L = L\bigcap M \subseteq M$ and we conclude that $L \subseteq M_0$. Thus $N_0 \subseteq M_0$. Since $M \subseteq N$ it follows that $M_0 \subseteq N_0$. Therefore $N_0 = M_0$.

Conversely, suppose that $N_0 = M_0$. Let L be a non-zero subcomodule of N. Then L contains a simple subcomodule L' by Proposition 3.2.8. Since $L' \subseteq N_0 = M_0$ we conclude that $L\bigcap M \neq (0)$. We have shown that N is an essential extension of M.

Part (b) follows by Zorn's Lemma. To show part (c) suppose N is injective and let I be a maximal essential extension of M in N. We show that I is injective. Observe that I has no essential extension in N other than itself. We claim that I has no essential extensions other than itself. For suppose that I' is an essential extension of I. Since N is injective the inclusion $\imath : I \longrightarrow N$ extends to a map of right C-modules $f : I' \longrightarrow N$. Since $\mathrm{Ker}(f)\bigcap I = (0)$ it follows that $\mathrm{Ker}(f) = (0)$. Therefore f is one-one which means $f(I')$ is an essential extension of I in N. Consequently $f(I') = I$ and thus $I' = I$.

Let I' be a subcomodule of N maximal with respect to $I'\bigcap I = (0)$. It suffices to show that N/I' is an essential extension of $(I + I')/I' \simeq I$. For then, since I has no essential extensions other than itself, $N/I' = (I+I')/I'$

which implies $N = I + I'$. Since this sum is direct I is injective by Corollary 3.5.3.

Suppose that L is a subcomodule of N which contains I' and $(L/I')\bigcap(I + I')/I' = (0)$. Then $L\bigcap I = (0)$. By the maximality of I' it follows that $L = I'$. Therefore N/I' is an essential extension $(I + I')/I'$ and I is injective. To show that N contains an injective hull of M we note that there is a maximal essential extension I of M in N by part (b). We have just shown that I is injective. Our proof of part (c) is complete. □

Theorem 3.5.7. *Suppose that C is a coalgebra over the field k and M is a right C-comodule. Then M has an injective hull.*

Proof. The comodule M is contained in some injective right C-comodule by part (c) of Proposition 3.5.4. Thus an injective hull for M exists by part (c) of Proposition 3.5.6. □

There are two very useful characterizations of injective hulls which have module analogs.

Theorem 3.5.8. *Suppose that C is a coalgebra over the field k and I is a right C-comodule. Then the following are equivalent:*

(a) *I is an injective right C-comodule.*
(b) *If N is a right C-comodule and I is a subcomodule of N then $I \bigoplus M = N$ for some subcomodule M of N.*
(c) *I has no essential extensions other than itself.*

Proof. We first show that part (a) implies part (b). Suppose that I is an injective right C-comodule which is a subcomodule of a right C-comodule N. Since I is injective the identity map I_I extends to a comodule map $F : N \longrightarrow I$. Thus $N = I \bigoplus \mathrm{Ker}(F)$ is a direct sum of right C-comodules and part (b) follows.

That part (b) implies part (c) is clear. Suppose that I has no essential extensions other than itself. Now I has an injective hull I' by Theorem 3.5.7 which is, by definition, and essential extension of I. Therefore $I = I'$ by assumption. Thus part (c) implies part (a) and our proof is complete. □

Corollary 3.5.9. *Suppose that C is a coalgebra over the field k and M is a right C-comodule. Then any two injective hulls of M are isomorphic.*

Proof. Suppose that I and I' are injective hulls of M. Then the inclusion map $f : M \longrightarrow I'$ extends to a map of right C-comodules $F : I \longrightarrow I'$.

Therefore $\mathrm{Ker}(F)\bigcap M = (0)$. Since I is an essential extension of M it follows that $\mathrm{Ker}(F) = (0)$. Thus F is one-one. Since $M \subseteq F(I)$ we conclude at this point that $F(I)$ is an injective hull of M in I'. Now I' is an essential extension of M and therefore I' is an essential extension of $F(I)$. By Theorem 3.5.8 we have $F(I) = I'$. $\square$

At this point we collect some of our results and apply them to right coideals of C.

Proposition 3.5.10. *Suppose that C is a coalgebra over the field k and regard (C, Δ) as a right C-comodule. Let N be a right coideal of C.*

(a) *N is injective if and only if $C = N\bigoplus M$ for some right coideal M of C.*
(b) *N is injective if and only if $N = C\leftharpoonup e$ for some idempotent $e \in C^*$.*
(c) *N has an injective hull in C.*
(d) *Suppose that $N = C\leftharpoonup e$ where $e \in C^*$ is an idempotent. Then $N_0 = C_0\leftharpoonup e$ and thus N an injective hull of $C_0\leftharpoonup e$.*

Suppose that N is an injective right coideal of C.

(e) *Write $C = N\bigoplus M$, where M is a right coideal of C. Then $N = C\leftharpoonup e$ for some idempotent $e \in C^*$ which satisfies $<e, M> = (0)$.*
(f) *$N\wedge N = N$.*

Proof. The right C-comodule (C, Δ) is injective by part (b) of Proposition 3.5.4. Thus part (a) follows by Corollary 3.5.3 and Theorem 3.5.8.

To show part (b) we let e be an idempotent of C^* and set $f = \epsilon - e$. Then $\epsilon = e + f$ and $ef = 0 = fe$ which imply that $C = (C\leftharpoonup e)\bigoplus(C\leftharpoonup f)$. Now $C\leftharpoonup c^*$ is a right coideal of C for all $c^* \in C^*$. Thus $C\leftharpoonup e$ is injective by part (a).

Conversely, suppose that N is injective and write $C = N\bigoplus M$ as the direct sum of right coideals of C. There is such a decomposition of C by part (a). Let $e \in C^*$ be the functional determined by $e|N = \epsilon|N$ and $e|M = 0$. Thus for $n \in N$ we compute $n = n\leftharpoonup e$ and thus $e^2(n) = e(n\leftharpoonup e) = e(n)$. For $m \in M$ observe that $e^2(m) = e(m\leftharpoonup e) = 0 = e(m)$. Thus $N = C\leftharpoonup e$ and e is an idempotent of C^*. We have completed the proof of parts (b) and (e). To prove part (f) we let the reader show that $N^\perp = (C\leftharpoonup e)^\perp = fC^*$, where f is the idempotent $f = \epsilon - e$. Thus

$$N\wedge N = (N^\perp N^\perp)^\perp = (fC^* fC^*)^\perp = (fC^*)^\perp = N$$

and part (f) is established.

Part (c) follows by part (c) of Proposition 3.5.6 since (C, Δ) is an injective right C-comodule. It remains to show part (d). Suppose that $N = C \leftharpoonup e$, where e is an idempotent of C^*. Then N is injective by part (b). Thus we need only show that $N_0 = C_0 \leftharpoonup e$ by part (a) of Proposition 3.5.6.

Since e is an idempotent $n \leftharpoonup e = n$ for all $n \in N$. Since $N_0 \subseteq C_0$ it now follows that $N_0 = N_0 \leftharpoonup e \subseteq C_0 \leftharpoonup e$. On the other hand $C_0 \leftharpoonup e \subseteq C \leftharpoonup e \bigcap C_0 = N_0$. We have shown $N_0 = C_0 \leftharpoonup e$. Thus part (d) follows and our proof is complete. □

There is a close connection between certain families of idempotents $\{e_i\}_{i \in \mathcal{I}}$ of C^* and direct sum decompositions of C into right coideals.

Definition 3.5.11. Let C be a coalgebra over the field k. An *orthogonal family of idempotents of* C^* is a family of idempotents $\{e_i\}_{i \in \mathcal{I}}$ of C^* such that $e_i e_j = 0$ whenever $i, j \in \mathcal{I}$ are different.

Suppose that $\{e_i\}_{i \in \mathcal{I}}$ is an orthogonal family of idempotents of C^*. Then $e = \sum_{i \in \mathcal{I}} e_i$ is a well-defined idempotent. For let $c \in C$. Then c is contained in a finite-dimensional subcoalgebra D of C by Theorem 2.2.3. Since any orthogonal set of non-zero idempotents of D^* is linearly independent, all but finitely many of the restrictions $e_i|D$'s are zero. Thus all but finitely many of the scalars $e_i(c)$ are zero, and therefore $\sum_{i \in \mathcal{I}} e_i(c)$ is a sum with only finitely many non-zero terms. Since $e|D$ is a finite sum of orthogonal idempotents it follows that $e|D$ is an idempotent. Since the D's we are describing span C, it now follows that e is an idempotent of C^*.

Definition 3.5.12. Suppose that C is a coalgebra over the field k. An *orthonormal family of idempotents of* C^* is an orthogonal family of idempotents $\{e_i\}_{i \in \mathcal{I}}$ of C^* such that $\sum_{i \in \mathcal{I}} e_i = \epsilon$.

Theorem 3.5.13. *Suppose that C is a coalgebra over the field k and regard (C, Δ) as a right C-comodule.*

(a) *Let $C = \bigoplus_{s \in S} I_s$ be the direct sum of an indexed family of right coideals of C. Then:*
   (i) *I_s is an injective right coideal of C for all $s \in S$.*
   (ii) *There exists an orthonormal set of idempotents $\{e_s\}_{s \in S}$ of C^* such that $I_s = C \leftharpoonup e_s$ for all $s \in S$.*
   (iii) *$C_0 = \bigoplus_{s \in S} (I_s)_0$.*

(b) *Let* $C_0 = \bigoplus_{s\in S} N_s$ *be the direct sum of an indexed family of simple right coideals of* C. *For all* $s \in S$ *let* I_s *be an injective hull of* N_s *in* C. *Then* $C = \bigoplus_{s\in S} I_s$.

Proof. We first show part (a). Suppose that $\{I_s\}_{s\in S}$ is a family of right coideals of C such that $C = \bigoplus_{s\in S} I_s$. Let $M_s = \bigoplus_{s'\neq s} I_{s'}$ for $s \in S$. Then I_s is an injective right C-comodule for all $s \in S$ by part (a) of Proposition 3.5.10 since $C = I_s \bigoplus M_s$.

By part (e) of Proposition 3.5.10 there is an idempotent $e_s \in C^*$ such that $I_s = C \leftharpoonup e_s$ and e_s vanishes on M_s. The reader is left with the small exercise of showing that $\{e_s\}_{s\in S}$ is an orthonormal set of idempotents of C^*.

By part (d) of Proposition 3.5.10 we have $(I_s)_0 = C_0 \leftharpoonup e_s$ for all $s \in S$. Since $\epsilon = \sum_{s\in S} e_s$ and C_0 is a subcoalgebra of C, the calculation

$$C_0 = C_0 \leftharpoonup \epsilon = \sum_{s\in S} C_0 \leftharpoonup e = \sum_{s\in S} (I_s)_0$$

shows that $C_0 = \bigoplus_{s\in S} (I_s)_0$. We have established part (a). Part (a)(i) also follows by part (b) of Proposition 3.5.4 and Corollary 3.5.3.

Suppose the hypothesis of part (b) is satisfied. We note that N_s has an injective hull in C by part (c) of Proposition 3.5.10. By Zorn's Lemma there is a maximal subset S' of S such that $I' = \sum_{s\in S'} I_s$ is direct. Suppose that $s \in S \backslash S'$. Then $I_s \bigcap I' \neq (0)$ by the maximality of S'. Since $N_s = (N_s)_0 = (I_s)_0$ by part (a) of Proposition 3.5.6 we have that $N_s \subseteq I'$. Therefore $C_0 \subseteq I'$.

Now I' is injective since it is the direct sum of injective right coideals of C by Corollary 3.5.3. Therefore $I' \bigoplus M = C$ for some right coideal M of C by Theorem 3.5.8. Since $C_0 \subseteq I'$ it follows that $M_0 = (0)$. Therefore $M = (0)$ and consequently $I' = C$. Since $C_0 = \bigoplus_{s\in S'} (I_s)_0 = \bigoplus_{s\in S'} N_s$ by part (a)(iii), necessarily $S = S'$. □

Suppose that $\{e_s\}_{s\in S}$ is an orthonormal family of idempotents of C^*. Since

$$e_s(c \leftharpoonup e_{s'}) = e_s e_{s'}(c) = \delta_{s,s'} e_s(c) = \delta_{s,s'} \epsilon(c \leftharpoonup e_s)$$

for all $c \in C$ we have as a corollary to Theorem 3.5.13:

Corollary 3.5.14. *Suppose that* C *is a coalgebra over* k *and that* $C_0 = \bigoplus_{s\in S} N_s$ *is the direct sum of simple right coideals of* C. *Then there exists an orthonormal family of idempotents* $\{e_s\}_{s\in S}$ *such that* $e_s|N_{s'} = \delta_{s,s'} \epsilon|N_s$ *for all* $s, s' \in S$. □

Since a simple subcoalgebra of C is the direct sum of simple right coideals of C, as a consequence of the Corollary 3.5.14, or by direct computation:

Corollary 3.5.15. *Suppose that C is a coalgebra over k and suppose that $C_0 = \bigoplus_{s\in S} D_s$ is the direct sum of subcoalgebras of C. Then there exists an orthonormal family of idempotents $\{e_s\}_{s\in S}$ such that $e_s|D_{s'} = \delta_{s,s'}\epsilon|D_s$ for all $s, s' \in S$.* □

Exercises

In the following exercises C is a coalgebra over the field k.

Exercise 3.5.1. Suppose that A is an algebra over the field k and I is a left A-module. Prove that the following are equivalent:

(a) I is injective.

(b) For every finitely generated left A-module N and module maps $\imath : M \longrightarrow N$ and $f : M \longrightarrow I$, where $\imath$ is one-one, there exists a module map $F : N \longrightarrow I$ such that $F\circ\imath = f$.

(c) For every cyclic left A-module N and all module maps $\imath : M \longrightarrow N$ and $f : M \longrightarrow I$, where $\imath$ is one-one, there exists a module map $F : N \longrightarrow I$ such that $F\circ\imath = f$.

Compare with Lemma 3.5.2.

Exercise 3.5.2. Prove the following:

Theorem 3.5.16. *Suppose that C is a coalgebra over the field k. Then the following are equivalent:*

(a) *All right C-comodules are injective.*
(b) *C is a cosemisimple coalgebra.*
(c) *All left C-comodules are injective.*

Exercise 3.5.3. Suppose that $C = D\bigoplus E$ is the direct sum of subcoalgebras, N is a right coideal of C and I is an injective hull of N in C. Show that if $N \subseteq D$ then $I \subseteq D$.

Exercise 3.5.4. Suppose that $g, h \in \mathrm{G}(C)$ and $C = kg\bigoplus kh\bigoplus V$ where V is a subspace of C and $\Delta(v) = g\otimes v + v\otimes h$ for all $v \in V$.

(a) Show that kg, where kg is considered a left coideal of C, has a unique injective hull in C which is kg.

Regard kg as a right coideal of C and for $\alpha \in V^*$ let R_α be the span of g and the elements $v + <\alpha, v>h$, where v runs over V. Show that:

(b) R_α is an injective hull of kg in C.

(c) $R_\alpha = R_{\alpha'}$ if and only if $\alpha = \alpha'$.

(d) Any injective hull of kg in C is R_α for some $\alpha \in k$.

Exercise 3.5.5. Suppose that $\{I_s\}_{s\in S}$ is a family of injective right C-comodules. Regard I_s as a rational left C^*-module and let $\prod_{s\in S} I_s$ be the direct product of left C^*-modules. Show that:

(a) $(\prod_{s\in S} I_s)_r$ is an injective right C-comodule.

(b) $\bigoplus_{s\in S} I_s$ is a direct summand of $\prod_{s\in S} I_s$.

Exercise 3.5.6. Suppose that (N, ρ) is right C-comodule and $\rho(N) \subseteq N\otimes D$ for some subcoalgebra D of C.

(a) Show that the C-subcomodules of N are the D-subcomodules of N.

(b) Let M be a C-subcomodule of N. Show that N is an essential extension of M as a C-comodule if and only if N is an essential extension of M as a D-comodule.

(c) Suppose that $s : C \longrightarrow C^{cop}$ is a one-one coalgebra map and that M, N are left coideals of C. Show that N is an essential extension of M if and only if the right coideal $s(N)$ of C is an essential extension of the right coideal $s(M)$ of C.

Exercise 3.5.7. Suppose that D is a subcoalgebra of C and $e{\rightharpoonup}C$ is an injective hull for D in C regarded as a left C-comodule, where e is an idempotent of C^*. Show that $C{\leftharpoonup}e$ is an injective hull for D in C regarded as a right C-comodule.

Exercise 3.5.8. Show that $\{e_s\}_{s\in S} \mapsto \{C{\leftharpoonup}e_s\}_{s\in S}$ describes a one-one correspondence between the collection of indexed orthonormal families of idempotents of C^* and the collection of indexed families of right coideals of C whose sum is a direct sum decomposition of C.

Exercise 3.5.9. Let $\mathcal{M}^C_{fd}$ (respectively ${}^C\mathcal{M}_{fd}$) be the category of finite-dimensional left (respectively right) C-comodules and maps of right C-comodules under function composition. Show that:

(a) $M \mapsto M^*$ and $f \mapsto f^*$ describe an anti-equivalence of categories $\mathcal{M}^C_{fd} \longrightarrow {}^C\mathcal{M}_{fd}$.

(b) M is an injective object of $\mathcal{M}^C_{fd}$ if and only if M^* is a projective object of ${}^C\mathcal{M}_{fd}$.

3.6 Coalgebras which are submodules of their dual algebras

We regard C as a left and right C^*-module under the rational actions defined by $c^*{\rightharpoonup}c = c_{(1)}{<}c^*, c_{(2)}{>}$ and $c{\leftharpoonup}c^* = {<}c^*, c_{(1)}{>}c_{(2)}$ respectively for all $c^* \in C^*$ and $c \in C$. We regard C^* as a left and right C^*-module under left and right multiplication respectively.

In this section we explore the implications of the existence of a one-one map of right C^*-modules $f : C \longrightarrow C^*$. The existence of such maps for Hopf algebras is discussed in Section 10.9. Recall that for any vector space V over k a linear map $f : V \longrightarrow V^*$ gives rise to bilinear forms $\beta, \beta' : V \times V \longrightarrow k$ determined by $\beta_\ell = f = \beta'_r$ respectively; see Exercise 12.4.1.

Definition 3.6.1. Suppose that C is a coalgebra over k. A *left coalgebra form on* C is a bilinear form $\beta : C \times C \longrightarrow k$ such that $\beta(c_{(1)}, d)c_{(2)} = d_{(1)}\beta(c, d_{(2)})$ for all $c, d \in C$.

Definition 3.6.2. A *left* (respectively *right*) *co-Frobenius coalgebra* over k is a coalgebra C over k with a one-one map of left (respectively right) C^*-modules $f : C \longrightarrow C^*$.

Coalgebra forms are basic building blocks for co-Frobenius coalgebras.

Lemma 3.6.3. *Let C be a coalgebra over the field k and $\beta : C \times C \longrightarrow k$ be a bilinear form. Then the following are equivalent:*

(a) $\beta_\ell : C \longrightarrow C^*$ *is a map of left C^*-modules.*
(b) β *is a left coalgebra form on C.*
(c) $\beta_r : C \longrightarrow C^*$ *is a map of right C^*-modules.*

Proof. Notice that the equivalence of parts (a) and (b) for C^{cop} and β^{op} is the equivalence of parts (c) and (b) for C. Thus to prove the lemma we need only establish the equivalence of parts (a) and (b).

Observe that

$$<\beta_\ell(c^*{\rightharpoonup}c), d> = <c^*, \beta(c_{(1)}, d)c_{(2)}>$$

and

$$<c^*\beta_\ell(c), d> = <c^*, d_{(1)}\beta(c, d_{(2)})>$$

for all $c, d \in C$ and $c^* \in C^*$. Thus parts (a) and (b) are equivalent. □

Regard (C, Δ) as a left C-comodule. By part (b) of Proposition 3.1.9 note that C^r is the sum of the finite-dimensional left ideals of C^*. There is a connection between left coalgebra forms on C and when C^r is a dense subspace of C^*.

Proposition 3.6.4. *Suppose that C is a right co-Frobenius coalgebra over the field k. Then:*

(a) *The sum of the finite-dimensional left ideals of C^* is a dense subspace of C^*.*
(b) *Suppose that $g \in \mathrm{G}(C)$ and $\lambda = \beta_\ell(g)$. Then $\lambda \neq 0$ and satisfies $\lambda c^* = <c^*, g>\lambda$ for all $c^* \in C^*$.*

Proof. Since C is right co-Frobenius there exists a left coalgebra form β on C such that β_r is one-one by Lemma 3.6.3. Since β_r is one-one $\mathrm{Im}(\beta_\ell)$ is a dense subspace of C^*. Since C is the sum of finite-dimensional left C^*-submodules it follows that $\mathrm{Im}(\beta_\ell)$ is the sum of finite-dimensional left ideals of C^* by Lemma 3.6.3. We have shown part (a).

To show part (b), suppose that $g \in \mathrm{G}(C)$ and let $\lambda = \beta_r(g)$. Since β_r is a map of right C^*-modules we calculate for $c^* \in C^*$ that

$$\lambda c^* = \beta_r(g)c^* = \beta_r(g \leftharpoonup c^*) = <c^*, g>\beta_r(g) = <c^*, g>\lambda.$$

Since β_r is one-one it follows that $\lambda \neq 0$. □

By Proposition 3.6.4 the next two results apply to right co-Frobenius coalgebras.

Theorem 3.6.5. *Suppose that C is a coalgebra over the field k and the sum of the finite-dimensional left ideals of C^* is a dense subspace of C^*. Let M and N be left C-comodules. Then:*

(a) *The intersection of the cofinite subcomodules of M is (0).*
(b) *Suppose that N is a subcomodule of M and $N \neq M$. Then there is a cofinite subcomodule P of M such that $N \subseteq P$ and $P \neq M$.*

Now suppose that M is finite-dimensional.

(c) *If N is an essential extension of M then N is finite-dimensional.*
(d) *If N is an injective hull of M then N is finite-dimensional.*

Proof. We first show part (a). Suppose that $\mathcal{L}$ is the set of all finite-dimensional left ideals of C^* and let $I = \sum_{L \in \mathcal{L}} L$. By assumption I is a dense subspace of C^*.

Regard M^* as a left C^*-module under the transpose of the right rational action on M and let $m^* \in M^*$. Then $f : C^* \longrightarrow M^*$ defined by $f(c^*) = c^* \succ m^*$ for all $c^* \in C^*$ is continuous. Thus

$$(I \succ m^*)^{\perp} = f(I)^{\perp} = f(\overline{I})^{\perp} = (C^* \succ m^*)^{\perp}$$

from which

$$(C^* \succ m^*)^{\perp} = (\sum_{L \in \mathcal{L}} L \succ m^*)^{\perp} = \bigcap_{L \in \mathcal{L}} (L \succ m^*)^{\perp}$$

follows. Therefore $(C^* \succ m^*)^{\perp}$ is the intersection of cofinite subcomodules of M since $L \succ m^*$ is finite-dimensional for all $L \in \mathcal{L}$. Since

$$(0) = (M^*)^{\perp} = (\sum_{m^* \in M^*} C^* \succ m^*)^{\perp} = \bigcap_{m^* \in M^*} (C^* \succ m^*)^{\perp}$$

the intersection of all cofinite subcomodules of M is (0).

Part (b) follows from part (a). For suppose that N is a subcomodule of M and $N \neq M$. By part (a) the intersection of the cofinite subcomodules of M/N is (0). Since $M/N \neq (0)$ it follows that M/N contains a cofinite subcomodule not equal to M/N. This subcomodule has the form L/N, where L is a subcomodule of M which satisfies $L \supseteq N$, L is a cofinite subcomodule of M and $L \neq M$.

Part (c) follows from part (a) as well. For suppose that N is an essential extension of M. Let $\mathcal{P}$ be the family of all cofinite subcomodules of N. Then $\mathcal{P}$ is closed under finite intersections. By part (a) we conclude that $\bigcap_{P \in \mathcal{P}} P = (0)$. Since M is finite-dimensional we conclude that $M \bigcap P = (0)$ for some $P \in \mathcal{P}$. Since N is an essential extension of M it now follows that $P = (0)$. Therefore (0) is a cofinite subspace of N which means that N is finite-dimensional.

Part (d) follows from part (c) since an injective hull of M is an essential extension of M. This completes our proof. □

Corollary 3.6.6. *Suppose that C is a coalgebra over the field k and the sum of the finite-dimensional left ideals of C^* is a dense subspace of C^*. Then:*

(a) *Let D be a finite-dimensional subcoalgebra of C. Then $D^{(\infty)}$ is finite-dimensional.*
(b) *Let D and E be finite-dimensional subcoalgebras of C. Then the wedge product $D \wedge E$ is finite-dimensional.*

Proof. Let D be a finite-dimensional subcoalgebra of C and regard D as a left coideal of C. Applying Proposition 3.5.10 to the coalgebra C^{cop} we see that D has an injective hull in C of the form $N = e{\rightharpoonup}C$ for some idempotent e of C^*. By part (d) of Theorem 3.6.5 it follows that N is finite-dimensional. By part (f) of Proposition 3.5.10 we see that N is closed under wedging. Parts (a) and (b) follow at this point. □

Definition 3.6.7. Let C be a coalgebra over k. A *right coalgebra form for* C is a left coalgebra form for C^{cop}.

When C has a left non-singular right coalgebra form then the results in this section apply to the coalgebra C^{cop}. The reader is left with the exercise of expressing them in terms of C.

Exercises

In the following exercises C is a coalgebra over the field k.

Exercise 3.6.1. Let S be a non-empty set and $C = k[S]$ be the grouplike coalgebra on the set S. Show that the left non-singular left coalgebra forms on C are in one-one correspondence with $\mathrm{Fun}(C, k^*)$, the set of all functions from C to k^*.

Exercise 3.6.2. Let $C = kg\bigoplus V$ be the coalgebra of Exercise 2.3.14. Show that C has a left non-singular left coalgebra form if and only if $\mathrm{Dim}\, V \leq 1$.

Exercise 3.6.3. Let C be a coalgebra with a left non-singular left coalgebra form. Show that any space of skew-primitives $P_{g,h}(C)$ is finite-dimensional for all $g, h \in \mathrm{G}(C)$.

Exercise 3.6.4. Suppose that C and C' are coalgebras with left coalgebra forms β and β' respectively. Show that β'' defined by $\beta''(c{\otimes}c', d{\otimes}d') = \beta(c, d)\beta(c', d')$ for all $c, d \in C$ and $c', d' \in C'$ is a left coalgebra form for the tensor product coalgebra $C{\otimes}C'$.

Exercise 3.6.5. Let $\beta : C{\times}C \longrightarrow k$ be a bilinear form. Show that:

(a) β is a left coalgebra form for C if and only if β is a right coalgebra form for C^{cop}.

(b) β is a left coalgebra form for C if and only if β^{op} is a right coalgebra form for C.

(c) β is a right coalgebra form for C if and only if β^{op} is a left coalgebra form for C^{cop}.

3.7 Indecomposable coalgebras

Just as finite-dimensional algebras can be written as a direct sum of indecomposable ideals, coalgebras can be written as the direct sum of indecomposable subcoalgebras. Our treatment of indecomposable coalgebras uses some of the results on injective comodules developed in Section 3.5. We begin our discussion with a generalization of Exercise 3.5.3.

Lemma 3.7.1. *Let C be a coalgebra over the field k. Suppose that $N \subseteq D$ and is a right coideal of C, I is an injective hull of N in C and $I \subseteq D+E$, where D, E are subcoalgebras such that the sum $D+E$ is direct. Then $I \subseteq D$.*

Proof. We first recall that $I = C \leftharpoonup e$ for some idempotent e of C^* by part (b) of Proposition 3.5.10. The calculation

$$C \leftharpoonup e = I \leftharpoonup e \subseteq (D \leftharpoonup e) + (E \leftharpoonup e) \subseteq C \leftharpoonup e$$

shows that $I = (D \leftharpoonup e) \bigoplus (E \leftharpoonup e)$ is the direct sum of right coideals of C. Since $N = N \leftharpoonup e \subseteq D \leftharpoonup e$ it follows that $N \bigcap (E \leftharpoonup e) = (0)$. Since I is an essential extension of N we conclude that $E \leftharpoonup e = (0)$. Thus $I = D \leftharpoonup e \subseteq D$. □

Definition 3.7.2. An *indecomposable coalgebra over k* is a coalgebra C over k such that whenever $C = D \bigoplus E$ is the direct sum of two subcoalgebras D and E of C either $D = (0)$ or $E = (0)$. An *indecomposable component of C* is a maximal indecomposable subcoalgebra of C.

Injective hulls of simple subcoalgebras of C generate indecomposable subcoalgebras of C. More generally:

Proposition 3.7.3. *Suppose that C is a coalgebra over the field k, N is a right coideal of C contained in a simple subcoalgebra of C and I is an injective hull of N in C. Then the subcoalgebra of C which I generates is indecomposable.*

Proof. Since I is a left C^*-submodule of $(C, \rightharpoonup)$ the subcoalgebra of C which I generates is $I \leftharpoonup C^*$. Write $I \leftharpoonup C^* = D \bigoplus E$ as the direct sum of subcoalgebras of C. By assumption N is contained in a simple subcoalgebra S of C. Since $N \subseteq S \bigcap (D \bigoplus E)$ it follows that $S \subseteq D \bigoplus E$ since S is simple. Thus by part (a) of Proposition 3.4.3 it follows that $S \subseteq D$ or $S \subseteq E$. We may assume without loss of generality that $S \subseteq D$. In this case $N \subseteq D$ and consequently $I \subseteq D$ by Lemma 3.7.1. Therefore $I \leftharpoonup C^* \subseteq D$ since D

is a subcoalgebra of C. Thus $E = (0)$ and $I \rightharpoonup C^*$ is an indecomposable subcoalgebra of C by definition. $\square$

Suppose that N is a simple right coideal of C. Then N is contained in a simple subcoalgebra of C by part (b) of Lemma 3.2.11. Thus an injective hull of N in C generates an indecomposable subcoalgebra of C by Proposition 3.7.3. As a coalgebra is the sum of injective hulls of simple right coideals by part (b) of Theorem 3.7.3, we have as a corollary to Proposition 3.7.3:

Corollary 3.7.4. *Suppose that C is a coalgebra over the field k. Then C is the sum of indecomposable subcoalgebras.* $\square$

We next show that indecomposable components exist in C and that C is the direct sum of all of them. A preliminary lemma will streamline our proof.

Lemma 3.7.5. *Suppose that C is a coalgebra over the field k.*

(a) *Let D, E and F be subcoalgebras of C, where D is indecomposable, $E + F$ is direct and $D \subseteq E \bigoplus F$. Then $D \subseteq E$ or $D \subseteq F$.*
(b) *Let $\{D_i\}_{i \in I}$ be a family of indecomposable subcoalgebras of C. If $D_i \bigcap D_j \neq (0)$ for all $i, j \in I$ then $\sum_{i \in I} D_i$ is indecomposable.*
(c) *Suppose that D is a non-zero indecomposable subcoalgebra of C. Then D is contained in a unique indecomposable component of C.*

Proof. First of all suppose that if D, E and F are subcoalgebras of C and $E + F$ is direct. Then $D \bigcap (E \bigoplus F) = (D \bigcap E) \bigoplus (D \bigcap F)$ by part (b) of Proposition 3.4.3. In particular if $D \subseteq E \bigoplus F$ then $D = (D \bigcap E) \bigoplus (D \bigcap F)$.

Assume the hypothesis of part (a). Then $D = (D \bigcap E) \bigoplus (D \bigcap F)$. Since D is indecomposable either $D \bigcap E = (0)$, in which case $D \subseteq F$, or $D \bigcap F = (0)$, in which case $D \subseteq E$. Thus part (a) follows.

Let $\{D_i\}_{i \in I}$ be a family of indecomposable subcoalgebras of C such that $D_i \bigcap D_j \neq (0)$ for all $i, j \in I$. Let $D = \sum_{i \in I} D_i$ and suppose $D = E \bigoplus F$ is the direct sum of subcoalgebras of C. Fix $i \in I$. Then $D_i \subseteq E$ or $D_i \subseteq F$ by part (a). Without loss of generality we may assume $D_i \subseteq E$. For $j \in I$ we cannot have $D_j \subseteq F$ since $D_i \bigcap D_j \neq (0)$ by assumption. Therefore $D_j \subseteq E$ for all $j \in I$ which means that $D \subseteq E$. Hence $F = (0)$ and we conclude that D is indecomposable. We have shown part (b). To show part (c) we note that the sum of all indecomposable subcoalgebras of C which contain D is indecomposable by part (b). $\square$

Theorem 3.7.6. *Suppose that C is a coalgebra over the field k. Then:*

(a) *C is the direct sum of its indecomposable components.*
(b) *Suppose that $C = \bigoplus_{i \in I} D_i$ is the direct sum of non-zero indecomposable subcoalgebras. Then the D_i's are the indecomposable components of C.*

Proof. We first show part (a). By Corollary 3.7.4 it follows that C is the sum of indecomposable subcoalgebras. By part (c) of Lemma 3.7.5 every indecomposable subcoalgebra of C is contained in an indecomposable component of C. Therefore C is the sum of its indecomposable components.

It remains to show directness. Suppose that $D_1, \ldots, D_r$ are distinct indecomposable components of C. We will show that $D_1 + \cdots + D_r$ is direct. By induction on r we may assume that $r > 1$ and that $D_1 + \cdots + D_{r-1}$ is direct. Suppose that $(D_1 + \cdots + D_{r-1}) \bigcap D_r \neq (0)$. Then the intersection contains a simple subcoalgebra S of C. Since $S \subseteq D_1 + \cdots + D_{r-1}$ it follows that $S \subseteq D_i$ for some $1 \leq i \leq r-1$ by part (a) of Proposition 3.4.3. Thus $D_i \bigcap D_r \neq (0)$. By part (b) of Lemma 3.7.5 we conclude that $D_i + D_r$ is indecomposable. Since D_i and D_r are maximal indecomposable subcoalgebras of C it follows that $D_i = D_i + D_r = D_r$, a contradiction. Thus $(D_1 + \cdots + D_{r-1}) \bigcap D_r = (0)$ after all which means $D_1 + \cdots + D_r$ is direct. As a consequence the sum of all indecomposable components of C is direct.

To show part (b) we suppose that D is an indecomposable component of C. We may write $D = \bigoplus_{i \in I} (D \bigcap D_i)$ by part (b) of Proposition 3.4.3. Since D is indecomposable $D = D \bigcap D_i \subseteq D_i$ for some $i \in I$. Since D is an indecomposable component of C and D_i is indecomposable $D = D_i$. Part (b) now follows by part (a). □

We discuss indecomposable coalgebras further in Section 4.8.

Exercises

In the following exercises C is a coalgebra over the field k.

Exercise 3.7.1. Let N be a simple right coideal of C. Show that:

(a) N is contained in a unique indecomposable component D of C.

(b) Any injective hull of N in C is contained in D.

Exercise 3.7.2. Suppose that $C \neq (0)$. Show that the following are equivalent:

(a) All non-zero subcoalgebras of C are indecomposable.

(b) All non-zero subcoalgebras of C are irreducible.

(c) C is irreducible.

Exercise 3.7.3. Suppose that C is cocommutative. Show that:

(a) If N is a simple right coideal (or equivalently a simple subcoalgebra) of C and I is an injective hull of N in C then I is an indecomposable component of C.

(b) N has a unique injective hull in C.

(c) The indecomposable subcoalgebras of C are irreducible subcoalgebras of C.

Exercise 3.7.4. Suppose that $C = kg\bigoplus kh\bigoplus V$ where $g, h \in \mathrm{G}(C)$, $V \neq (0)$ and $\Delta(v) = g\otimes v + v\otimes h$ for all $v \in V$.

(a) Regard kg as a left coideal of C. Show that the subcoalgebra generated by any injective hull of kg in C is kg.

(b) Regard kg as a right coideal of C. Show that the subcoalgebra generated by any injective hull of kg in C is C.

(c) Show that the indecomposable subcoalgebras of C are kg, kh and $kg\bigoplus kh\bigoplus U$, where U is a non-zero subspace of V. (In particular C is indecomposable.)

[Hint: See Exercise 3.5.4.]

Exercise 3.7.5. Find a coalgebra C with a simple right coideal N of C such that no injective hull of N in C generates an indecomposable component of C.

Definition 3.7.7. An *indecomposable algebra over* k is an A over k such that whenever $A = I\bigoplus J$ is the direct sum of ideals of A then $I = (0)$ or $J = (0)$.

Exercise 3.7.6. Suppose that C is finite-dimensional. Show that C is an indecomposable coalgebra if and only if C^* is an indecomposable algebra.

Exercise 3.7.7. Show that a coalgebra automorphism, or anti-automorphism, of C permutes the indecomposable components of C.

Chapter notes

The notions of rational module and comodule, and basic results about these objects, are well-developed in Sweedler's book [201]. Strongly rational modules are introduced and studied in a paper by Heyneman and the author [71]. Injective comodules were seen to be important in the theory of Hopf algebras early on in the work of Sweedler [203] and Sullivan [197]. Here we flesh out the theory of injective comodules in detail basing our discussion on the work of Lin [108]. His paper deals more generally with homological aspects of comodule theory, in particular with projective comodules. Section 3.6, which is about co-Frobenius coalgebras, is a slight reworking of important results of [108]. We refer the reader to Section 3.3 of the book by Dăscălescu, Năstăsescu, and Raianu [35] for an extensive discussion of co-Frobenius and quasi-co-Frobenius coalgebras.

Kaplansky showed that a coalgebra is the direct sum of indecomposable components in [85]. Our treatment of indecomposable coalgebras was inspired by Montgomery's work [133] and follows it rather closely. We connect injective hulls and indecomposable subcoalgebras.

Chapter 4

The coradical filtration and related structures

Let C be a coalgebra over the field k. Filtrations of C are certain chains of subspaces $V_0 \subseteq V_1 \subseteq V_2 \subseteq \dots$ of C which satisfy $\cup_{n=0}^{\infty} V_n = C$. Filtrations are important in that they provide a mechanism for inductive proofs and often provide insight into the nature of the coproduct of C. We discuss filtrations of coalgebras in Sections 4.1–4.3.

One of the most important filtrations of C is the coradical filtration whose terms are given by $V_n = C_n = \wedge^{n+1} C_0$ for all $n \geq 0$. The term $C_0 \wedge C_0$ of the coradical filtration contains useful information about C as we shall see. There is a very workable description of $C_0 \wedge C_0$ when C is pointed. Idempotents play an important role in the analysis of the coradical filtration.

If C is graded then C has a natural filtration. Conversely, a filtration of C give rise to an associated graded coalgebra. We discuss connections between filtered and graded structures on C in great detail in Section 4.4.

The sections after Section 4.4 concern special topics related to filtrations of coalgebras. The radical of the dual algebra C^* is characterized in various ways in Section 4.6; in particular in terms of the coradical C_0 of C. In Section 4.7 pointed and pointed irreducible coalgebras are associated with C. These are useful devices for proving results about C. In particular we use the associated pointed coalgebra to show that a coalgebra map $f : C \longrightarrow D$ is one-one if and only if the restrictions $f|S \wedge T$ are one-one for all simple subcoalgebras S and T of C.

The chapter closes with a resumption of our discussion of indecomposable coalgebras begun in Section 3.7 of Chapter 3. Our continued discussion is based on results from Section 4.2.

4.1 Filtrations of coalgebras

We begin with the definition of filtration of a coalgebra.

Definition 4.1.1. A *filtration of a coalgebra* C *over* k is a family of subspaces $\{V_n\}_{n=0}^{\infty}$ of C which satisfies $V_0 \subseteq V_1 \subseteq V_2 \subseteq \cdots \subseteq \cup_{n=0}^{\infty} V_n = C$ and $\Delta(V_n) \subseteq \sum_{\ell=0}^{n} V_{n-\ell} \otimes V_\ell$ for all $n \geq 0$.

The notion of filtration of a coalgebra is dual to the notion of decreasing filtration of an algebra. The reader is directed to Exercise 4.1.10.

Observe that the terms V_n in a filtration $\{V_n\}_{n=0}^{\infty}$ of C are subcoalgebras of C. By our next result the terms of a filtration must contain the coradical.

Proposition 4.1.2. *Suppose that $\{V_n\}_{n=0}^{\infty}$ is a filtration of a coalgebra C over the field k. Then any simple subcoalgebra of C is contained in V_0.*

Proof. Let S be a simple subcoalgebra of C. Since S is finite-dimensional $S \subseteq V_n$ for some $n \geq 0$. If $n \geq 1$ then $S \subseteq V_n \subseteq V_{n-1} \wedge V_0$ which means that $S \subseteq V_{n-1}$ or $S \subseteq V_0$ by part (a) of Proposition 2.4.3. Thus $S \subseteq V_0$ by induction on n. □

Filtrations of coalgebras give rise to filtrations of subcoalgebras, quotients, and tensor products in a natural way.

Lemma 4.1.3. *Suppose that C and D are coalgebras over the field k and that $\{V_n\}_{n=0}^{\infty}$ is a filtration of C. Then:*

(a) *If D is a subcoalgebra of C then $\{V_n \cap D\}_{n=0}^{\infty}$ is a filtration of D.*
(b) *If $f : C \longrightarrow D$ is a coalgebra map then $\{f(V_n)\}_{n=0}^{\infty}$ is a filtration of $f(C)$.*
(c) *If $\{W_n\}_{n=0}^{\infty}$ is a filtration of D then $\{\mathcal{V}_n\}_{n=0}^{\infty}$ is a filtration of $C \otimes D$, where $\mathcal{V}_n = \sum_{\ell=0}^{n} V_{n-\ell} \otimes W_\ell$ for all $n \geq 0$.*

Proof. Part (b) follows since f is a coalgebra map and part (c) is an easy calculation. To show part (a) we use Exercises 2.4.1 and 4.1.1 to calculate for $n \geq 1$ that

$$\begin{aligned}\Delta(V_n \cap D) &\subseteq (\sum_{\ell=0}^{n} V_{n-\ell} \otimes V_\ell) \cap (D \otimes D) \\ &= (\bigcap_{\ell=0}^{n-1} (V_{n-1-\ell} \otimes C + C \otimes V_\ell)) \cap (D \otimes D) \\ &= \bigcap_{\ell=0}^{n-1} ((V_{n-1-\ell} \otimes C + C \otimes V_\ell) \cap (D \otimes D)) \\ &= \bigcap_{\ell=0}^{n-1} ((V_{n-1-\ell} \cap D) \otimes D + D \otimes (V_\ell \cap D)) \\ &= \sum_{\ell=0}^{n} (V_{n-\ell} \cap D) \otimes (V_\ell \cap D).\end{aligned}$$

Since $V_0 \bigcap D \subseteq V_1 \bigcap D \subseteq V_2 \bigcap D \subseteq \ldots \subseteq \bigcup_{n=0}^{\infty}(V_n \bigcap D) = D$ the family $\{V_n \bigcap D\}_{n=0}^{\infty}$ is a filtration of D. □

Let $\{V_n\}_{n=0}^{\infty}$ be a filtration of C. We noted in the proof of Proposition 4.1.2 that $V_n \subseteq V_{n-1} \wedge V_0$ for $n \geq 1$. This suggests the following construction. Let D be any subcoalgebra of C. Define $D^{(0)} = D$ and $D^{(n)} = D^{(n-1)} \wedge D$ for $n \geq 1$. Thus $D^{(n)} = \wedge^{n+1} D$ for all $n \geq 0$. By part (e) of Proposition 2.4.2 each $D^{(n)}$ is a subcoalgebra of C and $D = D^{(1)} \subseteq D^{(2)} \subseteq \cdots$. Thus $D^{(\infty)} = \cup_{n=0}^{\infty} D^{(n)}$ is a subcoalgebra of C.

Proposition 4.1.4. *Let C be a coalgebra over the field k, D be a subcoalgebra of C and $\mathcal{D} = D^{(\infty)}$. Then:*

(a) *$\{D^{(n)}\}_{n=0}^{\infty}$ is a filtration of $\mathcal{D}$.*
(b) *Let $I = D^{\perp}$. Then $D^{(n)} = (I^{n+1})^{\perp}$ for all $n \geq 0$.*
(c) *$D_0 = \mathcal{D}_0$.*
(d) *$\mathcal{D}$ is saturated. Moreover $\mathcal{D}$ is the smallest saturated subcoalgebra of C containing D.*

Proof. We first show part (a). Since D is a subcoalgebra of C we have $\Delta(D^{(0)}) \subseteq D^{(0)} \otimes D^{(0)}$. Let $n \geq 1$. Since $D^{(n)} = D^{(n-\ell-1)} \wedge D^{(\ell)}$ for $0 \leq \ell < n$, we mimic the proof of part (a) of Lemma 4.1.3 to obtain

$$\Delta(D^{(n)}) \subseteq (\bigcap_{\ell=0}^{n-1} (D^{(n-\ell-1)} \otimes C + C \otimes D^{(\ell)})) \cap (D^{(n)} \otimes D^{(n)}) = \sum_{\ell=0}^{n} D^{(n-\ell)} \otimes D^{(\ell)}.$$

Thus $\{D^{(n)}\}_{n=0}^{\infty}$ is a filtration of $\mathcal{D}$.

Part (b) follows by part (c) of Proposition 2.4.2 and part (c) follows by Proposition 4.1.2.

It remains to show part (d). It is clear that we need only show $\mathcal{D}$ is saturated. Suppose that U and V are subspaces of $\mathcal{D}$ and let $c \in U \wedge V$. Let E be a finite-dimensional subcoalgebra of C which contains c. Since $E \cap \mathcal{D}$ is finite-dimensional it follows that $F = E \cap \mathcal{D} \subseteq D^{(n)}$ for some $n \geq 0$. Using part (a) of Proposition 2.4.3 we find that

$$c \in (U \wedge V) \cap E = (U \cap E) \wedge_E (V \cap E) \subseteq F \wedge F \subseteq D^{(2n+1)} \subseteq \mathcal{D}.$$

Therefore $U \wedge V \subseteq \mathcal{D}$ which means $\mathcal{D}$ is saturated. □

Let C_0 be the coradical of C and set $C_n = C_{n-1} \wedge C_0$ for $n \geq 1$. We will show that $C_0^{(\infty)} = C$.

Suppose that D is a finite-dimensional subcoalgebra of C. Since all subspaces of D^* are closed, $C^\perp$ is the intersection of the maximal ideals of C^* by part (d) of Proposition 2.3.7. Thus $D_0^\perp = \mathrm{Rad}(D^*)$. Since D is finite-dimensional, $\mathrm{Rad}(D^*)$ is nilpotent. Therefore $D_n = ((\mathrm{Rad}(D^*))^{n+1})^\perp = (0)^\perp = D$ for some $n \geq 0$. Since C is the sum of its finite-dimensional subcoalgebras $C_0^{(\infty)} = C$. By part (a) of Proposition 4.1.4:

Proposition 4.1.5. *Suppose that C is a coalgebra over the field k and $C_n = C_{n-1} \wedge C_0$ for all $n \geq 1$, where C_0 is the coradical of C. Then $\{C_n\}_{n=0}^\infty$ is a filtration of C.* □

Definition 4.1.6. The filtration of Proposition 4.1.5 is the *coradical filtration* of C.

One of the most important applications of Lemma 4.1.3 is to the coradical of subcoalgebras, quotients, and the tensor products of coalgebras. Part (a) of Lemma 4.1.3 provides a different proof of part (d) of Proposition 3.4.3 which is that $D_0 = D \cap C_0$ for all subcoalgebras D of C. As far as quotients and tensor products are concerned:

Proposition 4.1.7. *Let C and D be coalgebras over the field k. Then:*

(a) *Suppose that $f : C \longrightarrow D$ is an onto coalgebra map. Then $f(C_0) \supseteq D_0$. Furthermore, if F is a simple subcoalgebra of D then $F \subseteq f(E)$ for some simple subcoalgebra E of C.*

(b) *$(C \otimes D)_0 \subseteq C_0 \otimes D_0$. Furthermore, if E is a simple subcoalgebra of $C \otimes D$ then $E \subseteq F \otimes G$, where F is a simple subcoalgebra of C and G is a simple subcoalgebra of D.*

(c) *Suppose that* C *and* D *are pointed. Then* $C\otimes D$ *is pointed and* $\mathrm{G}(C\otimes D) = \{c\otimes d \,|\, c \in \mathrm{G}(C),\, d \in \mathrm{G}(D)\}$.

Proof. Suppose that $f : C \longrightarrow D$ is a coalgebra map. Then $\{f(C_n)\}_{n=0}^{\infty}$ is a filtration of $f(C)$ by part (b) of Lemma 4.1.3. Thus $f(C)_0 \subseteq f(C_0)$ by Proposition 4.1.2. Let $\mathcal{S}$ be the set of simple subcoalgebras of C. Since $f(C_0) = \sum_{S\in\mathcal{S}} f(S)$ by part (c) of Proposition 3.4.3, every simple subcoalgebra of $f(C)$ is contained in one of the $f(S)$'s by part (a) of the same. We have shown part (a).

To show part (b) we let $\mathcal{V}_n = \sum_{\ell=0}^{n} C_{n-\ell}\otimes D_\ell$ for $n \geq 0$. Then $\{\mathcal{V}_n\}_{n=0}^{\infty}$ is a filtration of $C\otimes D$ by part (c) of Lemma 4.1.3. Thus $(C\otimes D)_0 \subseteq C_0\otimes D_0$ by Proposition 4.1.2. Let $\mathcal{T}$ be set of simple subcoalgebras of D. We see $C_0\otimes D_0 = \sum_{S\in\mathcal{S},T\in\mathcal{T}} S\otimes T$ using Proposition 3.4.3 again as every simple subcoalgebra of $C\otimes D$ is contained in one of the $S\otimes T$'s. Part (c) follows from part (b). □

Suppose that k is a perfect field and A and B are finite-dimensional simple algebras over k. Then $A\otimes B$ is semisimple. Therefore the tensor product of finite-dimensional simple coalgebras over k is cosemisimple. Thus as a corollary to part (b) of Proposition 4.1.7:

Corollary 4.1.8. *Suppose that* C *and* D *are coalgebras over the field* k*. If* k *is perfect then* $(C\otimes D)_0 = C_0\otimes D_0$. □

It will be convenient to consider filtrations in a categorical setting.

Definition 4.1.9. A *filtered coalgebra over* k is a pair $(C, \mathcal{F})$, where C is a coalgebra over k and $\mathcal{F}$ is a filtration of C.

We will more informally write $C = \cup_{n=0}^{\infty} V_n$ for a filtered coalgebra $(C, \mathcal{F})$, where $\mathcal{F} = \{V_n\}_{n=0}^{\infty}$, or simply C for $(C, \mathcal{F})$ when the filtration is understood.

Definition 4.1.10. A *map of filtered coalgebras* $C = \cup_{n=0}^{\infty} V_n$ *and* $D = \cup_{n=0}^{\infty} W_n$ is a map $f : C \longrightarrow D$ of underlying coalgebras which satisfies $f(V_n) \subseteq W_n$ for all $n \geq 0$.

Filtered coalgebras over k and maps of filtered coalgebras form a category. Note that k is a filtered coalgebra in a unique way. Let $(C, \mathcal{F})$ and $(D, \mathcal{G})$ be filtered coalgebras over k with filtrations $\mathcal{F} = \{V_n\}_{n=0}^{\infty}$ and $\mathcal{G} = \{W_n\}_{n=0}^{\infty}$ respectively and set $\mathcal{F}\otimes\mathcal{G} = \{(\mathcal{V}\otimes\mathcal{W})_n\}_{n=0}^{\infty}$, where $(\mathcal{V}\otimes\mathcal{W})_n = \sum_{\ell=0}^{n} V_{n-\ell}\otimes W_\ell$ for all $n \geq 0$. Then $(C\otimes D, \mathcal{F}\otimes\mathcal{G})$ is a filtered coalgebra over k.

Exercises

In the following exercises C and D are coalgebras over the field k.

Exercise 4.1.1. Let V be a vector space over k with subspaces $V_0, \ldots, V_n$ for some $n \geq 1$ which satisfy $V_0 \subseteq V_1 \subseteq \cdots \subseteq V_n = V$. Show that

$$\bigcap_{\ell=0}^{n-1} (V_{n-1-\ell} \otimes V + V \otimes V_\ell) = \sum_{\ell=0}^{n} V_{n-\ell} \otimes V_\ell.$$

[Hint: There are subspaces $U_0, \ldots, U_n$ of V such that $U_0 \bigoplus \cdots \bigoplus U_i = V_i$ for all $0 \leq i \leq n$. Show that the intersection is the sum of the $U_s \otimes U_t$'s where $0 \leq s \leq n-1-\ell$ or $0 \leq t \leq \ell$ for all $0 \leq \ell \leq n$ and that this condition is met if and only if $s + t \leq n$.]

Exercise 4.1.2. Let $f : C \longrightarrow D$ be a coalgebra map. Show that:

(a) If S is a simple subcoalgebra of C and is cocommutative then $f(S)$ is a simple subcoalgebra of D. [Hint: We may regard the dual algebra $f(S)^*$ as a subalgebra of S^*. Since S is simple and cocommutative S^* is a finite dimensional field extension of k.]

(b) If C_0 is cocommutative then $f(C_0) \subseteq D_0$.

Exercise 4.1.3. Suppose that C or D is pointed. Show that:

(a) $S \otimes T$ is a simple subcoalgebra of $C \otimes D$ whenever S and T are simple subcoalgebras of C and D respectively and that all simple subcoalgebras of $C \otimes D$ have this form.

(b) $(C \otimes D)_0 = C_0 \otimes D_0$.

Exercise 4.1.4. Suppose that $C \neq (0)$. Show that the following are equivalent:

(a) C is cosemisimple.

(b) $f(D_0) = C_0$ for every coalgebra D over k and onto coalgebra map $f : D \longrightarrow C$.

Exercise 4.1.5. Suppose that $C = kg \bigoplus kh \bigoplus V$ is the coalgebra of Exercise 3.7.4. Find all of the filtrations of C.

Exercise 4.1.6. Suppose that k has characteristic $p > 0$ and is not perfect. Let $a \in k \backslash k^p$, let A be the p-dimensional algebra over k generated by x subject to the relation $x^p = a$ and let $C = A^*$. Show that:

(a) C is a simple coalgebra.

(b) $(C \otimes C)_0 \neq C_0 \otimes C_0$.

(c) $(C \otimes C)_0 \neq C \otimes C$.

[Hint: Show that $A = C^*$ is a field and that the commutative algebra $(C \otimes C)^* \simeq A \otimes A$ is not semisimple since $x \otimes 1 - 1 \otimes x$ is a non-zero nilpotent element of $A \otimes A$.]

Exercise 4.1.7. Let (M, ρ) be a left C-comodule and suppose that N is a subcomodule of M. Show that $N = M$ if and only if $C_0 \wedge N \subseteq N$. [Hint: Note that $C \wedge (0) = M$. Show that $C_n \wedge (0) \subseteq N$ for all $n \geq 0$.]

Exercise 4.1.8. Show that all subcoalgebras of C are saturated if and only if C is cosemisimple.

Definition 4.1.11. Let D be a subcoalgebra of a coalgebra C over the field k. Then D *generates* C if $D^{(\infty)} = C$.

Exercise 4.1.9. Show that a subcoalgebra D of C generates C if and only if $C_0 \subseteq D$.

Definition 4.1.12. A *decreasing filtration of an algebra A over the field k* is a family of ideals $\{I_n\}_{n=0}^{\infty}$ of A which satisfies $I_1 \supseteq I_2 \supseteq \cdots \supseteq \cap_{n=1}^{\infty} I_n = (0)$ and $I_m I_n \subseteq I_{m+n}$ for all $m, n \geq 0$.

Exercise 4.1.10. Let $\{V_n\}_{n=0}^{\infty}$ be a filtration of C. Show that $\{I_n\}_{n=1}^{\infty}$ is a decreasing filtration of C^* where $I_n = V_{n-1}^{\perp}$ for all $n \geq 1$.

Exercise 4.1.11. Suppose that $\{I_n\}_{n=1}^{\infty}$ is a decreasing filtration of C^* consisting of closed ideals and set $V_n = I_{n+1}^{\perp}$ for all $n \geq 0$. Show that $\{V_n\}_{n=0}^{\infty}$ is a filtration of C. [Hint: Let $n \geq 1$. Show that $\Delta(V_n) \subseteq (I_{n-\ell} \otimes I_{\ell+1})^{\perp}$ for all $0 \leq \ell < n$. See Exercises 2.3.2 and 4.1.1.]

Exercise 4.1.12. Suppose D is a subcoalgebra of C. Show that $D^{(n)} = D^{(\ell)} \wedge D^{(n-\ell-1)}$ for all $0 \leq \ell \leq n-1$. [Hint: See Propositions 2.3.6, 2.4.2, and 4.1.4.]

Exercise 4.1.13. Let D be a simple subcoalgebra of C. Show that $D^{(\infty)}$ is the irreducible component containing D.

Exercise 4.1.14. Suppose that e is an idempotent of C^* and $e \rightharpoonup C_0 = kg$ for some $g \in G(C)$. Suppose that $c \in C_n$ and $c = e \rightharpoonup c$. Show that $\Delta(c) = c \otimes g + \nu$, where $\nu \in C_{m-1} \otimes C_m$. (We set $C_{-1} = (0)$.)

4.2 The wedge product and the coradical filtration

In this short section we consider the coradical filtrations of subcoalgebras, quotients, and tensor products of coalgebras. Our results are consequences of more general statements about wedge powers.

Let C be a coalgebra over k and D, E be subcoalgebras of C with $D \subseteq E$. We set $D_E^{(0)} = D$ and $D_E^{(n)} = D_E^{(n-1)} \wedge_E D$ for $n \geq 1$.

Proposition 4.2.1. *Let C and F be coalgebras over the field k and let D be a subcoalgebra of C.*

(a) *Suppose that F is a subcoalgebra of C. Then $D^{(n)} \cap F = (D \cap F)_F^{(n)}$ for all $n \geq 0$.*
(b) *Suppose that $f : C \longrightarrow F$ is a coalgebra map. Then $f(D^{(n)}) \subseteq f(D)^{(n)}$ for all $n \geq 0$. If $\mathrm{Ker}(f) \subseteq D$ then $f(D^{(n)}) = f(D)^{(n)}$ for all $n \geq 0$.*
(c) *Suppose that E is a subcoalgebra of F and regard $D \otimes E$ as a subcoalgebra of $C \otimes F$. Then $(D \otimes E)^{(n)} = \sum_{\ell=0}^{n} D^{(n-\ell)} \otimes E^{(\ell)}$ for all $n \geq 0$.*

Proof. First of all suppose that F is a subcoalgebra of C. Then $D^{(0)} \cap F = D \cap F = D_F^{(0)}$ by definition. For $n \geq 1$ we calculate by part (a) of Proposition 2.4.3 that

$$\begin{aligned} D^{(n)} \cap F &= (D^{(n-1)} \wedge D) \cap F \\ &= (D^{(n-1)} \cap F) \wedge_F (D \cap F) \\ &= (D \cap F)_F^{(n-1)} \wedge_F (D \cap F) \\ &= (D \cap F)_F^{(n)} \end{aligned}$$

by induction on n. We have shown part (a).

Let $f : C \longrightarrow F$ be a coalgebra map. Then $f(D^{(0)}) = f(D) = f(D)^{(0)}$ by definition. By part (c) of Proposition 2.4.3 the relation $f(D^{(n)}) \subseteq f(D)^{(n)}$ for all $n \geq 0$ follows by induction on n. Suppose that $\mathrm{Ker}(f) \subseteq D$. Then $\mathrm{Ker}(f) \subseteq D^{(n)}$ for all $n \geq 0$. We have noted that $f(D^{(0)}) = f(D)^{(0)}$. For $n \geq 1$ we use part (d) of Proposition 2.4.3 and compute

$$\begin{aligned} f(D^{(n)}) &= f(D^{(n-1)} \wedge D) \\ &= f(D^{(n-1)}) \wedge f(D) \\ &= f(D)^{(n-1)} \wedge f(D) \\ &= f(D)^{(n)} \end{aligned}$$

by induction on n. Thus part (b) follows.

To show part (c), suppose that E is a subcoalgebra of F. Let $I = D^{\perp}$ and $J = E^{\perp}$. Then I is an ideal of C^* and J is an ideal of F^*. Observe that $\mathcal{I} = I \otimes F^* + C^* \otimes J$ is a dense subspace of $(D \otimes E)^{\perp}$ by part (a) of Exercise 2.3.2. Thus $(D \otimes E)^{(n)} = (\mathcal{I}^{n+1})^{\perp}$ for all $n \geq 0$ by part (c) of Proposition

2.4.2. For $n \geq 1$ we use Exercises 2.3.2, 2.4.1, and 4.1.1 to compute

$$
\begin{aligned}
(D\otimes E)^{(n)} &= ((I\otimes F^* + C^*\otimes J)^{n+1})^{\perp} \\
&= (I^{n+1}\otimes F^* + C^*\otimes J^{n+1} + \sum_{\ell=1}^{n} I^{n+1-\ell}\otimes J^{\ell})^{\perp} \\
&= (I^{n+1}\otimes F^* + C^*\otimes J^{n+1})^{\perp}\cap(\bigcap_{\ell=1}^{n}(I^{n+1-\ell}\otimes J^{\ell})^{\perp}) \\
&= (D^{(n)}\otimes E^{(n)})\cap\bigcap_{\ell=1}^{n}(D^{(n-\ell)}\otimes F + C\otimes E^{(\ell-1)}) \\
&= \bigcap_{\ell=1}^{n}(D^{(n-\ell)}\otimes E^{(n)} + D^{(n)}\otimes E^{(\ell-1)}) \\
&= \sum_{\ell=0}^{n} D^{(n-\ell)}\otimes E^{(\ell)}.
\end{aligned}
$$

□

Corollary 4.2.2. *Suppose that C and D are coalgebras over the field k. Then:*

(a) *If D is a subcoalgebra of C then $D_n = D\cap C_n$ for all $n \geq 0$.*
(b) *If $f : C \longrightarrow D$ is a coalgebra map and $f(C_0) \subseteq f(C)_0$ then $f(C_n) \subseteq f(C)_n$ for all $n \geq 0$. If f is also onto and $\mathrm{Ker}(f) \subseteq C_0$ then $f(C_n) = D_n$ for all $n \geq 0$.*
(c) *Suppose that $(C\otimes D)_0 = C_0\otimes D_0$. Then $(C\otimes D)_n = \sum_{m=0}^{n} C_{n-m}\otimes D_m$ for all $n \geq 0$.*

□

The hypothesis $(C\otimes D)_0 = C_0\otimes D_0$ of part (c) of Corollary 4.2.2 is met when k is perfect, as we noted in Corollary 4.1.8, or when C or D is pointed by Exercise 4.1.3.

Exercises

In the following exercises C and D are coalgebras over the field k.

Exercise 4.2.1. Show that $(C\otimes D)_n \subseteq \sum_{\ell=0}^{n} C_{n-\ell}\otimes D_\ell$ for all $n \geq 0$ in any case.

Exercise 4.2.2. Suppose that $f : C \longrightarrow D$ is a coalgebra map and that C_0 is cocommutative. Show that $f(C_n) \subseteq D_n$ for all $n \geq 0$. [Hint: See Exercise 4.1.2.]

Exercise 4.2.3. Suppose D is a subcoalgebra of C. Show that $D_C^{(n)} = D_{C^{cop}}^{(n)}$ for all $n \geq 0$.

Exercise 4.2.4. Let $f : C \longrightarrow D$ be a coalgebra map and let E be an irreducible subcoalgebra of C. Show that if $f(E_0)$ is a simple subcoalgebra of D then $f(E)$ is an irreducible subcoalgebra of D.

4.3 Idempotents and the coradical filtration

Let C be a coalgebra over k and $\mathcal{E}$ be an orthonormal family of idempotents of C^*. The family $\mathcal{E}$ can be used to decompose C into a direct sum of subspaces in terms of which a useful description of diagonalization can be given. We apply our results in particular to families $\mathcal{E}$ described in Corollary 3.5.15.

Suppose that C is pointed. The decomposition gives useful information about the terms C_n of the coradical filtration of C. In particular it gives a good description of C_1. We establish necessary and sufficient conditions for a coalgebra map $f : C \longrightarrow D$ to be one-one in the pointed case.

Let C be any coalgebra over k and let $\mathcal{E}$ be an orthonormal family of idempotents of C^*. For $e \in \mathcal{E}$ we set $\mathrm{L}_e = \mathrm{L}(e)$ and $\mathrm{R}_e = \mathrm{R}(e)$; see (2.15). Thus

$$\mathrm{L}_e(c) = e \rightharpoonup c \quad \text{and} \quad \mathrm{R}_e(c) = c \leftharpoonup e$$

for all $c \in C$. Observe that L_e and $\mathrm{R}_{e'}$ are commuting operators for all $e, e' \in \mathcal{E}$. For any subcoalgebra D of C note that $\mathrm{L}_e(D) \subseteq D$ and $\mathrm{R}_e(D) \subseteq D$ for all $e \in \mathcal{E}$. Since $\mathcal{E}$ is an orthogonal set of idempotents

$$\mathrm{L}_e \circ \mathrm{L}_{e'} = \delta_{e,e'} \mathrm{L}_e \quad \text{and} \quad \mathrm{R}_e \circ \mathrm{R}_{e'} = \delta_{e,e'} \mathrm{R}_e \tag{4.1}$$

for all $e, e' \in \mathcal{E}$. In particular L_e and R_e are idempotent operators for all $e \in \mathcal{E}$. Observe that

$$(\mathrm{I}_C \otimes \mathrm{L}_e) \circ \Delta = \Delta \circ \mathrm{L}_e \quad \text{and} \quad (\mathrm{R}_e \otimes \mathrm{I}_C) \circ \Delta = \Delta \circ \mathrm{R}_e \tag{4.2}$$

for all $e \in \mathcal{E}$.

Let D be a finite-dimensional subcoalgebra of C. We have noted in the discussion preceding Theorem 3.5.13 that all but finitely many of the e's in $\mathcal{E}$ vanish on D. Therefore all but finitely many of the L_e's and R_e's vanish on D. Consequently $\sum_{e \in \mathcal{E}} \mathrm{L}_e$ and $\sum_{e \in \mathcal{E}} \mathrm{R}_e$ are well-defined endomorphisms of C. Since $\sum_{e \in \mathcal{E}} e = \epsilon$ it follows that

$$\sum_{e \in \mathcal{E}} \mathrm{L}_e = \mathrm{I}_C = \sum_{e \in \mathcal{E}} \mathrm{R}_e. \tag{4.3}$$

Since

$$(\mathrm{L}_e \otimes \mathrm{R}_e) \circ \Delta = (\mathrm{I}_C \otimes \mathrm{R}_{e^2}) \circ \Delta = (\mathrm{I}_C \otimes \mathrm{R}_e) \circ \Delta$$

for all $e \in \mathcal{E}$ we conclude from (4.3) that

$$\sum_{e \in \mathcal{E}} (\mathrm{L}_e \otimes \mathrm{R}_e) \circ \Delta = \Delta. \tag{4.4}$$

Let $e \in \mathcal{E}$. For $c \in C$ set

$$^{e}c = c \leftharpoonup e = \mathrm{R}_e(c) \quad \text{and} \quad c^{e} = e \rightharpoonup c = \mathrm{L}_e(c).$$

For a subspace U of C set

$$^{e}U = U \leftharpoonup e = \mathrm{R}_e(U) \quad \text{and} \quad U^{e} = e \rightharpoonup U = \mathrm{L}_e(U).$$

We denote $^{e}(c^{e'}) = (^{e}c)^{e'}$ by $^{e}c^{e'}$ and likewise $^{e}(U^{e'}) = (^{e}U)^{e'}$ by $^{e}U^{e'}$. Let D be a subcoalgebra of C. Then

$$D = \oplus_{e,e' \in \mathcal{E}} {}^{e}D^{e'} \tag{4.5}$$

which follows by (4.1) and (4.3). Since $\Delta(D_n) \subseteq \sum_{m=0}^{n} D_{n-m} \otimes D_m$ for all $n \geq 0$ by Proposition 4.1.5, using (4.2) and (4.4) we compute

$$\Delta\, {}^{e}(D_n)^{e'} \subseteq \sum_{e'' \in \mathcal{E}} \sum_{m=0}^{n} {}^{e}(D_{n-m})^{e''} \otimes {}^{e''}(D_m)^{e'} \tag{4.6}$$

for all $e, e' \in \mathcal{E}$ and $n \geq 0$.

Now write $C_0 = \bigoplus_{i \in I} S_i$ as the direct sum of simple coalgebras of C. Let $\mathcal{E} = \{e_i\}_{i \in I}$ be an orthonormal family of idempotents of C^* which satisfies $e_i | S_j = \delta_{i,j} \epsilon | S_j$ for all $i, j \in I$. Such families exist by Corollary 3.5.15. It is easy to see that $e_i \rightharpoonup C_0 = S_i = C_0 \leftharpoonup e_i$ for all $i \in I$. Thus (4.6) in this case specializes to

$$\Delta\, {}^{e_i}(D_n)^{e_j} \subseteq S_i \otimes {}^{e_i}(D_n)^{e_j} + {}^{e_i}(D_n)^{e_j} \otimes S_j + \sum_{\ell \in I} \sum_{m=1}^{n-1} {}^{e_i}(D_{n-m})^{e_\ell} \otimes {}^{e_\ell}(D_m)^{e_j} \tag{4.7}$$

for all $i, j \in I$ and $n \geq 0$.

Now suppose that C is pointed. We specialize (4.7) even further. It is convenient to simplify notation when $\mathcal{E} = \{e_g\}_{g \in \mathrm{G}(C)}$ and satisfies $e_g(h) = \delta_{g,h}$ for all $g, h \in \mathrm{G}(C)$ by writing g for e_g when e_g is used as a superscript.

Proposition 4.3.1. *Suppose that C is a pointed coalgebra over the field k and $\mathcal{E} = \{e_g\}_{g \in \mathrm{G}(C)}$ is an orthonormal family of idempotents of C^* which satisfies $e_g(h) = \delta_{g,h}$ for all $g, h \in \mathrm{G}(C)$. Let $g, h \in \mathrm{G}(C)$. Then:*

(a) *If* $n \geq 1$ *and* $c \in {}^g(C_n)^h$ *then* $\Delta(c) = g \otimes c + c \otimes h + \nu$, *where* $\nu \in \sum_{\ell \in \mathrm{G}(C)} {}^g(C_{n-1})^\ell \otimes {}^\ell(C_{n-1})^h$.
(b) *If* $g \neq h$ *then* $k(g-h) \bigoplus {}^g(C_1)^h = \mathrm{P}_{g,h}(C)$.
(c) ${}^g(C_1)^g = \mathrm{P}_{g,g}(C) \bigoplus kg$.

Proof. Let $n \geq 1$ and $c \in {}^g(C_n)^h$. Then

$$\Delta(c) = g \otimes c' + c'' \otimes h + \nu', \tag{4.8}$$

by (4.7), where $\nu' \in {}^g(C_{n-1}) \otimes (C_{n-1})^h$. Applying $\epsilon \otimes \mathrm{I}_C$ to both sides of (4.8) we obtain $c = c' + \epsilon(c'')h + \nu''$, where $\nu'' \in (C_{n-1})^h$. Likewise, applying $\mathrm{I}_C \otimes \epsilon$ to both sides of (4.8) we obtain $c = g\epsilon(c') + c'' + \nu'''$, where $\nu''' \in {}^g(C_{n-1})$. Thus $\Delta(c) = g \otimes c + c \otimes h + \nu$, where $\nu = \nu' - g \otimes (\epsilon(c'')h + \nu'') - (\epsilon(c')g + \nu''') \otimes h \in {}^g(C_{n-1}) \otimes (C_{n-1})^h$. Since ${}^gc = c = c^h$, by (4.4) it follows that

$$\nu = \sum_{\ell \in \mathrm{G}(C)} (\mathrm{L}_{e_\ell} \otimes \mathrm{R}_{e_\ell})(\nu) \in \sum_{\ell \in \mathrm{G}(C)} {}^g(C_{n-1})^\ell \otimes {}^\ell(C_{n-1})^h.$$

We have shown part (a).

To show part (b) we first note that ${}^g(C_0)^h = \delta_{g,h}kg$. Suppose that $g \neq h$. Then ${}^g(C_1)^h \subseteq \mathrm{P}_{g,h}(C)$ by part (a). Since $g - h \in \mathrm{P}_{g,h}(C)$ we have $k(g-h) + {}^g(C_1)^h \subseteq \mathrm{P}_{g,h}(C)$. Since ${}^g(g-h)^h = 0$ the preceding sum is direct.

Let $c \in \mathrm{P}_{g,h}(C)$. Then $\Delta(c) = g \otimes c + c \otimes h$ by definition. Since ${}^gc = c \leftharpoonup e_g = c + e_g(c)h$ and $c^h = e_h \rightharpoonup c = c + e_h(c)g$ it follows that ${}^gc^h = c + \alpha g + \beta h$ for some $\alpha, \beta \in k$. Now $\epsilon(c) = 0$. Observe that $\epsilon({}^gc^h) = e_h e_g(c) = 0$ since $g \neq h$. Thus $\alpha + \beta = 0$ which means that $c = {}^gc^h - \alpha(g-h) \in k(g-h) + {}^g(C_1)^h$. We have shown that $\mathrm{P}_{g,h}(C) \subseteq k(g-h) + {}^g(C_1)^h$ which completes the proof of part (b).

The reader is left with the exercise of establishing part (c). Note that $\mathrm{P}_{g,h}(C) + kg$ is direct since $\epsilon(\mathrm{P}_{g,h}(C)) = (0)$. □

Suppose that C is pointed. By Corollary 3.5.15 there is an orthonormal family of idempotents of C^* which satisfies the hypothesis of Proposition 4.3.1. Thus:

Theorem 4.3.2. *Suppose* C *is a pointed coalgebra over the field* k. *There are subspaces* $V_{g,h} \subseteq \mathrm{P}_{g,h}(C)$ *such that* $C_1 = (\bigoplus_{g,h \in \mathrm{G}(C)} V_{g,h}) \bigoplus k[\mathrm{G}(C)]$. □

When C is pointed, to determine when a coalgebra map $f : C \longrightarrow D$ is one-one is a matter of checking f on skew-primitives.

Proposition 4.3.3. *Suppose that $f : C \longrightarrow D$ is a map of coalgebras over the field k, where C is pointed. Then the following are equivalent:*

(a) *f is one-one.*
(b) *The restriction $f|\mathrm{P}_{g,h}(C)$ is one-one for all $g, h \in \mathrm{G}(C)$.*
(c) *The restriction $f|kg\wedge kh$ is one-one for all $g, h \in \mathrm{G}(C)$.*

Proof. We first show that parts (b) and (c) are equivalent. Let $g, h \in \mathrm{G}(C)$. Then $kg\wedge kh = \mathrm{P}_{g,h}(C)\bigoplus kg$ by Exercise 2.4.5. Since $\mathrm{Ker}(f) \subseteq \mathrm{Ker}(\epsilon)$ it follows that $(\mathrm{Ker}(f))\cap(kg\wedge kh) \subseteq (\mathrm{Ker}(\epsilon))\cap(kg\wedge kh) = \mathrm{P}_{g,h}(C)$. Thus $f|kg\wedge kh$ is one-one if and only if $f|\mathrm{P}_{g,h}(C)$ is one-one.

Since part (a) implies part (b), to complete the proof of the proposition we need only show that part (b) implies part (a). Suppose that $f|\mathrm{P}_{g,h}(C)$ is one-one for all $g, h \in \mathrm{G}(C)$. We will show that $f|C_n$ is one-one, or equivalently that $\mathrm{Ker}(f)\cap C_n = (0)$, for all $n \geq 0$. Since $C = \cup_{n=0}^{\infty} C_n$ by Proposition 4.1.5 it will then follow that f is one-one.

Let $n = 0$ and $g, h \in \mathrm{G}(C)$ be distinct. Then $0 \neq g-h \in \mathrm{P}_{g,h}(C)$ means that $0 \neq f(g-h)$, or equivalently $f(g) \neq f(h)$. We have showed that the restriction $f|\mathrm{G}(C)$ is one-one. Since $\mathrm{G}(D)$ is independent by Lemma 2.1.12 the restriction $f|C_0$ is one-one.

As a convenience we will assume that f is onto. Thus $f(C_0) = D_0$ and D is pointed by part (a) of Proposition 4.1.7. Let $\mathcal{E}_D = \{e_{f(g)}\}_{g\in\mathrm{G}(C)}$ be an orthonormal family of idempotents of D^* which satisfies $e_{f(g)}(f(h)) = \delta_{g,h}$ for all $g, h \in \mathrm{G}(C)$. Again, such a family exists by Corollary 3.5.15. For $e \in \mathcal{E}$ let $e_g = e_{f(g)}\circ f$ and set $\mathcal{E}_C = \{e_g\}_{g\in\mathrm{G}(C)}$. Since f is a coalgebra map it is easy to see that $\mathcal{E}_C$ is an orthonormal family of idempotents of C^* which satisfies $e_g(h) = \delta_{g,h}$ for all $g, h \in \mathrm{G}(C)$. Now let $n \geq 1$. By (4.5) we have $C_n = \bigoplus_{g,h\in\mathrm{G}(C)} {}^g(C_n)^h$ and $D = \bigoplus_{g,h\in\mathrm{G}(C)} {}^{f(g)}(D)^{f(h)}$. For $g, h \in \mathrm{G}(C)$ and $c \in C$ by Exercise 4.3.1 we have $f({}^g(C_n)^h) \subseteq {}^{f(g)}(D)^{f(h)}$. Thus $f|C_n$ is one-one if and only if $f|{}^g(C_n)^h$ is one-one, or equivalently $(\mathrm{Ker}(f))\cap{}^g(C_n)^h = (0)$ for all $g, h \in \mathrm{G}(C)$.

We have shown that $f|C_0$ is one-one. Suppose that $n \geq 1$ and $f|C_{n-1}$ is one-one. Let $g, h \in \mathrm{G}(C)$ and $c \in (\mathrm{Ker}(f))\cap{}^g(C_n)^h$. Since $c \in {}^g(C_n)^h$ we can write $\Delta(c) = g\otimes c + c\otimes h + \nu$, where $\nu \in C_{n-1}\otimes C_{n-1}$, by part (a) of Proposition 4.3.1. Since $c \in \mathrm{Ker}(f)$ and f is a coalgebra map, using this expression for $\Delta(c)$ we compute

$$0 = \Delta(f(c)) = f(g)\otimes f(c) + f(c)\otimes f(h) + (f\otimes f)(\nu) = (f\otimes f)(\nu).$$

By assumption $f|C_{n-1}$ is one-one. Therefore $(f\otimes f)|C_{n-1}\otimes C_{n-1}$ is one-one and consequently $\nu = 0$. This means $c \in \mathrm{P}_{g,h}(C)$. By hypothesis $c = 0$.

We have shown that $(\mathrm{Ker}(f)) \cap {}^g(C_n)^h = (0)$. Thus $f|C_n$ is one-one. By induction on n it follows that $f|C_n$ is one-one for all $n \geq 0$ which concludes our proof. □

Corollary 4.3.4. *Let C be a pointed irreducible coalgebra over k with $\mathrm{G}(C) = \{g\}$ and let I be a coideal of C. Then $I = (0)$ if and only if $I \cap \mathrm{P}_g(C) = (0)$.*

Proof. Apply the preceding theorem to the coalgebra C/I and the projection $f : C \longrightarrow C/I$. □

Exercises

In the following exercises C is a coalgebra over the field k.

Exercise 4.3.1. Suppose that $f : C \longrightarrow D$ is an *onto* map of coalgebras over k such that the restriction $f|C_0$ is one-one and let $\mathcal{S}$ be the set of the simple subcoalgebras of C.

(a) Show that:

  (i) $f(S)$ is a simple subcoalgebra of D for all $S \in \mathcal{S}$.
  (ii) $D_0 = f(C_0) = \bigoplus_{S \in \mathcal{S}} f(S)$.
  (iii) $\{f(S)\}_{S \in \mathcal{S}}$ is the set of simple subcoalgebras of C.

Let $\mathcal{E}_D = \{e_{f(S)}\}_{S \in \mathcal{S}}$ be an orthonormal family of idempotents of D^* which satisfies $e_{f(S)}|f(T) = \delta_{S,T}\epsilon|f(S)$ for all $S, T \in \mathcal{S}$. Define $e_S \in C^*$ by $e_S = e_{f(S)} \circ f$ for all $S \in \mathcal{S}$ and let $\mathcal{E}_C = \{e_S\}_{S \in \mathcal{S}}$. Show that:

(b) $\mathcal{E}_C$ is an orthonormal family of idempotents of C^* which satisfies $e_S|T = \delta_{S,T}\epsilon|S$ for all $S \in \mathcal{S}$.

(c) $f({}^{e_S}c^{e_T}) = {}^{e_{f(S)}}f(c)^{e_{f(T)}}$ for all $S, T \in \mathcal{S}$. (Thus $f({}^{e_S}U^{e_T}) = {}^{e_{f(S)}}f(U)^{e_{f(T)}}$ for all subspaces U of C.)

(d) If E is a subcoalgebra of C then $f|E$ is one-one if and only if $f|{}^{e_S}E^{e_T}$ is one-one for all $S, T \in \mathcal{S}$.

4.4 Graded algebras and coalgebras

Examples of coalgebras graded by sets other than the natural numbers $\boldsymbol{N} = \{0, 1, 2, \ldots\}$ frequently arise in practice. For this reason we consider graded coalgebras in the context of vector spaces graded by a non-empty set and make very general definitions. However the results of this section concern objects graded by $\boldsymbol{N}$.

Definition 4.4.1. Let S be a non-empty set. An *S-graded vector space over k* is a vector space V over k with a designated direct sum decomposition $V = \bigoplus_{s \in S} V(s)$. An *$S$-graded subspace of V* is a subspace W of V such that $W = \bigoplus_{n=0}^{\infty}(W \cap V(s))$.

We will frequently denote a graded vector space $V = \bigoplus_{s \in S} V(s)$ by V to simplify notation when the grading is understood.

Definition 4.4.2. A *map of S-graded vector spaces V and W over k* is a linear map $f : V \longrightarrow W$ of underlying vector spaces which satisfies $f(V(s)) \subseteq W(s)$ for all $s \in S$.

Observe that the kernel of an S-graded vector space map is an S-graded subspace. It is easy to see that S-graded vector spaces over k and their maps under composition form a category which we refer to as the category of S-graded vector spaces over k.

$\mathcal{S}$-graded vector spaces arise very naturally in connection with cosemisimple coalgebras. Let C be a cosemisimple coalgebra over k and let $\mathcal{S}$ be the set of simple subcoalgebras of C. Then any left C-comodule (M, ρ) has an $\mathcal{S}$-graded vector space structure $M = \bigoplus_{S \in \mathcal{S}} M(S)$ where $M(S) = \rho^{-1}(S \otimes M)$. See Exercise 3.2.11. Maps of left C-comodules are easily seen to be maps of $\mathcal{S}$-graded vector spaces.

From this point on S is a monoid, which we usually take to be multiplicative with neutral element e. We regard k as an S-graded vector space over k where $k(e) = k$ and $k(s) = (0)$ for all $s \in S \backslash e$. If V and W are S-graded vector spaces over k then we turn the tensor product $V \otimes W$ over k into an S-graded vector space over k by defining

$$(V \otimes W)(s) = \sum_{s's''=s} V(s') \otimes W(s'')$$

for all $s \in S$.

Definition 4.4.3. Let S be a (multiplicative) monoid with neutral element e. An *S-graded algebra* (respectively *coalgebra*) *over k* is an S-graded vector space V over k together with a k-algebra (respectively coalgebra) structure on the underlying vector space V whose structure maps are S-graded linear maps.

Thus S-graded algebras and coalgebras are merely the algebras and coalgebras respectively in the category of S-graded vector spaces when S a monoid. An S-graded algebra over k can be described as an S-graded vector space $A = \bigoplus_{s\in S} A(s)$ over k together with a k-algebra structure on A such that

$$1 \in A(e)$$

and

$$A(s)A(s') \subseteq A(ss')$$

for all $s, s' \in S$. Likewise an S-graded coalgebra over k can be described as an S-graded vector space $C = \bigoplus_{s\in S} C(s)$ over k together with a k-coalgebra structure on C such that

$$\epsilon(C(s)) = (0)$$

for all $s \in S\backslash e$ and

$$\Delta(C(s)) \subseteq \sum_{s's''=s} C(s)\otimes C(s')$$

for all $s \in S$.

Maps of S-graded algebras and coalgebras over k are defined to be morphisms of algebras and coalgebras respectively in the category of S-graded vector spaces over k.

Definition 4.4.4. Let S be a monoid. A *map of S-graded algebras* (respectively *coalgebras*) *V and W over k* is an S-graded map $f : V \longrightarrow W$ which is also a map of the k-algebra (respectively coalgebra) structures of V and W.

Suppose that C is an S-graded coalgebra over k. Then subcoalgebras, left coideals, and the like of C are S-graded if they are S-graded subspaces of C. Similar definitions are made for S-graded ideals, left ideals, and the like of an S-graded algebra over k.

Note that any algebra A over k has an S-graded algebra structure determined by $A(e) = A$ and likewise any coalgebra C over k has an S-graded coalgebra structure determined by $C(e) = C$.

There are natural examples of S-graded algebras over k and, when S is finite, of S-graded coalgebras over k. These examples are built from S.

Example 4.4.5. Let $A = k[S]$ be the monoid algebra of S over k. Then $A = \bigoplus_{s\in S} A(s)$ is an S-graded algebra, where $A(s) = ks$ for all $s \in S$.

Example 4.4.6. Suppose S is finite and $C = k[S]^*$ is the dual coalgebra of the monoid algebra of S over k. Let $\{e_s\}_{s\in S}$ be the basis for C dual to the basis S for $k[S]$. Then $C = \bigoplus_{s\in S} C(s)$ is an S-graded coalgebra, where $C(s) = ke_s$ for all $s \in S$.

The assertion of the last example follows since $\epsilon(e_s) = \delta_{s,e}$ and $\Delta(e_s) = \sum_{s' s'' = s} e_s \otimes e_{s'}$ for all $s \in S$.

When $S = \boldsymbol{N}$ is the set of non-negative integers under addition we refer to S-graded objects simply as graded objects. Here we give two examples of graded coalgebras.

Example 4.4.7. Let C be the coalgebra over k with basis $\{c_0, c_1, \ldots\}$ with coproduct determined by $\Delta(c_n) = \sum_{\ell=0}^{n} c_{n-\ell} \otimes c_\ell$ and augmentation given by $\epsilon(c_n) = \delta_{n,0}$ for all $n \geq 0$. Then $C = \bigoplus_{n=0}^{\infty} C(n)$ is a graded coalgebra, where $C(n) = kc_n$ for all $n \geq 0$.

Example 4.4.8. Let $C = kg \bigoplus kh \bigoplus V$, where $\Delta(g) = g \otimes g$, $\Delta(h) = h \otimes h$ and $\Delta(v) = g \otimes v + v \otimes h$ for all $v \in V$. Give V any graded vector space structure $V = \bigoplus_{n=0}^{\infty} V(n)$. Then $C = \bigoplus_{n=0}^{\infty} C(n)$ is a graded coalgebra, where $C(n) = kg \bigoplus kh \bigoplus V(n)$ for all $n \geq 0$.

In the last example, which is graded, $C_0 \subseteq C(0)$. This is always the case for graded coalgebras since:

Proposition 4.4.9. *Suppose that $C = \bigoplus_{n=0}^{\infty} C(n)$ is a graded coalgebra over the field k. Then any simple subcoalgebra of C is contained in $C(0)$.*

Proof. Let

$$V_n = C(0) \oplus \cdots \oplus C(n) \tag{4.9}$$

for all $n \geq 0$. Then $\{V_n\}_{n=0}^{\infty}$ is a filtration of C. If S is a simple subcoalgebra of C then $S \subseteq V_0 = C(0)$ by Proposition 4.1.2. $\square$

A graded coalgebra $C = \bigoplus_{n=0}^{\infty} C(n)$ over k has a natural filtration $\{V_n\}_{n=0}^{\infty}$ defined by (4.9). The terms $C(0), C(1), \ldots$ of the grading can be recovered from the filtration since $C(n) \simeq V_n/V_{n-1}$ for all $n \geq 0$; we set $V_{-1} = (0)$.

Suppose that C is a coalgebra over k with filtration $\mathcal{F} = \{V_n\}_{n=0}^{\infty}$ and set $V_{-1} = (0)$. There is a way of defining a graded coalgebra structure on the graded vector space

$$\mathrm{gr}_{\mathcal{F}}(C) = \bigoplus_{n=0}^{\infty} (V_n/V_{n-1})$$

over k. When C is graded $C \simeq \mathrm{gr}_{\mathcal{F}}(C)$ as graded coalgebras where the terms of $\mathcal{F}$ are defined by (4.9).

First C is given a graded vector space structure $C = \bigoplus_{n=0}^{\infty} C(n)$ which satisfies (4.9) and

$$\epsilon(C(n)) = (0) \tag{4.10}$$

for all $n > 0$. The coproduct Δ is then replaced by a linear map $\delta : C \longrightarrow C{\otimes}C$ which makes (C, δ, ϵ) a graded coalgebra. Using an isomorphism of graded vector spaces $C \simeq \mathrm{gr}_{\mathcal{F}}(C)$ the graded coalgebra structure (C, δ, ϵ) is transferred to $\mathrm{gr}_{\mathcal{F}}(C)$. Now for the details.

We may as well assume that $C \neq (0)$. Then $C_0 \neq (0)$ and consequently $V_0 \neq (0)$ by Proposition 4.4.9. Since V_0 is a non-zero subcoalgebra of C it follows that $\epsilon(V_0) \neq (0)$ by Exercise 2.1.14. Therefore we can write $C = \bigoplus_{n=0}^{\infty} C(n)$ as a direct sum of subspaces which satisfy (4.9) and (4.10) for all $n \geq 0$. Since (4.10) is satisfied $\epsilon : C \longrightarrow k$ is a graded vector space map.

For $n \geq 0$ let $p_n : C{\otimes}C \longrightarrow (C{\otimes}C)(n)$ be the projection of the direct sum $C{\otimes}C = \bigoplus_{m=0}^{\infty}(C{\otimes}C)(m)$ onto the summand $(C{\otimes}C)(n)$. Define a linear map of graded vector spaces $\delta : C \longrightarrow C{\otimes}C$ by $\delta = \bigoplus_{n=0}^{\infty} p_n{\circ}(\Delta|C(n))$. Using the fact that

$$\Delta(C(n)) \subseteq \sum_{\ell=0}^{n} V_{n-\ell}{\otimes}V_\ell = \sum_{r+s\leq n} C(r){\otimes}C(s)$$

we conclude that

$$\epsilon{\circ}(\mathrm{I}_C{\otimes}\delta) = \mathrm{I}_C = \epsilon{\circ}(\delta{\otimes}\mathrm{I}_C)$$

and

$$\Delta(c) - \delta(c) \in \sum_{r<n}(C{\otimes}C)(r),$$

for all $n \geq 0$ and $c \in C(n)$, where $(C{\otimes}C)(-1) = (0)$. The last relation implies that the differences $(\mathrm{I}_C{\otimes}\Delta){\circ}\Delta(c) - (\mathrm{I}_C{\otimes}\delta){\circ}\delta(c)$ and $(\Delta{\otimes}\mathrm{I}_C){\circ}\Delta(c) - (\delta{\otimes}\mathrm{I}_C){\circ}\delta(c)$ lie in $\sum_{r<n}(C{\otimes}C{\otimes}C)(r)$ for all $c \in C(n)$. At this point it is easy to see that (C, δ, ϵ) is a graded coalgebra over k.

We next transfer the structure (C, δ, ϵ) to $\mathrm{gr}_{\mathcal{F}}(C)$. Since $V_{n-1}\bigoplus C(n) = V_n$, the inclusion $C(n) \longrightarrow V_n$ induces an isomorphism $\jmath_n : C(n) \longrightarrow V_n/V_{n-1}$. Let $\jmath : C \longrightarrow \mathrm{gr}_{\mathcal{F}}(C)$ be the isomorphism of graded vector spaces defined by $\jmath = \bigoplus_{n=0}^{\infty} \jmath_n$. There is a unique coalgebra structure on $(\mathrm{gr}_{\mathcal{F}}(C), \boldsymbol{\Delta}, \boldsymbol{\epsilon})$ such that $\jmath$ is an isomorphism of graded coalgebras.

The coalgebra structure $(\mathrm{gr}_{\mathcal{F}}(C), \boldsymbol{\Delta}, \boldsymbol{\epsilon})$ does not depend on the choice of vector space grading for C provided that (4.9) and (4.10) are satisfied for all $n \geq 0$. For $n \geq 0$ let $\pi_{n,\mathcal{F}} : V_n \longrightarrow \mathrm{gr}_{\mathcal{F}}(C)$ be the projection

$V_n \longrightarrow V_n/V_{n-1} = \overline{V_n}$ followed by the inclusion. Using Exercise 4.4.12 the reader can check that there is a linear map $\Pi_{n,\mathcal{F}}$ as indicated below which makes the diagram

$$\begin{array}{ccc} V_n & \xrightarrow{\quad \Delta|V_n \quad} & \sum_{i=0}^{n} V_{n-\ell}\otimes V_\ell \\ \Big\downarrow{\scriptstyle \pi_{n,\mathcal{F}}} & & \Big\downarrow{\scriptstyle \Pi_{n,\mathcal{F}}} \\ \overline{V_n} & \xrightarrow[\quad \mathbf{\Delta}|\overline{V_n} \quad]{} & \bigoplus_{\ell=0}^{n} \overline{V_{n-\ell}}\otimes\overline{V_\ell} \end{array}$$

commute. $\Pi_{n,\mathcal{F}}$ is determined by $\Pi_{n,\mathcal{F}}(v_{n-\ell}\otimes v_\ell) = \pi_{n-\ell,\mathcal{F}}(v_{n-\ell})\otimes\pi_{\ell,\mathcal{F}}(v_\ell)$ for all $0 \leq \ell \leq n$, $v_{n-\ell} \in V_{n-\ell}$ and $v_\ell \in V_\ell$.

Suppose that $C = \bigoplus_{n=0}^{\infty}$ is graded to start with and $\mathcal{F} = \{V_n\}_{n=0}^{\infty}$ is defined by (4.9). Then $\delta = \Delta$ and thus $(C, \Delta, \epsilon) \simeq (\mathrm{gr}_{\mathcal{F}}(C), \mathbf{\Delta}, \boldsymbol{\epsilon})$ as graded coalgebras.

Proposition 4.4.10. *Suppose that C and D are filtered coalgebras over the field k with filtrations $\mathcal{F} = \{V_n\}_{n=0}^{\infty}$ and $\mathcal{G} = \{W_n\}_{n=0}^{\infty}$ respectively.*

(a) *Suppose that $f : C \longrightarrow D$ is a map of filtered coalgebras. Then there is a map of graded coalgebras $\mathrm{gr}(f) : \mathrm{gr}_{\mathcal{F}}(C) \longrightarrow \mathrm{gr}_{\mathcal{G}}(D)$ determined by $\mathrm{gr}(f)\circ\pi_{n,\mathcal{F}} = \pi_{n,\mathcal{G}}\circ f$ for all $n \geq 0$.*
(b) *Let D be a subcoalgebra of C and $\imath : D \longrightarrow C$ be the inclusion. If $\mathcal{G} = \{D\cap V_n\}_{n=0}^{\infty}$ the inclusion $\imath : D \longrightarrow C$ is a map of filtered coalgebras and $\mathrm{gr}(\imath) : \mathrm{gr}_{\mathcal{G}}(D) \longrightarrow \mathrm{gr}_{\mathcal{F}}(C)$ is one-one.*
(c) *$\mathrm{gr}_{\mathcal{F}\otimes\mathcal{G}}(C\otimes D) \simeq \mathrm{gr}_{\mathcal{F}}(C)\otimes\mathrm{gr}_{\mathcal{G}}(D)$ as graded coalgebras.*

Proof. To show part (a) we take the direct sum $\mathrm{gr}(f) = \bigoplus_{n=0}^{\infty} f_n$ of the maps $f_n : V_n/V_{n-1} \longrightarrow W_n/W_{n-1}$ defined by $f_n(v+V_{n-1}) = f(v)+W_{n-1}$ for all $v \in V_n$ to form a map of graded vector spaces $\mathrm{gr}(f) : \mathrm{gr}_{\mathcal{F}}(C) \longrightarrow \mathrm{gr}_{\mathcal{G}}(D)$. We leave the reader with the exercise of showing that $\mathrm{gr}(f)$ satisfies the requirement of part (a). For the proof of part (b) we refer the reader to Exercise 4.4.12.

To show part (c) we let $(\mathcal{V}\otimes\mathcal{W})_n = \sum_{\ell=0}^{n} V_{n-\ell}\otimes W_\ell$ be the nth term of the filtration $\mathcal{F}\otimes\mathcal{G}$. Let $\imath_{\mathcal{F},\mathcal{G}}$ be the direct sum of the composite

$$(\mathcal{V}\otimes\mathcal{W})_n/(\mathcal{V}\otimes\mathcal{W})_{n-1} = (\sum_{\ell=0}^{n} V_{n-\ell}\otimes W_\ell)/(\sum_{\ell=0}^{n-1} V_{n-1-\ell}\otimes W_\ell)$$
$$\simeq \bigoplus_{\ell=0}^{n}(V_{n-\ell}/V_{n-1-\ell})\otimes(W_\ell/W_{\ell-1}),$$

where for $0 \leq \ell \leq n$, $v_{n-\ell} \in V_{n-\ell}$ and $w_\ell \in W_\ell$ the coset represented by $v_{n-\ell}\otimes w_\ell$ is sent to $(v_{n-\ell}+V_{n-1-\ell})\otimes(w_\ell+W_{\ell-1})$. See Exercise 4.4.12. The reader is left to check that $\imath_{\mathcal{F},\mathcal{G}} : \mathrm{gr}_{\mathcal{F}\otimes\mathcal{G}}(C\otimes D) \longrightarrow \mathrm{gr}_{\mathcal{F}}(C)\otimes\mathrm{gr}_{\mathcal{G}}(D)$ is an isomorphism of graded coalgebras. □

When $\mathcal{F}$ is the coradical filtration of C we write $\mathrm{gr}(C)$ for $\mathrm{gr}_{\mathcal{F}}(C)$ and $\pi_{n,C}$ for $\pi_{n,\mathcal{F}}$.

Proposition 4.4.11. *Suppose that C and D are coalgebras over the field k with filtrations $\mathcal{F}$ and $\mathcal{G}$ respectively.*

(a) *Let $f : C \longrightarrow D$ be a coalgebra map which satisfies $f(C_n) \subseteq D_n$ for all $n \geq 0$. Then there is a map of graded coalgebras $\mathrm{gr}(f) : \mathrm{gr}(C) \longrightarrow \mathrm{gr}(D)$ determined by $\mathrm{gr}(f)\circ\pi_{n,C} = \pi_{n,D}\circ f$ for all $n \geq 0$.*
(b) *Suppose that D is a subcoalgebra of C. Then the inclusion $\imath : D \longrightarrow C$ satisfies the hypothesis of part (a) and the map of graded coalgebras $\mathrm{gr}(\imath) : \mathrm{gr}(D) \longrightarrow \mathrm{gr}(C)$ is one-one.*
(c) $\mathrm{gr}(C\otimes D) \simeq \mathrm{gr}(C)\otimes\mathrm{gr}(D)$.

Proof. Let $\mathcal{F} = \{C_n\}_{n=0}^{\infty}$ and $\mathcal{G} = \{D_n\}_{n=0}^{\infty}$ be the coradical filtrations of C and D respectively. To say that a coalgebra map $f : C \longrightarrow D$ is a map of filtered coalgebras is to say that $f(C_n) \subseteq D_n$ for all $n \geq 0$. Thus part (a) follows by part (a) of Proposition 4.4.11. Suppose that D is a subcoalgebra of C. Then $D_n = D\cap C_n$ for all $n \geq 0$ by part (a) of Corollary 4.2.2. Thus part (b) follows by part (b) of Proposition 4.4.11. When $C_0\otimes D_0 = (C\otimes D)_0$ the coradical filtration of $C\otimes D$ is $\mathcal{F}\otimes\mathcal{G}$ by part (c) of Corollary 4.2.2. Thus part (c) follows from part (c) of Proposition 4.4.11. □

Suppose that $C = \bigoplus_{n=0}^{\infty} C(n)$ is a graded coalgebra over k. We identify $C(n)^*$ with the set of functionals on C^* which vanish on $\bigoplus_{m\neq n} C(m)$. Observe that then sum $\sum_{n=0}^{\infty} C(n)^*$ is direct and that $\mathrm{gr}(C)^\star = \bigoplus_{n=0}^{\infty} C(n)^*$ is a dense subspace of C^*. Since $\epsilon \in C(0)^*$ and $C(m)^*C(n)^* \subseteq C(m+n)^*$ for all $m, n \geq 0$ it follows that $\mathrm{gr}(C)^\star$ is a graded subalgebra of the dual algebra C^*.

Definition 4.4.12. Let $C = \bigoplus_{n=0}^{\infty} C(n)$ be a graded coalgebra over the field k. Then $\mathrm{gr}(C)^\star = \bigoplus_{n=0}^{\infty} C(n)^*$ is the *graded dual of* C.

Now suppose that A is an algebra over k with a decreasing filtration $\mathcal{I} = \{I_n\}_{n=1}^{\infty}$ and set $I_0 = A$. Then the graded vector space

$$\mathrm{gr}_{\mathcal{I}}(A) = \bigoplus_{n=0}^{\infty} I_n/I_{n+1}$$

is a graded algebra over k, where multiplication is defined by $(a+I_{n+1})(b+I_{m+1}) = (a+b) + I_{m+n+1}$ for all $a \in I_n$ and $b \in I_m$. If C is a filtered coalgebra with filtration $\mathcal{F} = \{V_n\}_{n=0}^{\infty}$ then C^* is a filtered algebra with filtration $\mathcal{I}(\mathcal{F}) = \{I_n\}_{n=1}^{\infty}$, where $I_n = V_{n-1}^{\perp}$ for all $n \geq 0$, by Exercise 4.1.10. The graded algebras $\mathrm{gr}_{\mathcal{F}}(C)^\star$ and $\mathrm{gr}_{\mathcal{I}(\mathcal{F})}(C^*)$ are one in the same.

Proposition 4.4.13. *Let C be a coalgebra with filtration $\mathcal{F}$ over the field k. Then $\mathrm{gr}_{\mathcal{F}}(C)^\star \simeq \mathrm{gr}_{\mathcal{I}(\mathcal{F})}(C^*)$ as graded algebras.* □

The reader is directed to Exercise 4.4.14 to see how a proof of the proposition may be constructed.

Definition 4.4.14. A graded coalgebra $C = \bigoplus_{n=0}^{\infty} C(n)$ over the field k is *coradically graded* if $C(0) \bigoplus \cdots \bigoplus C(n) = C_n$ for all $n \geq 0$ and is *strictly graded* if coradically graded and $\mathrm{Dim}(C(0)) = 1$.

Proposition 4.4.15. *Let C be a coalgebra over the field k. Then $\mathrm{gr}(C)$ is coradically graded.*

Proof. Any element of c is contained in a finite-dimensional subcoalgebra D of C by Theorem 2.2.3 and $C_n \cap D = D_n$ for all $n \geq 0$ by Corollary 4.2.2. Let $c \in \mathrm{gr}(C)$. Then $c \in \mathrm{gr}(D)$ for some finite-dimensional subcoalgebra D of C; we can identify $\mathrm{gr}(D)$ and $\mathrm{Im}(\mathrm{gr}(\imath))$, where $\imath : D \longrightarrow C$ is the inclusion, by part (b) of Proposition 4.4.11. Thus we may assume that C is finite-dimensional.

Suppose C is finite-dimensional. Then $\mathrm{gr}(C)^* \simeq \mathrm{gr}_{\mathcal{I}(\mathcal{F})}(C^*)$ by Proposition 4.4.13, where $\mathcal{F}$ is the coradical filtration of C. It is not hard to see that $\mathcal{I}(\mathcal{F})$ consists of powers of the radical of C^*. That $\mathrm{gr}(C)$ is coradically graded now follows by examining the preceding algebra isomorphism. □

Exercises

Unless otherwise stated, $C = \bigoplus_{n=0}^{\infty} C(n)$ is a graded coalgebra over k.

Exercise 4.4.1. Suppose that $V = \bigoplus_{s \in S} V(s)$ is an S-graded vector space over k. Show that:

(a) The sum and intersection of S-graded subspaces of V are S-graded subspaces of V.

(b) Every subspace of V is contained in a unique smallest graded subspace of V.

Exercise 4.4.2. Let $f : V = \bigoplus_{s\in S} V(s) \longrightarrow W = \bigoplus_{s\in S} W(s)$ be a map of S-graded vector spaces over k. Set $f(s) = f|V(s)$ for all $s \in S$. Show that:

(a) f is one-one if and only if $f(s)$ is one-one for all $s \in S$.

(b) f is onto if and only if $f(s)$ is onto for all $s \in S$.

We say that a linear endomorphism T of a vector space V over k *splits over* k if $V = \sum_{\lambda\in k} V_{(\lambda)}$, where

$$V_{(\lambda)} = \cup_{n=0}^{\infty}\mathrm{Ker}(T - \lambda \mathrm{I}_V)^n.$$

Note that $\sum_{\lambda\in k} V_{(\lambda)}$ is a direct sum for any linear endomorphism T of V.

Exercise 4.4.3. Let T be a split endomorphism of a vector space V over k and let S be the multiplicative submonoid of k generated by the eigenvalues of T. Then $V = \bigoplus_{\lambda\in S} V_{(\lambda)}$. Show that:

(a) If V is an algebra over k and T is an algebra endomorphism of V then $V = \bigoplus_{\lambda\in S} V_{(\lambda)}$ is an S-graded algebra structure on V.

(b) If V is a coalgebra over k and T is a coalgebra endomorphism of V then $V = \bigoplus_{\lambda\in S} V_{(\lambda)}$ is an S-graded coalgebra structure on V.

[Hint: For part (a) let $m : V\otimes V \longrightarrow V$ be the multiplication map. For $\lambda, \lambda' \in k$ note that

$$(T - \lambda\lambda'\mathrm{I}_V)\circ m = m\circ\mathcal{T},$$

where $\mathcal{T} = (T-\lambda\mathrm{I}_V)\otimes T + \lambda\mathrm{I}_V\otimes(T-\lambda'\mathrm{I}_V)$ and is the sum of commuting operators. Therefore $(T - \lambda\lambda'\mathrm{I}_V)^n\circ m = m\circ\mathcal{T}^n$ for all $n \geq 0$ and $\mathcal{T}^n$ can be computed by the Binomial Theorem. As for part (b) we note $\Delta\circ(T - \lambda\mathrm{I}_V) = \mathcal{T}\circ\Delta$, where $\mathcal{T} = T\otimes T - \lambda\mathrm{I}_V\otimes\mathrm{I}_V$ and is also the difference of commuting operators.]

Exercise 4.4.4. Suppose that $C(0) = kg$ where $g \in \mathrm{G}(C)$. Show that $C(1) \subseteq \mathrm{P}_g(C)$.

Exercise 4.4.5. Suppose that $C(0) = C_0$. Show that $C(0)\bigoplus\cdots\bigoplus C(n) \subseteq C_n$ for all $n \geq 0$.

Exercise 4.4.6. Suppose that C is coradically graded and D is a graded subcoalgebra of C. Show that D is coradically graded.

Exercise 4.4.7. Suppose that I is a graded coideal of C. Show that C/I has a unique graded coalgebra structure such that the projection $\pi : C \longrightarrow C/I$ satisfies $\pi(C(n)) = \pi(C)(n)$ for all $n \geq 0$.

Exercise 4.4.8. Suppose that $C(0) = kg$ where $g \in \mathrm{G}(C)$. Suppose I is a non-zero graded coideal of C and write $I = \bigoplus_{m=n}^{\infty} I \cap C(m)$, where $I \cap C(n) \neq (0)$. Show that $I \cap C(n) \subseteq \mathrm{P}_g(C)$.

Exercise 4.4.9. Suppose that U and V are graded subspaces of C. Show that:

(a) The left coideal, right coideal, and subcoalgebra of C generated by U is graded.

(b) $U \wedge V$ is graded.

(c) The unique coideal of C which is maximal among the coideals of C contained in U is graded.

[Hint: Let $\mathcal{A} = \mathrm{gr}(C)^{\star}$. Observe that there is a graded subspace I of $\mathcal{A}$ such that $I^{\perp} = U$. The product of graded subspaces of $\mathcal{A}$ is a graded subspace of $\mathcal{A}$.]

Exercise 4.4.10. Show that every finite-dimensional subspace of C is contained in a finite-dimensional graded subcoalgebra of C.

Exercise 4.4.11. Show that $P_{g,h}(C) = \bigoplus_{n=0}^{\infty}(P_{g,h}(C) \cap C(n))$ for all $g, h \in \mathrm{G}(C)$.

Exercise 4.4.12. Let V and W be vector spaces over the field k and $n \geq 0$. Suppose that $V_{-1} = (0) \subseteq V_0 \subseteq V_1 \subseteq \cdots \subseteq V_n$ and $W_{-1} = (0) \subseteq W_0 \subseteq W_1 \subseteq \cdots \subseteq W_n$ are subspaces of V and W respectively. For $0 \leq \ell \leq n$ let $\pi_{\ell,V} : V_\ell \longrightarrow V_\ell/V_{\ell-1}$ and $\pi_{\ell,W} : V_\ell \longrightarrow W_\ell/W_{\ell-1}$ be the projections. Show that:

(a) There is a linear map

$$f : \sum_{\ell=0}^{n} V_{n-\ell} \otimes W_\ell \longrightarrow \bigoplus_{\ell=0}^{n} (V_{n-\ell}/V_{n-\ell-1}) \otimes (W_\ell/W_{\ell-1})$$

determined by $f(v_{n-\ell} \otimes w_\ell) = \pi_{n-\ell,V}(v_{n-\ell}) \otimes \pi_{\ell,W}(w_\ell)$ for all $0 \leq \ell \leq n$, $v_{n-\ell} \in V_{n-\ell}$ and $w_\ell \in W_\ell$.

(b) $\mathrm{Ker}(f) = \sum_{\ell=0}^{n-1} V_{n-\ell-1} \otimes W_\ell$.

[Hint: Choose subspaces $U_0, \ldots, U_n$ and $\mathcal{U}_0, \ldots, \mathcal{U}_n$ of V and W respectively such that $U_0 \bigoplus \cdots \bigoplus U_\ell = V_\ell$ and $\mathcal{U}_0 \bigoplus \cdots \bigoplus \mathcal{U}_\ell = W_\ell$ for all $0 \leq \ell \leq n$. Note that $\sum_{\ell=0}^{n} V_{n-\ell} \otimes W_\ell = \bigoplus_{r+s \leq n} U_r \bigoplus \mathcal{U}_s$ and that $\bigoplus_{r+s=n} U_r \bigoplus \mathcal{U}_s \simeq \bigoplus_{\ell=0}^{n} (V_{n-\ell}/V_{n-\ell-1}) \otimes (W_\ell/W_{\ell-1})$.]

Exercise 4.4.13. Let C and D be filtered coalgebras with filtrations $\mathcal{F} = \{V_n\}_{n=0}^{\infty}$ and $\mathcal{G} = \{W_n\}_{n=0}^{\infty}$ respectively. Suppose that $f : C \longrightarrow D$ is a map of filtered coalgebras. Show that:

(a) $\mathrm{gr}(f) : \mathrm{gr}_{\mathcal{F}}(C) \longrightarrow \mathrm{gr}_{\mathcal{G}}(D)$ is one-one if and only if $f^{-1}(W_{n-1}) \cap V_n = V_{n-1}$ for all $n \geq 0$.

(b) $\mathrm{gr}(f) : \mathrm{gr}_{\mathcal{F}}(C) \longrightarrow \mathrm{gr}_{\mathcal{G}}(D)$ is onto if and only if $f(V_n) + W_{n-1} = W_n$ for all $n \geq 0$.

Exercise 4.4.14. Let U and V be subspaces of a vector space W over the field k and suppose that $U \subseteq V$.

(a) Show that $f : U^{\perp}/V^{\perp} \longrightarrow (V/U)^*$ defined by $f(\alpha + V^{\perp})(v + U) = \alpha(v)$ for all $\alpha \in U^{\perp}$ and $v \in V$ is a well-defined linear isomorphism.

(b) Use part (a) to construct a proof of Proposition 4.4.13.

This next exercise is used in the proof of Theorem 4.6.1 of the next section.

Exercise 4.4.15. Suppose that $\alpha \in C^*$. Show that:

(a) α has an inverse in C^* if and only if $\alpha|D$ has an inverse in D^* for all finite-dimensional subcoalgebras D of C.

(b) If α vanishes on C_0 then $\epsilon - \alpha$ has an inverse in C^*. [Hint: We may assume that C is finite-dimensional by part (a). Show that α^{m+1} vanishes on C_m for all $m \geq 0$ and thus α is nilpotent. Then show $\epsilon + \alpha + \alpha^2 + \cdots + \alpha^n$ is an inverse for $\epsilon - \alpha$, where $\alpha^{n+1} = 0$.]

4.5 The cofree pointed irreducible coalgebra on a vector space

Let $(\pi, T^{co}(V))$ be the cofree coalgebra on a vector space V over the field k of Section 2.7. By Exercise 2.7.1 there is a grouplike element $\mathbf{1}_V \in T^{co}(V)$ determined by $\pi(\mathbf{1}_V) = 0$. Let $\mathrm{Sh}(V) = T^{co}(V)_{k\mathbf{1}_V}$ be the irreducible component of $T^{co}(V)$ containing the simple subcoalgebra $k\mathbf{1}_V$ and let $\pi_{\mathrm{Sh}(V)} = \pi|\mathrm{Sh}(V)$. The pair $(\pi_{\mathrm{Sh}(V)}, \mathrm{Sh}(V))$ satisfies a universal mapping property.

Theorem 4.5.1. *Let V be a vector space over the field k. The pair $(\pi_{\mathrm{Sh}(V)}, \mathrm{Sh}(V))$ satisfies the following:*

(a) *$\mathrm{Sh}(V)$ is a pointed irreducible coalgebra over k, $\pi_{\mathrm{Sh}(V)} : \mathrm{Sh}(V) \longrightarrow V$ is a linear map, and $\pi_{\mathrm{Sh}(V)}(\mathbf{1}_V) = 0$, where $\mathrm{G}(\mathrm{Sh}(V)) = \{\mathbf{1}_V\}$.*

(b) *If C is a pointed irreducible coalgebra over k, $f : C \longrightarrow V$ is a linear map, and $f(g) = 0$, where $\mathrm{G}(C) = \{g\}$, then there is a coalgebra map $F : C \longrightarrow \mathrm{Sh}(V)$ determined by $\pi_{\mathrm{Sh}(V)} \circ F = f$.*

Proof. We need only show that part (b) holds. Assume the hypothesis of part (b). By Proposition 2.7.3 there is a coalgebra map $F : C \longrightarrow T^{co}(V)$

determined by $\pi \circ F = f$. We need only show that $F(g) = \mathbf{1}_V$ and $F(C) \subseteq \mathrm{Sh}(V)$.

Now $F(g) \in \mathrm{G}(T^{co}(V))$ since F is a coalgebra map. Since $\pi(F(g)) = f(g) = 0$, by part (a) of Exercise 2.7.1 we have $F(g) = \mathbf{1}_V$. Since C is irreducible with simple subcoalgebra kg, by Exercise 4.2.4 it follows that $F(C)$ is irreducible with simple subcoalgebra $F(kg) = k\mathbf{1}_V$. Thus $F(C) \subseteq T^{co}(V)_{k\mathbf{1}_V} = \mathrm{Sh}(V)$. □

Definition 4.5.2. Let V be a vector space over the field k. A *cofree pointed irreducible coalgebra on* V is a pair $(\pi_{\mathrm{Sh}(V)}, \mathrm{Sh}(V))$ which satisfies the conclusion of Theorem 4.5.1.

Every pointed irreducible coalgebra C over k can be embedded in a cofree one in a rather special way. For notational convenience we set $\mathrm{P}(C) = \mathrm{P}_{g,g}(C)$ where $\mathrm{G}(C) = \{g\}$.

Proposition 4.5.3. *Let C be a pointed irreducible coalgebra over the field k and $V = \mathrm{P}(C)$. Then there is a one-one coalgebra map $F : C \longrightarrow \mathrm{Sh}(V)$ with $F(\mathrm{P}(C)) = \mathrm{P}(\mathrm{Sh}(V))$.*

Proof. First of all note that $kg+\mathrm{P}(C)$ is a direct sum, where $\mathrm{G}(C) = \{g\}$. Thus there is a linear map $f : C \longrightarrow V$ which satisfies $f(g) = 0$ and $f(v) = v$ for all $v \in V$. By Theorem 4.5.1 there is a coalgebra map $F : C \longrightarrow \mathrm{Sh}(V)$ which satisfies $\pi_{\mathrm{Sh}(V)} \circ F = f$. Since F is a coalgebra map and $C, \mathrm{Sh}(V)$ are pointed irreducible, $F(g) = \mathbf{1}_V$ and thus $F(\mathrm{P}(C)) \subseteq \mathrm{P}(\mathrm{Sh}(V))$. Observe that $(\pi_{\mathrm{Sh}(V)}|\mathrm{P}(\mathrm{Sh}(V))) \circ (F|\mathrm{P}(C)) = \mathrm{I}_V$ since $\pi_{\mathrm{Sh}(V)}(F(v)) = f(v) = v$ for all $v \in V$. The first map of the composition is a linear isomorphism by part (b) of Exercise 2.7.1. Therefore $F|\mathrm{P}(C) : \mathrm{P}(C) \longrightarrow \mathrm{P}(\mathrm{Sh}(V))$ is a linear isomorphism. That F is one-one follows by part (b) of Proposition 4.3.3. □

We end this section with an explicit description of $\mathrm{Sh}(V)$. First of all assume that V is finite-dimensional. Then $\mathrm{Sh}(V) = (T(V^*)^o)_{\mathbf{1}_V}$ by Lemma 2.7.2. For a vector space X over k we let $X^{\otimes n} = X \otimes \cdots \otimes X$ (n tensor factors) for all $n \geq 1$.

Consider the ideal $\mathcal{M} = \bigoplus_{m=1}^{\infty} V^{*\otimes m}$ of $T(V^*)$. Observe that $\mathcal{M}^n = \bigoplus_{m=n}^{\infty} V^{*\otimes m}$ and is a cofinite ideal of $T(V^*)$ for all $n \geq 1$. Since V is finite-dimensional we will identify V with V^{**} via the linear isomorphism $\imath_V : V \longrightarrow V^{**}$ of Section 2.7 which is given by $\imath_V(v)(v^*) = v^*(v)$ for all $v \in V$ and $v^* \in V^*$. Observe that $\mathbf{1}_V \in \mathrm{G}(T^{co}(V)) = \mathrm{Alg}(T(V^*), k)$ is determined by $\mathbf{1}_V(v^*) = 0$ for all $v^* \in V^*$ as $0 = \pi(\mathbf{1}_V) = \imath^*(\mathbf{1}_V) = \mathbf{1}_V \circ \imath$, where

$\imath : V^* \longrightarrow T(V^*)$ is the inclusion. For $n \geq 1$ we regard $f \in V^{\otimes n} = V^{**\otimes n}$ as a function on $T(V^*)$ by setting $f = 0$ on $k1$ and $V^{*\otimes m}$ where $m \geq 1$ and $m \neq n$. Thus $\mathcal{M}^\perp = k\mathbf{1}_V$ and $(\mathcal{M}^n)^\perp = k\mathbf{1}_V \bigoplus V \bigoplus \cdots \bigoplus V^{\otimes(n-1)}$ for all $n \geq 2$. In any event $(\mathcal{M}^n)^\perp$ is a subcoalgebra of $T(V^*)^o$ by part (e) of Proposition 2.5.5. Note that $(\mathcal{M}^n)^\perp = \wedge^n k\mathbf{1}_V$ for all $n \geq 1$ by Exercise 2.5.21 and part (d) of Proposition 2.4.2. Thus the irreducible component $\mathrm{Sh}(V)$ of $k\mathbf{1}_V$ in $T(V^*)^o$ is

$$\mathrm{Sh}(V) = \mathbf{1}_V \oplus (\oplus_{n=1}^{\infty} V^{\otimes n}). \tag{4.11}$$

Thus as a vector space $\mathrm{Sh}(V) \simeq T(V)$.

The coalgebra structure of $\mathrm{Sh}(V)$ is particularly simple. Since $\mathbf{1}_V$ is grouplike

$$\Delta(\mathbf{1}_V) = \mathbf{1}_V \otimes \mathbf{1}_V \quad \text{and} \quad \epsilon(\mathbf{1}_V) = 1. \tag{4.12}$$

Let $n \geq 1$ and $v_1, \ldots, v_n \in V$. We will show that

$$\begin{aligned} \Delta(v_1 \otimes \cdots \otimes v_n) = \mathbf{1}_V \otimes (v_1 \otimes \cdots \otimes v_n) + (v_1 \otimes \cdots \otimes v_n) \otimes \mathbf{1}_V \\ + \sum_{i=1}^{n-1} (v_1 \otimes \cdots \otimes v_i) \otimes (v_{i+1} \otimes \cdots \otimes v_n) \end{aligned} \tag{4.13}$$

and

$$\epsilon(v_1 \otimes \cdots \otimes v_n) = 0. \tag{4.14}$$

Since $\epsilon_{T(V^*)^o}(v_1 \otimes \cdots \otimes v_n) = \langle v_1 \otimes \cdots \otimes v_n, 1 \rangle = 0$ the last equation follows. To establish (4.13) let $\{v_1, \ldots, v_\theta\}$ be a basis for V and let $\{v^1, \ldots, v^\theta\}$ be the dual basis for V^*. We need only establish (4.13) for $v_{i_1} \otimes \cdots \otimes v_{i_n}$, where $n \geq 1$ and $1 \leq i_1, \ldots, i_n \leq \theta$. Let $\mathcal{L} = v_{i_1} \otimes \cdots \otimes v_{i_n}$ and $\mathcal{R}$ be the right-hand side of (4.13) with v_j replaced by v_{i_j}. Note that $\mathcal{L}(1 \cdot 1) = \mathcal{L}(1) = 0 = \mathcal{R}(1 \otimes 1)$. Now let $m \geq 1$ and $1 \leq j_1, \ldots, j_m \leq \theta$. It is easy to see that

$$\begin{aligned} \mathcal{L}(1(v^{j_1} \otimes \cdots \otimes v^{j_m})) &= \mathcal{R}(1 \otimes (v^{j_1} \otimes \cdots \otimes v^{j_m})) \\ \mathcal{L}((v^{j_1} \otimes \cdots \otimes v^{j_m})1) &= \mathcal{R}((v^{j_1} \otimes \cdots \otimes v^{j_m}) \otimes 1), \quad \text{and} \\ \mathcal{L}((v^{j_1} \otimes \cdots \otimes v^{j_u})(v^{j_{u+1}} \otimes \cdots \otimes v^{j_m})) &= \mathcal{R}((v^{j_1} \otimes \cdots \otimes v^{j_u}) \otimes (v^{j_{u+1}} \otimes \cdots \otimes v^{j_m})) \end{aligned}$$

for all $1 \leq u < m$. Thus (4.13) holds as needed.

Finally note that $\pi_{\mathrm{Sh}(V)} = \imath^* | \mathrm{Sh}(V)$, where again $\imath : V^* \longrightarrow T(V^*)$ is the inclusion. Thus

$$\pi(\mathbf{1}_V) = 0, \quad \text{and} \quad \pi(v_1 \otimes \cdots \otimes v_n) = \delta_{n,1} v_1 \tag{4.15}$$

for all $n \geq 1$ and $v_1, \ldots, v_n \in V$. As a consequence π is the projection of $\mathrm{Sh}(V) = \mathbf{1}_V \bigoplus (\bigoplus_{n=0}^{\infty} V^{\otimes n})$ onto the term $V^{\otimes 1} = V$.

Now suppose V is any vector space over k. Then $\mathrm{Sh}(V)$ defined by (4.11) is a coalgebra over k with coproduct and counit defined by (4.12), (4.13, and (4.14). Let $\pi_{\mathrm{Sh}(V)} : \mathrm{Sh}(V) \longrightarrow V$ be defined by (4.15). We will show that $(\pi_{\mathrm{Sh}(V)}, \mathrm{Sh}(V))$ is indeed a cofree pointed irreducible coalgebra on V. For a subspace U of V and $n \geq 1$ we regard $U^{\otimes n}$ as a subspace of $V^{\otimes n}$.

Let U be a finite-dimensional subspace of V. Then $\mathrm{sh}(U) = \mathbf{1}_V \bigoplus (\bigoplus_{n=1}^{\infty} U^{\otimes n})$ is a subcoalgebra of $\mathrm{Sh}(V)$ and $(\pi_{\mathrm{sh}(U)}, \mathrm{sh}(U))$ is a cofree coalgebra on U, where $\pi_{\mathrm{sh}}(U) = \pi_{\mathrm{Sh}(V)}|\mathrm{sh}(U)$. Let $\mathcal{F}$ be the set of all finite-dimensional subspaces of V. Then $\mathrm{Sh}(V) = \sum_{U \in \mathcal{F}} \mathrm{sh}(U)$ and for $U, U' \in \mathcal{F}$ such that $U \subseteq U'$ we have that $\mathrm{sh}(U)$ is a subcoalgebra of $\mathrm{sh}(U')$ and $\pi_{\mathrm{sh}(U)} = \pi_{\mathrm{sh}(U')}|U$. By Exercise 4.5.4 it follows that $(\pi_{\mathrm{Sh}(V)}, \mathrm{Sh}(V))$ is a cofree pointed irreducible coalgebra on V.

Let $\mathrm{Sh}(V)(0) = k\mathbf{1}_V$ and $\mathrm{Sh}(V)(n) = V^{\otimes n}$ for $n \geq 1$. Note that $\mathrm{Sh}(V) = \bigoplus_{n=0}^{\infty} \mathrm{Sh}(V)(n)$ is a graded coalgebra. When V is finite-dimensional $\wedge^n k\mathbf{1}_V = k\mathbf{1}_V \bigoplus V \bigoplus \cdots \bigoplus V^{\otimes(n-1)}$ for all $n \geq 1$. In this case the terms of the coradical filtration are given by $\mathrm{Sh}(V)_n = \mathrm{Sh}(V)(0) \bigoplus \cdots \bigoplus \mathrm{Sh}(V)(n)$ for all $n \geq 0$.

Proposition 4.5.4. *Let V be any vector space over the field k. Then $\mathrm{Sh}(V) = \bigoplus_{n=0}^{\infty} \mathrm{Sh}(V)(n)$ is a coradically graded, hence a strictly graded, coalgebra over k.*

Proof. We need only show that $\mathrm{Sh}(V)_n = \bigoplus_{m=0}^{n} \mathrm{Sh}(V)(m)$ for all $n \geq 0$. We have noted this is the case when V is finite-dimensional.

Let $n \geq 0$. By definition of the coradical filtration $\bigoplus_{m=0}^{n} \mathrm{Sh}(V)(m) \subseteq \mathrm{Sh}(V)_n$. Suppose $c \in \mathrm{Sh}(V)_n$. Now $c \in \mathrm{sh}(U)$ for some finite-dimensional subspace U of V. Using part (a) of Corollary 4.2.2 we see that

$$c \in \mathrm{sh}(U) \cap \mathrm{Sh}(V)_n = \mathrm{sh}(U)_n = \bigoplus_{m=0}^{n} \mathrm{sh}(U)(m) \subseteq \bigoplus_{m=0}^{n} \mathrm{Sh}(V)(m).$$

Thus $\mathrm{Sh}(V)_n = \bigoplus_{m=0}^{n} \mathrm{Sh}(V)(m)$ for all $n \geq 0$. □

There is a cocommutative analog of $(\pi_{\mathrm{Sh}(V)}, \mathrm{Sh}(V))$. Let $\mathrm{B}(V)$ be the sum of all cocommutative subcoalgebras of $\mathrm{Sh}(V)$. Then $\mathrm{B}(V)$ is pointed irreducible, $\mathrm{G}(\mathrm{B}(V)) = \{\mathbf{1}_V\}$, and $\mathrm{P}(\mathrm{B}(V)) = \mathrm{P}(\mathrm{Sh}(V)) = V$. Let $\pi_{\mathrm{B}(V)} = \pi_{\mathrm{Sh}(V)}|\mathrm{B}(V)$.

Theorem 4.5.5. *Let V be a vector space over the field k. The pair $(\pi_{\mathrm{B}(V)}, \mathrm{B}(V))$ satisfies the following:*

(a) $\mathrm{B}(V)$ *is a cocommutative pointed irreducible coalgebra over* k, *and* $\pi_{\mathrm{B}(V)} : \mathrm{B}(V) \longrightarrow V$ *is a linear map, and* $\pi_{\mathrm{B}(V)}(\mathbf{1}_V) = 0$, *where* $\mathrm{G}(\mathrm{B}(V)) = \{\mathbf{1}_V\}$.

(b) *Whenever* C *is a cocommutative pointed irreducible coalgebra over* k, $f : C \longrightarrow V$ *is a linear map, and* $f(g) = 0$, *where* $\mathrm{G}(C) = \{g\}$, *there is a coalgebra map* $F : C \longrightarrow \mathrm{B}(V)$ *determined by* $\pi_{\mathrm{B}(V)} \circ F = f$.

□

Exercises

Throughout these exercises U, V are vector space over the field k.

Exercise 4.5.1. Prove Theorem 4.5.5.

Exercise 4.5.2. Show that $T^{co}(\{0\}) = k$, where $\mathrm{G}(k) = \{1\}$.

Exercise 4.5.3. Show that $\mathrm{Sh}(\{0\}) = k$, where $\mathrm{G}(k) = \{1\}$.

Exercise 4.5.4. Let D be a coalgebra over the field k. Suppose $\mathcal{F}$ is a family of subspaces of V such that:

(a) $\sum_{U \in \mathcal{F}} U = V$.

(b) For a finite-dimensional subspace $U' \subseteq V$ there is a $U \in \mathcal{F}$ such that $U' \subseteq U$.

(c) There is a family $\{(\pi_{\mathrm{sh}(U)}, \mathrm{sh}(U))\}_{U \in \mathcal{F}}$, where $(\pi_{\mathrm{sh}(U)}, \mathrm{sh}(U))$ is a cofree pointed irreducible coalgebra on U, $\mathrm{sh}(U)$ is a subcoalgebra of D, and $U, U' \in \mathcal{F}$ with $U \subseteq U'$ implies $\mathrm{sh}(U) \subseteq \mathrm{sh}(U')$ and $\pi_{\mathrm{sh}(U)} = \pi_{\mathrm{sh}(U')}|U$.

Show that $(\pi, \mathrm{sh}(V))$ is a cofree pointed irreducible coalgebra on V, where $\mathrm{sh}(V) = \sum_{U \in \mathcal{F}} \mathrm{sh}(U)$ and $\pi : \mathrm{sh}(V) \longrightarrow V$ is determined by $\pi|\mathrm{sh}(U) = \pi_{\mathrm{sh}(U)}$ for all $U \in \mathcal{F}$. [Hint: Note that $\mathrm{sh}(V)$ is pointed irreducible by part (a) of Proposition 3.4.3.]

Exercise 4.5.5. Show that:

(a) If $f : U \longrightarrow V$ is linear then there is a coalgebra map $F : \mathrm{Sh}(U) \longrightarrow \mathrm{Sh}(V)$ determined by $\pi_{\mathrm{Sh}(V)} \circ F = f \circ \pi_{\mathrm{Sh}(U)}$.

(b) The assignments $V \mapsto \mathrm{Sh}(V)$ and $f \mapsto F$ determine a functor from k-Vec to k-PICoalg, where the latter is the full subcategory of k-Coalg whose objects are pointed irreducible coalgebras over k.

Exercise 4.5.6. Let $f : U \longrightarrow V$ be linear and $F : \mathrm{Sh}(U) \longrightarrow \mathrm{Sh}(V)$ be the coalgebra map of part (a) of Exercise 4.5.5. Show that:

(a) f one-one implies F is one-one.

(b) f onto implies F is onto.

[Hint: If $f : U \longrightarrow V$ and $g : V \longrightarrow U$ are functions which satisfy $g \circ f = \mathrm{I}_U$ then f is one-one and g is onto.]

Exercise 4.5.7. Let $(\pi_{\mathrm{Sh}(V)}, \mathrm{Sh}(V))$ be a pointed irreducible coalgebra on V. For a subspace $U \subseteq V$ let $F_{U,V} : \mathrm{Sh}(U) \longrightarrow \mathrm{Sh}(V)$ be the coalgebra map of part (a) of Exercise 4.5.5 determined by $\pi_{\mathrm{Sh}(V)} \circ F_{U,V} = f_{U,V} \circ \pi_{\mathrm{Sh}(U)}$, where $f_{U,V} : U \longrightarrow V$ is the inclusion. Let $\mathcal{F}$ be a family of subspaces of V which satisfies (a)–(c) of Exercise 4.5.4. Show that $\mathrm{Sh}(V) = \sum_{U \in \mathcal{F}} \mathrm{Im}(F_{U,V})$. [Hint: Show that $(\pi, \mathrm{Sh}(V)')$ is a cofree pointed irreducible coalgebra on V, where $\mathrm{Sh}(V)' = \sum_{U \in \mathcal{F}} \mathrm{Im}(F_{U,V})$ and $\pi = \pi_{\mathrm{Sh}(V)} | \mathrm{Sh}(V)'$.]

4.6 The radical of the dual algebra

We have noted that $\mathrm{Rad}(C^*) = C_0^\perp$ when C is a finite-dimensional coalgebra over k. The radical of C^* can always be described this way. There are characterizations of $\mathrm{Rad}(C^*)$ in terms of cofinite maximal left, right or two-sided ideals of C^*.

Theorem 4.6.1. *Suppose that C is a coalgebra over the field k. Then $\mathrm{Rad}(C^*)$ is:*

(a) *$C_0^\perp$, where C_0 is the coradical of C;*
(b) *The intersection of the cofinite maximal ideals of C^*;*
(c) *The intersection of the closed cofinite maximal ideals of C^*;*
(d) *The intersection of the cofinite maximal left ideals of C^*;*
(e) *The intersection of the closed cofinite maximal left ideals of C^*.*

Proof. Suppose that A is an algebra over k. Then $\mathrm{Rad}(A)$ is the intersection of the maximal left ideals of A and is the intersection of the maximal ideals of A. Thus a cofinite maximal ideal of A is the intersection of the cofinite maximal left ideals of A which contain it. Recall that the intersection of closed subspaces of C^* is closed and that subspaces of C^* which contain a cofinite closed subspace of C^* are closed. Note that $C_0^\perp = \bigcap_D D^\perp$, where D runs over the simple subcoalgebras of C, by part (c) of Proposition 3.4.3. If D is a simple subcoalgebra of C then $D^\perp$ is a closed cofinite maximal ideal of C^* by part (d) of Proposition 2.3.7. Thus $\mathrm{Rad}(C^*) \subseteq C_0^\perp$ and if I is one of the intersections described in parts (b) – (e) then $\mathrm{Rad}(C^*) \subseteq I \subseteq C_0^\perp$. To complete the proof we need only show

$C_0^\perp \subseteq \mathrm{Rad}(C^*)$.

Let L be a maximal left ideal of C^*. To show that $C_0^\perp \subseteq \mathrm{Rad}(C^*)$ it suffices to show that $C_0^\perp \subseteq L$.

Suppose that $C_0^\perp \not\subseteq L$. Then $C^* = C_0^\perp + L$ since L is a maximal left ideal of C^*. Therefore $\epsilon = \alpha + \ell$, where $\alpha \in C_0^\perp$ and $\ell \in L$. By Exercise 4.4.15 we conclude that $\ell = \epsilon - \alpha$ has an inverse in C^*, a contradiction. Therefore $C_0^\perp \subseteq L$ and the theorem is proved. □

Since $\mathrm{Rad}(C^*) = \mathrm{Rad}(\mathcal{C}^*)$, where $\mathcal{C} = C^{cop}$, the word "left" can be replaced by "right" in the statement of Theorem 4.6.1 and the list of equivalent descriptions of $\mathrm{Rad}(C^*)$ is thereby extended.

In light of Theorem 4.6.1 there is a natural connection between the maximal ideals of C^* and the maximal ideals of C_0^*. More generally:

Corollary 4.6.2. *Let C be a coalgebra over the field k and $\pi : C^* \longrightarrow C_0^*$ be the restriction map. Then the association $I \mapsto \pi(I)$ is a one-one correspondence between the set of left (respectively right, two-sided) ideals of C^* which contain* $\mathrm{Rad}(C^*)$ *and the left (respectively right, two-sided) ideals of C_0^*. Under this correspondence I is a closed (respectively cofinite) subspace of C^* if and only if $\pi(I)$ is a closed (respectively cofinite) subspace of C_0^*.*

Proof. Observe that $\pi = \imath^*$ where $\imath : C_0 \longrightarrow C$ is the inclusion map. Therefore π is a continuous onto map of algebras since $\imath$ is a one-one map of coalgebras. Observe that $\mathrm{Ker}(\pi) = C_0^\perp = \mathrm{Rad}(C^*)$ by Theorem 4.6.1. Since π is continuous, $\pi(I)$ is a closed subspace of C_0^* whenever I is a closed subspace of C^* and $\pi^{-1}(J)$ is a closed subspace of C^* whenever J is a closed subspace of C_0^* by part (b) and part (a) of Theorem 1.3.10 respectively. Since $[C^*{:}\pi^{-1}(J)] = [C_0^*{:}J]$ for all subspaces J of C_0^* our proof is complete. □

Exercises

In the following exercises C is a coalgebra over the field k.

Exercise 4.6.1. Suppose injective hulls of finite-dimensional right coideals of C are finite-dimensional and that the coradical C_0 of C is a reflexive coalgebra. Show that C^* is almost left noetherian. [Hint: Let L be a cofinite left ideal of C^*. Show that L contains a cofinite ideal I of C^*. Show that $J = \mathrm{Rad}(I)$ is a cofinite closed ideal of C^* which contains an idempotent f such that C^*f is a cofinite left ideal of C^* and thus $f \in I$ as $J^n \subseteq I$ for some $n \geq 1$. See Proposition 3.5.10.]

4.7 Free pointed coalgebras associated to coalgebras

Let C be a coalgebra over k. In this section we associate a pointed coalgebra and a pointed irreducible coalgebra over k to C. Using the associated pointed coalgebra we formulate and prove a version of Proposition 4.3.3 which holds for C.

Throughout this section $\mathcal{S}$ is the set of the simple subcoalgebras of C. Recall that $C_0 = \bigoplus_{S\in\mathcal{S}} S$ by part (c) of Proposition 3.4.3. Let $I = \bigoplus_{S\in\mathcal{S}} S^+$ where $S^+ = \mathrm{Ker}(\epsilon)\cap S$. Then I is a coideal of C since it is the sum of coideals of C. Give $\mathrm{p}(C) = C/I$ the quotient coalgebra structure and let $\pi_C : C \longrightarrow \mathrm{p}(C)$ be the projection. Since π_C is an onto coalgebra map and $\mathrm{Ker}(\pi_C) \subseteq C_0$ it follows that $\pi_C(C_n) = \pi(C)_n$ for all $n \geq 0$ by part (b) of Corollary 4.2.2. Observe that $\pi_C(S)$ is one-dimensional for all $S \in \mathcal{S}$. Since π_C is onto $\mathrm{p}(C)$ is a pointed coalgebra by part (a) of Proposition 4.1.7. The triple $(C, \pi_C, \mathrm{p}(C))$ satisfies a universal mapping property.

Theorem 4.7.1. *Suppose that C is a coalgebra over the field k, $\mathcal{S}$ is the set of simple subcoalgebras of C and $(C, \pi_C, \mathrm{p}(C))$ is the triple described above. Then:*

(a) *$\mathrm{p}(C)$ is a pointed coalgebra and $\pi_C : C \longrightarrow \mathrm{p}(C)$ is a coalgebra map such that $\pi_C(S)$ is one-dimensional for all $S \in \mathcal{S}$.*
(b) *Suppose that $f : C \longrightarrow D$ is a coalgebra map, where D is pointed, and $f(S)$ is one-dimensional for all $S \in \mathcal{S}$. Then there is a coalgebra map $F : \mathrm{p}(C) \longrightarrow D$ determined by $F\circ\pi_C = f$.*

Proof. We need only establish part (b). Assume the hypothesis of part (b) holds and let $S \in \mathcal{S}$. Since $S \neq (0)$ we have $f(S^+) = f(S)^+ \neq f(S)$. Since $f(S)$ is one-dimensional $f(S^+) = (0)$. Thus $f(I_C) = (0)$. By part (b) of Theorem 2.1.21 there is a coalgebra map $F : \mathrm{p}(C) \longrightarrow D$ which is determined by $F\circ\pi_C = f$. This equation determines F as π_C is onto. □

Definition 4.7.2. Let C be a coalgebra over k. A *free pointed coalgebra on* C is a triple $(C, \pi_C, \mathrm{p}(C))$ which satisfies the conditions of Theorem 4.7.1.

If $(C, \pi_C, \mathrm{p}(C))$ and $(C, \pi'_C, \mathrm{p}(C)')$ are free pointed coalgebras on C then there exists a coalgebra isomorphism $F : \mathrm{p}(C) \longrightarrow \mathrm{p}(C)'$ such that $F\circ\pi_C = \pi'_C$. Therefore

$$\pi_C \quad \text{is onto,} \quad \mathrm{Ker}(\pi_C) = \bigoplus_{S\in\mathcal{S}} S^+, \quad \text{and} \quad \pi_C(C_n) = \mathrm{p}(C)_n \qquad (4.16)$$

for all $n \geq 0$ since these statements are true for the particular free pointed coalgebra on C which we constructed.

Corollary 4.7.3. *Suppose that $f : C \longrightarrow D$ is a map of coalgebras over the field k such that $f(S)$ is a simple subcoalgebra of D whenever S is a simple subcoalgebra of C. Let $(C, \pi_C, \mathrm{p}(C))$ and $(D, \pi_D, \mathrm{p}(D))$ be free pointed coalgebras on C and D respectively. Then:*

(a) *There exists a coalgebra map $F : \mathrm{p}(C) \longrightarrow \mathrm{p}(D)$ determined by $F \circ \pi_C = \pi_D \circ f$.*

(b) *Suppose that $f|C_0$ is one-one. Then f is one-one if and only if F is one-one.*

Proof. We first show part (a). Let $S \in \mathcal{S}$. Then $f(S)$ is a simple subcoalgebra of D which means that $\pi_D(f(S))$ is one-dimensional. Thus the coalgebra map $\pi_D \circ f : C \longrightarrow \mathrm{p}(D)$ satisfies the hypothesis of part (b) of Theorem 4.7.1. Consequently there is a coalgebra map $F : \mathrm{p}(C) \longrightarrow \mathrm{p}(D)$ determined by $F \circ \pi_C = \pi_D \circ f$.

To show part (b), assume that $f|C_0$ is one-one. Suppose first of all that F is one-one and let $c \in \mathrm{Ker}(f)$. Then $F(\pi_C(c)) = \pi_D(f(c)) = 0$ which means that $\pi_C(c) \in \mathrm{Ker}(F) = (0)$. Therefore $c \in \mathrm{Ker}(\pi_C) \subseteq C_0$. Since $f|C_0$ is one-one $c = 0$. We have shown that f is one-one.

Conversely, suppose that f is one-one and $\pi_C(c) \in \mathrm{Ker}(F)$, where $c \in C$. Since π_C is onto all elements of $\mathrm{Ker}(F)$ have this form. Since $0 = F(\pi_C(c)) = \pi_D(f(c))$ we conclude that $f(c) \in (\mathrm{Ker}(\pi_D)) \cap f(C)$. To complete the proof of part (b) we need only show that $(\mathrm{Ker}(\pi_D)) \cap f(C) = f(\mathrm{Ker}(\pi_C))$. For then, since f is one-one, $c \in \mathrm{Ker}(\pi_C)$ which means $\pi_C(c) = 0$.

We now show that $(\mathrm{Ker}(\pi_D)) \cap f(C) = f(\mathrm{Ker}(\pi_C))$. Since f is one-one $f(C_0) = \bigoplus_{S \in \mathcal{S}} f(S)$ and $f(S)$ is a simple subcoalgebra of D for all $S \in \mathcal{S}$. Thus

$$f(\mathrm{Ker}(\pi_C)) = \bigoplus_{S \in \mathcal{S}} f(S^+) = \bigoplus_{S \in \mathcal{S}} f(S)^+$$

and consequently $f(\mathrm{Ker}(\pi_C)) \subseteq (\mathrm{Ker}(\pi_D)) \cap f(C)$. Since $f(C_0) \supseteq f(C)_0$ by part (a) of Proposition 4.1.7 it follows that

$$\begin{aligned}(\mathrm{Ker}(\pi_D)) \cap f(C) &\subseteq (\mathrm{Ker}(\pi_D)) \cap D_0 \cap f(C) \\ &= (\mathrm{Ker}(\pi_D)) \cap f(C)_0 \\ &\subseteq (\mathrm{Ker}(\pi_D)) \cap f(C_0).\end{aligned}$$

Thus

$$(\mathrm{Ker}(\pi_D)) \cap f(C) \subseteq \bigoplus_{S \in \mathcal{S}} f(S)^+ = f(\mathrm{Ker}(\pi_C)).$$

We have shown that $(\mathrm{Ker}(\pi_D))\cap f(C) = f(\mathrm{Ker}(\pi_C))$ and the proof of part (b) is complete. □

The equivalence of parts (a) and (c) of Proposition 4.3.3 is a special case of a more general result.

Theorem 4.7.4. *Suppose that $f : C \longrightarrow D$ is a map of coalgebras over the field k. Then the following are equivalent:*

(a) *f is one-one.*
(a) *$f|S\wedge T$ is one-one for all simple subcoalgebras S and T of C.*

Proof. Without loss of generality we may assume that f is onto. We need only show that part (b) implies part (a).

Suppose that $f|S\wedge T$ is one-one for all $S, T \in \mathcal{S}$. Let $S, T \in \mathcal{S}$. Since $S, T \subseteq S\wedge T$ it follows that $f(S)$ and $f(T)$ are simple subcoalgebras of D and that $f(S) \neq f(T)$ whenever $S \neq T$. Consequently $f(C_0) = \bigoplus_{S\in\mathcal{S}} f(S)$ by part (c) of Proposition 3.4.3. Thus $f|C_0$ is one-one. Since f is onto $f(C_0) \supseteq f(C)_0$ by part (a) of Proposition 4.1.7. Thus $D_0 = f(C_0) = \bigoplus_{S\in\mathcal{S}} f(S)$. By part (a) of Proposition 3.4.3 the set of simple subcoalgebras of D is $\{f(S)\}_{S\in\mathcal{S}}$. Note that $f(\mathrm{Ker}(\pi_C)) = \mathrm{Ker}(\pi_D)$ as well.

Now let $(C, \pi_C, \mathrm{p}(C))$ and $(D, \pi_D, \mathrm{p}(D))$ be free pointed coalgebras on C and D respectively. Since $f|C_0$ is one-one there is a coalgebra map $F : \mathrm{p}(C) \longrightarrow \mathrm{p}(D)$ such that $F\circ\pi_C = \pi_D\circ f$ by part (a) of Corollary 4.7.3.

We need to relate the simple subcoalgebras of C and $\mathrm{G}(\mathrm{p}(C))$ in order to proceed with the proof. First of all note that $S \mapsto \pi_C(S)$ determines a one-one correspondence of the set of simple subcoalgebras $\mathcal{S}$ of C and the set of simple subcoalgebras of $\mathrm{p}(C)$.

Now $\pi_C(S)$ one-dimensional. Thus $\pi_C(S) = kg$ for some unique group-like element g of $\mathrm{p}(C)$ which we denote by $g(S)$. Thus $S \mapsto g(S)$ describes a one-one correspondence between $\mathcal{S}$ and $\mathrm{G}(\mathrm{p}(C))$.

The proof will be completed once we establish that $F|\pi_C(S\wedge T)$ is one-one and that $\pi_C(S\wedge T) = kg(S)\wedge kg(T)$. For then $F|\ kg(S)\wedge kg(T)$ is one-one which means that F is one-one by Proposition 4.3.3. Thus f is one-one follows by part (b) of Corollary 4.7.3.

To establish these two facts we utilize several families of idempotents. By Corollary 3.5.15 there is an orthonormal family of idempotents $\mathcal{E}_D = \{e_{f(S)}\}_{S\in\mathcal{S}}$ which satisfies $e_{f(S)}|f(T) = \delta_{S,T}\epsilon|f(S)$ for all $S, T \in \mathcal{S}$. For $S \in \mathcal{S}$ set $e_S = e_{f(S)}\circ f$ and let $\mathcal{E}_C = \{e_S\}_{S\in\mathcal{S}}$. Observe that $e_S|T = \delta_{S,T}\epsilon|S$ for all $S, T \in \mathcal{S}$. For $S \in \mathcal{S}$ note that $e_S(\mathrm{Ker}(\pi_C)) = (0)$. Therefore there is an $e_{g(S)} \in \mathrm{p}(C)^*$ uniquely determined by $e_{g(S)}\circ\pi_C = e_S$. Set

$\mathcal{E}_{\mathrm{p}(C)} = \{e_{g(S)}\}_{S\in\mathcal{S}}$. Observe that $e_{g(S)}|g(T) = \delta_{S,T}$ for all $S, T \in \mathcal{S}$. By Exercise 4.3.1 we see that $\mathcal{E}_{\mathrm{p}(C)}$ and $\mathcal{E}_C$ are orthonormal families of idempotents for C and $\mathrm{p}(C)$ respectively.

We now show that $\mathrm{P}_{g(S),g(T)}(\mathrm{p}(C)) \subseteq \pi_C(S\wedge T)$. Let $c \in S\wedge T$. Since $\Delta(c) \in S\otimes C_1 + C_1\otimes T$, and $c = \sum_{S',S''\in\mathcal{S}} {}^{e_{S'}}c^{e_{S''}}$ by (4.3), it follows by (4.7) that

$$S\wedge T = {}^{e_S}C_1^{e_T} + S + T. \tag{4.17}$$

Applying π_C to both sides of (4.17) and using part (c) of Exercise 4.3.1 we obtain $\pi_C(S\wedge T) = {}^{g(S)}(\mathrm{p}(C)_1)^{g(T)} + kg(S) + kg(T) = kg(S)\wedge g(T)$.

We next show that $F|\pi_C(S\wedge T)$ is one-one. Let $c \in S\wedge T$ and assume that $F(\pi_C(c)) = 0$. From the equations $\pi_D(f(c)) = F(\pi_C(c)) = 0$ we conclude that $f(c) \in \mathrm{Ker}(\pi_D) = f(\mathrm{Ker}(\pi_C))$. Therefore $f(c) = f(c')$ for some $c' \in \mathrm{Ker}(\pi_C) \subseteq C_0$. From (4.17) we conclude that ${}^{e_{S'}}c^{e_{S'}} = 0$ whenever $S' \in \mathcal{S}$ and is not one of S or T. By part (c) of Exercise 4.3.1 we calculate for such an S' that

$$f({}^{e_{S'}}c'^{e_{S'}}) = {}^{e_{f(S')}}f(c')^{e_{f(S')}} = {}^{e_{f(S')}}f(c)^{e_{f(S')}} = f({}^{e_{S'}}c^{e_{S'}}) = 0.$$

Since $f|C_0$ is one-one and $c' \in C_0$ we conclude that ${}^{e_{S'}}c'^{e_{S'}} = 0$. Thus $c' \in S + T \subseteq S\wedge T$. Since $c \in S\wedge T$, and $f|S\wedge T$ is one-one by assumption, it follows that $c = c'$. Therefore $\pi_C(c) = \pi_C(c') = 0$ and hence $F|\pi_C(S\wedge T)$ is one-one. □

Corollary 4.7.5. *Let C be a pointed irreducible coalgebra over the field k and suppose I is a coideal of C. Then $I = (0)$ if and only if $I\cap\mathrm{P}(C) = (0)$.*

Proof. Apply Theorem 4.7.4 to the projection $\pi : C \longrightarrow C/I$. □

We end this section by describing a pointed irreducible coalgebra $\mathrm{pi}(C)$ which is associated with C in the same manner as the pointed coalgebra $\mathrm{p}(C)$ is associated with C. Give $\mathrm{pi}(C) = C/C_0^+$ the quotient coalgebra structure and let $\pi_C : C \longrightarrow \mathrm{pi}(C)$ be the projection. Since π_C is an onto coalgebra map $\pi_C(C_0) \supseteq \mathrm{pi}(C)_0$ by part (a) of Proposition 4.1.7. Since $\pi_C(C_0)$ is one-dimensional $\mathrm{pi}(C)$ is pointed irreducible. The triple $(C, \pi_C, \mathrm{pi}(C))$ satisfies a universal mapping property which we leave to the reader to verify.

Theorem 4.7.6. *Suppose that C is a coalgebra over the field k and $(C, \pi_C, \mathrm{pi}(C))$ is the triple described above. Then:*

(a) *$\mathrm{pi}(C)$ is a pointed irreducible coalgebra and $\pi_C : C \longrightarrow \mathrm{pi}(C)$ is a coalgebra map such that $\pi_C(C_0)$ is one-dimensional.*

(b) *Suppose that* $f : C \longrightarrow D$ *is a coalgebra map, where* D *is pointed irreducible, and* $f(C_0)$ *is one-dimensional. Then there is a coalgebra map* $F : \mathrm{pi}(C) \longrightarrow D$ *determined by* $F \circ \pi_C = f$.

□

Definition 4.7.7. Let C be a coalgebra over k. A *free pointed irreducible coalgebra on* C is a triple $(C, \pi_C, \mathrm{pi}(C))$ which satisfies the conditions of Theorem 4.7.6.

Let $(C, \pi_C, \mathrm{pi}(C))$ be a free pointed irreducible coalgebra on C. The reasons which justify (4.16) applied to $\mathrm{pi}(C)$ yield

$$\pi_C \quad \text{is onto,} \quad \mathrm{Ker}(\pi_C) = C_0^+, \quad \text{and} \quad \pi_C(C_n) = \mathrm{pi}(C)_n \tag{4.18}$$

for all $n \geq 0$.

Exercises

In the following exercises C and D are coalgebras over the field k.

Definition 4.7.8. A *locally finite coalgebra over* k is a coalgebra C over k such that $U \wedge V$ is finite-dimensional for all finite-dimensional subspaces U and V of C.

Exercise 4.7.1. Suppose that $f : C \longrightarrow D$ is an onto coalgebra map and $\mathrm{Ker}(f)$ is finite-dimensional. Show that C is locally finite if and only if D is locally finite.

Exercise 4.7.2. Prove the following:

Lemma 4.7.9. *Suppose that* C *is a pointed irreducible coalgebra over the field* k. *Then* C *is locally finite if and only if* C_1 *is finite-dimensional.* □

[Hint: $\mathrm{G}(C) = \{g\}$ and $C_0 = kg$. Define $\delta : C \longrightarrow C \otimes C$ by $\delta(c) = \Delta(c) - (c \otimes g + c \otimes g)$ for all $c \in C$. Then $\mathrm{Ker}(\delta) \subseteq C_1$ and $\delta(C_n) \subseteq C_{n-1} \otimes C_{n-1}$ for all $n \geq 1$ by part (a) of Proposition 4.3.1.]

Exercise 4.7.3. Prove the following:

Theorem 4.7.10. *Let* C *be a coalgebra over the field* k. *Then the following are equivalent:*

(a) *C is locally finite.*

(b) $D \wedge D$ *is finite-dimensional for all finite-dimensional subcoalgebras* D *of* C_0.

□

[Hint: Suppose that U, V are finite-dimensional subspaces of C. Then U and V are contained in a finite-dimensional subcoalgebra E of C and thus $U, V \subseteq \mathcal{D} = D^{(\infty)}$, where $D = E_0$. Note that $U \wedge V = U \wedge_{\mathcal{D}} V \subseteq \mathcal{D}$. Thus to show that the condition of part (b) implies C is locally finite we may assume that $C = \mathcal{D}$ and C_1 is finite-dimensional. Let $\pi_C : C \longrightarrow \mathrm{pi}(C)$ be the map of Theorem 4.7.6 and see Exercise 4.7.2.]

Exercise 4.7.4. Suppose C is a graded coalgebra. Show that $\mathrm{p}(C)$ (respectively $\mathrm{pi}(C)$) is a graded coalgebra and $\pi_C : C \longrightarrow \mathrm{p}(C)$ (respectively $\pi_C : C \longrightarrow \mathrm{pi}(C)$) is a map of graded coalgebras which satisfies $\pi_C(C(n)) = \pi_C(C)(n)$ for all $n \geq 0$.

4.8 Linked simple subcoalgebras

The simple subcoalgebras of an indecomposable coalgebra over k are "linked" by a relationship given by the wedge product. Here we continue the discussion of indecomposable subcoalgebras of Section 3.7 using some results from Section 4.2.

Let C be a coalgebra over the field k and suppose that D and E are subcoalgebras of C. Then $D + E \subseteq D \wedge E$ and $D + E \subseteq E \wedge D$. Considering whether or not one of these inclusions is proper leads to an interesting circle of ideas related to the indecomposable subcoalgebras of C.

Definition 4.8.1. Suppose that D, E are subcoalgebras of a coalgebra C over k. Then D *and* E *are directly linked in* C if $D + E$ is a proper subset of $D \wedge E + E \wedge D$.

Suppose D, E and F are subcoalgebras of C and $D, E \subseteq F$. Since $D \wedge_F E \subseteq D \wedge_C E$ it follows that if D and E directly linked in F then D and E are directly linked in C.

To say that D and E are *not* directly linked in C is to say that $D \wedge E = D + E = E \wedge D$. We are interested in whether or not simple subcoalgebras of C are linked in the following sense.

Definition 4.8.2. Suppose that S, T are simple subcoalgebras of a coalgebra C over k. Then S *and* T *are linked in* C if there is an $n \geq 0$ and simple subcoalgebras $S_0, \ldots, S_n$ of C such that $S = S_0, T = S_n$ and S_i and S_{i+1} are directly linked in C for $0 \leq i < n$.

Observe that $S \sim T$ if and only if S and T are linked in C is an equivalence relation on the set of the simple subcoalgebras of C. As we see by the next lemma, simple subcoalgebras of C which lie in different direct summands are not linked in C.

Lemma 4.8.3. *Suppose that C is a coalgebra over the field k and $C = D\bigoplus E$ is the direct sum of subcoalgebras D and E. Let S and T be subcoalgebras of C. Then:*

(a) *If $S \subseteq D$ and $T \subseteq E$ then S and T are not directly linked in C.*
(b) *If S and T are simple and linked in C then $S, T \subseteq D$ or $S, T \subseteq E$.*

Proof. Part (b) follows from part (a). For any simple subcoalgebra of C is contained in D or E by part (a) of Proposition 3.4.3. To show part (a) we note that $S \wedge T = D' \bigoplus E'$, where D' and E' are subcoalgebras of C which satisfy $D' \subseteq D$ and $E' \subseteq E$, by part (b) of Proposition 3.4.3. By Exercise 1.2.8 the calculation

$$\Delta(D') \subseteq (D' \otimes D') \cap (S \otimes (D \oplus E) + (D \oplus E) \otimes T) = (D' \cap S) \otimes D'$$

shows that $D' \subseteq S \cap D'$. Therefore $D' \subseteq S$ and likewise $E' \subseteq T$. We have shown that $S \wedge T \subseteq S + T$. Since $S + T \subseteq S \wedge T$ in any case it follows that $S \wedge T = S + T$. Since $C = E \bigoplus D$ we have $T \wedge S = T + S$ as well. Therefore S and T are not directly linked. □

Definition 4.8.4. Let C be a coalgebra over k. A *link indecomposable subcoalgebra of C* is a subcoalgebra D of C such that whenever S, T are simple subcoalgebras of D Definition 4.8.2 is satisfied for some simple subcoalgebras $S_0, \ldots, S_n \subseteq D$. A *link indecomposable component of C* is maximal link indecomposable subcoalgebra of C.

Suppose that D and E are subcoalgebras of C and $D_0 = E_0$. Then D is link indecomposable if and only if E is link indecomposable. Observe that any simple subcoalgebra of C is link indecomposable by definition. By Zorn's Lemma any link indecomposable subcoalgebra of C is contained in a link indecomposable component of C.

The concepts of link indecomposable and indecomposable are very closely related.

Lemma 4.8.5. *Let C be a coalgebra over the field k and suppose that D is a subcoalgebra of C. Then:*

(a) *If D is link indecomposable then D is contained in an indecomposable subcoalgebra $\mathcal{D}$ of C which satisfies $D_0 = \mathcal{D}_0$.*
(b) *If D is indecomposable then D is link indecomposable.*

Proof. We first show part (a). Suppose that D is link indecomposable and set $\mathcal{D} = D^{(\infty)}$. Then $\mathcal{D}$ is closed under the wedge product; more precisely $U \wedge_{\mathcal{D}} V = U \wedge V$ for all subspaces U and V of $\mathcal{D}$ by part (a) of Proposition 2.4.3.

Suppose $\mathcal{D} = E \bigoplus F$ is the direct sum of non-zero subcoalgebras of $\mathcal{D}$. Let $S, T \subseteq \mathcal{D}$ be simple subcoalgebras. If S and T are directly linked in C we deduce from part (a) of Lemma 4.8.3 that $S, T \subseteq E$ or $S, T \subseteq F$. Since D is link indecomposable and $D_0 = \mathcal{D}_0$, all simple subcoalgebras of $\mathcal{D}$ lie on E or they all lie in F. We have shown that $\mathcal{D}$ is indecomposable.

To show part (b) we may assume that $D = C$. It suffices to show that C is the direct sum of its link indecomposable components. For then, since C is indecomposable, C must be one of its link indecomposable components.

Let $D_1, \ldots, D_r$ be distinct link indecomposable components of C. We first show that $D_1 + \cdots + D_r$ is direct. We may assume $r > 1$. Suppose $(D_1 + \cdots + D_{r-1}) \cap D_r \neq (0)$. Since the intersection is not zero it contains a simple subcoalgebra S. Since $S \subseteq D_1 + \cdots + D_{r-1}$ it follows that $S \subseteq D_j$ for some $1 \leq j \leq r-1$ by part (a) of Proposition 3.4.3. By the same result a simple subcoalgebra of $D_j + D_r$ is contained in either D_j or D_r. Since S is a simple subcoalgebra of both D_j and D_r it follows that $D_j + D_r$ is link indecomposable. By maximality we have $D_j = D_j + D_r = D_r$, a contradiction. Thus $(D_1 + \cdots + D_{r-1}) \cap D_r = (0)$. By induction on r the sum $D_1 + \cdots + D_r$ is direct. Thus the sum of all link indecomposable components of C is direct.

To complete the proof of part (b) we need to show that sum of the link indecomposable components of C spans C. Since every link indecomposable subcoalgebra of C is contained in a link indecomposable component of C, we may assume that C is finite-dimensional. By Theorem 3.7.6 we may further assume that C is indecomposable.

Suppose that C is finite-dimensional and indecomposable. Let $\mathcal{M}$ and $\mathcal{N}$ be maximal ideals of C^*. Then $S = \mathcal{M}^\perp$ and $T = \mathcal{N}^\perp$ are simple subcoalgebras of C. Now $(\mathcal{M}\mathcal{N})^\perp = (S^\perp T^\perp)^\perp = S \wedge T$ by part (a) of Proposition 2.4.2. Thus $(\mathcal{M}\mathcal{N})^\perp = S \wedge T$ and $(\mathcal{N}\mathcal{M})^\perp = T \wedge S$. If S and T are not directly linked then $(\mathcal{M}\mathcal{N})^\perp = (\mathcal{N}\mathcal{M})^\perp$, or equivalently $\mathcal{M}\mathcal{N} = \mathcal{N}\mathcal{M}$.

Suppose that the simple subcoalgebras of C fall into more than one equivalence class under the link relation. Then the maximal ideals of C^* can be partitioned into two non-empty sets $\{\mathcal{M}_1, \ldots, \mathcal{M}_r\}$ and $\{\mathcal{N}_1, \ldots, \mathcal{N}_s\}$ where the $\mathcal{M}_i$'s commute with the $\mathcal{N}_j$'s. Let $I = \mathcal{M}_1 \cdots \mathcal{M}_r$ and $J = \mathcal{N}_1 \cdots \mathcal{N}_s$. Since maximal ideals are prime it follows that I^n and J^n are comaximal ideals of C^* for all $n \geq 0$. Since the $\mathcal{M}_i$'s commute with the $\mathcal{N}_j$'s it follows that I and J commute. Thus I^n and J^n commute for all $n \geq 0$. Since $IJ \subseteq \mathrm{Rad}(C^*)$ and C is finite-dimensional $I^n J^n = (IJ)^n = (0)$ for some $n \geq 1$. Since $I^n + J^n = C^*$ we conclude that $I^n \cap J^n = I^n J^n = (0)$, and thus $C = D \bigoplus E$, where $D = (I^n)^\perp$ and $E = (J^n)^\perp$ are non-zero subcoalgebras of C. This is a contradiction since C is indecomposable. Therefore C is link indecomposable. We have completed the proof of part (b). □

Theorem 4.8.6. *Suppose that C is a coalgebra over the field k.*

(a) *The link indecomposable components of C are the indecomposable components of C. Thus C is the direct sum of its link indecomposable components.*
(b) *Suppose that $C = \bigoplus_{i \in I} D_i$ is the direct sum of non-zero link indecomposable subcoalgebras of C. Then the D_i's are the link indecomposable components (hence the indecomposable components) of C.*
(c) *Two simple subcoalgebras of C are linked in C if and only if they belong to the same indecomposable component of C.*
(d) *C is indecomposable if and only if any two simple subcoalgebras of C are linked.*

Proof. Part (a) follows by Lemma 4.8.5 and part (a) of Theorem 3.7.6. To show part (b) we recall that any link indecomposable subcoalgebra of C is contained in a link indecomposable component of C. Therefore each D_i is contained in an indecomposable component of C by part (a). As C is the direct sum of its indecomposable components, each D_i is a direct summand of an indecomposable component of C. Our conclusion: each D_i is an indecomposable component of C. Since the link indecomposable components of C are the indecomposable components of C, we have established part (b). Part (c) follows by part (a) and Lemma 4.8.3. Part (d) is a direct consequence of part (c). □

Quite a bit of information about the indecomposable components of C is contained in C_1.

Corollary 4.8.7. *Let C be a non-zero coalgebra over the field k and write*

$C = \bigoplus_{i\in I} D_i$ as the direct sum of its indecomposable components.

(a) *Suppose that D is a subcoalgebra of C and $C_1 \subseteq D$. Then the $D\cap D_i$'s are the indecomposable components of D.*

(b) *C is indecomposable if and only if C_1 is indecomposable.*

Proof. Let D be any subcoalgebra of C. Then $D = \bigoplus_{i\in I}(D\cap D_i)$ by part (b) of Proposition 3.4.3. Now suppose that $C_1 \subseteq$D. Since $D_i \neq (0)$ and $C_0 \subseteq D$ it follows that $(0) \neq D_i\cap C_0 \subseteq D\cap D_i$.

To show that $D\cap D_i$ is an indecomposable component of D it suffices to show that $D\cap D_i$ a link indecomposable subcoalgebra of D by part (b) of Theorem 4.8.6. Suppose that S and T are simple subcoalgebras of D. Since $S\wedge T \subseteq C_0\wedge C_0 = C_1 \subseteq D$ we conclude that $S\wedge T = S\wedge_D T$ by part (a) of Proposition 2.4.3. Therefore S and T are directly linked in C if and only if they are directly linked in D. Consequently S and T are linked in C if and only if they are linked in D. Thus any two simple subcoalgebras $S, T \subseteq D\cap D_i$ are linked as subcoalgebras of D. We have established part (a). Part (b) is an immediate consequence of part (a). □

Exercises

Throughout the following exercises C and D are coalgebras over the field k.

Exercise 4.8.1. Find a coalgebra C with a link indecomposable subcoalgebra D which is not indecomposable.

Exercise 4.8.2. Show that any irreducible component of C is contained in an indecomposable component of C.

Exercise 4.8.3. Show that any indecomposable component of C is saturated (recall meaning closed under the wedge product of subspaces).

Exercise 4.8.4. Suppose that S and T are subcoalgebras of C and $S\cap T = (0)$. Show that $S\wedge T = T\wedge S$ if and only if $S\wedge T = S+T = T\wedge S$. (Note that $S\cap T = (0)$ if S and T are distinct simple subcoalgebras of C.)

Exercise 4.8.5. Suppose that C is cocommutative.

(a) Show that simple subcoalgebras S and T are linked in C if and only if $S = T$.

(b) Show that the irreducible components of C are the indecomposable components of C.

(c) Prove the following theorem:

Theorem 4.8.8. *Suppose C is a cocommutative coalgebra. Then C is the direct sum of its irreducible components.*

Exercise 4.8.6. Find a coalgebra C whose irreducible components are not its indecomposable components.

Exercise 4.8.7. Suppose that $\{D_i\}_{i \in I}$ is a family of link indecomposable subcoalgebras of C. Show that $\sum_{i \in I} D_i$ is link indecomposable if $\bigcap_{i \in I} D_i \neq (0)$.

Exercise 4.8.8. Use Exercise 4.8.7 to show that every non-zero link indecomposable subcoalgebra of C is contained in a unique link indecomposable component of C. (Thus Zorn's Lemma is not needed to establish the existence of link indecomposable components.)

Exercise 4.8.9. Let $C = kg \bigoplus kh \bigoplus V$ be the coalgebra of Exercise 3.7.4. Show that C is indecomposable by showing that $C_0 = kg \bigoplus kh$ and kg and kh are directly linked.

Exercise 4.8.10. Show that subcoalgebras and quotients of indecomposable coalgebras are not necessarily indocomposable.

Exercise 4.8.11. Suppose that $g, h \in \mathrm{G}(C)$ are distinct grouplike elements of C. Show that kg and kh are directly linked if and only if there is some $x \notin k(g-h)$ such that either $\Delta(x) = g \otimes x + x \otimes h$ or $\Delta(x) = h \otimes x + x \otimes g$.

Exercise 4.8.12. Suppose that C and D are pointed and indecomposable. Show that $C \otimes D$ is pointed and indecomposable. [Hint: Recall that $\mathrm{G}(C \otimes D) = \{c \otimes d \mid c \in \mathrm{G}(C),\, d \in \mathrm{G}(D)\}$. See part (c) of Theorem 4.8.6 and Exercise 4.8.11.]

Exercise 4.8.13. Suppose that C and D are pointed and that $\{C_i\}_{i \in I}$ and $\{D_j\}_{j \in J}$ are the sets of indecomposable components of C and D respectively. Show that $\{C_i \otimes D_j\}_{i \in I,\, j \in J}$ is the set of indecomposable components of $C \otimes D$.

Exercise 4.8.14. Suppose that F is a field extension of k. Show that:

(a) If the F-coalgebra $C \otimes F$ is indecomposable then C is indecomposable.

(b) If C is indecomposable then it may not be the case that $C \otimes F$ is indecomposable. [Hint: See Exercise 2.5.9.]

Exercise 4.8.15. Suppose that D is a finite-dimensional subcoalgebra of C. Show that $D \subseteq S_1 \wedge \cdots \wedge S_n$ where $S_1, \ldots, S_n$ are simple subcoalgebras (not necessarily distinct) of C.

Chapter notes

The coradical and coradical filtration play a central role in the theory of coalgebras and most of the basic results about them are found in Sweedler's book [201]. The connection between the filtration and idempotents is described in [165]. Theorem 4.3.2 is the Taft-Wilson Theorem, a basic result in the study of pointed Hopf algebras. This result is found in [209].

Let $f : C \longrightarrow D$ be a coalgebra map. Necessary and sufficient conditions for f to be one-one emerged rather gradually. All involve the restriction $f|C_1$. Sweedler showed that when C is pointed irreducible then f is one-one if and only if the restriction $f|P(C)$ is one-one in Lemma 11.0.1 of [201]. It was shown by Heyneman and the author in Proposition 2.4.2 of [71] that f is one-one if and only if $f|C_1$ is one-one by reduction to the pointed irreducible case. When C is pointed, that f is one-one if and only if $f|\mathrm{P}_{g,h}(C)$ is one-one for all $g, h \in \mathrm{G}(C)$ was established by Takeuchi in [219] and in [170]. Theorem 4.7.4, which generalizes all of these results on when f is one-one, reduces to the pointed case.

The characterization of the radical of C^* as $C_0^\perp$ is found in [71]. Also see Proposition 5.2.9 in Montgomery's book [133]. The pointed irreducible coalgebra $\mathrm{pi}(C)$ of Section 4.7 is the coalgebra R of [71]. Section 4.8 for the most part is a slightly different presentation of some of Montgomery's results found in [134]. The reader is encouraged to consult this paper for a discussion of the contributions of others to the study of linkage, in particular Chin and Musson [32], Goodearl and Warfield [61], Shudo and Miyamoto [193], and Xu and Fong [225].

The theory of pointed Hopf algebras is well developed at this stage. Important structures in the analysis of pointed Hopf algebras are the associated graded pointed Hopf algebras of Section 15.5. These are special cases of the graded coalgebras associated to coalgebras developed in Section 4.4. See the paper by Heyneman and Sweedler [72] also.

Chapter 5

Bialgebras

Bialgebras are vector spaces with a coalgebra structure and an algebra structure which are compatible in a natural way. Hopf algebras are bialgebras with additional structure. In this chapter we define bialgebra and establish some of the basic properties bialgebras possess. Examples of bialgebras include monoid algebras and universal enveloping algebras of Lie algebras. There are free and cofree bialgebras of various types. We consider a few in great detail.

Free bialgebras on coalgebras are basic; indeed every bialgebra is the quotient of one of these. For this reason construction of many examples of bialgebras starts with a free bialgebra on a coalgebra. We work out an example in great detail and explain useful techniques for construction in the process.

5.1 Basic definitions and results

There are two natural compatibilities between an algebra structure and a coalgebra structure on a vector space over k.

Lemma 5.1.1. *Let A be a vector space over the field k which has an algebra structure (A, m, η) and a coalgebra structure (A, Δ, ϵ). Give $A{\otimes}A$ the tensor product algebra and coalgebra structures. Then the following are equivalent:*

(a) *m and η are coalgebra maps.*
(b) *Δ and ϵ are algebra maps.*

Proof. Observe that m is a coalgebra map if and only if $\Delta{\circ}m = (m{\otimes}m){\circ}\Delta_{A\otimes A}$ and $\epsilon{\circ}m = \epsilon_{A\otimes A}$; that is if and only if

$$\Delta(ab) = a_{(1)}b_{(1)}{\otimes}a_{(2)}b_{(2)} = \Delta(a)\Delta(b)$$

and

$$\epsilon(ab) = \epsilon(a)\epsilon(b)$$

for all $a, b \in A$. Note that η is a coalgebra map if and only if $\Delta \circ \eta = (\eta \otimes \eta) \circ \Delta_k$ and $\epsilon \circ \eta = \epsilon_k$; that is if and only if

$$\Delta(1) = 1 \otimes 1 \quad \text{and} \quad \epsilon(1) = 1.$$

□

Definition 5.1.2. A *bialgebra over the field* k, or a k*-bialgebra*, is a tuple $(A, m, \eta, \Delta, \epsilon)$, where (A, m, η) is an algebra and (A, Δ, ϵ) is a coalgebra over k, such that either of the equivalent conditions of Lemma 5.1.1 are satisfied.

Ordinarily we denote a bialgebra $(A, m, \eta, \Delta, \epsilon)$ by its underlying vector space A. Bialgebras can also be defined as algebras or coalgebras in certain categorical settings. See Exercise 5.1.1.

Definition 5.1.3. A *commutative bialgebra over* k is a bialgebra A over k whose underlying algebra structure is commutative. A *cocommutative bialgebra over* k is a bialgebra A over k whose underlying coalgebra structure is cocommutative.

Cocommutative bialgebras arise in the contexts of monoids and Lie algebras.

Example 5.1.4. Let S be a monoid. The monoid algebra $k[S]$ of S over k, regarded as the grouplike coalgebra on the set S over k, is a cocommutative bialgebra over k.

Recall that the grouplike coalgebra structure on $k[S]$ is determined by $\Delta(s) = s \otimes s$ and $\epsilon(s) = 1$ for $s \in S$. Note that $k[S]$ is a commutative bialgebra if and only if S is a commutative monoid.

The universal enveloping algebra $(\iota, \mathrm{U}(L))$ of a Lie algebra L over k has a cocommutative bialgebra structure where $\Delta(\ell) = \ell \otimes 1 + 1 \otimes \ell$ and $\epsilon(\ell) = 0$ for all $\ell \in L$.

Definition 5.1.5. Let A be a bialgebra over the field k. A *primitive element of* A is an $a \in A$ which satisfies $\Delta(a) = a \otimes 1 + 1 \otimes a$. The set of all primitive elements of A is denoted by $\mathrm{P}(A)$.

Note that $\mathrm{P}(A)$ is a coideal of A. By definition L consists of primitive elements of the bialgebra $\mathrm{U}(L)$. We postpone further discussion of the enveloping algebra until Section 5.4.

A bialgebra $(A, m, \eta, \Delta, \epsilon)$ over k gives rise to three other bialgebras by twisting the product and/or the coproduct. These bialgebras are $(A, m^{op}, \eta, \Delta, \epsilon)$, $(A, m, \eta, \Delta^{cop}, \epsilon)$, and $(A, m^{op}, \eta, \Delta^{cop}, \epsilon)$, which we denote by A^{op}, A^{cop}, and $A^{op\,cop}$ respectively. See Exercise 5.1.2.

The notion of "opposite bialgebra" takes both the product and coproduct into account.

Definition 5.1.6. Let A be a bialgebra over the field k. Then $A^{op\,cop}$ is the *opposite bialgebra.*

The notion of sub-bialgebra also takes both the product and coproduct into account.

Definition 5.1.7. Let A be a bialgebra over the field k. A *sub-bialgebra of* A is a subspace B of A which is simultaneously a subalgebra and a subcoalgebra of A.

Since the intersection of a family of subcoalgebras (respectively subalgebras) of a bialgebra A over k is a subcoalgebra (respectively a subalgebra) of A, the intersection of a family of sub-bialgebras of A is a sub-bialgebra of A. Hence for a subspace V of A there is a unique sub-bialgebra B of A which is minimal with respect to the property that B contains V.

Definition 5.1.8. Let A be a bialgebra over the field k and let V be a subspace of A. The *sub-bialgebra of A generated by V* is the sub-bialgebra B of A described above.

At this point the definition of bialgebra map should come as no surprise.

Definition 5.1.9. Let A and B be bialgebras over k. A *bialgebra map* $f : A \longrightarrow B$ is a map of underlying vector spaces which is both an algebra map and a coalgebra map. A *bialgebra isomorphism* is a bijective bialgebra map.

The kernel of a bialgebra map is an ideal and a coideal. These subspaces are therefore fundamental to the study of bialgebras.

Definition 5.1.10. Let A be a bialgebra over k. A *bi-ideal of a bialgebra* A over k is a subspace of A which is both an ideal and a coideal of A.

Observe that the sum of bi-ideals A is a bi-ideal of A. The kernel of a bialgebra map is a bi-ideal by definition. If I is a bi-ideal of A then the quotient algebra and coalgebra structures endow A/I with a unique bialgebra structure such that the projection $\pi : A \longrightarrow A/I$ is a bialgebra map. We leave it to the reader at this point to formulate and prove a fundamental homomorphism theorem for bialgebras; that is formulate and prove an appropriate analog of Theorem 2.1.21.

The fact that the coproduct and counit of A are multiplicative has interesting implications for multiplying and wedging certain types of subspaces of A.

Proposition 5.1.11. *Let A be a bialgebra over the field k.*

(a) *Suppose that C is a subcoalgebra of A and that I is a left coideal (respectively a right coideal, coideal, bi-ideal, subcoalgebra) of A. Then CI, IC, and CIC are left coideals (respectively right coideals, coideals, bi-ideals, subcoalgebras) of A.*

(b) *The left, right, or two-sided ideal of A generated by a left coideal (respectively a right coideal, coideal, bi-ideal, subcoalgebra) of A is a left coideal (respectively a right coideal, coideal, bi-ideal, subcoalgebra) of A.*

(c) *The sub-bialgebra of A generated by a subcoalgebra C of A is the subalgebra $<C> = k1 + C + C^2 + \cdots$ of A generated by C.*

(d) *Let I and J be left (respectively right, two-sided) ideals of A. Then $I \wedge J$ is a left (respectively right, two-sided) ideal of A.*

Proof. Part (b) follows from part (a) with $C = A$. Assuming part (a), the product of two subcoalgebras of A is a subcoalgebra of A. As the sum of subcoalgebras of A is again a subcoalgebra of A, part (c) follows from part (a).

Let U, V be subspaces of A. Since Δ and ϵ are multiplicative we have $\Delta(UV) \subseteq \Delta(U)\Delta(V)$ and $\epsilon(UV) \subseteq \epsilon(U)\epsilon(V)$. Part (a) follows from these equations. Since $I \wedge J = \Delta^{-1}(I \otimes A + A \otimes J)$ and Δ is multiplicative, part (d) follows. □

Part (c) of the preceding proposition has an interesting implication for the coradical of a sub-bialgebra of A generated by a subcoalgebra.

Corollary 5.1.12. *Let A be a bialgebra over the field k and suppose that C, D are subcoalgebras of A. Then:*

(a) $(CD)_0 \subseteq C_0 D_0$.

(b) $<C>_0 \subseteq <C_0>$.

(c) *If C generates A as an algebra then $A_0 \subseteq <C_0>$.*

Proof. To show part (a) we first note that CD is a subcoalgebra of A by part (a) of Proposition 5.1.11. Since the multiplication map $m : A\otimes A \longrightarrow A$ is a coalgebra map, and $C\otimes D$ is a subcoalgebra of $A\otimes A$, the restriction $m' = m|(C\otimes D)$ is an (onto) coalgebra map $m' : C\otimes D \longrightarrow CD$. We now apply parts (a) and (b) of Proposition 4.1.7 to conclude

$$(CD)_0 = m'(C\otimes D)_0 \subseteq m'((C\otimes D)_0) \subseteq m'(C_0\otimes D_0) = C_0D_0.$$

We have established part (a). To show part (b) use part (a) of Proposition 3.4.3 and part (a) to calculate

$$\begin{aligned} <C>_0 &= (k1 + C + C^2 + \cdots)_0 \\ &= (k1)_0 + C_0 + (C^2)_0 + \cdots \\ &\subseteq k1 + C_0 + (C_0)^2 + \cdots \\ &= <C_0>. \end{aligned}$$

Part (c) follows from part (b). □

There are important types of bialgebras defined in terms of their coradicals.

Definition 5.1.13. A *pointed* (respectively an *irreducible*) *bialgebra over* k is a bialgebra over k whose underlying coalgebra structure is pointed (respectively irreducible).

Thus A is a pointed bialgebra if and only if $A_0 = k[\mathrm{G}(A)]$. Since $k1$ is always a simple subcoalgebra of A it follows that A is irreducible if and only if $A_0 = k1$. In particular irreducible bialgebras over k are pointed irreducible as coalgebras.

The following characterization of the coradical of certain finitely generated bialgebras is useful in applications.

Corollary 5.1.14. *Suppose A is a bialgebra over k generated by $\mathcal{S}\cup\mathcal{P}$, where $\mathcal{S} \subseteq \mathrm{G}(A)$ and $\mathcal{P}$ consists of skew-primitives x which satisfy $\Delta(x) = s\otimes x + x\otimes s'$ for some $s, s' \in \mathcal{S}$. Then:*

(a) *A is pointed.*
(b) *$\mathrm{G}(A)$ is the multiplicative submonoid of A generated by $\mathcal{S}$.*

Proof. Let C be the span of $\mathcal{S}\cup\mathcal{P}$. Then $k[\mathcal{S}]$, C are the first two terms of a coalgebra filtration of C. Therefore $C_0 \subseteq k[\mathcal{S}]$ by Proposition 4.1.2. Since $k[\mathcal{S}] \subseteq C_0$ it follows that $C_0 = k[\mathcal{S}]$. By part (c) of Corollary 5.1.12 we are done. □

We next consider the grouplike and primitive elements of A and left, right multiplication by them. Recall that an associative algebra $\mathcal{A}$ has an associated Lie algebra structure defined by $[a, b] = ab - ba$ for all $a, b \in \mathcal{A}$. For subsets S, T of $\mathcal{A}$ we define $[S, T] = \{[s, t] \mid s \in S, t \in T\}$.

Proposition 5.1.15. *Let A be a bialgebra over the field k. Then:*

(a) *$\mathrm{G}(A)$ is a submonoid of A under multiplication. If $g \in \mathrm{G}(A)$ has a multiplicative inverse in A then $g^{-1} \in \mathrm{G}(A)$.*
(b) *Let $g \in \mathrm{G}(A)$. Then $\ell(g), \mathrm{r}(g) : A \longrightarrow A$ defined by (2.16) are coalgebra maps.*
(c) *Let $a, g, h \in \mathrm{G}(A)$. Then*
$$\ell(a)(\mathrm{P}_{g,h}(A)) \subseteq \mathrm{P}_{ag,ah}(A) \qquad \text{and} \qquad \mathrm{r}(a)(\mathrm{P}_{g,h}(A)) \subseteq \mathrm{P}_{ga,ha}(A).$$
(d) *Let $L = \mathrm{P}(A)$ be the coideal of primitive elements of A. Then $[L, L] \subseteq L$; thus L is a Lie subalgebra of the Lie algebra associated to A.*
(e) *Suppose that $a \in \mathrm{P}(A)$. Then $\ell(a)$ and $\mathrm{r}(a)$ are coderivations of A.*

Proof. Suppose that $g \in \mathrm{G}(A)$ has an inverse in A. The calculation
$$\Delta(g^{-1}) = (\Delta(g))^{-1} = (g \otimes g)^{-1} = g^{-1} \otimes g^{-1},$$
which follows since Δ is an algebra map, and the calculation
$$\epsilon(g^{-1}) = (\epsilon(g))^{-1} = 1,$$
which follows since ϵ is an algebra map, show that $g^{-1} \in \mathrm{G}(A)$. The remainder of the proof is an elementary exercise in definitions and is left to the reader. Coderivations are discussed in Exercise 2.3.11; see Definition 2.3.12. □

In later chapters we will develop the structure theory of pointed Hopf algebras. Here we take first steps in that direction by considering the multiplicative nature of the irreducible components of A. Our discussion is based on the material of Section 3.4.

Let A be any bialgebra over the field k and consider its underlying coalgebra structure. Let C be a simple subcoalgebra C of A. Since C is irreducible, by part (a) of Corollary 3.4.7 the subcoalgebra C is contained in a unique irreducible component of A which we denote by A_C. By the discussion preceding this corollary A_C is the sum of all subcoalgebras D of A such that $D_0 = C$. By Part (b) of Corollary 3.4.7 the sum of the irreducible components of the coalgebra A is direct. If A is cocommutative then A is the direct sum of its irreducible components by Theorem 4.8.8. See Exercise 4.8.5.

Proposition 5.1.16. *Let A be a bialgebra over the field k and for a simple subcoalgebra C of A let A_C denote the irreducible component of A which contains C.*

(a) *Suppose that C, D and the product CD are simple subcoalgebras of A. Then $A_C A_D \subseteq A_{CD}$.*
(b) *$A_{kg}A_{kg'} \subseteq A_{kgg'}$ for all $g, g' \in \mathrm{G}(A)$.*
(c) *Suppose that $g \in \mathrm{G}(A)$ is invertible and let C be a simple subcoalgebra of A. Then gC, Cg are simple subcoalgebras of A, $gA_C = A_{kg}A_C = A_{gC}$ and $(A_C)g = A_C A_{kg} = A_{Cg}$.*
(d) *Let $B = A_{k1}$. Then B is a sub-bialgebra of A and $BA_C = A_C = A_C B$ for all simple subcoalgebras C of A.*

Proof. Let C and D be subcoalgebras of A. By part (a) of Corollary 5.1.12 it follows that $(A_C A_D)_0 \subseteq (A_C)_0(A_D)_0 = CD$. Therefore if CD is a simple subcoalgebra of A also, the product $A_C A_D$ is contained in A_{CD}. We have established part (a). Part (b) follows from part (a) since $\mathrm{G}(A)$ is closed under multiplication by part (a) of Proposition 5.1.15.

To establish part (c) let f be any coalgebra automorphism of A. Since f takes subcoalgebras of A to subcoalgebras of A, and f is bijective, $f(C)$ is a simple subcoalgebra of A and $f(A_C) = A_{f(C)}$. Now suppose that $g \in \mathrm{G}(A)$ is invertible. Then $\ell(g)$ and $\mathrm{r}(g)$ are coalgebra automorphisms of A by parts (a) and (b) of Proposition 5.1.15. Thus part (c) follows from part (a) with $f = \ell(g), \mathrm{r}(g)$. Part (d) follows from part (c) with $g = 1$. □

We return to comment made after the definition of bialgebra, namely that bialgebras can be regarded as algebras or as coalgebras in certain categorical contexts. This comment foreshadows Chapter 11 and will be explained in Exercise 5.1.1 below. First two definitions.

Consider the category k-Coalg whose objects are coalgebras over k and whose morphisms are coalgebra maps under composition. For objects C, D we regard the tensor product $C \otimes D$ as an object of k-Coalg with the tensor product coalgebra structure. Observe that k is an object of k-Coalg where $\Delta(1) = 1 \otimes 1$ and $\epsilon(1) = 1$. It is easy to see that the canonical identifications $C \otimes k \simeq C$ and $k \otimes C \simeq C$ are in fact morphisms and that n-fold tensors products are objects of k-Coalg.

Definition 5.1.17. An *algebra in* k-Coalg is a triple (A, m, η), where A is an object of k-Coalg and $m : A \otimes A \longrightarrow A$ and $\eta : k \longrightarrow A$ are morphisms which satisfy (2.1) and (2.2).

Now consider the category k-Alg whose objects are algebras over k and whose morphisms are algebra maps under composition. For objects A, B we regard the tensor product $A \otimes B$ as an object of k-Alg with the tensor product coalgebra structure. Observe that k is an object of k-Alg. It is clear that the canonical identifications $A \otimes k \simeq A$ and $k \otimes A \simeq A$ are in fact morphisms and that n-fold tensors products are objects of k-Alg.

Definition 5.1.18. A *coalgebra in* k-Alg is a triple (C, Δ, ϵ), where C is an object of k-Coalg and $\Delta : C \longrightarrow C \otimes C$ and $\epsilon : C \longrightarrow k$ are morphisms which satisfy (2.3) and (2.4).

Exercises

In the following exercises A and B are bialgebras over the field k.

Exercise 5.1.1. Bialgebras can be considered as algebras or as coalgebras in certain categorical settings.

(a) Let (A, m, η) be an object of k-Alg, and suppose that $\Delta : A \longrightarrow A \otimes A$ and $\epsilon : A \longrightarrow k$ are linear maps. Show that $(A, m, \eta, \Delta, \epsilon)$ is a bialgebra over k if and only if $((A, m, \eta), \Delta, \epsilon)$ is a coalgebra in k-Alg.

(b) Let (C, Δ, ϵ) be an object of k-Coalg, and suppose that $m : C \otimes C \longrightarrow C$ and $\eta : k \longrightarrow C$ are linear maps. Show that $(C, m, \eta, \Delta, \epsilon)$ is a bialgebra over k if and only if $((C, \Delta, \epsilon), m, \eta)$ is an algebra in k-Coalg.

Exercise 5.1.2. Suppose $(A, m, \eta, \Delta, \epsilon)$ is a bialgebra over the field k and let $\tau : A \otimes A \longrightarrow A \otimes A$ be the "twist" map defined by $\tau(a \otimes b) = b \otimes a$ for all $a, b \in A$.

(a) Show that τ is an algebra and a coalgebra automorphism of $A \otimes A$, where the latter is given the tensor product algebra and coalgebra structure.

(b) Using the fact that $m^{op} = m \circ \tau$ and $\Delta^{op} = \tau \circ \Delta$, show that $(A, m^{op}, \eta, \Delta, \epsilon)$ and $(A, m, \eta, \Delta^{cop}, \epsilon)$ are bialgebras over k, and consequently $(A, m^{op}, \eta, \Delta^{cop}, \epsilon)$ is a bialgebra over k.

Exercise 5.1.3. Formulate and prove a fundamental homomorphism theorem for bialgebras. See Theorem 2.1.21.

Exercise 5.1.4. Show that:

(a) The composition of bialgebra maps is a bialgebra map;

(b) The inverse of a bialgebra isomorphism is a bialgebra isomorphism;

(c) The identity map of a bialgebra is a bialgebra map.

(In particular bialgebras over k and their bialgebra maps form a category. See Exercise 2.1.34.)

Definition 5.1.19. The category whose objects are bialgebras over the field k and whose morphisms are bialgebra maps under composition is denoted by k-Bialg.

Exercise 5.1.5. Show that:

(a) The subspace $\mathrm{Cc}(A)$ of the cocommutative elements of A is $\mathrm{Ker}(\Delta - \Delta^{cop})$ and is therefore a subalgebra of A. [Hint: If $f, g : \mathcal{A} \longrightarrow \mathcal{B}$ are algebra maps then $\mathrm{Ker}(f - g)$ is a subalgebra of $\mathcal{A}$.]

(b) The sum of all cocommutative subcoalgebras of A, which we denote by A_{cocom}, satisfies $A_{cocom} \subseteq \mathrm{Cc}(A)$ and is a sub-bialgebra of A.

(c) If $f : A \longrightarrow B$ is a bialgebra map then $f(A_{cocom}) \subseteq B_{cocom}$ and the restriction $f_{cocom} = f|A_{cocom}$ is a bialgebra map $f_{cocom} : A_{cocom} \longrightarrow B_{cocom}$.

Apropos of the last exercise, note that the assignments $A \mapsto A_{cocom}$ and $f \mapsto f_{cocom}$ determine a functor from k-Bialg to k-CcBialg, where the latter is the full subcategory of k-Bialg whose objects are the cocommutative bialgebras over k.

Exercise 5.1.6. Let $k[X]$ be the algebra of all polynomials in indeterminate X over k. Show that $k[X]$ has a bialgebra structure determined by $\Delta(X) = 1 \otimes X + X \otimes 1$ and $\epsilon(X) = 0$. [Hint: Show that the assignments $X \mapsto 1 \otimes X + X \otimes 1$ and $X \mapsto 0$ determine algebra maps $\Delta : k[X] \longrightarrow k[X] \otimes k[X]$ and $\epsilon : k[X] \longrightarrow k$. To show that $(k[X], \Delta, \epsilon)$ is a coalgebra observe that the compositions of algebra maps

$$(\Delta \otimes \mathrm{I}_{k[X]}) \circ \Delta,\ (\mathrm{I}_{k[X]} \otimes \Delta) \circ \Delta : k[X] \longrightarrow k[X] \otimes k[X] \otimes k[X]$$

agree on the algebra generator X as do the algebra maps

$$(\epsilon \otimes \mathrm{I}_{k[X]}) \circ \Delta,\ \mathrm{I}_{k[X]},\ (\mathrm{I}_{k[X]} \otimes \epsilon) \circ \Delta : k[X] \longrightarrow k.]$$

Exercise 5.1.7. Let $A = k[X]$ be the bialgebra of Exercise 5.1.6.

(a) Show that $\mathrm{G}(A) = \{1\}$.

(b) Suppose that the characteristic of k is zero. Show that $\mathrm{P}(A)$ has basis $\{X\}$.

(c) Suppose that the characteristic of k is $p > 0$. Show that $\mathrm{P}(A)$ has basis $\{X^{p^n} \mid n \geq 0\}$. [Hint: Since the characteristic of k is $p > 0$, for all commuting a, b in an algebra over k the equation $(a + b)^p = a^p + b^p$ holds.]

Exercise 5.1.8. Let S be a monoid and $H = k[S]$ be the monoid algebra of S over k. Suppose that K is a sub-bialgebra of H. Show that $K = k[T]$ for some sub-monoid T of S. [Hint: See Exercise 2.1.1 and Lemma 2.1.12.]

Exercise 5.1.9. Let C be the coalgebra over k with basis $\{c_0, c_1, c_2, \ldots\}$ whose coalgebra structure is determined by $\Delta(c_n) = \sum_{\ell=0}^{n} c_{n-\ell} \otimes c_\ell$ and $\epsilon(c_n) = \delta_{0,n}$ for all $n \geq 0$. See Example 2.1.13 and the ensuing discussion.

(a) Show that C is a bialgebra with this coalgebra structure and multiplication defined in terms of the usual binomial symbols by

$$c_m c_n = \binom{m+n}{n} \cdot c_{m+n}$$

for all $m, n \geq 0$; in particular the unity 1 of C is c_0.

(b) Show that $\mathrm{G}(C) = \{1\}$ and that $\{c_1\}$ is a basis for $\mathrm{P}(C)$.

(c) Show that C and the bialgebra of Exercise 5.1.6 are isomorphic if and only if the characteristic of k is zero.

We will make the following definition of simple bialgebra.

Definition 5.1.20. A *simple bialgebra over* k is a bialgebra over a field k which has exactly two bi-ideals, namely (0) and $\mathrm{Ker}(\epsilon)$.

Exercise 5.1.10. Suppose that $x \in \mathrm{P}(A)$ and $x \neq 0$. Show that:

(a) The subalgebra $k[x]$ of A generated by x is a cocommutative sub-bialgebra of A. [Hint: To show that $k[x]$ is cocommutative, let τ be the algebra automorphism of $k[x]$ defined by $\tau(a \otimes b) = b \otimes a$ for all $a, b \in k[x]$. Note that the algebra maps $\Delta^{cop} = \tau \circ \Delta$, $\Delta : k[x] \longrightarrow k[x] \otimes k[x]$ agree on the algebra generator x for $k[x]$.]

(b) $\Delta(x^n) = \sum_{\ell=0}^{n} \binom{n}{\ell} \cdot x^{n-\ell} \otimes x^{\ell}$ and $\epsilon(x^n) = \delta_{n,0}$ for all $n \geq 0$. [Hint: $\Delta(x^n) = (\Delta(x))^n$ for all $n \geq 0$.]

(c) If the characteristic of k is $p > 0$ then $x^{p^n} \in \mathrm{P}(A)$ for all $n \geq 0$. [Hint: See part (c) of Exercise 5.1.7.]

Suppose that k has characteristic zero. Show that:

(d) x is an indeterminate over k. (Thus the bialgebra $k[x]$ is the bialgebra of Exercise 5.1.6 in this case.)

(e) $k[x]$ is a simple bialgebra. [Hint: Let I be a bi-ideal of $k[x]$ and consider the quotient bialgebra $k[x]/I$.]

Exercise 5.1.11. Show that the algebra $k[X]$ of all polynomials in indeterminate X over k has a bialgebra structure determined by $\Delta(X) = X \otimes X$ and $\epsilon(X) = 1$, and show that $k[X]$ with this bialgebra structure is the bialgebra of Example 5.1.4, where S is the free monoid on a singleton set.

Exercise 5.1.12. Let $f : A \longrightarrow B$ be a bialgebra map. Show that:

(a) If J is a bi-ideal (respectively a sub-bialgebra) of B then $f^{-1}(J)$ is a bi-ideal (respectively a sub-bialgebra) of A;

(b) If I is a sub-bialgebra of A then $f(I)$ is a sub-bialgebra of B;

(c) If I is a bi-ideal of A then $f(I)$ is a bi-ideal of $f(A)$.

See Exercise 2.1.28.

Exercise 5.1.13. Let K be a field extension of k. Recall that $K \otimes A$ is a K-algebra where $(\alpha \otimes a)(\beta \otimes b) = \alpha\beta \otimes ab$ for all $\alpha, \beta \in K$ and $a, b \in A$. Show that:

(a) $K \otimes A$ with the algebra structure just described and the coalgebra structure of Exercise 2.1.36 is a K-bialgebra.

(b) If $f : A \longrightarrow B$ is a map of k-bialgebras then $\mathrm{I}_K \otimes f : K \otimes A \longrightarrow K \otimes B$ is a map of K-bialgebras.

(c) F : k-Bialg $\longrightarrow$ K-Bialg given by $F(A) = K \otimes A$ and $F(f) = \mathrm{I}_K \otimes f$ is a functor.

Exercise 5.1.14. The tensor product of bialgebras A and B over k has a natural bialgebra structure. Show that:

(a) The tensor product of vector spaces $A \otimes B$ with the tensor product coalgebra and algebra structures is a bialgebra over k. See Exercise 2.1.19.

(b) $(A \otimes B)^{op} = A^{op} \otimes B^{op}$, $(A \otimes B)^{cop} = A^{cop} \otimes B^{cop}$, and $(A \otimes B)^{op\,cop} = A^{op\,cop} \otimes B^{op\,cop}$.

(c) $\pi_A : A \otimes B \longrightarrow A$ and $\pi_B : A \otimes B \longrightarrow B$ defined by $\pi_A(a \otimes b) = a\epsilon(b)$ and $\pi_B(a \otimes b) = \epsilon(a)b$ for all $a \in A$ and $b \in B$ respectively are bialgebra maps.

(d) If $f : A \longrightarrow A'$ and $g : B \longrightarrow B'$ are bialgebra maps then the tensor product map $f \otimes g : A \otimes B \longrightarrow A' \otimes B'$ is a bialgebra map.

Definition 5.1.21. The bialgebra structure described in part (a) of Exercise 5.1.14 is the *tensor product bialgebra structure on* $A \otimes B$.

Exercise 5.1.15. Let C also be a bialgebra over k. Show that the linear isomorphisms $(A \otimes B) \otimes C \simeq A \otimes (B \otimes C)$ and $A \otimes B \simeq B \otimes A$ described by $(a \otimes b) \otimes c \mapsto a \otimes (b \otimes c)$ and $a \otimes b \mapsto b \otimes a$ for all $a \in A$, $b \in B$, and $c \in C$ respectively are in fact bialgebra isomorphisms. See Exercise 2.1.20.

Exercise 5.1.16. Suppose that A is finite-dimensional and that $\{a_i\}_{i \in I}$ is a basis for A. Write the structure constants for A as follows:

$$a_i a_j = \sum_{\ell \in I} m_{i,j}^{\ell} a_\ell, \qquad \eta(1) = \sum_{\ell \in I} \eta^{\ell} a_i$$

for all $i, j \in I$, and

$$\Delta(a_i) = \sum_{j,\ell \in I} \Delta_i^{j,\ell}(a_j \otimes a_\ell), \qquad \epsilon(a_i) = \epsilon_i$$

for all $i \in I$. Express the bialgebra axioms for A in terms of the structure constants using the Einstein summation convention. See Exercises 2.1.22 and 2.5.18.

Exercise 5.1.17. Suppose that C is a subcoalgebra of A which generates A as an algebra. Show that:

(a) A is cocommutative if and only if C is cocommutative.

(b) A is pointed if and only if C is pointed.

[Hint: For part (a), recall that Δ and Δ^{op} are algebra maps. See Exercise 5.1.2.]

Exercise 5.1.18. Let V be a subspace of A. Show that the sub-bialgebra of A generated by V is $<C>$, where C is the subcoalgebra of A generated by V.

Definition 5.1.22. A bialgebra A over the field k is *finitely generated as bialgebra* if there is a finite-dimensional subspace V of A such that the smallest sub-bialgebra which contains V is A itself.

Exercise 5.1.19. Prove that the following are equivalent:

(a) A is finitely generated as a bialgebra.

(b) A is finitely generated as an algebra.

Exercise 5.1.20. Suppose that A is pointed, cocommutative, $\mathrm{G}(A)$ is a group under multiplication, and consider the sub-bialgebras $H = k[\mathrm{G}(A)]$ and $B = A_{k1}$ of A. Show that:

(a) The map $H\otimes B \longrightarrow A$ defined by $h\otimes b \mapsto hb$ for all $h \in H$ and $b \in B$ is an isomorphism of coalgebras.

(b) If A is commutative then the coalgebra isomorphism of part (a) is a bialgebra isomorphism.

Exercise 5.1.21. Let U, V be subspaces of A and suppose that C is a subcoalgebra of A. Show that:

(a) $C(U\wedge V) \subseteq CU\wedge CV$ and $(U\wedge V)C \subseteq UC\wedge VC$.

(b) If $g \in \mathrm{G}(A)$ then $g(U\wedge V) \subseteq gU\wedge gV$ and $(U\wedge V)g \subseteq Ug\wedge Vg$.

(c) If $UC \subseteq U$ then $U^{(n)}C \subseteq U^{(n)}$ for all $n \geq 0$.

(d) If $CU \subseteq U$ then $CU^{(n)} \subseteq U^{(n)}$ for all $n \geq 0$.

Exercise 5.1.22. Complete the proof of Proposition 5.1.15.

5.2 The dual bialgebra

The discussion of this section draws heavily from the material from Section 2.6 which we shall use without particular reference. Let A be a bialgebra over k. The underlying algebra structure on A accounts for the dual coalgebra A^o. Likewise the underlying coalgebra structure on A accounts for the dual algebra structure on A^*. A very natural question to ask at this point is whether or not A^o is a subalgebra of A^*.

Proposition 5.2.1. *Let A be a bialgebra over the field k. Then:*

(a) *The dual coalgebra A^o is also a subalgebra of the dual algebra A^*.*
(b) *A^o is a bialgebra over k with the dual coalgebra structure and the subalgebra structure of the dual algebra A^*.*

Proof. Since ϵ is an algebra map Δ_ϵ exists and $\Delta_\epsilon = \epsilon \otimes \epsilon$. Let $a^o, b^o \in A^o$. The calculation

$$\begin{aligned}(a^o_{(1)}b^o_{(1)})(a)(a^o_{(2)}b^o_{(2)})(b) &= a^o_{(1)}(a_{(1)})b^o_{(1)}(a_{(2)})a^o_{(2)}(b_{(1)})b^o_{(2)}(b_{(2)}) \\ &= a^o(a_{(1)}b_{(1)})b^o(a_{(2)}b_{(2)}) \\ &= a^o((ab)_{(1)})b^o((ab)_{(2)}) \\ &= (a^o b^o)(ab)\end{aligned}$$

for all $a, b \in A$ shows that $\Delta_{a^o b^o}$ exists and $\Delta_{a^o b^o} = a^o_{(1)}b^o_{(1)} \otimes a^o_{(2)}b^o_{(2)}$. Therefore $a^o b^o \in A^o$.

We have shown that A^o is a subalgebra of A^*. It is a small step to complete the proof at this point. □

Definition 5.2.2. Let A be a bialgebra A over a field k. The *dual bialgebra of* A is the dual coalgebra A^o regarded as a subalgebra of the dual algebra A^*.

Observe that A^o is commutative if A is cocommutative, and that A^o is cocommutative if A is commutative. See Exercise 5.2.6. In the previous section we noted that twisting the product and/or coproduct of A gives rise to the bialgebras A^{op}, A^{cop}, and $A^{op\ cop}$ over k. Therefore $(A^o)^{op} = (A^{cop})^o, (A^o)^{cop} = (A^{op})^o$, and $(A^o)^{op\ cop} = (A^{cop\ op})^o$ are also bialgebras over k; thus A formally gives rise to eight bialgebras, including itself.

Now suppose that $f : A \longrightarrow B$ is a bialgebra map and set $f^o = f^*|B^o$. Since f is an algebra map, $f^o(B^o) \subseteq A^o$ and $f^o : B^o \longrightarrow A^o$ is a coalgebra map by Proposition 2.5.4. Since f is a coalgebra map, $f^* : B^* \longrightarrow A^*$

is an algebra map by Proposition 2.3.3. Therefore $f^o : B^o \longrightarrow A^o$ is an algebra map as well. We have shown:

Proposition 5.2.3. *Let $f : A \longrightarrow B$ be a bialgebra map and let $f^o = f^*|B^o$. Then $f^o(B^o) \subseteq A^o$ and $f^o : B^o \longrightarrow A^o$ is a bialgebra map.* □

We denote $(A^o)^o$ by A^{oo} and for a bialgebra map $f : A \longrightarrow B$ we denote $(f^o)^o$ by f^{oo}. There is a natural relationship between A and its double dual A^{oo}.

Proposition 5.2.4. *Suppose A is a bialgebra over the field k and suppose $\jmath_A : A \longrightarrow (A^o)^*$ is the linear map defined by $\jmath(a)(a^o) = a^o(a)$ for all $a \in A$ and $a^o \in A^o$. Then:*

(a) $\mathrm{Im}(\jmath_A) \subseteq A^{oo}$ *and $\jmath_A : A \longrightarrow A^{oo}$ is a bialgebra map.*
(b) $\mathrm{Ker}(\jmath_A)$ *is the intersection of the cofinite ideals of A.*
(c) $\mathrm{Im}(\jmath_A)$ *is a dense subspace of $(A^o)^*$.*

Proof. Regarding A as an algebra over k, we note that $\jmath_A : A \longrightarrow (A^o)^*$ is an algebra map and that parts (b), (c) hold by Proposition 2.6.1. We show part (a).

By Proposition 2.5.4 the inclusion $\imath : A^o \longrightarrow A^*$, which is an algebra map, induces a coalgebra map $\imath^o : (A^*)^o \longrightarrow A^{oo}$, defined by $\imath^o(\alpha) = \alpha \circ \imath = \alpha|A^o$ for all $\alpha \in (A^*)^o$. Regard A as a coalgebra. Then $\imath_A : A \longrightarrow (A^*)^o$, defined by $\imath(a)(a^o) = a^o(a)$ for all $a \in A$ and $a^o \in A^o$, is a coalgebra map by part (a) of Proposition 2.6.2. Since $\jmath_A = \imath^o \circ \imath_A$, it follows that $\mathrm{Im}(\jmath_A) \subseteq A^{oo}$ and that $\jmath_A$ is a coalgebra map also. □

We end this section by relating the bialgebra structures of A and A^o.

Proposition 5.2.5. *Let A be a bialgebra over the field k. Then:*

(a) *If U is a subspace of A which is a coideal (respectively subcoalgebra, left coideal, right coideal) of A then $U^\perp \cap A^o$ is a subalgebra (respectively ideal, left ideal, right ideal) of A^o.*
(b) *If I is a subspace of A which is a subalgebra (respectively ideal, left ideal, right ideal) of A then $I^\perp \cap A^o$ is a coideal (respectively subcoalgebra, left coideal, right coideal) of A^o.*
(c) *If V is a subspace of A^o which is a coideal (respectively subcoalgebra, left coideal, right coideal) of A^o then $V^\perp$ is a subalgebra (respectively ideal, left ideal, right ideal) of A.*

(d) *Suppose that A is a proper algebra. If J is a subspace of A^o which is a subalgebra (respectively ideal, left ideal, right ideal) of A^o then $J^\perp$ coideal (respectively subcoalgebra, left coideal, right coideal) of A.*

Proof. Since A^o is a subalgebra of the dual algebra A^* of the underlying coalgebra structure of A, part (a) follows by part (b) of Proposition 2.3.7. Part (b) follows by part (a) of Proposition 2.6.4 and part (c) follows by part (a) of the same.

Suppose that A is proper and let $\jmath : A \longrightarrow (A^o)^o$ be the bialgebra map of Proposition 5.2.4. For a subspace J of A^o let $J^{\perp} = \{\alpha \in (A^o)^* \mid \alpha(J) = (0)\}$. Since $J^\perp = \jmath^{-1}(J^{\perp})$, part (d) follows by part (b) and Exercise 2.1.8. ∎

Exercises

Throughout the following exercises A and B are bialgebras over k.

Exercise 5.2.1. Show that $\mathrm{G}(A^o) = \mathrm{Alg}\,(A, k)$ is multiplicative submonoid of A^*.

Exercise 5.2.2. Show that $(A{\otimes}B)^o \simeq A^o{\otimes}B^o$ as bialgebras, where the tensor products have the tensor product bialgebra structures.

Exercise 5.2.3. Suppose that A is finite-dimensional. Show that A is simple if and only if the only sub-bialgebras of A^* are $k\epsilon$ and A^*.

Exercise 5.2.4. Here we explore double duals of bialgebras in more detail.

(a) Let $f : A \longrightarrow B$ be a bialgebra map. Show that $f^{oo} : A^{oo} \longrightarrow B^{oo}$ is a bialgebra map.

(b) Show that $F : k\text{-Bialg} \longrightarrow k\text{-Bialg}$ given by $F(A) = A^o$ and $F(f) = f^o$ is a contravariant functor.

(c) Show that $G : k\text{-Bialg} \longrightarrow k\text{-Bialg}$ given by $G(A) = A^{oo}$ and $G(f) = f^{oo}$ is a functor.

(d) Establish the adjoint relation

$$\mathrm{Hom}_{\mathrm{Bialg}}(A, F(B)) \simeq \mathrm{Hom}_{\mathrm{Bialg}}(B, F(A))$$

given by $f \mapsto f^o \circ \jmath_B$.

Let $k\text{-Bialg}_{fd}$ be the full sub-category of k-Bialg whose objects are finite-dimensional bialgebras over k.

(e) Show that F restricts to an anti-categorical equivalence of $k\text{-Bialg}_{fd}$ with itself.

See Exercise 2.5.20.

Exercise 5.2.5. Let $A = k[X]$ be the bialgebra of Exercise 5.1.6; thus $\Delta(X) = 1\otimes X + X\otimes 1$ and $\epsilon(X) = 0$. Let $\jmath_A : A \longrightarrow A^{oo}$ be the bialgebra map of Proposition 5.2.4; thus $\jmath_A(a)(a^o) = a^o(a)$ for all $a \in A$ and $a^o \in A^o$.

(a) Show that $\jmath_A$ is one-one.

(b) Let $\mathcal{H} = k[\mathrm{G}(A^o)]$ and $\mathcal{B} = (A^o)_{k\epsilon}$ be the irreducible component of A^o which contains $k\epsilon$. Show that multiplication $\mathcal{H}\otimes\mathcal{B} \longrightarrow A^o$ is an isomorphism of bialgebras. [Hint: See Exercise 5.1.20.]

(c) Show that $\jmath_A$ is not onto.

Exercise 5.2.6. Show that:

(a) A commutative (respectively cocommutative) implies that A^o is cocommutative (respectively commutative).

(b) If A^o is a dense subspace of A^* then A^o cocommutative (respectively commutative) implies that A is commutative (respectively cocommutative).

5.3 The free bialgebra on a coalgebra and related constructions

Recall that the tensor algebra, or the free algebra, on a vector space V over k is a pair $(\iota, T(V))$ which satisfies the following universal mapping property:

(FA.1) $T(V)$ is an algebra over k and $\iota : V \longrightarrow T(V)$ is linear map.
(FA.2) If A is an algebra over k and $f : V \longrightarrow A$ is a linear map there is an algebra map $F : T(V) \longrightarrow A$ determined by $F\circ\iota = f$.

Any two free algebras on V are isomorphic. More precisely, suppose that $(\iota, T(V))$ and $(\iota', T(V)')$ are pairs which satisfy the universal mapping property described above. Then there is an isomorphism of algebras $F : T(V) \longrightarrow T(V)'$ determined by $F\circ\iota = \iota'$. See Exercise 5.3.1. Since any two free algebras on a vector space V over k are isomorphic, we shall refer to a free algebra as the free algebra. Generally objects satisfying a particular universal mapping property are isomorphic and we do not distinguish them.

There is a commutative analog of the free algebra on a vector space. Let V be a vector space over k and suppose that $(\iota, T(V))$ is the free algebra on V. Let I be the ideal of $T(V)$ generated by $[T(V), T(V)]$. Then the quotient algebra $S(V) = T(V)/I$ is commutative. Let $\jmath : V \longrightarrow S(V)$ be $\jmath = \pi\circ\iota$, where $\pi : T(V) \longrightarrow T(V)/I$ is the projection. The pair $(\jmath, S(V))$

satisfies (FA.1) and (FA.2) with "algebra over k" replaced by "commutative algebra over k". Any such pair is called a free commutative algebra on V. As in the case of free algebras on V, if $(\jmath, S(V))$ and $(\jmath', S(V)')$ are free commutative algebras over k then there exists an algebra isomorphism $F : S(V) \longrightarrow S(V)'$ determined by $F\circ\jmath = \jmath'$. See Exercises 5.3.1 and 5.3.3.

The algebra structure on $T(V)$ extends to a bialgebra structure and $S(V)$ is a quotient bialgebra. See Exercise 5.3.9.

Let (C, δ, e) be a coalgebra over the field k, and let $(\iota, T(C))$ be the tensor algebra on the vector space C. Thus the linear maps $e : C \longrightarrow k$ and $(\iota\otimes\iota)\circ\delta : C \longrightarrow T(C)\otimes T(C)$ give rise to algebra maps $\epsilon : T(C) \longrightarrow k$ and $\Delta : T(C) \longrightarrow T(C)\otimes T(C)$ which are determined by $\epsilon\circ\iota = e$ and $\Delta\circ\iota = (\iota\otimes\iota)\circ\delta$ respectively. To show that $T(C)$ is a bialgebra with these structures is a matter of showing that $(T(C), \Delta, \epsilon)$ is a coalgebra. It is an easy exercise to show that the algebra maps $(\mathrm{I}_{T(C)}\otimes\Delta)\circ\Delta$ and $(\Delta\otimes\mathrm{I}_{T(C)})\circ\Delta$ agree on $\iota(C)$. Since $\iota(C)$ generates $T(C)$ as an algebra over k it follows that $(\mathrm{I}_{T(C)}\otimes\Delta)\circ\Delta = (\Delta\otimes\mathrm{I}_{T(C)})\circ\Delta$. For the same reasons $(\epsilon\otimes\mathrm{I}_{T(C)})\circ\Delta = \mathrm{I}_{T(C)} = (\mathrm{I}_{T(C)}\otimes\epsilon)\circ\Delta$.

The pair $(\iota, T(C))$ satisfies a universal mapping property.

Theorem 5.3.1. *Suppose that C is a coalgebra over the field k. Then the pair $(\iota, T(C))$ defined above satisfies:*

(a) *$T(C)$ is a bialgebra over k and $\iota : C \longrightarrow T(C)$ is a coalgebra map.*
(b) *If A is a bialgebra over k and $f : C \longrightarrow A$ is a coalgebra map there exists a bialgebra map $F : T(C) \longrightarrow A$ determined by $F\circ\iota = f$.*

Proof. In light of our comments preceding the statement of the theorem, we need only justify part (b). Suppose that A is a bialgebra over k and suppose that $f : C \longrightarrow A$ is a coalgebra map. By the universal mapping property of the tensor algebra of the vector space C there is an algebra map $F : T(C) \longrightarrow A$ determined by $F\circ\iota = f$. To see that F is a coalgebra map we observe that the algebra maps $(F\otimes F)\circ\Delta$ and $\Delta_A\circ F$ agree on $\iota(C)$ which generates $T(C)$ as an algebra over k. Thus $(F\otimes F)\circ\Delta = \Delta_A\circ F$; likewise $\epsilon_A\circ F = \epsilon$. □

Definition 5.3.2. Let C be a coalgebra over the field k. A *free bialgebra on the coalgebra C*, or a *tensor bialgebra on the coalgebra C, over k*, is a pair $(\iota, T(C))$ which satisfies the conclusion of Theorem 5.3.1.

For uniqueness of the free bialgebra on a coalgebra C over k see Exercise 5.3.4. Note that ι is one-one. Thus C can be regarded as a subcoalgebra of

$T(C)$; see Exercise 5.3.5.

Every bialgebra is a quotient of a free bialgebra.

Corollary 5.3.3. *Let A be a bialgebra over the field k. Then there is a coalgebra C over k and an onto bialgebra map $F : T(C) \longrightarrow A$.*

Proof. Let C be a subcoalgebra of A which generates A as an algebra, for example $C = A$, and let $f : C \longrightarrow A$ be the inclusion. Then f is a coalgebra map. Thus there exists a bialgebra map $F : T(C) \longrightarrow A$ such that $F \circ \iota = f$ by Theorem 5.3.1. Since F is an algebra map and $C = \mathrm{Im}(f) \subseteq \mathrm{Im}(F)$, necessarily F is onto. □

Suppose that A is a bialgebra over k which is finitely generated as an algebra. Let V be a finite-dimensional subspace of A which generates A as an algebra over k. Then V is contained in a finite-dimensional subcoalgebra C of A by Theorem 2.2.3. We may take the subcoalgebra C to be the coalgebra of Corollary 5.3.3. Better yet, we may take the coalgebra of the preceding corollary to be a comatrix coalgebra.

By Corollary 2.2.2 there is coalgebra map $\pi : \mathrm{C}_n(k) \longrightarrow C$ which is onto for some $n > 0$. Regard π as a coalgebra map to A. Since the image of a bialgebra map $f : A \longrightarrow B$ is a sub-bialgebra of B, it now follows by Theorem 5.3.1 that:

Corollary 5.3.4. *Let A be a bialgebra over the field k which is finitely generated as an algebra over k. Then A is the homomorphic image of the free bialgebra $T(\mathrm{C}_n(k))$ on the comatrix coalgebra $\mathrm{C}_n(k)$ for some $n > 0$.*

□

By Corollary 5.3.3 every bialgebra over k is the quotient of a free bialgebra on a coalgebra C over k. Many bialgebras are constructed explicitly as quotients. We go through a typical construction in great detail, demonstrating some useful techniques along the way.

We will construct a 4-dimensional bialgebra A over k which is completely determined by the following data: A is generated an algebra over k by g, x subject to the relations

$$g^2 = 1, \qquad x^2 = 0, \qquad xg = -gx \tag{5.1}$$

and the coalgebra structure of A is determined by

$$\Delta(g) = g \otimes g, \qquad \Delta(x) = x \otimes g + 1 \otimes x. \tag{5.2}$$

Observe the relation $g^2 = 1$ implies that $g \neq 0$. Thus $\Delta(g) = g \otimes g$ implies that $\epsilon(g) = 1$. It now follows that $\epsilon(x) = 0$. Since Δ and ϵ are algebra maps, their affect on the generators g, x determine them.

To begin we let C be a 3-dimensional vector space k with basis of formal symbols $\{G, U, X\}$. We endow C with the coalgebra structure (C, δ, e) determined by

$$\delta(G) = G \otimes G, \qquad \delta(U) = U \otimes U, \qquad \delta(X) = X \otimes G + U \otimes X.$$

Necessarily $e(G) = e(U) = 1$ and $e(X) = 0$. Let $(\iota, T(C))$ be the tensor bialgebra on the coalgebra C and denote the coalgebra structure on $T(C)$ by $(T(C), \boldsymbol{\Delta}, \epsilon)$. By Exercise 5.3.5 the coalgebra map ι is one-one. Thus we may assume that C is a subcoalgebra of $T(C)$, which we will do for notational convenience.

Let I be the ideal of $T(C)$ generated by

$$U - 1, \qquad G^2 - 1, \qquad X^2, \qquad XG + GX.$$

Consider the quotient algebra $T(C)/I$ and let $\pi : T(C) \longrightarrow T(C)/I$ be the projection. Note that $1 = \pi(U)$ and set $g = \pi(G)$ and $x = \pi(X)$. Observe that g, x generate $T(C)/I$ as an algebra over k and they satisfy the relations of (5.1).

Suppose that I is in fact a bi-ideal of $T(C)$. Then $T(C)/I$ is a bialgebra and π is a bialgebra map. Thus the coproduct of $T(C)/I$ satisfies (5.2). If, in addition, $\mathrm{Dim}(T(C)/I) = 4$ then $A = T(C)/I$ is the desired bialgebra. We will show that I is a bi-ideal of $T(C)$ and then show that $\mathrm{Dim}(T(C)/I) = 4$.

Since I is an ideal of $T(C)$, to show that I is a bi-ideal of $T(C)$ we need only show that I is a coideal of $T(C)$. Establishing that I is a coideal of $T(C)$ is one of the more interesting steps of the construction.

Since $\epsilon(U) = \epsilon(G) = 1$, $\epsilon(X) = 0$, and ϵ is an algebra map, it follows that $\epsilon(I) = (0)$. To show that I is a coideal of $T(C)$ it remains to show that $\boldsymbol{\Delta}(I) \subseteq I \otimes T(C) + T(C) \otimes I$.

Now the composite $(\pi \otimes \pi) \circ \boldsymbol{\Delta} : T(C) \longrightarrow T(C)/I \otimes T(C)/I$ is an algebra map, since $\boldsymbol{\Delta}$ and π are algebra maps, and

$$\mathrm{Ker}((\pi \otimes \pi) \circ \boldsymbol{\Delta}) = \boldsymbol{\Delta}^{-1}(\mathrm{Ker}(\pi \otimes \pi)) = \boldsymbol{\Delta}^{-1}(I \otimes T(C) + T(C) \otimes I)$$

is an ideal of $T(C)$. Therefore to show that I is a coideal of $T(C)$ we need only show that $(\pi \otimes \pi) \circ \boldsymbol{\Delta}$ vanishes on the generators of I. The calculations

$$\begin{aligned} ((\pi \otimes \pi) \circ \boldsymbol{\Delta})(U - 1) &= (\pi \otimes \pi)(\boldsymbol{\Delta}(U) - \boldsymbol{\Delta}(1)) \\ &= (\pi \otimes \pi)(U \otimes U - 1 \otimes 1) \\ &= 1 \otimes 1 - 1 \otimes 1 \\ &= 0, \end{aligned}$$

$$\begin{aligned}((\pi\otimes\pi)\circ\boldsymbol{\Delta})(G^2-1) &= (\pi\otimes\pi)(\boldsymbol{\Delta}(G^2)-\boldsymbol{\Delta}(1))\\ &= (\pi\otimes\pi)((G\otimes G)^2-1\otimes 1)\\ &= g^2\otimes g^2-1\otimes 1\\ &= 1\otimes 1-1\otimes 1\\ &= 0,\end{aligned}$$

$$\begin{aligned}((\pi\otimes\pi)\circ\boldsymbol{\Delta})(X^2) &= (\pi\otimes\pi)(\boldsymbol{\Delta}(X^2))\\ &= (\pi\otimes\pi)((X\otimes G+U\otimes X)^2)\\ &= (x\otimes g+1\otimes x)^2\\ &= x^2\otimes g^2+x\otimes(gx+xg)+1\otimes x^2\\ &= 0,\end{aligned}$$

and

$$\begin{aligned}&((\pi\otimes\pi)\circ\boldsymbol{\Delta})(XG+GX)\\ &= (\pi\otimes\pi)(\boldsymbol{\Delta}(XG)+\boldsymbol{\Delta}(GX))\\ &= (\pi\otimes\pi)((X\otimes G+U\otimes X)(G\otimes G)+(G\otimes G)(X\otimes G+U\otimes X))\\ &= (x\otimes g+1\otimes x)(g\otimes g)+(g\otimes g)(x\otimes g+1\otimes x)\\ &= (xg+gx)\otimes g^2+g\otimes(xg+gx)\\ &= 0\end{aligned}$$

show that $(\pi\otimes\pi)\circ\boldsymbol{\Delta}$ vanishes on the generators of I. Therefore I is a coideal of $T(C)$ and consequently $T(C)/I$ is a bialgebra.

It remains to show that $\mathrm{Dim}(T(C)/I)=4$. Since $\{g,x\}$ generates $T(C)/I$, the relations of (5.1) imply that $\mathrm{Dim}(T(C)/I)\leq 4$. Thus we need only show that $\mathrm{Dim}(T(C)/I)\geq 4$. To this end we need only find a 4-dimensional algebra $\mathcal{A}$ over k and an onto algebra map $F:T(C)\longrightarrow\mathcal{A}$ such that $I\subseteq\mathrm{Ker}(f)$.

Let $\mathcal{A}$ be the vector space over k with basis $\{(i,j)\,|\,i,j\in\boldsymbol{Z}_2\}$. It is easy to see that $\mathcal{A}$ is an algebra over k where

$$(i,j)\cdot(i',j')=(-1)^{ji'}\delta_{0,jj'}(i+i',j+j')$$

for all $i,j,i',j'\in\boldsymbol{Z}_2$. Regarding $T(C)$ as the free algebra over the vector space C, the assignments $U\mapsto 1$, $G\mapsto(1,0)$ and $X\mapsto(0,1)$ define a linear map $f:C\longrightarrow\mathcal{A}$ which determines an algebra map $F:T(C)\longrightarrow\mathcal{A}$ with the desired property.

There is another method of showing that $\mathrm{Dim}(T(C)/I)=4$ which does not involve an explicit construction of a 4-dimensional algebra $\mathcal{A}$ over k

generated by g, x which satisfy the relations (5.1). This method uses the Diamond Lemma of [21]. This important method will be illustrated in detail in Section 7.3.

The bialgebra we have constructed is the first in a series of fundamental examples which will be discussed in Section 7.3. We will give it a name for easy reference.

Definition 5.3.5. The 4-dimensional bialgebra over k with generators g, x whose structure is determined by (5.1) and (5.2) is denoted $H_{2,-1}(k)$.

We close with a brief discussion of the free commutative bialgebra on a coalgebra C over k. Justification of the following assertions is outlined in Exercises 5.3.8 and 5.3.12.

Let $(\iota, T(C))$ be the free bialgebra on the coalgebra C and let I be the ideal of $T(C)$ generated by $[\iota(C), \iota(C)]$. Then I is a bi-ideal of $T(C)$, the quotient $T(C)/I$ is a commutative bialgebra over k, and the composition $\jmath = \pi \circ \iota$, where $\pi : T(C) \longrightarrow T(C)/I$ is the projection, is a coalgebra map $\jmath : C \longrightarrow T(C)/I$. Set $S(C) = T(C)/I$. Then the pair $(\jmath, S(C))$ satisfies the following universal mapping property:

Theorem 5.3.6. *Suppose that C is a coalgebra over the field k. Then the pair $(\jmath, S(C))$ defined above satisfies:*

(a) *$S(C)$ is a commutative bialgebra over k and $\jmath : C \longrightarrow S(C)$ is a coalgebra map.*
(b) *If A is a commutative bialgebra over k and $f : C \longrightarrow A$ is a coalgebra map there exists a bialgebra map $F : S(C) \longrightarrow A$ determined by $F \circ \jmath = f$.*

□

Definition 5.3.7. Let C be a coalgebra over the field k. A *free commutative bialgebra on the coalgebra C* is any pair $(\jmath, S(C))$ which satisfies the conclusion of Theorem 5.3.6.

The map $\jmath$ is one-one which means that we may regard C as a subcoalgebra of $S(C)$. See Exercise 5.3.3.

The proofs of Corollaries 5.3.3 and 5.3.4 can be modified to give:

Corollary 5.3.8. *Let A be a commutative bialgebra over the field k. Then there is a coalgebra over k and an onto bialgebra map $F : S(C) \longrightarrow A$.* □

Corollary 5.3.9. *Let A be a commutative bialgebra over the field k which is finitely generated as an algebra over k. Then A is the homomorphic image of the free commutative bialgebra $S(\mathrm{C}_n(k))$ on the comatrix coalgebra $\mathrm{C}_n(k)$ for some $n > 0$.* □

There are other constructions directly related to the free bialgebra on a coalgebra, namely localization and the coproduct of bialgebras. We discuss these in Exercises 5.3.22, 5.3.23, and 5.3.24 respectively.

Exercises

Throughout the following exercises A and B are bialgebras over the field k, U and V are vector spaces over k, $(\iota, T(V))$ or $(\iota_V, T(V))$ denotes the free algebra on V over k, C is a coalgebra over k, $(\iota, T(C))$ or $(\iota_C, T(C))$ denotes the free bialgebra on the coalgebra C over k, and $(\iota, S(C))$ or $(\iota_C, S(C))$ denotes the free commutative bialgebra on the coalgebra C over k.

Exercise 5.3.1. Suppose $(\iota', T(V)')$ is also a free algebra on the vector space V over. Show that:

(a) There are algebra maps $F : T(V) \longrightarrow T(V)'$ and $F' : T(V)' \longrightarrow T(V)$ determined by $F\circ\iota = \iota'$ and $F'\circ\iota' = \iota$ respectively.

(b) $F'\circ F = \mathrm{I}_{T(V)}$ and $F\circ F' = \mathrm{I}_{T(V)'}$.

(c) ι is one-one. [Hint: This is the case for *at least one realization* of the free algebra on V.]

(d) $\iota(V)$ generates $T(V)$ as an algebra. [Hint: Let B be the subalgebra of $T(V)$ generated by $\iota(V)$. Show that the pair (ι, B) satisfies the universal mapping property of the free algebra on V.]

Exercise 5.3.2. Show that:

(a) If $f : V \longrightarrow W$ be a linear map there is an algebra map $F : T(V) \longrightarrow T(W)$ determined by $F\circ\iota_V = \iota_W\circ f$.

(b) The correspondences $V \mapsto T(V)$ and $f \mapsto F$ determine a functor from k-Vec to k-Alg.

Continuing with part (a), show that:

(c) If f is one then F is one-one.

(d) If f is onto then F is onto.

[Hint: For part (c) observe f one-one means there is a linear map $g : W \longrightarrow V$ which satisfies $g\circ f = \mathrm{I}_V$. See part (b).]

Exercise 5.3.3. Formulate analogs of Exercises 5.3.1 and 5.3.2 for the free commutative algebra on a vector space V and do them.

Exercise 5.3.4. Suppose that $(\iota', T(C)')$ is another free bialgebra on the coalgebra C over k.

(a) Show that there are bialgebra maps $F : T(C) \longrightarrow T(C)'$, $F' : T(C)' \longrightarrow T(C)$ determined by $F\circ\iota = \iota'$ and $F'\circ\iota' = \iota$ respectively.

(b) Show that $F'\circ F = \mathrm{I}_{T(C)}$ and $F\circ F' = \mathrm{I}_{T(C)'}$.

[Hint: See parts (a) and (b) of Exercise 5.3.1.]

Exercise 5.3.5. For $(\imath, T(C))$ show that:

(a) ι is one-one. [Hint: This is the case for *at least one realization* of the free bialgebra on the coalgebra C.]

(b) Show that $\iota(C)$ generates $T(C)$ as an algebra.

[Hint: See parts (c) and (b) of Exercise 5.3.1.]

Exercise 5.3.6. Show that:

(a) If $f : C \longrightarrow D$ is a coalgebra map then there is a bialgebra map $F : T(C) \longrightarrow T(D)$ determined by $F\circ\iota_C = \iota_D\circ F$.

(b) Show that the associations $C \mapsto T(C)$ and $f \mapsto F$ determine a covariant functor from k-Coalg to k-Bialg.

Exercise 5.3.7. Show that:

(a) If C is pointed then $T(C)$ is pointed.

(b) If C is cocommutative then $T(C)$ is cocommutative. [Hint: See Exercises 5.1.17 and 5.3.5.]

(c) If C is a sum of comatrix coalgebras then $T(C)$ is a sum of comatrix coalgebras.

Exercise 5.3.8. Show that:

(a) $\epsilon([a, b]) = 0$ and $\Delta([a, b]) = [a_{(1)}, b_{(1)}]\otimes a_{(2)}b_{(2)} + b_{(1)}a_{(1)}\otimes[a_{(2)}, b_{(2)}]$ for $a, b \in A$;

(b) If C and D are subcoalgebras of A then the span of $[C, D]$ is a coideal of A.

Exercise 5.3.9. Consider $(\iota, T(V))$. Since ι is one-one we may regard V as a subspace of $T(V)$; see part (c) of Exercise 5.3.1.

(a) Show that the algebra $T(V)$ has a bialgebra structure determined by $\Delta(v) = v\otimes 1 + 1\otimes v$ for all $v \in V$. [Hint: The linear maps $V \longrightarrow T(V)\otimes T(V)$ and $V \longrightarrow k$ determined by $v \mapsto v\otimes 1 + 1\otimes v$ and $v \mapsto 0$ respectively can be extended to algebra maps $\Delta : T(V) \longrightarrow T(V)\otimes T(V)$ and $\epsilon : T(V) \longrightarrow k$.]

(b) Show that $T(V)$ is an irreducible bialgebra.

(c) Show that the ideal of $T(V)$ generated by $[T(V), T(V)]$ is a bi-ideal of $T(V)$; thus $S(V) = T(V)/I$ is a commutative irreducible bialgebra with the quotient structure.

Exercise 5.3.10. Regard $T(V)$ as a bialgebra where $V \subseteq \mathrm{P}(T(V))$; see Exercise 5.3.9.

(a) Let $f : V \longrightarrow W$ be a linear map. Show that there is a bialgebra map $F : T(V) \longrightarrow T(W)$ determined by $f \circ \iota_V = \iota_W \circ f$.

(b) Show that the assignments $V \mapsto T(V)$ and $f \mapsto F$ determine a functor from k-Vec to k-Bialg.

Exercise 5.3.11. Let $C = V \oplus kg$ be the direct sum of a vector space V and a one-dimensional vector space over k.

(a) Show C has a coalgebra structure (C, δ, e) determined by $\delta(v) = g \otimes v + v \otimes g$ for all $v \in V$ and $\delta(g) = g \otimes g$.

(b) Let $(\imath, T(C))$ be the free coalgebra on C and let I be the ideal of $T(C)$ generated by $g - 1$. Show that I is a bi-ideal of $T(C)$ and that the quotient bialgebra $T(C)/I$ is isomorphic to the bialgebra $T(V)$ of Exercise 5.3.10.

Exercise 5.3.12. Show that:

(a) The ideal I of $T(C)$ generated by $[C, C]$ is a coideal of $T(C)$; thus I is a bi-ideal of $T(C)$.

(b) Show $S(C) = T(C)/I$ is a commutative bialgebra over k and $\jmath : C \longrightarrow S(C)$ defined by $\jmath(c) = \iota(c) + I$ is a bialgebra map.

(c) Show that the pair $(\jmath, S(C))$ is a free commutative bialgebra on the coalgebra C.

Exercise 5.3.13. Suppose $(\iota', S(C)')$ is also a free commutative bialgebra on the coalgebra C over k.

(a) Show there are bialgebra maps $F : S(C) \longrightarrow S(C)'$ and $F' : S(C)' \longrightarrow S(C)$ determined by $F \circ \iota = \iota'$ and $F' \circ \iota' = \iota$ respectively.

(b) Show that $F' \circ F = \mathrm{I}_{S(C)}$ and $F \circ F' = \mathrm{I}_{S(C)'}$.

[Hint: See Exercise 5.3.4. Evidently Exercise 5.3.4 and this exercise fit into a more general context.]

Exercise 5.3.14. Show that:

(a) If $f : C \longrightarrow D$ is a coalgebra map there is a bialgebra map $F : S(C) \longrightarrow S(D)$ determined by $F \circ \iota_C = \iota_D \circ F$.

(b) The associations $C \mapsto S(C)$ and $f \mapsto F$ determine a functor from k-Coalg to k-CmBialg, where the latter is the full subcategory of k-Bialg whose objects are the commutative bialgebras over k.

[Hint: See Exercise 5.3.6 and the comment in the hint for Exercise 5.3.13.]

Exercise 5.3.15. For the free commutative bialgebra $(\iota, S(C))$ on the coalgebra C over k show that parts (a)–(b) of Exercises 5.3.5 and 5.3.7 hold with $(\iota, S(C))$ replacing $(\iota, T(C))$.

Exercise 5.3.16. Suppose that $g, h, g', h' \in \mathrm{G}(A)$ commute with each other and centralize $\mathrm{P}_{g,h}(A)$ and $\mathrm{P}_{g',h'}(A)$. Show that $[\mathrm{P}_{g,h}(A), \mathrm{P}_{g',h'}(A)] \subseteq \mathrm{P}_{gg',hh'}(A)$. Thus part (d) of Proposition 5.1.15 can be generalized.

Exercise 5.3.17. Prove that there is a pair (π, A_{com}) which satisfies the following universal mapping property:

(a) A_{com} is a commutative bialgebra over k and $\pi : A \longrightarrow A_{com}$ is a bialgebra map.

(b) If B is a commutative bialgebra over k and $f : A \longrightarrow B$ is a bialgebra map, then there exists a map of (commutative) bialgebras $F : A_{com} \longrightarrow B$ determined by $F \circ \pi = f$.

[Hint: Consider the quotient A/I, where I is the ideal of A generated by $[A, A]$. See Exercise 5.3.8.]

Exercise 5.3.18. Let (π_A, A_{com}) denote the pair described in Exercise 5.3.17.

(a) Using Exercises 5.3.6 and 5.3.14 as models, show that a bialgebra map $f : A \longrightarrow B$ gives rise to a bialgebra map $F : A_{com} \longrightarrow B_{com}$ which is determined by $f \circ \pi_A = \pi_B \circ F$.

(b) Show that the correspondences $A \mapsto A_{com}$ and $f \mapsto F$ describe a covariant functor from k-Bialg to k-CmBialg.

Exercise 5.3.19. Suppose that k has characteristic $p > 0$ and let $n \geq 1$. Let $\alpha \in k$. Show that there is a bialgebra A over k of dimension p^n which satisfies the following data: A is generated as an algebra over k by x subject to the relation $x^{p^n} = \alpha x$ and whose coalgebra structure is determined by $\Delta(x) = 1 \otimes x + x \otimes 1$.

Exercise 5.3.20. Let $A = H_{2,-1}(k)$ be the bialgebra of Definition 5.3.5. Show that A, A^{op}, A^{cop}, $A^{op\,cop}$ and A^o are all isomorphic as bialgebras.

Exercise 5.3.21. Here is an apparent variation on the bialgebra $H_{2,-1}(k)$ described in Definition 5.3.5.

(a) Show that there is a 4-dimensional bialgebra A over k determined by the following data: A is generated as an algebra over k by g, x subject to the relations

$$g^2 = 1, \qquad x^2 = x, \qquad xg + gx = 1 + g$$

and whose coalgebra structure is determined by

$$\Delta(g) = g\otimes g, \qquad \Delta(x) = x\otimes g + 1\otimes x.$$

(b) Show that A, A^{op}, A^{cop}, $A^{op\,cop}$ and A^{o} are all isomorphic as bialgebras.

(c) Show that A and $H_{2,-1}(k)$ are isomorphic bialgebras if and only if the characteristic of k is not 2.

We explore localization for bialgebras in this exercise and the next.

Exercise 5.3.22. It is possible to formally invert grouplike elements which belong to a bialgebra. Let $S \subseteq \mathrm{G}(A)$ be a non-empty subset, let $C = A\oplus k[S]$ be the direct sum of the underlying coalgebra of A and the grouplike coalgebra $k[S]$ of S over k, and let $(\iota, T(C))$ be the free bialgebra on C. For $a \in A$ write $a = a\oplus 0$ and for $s \in S$ write $s^{-1} = 0\oplus s$. Finally, for $a, b \in A$ set $\{a, b\} = \iota(a)\iota(b) - \iota(ab)$.

(a) Show that

$$\Delta(\{a,b\}) = \{a_{(1)}, b_{(1)}\}\otimes\iota(a_{(2)})\iota(b_{(2)}) + \iota(a_{(1)})\iota(b_{(1)})\otimes\{a_{(2)}, b_{(2)}\}$$

and

$$\epsilon(\{a,b\}) = 0$$

for all $a, b \in A$.

(b) Let I be the ideal of $T(C)$ generated by the set of $\{a, b\}$'s, where $a, b \in A$, and the set of differences $\iota(s)\iota(s^{-1}) - 1, \iota(s^{-1})\iota(s) - 1$, where $s \in S$. Show that I is a bi-ideal of $T(C)$ and that $\jmath : A \longrightarrow T(C)/I$, defined by $\jmath(a) = \iota(a\oplus 0) + I$, for all $a \in A$ is a bialgebra map.

Exercise 5.3.23. We continue with Exercise 5.3.22. Let $A\{S^{-1}\} = T(C)/I$. Show that the pair $(\jmath, A\{S^{-1}\})$ satisfies the following universal mapping property:

(a) $A\{S^{-1}\}$ is a bialgebra over k, $\jmath : A \longrightarrow A\{S^{-1}\}$ is a bialgebra map, and $\jmath(S)$ consists of invertible elements;

(b) If $f : A \longrightarrow B$ is a bialgebra map and $f(S)$ consists of invertible elements, then there is a bialgebra map $F : A\{S^{-1}\} \longrightarrow B$ determined by $F\circ\jmath = f$.

Exercise 5.3.24. The categorical coproduct $(\imath_A, \imath_B, A\coprod B)$ of A and B exists. Let $A\oplus B$ be the direct sum of the underlying coalgebras of A and B respectively and let $(\iota, T(A\oplus B))$ be the free bialgebra on the coalgebra $A\oplus B$. For $a\oplus b, a'\oplus b' \in A\oplus B$ set

$$\{a\oplus b, a'\oplus b'\} = \iota(a\oplus b)\iota(a'\oplus b') - \iota(aa'\oplus bb').$$

Finally, let C be a bialgebra over k and suppose that $f_A : A \longrightarrow C$, $f_B : B \longrightarrow C$ are bialgebra maps.

(a) Show that $f_A + f_B : A\oplus B \longrightarrow C$, defined by $(f_A + f_B)(a\oplus b) = f_A(a) + f_B(b)$ for all $a\oplus b \in A\oplus B$, is a coalgebra map. [Hint: See Exercise 2.1.35.]

(b) Show that

$$\begin{aligned}&\Delta(\{a\oplus b, a'\oplus b'\})\\&= \{a_{(1)}\oplus 0, a'_{(1)}\oplus 0\}\otimes\iota(a_{(2)}\oplus 0)\iota(a'_{(2)}\oplus 0) + \iota(a_{(1)}a'_{(1)}\oplus 0)\otimes\{a_{(2)}\oplus 0, a'_{(2)}\oplus 0\}\\&\quad+\{0\oplus b_{(1)}, 0\oplus b'_{(1)}\}\otimes\iota(0\oplus b_{(2)})\iota(0\oplus b'_{(2)}) + \iota(0\oplus b_{(1)}b'_{(1)})\otimes\{0\oplus b_{(2)}, 0\oplus b'_{(2)}\}\\&\quad+\iota(a_{(1)}\oplus 0)\iota(0\oplus b'_{(1)})\otimes\iota(a_{(2)}\oplus 0)\iota(0\oplus b'_{(2)})\\&\quad+\iota(0\oplus b_{(1)})\iota(a'_{(1)}\oplus 0)\otimes\iota(0\oplus b_{(2)})\iota(a'_{(2)}\oplus 0)\end{aligned}$$

and

$$\epsilon(\{a\oplus b, a'\oplus b'\}) = \epsilon(a)\epsilon(b') + \epsilon(b)\epsilon(a')$$

for all $a\oplus b, a'\oplus b' \in A\oplus B$.

Let I be the ideal of $T(A\oplus B)$ generated by $\iota(1_A\oplus 0)-1_{T(A\oplus B)}$, $\iota(0\oplus 1_B)-1_{T(A\oplus B)}$, the $\{a\oplus 0, a'\oplus 0\}$'s where $a, a' \in A$, and the $\{0\oplus b, 0\oplus b'\}$'s where $b, b' \in B$.

(c) Show that I is a bi-ideal of $T(A\oplus B)$ and the maps $\imath_A : A \longrightarrow T(A\oplus B)/I$ and $\imath_B : B \longrightarrow T(A\oplus B)/I$, defined by $\imath_A(a) = \iota(a\oplus 0)+I$ and $\imath_B(b) = \iota(0\oplus b)+I$ respectively for all $a \in A$ and $b \in B$, are bialgebra maps.

(d) Let $F' : T(A\oplus B) \longrightarrow C$ be the bialgebra map determined by $F'\circ\iota = f_A+f_B$. Show that $F'(I) = (0)$ and the bialgebra map $F : T(A\oplus B)/I \longrightarrow C$, which is defined by $F(a\oplus b) = F'(a\oplus b) + I$ for all $a\oplus b \in A\oplus B$, is determined by $F\circ\imath_A = f_A$ and $F\circ\imath_B = f_B$.

(e) Let $A\coprod B = T(A\oplus B)/I$. Show that $(\imath_A, \imath_B, A\coprod B)$ is the coproduct of A and B in the category k-Bialg.

5.4 The universal enveloping algebra

Let L be a Lie algebra over the field k. A universal enveloping algebra of L over k is a pair $(\iota, U(L))$ which satisfies the following universal mapping property:

(UL.1) $U(L)$ is an associative algebra over k and $\iota : L \longrightarrow U(L)$ is a Lie algebra map of L to the Lie algebra associated to $U(L)$.
(UL.2) If A is an associative algebra over k and $f : L \longrightarrow A$ is a map of the Lie algebra L to the Lie algebra associated to A, then there is a map of associative algebras $F : U(L) \longrightarrow A$ determined by $F\circ\iota = f$.

Observe that the universal mapping property implies that $\mathrm{Im}(\iota)$ generates $U(L)$ as an associative algebra and that any universal enveloping algebras of L over k are isomorphic. See Exercises 5.3.1, 5.3.4 and 5.3.5.

There is a very natural coalgebra structure on $U(L)$ which makes the algebra $U(L)$ a bialgebra. The details are very easy after our discussion of free objects in Section 5.3.

Think of $U(L)$ as a Lie algebra with the Lie structure derived from the associative algebra structure of $U(L)$. To determine a coalgebra structure for $U(L)$ we note that the zero map $0 : L \longrightarrow k$ and $\delta : L \longrightarrow U(L){\otimes}U(L)$, the latter defined by $\delta(\ell) = \iota(\ell){\otimes}1 + 1{\otimes}\iota(\ell)$ for all $\ell \in L$, are Lie algebra maps. Therefore there are associative algebra maps $\epsilon : U(L) \longrightarrow k$ and $\Delta : U(L) \longrightarrow U(L){\otimes}U(L)$ determined by $\epsilon{\circ}\iota = 0$ and $\Delta{\circ}\iota = \delta$.

To show that the algebra $U(L)$ together with $(U(L), \Delta, \epsilon)$ is a bialgebra we need only show that $(U(L), \Delta, \epsilon)$ is a coalgebra. To this end we note that $(\mathrm{I}_{U(L)}{\otimes}\Delta){\circ}\Delta$ and $(\Delta{\otimes}\mathrm{I}_{U(L)}){\circ}\Delta$ are algebra maps. An easy calculation shows that both agree on $\mathrm{Im}(\iota)$ which generates $U(L)$ as an algebra. Therefore $(\mathrm{I}_{U(L)}{\otimes}\Delta){\circ}\Delta = (\Delta{\otimes}\mathrm{I}_{U(L)}){\circ}\Delta$. Likewise $(\mathrm{I}_{U(L)}{\otimes}\epsilon){\circ}\Delta$, $\mathrm{I}_{U(L)}$, and $(\epsilon{\otimes}\mathrm{I}_{U(L)}){\circ}\Delta$ are associative algebra maps which agree on $\mathrm{Im}(\iota)$. Thus $(\mathrm{I}_{U(L)}{\otimes}\epsilon){\circ}\Delta = \mathrm{I}_{U(L)} = (\epsilon{\otimes}\mathrm{I}_{U(L)}){\circ}\Delta$.

By virtue of the bialgebra structure on $U(L)$ the set $\iota(L)$ consists of primitive elements.

Exercises

Exercise 5.4.1. Let L be a Lie algebra over k. The bialgebra $U(L)$ can be constructed directly from the tensor algebra $(\iota, T(L))$ of the vector space L.

(a) Show that the algebra maps $\Delta : T(L) \longrightarrow T(L){\otimes}T(L)$ and $\epsilon : T(L) \longrightarrow k$ which arise from the linear maps $\delta : L \longrightarrow T(L){\otimes}T(L)$ and the zero map $0 : L \longrightarrow k$ respectively, where $\delta(\ell) = \iota(\ell){\otimes}1 + 1{\otimes}\iota(\ell)$ for all $\ell \in L$, give the tensor algebra $T(L)$ a bialgebra structure.

(b) For $\ell, \ell' \in L$ show that $d(\ell, \ell') = [\ell, \ell']_a - \iota([\ell, \ell'])$ spans a coideal, where $[\ell, \ell']_a = \iota(\ell)\iota(\ell') - \iota(\ell')\iota(\ell)$ is the Lie product of $\iota(\ell)$ and $\iota(\ell')$ derived from the associative algebra structure on $T(L)$.

(c) Let I be the ideal of $T(L)$ generated by the differences $d(\ell, \ell')$, where $\ell, \ell' \in L$. Show that I is a bi-ideal of $T(L)$.

(d) Set $U(L) = T(L)/I$ and let $\jmath : L \longrightarrow U(L)$ be the linear map defined by $\jmath(\ell) = \iota(\ell) + I$ for all $\ell \in L$. Show that the pair $(\jmath, U(L))$, where $U(L)$ is regarded as an algebra over k, is a universal enveloping algebra of L over k, and the bialgebra structure on the quotient $U(L)$ is that described above.

Exercise 5.4.2. Let $A = k[X_1, \ldots, X_r]$ be the polynomial algebra in indeterminates $X_1, \ldots, X_r$ over k.

(a) Show that there is a unique bialgebra structure on A with $X_1, \ldots, X_r \in \mathrm{P}(A)$.

(b) Show that $A \simeq U(L)$ as bialgebras, where L is an r-dimensional abelian Lie algebra.

Exercise 5.4.3. The universal enveloping algebra construction is a special case of the free bialgebra on a coalgebra construction. Let L be a Lie algebra over k and let $C = L \oplus kg$ be the direct sum of the vector space L and a one-dimensional vector space kg.

(a) Show that there is a unique coalgebra structure on (C, δ, e) determined by $\delta(\ell) = \ell \otimes g + g \otimes \ell$ for all $\ell \in L$ and $\Delta(g) = g \otimes g$.

Let $(\iota, T(C))$ be the free bialgebra on the coalgebra C and let I be the ideal of $T(C)$ generated by $g - 1$ and the differences $d(\ell, \ell')$ of Exercise 5.4.1.

(b) Show that I is a bi-ideal of $T(C)$.

(c) Show that the pair $(\jmath, T(C)/I)$, where $\jmath(\ell) = \iota(\ell) + I$ for all $\ell \in L$, is a universal enveloping algebra for L and the quotient bialgebra structure on $T(C)/I$ is that described in for the enveloping algebra.

5.5 The cofree bialgebra on an algebra

The constructions in this section are based on the cofree coalgebra on a vector space over k of Section 2.7. Let (A, m, η) be an associative algebra over k and let $(\pi, T^{co}(A))$ be the cofree coalgebra on the vector space A. Consider the tensor product coalgebra $C = T^{co}(A) \otimes T^{co}(A)$ and the linear map $f : C \longrightarrow A$ defined by $f = m \circ (\pi \otimes \pi)$. By the universal mapping property of $(\pi, T^{co}(A))$ there exists a coalgebra map $\boldsymbol{m} : C \longrightarrow T^{co}(A)$ determined by $\pi \circ \boldsymbol{m} = f = m \circ (\pi \otimes \pi)$. Now let $C = k$ and $f = \eta$. Then there exists a coalgebra map $\boldsymbol{\eta} : k \longrightarrow T^{co}(A)$ which is determined by $\pi \circ \boldsymbol{\eta} = f = \eta$. It is an easy exercise in the universal mapping property of the cofree coalgebra to show that $(T^{co}(A), \boldsymbol{m}, \boldsymbol{\eta})$ is an associative algebra over k. See Exercise 5.5.4. Therefore $T^{co}(A)$ with its coalgebra structure and this particular algebra structure is a bialgebra over k. The pair $(\pi, T^{co}(A))$ satisfies the following universal mapping property. Observe that $\pi : T^{co}(A) \longrightarrow A$ is an algebra map.

Theorem 5.5.1. *Suppose that A is an algebra over the field k. Then pair $(\pi, T^{co}(A))$ defined above satisfies the following:*

(a) *$T^{co}(A)$ is a bialgebra over k and $\pi : T^{co}(A) \longrightarrow A$ is an algebra map.*
(b) *If B is a bialgebra over k and $f : B \longrightarrow A$ is an algebra map there exists a bialgebra map $F : B \longrightarrow T^{co}(A)$ determined by $\pi \circ F = f$.*

Proof. Let $(\pi, T^{co}(A))$ be the pair described before the statement of the theorem. We need only establish part (b). Let (A, m_A, η_A) and (B, m_B, η_B) denote the algebra structures of A and B respectively. Tensor products will have the tensor product algebra and coalgebra structures.

Suppose that $f : B \longrightarrow A$ is an algebra map. Regarding $(\pi, T^{co}(A))$ as the cofree coalgebra on the underlying vector space of A, it follows that there exists a coalgebra map $F : B \longrightarrow T^{co}(A)$ determined by $\pi \circ F = f$. We need only show that F is an algebra map; that is $F \circ m_B = \boldsymbol{m} \circ (F \otimes F)$ and $F \circ \eta_B = \boldsymbol{\eta}$.

Note that $F \circ m_B, \boldsymbol{m} \circ (F \otimes F) : B \otimes B \longrightarrow T^{co}(A)$ are compositions of coalgebra maps and are therefore coalgebra maps. Since f is an algebra map

$$\pi \circ (F \circ m_B) = (\pi \circ F) \circ m_B = f \circ m_B = m_A \circ (f \otimes f).$$

Since π is an algebra map

$$\pi \circ (\boldsymbol{m} \circ (F \otimes F)) = (\pi \circ \boldsymbol{m}) \circ (F \otimes F) = m_A \circ (\pi \otimes \pi) \circ (F \otimes F) = m_A \circ (f \otimes f).$$

Therefore $F \circ m_B = \boldsymbol{m} \circ (F \otimes F)$.

Using the fact that F, η_B, and $\boldsymbol{\eta}$ are coalgebra maps we conclude that $F \circ \eta_B, \boldsymbol{\eta} : k \longrightarrow T^{co}(A)$ are coalgebra maps. Since f is an algebra map

$$\pi \circ (F \circ \eta_B) = (\pi \circ F) \circ \eta_B = f \circ \eta_B = \eta_A = \pi \circ \boldsymbol{\eta}.$$

Therefore $\pi \circ (F \circ \eta_B) = \pi \circ \boldsymbol{\eta}$ from which $F \circ \eta_B = \boldsymbol{\eta}$ follows. □

Definition 5.5.2. Let A be an algebra over the field k. A *cofree bialgebra on the algebra* A is a pair $(\pi, T^{co}(A))$ which satisfies the conclusion of Theorem 5.5.1.

Any pair of cofree bialgebras on an associative algebra over k are isomorphic for the same reason that any pair of cofree coalgebras on a vector space over k are isomorphic. See Exercise 5.5.2.

Let A be an associative algebra over k and let $(\pi_A, T^{co}(A))$ be the cofree bialgebra on A. By part (b) of Exercise 5.1.5 the sum $S^{co}(A) = T^{co}(A)_{com}$ of all commutative subcoalgebras of $T^{co}(A)$ is a sub-bialgebra of $T^{co}(A)$. Let $\pi : S^{co}(A) \longrightarrow A$ be the restriction $\pi_A | S^{co}(A)$. Using part (c) of Exercise 5.1.5 we deduce:

Theorem 5.5.3. *Suppose that A is an algebra over the field k. Then pair $(\pi, S^{co}(A))$ defined above satisfies the following:*

(a) *$S^{co}(A)$ is a cocommutative bialgebra over k and $\pi : S^{co}(A) \longrightarrow A$ is an algebra map;*

(b) *If B is a cocommutative bialgebra over k and $f : B \longrightarrow A$ is an algebra map there exists a bialgebra map $F : B \longrightarrow S^{co}(A)$ determined by $\pi \circ F = f$.*

□

Definition 5.5.4. Let A be an algebra over the field k. A *cofree cocommutative bialgebra on the algebra A* is a pair $(\pi, S^{co}(A))$ which satisfies the conclusion of Theorem 5.5.3.

Just as every bialgebra is a quotient of a free bialgebra on a coalgebra, every bialgebra is a sub-bialgebra of a cofree bialgebra on an algebra.

Corollary 5.5.5. *Let A be a bialgebra over the field k. Then there is an algebra B over k and a one-one bialgebra map $\jmath : A \longrightarrow T^{co}(B)$.*

Proof. Let I be an ideal of A which contains no coideals of A other than (0), for example $I = (0)$. Let $B = A/I$ be the quotient algebra and $f : A \longrightarrow B$ be the projection. By the universal mapping property of the cofree bialgebra $(\pi_B, T^{co}(B))$ there is a bialgebra map $F : A \longrightarrow T^{co}(B)$ such that $\pi \circ F = f$. Since $\mathrm{Ker}(F)$ is a coideal of A and $\mathrm{Ker}(F) \subseteq \mathrm{Ker}(f) = I$, it follows that $\mathrm{Ker}(F) = (0)$. Thus F is one-one. □

We explore when the algebra B of the corollary can be taken to be a matrix algebra in Exercise 5.5.8.

Exercises

In the following exercises A and B are bialgebras over k.

Exercise 5.5.1. Prove Theorem 5.5.3.

Exercise 5.5.2. Let A be an algebra over k and suppose that $(\pi_A, T^{co}(A))$ is a cofree bialgebra on the algebra A over k. Show that:

(a) $\pi_A : T^{co}(A) \longrightarrow A$ is onto.

(b) If $f : A \longrightarrow B$ is an algebra map then there exists a bialgebra map $F : T^{co}(A) \longrightarrow T^{co}(B)$ determined by $\pi_B \circ F = f \circ \pi_A$.

[Hint: See Exercise 2.7.2.]

Exercise 5.5.3. For an algebra A over k let $(\pi_A, T^{co}(A))$ be the cofree bialgebra on A. Show that:

(a) If $f : A \longrightarrow B$ is an algebra map then there exists a bialgebra map $F : T^{co}(A) \longrightarrow T^{co}(B)$ determined by $F\circ\pi_A = \pi_B\circ f$.

(b) The correspondences $A \mapsto T^{co}(A)$ and $f \mapsto F$ determine a functor from k-Alg to k-Bialg.

Continuing with part (a) show that:

(c) If f is one then F is one-one.

(d) If f is onto then F is onto.

Exercise 5.5.4. Let (A, m, η) be an algebra over k, let $(\pi_A, T^{co}(A))$ be the cofree coalgebra on the vector space A, and $\boldsymbol{m} : T^{co}(A)\otimes T^{co}(A) \longrightarrow T^{co}(A)$ and $\boldsymbol{\eta} : k \longrightarrow T^{co}(A)$ be the coalgebra maps determined by

$$\pi_A\circ\boldsymbol{m} = m\circ(\pi_A\otimes\pi_A) \qquad \text{and} \qquad \pi_A\circ\boldsymbol{\eta} = \eta$$

respectively.

(a) Show that

$$\pi_A\circ\boldsymbol{m}\circ(\boldsymbol{m}\otimes \mathrm{I}_{T^{co}(A)}) = (m\circ(m\otimes\mathrm{I}_A))\circ(\pi_A\otimes\pi_A\otimes\pi_A)$$

and

$$\pi_A\circ\boldsymbol{m}\circ(\mathrm{I}_{T^{co}(A)}\otimes\boldsymbol{m}) = (m\circ(\mathrm{I}_A\otimes m))\circ(\pi_A\otimes\pi_A\otimes\pi_A).$$

(b) Let $c : A \longrightarrow A\otimes k$ and $\boldsymbol{c} : T^{co}(A) \longrightarrow T^{co}(A)\otimes k$ be the maps defined by $x \mapsto x\otimes 1$. Show that

$$\pi_A\circ(\boldsymbol{m}\circ(\mathrm{I}_A\otimes\boldsymbol{\eta})\circ\boldsymbol{c}) = (m\circ(\mathrm{I}_A\otimes\eta)\circ c)\circ\pi_A = \pi_A\circ\mathrm{I}_{T^{co}(A)}$$

and

$$\pi\circ(\boldsymbol{m}\circ(\boldsymbol{\eta}\otimes\mathrm{I}_A)\circ\boldsymbol{c}) = (m\circ(\eta\otimes\mathrm{I}_A)\circ c)\circ\pi_A = \pi_A\circ\mathrm{I}_{T^{co}(A)}.$$

(c) Use parts (a) and (b) to show that $(T^{co}(A), \boldsymbol{m}, \boldsymbol{\eta})$ is an algebra over k.

By Lemma 5.1.1 we conclude that the coalgebra $T^{co}(A)$ together with the algebra structure $(T^{co}(A), \boldsymbol{m}, \boldsymbol{\eta})$ is a bialgebra over k.

Exercise 5.5.5. Let A be an algebra over k and suppose that $(\pi, T^{co}(A))$ is the cofree bialgebra on A. Show that $T^{co}(A)$ is commutative if and only if A is commutative. [Hint: See part (a) of Exercise 5.5.2.]

Exercise 5.5.6. Let $V = kv$ be a one-dimensional vector space over k and let $k[X]$ be the algebra of polynomials over k in indeterminate X. Show that:

(a) $(\pi, k[X]^o)$ is the cofree coalgebra on V, where $\pi(\alpha) = \alpha(X)v$ for all $\alpha \in k[X]^o$.

(b) The bialgebra structure of Exercise 5.5.4 on $k[X]^o$ is the dual bialgebra structure.

Exercise 5.5.7. Let $(\pi_V, C(V))$ be the cocommutative cofree coalgebra on the vector space V over k, and regard $C(V)$ as a bialgebra with the structure of Exercise 5.1.5.

(a) For vector spaces U and V over k, show that $(\pi, C(U)\otimes C(V))$ is a cofree cocommutative coalgebra on $U\oplus V$, where $\pi = \pi_U\otimes\epsilon + \epsilon\otimes\pi_V$.

(b) Show that the isomorphism of coalgebras $F : C(U)\otimes C(V) \longrightarrow C(U\oplus V)$ determined by $\pi = \pi_{U\oplus V}\circ F$ is in fact an isomorphism of bialgebras.

Suppose V has basis $\{v_1, \ldots, v_n\}$.

(c) Show that $(\pi, k[X_1]^o\otimes\cdots\otimes k[X_n]^o)$ is a cofree cocommutative coalgebra on V, where $\pi(\alpha_1\otimes\cdots\otimes\alpha_n) = \alpha_1(X_1)v_1 + \cdots + \alpha_n(X_n)v_n$ for all $\alpha_i \in k[X_i]$.

(d) Show that $C(V) \simeq k[X_1]^o\otimes\cdots\otimes k[X_n]^o$ as bialgebras, where the right-hand side has the tensor product bialgebra structure on the dual bialgebras.

[Hint: For parts (c) and (d) see Exercise 5.5.6.]

Exercise 5.5.8. Let A be an algebra over k and suppose that I is a cofinite ideal of A such that (0) is the only coideal of A. Show that there is a one-one bialgebra map $F : A \longrightarrow T^{co}(\mathrm{M}_n(k))$ for some $n > 0$. [Hint: If (A/I) $= n$ construct an algebra map $f : A \longrightarrow \mathrm{End}(A/I) \simeq \mathrm{M}_n(k)$.]

5.6 Filtrations and gradings of bialgebras

We begin with filtered algebras.

Definition 5.6.1. A *filtration of an algebra A over the field k* is a family of subspaces $\mathcal{F} = \{V_n\}_{n=0}^{\infty}$ of C which satisfies $V_0 \subseteq V_1 \subseteq V_2 \subseteq \cdots \subseteq \cup_{n=0}^{\infty} V_n = A$ and $V_mV_n \subseteq V_{m+n}$ for all $m, n \geq 0$. The pair $(A, \mathcal{F})$ is a *filtered algebra.*

As usual we will tend to denote this new object by its underlying vector space. Apropos of the definition, $1 \in V_0$. See Exercise 5.6.1.

Suppose (A, m, η) is an algebra over the field k with filtration $\mathcal{F} = \{V_n\}_{n=0}^{\infty}$ and set $V_{-1} = (0)$. Then $\mathrm{gr}_{\mathcal{F}}(A) = \bigoplus_{n=0}^{\infty} V_n/V_{n-1}$ has an algebra structure which gives $\mathrm{gr}_{\mathcal{F}}(A)$ the structure of a graded algebra $(\mathrm{gr}_{\mathcal{F}}(A), \boldsymbol{m}, \eta)$. Let $a \in V_m$ and $b \in V_n$. Then

$$(a + V_{m-1})(b + V_{n-1}) = ab + V_{m+n-1} \tag{5.3}$$

is a well-defined product which gives $\mathrm{gr}_{\mathcal{F}}(A)$ the structure of a graded algebra.

Suppose further that A is a bialgebra, $\mathcal{F}$ is also a coalgebra filtration, and $\mathrm{gr}_{\mathcal{F}}(A)$ has the graded coalgebra structure described at the end of Section 4.4. We will show that $\mathrm{gr}_{\mathcal{F}}(A)$ is a graded bialgebra with the product just described. First of all $(A, \mathcal{F})$ is a filtered bialgebra.

Definition 5.6.2. A *filtration of a bialgebra A over the field k* is a family of subspaces $\mathcal{F} = \{V_n\}_{n=0}^{\infty}$ of A which is an algebra and coalgebra filtration of A. The pair $(A, \mathcal{F})$ is a *filtered bialgebra.*

Definition 5.6.3. A *graded bialgebra over the field k* is a bialgebra A over k with a designated direct sum decomposition $A = \bigoplus_{n=0}^{\infty} A(n)$ which gives A the structure of a graded algebra and a graded coalgebra over k.

We will show that $\mathrm{gr}_{\mathcal{F}}(A)$ is a graded bialgebra over k.

To do this we continue the discussion begun after the proof of Proposition 4.4.9 and freely use notation introduced there, replacing C by A. The product $\boldsymbol{m}$ is determined by the fact that the diagrams

$$\begin{array}{ccc} V_\ell \otimes V_n & \xrightarrow{\quad m \quad} & V_{\ell+n} \\ \Big\downarrow {\scriptstyle \pi_{\ell,\mathcal{F}} \otimes \pi_{n,\mathcal{F}}} & & \Big\downarrow {\scriptstyle \pi_{\ell+n,\mathcal{F}}} \\ \overline{V}_\ell \otimes \overline{V}_n & \xrightarrow[\quad \boldsymbol{m} \quad]{} & \overline{V}_{\ell+n} \end{array}$$

commute for all $\ell, n \geq 0$, where $\overline{V}_n = V_v/V_{n-1}$ for all $n \geq 0$. By slight abuse of notion we are letting m and $\boldsymbol{m}$ denote restrictions of the multiplications of A and $\mathrm{gr}_{\mathcal{F}}$ respectively.

Write $A = \oplus_{n=0}^{\infty} A(n)$, where $V_n = A(0) \oplus \cdots \oplus A(n)$ for all $n \geq 0$. The linear isomorphism $\jmath : A \longrightarrow \mathrm{gr}_{\mathcal{F}}(A)$ is an isomorphism of coalgebras, where A has the coalgebra structure (A, δ, ϵ).

We define a product "$\cdot$" on A as follows. Let $a \in A(m)$ and $b \in A(n)$. Then $ab \in V_{m+n} = A(m+n) \oplus V_{m+n-1}$. The product $a{\cdot}b$ is the first summand in the decomposition of ab. Thus $a{\cdot}b$ is determined by the equation $ab = a{\cdot}b + \nu$, where $a{\cdot}b \in A(m+n)$ and $\nu \in V_{m+n-1}$.

Observe that $\jmath(a{\cdot}b) = \jmath(a)\jmath(b)$. Since $\jmath$ is a linear isomorphism our product gives $\mathrm{gr}_{\mathcal{F}}(A)$ the structure of a graded algebra over k such that $\jmath$ is an isomorphism. To complete the proof that $\mathrm{gr}_{\mathcal{F}}(A)$ is a bialgebra we need only show that the coalgebra structure maps of (C, δ, ϵ) are algebra maps.

Note that $1 \in A_{(0)}$ and $a{\cdot}b = ab$ for all $a, b \in A(0)$. Since $\epsilon(A(n)) = (0)$ for $n > 0$ it follows that ϵ is multiplicative.

To show that δ is multiplicative, we set

$$\mathcal{A}(n) = \sum_{r+s=n} A(r) \otimes A(s)$$

and

$$\mathcal{V}_n = \sum_{r+s \leq n} A(r) \otimes A(s) = \sum_{\ell=0}^{n} V_{n-\ell} \otimes V_\ell$$

for $n \geq 0$. Observe that $\mathcal{V}_0 \subseteq \mathcal{V}_1 \subseteq \cdots \subseteq \cup_{n=0}^{\infty} \mathcal{V}_n = A \otimes A$, $\mathcal{V}_n = \mathcal{A}(n) \oplus \mathcal{V}_{n-1}$ for all $n \geq 0$, and $A \otimes A = \bigoplus_{n=0}^{\infty} \mathcal{A}(n)$.

That $\delta(1) = 1 \otimes 1$ is easy to see. To show that δ is multiplicative, let $a \in A(m)$ and $b \in A(m)$. Then $\Delta(ab) \in \mathcal{V}_{m+n}$ and $\delta(a{\cdot}b) \in \mathcal{A}(m+n)$. Let $\ell \geq 0$. For $x, y \in A \otimes A$ we write $x \sim_\ell y$ if and only if $x - y \in \mathcal{V}_{\ell-1}$. Observe that $\sim_\ell$ is an equivalence relation.

Write $\sim$ for $\sim_{m+n}$. The calculation

$$\delta(a{\cdot}b) \sim \delta(a{\cdot}b) \sim \Delta(ab) = \Delta(a)\Delta(b) \sim \delta(a)\Delta(b) \sim \delta(a)\delta(b) \sim \delta(a){\cdot}\delta(b)$$

shows that $\delta(a{\cdot}b) \sim \delta(a){\cdot}\delta(b)$. Since $\delta(a{\cdot}b), \delta(a){\cdot}\delta(b) \in \mathcal{A}(m+n)$ it follows that $\delta(a{\cdot}b) = \delta(a){\cdot}\delta(b)$. We have shown:

Theorem 5.6.4. *Suppose that $(A, \mathcal{F})$ is a filtered bialgebra over the field k. Then $\mathrm{gr}_{\mathcal{F}}(A)$ is a graded bialgebra over k with the coalgebra structure described at the end of Section 4.4 and the algebra structure described in (5.3).* □

A very important result.

Lemma 5.6.5. *Let A be a bialgebra over k. The coradical filtration of A is a bialgebra filtration of A if and only if A_0 is a subalgebra of A.*

Proof. Suppose A_0 is a subalgebra of A. We need only show that $A_m A_n \subseteq A_{m+n}$ for all $m, n \geq 0$. We do this by induction on $m+n$. If $m+n = 0$ then $m = n = 0$ and $A_m A_n \subseteq A_{m+n}$ by assumption.

Suppose $m+n > 0$ and $A_{m'} A_{n'} \subseteq A_{m'+n'}$ whenever $m'+n' < m+n$. Then

$$\begin{aligned}
\Delta(A_m A_n) &= \Delta(A_m)\Delta(A_n) \\
&\subseteq (A_0 \otimes A_m + A_m \otimes A_{m-1})(A_0 \otimes A_n + A_n \otimes A_{n-1}) \\
&\subseteq A_0 A_0 \otimes A_m A_n + A_0 A_n \otimes A_m A_{n-1} \\
&\qquad + A_m A_0 \otimes A_{m-1} A_n + A_m A_n \otimes A_{m-1} A_{n-1} \\
&\subseteq A_0 \otimes A + A \otimes A_{m+n-1}
\end{aligned}$$

which means $A_m A_n \subseteq A_0 \wedge A_{m+n-1} = A_{m+n}$. □

When C is a coalgebra over k, recall that $\mathrm{gr}(C) = \mathrm{gr}_{\mathcal{F}}(C)$ when $\mathcal{F}$ is the coradical filtration on C. As a consequence of the preceding lemma and theorem:

Proposition 5.6.6. *Let A be a bialgebra over the field k whose coradical A_0 is a subalgebra of A. Then $\mathrm{gr}(A)$ is a graded bialgebra with the coalgebra structure described at the end of Section 4.4 and the algebra structure described in (5.3).* □

For the record:

Definition 5.6.7. Let S be a (multiplicative) monoid. An *S-graded bialgebra over the field k* is a bialgebra A over k with an S-graded vector space structure over k which gives A an S-graded algebra and an S-graded coalgebra structure over k.

Exercises

Exercise 5.6.1. Let $\mathcal{F} = \{V_n\}_{n=0}^{\infty}$ be a filtration on an algebra A over k. Show that $1 \in V_0$.

Exercise 5.6.2. Justify the sequence of equivalences "$\sim$" in the proof of Theorem 5.6.4, noting in particular how the equivalence relation "$\sim_\ell$" relates left and right multiplication.

5.7 Representations of bialgebras

Generally there is no natural way to give the tensor product of two representations of an algebra or a coalgebra the structure of a representation for the algebra or coalgebra. However, such tensor product structures exist for algebras and coalgebras which are ingredients of a bialgebra.

Let A be a bialgebra over k and suppose that M and N are left A-modules. Then the tensor product $M \otimes N$ is a left A-module where

$$a \cdot (m \otimes n) = a_{(1)} \cdot m \otimes a_{(2)} \cdot n \tag{5.4}$$

for all $a \in A, m \in M$ and $n \in N$. If M and N are left A-comodules then $M \otimes N$ is a left A-comodule where

$$\rho(m \otimes n) = m_{(-1)} n_{(-1)} \otimes (m_{(0)} \otimes n_{(0)}) \tag{5.5}$$

for all $m \in M$ and $n \in N$. Observe that any vector space V over k is a left A-module via

$$a{\cdot}v = \epsilon(h)v \tag{5.6}$$

for all $a \in A$ and $v \in V$ and is a left A-comodule via

$$\rho(v) = 1{\otimes}v \tag{5.7}$$

for all $v \in V$.

Suppose that M and N are right A-modules. Then $M{\otimes}N$ is a right A-module where

$$(m{\otimes}n){\cdot}a = m{\cdot}a_{(1)}{\otimes}n{\cdot}a_{(2)} \tag{5.8}$$

for all $a \in A$, $m \in M$, and $n \in N$. Now suppose that M and N are right A-comodules. Then $M{\otimes}N$ is a right A-comodule where

$$\rho(m{\otimes}n) = (m_{(0)}{\otimes}n_{(0)}){\otimes}m_{(1)}n_{(1)} \tag{5.9}$$

for all $m \in M$ and $n \in N$. Observe that any vector space V over k is a right A-module via

$$v{\cdot}a = \epsilon(h)v \tag{5.10}$$

for all $v \in V$ and $a \in A$ and is a right A-comodule via

$$\rho(v) = v{\otimes}1 \tag{5.11}$$

for all $v \in V$.

Definition 5.7.1. Let A be a bialgebra over k. The module structure on the tensor product of two left (respectively right) A-modules M and N defined by (5.4) (respectively (5.8)) is the *tensor product left* (respectively *right*) *A-module structure.*

Definition 5.7.2. Let A be a bialgebra over k. The module action described by (5.6) (respectively (5.10)) is the *trivial left* (respectively *right*) *A-module action.*

Definition 5.7.3. Let A be a bialgebra over k. The comodule structure on the tensor product of two left (respectively right) A-comodules M and N defined by (5.5) (respectively (5.9)) is the *tensor product left* (respectively *right*) *A-comodule structure.*

Definition 5.7.4. Let A be a bialgebra over k. The comodule action described by (5.7) (respectively (5.11)) is the *trivial left* (respectively *right*) *A-comodule action.*

Chapter notes

A good part of the basic material on bialgebras is drawn from Sweedler's book [201] and some is found in Section 1.5 of the book by Lambe and the author [99]. Our treatment of the cofree coalgebra on a vector space is slightly different from that found in Section 6.4 of [201]. It is less categorical and more computational. Since many Hopf algebras of interest are pointed and generated by skew primitives, Corollary 5.1.14 is useful in application. For an important class of such examples, see Section 15.6.

Chapter 6

The convolution algebra

The definition of a Hopf algebra involves the convolution algebra, the subject of this very short chapter. The convolution algebra is constructed from a coalgebra (C, Δ, ϵ) and an algebra (A, m, η) over the field k.

6.1 Definition and basic properties

Definition 6.1.1. The *convolution algebra of a coalgebra C and an algebra A over k* is the linear space $\mathrm{Hom}(C, A)$ with product defined by

$$f{*}g = m{\circ}(f{\otimes}g){\circ}\Delta \tag{6.1}$$

for all $f, g \in \mathrm{Hom}(C, A)$ and multiplicative identity $\eta{\circ}\epsilon$.

Thus

$$(f{*}g)(c) = f(c_{(1)})g(c_{(2)})$$

for all $f, g \in \mathrm{Hom}(C, A)$ and $c \in C$. When $A = k$ the convolution algebra $\mathrm{Hom}(C, k) = C^*$ is the dual algebra of C. When $C = k$ observe that $\mathrm{Hom}(k, A) = A$.

Let $\pi : C \longrightarrow D$ be a coalgebra map and $\jmath : A \longrightarrow B$ be an algebra map. By an easy exercise in definitions the map

$$\pi^* : \mathrm{Hom}(D, A) \longrightarrow \mathrm{Hom}(C, A) \tag{6.2}$$

defined by $\pi^*(f) = f{\circ}\pi$ for all $f \in \mathrm{Hom}(D, A)$, and the map

$$\jmath_* : \mathrm{Hom}(C, A) \longrightarrow \mathrm{Hom}(C, B), \tag{6.3}$$

defined by $\jmath_*(f) = \jmath{\circ}f$ for all $f \in \mathrm{Hom}(C, A)$, are seen to be algebra maps.

Consider the linear map

$$\iota : C^*{\otimes}A \longrightarrow \mathrm{Hom}(C, A)$$

defined by $\iota(c^*\otimes a)(c) = \langle c^*, c\rangle a$ for all $c^* \in C^*$ and $a \in A$. By Exercise 1.2.4 it follows that ι is one-one and $\mathrm{Im}(\iota) = \mathrm{Hom}_f(C, A)$ is the subspace of $\mathrm{Hom}(C, A)$ consisting of all linear maps with finite rank. Note that ι is an algebra map, where $C^*\otimes A$ is given the tensor product algebra structure. As a consequence $\mathrm{Hom}_f(C, A)$ is a subalgebra of $\mathrm{Hom}(C, A)$. Observe that ι is an algebra isomorphism if and only if either C or A is finite-dimensional.

We examine the inverses (when they exist) of algebra maps and coalgebra maps in certain convolution coalgebras.

Proposition 6.1.2. *Let C be a bialgebra and A be an algebra over the field k. Suppose that $f \in \mathrm{Hom}(C, A)$ has a convolution inverse f^{-1}.*

(a) *If $f : C \longrightarrow A$ is an algebra map then $f^{-1} : C \longrightarrow A^{op}$ is an algebra map.*
(b) *If $f : C \longrightarrow A^{op}$ is an algebra map then $f^{-1} : C \longrightarrow A$ is an algebra map.*

Proof. Parts (a) and (b) are easily seen to be equivalent. To show part (a), let $\mathcal{C} = C\otimes C$ have the tensor product coalgebra structure. Since C is a bialgebra multiplication $m_C : \mathcal{C} \longrightarrow C$ is a coalgebra map. Thus $m_C^*(f)$ has an inverse $m_C^*(f^{-1})$ in the convolution algebra $\mathrm{Hom}(\mathcal{C}, A)$ by (6.2). We show that $\ell : C\otimes C \longrightarrow A$ defined by $\ell(c\otimes d) = f^{-1}(d)f^{-1}(c)$ is a left convolution inverse for $m_C^*(f)$ as well. This will mean that $f^{-1}\circ m_C = m_C^*(f^{-1}) = \ell$ and part (a) will be established.

Let $c, d \in C$. The calculation

$$\begin{aligned}(\ell * m_C^*(f))(c\otimes d) &= \ell(c_{(1)}\otimes d_{(1)})m_C^*(f)(c_{(2)}\otimes d_{(2)}) \\ &= f^{-1}(d_{(1)})f^{-1}(c_{(1)})f(c_{(2)}d_{(2)}) \\ &= f^{-1}(d_{(1)})f^{-1}(c_{(1)})f(c_{(2)})f(d_{(2)}) \\ &= f^{-1}(d_{(1)})\epsilon(c)1f(d_{(2)}) \\ &= \epsilon(c)\epsilon(d)1 \\ &= \epsilon_{C\otimes C}(c\otimes d)1\end{aligned}$$

shows that ℓ is a left convolution inverse for $m_C^*(f)$. □

The preceding proposition has an analog for coalgebra maps.

Proposition 6.1.3. *Let A be a bialgebra and C be a coalgebra over k. Suppose that $f \in \mathrm{Hom}(C, A)$ has a convolution inverse f^{-1}.*

(a) *If $f : C \longrightarrow A$ is a coalgebra map then $f^{-1} : C \longrightarrow A^{cop}$ is a coalgebra map.*

(b) *If $f : C \longrightarrow A^{cop}$ is a coalgebra map then $f^{-1} : C \longrightarrow A$ is a coalgebra map.*

Proof. We sketch a proof. Again, parts (a) and (b) are easily seen to be equivalent. Thus only part (a) needs to be established.

Give $\mathcal{A} = A \otimes A$ the tensor product algebra structure. Since A is a bialgebra the coproduct $\Delta_A : A \longrightarrow \mathcal{A}$ is an algebra map. Therefore $(\Delta_A)_*(f)$ and $(\Delta_A)_*(f^{-1})$ are inverses in the convolution algebra $\mathrm{Hom}(C, \mathcal{A})$ by (6.3). Define $\ell : C \longrightarrow A \otimes A$ by $\ell(c) = f^{-1}(c_{(2)}) \otimes f^{-1}(c_{(1)})$ for all $c \in C$. It is easy to see that ℓ is a left inverse for $(\Delta_A)_*(f)$ also. The reader is left with the remaining details. □

Some of the following exercises are generalizations of exercises on the dual algebra.

Exercises

In these exercises C is a coalgebra and A is an algebra over k.

Exercise 6.1.1. Let $C = k[S]$ be the grouplike coalgebra of Example 2.1.9 on a non-empty set S. Show that the convolution algebra $\mathrm{Hom}(C, A)$ is isomorphic to the k-algebra of all functions $f : S \longrightarrow A$ under pointwise operations. See Exercise 2.3.23.

Exercise 6.1.2. Let $C = \mathrm{C}_n(k)$ be the comatrix coalgebra of Definition 2.1.16 on the set $S = \{1, \ldots, n\}$. Let A be any algebra over k. Show that the convolution algebra $\mathrm{Hom}(C, A) \simeq \mathrm{M}_n(A)$, where the latter is the k-algebra of all $n \times n$ matrices with coefficients in A with the usual matrix operations. See Exercise 2.3.24.

Exercise 6.1.3. Let A be an algebra over k and let $A[[X]]$ be the set of all formal power series $\sum_{n=0}^{\infty} a_n X^n$, where $a_0, a_1, a_2, \ldots \in A$.

(a) Show that $A[[X]]$ is an algebra over k, where

$$\sum_{n=0}^{\infty} a_n X^n + \sum_{n=0}^{\infty} b_n X^n = \sum_{n=0}^{\infty} (a_n + b_n) X^n,$$

$$\alpha \left(\sum_{n=0}^{\infty} a_n X^n \right) = \sum_{n=0}^{\infty} \alpha a_n X^n, \text{ and}$$

$$\left(\sum_{n=0}^{\infty} a_n X^n \right) \left(\sum_{n=0}^{\infty} b_n X^n \right) = \sum_{n=0}^{\infty} \left(\sum_{\ell=0}^{n} a_{n-\ell} b_\ell \right) X^n$$

for all $\sum_{n=0}^{\infty} a_n X^n, \sum_{n=0}^{\infty} a_n X^n \in A[[X]]$, and $\alpha \in k$.

Let $C = \mathrm{span}\{c_0, c_1, c_2, \ldots\}$ be the coalgebra of Definition 2.1.14.

(b) Show that $\mathrm{Hom}(C, A) \simeq A[[X]]$ as algebras.

(c) Let

$$\mathcal{A} = \{f \in \mathrm{Hom}(C, A) \mid f(c_m) = f(c_{m+1}) = \cdots = 0 \text{ for some } m \geq 0\}.$$

Show that $\mathcal{A}$ is a subalgebra of $\mathrm{Hom}(C, A)$ and is isomorphic to the k-algebra of polynomials in one indeterminate with coefficients in A.

Exercise 6.1.4. Suppose that C is a bialgebra over k and that A is commutative. Show that the subset $\mathrm{Alg}(C, A)$ of $\mathrm{Hom}(C, A)$ consisting of the k-algebra maps is closed under convolution multiplication and is a submonoid of the monoid $\mathrm{Hom}(C, A)$ with convolution multiplication.

Exercise 6.1.5. Suppose that C is cocommutative and that A is a bialgebra over k. Show that the subset $\mathrm{Coalg}(C, A)$ of $\mathrm{Hom}(C, A)$ consisting of the k-coalgebra maps is closed under convolution multiplication and is a submonoid of the monoid $\mathrm{Hom}(C, A)$ with convolution multiplication.

Exercise 6.1.6. Let $\jmath : A^*{\otimes}C \longrightarrow \mathrm{Hom}(C, A)^*$ be the linear map defined by $\jmath(a^*{\otimes}c)(f) = a^*(f(c))$ for all $a^* \in A^*$, $c \in C$, and $f \in \mathrm{Hom}(C, A)$. Show that:

(a) $\jmath$ is one-one. [Hint: See Corollary 1.2.4.]

(b) $\jmath(A^o{\otimes}C) \subseteq \mathrm{Hom}(C, A)^o$ and that the restriction $f = \jmath|(A^o{\otimes}C)$ is a coalgebra map $f : A^o{\otimes}C \longrightarrow \mathrm{Hom}(C, A)^o$.

(c) f is an isomorphism when C is finite-dimensional.

Exercise 6.1.7. Suppose A is a bialgebra. Let $f \in \mathrm{Hom}(C, A)$ be a coalgebra map with convolution inverse. Show that $f^*|A^o : A^o \longrightarrow C^*$ is an algebra map with convolution inverse which is an algebra map $(f^{-1})^*|A^o : A^o \longrightarrow C^{*\,op}$.

Exercise 6.1.8. Suppose C is a bialgebra over k and $f \in \mathrm{Hom}(C, A)$ is an algebra map with a convolution inverse. Let $f^o = f^*|A^o$ and $(f^{-1})^o = (f^{-1})^*|A^o$. Show that $\mathrm{Im}(f^o), \mathrm{Im}((f^{-1})^o) \subseteq C^o$ and $f^o : A^o \longrightarrow C^o$, $(f^{-1})^o : A^o \longrightarrow C^{o\,cop}$ are coalgebra maps which are convolution inverses.

Exercise 6.1.9. Complete the proof of Proposition 6.1.3.

6.2 Invertible elements in the convolution algebra

Power series in elements of $\mathrm{Hom}(C, A)$ which vanish on the coradical C_0 of C are meaningful and prove to be useful in connection with the question of invertibility of elements of $\mathrm{Hom}(C, A)$. Let $f \in \mathrm{Hom}(C, A)$ and suppose that $f(C_0) = (0)$. Since $\{C_n\}_{n=0}^{\infty}$ is a filtration of C it follows by induction on n that $f^n(C_{n-1}) = (0)$ for all $n \geq 1$. Since all functions in $\mathrm{Hom}(C, A)$ which vanish on a given subcoalgebra of C form an ideal of $\mathrm{Hom}(C, A)$, the last equation implies that $f^m(C_{n-1}) = (0)$ for all $m \geq n > 0$.

Let $\alpha_0, \alpha_1, \alpha_2, \ldots \in k$ be any sequence of scalars. Now let $c \in C$. Since $c \in C_{r-1}$ for some $r \geq 1$ the expression $\sum_{n=0}^{\infty} \alpha_n f^n(c)$ is meaningful since all but finitely many of the summands are zero. At this point it is not hard to see that $\sum_{n=0}^{\infty} \alpha_n f^n \in \mathrm{Hom}(C, A)$, where

$$\left(\sum_{n=0}^{\infty} \alpha_n f^n\right)(c) = \sum_{n=0}^{\infty} \alpha_n f^n(c)$$

for all $c \in C$. We have all but shown that:

Lemma 6.2.1. *Let C be a coalgebra and A be an algebra over the field k. Suppose that $f \in \mathrm{Hom}(C, A)$ vanishes on the coradical C_0 of C. Then the function $\pi(f) : k[[X]] \longrightarrow \mathrm{Hom}(C, A)$ defined by $\pi(f)(\sum_{n=0}^{\infty} \alpha_n X^n) = \sum_{n=0}^{\infty} \alpha_n f^n$ is an algebra map.* □

It will be important for us to know when $f \in \mathrm{Hom}(C, A)$ is invertible.

Proposition 6.2.2. *Let C be a coalgebra and let A be an algebra over the field k. For $f \in \mathrm{Hom}(C, A)$ the following are equivalent:*

(a) *f has a left inverse (respectively has a right inverse, is invertible) in the convolution algebra* $\mathrm{Hom}(C, A)$.
(b) *The restriction $f|D$ has a left inverse (respectively has a right inverse, is invertible) in the convolution algebra* $\mathrm{Hom}(D, A)$ *for all simple subcoalgebras D of C.*
(c) *The restriction $f|C_0$ has a left inverse (respectively has a right inverse, is invertible) in the convolution algebra* $\mathrm{Hom}(C_0, A)$.

Proof. Let $f \in \mathrm{Hom}(C, A)$. A right inverse for f in $\mathrm{Hom}(C, A)$ is a left inverse for f in $\mathrm{Hom}(C^{cop}, A^{op})$. An element of any algebra has an inverse if and only if it has a left inverse and a right inverse. Thus to prove the proposition we need only establish the equivalences for left inverses.

First of all, suppose that D is any subcoalgebra of C and $\pi : D \longrightarrow C$ be the inclusion. Then $\pi^* : \mathrm{Hom}(C, A) \longrightarrow \mathrm{Hom}(D, A)$ is an algebra map by (6.2). Since $\pi^*(f) = f|D$ for all $f \in \mathrm{Hom}(C, A)$, part (a) implies part (b).

Let $\{D_i\}_{i\in\mathcal{I}}$ be the set of simple subcoalgebras of C. Then $C_0 = \oplus_{i\in\mathcal{I}} D_i$ by part (c) of Proposition 3.4.3. Thus if $g_i : D_i \longrightarrow A$ is a left inverse for $f|_{D_i} : D \longrightarrow A$ the linear map $g = \bigoplus_{i\in\mathcal{I}} g_i : C_0 \longrightarrow A$ is a left inverse for $f|C_0$. Therefore part (b) implies part (c).

Suppose that the restriction $f|C_0 : C_0 \longrightarrow A$ has a left inverse g' in the convolution algebra $\text{Hom}(C_0, A)$. Extend g' to a linear map $g \in \text{Hom}(C, A)$. Since

$$(g{*}f)|C_0 = (g|C_0){*}(f|C_0) = g'{*}(f|C_0) = \eta\circ\epsilon|C_0$$

it follows that $\eta\circ\epsilon - g{*}f$ vanishes on C_0. Since $1 - X$ has an inverse in the formal power series algebra $k[[X]]$, by Lemma 6.2.1 we conclude that $g{*}f = \eta\circ\epsilon - (\eta\circ\epsilon - g{*}f)$ has an inverse in the convolution algebra $\text{Hom}(C, A)$. In particular f has a left inverse in $\text{Hom}(C, A)$. We have shown that part (c) implies part (a) and our proof is complete. □

A *locally finite algebra over* k is an algebra A over k such that its finite-dimensional subspaces generate finite-dimensional subalgebras.

Proposition 6.2.3. *Let C be a coalgebra over k, let A be a locally finite algebra over k, and suppose $f \in \text{Hom}(C, A)$ has a left or right inverse. Then f has an inverse.*

Proof. Let $g \in \text{Hom}(C, A)$. Then $f{*}g = \eta\circ\epsilon$ in $\text{Hom}(C, A)$ if and only if $g{*}f = \eta\circ\epsilon$ in $\text{Hom}(C^{cop}, A^{op})$. Thus we need only show that if g is a left inverse for f in $\text{Hom}(C, A)$ then g is also a right inverse.

Suppose $g{*}f = \eta\circ\epsilon$. Let D be a finite-dimensional subcoalgebra of C. Then $f(D)$ and $g(D)$ are finite-dimensional. Therefore there is a finite-dimensional subalgebra B of A with $f(D), g(D) \subseteq B$. Thus f, g belong to the subalgebra $\mathcal{B} = \{h \in \text{Hom}(C, A) \mid h(D) \subseteq B\}$ of $\text{Hom}(C, A)$. The restriction $\pi : \mathcal{B} \longrightarrow \text{Hom}(D, B)$ is a well-defined algebra homomorphism. This means $\pi(g)$ is a left inverse of $\pi(f)$. Since $\text{Hom}(D, B)$ is finite-dimensional it follows that $\pi(g)$ is a right inverse for $\pi(f)$. Therefore $f(d_{(1)})g(d_{(2)}) = \epsilon(d)1$ for all $d \in D$. This last equation holds on a subspace of C. Since C is the union of its finite-dimensional subcoalgebras by Theorem 2.1.21, it follows that g is a right inverse for f. □

We end this section with a few elementary exercises on the convolution algebra.

Exercises

Throughout these exercises C is a coalgebra and A is an algebra over k.

Exercise 6.2.1. Let C be a bialgebra over k and suppose that $f : C \longrightarrow A$ and $g : C \longrightarrow A^{op}$ are algebra maps. Suppose that V is a subspace of C which generates C as an algebra. Show that $f{*}g = \eta{\circ}\epsilon$ if and only if $(f{*}g)|V = (\eta{\circ}\epsilon)|V$.

Exercise 6.2.2. Let $(\iota, T(C))$ be the free bialgebra on C and suppose that $f : T(C) \longrightarrow A$ is an algebra map. Show that f has a left (respectively right, two sided) inverse in the convolution algebra $\mathrm{Hom}(T(C), A)$ if and only if $f{\circ}\iota$ has a left (respectively right, two sided) inverse in the convolution algebra $\mathrm{Hom}(C, A)$.

Chapter notes

The discussion of the convolution algebra $\mathrm{Hom}(C, A)$ in Sweedler's book [201] is primarily for the case when C or A is a Hopf algebra. Takeuchi extends this discussion in [210] and we follow his ideas here. Part (a) of Proposition 6.1.2 is [210, Lemma 25] and Proposition 6.2.2 is a slight generalization of [210, Lemma 14]. Our Lemma 6.2.1 is a formal treatment of the idea of the proof of [210, Lemma 14].

The results of Section 6.1 are found in a result of Heyneman and Sweedler [72, 1.5.2 Proposition] when the bialgebra is a Hopf algebra. The mathematics and all but the last proposition of Section 6.2 are found in [72, Section 3] in the context of pointed irreducible Hopf algebras.

Chapter 7

Hopf algebras

A Hopf algebra over the field k is a bialgebra A over k with a linear endomorphism S which can be thought of as analogous to the inverse map for groups. It may be useful to think of the relationship of bialgebras to Hopf algebras as similar to the relationship of semigroups to groups. There is a richness in the theory for groups which is lacking for semigroups. Throughout this chapter A is a bialgebra over k.

7.1 Definition of Hopf algebra, the antipode

Let $(A, m, \eta, \Delta, \epsilon)$ be a bialgebra over a field k. To define Hopf algebra we use the underlying coalgebra and algebra structures A_c and A_m respectively of A and regard the linear space $\mathrm{End}(A)$ as the convolution algebra $\mathrm{Hom}(A_c, A_m)$.

Definition 7.1.1. A *Hopf algebra over the field k* is a bialgebra A over k such that the identity map I_A has an inverse S in the convolution algebra $\mathrm{End}(A)$. In this case S is an *antipode of* A.

Since any element in an algebra has at most one inverse, the bialgebra A has at most one antipode. In terms of compositions, an antipode S for A is defined by

$$m\circ(S\otimes\mathrm{I}_A)\circ\Delta = \eta\circ\epsilon = m\circ(\mathrm{I}_A\otimes S)\circ\Delta$$

which translates in the coproduct notation to

$$S(a_{(1)})a_{(2)} = \epsilon(a)1 = a_{(1)}S(a_{(2)}) \tag{7.1}$$

for all $a \in A$.

From these equations it is clear that if A is a Hopf algebra with antipode S then the bialgebra $A^{cop\ op}$ is a Hopf algebra with antipode S as well. We will generally denote Hopf algebras by H or K instead of A or B and will at times denote the antipode of a Hopf algebra H by S_H. Here is the first of many examples.

Example 7.1.2. Let G be a group and $H = k[G]$ be the monoid algebra of G over k. Then H is a Hopf algebra with antipode S given by $S(g) = g^{-1}$ for all $g \in G$.

The existence of an antipode means that algebra maps and coalgebra maps of certain convolution algebras have inverses.

Lemma 7.1.3. *Let H be a Hopf algebra with antipode S over the field k.*

(a) *Let A be an algebra over k. Then an algebra map $f : H \longrightarrow A$ has an inverse in the convolution algebra* $\mathrm{Hom}(H, A)$ *which is $f \circ S$.*
(b) *Let C be a coalgebra over k. A coalgebra map $f : C \longrightarrow H$ has an inverse in the convolution algebra* $\mathrm{Hom}(C, H)$ *which is $S \circ f$.*
(c) *Suppose that H' is also a Hopf algebra over k with antipode S' and $f : H \longrightarrow H'$ is a bialgebra map. Then $f \circ S = S' \circ f$.*

Proof. We first show part (a). Since $\mathrm{I} = \mathrm{I}_H$ and S are inverses in the convolution $\mathrm{End}(H)$, it follows by (6.3) that $f_*(\mathrm{I}) = f$ and $f_*(S) = f \circ S$ are inverses in the convolution algebra $\mathrm{Hom}(H, A)$. As for part (b), by (6.2) it follows that $f^*(\mathrm{I}) = f$ and $f^*(S) = S \circ f$ are inverses in the convolution algebra $\mathrm{Hom}(C, H)$.

Now let $f : H \longrightarrow H'$ be a map of bialgebras. Part (c) follows from parts (a) and (b) since both $S' \circ f$ and $f \circ S$ are inverses for f in the convolution algebra $\mathrm{Hom}(H, H')$. □

Definition 7.1.4. Let H and H' be Hopf algebras over k with antipodes S and S' respectively. A *Hopf algebra map* $f : H \longrightarrow H'$ is a bialgebra map f which satisfies $f \circ S = S' \circ f$. A *Hopf algebra isomorphism* is a bijective Hopf algebra map.

By part (c) of the preceding lemma bialgebra maps between Hopf algebras are automatically Hopf algebra maps.

Suppose that H is a Hopf algebra with antipode S over k and B is a sub-bialgebra over k such that $S(B) \subseteq B$. Then B is a Hopf algebra whose antipode is the restriction $S|B$.

Definition 7.1.5. A *Hopf subalgebra of a Hopf algebra* H with antipode S over the field k is a sub-bialgebra B of H such that $S(B) \subseteq B$.

Proposition 7.1.6. *Let $f : H \longrightarrow H'$ be a map of Hopf algebras over the field k. If B is a Hopf subalgebra of H then $f(B)$ is a Hopf subalgebra of H'.* □

We have noted that the intersection of sub-bialgebras of H is a sub-bialgebra. Thus it follows that the intersection of Hopf subalgebras of H is again a Hopf subalgebra of H. Therefore for every subspace V of H there is a unique Hopf subalgebra K of H which is minimal with respect to the property that K contains V.

Definition 7.1.7. Let H be a Hopf algebra over k and suppose V is a subspace of H. The Hopf subalgebra K of H described above is the *Hopf subalgebra of H generated by V.*

Definition 7.1.8. Suppose that H is a Hopf algebra with antipode S over k. A *Hopf ideal of* H is a bi-ideal I of H such that $S(I) \subseteq I$.

Suppose that $f : H \longrightarrow K$ is a Hopf algebra map. Then $\mathrm{Ker}(f)$ is a Hopf ideal of H. Let I be any Hopf ideal of H. Then there is a unique Hopf algebra structure on the quotient space H/I such that the projection $\pi : H \longrightarrow H/I$ is a Hopf algebra map.

Some fundamental properties of the antipode are described in our next proposition.

Proposition 7.1.9. *Suppose that H is a Hopf algebra with antipode S over the field k. Then:*

(a) $S : H \longrightarrow H^{op}$ *is an algebra map; that is* $S(1) = 1$ *and* $S(ab) = S(b)S(a)$ *for all* $a, b \in H$.

(b) $S : H \longrightarrow H^{cop}$ *is a coalgebra map; that is* $\epsilon(S(a)) = \epsilon(a)$ *and* $\Delta(S(a)) = S(a_{(2)}) \otimes S(a_{(1)})$ *for all* $a \in H$.

(c) $S : H \longrightarrow H^{op\ cop}$ *is a Hopf algebra map and* $\mathrm{Ker}(S)$ *is a Hopf ideal of* H.

(d) $\mathrm{G}(H)$ *is a multiplicative subgroup of* H *and* $S(g) = g^{-1}$ *for* $g \in \mathrm{G}(H)$.

(e) $S(\mathrm{P}_{g,h}(H)) = \mathrm{P}_{h^{-1},g^{-1}}(H)$ *for* $g, h \in \mathrm{G}(H)$. *In particular* $S(x) = -g^{-1}xh^{-1}$ *for all* $x \in \mathrm{P}_{g,h}(H)$ *and S restricts to a linear isomorphism*

$$\mathrm{P}_{g,h}(H) \simeq \mathrm{P}_{h^{-1},g^{-1}}(H).$$

Proof. Since $\mathrm{I}_H : H \longrightarrow H$ is both an algebra map and a coalgebra map, parts (a) and (b) follows by Propositions 6.1.2 and 6.1.3 respectively. Part (c) follows from parts (a) and (b). Parts (d) and (e) are left as exercises. □

We have observed that if H is a Hopf algebra with antipode S then $H^{op\ cop}$ is a Hopf algebra also with antipode S. If H^{op} or H^{cop} is a Hopf algebra then S is a linear automorphism of H, and vice versa.

Proposition 7.1.10. *Suppose that H is a Hopf algebra with antipode S over the field k. Then the following are equivalent:*

(a) *H^{op} is a Hopf algebra.*
(b) *H^{cop} is a Hopf algebra.*
(c) *S is bijective.*

If S is bijective then S^{-1} is the antipode of H^{op} and H^{cop}.

Proof. Since $H^{cop} = (H^{op})^{op\ cop}$ and $H^{op} = (H^{cop})^{op\ cop}$ parts (a) and (b) are equivalent. Suppose that H^{op} is a Hopf algebra with antipode T. Then $h_{(2)}T(h_{(1)}) = \epsilon(h)1 = T(h_{(2)})h_{(1)}$ for $h \in H$. Applying S to the left-hand equation results in

$$(S \circ T)(h_{(1)})S(h_{(2)}) = \epsilon(h)1$$

by part (a) of Proposition 7.1.9. Replacing h with $S(h)$ in the right-hand equation, by part (b) of the same

$$(T \circ S)(h_{(1)})S(h_{(2)}) = \epsilon(h)1.$$

Therefore $S \circ T$ and $T \circ S$ are both left inverses of S in the convolution algebra $\mathrm{End}(H)$. We have shown that $S \circ T = \mathrm{I}_H = T \circ S$ which establishes that part (a) implies part (c).

Conversely, suppose that S is a bijection and S^{-1} is its linear inverse. Since $S : H \longrightarrow H^{op}$ is an algebra map $S^{-1} : H^{op} \longrightarrow H$ is an algebra map as well. Thus applying S^{-1} to the two equations $S(h_{(1)})h_{(2)} = \epsilon(h)1 = h_{(1)}S(h_{(2)})$, and noting that $S(1) = 1$, we conclude that $S^{-1}(h_{(2)})h_{(1)} = \epsilon(h)1 = h_{(2)}S^{-1}(h_{(1)})$. This means that S^{-1} is an antipode for H^{op}. Thus part (c) implies part (a) and the proof of the proposition is complete. □

Corollary 7.1.11. *Suppose that H is a commutative or cocommutative Hopf algebra over the field k with antipode S. Then $S^2 = \mathrm{I}_H$.* □

Definition 7.1.12. An *involutory Hopf algebra over k* is a Hopf algebra H with antipode S over k such that $S^2 = \mathrm{I}_H$.

Suppose that H is a Hopf algebra with antipode S and K is a Hopf subalgebra of H. Then $S(K)$ is a Hopf subalgebra of H by Proposition 7.1.6 and part (c) of Proposition 7.1.9. In particular $S^n(H)$ is a Hopf subalgebra of H for all $n \geq 0$.

We consider other conditions under which S is bijective. The basis for our results is the following lemma.

Lemma 7.1.13. *Let H be a Hopf algebra with antipode S over the field k, let $K = S(H)$, and suppose that $S|K : K \longrightarrow K$ is bijective. Then S is bijective.*

Proof. Since $S : H \longrightarrow K$ is onto and $S|K : K \longrightarrow K$ is bijective it follows that $H = \mathrm{Ker}(S) \oplus K$. Let $\pi : H \longrightarrow K$ be the projection. Since $\mathrm{Ker}(S) = \mathrm{Ker}(\pi)$ and is a Hopf ideal of H it follows that $\epsilon(\mathrm{Ker}(S)) = (0)$ and $\Delta(\mathrm{Ker}(S)) \subseteq \mathrm{Ker}(S) \otimes H + H \otimes \mathrm{Ker}(S) = \mathrm{Ker}(\pi) \otimes H + H \otimes \mathrm{Ker}(S)$. Therefore $(\pi * S)(h) = 0 = \epsilon(h)1$ for $h \in \mathrm{Ker}(S)$. For $h \in K$ we compute $(\pi * S)(h) = \pi(h_{(1)})S(h_{(2)}) = h_{(1)}S(h_{(2)}) = \epsilon(h)1$.

We have shown that π is a left inverse of S in the convolution algebra $\mathrm{End}(H)$. Thus $\pi = \mathrm{I}_H$ which means $K = H$ and $S = S|K$; thus S is bijective. □

The antipode of a finite-dimensional Hopf algebra over k is bijective.

Theorem 7.1.14. *Suppose that H is a Hopf algebra with antipode S over the field k.*

(a) *S is bijective if and only if the restriction $S|S^n(H) : S^n(H) \longrightarrow S^n(H)$ is bijective for some $n \geq 0$.*
(b) *If H is finite-dimensional then S is a bijective. (In this case H^{op} and H^{cop} are Hopf algebras with antipode S^{-1}.)*
(c) *If $S^n(H) = S^{n+1}(H)$ and $\mathrm{Ker}(S^n) = \mathrm{Ker}(S^{n+1})$ for some $n \geq 0$ then S is bijective.*

Proof. Part (b) follows by part (a) and Proposition 7.1.10. Assume the hypothesis of part (c). Then $S|S^n(H) : S^n(H) \longrightarrow S^n(H)$ is bijective. Thus part (c) follows by part (a) as well.

To show part (a) we need only show that if $S|S^n(H) : S^n(H) \longrightarrow S^n(H)$ is an isomorphism for some $n \geq 0$ then S is a linear automorphism of H. By induction on n we are reduced to the case $n = 1$. But then with $K = S(H)$ Lemma 7.1.13 implies S is bijective. □

Suppose that the chain $\mathrm{Im}(S) \supseteq \mathrm{Im}(S^2) \supseteq \mathrm{Im}(S^2) \supseteq \cdots$ terminates and H satisfies the ascending chain condition on two-sided ideals. Then S is bijective by part (c) of the preceding result. In particular if H is finitely generated as an algebra and S is onto then S is bijective.

Exercises

Exercise 7.1.1. Formulate and prove a fundamental homomorphism theorem for Hopf algebras. (See Theorem 2.1.21.)

Exercise 7.1.2. Suppose that B is a finite-dimensional subalgebra of an algebra A over the field k. Assume that $a \in B$ has a multiplicative inverse a^{-1} in A. Show that $a^{-1} \in B$ in two different ways:

(a) Show that $\ell(a), \mathrm{r}(a) : B \longrightarrow B$ defined by (2.16) are one-one, and hence onto.

(b) Consider the sequence $\{1, a, a^2, \ldots\}$. Since B is finite-dimensional, show that there must be a non-trivial dependency relation in this set. Show that one of minimal length must have the form $\alpha_n a^n + \cdots + \alpha_0 1 = 0$, where $\alpha_n \neq 0 \neq \alpha_0$, and thus a^{-1} is a linear combination of powers of a.

Exercise 7.1.3. Let G be a group and $H = k[G]$ be the group algebra of G over k. Suppose that K is a Hopf subalgebra of H. Show that $K = k[L]$ for some subgroup L of G. [Hint: See Exercise 5.1.8 and Lemma 2.1.12.]

Exercise 7.1.4. Let L be a Lie algebra over k. Show that $U(L)$ is a Hopf algebra, where $(\iota, U(L))$ is the universal enveloping algebra of L. [Hint: First show that the bialgebra $T(L)$ of Exercise 5.4.1 is a Hopf algebra with antipode which is the algebra map $S : T(L) \longrightarrow T(L)^{op}$ arising from the linear map $\varsigma : L \longrightarrow T(L)^{op}$ defined by $\varsigma(\ell) = \iota(-\ell)$. See Exercise 6.2.1. Then show that S determines an antipode for $U(L)$.]

Exercise 7.1.5. Let A, B be Hopf algebras over the field k. Show that the tensor product bialgebra $A \otimes B$ is a Hopf algebra with antipode $S_{A \otimes B} = S_A \otimes S_B$. See Exercise 5.1.14.

Definition 7.1.15. The Hopf algebra structure described in Exercise 7.1.5 is the *tensor product Hopf algebra structure on* $A \otimes B$.

Exercise 7.1.6. Let H be a finite-dimensional Hopf algebra with basis $\{h_i\}_{i \in I}$. Write the axioms for H in terms of structure constants. For a vector space U over k with basis $\{u_i\}_{i \in I}$ and $T \in \mathrm{End}(U)$ write $T(u_j) = \sum_{i \in I} T^i_j v_i$, or $T(u_j) = T^i_j v_i$ using the Einstein summation convention. See Exercise 5.1.16.

Exercise 7.1.7. Show that:

(a) The composition of Hopf algebra maps is a Hopf algebra map.

(b) The inverse of a Hopf algebra isomorphism is a Hopf algebra isomorphism.

(c) The identity map of a Hopf algebra is a Hopf algebra map.

(In particular Hopf algebras over k and their maps form a category. See Exercise 5.1.4.)

Definition 7.1.16. The category of Hopf algebras over k and their maps under composition form a category denoted k-HopfAlg.

Exercise 7.1.8. Let K be a field extension of k and suppose that H is a Hopf algebra with antipode S over k. Show that:

(a) The bialgebra $K \otimes H$ of Exercise 5.1.13 is a Hopf algebra over K with antipode $\mathrm{I}_K \otimes S$.

(b) If $f : H \longrightarrow H'$ is a map of k-Hopf algebras then $\mathrm{I}_K \otimes f : K \otimes H \longrightarrow K \otimes H'$ is a map of K-Hopf algebras.

(c) $F : k\text{-HopfAlg} \longrightarrow K\text{-HopfAlg}$ is a functor, where $F(H) = K \otimes H$ and $F(f) = \mathrm{I}_K \otimes f$.

7.2 Q-binomial symbols

The simplest non-commutative non-cocommutative Hopf algebras involve q-binomial symbols in their analysis. These symbols arise when one considers the commutation relation $xa = qax$, where $q \in k$ is not zero.

Suppose that $q \in k$ is not zero and consider the algebra A over k generated by symbols x and a subject to the relation $xa = qax$. Then A has linear basis $\{a^i x^j \mid i, j \geq 0\}$. This is easy to see by simply showing that the vector space with basis $\{(i,j) \mid i, j \geq 0\}$ has an associative algebra structure defined by $(i,j) \cdot (i',j') = q^{ji'}(i+i', j+j')$.

Let $n \geq 0$. For $0 \leq m \leq n$ there is a scalar $\binom{n}{m}_q$ determined by the equation

$$(a+x)^n = \sum_{m=0}^{n} \binom{n}{m}_q a^{n-m} x^m. \tag{7.2}$$

Thus if a and x belong to *any* algebra over k and $xa = qax$ the equation of (7.2) holds. It is convenient to set $\binom{n}{m}_q = 0$ when $m < 0$ or $n < m$.

The coefficients of (7.2), which are referred to as q-binomial symbols, are determined recursively by

$$\binom{0}{0}_q = 1$$

and

$$\binom{n}{m}_q = q^m \binom{n-1}{m}_q + \binom{n-1}{m-1}_q$$

for $0 \leq m \leq n$. Using the recursive definition one sees that

$$\binom{n}{0}_q = 1 = \binom{n}{n}_q \tag{7.3}$$

for all $n \geq 0$. We shall show how q-binomial symbols and binomial symbols are related, as well as show that the computation of q-binomial symbols usually involves a q-analog of factorial.

Set $(0)_q = 0$ and for $n \geq 1$ set $(n)_q = 1 + q + \cdots + q^{n-1}$. Then $(n)_q = n$ when $q = 1$ and $(n)_q = \dfrac{1-q^n}{1-q}$ when $q \neq 1$. Consequently if $q \neq 1$ then $(n)_q = 0$ if and only if q is an nth root of unity. Set $(0)_q! = 1$ and $(n)_q! = (n)_q(n-1)_q!$ for $n \geq 1$. The scalars $(n)_q!$ are referred to as q-factorials.

Now let $N \geq 1$ be a fixed positive integer. Let n_D and n_R be the integers determined by $n = n_D N + n_R$, where $0 \leq n_R < N$. The following proposition describes the q-binomial symbols in terms of q-factorials and ordinary binomial symbols.

Proposition 7.2.1. *Suppose that $n \geq 1$ and $q \in k$ is neither 0 nor 1.*

(a) *If $(n-1)_q! \neq 0$, in particular if q is not a root of unity, then*

$$\binom{n}{m}_q = \frac{(n)_q!}{(m)_q!(n-m)_q!}$$

for all $0 \leq m \leq n$.

(b) *Suppose that q is a primitive n^{th} root of unity. Then $\binom{n}{m}_q = 0$ for all $1 \leq m < n$.*

(c) *Suppose that q is a primitive N^{th} root of unity. Then*

$$\binom{n}{m}_q = \binom{n_R}{m_R}_q \binom{n_D}{m_D}$$

for all $0 \leq m \leq n$.

Proof. Part (a) follows by the recursive definition of q-binomial symbols. Part (b) follows by part (a) since $(n)_q = 0$ when q is a primitive n^{th} root of unity.

Suppose that q is a primitive N^{th} root of unity and consider the algebra A over k defined above. Then $xa^N = q^N a^N x = a^N x$. In particular a^N and x^N commute. By part (b) we compute

$$\begin{aligned}(a+x)^n &= (a+x)^{Nn_D+n_R}\\ &= ((a+x)^N)^{n_D}(a+x)^{n_R}\\ &= \left(\sum_{m=0}^{N}\binom{N}{m}_q x^{N-m}a^m\right)^{n_D}(a+x)^{n_R}\\ &= (a^N+x^N)^{n_D}(a+x)^{n_R}\end{aligned}$$

from which we deduce the formula of part (c). □

We highlight an important calculation made in the proof of the preceding proposition.

Corollary 7.2.2. *Let $n \geq 1$ and suppose that A is an algebra over the field k with elements a, x which satisfy $xa = qax$, where $q \in k$ is a primitive n^{th} root of unity. Then $(a+x)^n = a^n + x^n$.* □

Exercises

In the following exercises $q \in k$ and is not zero.

Exercise 7.2.1. Show that:

(a) $\binom{n+1}{m}_q = \binom{n}{m-1}_q + \binom{n}{m}_q q^m = \binom{n}{m-1}_q q^{n+1-m} + \binom{n}{m}_q$

for all $n \geq 0$ and $0 \leq m \leq n+1$.

(b) $\binom{n}{m}_q = \binom{n}{n-m}_q$

for all $n \geq 0$ and $0 \leq m \leq n$. Use (7.2) for establish part (a).

Exercise 7.2.2. Show for all $n, m \in \boldsymbol{Z}$ that

(a) $\binom{n}{m}_q = \binom{n}{m}_{q^{-1}} q^{m(n-m)}$

and

(b) $\binom{n}{m}_{q^2} q^{m(m-n)} = \binom{n}{m}_{q^{-2}} q^{-n(m-n)}$.

[Hint: For part (a) consider $(a+x)^n = (x+a)^n$, where $xa = qax$. One approach to part (b) is to consider $(a^2+x)^n = (x+a^2)^n$.]

Exercise 7.2.3. Show for all $n \geq 1$ that:

(a) $\sum_{\ell=0}^{n}(-1)^{\ell}q^{(\ell-1)\ell/2}\binom{n}{\ell}_q = 0$;

(b) $\sum_{\ell=0}^{n}(-1)^{\ell}q^{(\ell+1)\ell/2}q^{-n\ell}\binom{n}{\ell}_q = 0$;

and

(c) $\sum_{\ell=0}^{n}(-1)^{\ell}q^{(\ell-1)\ell/2}q^{(1-n)\ell}\binom{n}{\ell}_q = 0$.

[Hint: To establish part (b), consider part (a) in light of part (b) of Exercise 7.2.1.]

Exercise 7.2.4. Show that:

(a) $\binom{n}{\ell}_q\binom{n-\ell}{m}_q = \binom{n}{\ell+m}_q\binom{\ell+m}{m}_q$ for all $n, \ell, m \geq 0$, where $n \geq \ell + m$.

(b) $\binom{n}{m}_q\binom{n-m}{r}_q\binom{m}{s}_q = \binom{n}{r+s}_q\binom{r+s}{r}_q\binom{n-(r+s)}{m-s}_q$ for all $n,\ m,\ r,\ s \geq 0$, where $n \geq m$, $n-m \geq r$, and $m \geq s$.

[Hint: To establish part (a) consider the k-algebra generated by x_1, x_2, x_3 subject to the relations $x_jx_i = qx_ix_j$ for all $1 \leq i < j \leq 3$ and calculate $((x_1+x_2)+x_3)^n = ((x_1+x_2)+x_3)^n$.]

7.3 Two families of examples

The Taft algebras $H_{n,q}$ and the Hopf algebra $U_{\mathsf{q}}(sl_2)'$ from quantum groups are important examples of Hopf algebras which are neither commutative nor cocommutative. We construct them from a generator and relation description following the method outlined at the end of Section 5.3. Since we are also constructing an antipode we will need to add another aspect to this method. First the Taft algebras.

Let $n \geq 1$ and suppose that k contains a primitive n^{th} root of unity q. The Taft algebra $H_{n,q}$ is a Hopf algebra described as follows. As an algebra $H_{n,q}$ is generated by a and x subject to the relations

$$a^n = 1, \qquad x^n = 0, \qquad \text{and} \qquad xa = qax. \tag{7.4}$$

The coalgebra structure of $H_{n,q}$ is determined by

$$\Delta(a) = a \otimes a \qquad \text{and} \qquad \Delta(x) = x \otimes a + 1 \otimes x.$$

The antipode S of $H_{n,q}$ is determined by $S(a) = a^{n-1} = a^{-1}$ and $S(x) = -xa^{-1}$. Note that $S^2(a) = a$ and $S^2(x) = axa^{-1} = q^{-1}x$. Thus S^2 is a diagonalizable operator. $H_{n,q}$ has basis $\{a^i x^j \,|\, 0 \le i, j < n\}$ over k; thus $\mathrm{Dim}(H_{n,q}) = n^2$.

First of all note that there is an algebra $\mathcal{A}$ over k generated by a, x satisfying (7.4) of dimension n^2. To see this let $\mathcal{A}$ be the vector space with basis $\{(i,j)\}_{0 \le i,j < n}$ and define a product on $\mathcal{A}$ by

$$(i,j)\cdot(i',j') = \begin{cases} 0 & : j + j' \ge n \\ q^{ji'}(i \oplus i', j + j') & : j + j' < n, \end{cases}$$

where addition in the first component takes place in $\boldsymbol{Z}_n$. Then $\mathcal{A}$ is an associative algebra. Set $a = (1,0)$ and $x = (0,1)$.

To construct $H_{n,q}$ we start with the three-dimensional coalgebra (C, δ, e) over k described after Corollary 5.3.4 of Section 5.3 and set $A = G$. Let $(T(C), \boldsymbol{\Delta}, \boldsymbol{\epsilon})$ be the coalgebra structure of the free bialgebra on C and let I be the ideal of $T(C)$ generated by

$$U - 1, \qquad A^n - 1, \qquad XA - qAX, \qquad \text{and} \quad X^n.$$

Let $\pi : T(C) \longrightarrow T(C)/I$ be the projection, $a = \pi(A)$, and $x = \pi(X)$. We show that I is a bi-ideal of $T(C)$. Thus $H_{n,q} = T(C)/I$ is a bialgebra and π is a bialgebra map.

It is easy to see that $\boldsymbol{\epsilon}(I) = (0)$. To show that $\boldsymbol{\Delta}(I) \subseteq I \otimes T(C) + T(C) \otimes I$ we need only show that the algebra map $(\pi \otimes \pi) \circ \boldsymbol{\Delta}$ vanishes on the generators of I. In light of our discussion following Corollary 5.3.4 we need only consider X^n. Note that

$$(\pi \otimes \pi) \circ \boldsymbol{\Delta}(X^n) = ((\pi \otimes \pi) \circ \boldsymbol{\Delta}(X))^n = (x \otimes a + 1 \otimes x)^n$$

and $(1 \otimes x)(x \otimes a) = q(x \otimes a)(1 \otimes x)$. Thus $(x \otimes a + 1 \otimes x)^n = x^n \otimes a^n + 1 \otimes x^n = 0$ by Corollary 7.2.2 from which we conclude that $(\pi \otimes \pi) \circ \boldsymbol{\Delta}(X^n) = 0$. Therefore I is a coideal of $T(C)$ and the quotient $T(C)/I$ is a bialgebra over k.

To show that $H_{n,q}$ has an antipode we consider the algebra map $\boldsymbol{S} : T(C) \longrightarrow T(C)^{op}$ determined by $\boldsymbol{S}(U) = U$, $\boldsymbol{S}(A) = A^{n-1}$, and $\boldsymbol{S}(X) = -XA^{n-1}$. It is easy to see that $\pi \circ \boldsymbol{S}$ vanishes on the generators of I. Therefore $\boldsymbol{S}$ induces an algebra map $S : T(C)/I \longrightarrow T(C)^{op}/I$, that is $S : H_{n,q} \longrightarrow H_{n,q}^{op}$, which is determined by $S(a) = a^{-1}$ and $S(x) = -xa^{-1}$. By Exercise 7.3.2 it follows that S is an antipode for H.

To complete our description of $H_{n,q}$ we need to prove the basis assertion. Let $\pi' : T(C) \longrightarrow \mathcal{A}$ be the algebra map determined by $\pi'(U) = 1$, $\pi'(A) =$

a, and $\pi'(X) = x$. Then $I \subseteq \text{Ker}(\pi')$ since (7.4) is satisfied by a and x. Since $\text{Dim}(T(C)/I) \leq n^2 = \text{Dim}(\mathcal{A}) = \text{Dim}(T(C)/\text{Ker}(\pi)')$, it follows that $I = \text{Ker}(\pi')$ and therefore π' induces an isomorphism of algebras $T(C)/I \simeq \mathcal{A}$. Thus $\text{Dim}(H_{n,q}) = n^2$ and the basis assertion for $H_{n,q}$ has been established. We have constructed the Taft algebra.

There is an important point implicit in the preceding calculation which we emphasize.

Lemma 7.3.1. *Let A be a bialgebra over the field k. Suppose that $a, x \in A$ satisfy $xa = qax$ for some non-zero $q \in k$, $\Delta(a) = a \otimes a$, and $\Delta(x) = x \otimes a + 1 \otimes x$. Then $\Delta(x^n) = \sum_{m=0}^{n} \binom{n}{m}_q x^{n-m} \otimes a^{n-m} x^m$ for all $n \geq 0$.*

Proof. $\Delta(x) = x \otimes a + 1 \otimes x = \mathcal{A} + \mathcal{X}$, where $\mathcal{A} = x \otimes a$ and $\mathcal{X} = 1 \otimes x$. Since $\mathcal{X}\mathcal{A} = q\mathcal{A}\mathcal{X}$ the lemma follows by the comment after (7.2). □

Using Lemma 7.3.1 one can write down the coproduct of basis elements of $H_{n,q}$.

We now turn to a family of examples which account for the finite-dimensional Hopf algebras $U_{\mathsf{q}}(sl_2)' = U_{n,q}$, where $q = \mathsf{q}^2$. The construction follows the lines of that for the Taft algebra. We will use the Diamond Lemma of [21] to establish a linear basis for our examples, an important technique in practice.

Let n be a positive integer and suppose that k contains a primitive n^{th} of unity q. There is a Hopf algebra $U_{n,q}$ described as follows. As an algebra $U = U_{n,q}$ is generated by a, x, and y subject to the relations

$$a^n = 1, \quad x^n = 0, \quad y^n = 0, \quad xa = qax, \quad ya = q^{-1}ay, \quad \text{and}$$

$$yx - q^{-1}xy = a^2 - 1.$$

The coalgebra structure of U is determined by

$$\Delta(a) = a \otimes a, \quad \Delta(x) = x \otimes a + 1 \otimes x, \quad \text{and} \quad \Delta(y) = y \otimes a + 1 \otimes y.$$

The antipode of U is the algebra map $S : U \longrightarrow U^{op}$ determined by

$$S(a) = a^{-1}, \qquad S(x) = -q^{-1}a^{-1}x, \qquad \text{and} \quad S(y) = -qa^{-1}y.$$

Note that $S^2(a) = a$, $S^2(x) = q^{-1}x$, $S^2(y) = qy$, and $S^2(h) = aha^{-1}$ for all $h \in U$. In particular S^2 is an inner automorphism. The Hopf algebra $U_{n,q}$ has basis $\{a^k x^\ell y^m \mid 0 \leq k, \ell, m < n\}$ and therefore $\text{Dim}(U_{n,q}) = n^3$.

The Hopf algebra $U_{n,q}$ is constructed in the same manner as is $H_{n,q}$. We start with the 4-dimensional coalgebra C over k with basis $\{A, B, X, Y\}$ whose structure is determined by

$$\Delta(A) = A \otimes A, \quad \Delta(B) = B \otimes B,$$

$$\Delta(X) = X \otimes A + B \otimes X, \quad \text{and} \quad \Delta(Y) = Y \otimes A + B \otimes Y,$$

and we let I be the ideal of $T(C)$ generated by

$$A^n - 1, \quad B - 1, \quad X^n, \quad Y^n,$$

$$XA - qAX, \quad YA - q^{-1}AY, \quad \text{and} \quad YX - q^{-1}XY - A^2 + 1.$$

That $U = T(C)/I$ is a Hopf algebra is left as an exercise for the reader.

Let $a = A + I$, $x = X + I$, and $y = Y + I$. We will show that ß$= \{a^k x^\ell y^m \,|\, 0 \leq k, \ell, m < n\}$ is a linear basis for $U_{n,q}$ by applying the Diamond Lemma to $T(C) = k\{A, B, X, Y\}$ with $A < B < X < Y$ and the substitution rules

$$A^n \leftarrow 1, \quad B \leftarrow 1, \quad X^n \leftarrow 0, \quad Y^n \leftarrow 0,$$

$$XA \leftarrow qAX, \quad YA \leftarrow q^{-1}AY \quad \text{and}$$

$$YX \leftarrow q^{-1}XY + A^2 - 1.$$

The symbolism $\mathcal{L} \leftarrow \mathcal{R}$ means that $\mathcal{L}$ is replaced by $\mathcal{R}$. Monomials in A, B, X, and Y are ordered lexicographically. A substitution rule is a monomial $\mathcal{L}$ to be replaced by a linear combination of monomials $\mathcal{R}$ each of which proceed $\mathcal{L}$ in the lexicographical order. The substitution rules are a different way of expressing the relations defining I.

Loosely speaking, an ambiguity is monomial to which more than one substitution rule can be applied. An *overlap ambiguity* is a monomial LMR, where $L, M, R \neq 1$ and LM, MR are left-hand sides of substitution rules. An *inclusion ambiguity* is a monomial LMR, where LMR and M are left-hand sides of substitution rules. If the two expressions resulting from the initial substitutions can be reduced to the same expression, the ambiguity is said to be resolvable. If all ambiguities are resolvable then the cosets of $T(C)/I$ represented by monomials which do not contain a left-hand side of a substitution rule form a basis for the quotient. In practice resolution of ambiguities results in determining more substitution rules and the resolution process is repeated with the expanded list of substitution rules.

There are ten ambiguities to resolve, all of them overlap ambiguities in this case:

$$(A^\ell A^{n-\ell})A^\ell = A^\ell(A^{n-\ell}A^\ell),$$
$$(X^\ell X^{n-\ell})X^\ell = X^\ell(X^{n-\ell}X^\ell), \qquad (Y^\ell Y^{n-\ell})Y^\ell = Y^\ell(Y^{n-\ell}Y^\ell),$$
$$(XA)A^{n-1} = X(AA^{n-1}), \qquad (YA)A^{n-1} = Y(AA^{n-1}),$$
$$(X^{n-1}X)A = X^{n-1}(XA), \quad (Y^{n-1}Y)A = Y^{n-1}(YA),$$
$$(YX)A = Y(XA),$$
$$(YX)X^{n-1} = Y(XX^{n-1}) \qquad \text{and} \qquad (Y^{n-1}Y)X = Y^{n-1}(YX).$$

The first eight ambiguities are easy to resolve. We demonstrate how to resolve the fourth which is $(XA)A^{n-1} = X(A^{n-1}A)$. Let $S1$ denote the rule $A^n \leftarrow 1$ and $S5$ denote the rule $XA \leftarrow qAX$. We use the symbolism $\stackrel{S}{\Longrightarrow}$ to denote the application of substitution rule S. Thus

$$\begin{aligned}(XA)A^{n-1} &\stackrel{S5}{\Longrightarrow} (qAX)A^{n-1} = qA(XA)A^{n-2} \stackrel{S5}{\Longrightarrow} \cdots \\ &\stackrel{S5}{\Longrightarrow} q^n(A^n)X \stackrel{S1}{\Longrightarrow} q^n(1)X = X\end{aligned}$$

and

$$X(AA^{n-1}) = X(A^n) \stackrel{S1}{\Longrightarrow} X(1) = X.$$

To resolve the last two we first note by induction on $\ell \geq 1$ that the monomial YX^ℓ reduces to

$$q^{-\ell}X^\ell Y + (1+q+\ldots+q^{(\ell-1)})A^2X^{\ell-1} - (1+q^{-1}+\ldots+q^{-(\ell-1)})X^{\ell-1}$$

under application of the last substitution rule and the monomial $Y^\ell X$ reduces to

$$\begin{aligned}q^{-\ell}XY^\ell + q^{-(\ell-1)}(1+q^{-1}+\ldots+q^{-(\ell-1)})A^2Y^{\ell-1} \\ -(1+q^{-1}+\ldots+q^{-(\ell-1)})Y^{\ell-1}\end{aligned}$$

for $\ell \geq 1$ by application of the same. These reductions give immediate resolutions of the last two ambiguities. We have shown that ß is a basis for $U_{n,q}$ over k.

Observe that $U_{\mathsf{q}}(sl_2)' = U_{n,q}$, where $\mathsf{q}^2 = q$. For if we set $e = xa^{-1}$, $f = y/(\mathsf{q}-\mathsf{q}^{-1})$, and $k = a^{-1}$, the algebra structure of $U_{n,q}$ is defined by the relations

$$k^n = 1, \quad ke = \mathsf{q}^2ek, \quad kf = \mathsf{q}^{-2}fk, \quad e^n = 0, \quad f^n = 0 \quad \text{and}$$
$$ef - fe = \frac{k-k^{-1}}{\mathsf{q}-\mathsf{q}^{-1}},$$

and the coalgebra structure of $U_{(n,q)}$ is determined by

$$\Delta(k) = k \otimes k, \quad \Delta(e) = e\otimes 1 + k\otimes e \quad \text{and} \quad \Delta(f) = f\otimes k^{-1} + 1\otimes f.$$

Exercises

Exercise 7.3.1. Complete the proof that $U_{n,q} = T(C)/I$ is a Hopf algebra.

Exercise 7.3.2. Let A be a bialgebra over the field k and $S : A \longrightarrow A^{op}$ be an algebra map. Show that

$$L = \{a \in A \,|\, a_{(1)}S(a_{(2)}) = \epsilon(a)1\}$$

and

$$R = \{a \in A \,|\, S(a_{(1)})a_{(2)} = \epsilon(a)1\}$$

are subalgebras of A.

Exercise 7.3.3. Let n, N and ν be positive integers such that n divides N and $1 \leq \nu < N$. Suppose that $q \in k$ is a primitive n^{th} of unity and let r the order of q^ν. Use the methods of this section to construct a Hopf algebra $H = H_{n,\,q,\,N,\,\nu}$ over k satisfying the following: as an algebra H is generated by a and x such that

$$a^N = 1, \qquad x^r = 0, \qquad \text{and} \qquad xa = qax;$$

the coproduct of H is determined by

$$\Delta(a) = a{\otimes}a \qquad \text{and} \qquad \Delta(x) = x{\otimes}a^\nu + 1{\otimes}x;$$

and $\mathrm{Dim}(H_{n,\,q,\,N,\,\nu}) = rN$.

7.4 The dual Hopf algebra

When H is a Hopf algebra then the bialgebra H^o is a Hopf algebra as well.

Theorem 7.4.1. *Suppose that H is a Hopf algebra with antipode S over the field k. Then $S^*(H^o) \subseteq H^o$ and the restriction $S^o = S^*|H^o$ is an antipode for H^o.*

Proof. Let $a^o \in H^o$. The calculation

$$\begin{aligned} S^*(a^o_{(2)})(a)S^*(a^o_{(1)})(b) &= a^o_{(2)}(S(a))a^o_{(1)}(S(b)) \\ &= a^o(S(b)S(a)) \\ &= a^o(S(ab)) \\ &= S^*(a^o)(ab) \end{aligned}$$

shows that $\Delta_{S^*(a^o)}$ exists and that $\Delta_{S^*(a^o)} = S^*(a^o_{(2)}){\otimes}S^*(a^o_{(1)})$. Therefore $S^*(a^o) \in H^o$. We leave the calculation that the restriction $S^o = S^*|H^o$ is an antipode for H^o as an exercise for the reader. □

Definition 7.4.2. Let H be a Hopf algebra with antipode S over the field k. The *dual Hopf algebra of* H is the dual bialgebra H^o with antipode $S^o = S^*|H^o$.

When A is a bialgebra over k the relationship between the bialgebras A and A^o is given by Proposition 5.2.5. There is not much more to add when A is a Hopf algebra.

Proposition 7.4.3. *Suppose that H is a Hopf algebra over the field k. Then:*

(a) *If I is a Hopf ideal (respectively Hopf subalgebra) of H then $I^{\perp} \cap H^o$ is a Hopf subalgebra (respectively Hopf ideal) of H^o.*
(b) *Suppose that H^o is a dense subspace of H^*. If I is a Hopf ideal (respectively Hopf subalgebra) of H^o then $I^{\perp}$ is a Hopf subalgebra (respectively Hopf ideal) of H.*

Proof. Let U be a subspace of H. If $S(U) \subseteq U$ then $S^*(U^{\perp}) \subseteq U^{\perp}$. Conversely, if U is a subspace of H^* and $S^*(U) \subseteq U$ then $S(U^{\perp}) \subseteq U^{\perp}$. At this point we apply Proposition 5.2.5. □

Suppose that H is a finite-dimensional Hopf algebra over k. Then H formally gives rise to eight others. We have shown that $H^{op\ cop}$, H^{op} and H^{cop} are Hopf algebras. We have noted that $H \simeq H^{op\ cop}$, and $H^{op} \simeq H^{cop}$ as Hopf algebras. Now H^* is a Hopf algebra by the preceding result. Thus a finite-dimensional Hopf algebra gives rise to eight, including itself, by twisting structures on H and H^*. Among these eight there are four potentially different isomorphism types. Observe that $(H^*)^{op\ cop} = (H^{cop\ op})^*$, $(H^*)^{op} = (H^{cop})^*$, and $(H^*)^{cop} = (H^{op})^*$.

Exercises

Exercise 7.4.1. Complete the proof of Theorem 7.4.1.

Exercise 7.4.2. Let A and B be Hopf algebras over the field k. Show that the linear isomorphism $(A \otimes B)^o \simeq A^o \otimes B^o$ of Exercise 2.5.13 is a Hopf algebra isomorphism, where the tensor products have the tensor product bialgebra structures. See Exercises 5.2.2 and 7.1.5.

Exercise 7.4.3. Let $H = H_{n,q}$ be a Taft algebra.

(a) Show that H^{op} and H^{cop} are Taft algebras.

The Taft algebra is self dual. Here is an outline of a proof.

Define $A, X \in H^*$ by $A(a^\ell x^m) = q^\ell \delta_{m,0}$ and $X(a^\ell x^m) = \delta_{m,1}$ for all $0 \leq \ell, m < n$.

(b) Show that $\Delta_{H^*}(A) = A \otimes A$, $\Delta_{H^*}(X) = X \otimes A + \epsilon \otimes X$, and $XA = qAX$.

(c) Find a formula for $A^\ell X^m(a^u x^v)$ for all $0 \leq \ell, m, u, v < n$ and use it to establish $A^n = \epsilon$, $X^n = 0$, and that $\{A^\ell X^m \,|\, 0 \leq \ell, m < n\}$ is a basis for H^*.

(d) Show that $f : H \longrightarrow H^*$ defined by $f(a^\ell x^m) = A^\ell X^m$ for all $0 \leq \ell, m < n$ is an isomorphism of Hopf algebras.

Exercise 7.4.4. Show for the Hopf algebra H of Exercise 7.3.3 that $H \simeq H^*$ if and only if k has a primitive N^{th} root of unity ω such that $q = \omega^\nu$.

Let G be a group. Recall that a *character of* G, more precisely a *k-valued character of* G, is a group map $\chi : G \longrightarrow k \backslash 0$ from G to the multiplicative group of units of k. The set of k-valued characters of G is informally denoted $\widehat{G}$ and is a group under pointwise multiplication.

Exercise 7.4.5. Let G be a group and $H = k[G]$ be the group algebra of G over k. Show that $\eta \in \mathrm{G}(H^o)$ implies $\eta|G \in \widehat{G}$ and the map $\mathrm{G}(H^o) \longrightarrow \widehat{G}$ given by $\eta \mapsto \eta|G$ is a group isomorphism.

Exercise 7.4.6. We add to the previous exercise where G is finite abelian and k is algebraically closed of characteristic 0. Let M be a left H-module and for $\chi \in \widehat{G}$ let

$$M_{(\chi)} = \{m \in M \,|\, g{\cdot}m = \chi(g)m \text{ for all } g \in G\}. \tag{7.5}$$

Show that $M = \bigoplus_{\chi \in \widehat{G}} M_{(\chi)}$ when M is finite-dimensional. [Hint: For $g \in G$ let ℓ_g be the endomorphism of M defined by $\ell_g(m) = g{\cdot}m$ for all $m \in M$. Then $\{\ell_g\}_{g \in G}$ is a family of commuting diagonalizable endomorphisms of M. If $\mathcal{T}$ is any family of diagonalizable commuting endomorphisms of a finite-dimensional vector space V over k then V has a basis whose vectors are eigenvectors for all $T \in \mathcal{T}$.]

7.5 The free Hopf algebra on a coalgebra

We continue the discussion of free objects on a coalgebra begun in Section 5.3. In this section we construct the free Hopf algebra on a coalgebra. We will make an intermediate free construction which in essence attempts to turn a coalgebra map $\varsigma : C \longrightarrow C^{cop}$ into an antipode.

Suppose A is a Hopf algebra over the field k with antipode S and C is a subcoalgebra of A. In Section 5.3.1 we showed that C generates A as a bialgebra if and only if $A = k1 + C + C^2 + \cdots$, that is C generates A as an algebra. What does it mean to say that C generates A *as a Hopf algebra*? Let B be the Hopf subalgebra of A which C generates. Since $S(B) \subseteq B$ it follows that $\mathcal{C} = C + S(C) + S^2(C) \ldots \subseteq B$. Notice that $\mathcal{C}$ is a subcoalgebra of B and $S(\mathcal{C}) \subseteq \mathcal{C}$. It is easy to see that the subalgebra generated by $\mathcal{C}$ is a sub-bialgebra of B which contains C and is invariant under S. Therefore $B = k1 + \mathcal{C} + \mathcal{C}^2 + \cdots$. We conclude that C generates A as a Hopf algebra if $\mathcal{C}$ generates A as an algebra. Observe that the restriction $\varsigma = S|\mathcal{C}$ is a coalgebra map $\varsigma : \mathcal{C} \longrightarrow \mathcal{C}^{cop}$.

In light of the preceding discussion we are motivated to construct a free Hopf algebra $(\iota, \mathrm{H}_\varsigma(C))$ on a pair (C, ς), where $\varsigma : C \longrightarrow C^{cop}$ is a coalgebra map, as an intermediate step in the construction of the free Hopf algebra on a coalgebra.

Theorem 7.5.1. *Let C be a coalgebra over the field k and $\varsigma : C \longrightarrow C^{cop}$ be a coalgebra map. Then there exists a pair $(\iota, \mathrm{H}_\varsigma(C))$ such that:*

(a) $\mathrm{H}_\varsigma(C)$ *is a Hopf algebra over k with antipode S and $\iota : C \longrightarrow \mathrm{H}_\varsigma(C)$ is a coalgebra map such that $\iota \circ \varsigma = S \circ \iota$.*
(b) *For any pair (f, A), where A is a Hopf algebra with antipode S_A over k and $f : C \longrightarrow A$ is a coalgebra map satisfying $f \circ \varsigma = S_A \circ f$, there exists a Hopf algebra map $F : \mathrm{H}_\varsigma(C) \longrightarrow A$ determined by $F \circ \iota = f$.*

Proof. We will sketch a proof. Some major details are carefully developed in Exercise 7.5.1 at the end of this section. The reader should find it worthwhile to spend time on them.

Let $(\jmath, T(C))$ be the free bialgebra on C. Then $\jmath : C \longrightarrow T(C)$ is a coalgebra map and $\mathrm{Im}(\jmath)$ generates $T(C)$ as an algebra. Observe that the composite $\jmath \circ \varsigma : C \longrightarrow T(C)^{op\,cop}$ is a coalgebra map. Therefore there is a bialgebra map $\mathsf{S} : T(C) \longrightarrow T(C)^{op\,cop}$ determined by $\mathsf{S} \circ \jmath = \jmath \circ \varsigma$. This last equation implies that the subcoalgebra $D = \mathrm{Im}(\jmath)$ is invariant under S.

Now let I be the ideal of $T(C)$ generated by the differences

$$\mathsf{S}(d_{(1)})d_{(2)} - \epsilon(d)1 \qquad \text{and} \qquad d_{(1)}\mathsf{S}(d_{(2)}) - \epsilon(d)1 \tag{7.6}$$

for all $d \in D$. Then I is a bi-ideal of $T(C)$. Since $\mathsf{S}(D) \subseteq D$ it follows that $\mathsf{S}(I) \subseteq I$. Therefore the quotient $T(C)/I$ has a unique bialgebra structure such that the projection $\pi : T(C) \longrightarrow T(C)/I$ is a bialgebra map and S lifts to a bialgebra map $S : T(C)/I \longrightarrow (T(C)/I)^{op\,cop}$ determined by $\pi \circ \mathsf{S} = S \circ \pi$.

Let $\mathrm{H}_\varsigma(C) = T(C)/I$ and $\iota = \pi\circ\jmath$. We first show that the bialgebra $\mathrm{H}_\varsigma(C)$ is a Hopf algebra with antipode S. For $d \in \mathcal{D} = \pi(D)$ observe that

$$S(d_{(1)})d_{(2)} - \epsilon(d)1 = 0 = d_{(1)}S(d_{(2)}) - \epsilon(d)1.$$

Since $\mathcal{D}$ generates $\mathrm{H}_\varsigma(C)$ as an algebra it follows that S is an antipode for $\mathrm{H}_\varsigma(C)$. Since

$$\pi\circ\jmath\circ\varsigma = \pi\circ\mathrm{S}\circ\jmath = S\circ\pi\circ\jmath$$

it follows that $\iota\circ\varsigma = S\circ\iota$. Thus part (a) is satisfied with our choice for $\mathrm{H}_\varsigma(C)$ and ι.

To show part (b) we will suppose that $f : C \longrightarrow A$ is a coalgebra map satisfying $f\circ\varsigma = S_A\circ f$. By the universal mapping property of the tensor bialgebra there is a bialgebra map $\mathcal{F} : T(C) \longrightarrow A$ determined by $\mathcal{F}\circ\jmath = f$. We will leave it to the reader to show that $\mathcal{F}$ vanishes on the differences described in (7.6). Therefore $\mathcal{F}$ lifts to a bialgebra map $F : T(C)/I \longrightarrow A$ such that $F\circ\pi = \mathcal{F}$. The calculation

$$F\circ\pi\circ\jmath = \mathcal{F}\circ\jmath = f$$

shows that $F\circ\iota = f$. Uniqueness of such an F is clear since $\mathrm{Im}(\iota)$ generates $\mathrm{H}_\varsigma(C)$ as an algebra. This completes our proof. □

Theorem 7.5.2. *Suppose that C is coalgebra over the field k. Then there exists a pair $(\jmath, \mathrm{H}(C))$ such that:*

(a) *$\mathrm{H}(C)$ is a Hopf algebra over k and $\jmath : C \longrightarrow \mathrm{H}(C)$ is a coalgebra map.*
(b) *For any pair (f, A), where A is a Hopf algebra over k and $f : C \longrightarrow A$ is a coalgebra map, there exists a Hopf algebra map $F : \mathrm{H}(C) \longrightarrow A$ determined by $F\circ\jmath = f$.*

Proof. Let $\mathcal{C} = C \oplus C^{cop} \oplus C \oplus C^{cop} \oplus \cdots$ and $\varsigma \in \mathrm{End}(\mathcal{C})$ be defined by $\varsigma((c_1, c_2, c_3, c_4, \ldots)) = (0, c_1, c_2, c_3, c_4, \ldots)$. Then $\varsigma : \mathcal{C} \longrightarrow \mathcal{C}^{cop}$ is a one-one coalgebra map. Identifying C with the first summand of $\mathcal{C}$ we have

$$\mathcal{C} = \oplus_{i=0}^{\infty}\varsigma^i(C).$$

Let $(\iota, \mathrm{H}_\varsigma(\mathcal{C}))$ be the universal pair of the previous theorem. Consider $(\jmath, \mathrm{H}(C))$, where $\mathrm{H}(C) = \mathrm{H}_\varsigma(\mathcal{C})$ and $\jmath = \iota|C$. This latter pair satisfies part (a). Now suppose that $f : C \longrightarrow A$ is a coalgebra map, where A is a Hopf algebra over k. Let S_A be the antipode of A. The map $\mathcal{F} : \mathcal{C} \longrightarrow A$ defined by $\mathcal{F}(\varsigma^\ell(c)) = S_A^\ell(f(c))$ for $\ell \geq 0$ and $c \in C$ is a coalgebra map and satisfies $\mathcal{F}\circ\varsigma = S_A\circ\mathcal{F}$. Therefore there exists a map of Hopf algebras $F : \mathrm{H}_\varsigma(\mathcal{C}) \longrightarrow A$ which satisfies $F\circ\iota = \mathcal{F}$. The last equation implies $F\circ\jmath = f$. This condition defines F since $\mathrm{Im}(\jmath)$ generates $\mathrm{H}_\varsigma(\mathcal{C})$ as a Hopf algebra. □

Definition 7.5.3. Let C be a coalgebra over the field k. A *free Hopf algebra on the coalgebra C over k* is any pair $(\iota, \mathrm{H}(C))$ which satisfies the conclusion of Theorem 7.5.2.

Corollary 7.5.4. *Suppose that A is a Hopf algebra over the field k which is finitely generated as a Hopf algebra. Then A is the homomorphic image of $\mathrm{H}(\mathrm{C}_n(k))$ for some $n \geq 1$.*

Proof. The Hopf algebra generators are contained in a finite-dimensional subcoalgebra C of A by Theorem 2.2.3. By Corollary 2.2.2 there is a coalgebra map $f : \mathrm{C}_n(k) \longrightarrow C$ for some $n \geq 1$ which is onto. By the previous theorem there is a Hopf algebra map $F : \mathrm{H}(\mathrm{C}_n(k)) \longrightarrow A$ such that $F{\circ}\iota = f$. In particular $\mathrm{Im}(F) \supseteq C$. Since $\mathrm{Im}(F)$ is a Hopf subalgebra of A it follows that $\mathrm{Im}(F) = A$. □

Exercises

Exercise 7.5.1. Let A be a bialgebra over the field k and $S : A \longrightarrow A^{op\,cop}$ be a bialgebra map. Define $\ell, r \in \mathrm{End}(A)$ by

$$\ell(a) = S(a_{(1)})a_{(2)} - \epsilon(a)1 \qquad \text{and} \qquad r(a) = a_{(1)}S(a_{(2)}) - \epsilon(a)1$$

for $a \in A$.

(a) Show that ℓ satisfies:

(1) $\epsilon(\ell(a)) = 0$ and $\Delta(\ell(a)) = \ell(a_{(2)}){\otimes}S(a_{(1)})a_{(3)} + 1{\otimes}\ell(a)$ for all $a \in A$.
(2) $\ell(1) = 0$ and $\ell(ab) = S(b_{(1)})\ell(a)b_{(2)} + \epsilon(a)\ell(b)$ for $a, b \in A$.

(b) Show that r satisfies:

(3) $\epsilon(r(a)) = 0$ and $\Delta(r(a)) = a_{(1)}S(a_{(2)}){\otimes}r(a_{(2)}) + r(a){\otimes}1$ for all $a \in A$.
(4) $r(1) = 0$ and $r(ab) = a_{(1)}r(b)S(a_{(2)}) + \epsilon(b)r(a)$ for $a, b \in A$.

(c) Show that $S{\circ}\ell = r{\circ}S$ and $S{\circ}r = \ell{\circ}S$.

(d) Show that the formulas of part (a) follow from those of part (b) and vice versa when A is replaced by $A^{op\,cop}$. Can you draw a similar conclusion for part (c)?

Let C be a subcoalgebra of A and let I_ℓ be the ideal generated by the $\ell(c)$'s and let I_r be the ideal generated by the $r(c)$'s where c runs over C.

(e) Show that I_ℓ and I_r are bi-ideals of A.

(f) Show that $I = I_\ell + I_r$ is a bi-ideal of A invariant under S when C is invariant under S.

(g) Suppose that C is invariant under S and generates A as an algebra. Show that the quotient A/I is a Hopf algebra with antipode S' defined by $S'(a+I) = S(a)+I$ for $a \in A$.

Exercise 7.5.2. Suppose that $f, g : A \longrightarrow B$ are maps of Hopf algebras over the field k and C is a subcoalgebra of A which generates A as a Hopf algebra. Show that $f|C = g|C$ if and only if $f = g$.

7.6 When a bialgebra is a Hopf algebra

Just as a non-empty finite multiplicatively closed subset of a group is a subgroup:

Proposition 7.6.1. *Let H be a Hopf algebra over the field k. Then any finite-dimensional sub-bialgebra of H is a Hopf subalgebra of H.*

Proof. Let B be a finite-dimensional sub-bialgebra of H and let S be the antipode of H. Since B is a subcoalgebra of H, the restriction $S|B$ is an inverse for I_B in the convolution algebra $\mathrm{Hom}(B, H)$. Observe that the convolution algebra $\mathrm{Hom}(B, B)$ is a subalgebra of $\mathrm{Hom}(B, H)$.

Let ℓ be the endomorphism of the algebra $\mathrm{Hom}(B, B)$ which is left multiplication by I_B. Then ℓ is one-one since I_B has a left inverse in $\mathrm{Hom}(B, H)$. Since $\mathrm{Hom}(B, B)$ is finite-dimensional ℓ is onto. Therefore I_B has a right inverse, hence an inverse, in the convolution algebra $\mathrm{Hom}(B, B)$. □

Lemma 7.6.2. *Let A be a bialgebra over the field k and suppose that B is a sub-bialgebra of A with an antipode. If $A_0 \subseteq B$ then A is a Hopf algebra.*

Proof. Let S be an antipode for B. Then I_{A_0} and $S|A_0$ are inverses in the convolution algebra $\mathrm{Hom}(A_0, A)$. As $\mathrm{I}_A|A_0 = \mathrm{I}_{A_0}$ it follows by Proposition 6.2.2 that I_A has an inverse in the convolution algebra $\mathrm{Hom}(A, A) = \mathrm{End}(A)$. □

Invertibility of grouplike elements is necessary and sufficient for a pointed bialgebra to be a Hopf algebra.

Proposition 7.6.3. *Let A be a pointed bialgebra over the field k. Then A is a Hopf algebra if and only if the set $\mathrm{G}(A)$ consists of invertible elements (in which case $\mathrm{G}(A)$ is a group under multiplication). In particular pointed irreducible bialgebras are Hopf algebras.*

Proof. Suppose that A is a Hopf algebra. Then $\mathrm{G}(A)$ is a group under multiplication by part (d) of Proposition 7.1.9. Conversely, suppose that $\mathrm{G}(A)$ consists of elements which have multiplicative inverses in H. Then $\mathrm{G}(A)$ is a multiplicative group by part (a) of Proposition 5.1.15. Therefore $A_0 = k[\mathrm{G}(A)]$ is a sub-bialgebra of A which has an antipode. By Lemma 7.6.2 we conclude that A is a Hopf algebra. □

Corollary 7.6.4. *The antipode S of a pointed Hopf algebra A over a field k is bijective. In particular the antipode of a pointed irreducible Hopf algebra is bijective.*

Proof. By Proposition 7.6.3 the set $\mathrm{G}(A)$ consists of invertible elements. Therefore $\mathrm{G}(A^{op}) = \mathrm{G}(A)$ consists of elements invertible in A^{op}. By the proposition again A^{op} is a Hopf algebra. Therefore S is bijective by Proposition 7.1.10. □

Corollary 7.6.5. *The universal enveloping algebra $U(L)$ of a Lie algebra L over the field k is a pointed irreducible Hopf algebra with antipode S determined by $S(\ell) = -\ell$ for all $\ell \in L$.*

Proof. As an algebra $U(L)$ is generated by L which consists of primitive elements. Thus $U(L)_0 = k1$ by Corollary 5.1.14 which means $U(L)$ is pointed irreducible and is therefore a Hopf algebra by Proposition 7.6.3. Let S be the antipode of $U(L)$ and $\ell \in L$. Then $S : U(L) \longrightarrow U(L)^{op}$ is an algebra homomorphism and $S(\ell) = -\ell$ by parts (a) and (e) of Proposition 7.1.9. □

Corollary 7.6.6. *The irreducible component A_{k1} of the subcoalgebra $k1$ of a bialgebra A over the field k is a pointed irreducible Hopf algebra over k.*

Proof. $(A_{k1})_0 = k1$ by definition and A_{k1} is a sub-bialgebra of A by part (d) of Proposition 5.1.16. Now the corollary follows by Proposition 7.6.3. □

Corollary 7.6.7. *Let A be a bialgebra over k generated by $\mathcal{S} \cup \mathcal{P}$, where $\mathcal{S} \subseteq \mathrm{G}(A)$, $\mathcal{P}$ consists of skew-primitives x which satisfy $\Delta(x) = x \otimes s + s' \otimes x$ for some $s, s' \in \mathcal{S}$, and the elements of $\mathcal{S}$ are invertible. Then A is a pointed Hopf algebra.*

Proof. By part (b) of Corollary 5.1.14 the set $\mathrm{G}(A)$ consists of invertible elements. Proposition 7.6.3 applies. □

For the remainder of this section we consider the relationship between the existence of an antipode and the comatrix identities. Suppose that A is a bialgebra over the field k. We describe a necessary and sufficient condition for A to be a Hopf algebra in terms of invertibility of the inclusion map $\iota_C : C \longrightarrow A$ in the convolution algebra $\mathrm{Hom}(C, A)$, where C belongs to a certain class of subcoalgebras of A. Our discussion is based on matrices with coefficients in A whose entries satisfy the comatrix identities.

Let V be a vector space over k and for $n > 0$ let $\mathrm{M}_n(V)$ be the set of $n \times n$ matrices with coefficients in V. Observe that $\mathrm{M}_n(V)$ is a vector space over k with the usual rules for matrix addition and scalar multiplication. For $\mathcal{A} \in \mathrm{M}_n(V)$ we denote the entry in the ith row and j^{th} column of $\mathcal{A}$ by $v_{i,j}$ and we write $\mathcal{A} = (v_{i,j})$. When V is an algebra over k the vector space $\mathrm{M}_n(V)$ is an algebra over k with the usual rule for matrix multiplication.

Lemma 7.6.8. *Let A be a bialgebra over the field k, suppose that C is a non-zero finite-dimensional subcoalgebra of A of dimension n, and let $\iota_C : C \longrightarrow A$ be the inclusion map. Then the following are equivalent:*

(a) *For all positive integers r, every $\mathcal{A} \in \mathrm{M}_r(C)$ whose entries satisfy the comatrix identities has a right inverse in $\mathrm{M}_r(A)$.*

(b) *There exists a basis $\{v_1, \ldots, v_n\}$ for C such that $\epsilon(v_i) = \delta_{i,1}$ for all $1 \leq i \leq n$ and $\mathcal{A} = (a_{i,j}) \in \mathrm{M}_n(C)$ has a right inverse in $\mathrm{M}_n(A)$, where the entries of $\mathcal{A}$ are determined by*

$$\Delta(v_i) = \sum_{j=1}^{n} a_{i,j} \otimes v_j$$

for all $1 \leq i \leq n$.

(c) *The inclusion ι_C has a right inverse in the convolution algebra $\mathrm{Hom}(C, A)$.*

Proof. We first show part (a) implies part (b). Since $C \neq (0)$ necessarily $\epsilon(C) \neq (0)$. Thus C has a basis $\{v_1, \ldots, v_n\}$ such that $\epsilon(v_i) = \delta_{i,1}$ for all $1 \leq i \leq n$. Since the entries of any matrix $\mathcal{A}$ satisfying the conditions of part (b) satisfy the comatrix identities, part (a) implies part (b).

We next show that part (b) implies part (c). Let $\{v_1, \ldots, v_n\}$ and $\mathcal{A}$ satisfy the conclusion of part (b). For $1 \leq i \leq r$ the equation $v_i = \sum_{j=1}^{n} a_{i,j}\epsilon(v_j) = a_{i,1}$ implies $v_i = a_{i,1}$. Let $\mathcal{B} = (b_{i,j})$ be a right inverse for $\mathcal{A}$ in $\mathrm{M}_n(A)$ and define a linear map $t : C \longrightarrow A$ by $t(v_j) = b_{j,1}$ for $1 \leq j \leq n$. For $c = v_i$ we compute

$$c_{(1)}t(c_{(2)}) = \sum_{j=1}^{n} a_{i,j}t(v_j) = \sum_{j=1}^{n} a_{i,j}b_{j,1} = \delta_{i,1}1 = \epsilon(c)1.$$

Therefore t is a right inverse for ι_C in the convolution algebra $\mathrm{Hom}(C, A)$. We have shown that part (b) implies part (c).

To complete the proof of the lemma, we need only show that part (c) implies part (a). Suppose that $t : C \longrightarrow A$ is a right inverse for ι_C in the convolution algebra $\mathrm{Hom}(C, A)$. Then for all $r > 0$, any $\mathcal{A} = (a_{i,j}) \in \mathrm{M}_r(C)$ which satisfies the comatrix identities has a right inverse $\mathcal{B} = (b_{i,j}) \in \mathrm{M}_r(A)$, where $b_{i,j} = t(a_{i,j})$. □

Applying the preceding lemma to the bialgebra $A^{op\,cop}$ yields the variation of the lemma where "left" is replaced by "right". The entries of the matrix $\mathcal{A}$ of part (b) are defined by $\Delta(v_j) = \sum_{i=1}^{n} v_i \otimes a_{i,j}$ in order for the comatrix identities to be satisfied.

Proposition 7.6.9. *Suppose that A is a bialgebra over the field k. Then the following are equivalent:*

(a) *For all $r > 0$, every $\mathcal{A} \in \mathrm{M}_n(A)$ whose coefficients satisfy the comatrix identities has an inverse in $\mathrm{M}_n(A)$.*
(b) *A is a Hopf algebra.*

Proof. We first prove that part (a) implies part (b). Assume the hypothesis of part (a) and let C be a finite-dimensional subcoalgebra of A. Then the inclusion $\iota_C : C \longrightarrow A$ has an inverse t_C in the convolution algebra $\mathrm{Hom}(C, A)$ by the previous lemma and subsequent remark. By uniqueness of inverses, if D is a subcoalgebra of C then $t_D = t_C|D$. By Theorem 2.2.3 any finite-dimensional subspace of A is contained in a finite-dimensional subcoalgebra of A. Therefore there exists a linear endomorphism S of A such that $S|C = t_C$ for all finite-dimensional subcoalgebras C of A. It is clear that S is an antipode for A. We have shown that part (a) implies part (b).

To show part (b) implies part (a) we suppose that $r > 0$ and $\mathcal{A} \in \mathrm{M}_r(A)$ satisfies the comatrix identities. Then the coefficients of $\mathcal{A}$ span a finite-dimensional subcoalgebra C of A. If A has an antipode S, then $S|C$ is an inverse for the inclusion $\iota_C : C \longrightarrow A$. Therefore part (b) implies part (a) by the previous lemma and subsequent remark also. Our proof of the proposition is complete. □

By Lemma 7.6.8, the following comment regarding $A^{op\,cop}$, and Proposition 6.2.2, it follows that a bialgebra A has an antipode if and only if all $\mathcal{A} \in \mathrm{M}_n(C)$ which satisfy the comatrix identities have an inverse in $\mathrm{M}_n(A)$ where C is a *simple* subcoalgebra of A. When the entries of $\mathcal{A}$ commute then the determinant of $\mathcal{A}$ plays a key role.

Recall that an $n \times n$ matrix $\mathcal{A}$ with coefficients in a commutative algebra A is invertible if and only if its determinant $\mathrm{Det}(\mathcal{A})$ is an invertible element of A. If A is a commutative bialgebra then $\mathrm{Det}(\mathcal{A})$ is a grouplike element.

Lemma 7.6.10. *Suppose that A is a bialgebra over the field k, let $n > 0$ and $\mathcal{A} \in \mathrm{M}_n(A)$. Suppose that the entries of $\mathcal{A}$ generate a commutative subalgebra of A and satisfy the comatrix identities. Then* $\mathrm{Det}(\mathcal{A}) \in \mathrm{G}(A)$.

Proof. Write $\mathcal{A} = (a_{i,j})$ as usual. S_n and A_n denote the groups of permutations and even permutations respectively on $\{1, \ldots, n\}$. We first note

$$\epsilon\left(\mathrm{Det}(\mathcal{A})\right) = \epsilon\left(\sum_{\sigma \in S_n} \mathrm{sgn}(\sigma)\, a_{1,\sigma(1)} \cdots a_{n,\sigma(n)}\right) = \epsilon(a_{1,1}) \cdots \epsilon(a_{n,n}) = 1$$

and

$$\begin{aligned}
&\Delta(\mathrm{Det}(\mathcal{A})) \\
&= \sum_{\sigma \in S_n}\left(\sum_{1 \leq j_1, \ldots, j_n \leq n} \mathrm{sgn}(\sigma)\, a_{1,j_1} \cdots a_{n,j_n} \otimes a_{j_1,\sigma(1)} \cdots a_{j_n,\sigma(n)}\right) \\
&= \sum_{1 \leq j_1, \ldots, j_n \leq n}\left(a_{1,j_1} \cdots a_{n,j_n} \otimes \left(\sum_{\sigma \in S_n} \mathrm{sgn}(\sigma)\, a_{j_1,\sigma(1)} \cdots a_{j_n,\sigma(n)}\right)\right).
\end{aligned}$$

Now let $j_1, \ldots, j_n$ be fixed, suppose that $1 \leq u < v \leq n$ and $j_u = j_v$. Using the transposition $\tau = (u\, v)$ we calculate

$$\begin{aligned}
&\sum_{\sigma \in S_n} \mathrm{sgn}(\sigma)\, a_{j_1,\sigma(1)} \cdots a_{j_n,\sigma(n)} \\
&= \sum_{\sigma \in A_n} \mathrm{sgn}(\sigma)\, a_{j_1,\sigma(1)} \cdots a_{j_n,\sigma(n)} + \sum_{\sigma \in A_n} \mathrm{sgn}(\sigma\tau)\, a_{j_1,\sigma\circ\tau(1)} \cdots a_{j_n,\sigma\circ\tau(n)} \\
&= \sum_{\sigma \in S_n} \mathrm{sgn}(\sigma)(a_{j_1,\sigma(1)} \cdots a_{j_n,\sigma(n)} - a_{j_1,\sigma(1)} \cdots a_{j_n,\sigma(n)}) \\
&= 0.
\end{aligned}$$

Thus

$$\begin{aligned}
&\Delta(\mathrm{Det}(\mathcal{A}))\\
&= \sum_{\sigma,\tau\in S_n} \mathrm{sgn}(\sigma)\, a_{1,\tau(1)}\cdots a_{n,\tau(n)} \otimes a_{\tau(1),\sigma(1)}\cdots a_{\tau(n),\sigma(n)}\\
&= \sum_{\tau\in S_n} \Bigg(\mathrm{sgn}(\tau)\, a_{1,\tau(1)}\cdots a_{n,\tau(n)} \otimes \\
&\qquad\qquad \Bigg(\sum_{\sigma\in S_n} \mathrm{sgn}(\sigma\circ\tau^{-1})\, a_{\tau(1),\sigma\circ\tau^{-1}(\tau(1))}\cdots a_{\tau(n),\sigma\circ\tau^{-1}(\tau(n))}\Bigg)\Bigg)\\
&= \mathrm{Det}(\mathcal{A})\otimes\mathrm{Det}(\mathcal{A}).
\end{aligned}$$

□

We have seen that the set of grouplike elements of a Hopf algebra form a multiplicative group. By the preceding lemma and Proposition 7.6.9:

Corollary 7.6.11. *Let A be a commutative bialgebra over the field k. Then A is a Hopf algebra if and only if all grouplike elements of A have a multiplicative inverse in A.* □

Exercises

Throughout the following exercises A is a bialgebra over k.

Exercise 7.6.1. Show that there is a unique subcoalgebra C of A maximal with respect to the property that the inclusion $\iota_C : C \longrightarrow A$ has an inverse in the convolution algebra $\mathrm{Hom}(C, A)$.

Exercise 7.6.2. Suppose that A is a commutative bialgebra over k and S is a non-empty subset of $\mathrm{G}(A)$. Let $X_S = \{X_s \mid s \in S\}$ be a set of symbols indexed by S and $k[X_S]$ be the polynomial algebra over k with the bialgebra structure determined by letting each X_s be a grouplike element.

(a) Show that $A[S^{-1}] = (A\otimes k[X_S])/I$ is a commutative bialgebra, where I is the ideal of $A\otimes k[X_S]$ generated by the differences $s\otimes X_s - 1\otimes 1$ for $s \in S$.

(b) Let $\iota : A \longrightarrow A[S^{-1}]$ be the bialgebra map defined by $\iota(a) = a\otimes 1 + I$. Show that the pair $(\iota, A[S^{-1}])$ satisfies the following universal property:

  (i) $\iota : A \longrightarrow A[S^{-1}]$ is a bialgebra map and $\iota(s)$ is invertible for all $s \in S$, and

  (ii) if $f : A \longrightarrow B$ is a map of commutative bialgebras such that $f(s)$ is invertible for all $s \in S$ then there exists a bialgebra map $F : A[S^{-1}] \longrightarrow B$ determined by $F\circ\iota = f$.

When $S = \{s\}$ is a singleton set we set $A[s^{-1}] = A[S^{-1}]$.

Exercise 7.6.3. Let $n > 0$ and $\{e_{i,j}\}_{1 \leq i,j \leq n}$ be the standard basis for $\mathrm{C}_n(k)$. Set $\mathcal{A} = (e_{i,j}) \in \mathrm{M}_n(\mathrm{C}_n(k))$. Show that $S(\mathrm{C}_n(k))[(\mathrm{Det}(\mathcal{A}))^{-1}]$ is a Hopf algebra, and that every finitely generated commutative Hopf algebra is the quotient of such a Hopf algebra.

7.7 Two-cocycles, pairings, and skew pairings of bialgebras

Two-cocycles provide a way of altering the product of a bialgebra to produce another bialgebra. Skew pairings give rise to 2-cocycles. Skew pairings are special types of pairings of bialgebras. This section, and the next, justifiably belong in the chapter on bialgebras, Chapter 5. Given the length of that chapter, plus the fact that the material of this section and the next are usually applied to Hopf algebras, we include Sections 7.7 and 7.8 in this one.

Let C, D be coalgebras over the field k, let $\beta : C \times D \longrightarrow k$ be a bilinear form, and let $\beta_{lin} : C \otimes D \longrightarrow k$ be the associated linear form. Recall $\beta_{lin}(c \otimes d) = \beta(c, d)$ for all $c \in C$ and $d \in D$. We say that β is *convolution invertible* if β_{lin} is invertible in the dual algebra $(C \otimes D)^*$. Thus β is convolution invertible if and only if there is a bilinear form $\beta' : C \times D \longrightarrow k$ which satisfies $\beta(c_{(1)}, d_{(1)})\beta'(c_{(2)}, d_{(2)}) = \epsilon(c)\epsilon(d) = \beta'(c_{(1)}, d_{(1)})\beta(c_{(2)}, d_{(2)})$ for all $c \in C$ and $d \in D$. When β is convolution invertible there is a unique solution β' to the two preceding equations. In this case we write β^{-1} for β' and call β^{-1} the inverse of β.

Definition 7.7.1. Let A be a bialgebra over the field k. A *left 2-cocycle for* A is a convolution invertible bilinear form $\sigma : A \times A \longrightarrow k$ which satisfies

$$\sigma(a_{(1)}, b_{(1)})\sigma(a_{(2)}b_{(2)}, c) = \sigma(b_{(1)}, c_{(1)})\sigma(a, b_{(2)}c_{(2)})$$

for all $a, b, c \in A$. A *right 2-cocycle for* A is a left 2-cocycle for A^{cop}.

Thus a bilinear form $\sigma : A \times A \longrightarrow k$ is a right 2-cocycle if it is convolution invertible and

$$\sigma(a_{(2)}, b_{(2)})\sigma(a_{(1)}b_{(1)}, c) = \sigma(b_{(2)}, c_{(2)})\sigma(a, b_{(1)}c_{(1)})$$

for all $a, b, c \in A$.

A left or right 2-cocycle σ for A is normal if $\sigma(a, 1) = \epsilon(a) = \sigma(a, 1)$ for all $a \in A$. Some basic properties of left 2-cocycles are:

Lemma 7.7.2. *Suppose A is a bialgebra over the field k and suppose that $\sigma : A{\times}A \longrightarrow k$ is a left 2-cocycle for A. Then:*

(a) *$\sigma(1,1)$ and $\sigma^{-1}(1,1)$ are multiplicative inverses.*
(b) *$\sigma(a,1) = \sigma(1,1)\epsilon(a) = \sigma(1,a)$ for all $a \in A$.*
(c) *σ^{-1} is a right 2-cocycle for A.*
(d) *$\sigma^{-1}(a,1) = \sigma^{-1}(1,1)\epsilon(a) = \sigma^{-1}(1,a)$ for all $a \in A$.*

Proof. Part (a) follows directly from the definition of convolution invertible. To show part (b) we define $\ell, r \in A^*$ by $\ell(a) = \sigma(a,1)$ and $r(a) = \sigma^{-1}(a,1)$ for all $a \in A$. Then ℓ and r are inverses in the dual algebra A^*. With $b = c = 1$ the left 2-cocycle condition reads $\ell^2 = \sigma(1,1)\ell$. Thus $\ell = \ell^2{*}r = \sigma(1,1)\ell{*}r = \sigma(1,1)\epsilon$, or equivalently $\sigma(a,1) = \sigma(1,1)\epsilon(a)$ for all $a \in A$. A similar calculation gives $\sigma(1,a) = \sigma(1,1)\epsilon(a)$ for all $a \in A$.

Part (d) follows from parts (b) and (c). It remains to show part (c). Define $\ell, r, L, R \in (A{\otimes}A{\otimes}A)^*$ by

$$\begin{aligned}
\ell(a{\otimes}b{\otimes}c) &= \sigma(a_{(1)}, b_{(1)})\sigma(a_{(2)}b_{(2)}, c),\\
L(a{\otimes}b{\otimes}c) &= \sigma^{-1}(a_{(2)}, b_{(2)})\sigma^{-1}(a_{(1)}b_{(1)}, c),\\
r(a{\otimes}b{\otimes}c) &= \sigma(b_{(1)}, c_{(1)})\sigma(a, b_{(2)}c_{(2)}),\\
R(a{\otimes}b{\otimes}c) &= \sigma^{-1}(b_{(2)}, c_{(2)})\sigma^{-1}(a, b_{(1)}c_{(1)})
\end{aligned}$$

for all $a, b, c \in A$. We need only show that $L = R$. This follows as $\ell = r$ and $L{*}\ell = \epsilon = r{*}R$ in the convolution algebra $(A{\otimes}A{\otimes}A)^*$. □

We may use a left 2-cocycle σ for a bialgebra $(A, m, \eta, \Delta, \epsilon)$ to create a new bialgebra. Regarding k as a subalgebra of A and identifying σ and σ_{lin}, we may regard $\sigma \in \mathrm{Hom}(A{\otimes}A, A)$.

Proposition 7.7.3. *Let $(A, m, \eta, \Delta, \epsilon)$ be a bialgebra over the field k and suppose that σ is a left 2-cocycle for A. Let $m_\sigma = \sigma{*}m{*}\sigma^{-1}$ in the convolution algebra $\mathrm{Hom}(A{\otimes}A, A)$. Then:*

(a) *$(A, m_\sigma, \eta, \Delta, \epsilon)$ is a bialgebra over k.*
(b) *Suppose that A has antipode S. Then A_σ has an antipode S_σ given by $S_\sigma(a) = \sigma(a_{(1)}, S(a_{(2)}))S(a_{(3)})\sigma^{-1}(S(a_{(4)}), a_{(5)})$ for all $a \in A$.*

Proof. For $a, b \in A$ the product $a{\cdot}b$ in A_σ is given by

$$a{\cdot}b = \sigma(a_{(1)}, b_{(1)})a_{(2)}b_{(2)}\sigma^{-1}(a_{(3)}, b_{(3)}) \tag{7.7}$$

for all $a, b \in A$. We first show that (A, m_σ, η) is an algebra over k. For $a \in A$ the equations $a{\cdot}1 = a = 1{\cdot}a$ follow by parts (a), (b), and (d) of Lemma 7.7.2. For $a, b, c \in A$ the equation $(a{\cdot}b){\cdot}c = a{\cdot}(b{\cdot}c)$ follows from the fact that σ is a left 2-cocycle for A and σ^{-1} is a left 2-cocycle for A^{cop}. The fact that Δ and ϵ are algebra maps with respect to A_σ follows from the definition of m_σ basically. We have sketched a proof of part (a).

Suppose that A has antipode S and let $a \in A$. Then

$$S_\sigma(a_{(1)}){\cdot}a_{(2)} = \sigma(a_{(1)}, S(a_{(2)}))\sigma^{-1}(S(a_{(3)}), a_{(4)})1 = a_{(1)}{\cdot}S_\sigma(a_{(2)})$$

follows by straightforward calculations. To complete the proof we need only show that $\sigma(a_{(1)}, S(a_{(2)}))\sigma^{-1}(S(a_{(3)}), a_{(4)}) = \epsilon(a)$. Using parts (b) and (c) of Lemma 7.7.2 we see that

$$\begin{aligned}
&\sigma(a_{(1)}, S(a_{(2)}))\sigma^{-1}(S(a_{(3)}), a_{(4)}) \\
&= \sigma(a_{(1)}, S(a_{(3)}))\epsilon(a_{(2)})\epsilon(a_{(6)})\sigma^{-1}(S(a_{(4)}), a_{(5)}) \\
&= \sigma(a_{(1)}, S(a_{(3)}))\sigma^{-1}(1,1)\sigma(\epsilon(a_{(2)})1, a_{(6)})\sigma^{-1}(S(a_{(4)}), a_{(5)}) \\
&= \sigma^{-1}(1,1)\sigma(a_{(1)}, S(a_{(4)}))\sigma(a_{(2)}S(a_{(3)}), a_{(7)})\sigma^{-1}(S(a_{(5)}), a_{(6)}) \\
&= \sigma^{-1}(1,1)\sigma(a_{(1)(1)}, S(a_{(2)})_{(1)})\sigma(a_{(1)(2)}S(a_{(2)})_{(2)}, a_{(5)})\sigma^{-1}(S(a_{(3)}), a_{(4)}) \\
&= \sigma^{-1}(1,1)\sigma(S(a_{(2)})_{(2)}, a_{(4)(2)})\sigma(a_{(1)}, S(a_{(2)})_{(3)}a_{(5)})\sigma^{-1}(S(a_{(2)})_{(1)}, a_{(4)(1)}) \\
&= \sigma^{-1}(1,1)\sigma(a_{(1)}, S(a_{(2)})a_{(3)}) \\
&= \sigma^{-1}(1,1)\sigma(a_{(1)}, \epsilon(a_{(2)})1) \\
&= \epsilon(a).
\end{aligned}$$

We have shown part (b). □

Definition 7.7.4. The bialgebra of part (a) of Proposition 7.7.3 is denoted by A_σ.

Let A and B be bialgebras over k and $\tau : A{\times}B \longrightarrow k$ be a bilinear form. Consider the following axioms:

(P.1) $\tau(aa', b) = \tau(a, b_{(1)})\tau(a', b_{(2)})$ for all $a, a' \in A$ and $b \in B$.
(P.2) $\tau(1, b) = \epsilon(b)$ for all $b \in B$.
(P.3) $\tau(a, bb') = \tau(a_{(1)}, b)\tau(a_{(2)}, b')$ for all $a \in A$ and $b, b' \in B$.
(P.4) $\tau(a, 1) = \epsilon(a)$ for all $a \in A$.

There are equivalent formulations of the axioms (P.1)–(P.4) in terms of $\tau_\ell : A \longrightarrow B^*$ and $\tau_r : A \longrightarrow A^*$.

Proposition 7.7.5. *Let A and B be bialgebras over the field k and suppose that $\tau : A{\times}B \longrightarrow k$ is a bilinear form. Then:*

(a) *The following are equivalent:*

(1) *(P.1) and (P.2) hold for τ.*

(2) $\tau_\ell : A \longrightarrow B^*$ *is an algebra map.*

(3) $\mathrm{Im}(\tau_r) \subseteq A^o$ *and* $\tau_r : B \longrightarrow A^o$ *is a coalgebra map.*

(b) *The following are equivalent:*

(1) *(P.3) and (P.3) hold for* τ.

(2) $\tau_r : B \longrightarrow A^*$ *is an algebra map.*

(3) $\mathrm{Im}(\tau_\ell) \subseteq B^o$ *and* $\tau_\ell : A \longrightarrow B^o$ *is a coalgebra map.*

Proof. We demonstrate part (a) and relegate the proof of part (b) to Exercise 7.7.5. Observe that (P.1) holds if and only if $\tau_\ell(aa')(b) = (\tau_\ell(a)(b_{(1)}))(\tau_\ell(a')(b_{(2)}))$ for all $a, a' \in A$ and $b \in b$ or equivalently $\tau_\ell(aa') = \tau_\ell(a)\tau_\ell(a')$ in the dual algebra B^*. Likewise (P.2) holds if and only if $\tau_\ell(1)(b) = \epsilon(b)$ for all $b \in B$ or equivalently $\tau_\ell(1) = 1_{B^*}$. We have shown that (1) and (2) of part (a) are equivalent.

Note (P.1) holds if and only if $\tau_r(b)(aa') = (\tau_r(b_{(1)})(a))(\tau_r(b_{(2)})(a'))$ for all $b \in B$ and $a, a' \in A$ or equivalently $\Delta_{\tau_r(b)}$ exists (thus $\tau_r(b) \in A^o$) and $\Delta_{\tau_r(b)} = \tau_r(b_{(1)}) \otimes \tau_r(b_{(2)})$ for all $b \in B$. (P.2) is equivalent to $\tau_r(b)(1) = \epsilon(b)$ for all $b \in B$ which is the same as $\epsilon_{A^o} \circ \tau_r = \epsilon_B$ when $\mathrm{Im}(\tau_r) \subseteq A^o$. We have shown (1) and (3) of part (a) are equivalent. □

Definition 7.7.6. A *pairing of bialgebras A and B over k* is a bilinear form $\tau : A \times B \longrightarrow k$ which satisfies any of the equivalent conditions of part (a) and any of the equivalent conditions of part (b) of Proposition 7.7.5. A *pairing of a bialgebra A with itself over k* is a pairing of A and A.

Definition 7.7.7. A *skew pairing of bialgebras A and B over k* is a pairing of A^{cop} and B. A *skew pairing of a bialgebra A with itself over k* is a pairing of A^{cop} and A.

Since skew pairings are rather important, we write down the analogs of (P.1)–(P.4) for them:

(SP.1) $\tau(aa', b) = \tau(a, b_{(1)})\tau(a', b_{(2)})$ for all $a, a' \in A$ and $b \in B$.
(SP.2) $\tau(1, b) = \epsilon(b)$ for all $b \in B$.
(SP.3) $\tau(a, bb') = \tau(a_{(2)}, b)\tau(a_{(1)}, b')$ for all $a \in A$ and $b, b' \in B$.
(SP.4) $\tau(a, 1) = \epsilon(a)$ for all $a \in A$.

Even though part of the next result is redundant, we record what it means for τ to be a skew pairing in terms of τ_ℓ and τ_r as a matter of convenience.

Proposition 7.7.8. *Let A and B be bialgebras over the field k and suppose that* $\tau : A \times B \longrightarrow k$ *is a bilinear form. Then:*

(a) *The following are equivalent:*

(1) *(SP.1) and (SP.2) hold for* τ.
(2) $\tau_\ell : A \longrightarrow B^*$ *is an algebra map.*
(3) $\mathrm{Im}(\tau_r) \subseteq A^o$ *and* $\tau_r : B \longrightarrow A^o$ *is a coalgebra map.*

(b) *The following are equivalent:*

(1) *(SP.3) and (SP.4) holds for* τ.
(2) $\tau_r : B \longrightarrow A^{*\,op}$ *is an algebra map.*
(3) $\mathrm{Im}(\tau_\ell) \subseteq B^o$ *and* $\tau_\ell : A \longrightarrow B^{o\,cop}$ *is a coalgebra map.*

□

When a pairing is convolution invertible will be important.

Proposition 7.7.9. *Let A and B be bialgebras over the field k and suppose that $\tau : A \times B \longrightarrow k$ is a pairing of A and B. Then:*

(a) *If τ is convolution invertible then τ^{-1} is a pairing of A^{cop} and B^{cop}, that is a skew pairing of A and B^{cop}.*
(b) *Suppose that A has antipode S. Then τ is convolution invertible and $\tau^{-1}(a,b) = \tau(S(a),b)$ for all $a \in A$ and $b \in B$.*
(c) *Suppose that B has antipode S. Then τ is convolution invertible and $\tau^{-1}(a,b) = \tau(a,S(b))$ for all $a \in A$ and $b \in B$.*
(d) *Suppose that A^{cop} and B have antipodes S_A^{cop} and S_B respectively. Then $\tau(a,b) = \tau(S_A^{cop}(a), S_B(b))$ for all $a \in A$ and $b \in B$.*

Proof. Since $\tau_\ell : A \longrightarrow B^*$ and $\tau_r : B \longrightarrow A^*$ are algebra maps by Proposition 7.7.5, and τ_ℓ, τ_r are convolution invertible with inverses $(\tau^{-1})_\ell$ and $(\tau^{-1})_r$ respectively, by part (a) of Proposition 6.1.2 we conclude that $(\tau^{-1})_\ell : A \longrightarrow B^{*\,op}$ and $(\tau^{-1})_r : B \longrightarrow A^{*\,op}$ are algebra maps. As $B^{*\,op} = (B^{cop})^*$ and $A^{*\,op} = (A^{cop})^*$ part (a) follows by Proposition 7.7.5 again.

Suppose that A has antipode S and define $\tau' : A \times B \longrightarrow k$ by $\tau'(a,b) = \tau(S(a),b)$ for all $a \in A$ and $b \in B$. The calculation

$$\begin{aligned}
\tau(a_{(1)}, b_{(1)})\tau'(a_{(2)}, b_{(2)}) &= \tau(a_{(1)}, b_{(1)})\tau(S(a_{(2)}), b_{(2)}) \\
&= \tau(a_{(1)}S(a_{(2)}), b) \\
&= \tau(\epsilon(a)1, b) \\
&= \epsilon(a)\epsilon(b)
\end{aligned}$$

for all $a \in A$ and $b \in B$ shows that τ' is a right inverse for τ. By a similar calculation τ' is a left inverse for τ and thus part (b) is established. Part

(c) follows in a similar manner. As for part (d), one needs only show that $\tau'' : A \times B \longrightarrow k$ defined by $\tau''(a,b) = \tau(S_A^{cop}(a), S_B(b))$ for all $a \in A$ and $b \in B$ is a left inverse for τ^{-1}. □

The main result of this section is easy to prove and important in practice.

Proposition 7.7.10. *Suppose that A and B are bialgebras over the field k and suppose $\tau : A \times B \longrightarrow k$ is a convolution invertible skew pairing. Consider the map $\sigma_\tau : (A \otimes B) \otimes (A \otimes B) \longrightarrow k$ defined by*

$$\sigma_\tau(a \otimes b, a' \otimes b') = \epsilon(a)\tau(a', b)\epsilon(b')$$

for all $a, a' \in A$ and $b, b' \in B$. Then σ_τ is a left 2-cocycle for $A \otimes B$ whose inverse is given by

$$\sigma_\tau^{-1}(a \otimes b, a' \otimes b') = \epsilon(a)\tau^{-1}(a', b)\epsilon(b')$$

for all $a, a' \in A$ and $b, b' \in B$. □

The product of the bialgebra $(A \otimes B)_{\sigma_\tau}$ of the preceding proposition is

$$(a \otimes b) \cdot (a' \otimes b') = a\tau(a'_{(1)}, b_{(1)})a'_{(2)} \otimes b_{(2)}\tau^{-1}(a'_{(3)}, b_{(3)})b' \tag{7.8}$$

for all $a, a' \in A$ and $b, b' \in B$.

Exercises

Throughout the following exercises, unless otherwise stated, A is an algebra over k and C is a coalgebra over k.

Exercise 7.7.1. Prove Proposition 7.7.10.

The notion of pairing of bialgebras can be generalized in two ways, according to Definition 2.6.11 or

Definition 7.7.11. Let C be a coalgebra over k and A be an algebra over k. A *pairing of C and A* is a bilinear form $\tau : C \times A \longrightarrow k$ which satisfies $\tau(c, aa') = \tau(c_{(1)}, a)\tau(c_{(2)}, a')$ and $\tau(c, 1) = \epsilon(c)$ for all $c \in C$ and $a, a' \in A$.

Exercise 7.7.2. Let $\tau : A \times C \longrightarrow k$ be a bilinear form. Show that the following are equivalent:

(a) $\tau_\ell : A \longrightarrow C^*$ is an algebra map.

(b) $\mathrm{Im}(\tau_r) \subseteq A^o$ and $\tau_r : C \longrightarrow A^o$ is a coalgebra map.

(c) τ is a pairing of A and C.

Exercise 7.7.3. Let $\tau : C{\times}A \longrightarrow k$ be a bilinear form. Show that the following are equivalent:

(a) $\tau_r : A \longrightarrow C^*$ is an algebra map.

(b) $\mathrm{Im}(\tau_\ell) \subseteq A^o$ and $\tau_\ell : C \longrightarrow A^o$ is a coalgebra map.

(c) τ is a pairing of C and A.

[Hint: Note that $\tau^{op} : A{\times}C \longrightarrow k$, where $\tau^{op}(a,c) = \tau(c,a)$ for all $a \in A$ and $c \in C$, is a pairing of A and C if and only if τ is a pairing of C and A.]

Exercise 7.7.4. Define pairing of bialgebras in terms of pairings of algebras and coalgebras and pairings of coalgebras and algebras. Show that Lemma 7.7.5 follows from Exercises 7.7.2 and 7.7.3.

Exercise 7.7.5. Prove part (b) of Proposition 7.7.5.

Exercise 7.7.6. Prove Proposition 7.7.8.

7.8 Twists of bialgebras

The purpose of this section is to describe the dual notion of left 2-cocycle σ for a bialgebra A over the field k and to describe a construction dual to A_σ. To do this we start with a finite-dimensional bialgebra $(A, m, \eta, \Delta, \epsilon)$ over k and a bilinear form $\sigma : A{\times}A \longrightarrow k$.

Let $\sigma_{lin} : A{\otimes}A \longrightarrow k$ be the associated linear map and let $\overline{1} \in k^*$ be defined by $\overline{1}(1) = 1$. Set $J = \sigma^*_{lin}(\overline{1}) \in (A{\otimes}A)^*$. Since A is finite-dimensional, we make the usual identification $(A{\otimes}A)^* = A^*{\otimes}A^*$ and write $J = \sum_{i=1}^r J_i{\otimes}J^i$ which is expressed $J = J_i{\otimes}J^i$ using the Einstein summation convention. Thus $\sigma(a,b) = J(a{\otimes}b) = J_i(a)J^i(b)$ for all $a \in A$ and $b \in B$.

Observe that σ has an inverse if and only if J has an inverse in the algebra $A^*{\otimes}A^*$ in which case $\sigma^{-1}(a,b) = J^{-1}(a{\otimes}b)$ for all $a \in A$ and $b \in B$. Since

$$\begin{aligned}
&\sigma(a_{(1)}, b_{(1)})\sigma(a_{(2)}b_{(2)}, c)\\
&= J_i(a_{(1)})J^i(b_{(1)})J_j(a_{(2)}b_{(2)})J^j(c)\\
&= J_i(a_{(1)})J^i(b_{(1)})J_{j\,(1)}(a_{(2)})J_{j\,(2)}(b_{(2)})J^j(c)\\
&= (J_iJ_{j\,(1)}{\otimes}J^iJ_{j\,(2)}{\otimes}J^j)(a{\otimes}b{\otimes}c)
\end{aligned}$$

and likewise

$$\begin{aligned}&\sigma(b_{(1)},c_{(1)})\sigma(a,b_{(2)}c_{(2)})\\&=J_i(b_{(1)})J^i(c_{(1)})J_j(a)J^j(b_{(2)}c_{(2)})\\&=J_i(b_{(1)})J^i(c_{(1)})J_j(a){J^j}_{(1)}(b_{(2)}){J^j}_{(2)}(c_{(2)})\\&=(J_j\otimes J_i{J^j}_{(1)}\otimes J^i{J^j}_{(2)})(a\otimes b\otimes c)\end{aligned}$$

it follows that σ is a left 2-cocycle for A if and only if J is invertible and

$$(J\otimes 1)\left((\Delta\otimes \mathrm{I}_{A^*})(J)\right)=(1\otimes J)\left((\mathrm{I}_{A^*}\otimes\Delta)(J)\right). \tag{7.9}$$

Thus the solution J to (7.9) is a left twist for A^*.

Definition 7.8.1. Let A be a bialgebra over the field k. A *left twist for* A is an invertible $J\in A\otimes A$ which satisfies $(J\otimes 1)((\Delta\otimes \mathrm{I}_A)(J))=(1\otimes J)((\mathrm{I}_A\otimes\Delta)(J))$. A *right twist for* A is a $J\in A\otimes A$ which is a left twist for A^{op}.

Suppose $J\in A\otimes A$ is invertible. Write $J=J_i\otimes J^i$ using the Einstein summation convention. Then J is a left twist for A if and only if

$$J\Delta(J_i)\otimes J^i=J_i\otimes J\Delta(J^i) \tag{7.10}$$

and J is a right twist for A if and only if

$$((\Delta\otimes \mathrm{I}_A)(J))(J\otimes 1)=((\mathrm{I}_A\otimes\Delta)(J))(1\otimes J);$$

that is

$$\Delta(J_i)J\otimes J^i=J_i\otimes\Delta(J^i)J. \tag{7.11}$$

A right twist is also called a Drinfel'd twist. See the chapter notes.

Continuing our discussion, suppose that σ is a left 2-cocycle for A. Then $A_\sigma=(A,m_\sigma,\eta,\Delta,\epsilon)$ is a bialgebra over k by part (a) of Proposition 7.7.3. Its dual is the bialgebra $(A_\sigma)^*=(A^*,\Delta^*,\epsilon^*,(m_\sigma)^*,\eta^*)$. We wish to determine $\Delta_{(A_\sigma)^*}=(m_\sigma)^*$. Let $p\in A^*$. Since

$$\begin{aligned}\Delta_{(A_\sigma)^*}(p)(a\otimes b)&=p(m_\sigma(a\otimes b))\\&=p(\sigma(a_{(1)},b_{(1)})a_{(2)}b_{(2)}\sigma^{-1}(a_{(3)},b_{(3)}))\\&=\sigma(a_{(1)},b_{(1)})p_{(1)}(a_{(2)})p_{(2)}(b_{(2)})\sigma^{-1}(a_{(3)},b_{(3)})\\&=J_i(a_{(1)})J^i(b_{(1)})p_{(1)}(a_{(2)})p_{(2)}(b_{(2)})(J^{-1})_j(a_{(3)})(J^{-1})^j(b_{(3)})\\&=\left(J_ip_{(1)}(J^{-1})_j\otimes J^ip_{(2)}(J^{-1})^j\right)(a\otimes b)\end{aligned}$$

for all $a,b\in A$ we conclude that $\Delta_{(A_\sigma)^*}(p)=J\Delta_{A^*}(p)J^{-1}$.

If $(A, m, \eta, \Delta, \epsilon)$ is a bialgebra and $J \in A{\otimes}A$ is a left twist for A, we will show that $(A, m, \eta, \Delta^J, \epsilon)$ is a bialgebra over k, where $\Delta^J(a) = J\Delta(a)J^{-1}$ for all $a \in A$. First an analog of Lemma 7.7.2.

Lemma 7.8.2. *Let A be a bialgebra over the field k and suppose that J is a left twist for A. Write $J = J_i{\otimes}J^i$ and $J^{-1} = (J^{-1})_j{\otimes}(J^{-1})^j$. Then:*

(a) $\epsilon(J_i)J^i$ *and* $\epsilon((J^{-1})_j)(J^{-1})^j$ *are multiplicative inverses.*
(b) $\epsilon(J_i)J^i = J_i\epsilon(J^i) = \epsilon(J_i)\epsilon(J^i)1$.
(c) J^{-1} *is a right twist for the bialgebra A.*
(d) $\epsilon((J^{-1})_i)(J^{-1})^i = (J^{-1})_i\epsilon((J^{-1})^i) = \epsilon((J^{-1})_i)\epsilon((J^{-1})^i)1$.

Proof. Part (c) is easy to see on computing the inverse of both sides of (7.10). Part (d) follows by parts (b) and (c). As for part (a), we apply the algebra map $\epsilon{\otimes}\mathrm{I}_A : A{\otimes}A \longrightarrow k$ to both sides of the equations which express that J and J^{-1} are inverses. It remains to show part (b).

Let $u = \epsilon(J_i)J^i$ and $v = J_i\epsilon(J^i)$. Then applying the algebra map $\mathrm{I}_A{\otimes}\epsilon{\otimes}\mathrm{I}_A : A{\otimes}A{\otimes}A \longrightarrow A{\otimes}A$ to both sides of (7.10) we obtain $(v{\otimes}1)J = (1{\otimes}u)J$. Since J is invertible $v{\otimes}1 = 1{\otimes}u$ from which we deduce $\epsilon(v) = \epsilon(u)$ and $v = \epsilon(u)1 = \epsilon(v)1 = u$. □

Proposition 7.8.3. *Let $(A, m, \eta, \Delta, \epsilon)$ be a bialgebra over the field k and suppose that $J = J_i{\otimes}J^i \in A{\otimes}A$ is a left twist for A. Then:*

(a) $(A, m, \eta, \Delta^J, \epsilon)$ *is a bialgebra over k.*
(b) *Suppose that A has antipode S. Then A^J has antipode S^J defined by* $S^J(a) = J_iS(J^i)S(a)S((J^{-1})_j)(J^{-1})^j$ *for all $a \in A$.*

Proof. Since Δ is an algebra map it follows that Δ^J is an algebra map. That $(\epsilon{\otimes}\mathrm{I}_A){\circ}\Delta^J = \mathrm{I}_A$ follows by parts (a) and (b) of Lemma 7.8.2 and that ϵ is an algebra map. That $(\mathrm{I}_A{\otimes}\epsilon){\circ}\Delta^J = \mathrm{I}_A$ follows by parts (b), (c), and (d) of the same. The fact that J is a left twist for A and J^{-1} is a right twist for A accounts for the coassociativity of Δ^J. This completes our proof of part (a). To show part (b) we note that

$$\begin{aligned}
&\left(m\circ(S^J\otimes \mathrm{I}_A)\circ\Delta^J\right)(a)\\
&= S^J(J_i a_{(1)}(J^{-1})_j)(J^i a_{(2)}(J^{-1})^j)\\
&= J_\ell S(J^\ell)S(J_i a_{(1)}(J^{-1})_j)S((J^{-1})_m)(J^{-1})^m J^i a_{(2)}(J^{-1})^j\\
&= J_\ell S(J^\ell)S((J^{-1})_j)S(a_{(1)})S(J_i)S((J^{-1})_m)(J^{-1})^m J^i a_{(2)}(J^{-1})^j\\
&= J_\ell S(J^\ell)S((J^{-1})_j)S(a_{(1)})S((J^{-1})_m J_i)(J^{-1})^m J^i a_{(2)}(J^{-1})^j\\
&= J_\ell S(J^\ell)S((J^{-1})_j)S(a_{(1)})a_{(2)}(J^{-1})^j\\
&= \epsilon(a)J_\ell S(J^\ell)S((J^{-1})_j)(J^{-1})^j,
\end{aligned}$$

and thus $\left(m\circ(S^J\otimes \mathrm{I}_A)\circ\Delta^J\right)(a) = \epsilon(a)J_\ell S(J^\ell)S((J^{-1})_j)(J^{-1})^j$, which is also equal to $\left(m\circ(\mathrm{I}_A\otimes S^J)\circ\Delta^J\right)(a)$ by a similar calculation. Now $(\mathrm{I}_A\otimes J^{-1})(J\Delta(J_i)\otimes J^i) = J_i\otimes\Delta(J^i)$ by (7.10). We apply $\mathrm{I}_A\otimes S\otimes \mathrm{I}_A$ to both sides of this equation and then $m\circ(m\otimes \mathrm{I}_A)$ to both sides of the result to obtain $J_\ell S(J^\ell)S((J^{-1})_j)(J^{-1})^j = 1$. □

Definition 7.8.4. The bialgebra of part (a) of Proposition 7.8.3 is denoted by A^J.

To complete the duality picture for left 2-cocycles and left twists:

Proposition 7.8.5. *Let A be a bialgebra over the field k and suppose that J is a left twist for A. Define $\sigma : A^o\times A^o \longrightarrow k$ by $\sigma(a^o, b^o) = (a^o\otimes b^o)(J)$ for all $a^o, b^o \in A^o$. Then:*

(a) *σ is a left 2-cocycle for A^o.*
(b) *$(A^J)^o = (A^o)_\sigma$ as bialgebras.*

□

Exercises

Exercise 7.8.1. Prove Proposition 7.8.5.

Exercise 7.8.2. State and prove the analog of Proposition 7.8.5, where "left twist" is replaced by "right twist".

7.9 Filtrations and gradings on Hopf algebras

Here we add a bit to the material of Section 5.6.

Definition 7.9.1. A *filtration of a Hopf algebra H with antipode S over the field k* is a bialgebra filtration $\mathcal{F} = \{V_n\}_{n=0}^{\infty}$ such that $S(V_n) \subseteq V_n$ for all $n \geq 0$. In this case the pair $(H, \mathcal{F})$ is a *filtered Hopf algebra over k*.

Our main result is:

Proposition 7.9.2. *Let $(H, \mathcal{F})$ be a filtered Hopf algebra over the field k. Then $\mathrm{gr}_{\mathcal{F}}(H)$ is a graded Hopf algebra over k with antipode $\boldsymbol{S}$ which is determined by $\boldsymbol{S}(h + V_{n-1}) = S(h) + V_{n-1}$ for all $n \geq 0$ and $h \in V_n$, where $\mathcal{F} = \{V_n\}_{n=0}^{\infty}$ and S is the antipode of H.*

Proof. Let $n \geq 0$. We follow the notation of Sections 4.4 and 5.6. Using the commutative diagrams following Proposition 4.4.9 and Definition 5.6.3 we observe that

$$
\begin{array}{ccccccc}
V_n & \xrightarrow{\ \Delta\ } & \sum_{\ell=0}^{n} V_{n-\ell}\otimes V_\ell & \xrightarrow{\ \mathrm{I}\otimes S\ } & \sum_{\ell=0}^{n} V_{n-\ell}\otimes V_\ell & \xrightarrow{\ m\ } & V_n \\
\Big\downarrow {\scriptstyle \pi_{n,\mathcal{F}}} & & \Big\downarrow {\scriptstyle \Pi_{n,\mathcal{F}}} & & \Big\downarrow {\scriptstyle \Pi_{n,\mathcal{F}}} & & \Big\downarrow {\scriptstyle \pi_{n,\mathcal{F}}} \\
\overline{V}_n & \xrightarrow[\ \boldsymbol{\Delta}\ ]{} & \bigoplus_{\ell=0}^{n} \overline{V}_{n-\ell}\otimes\overline{V}_\ell & \xrightarrow[\ \mathrm{I}\otimes\boldsymbol{S}\ ]{} & \bigoplus_{\ell=0}^{n} \overline{V}_{n-\ell}\otimes\overline{V}_\ell & \xrightarrow[\ \boldsymbol{m}\ ]{} & \overline{V}_n
\end{array}
$$

is the composition of commutative diagrams which means that the diagram

$$
\begin{array}{ccc}
V_n & \xrightarrow{\ m\circ(\mathrm{I}\otimes S)\circ\Delta\ } & V_n \\
\Big\downarrow {\scriptstyle \pi_{n,\mathcal{F}}} & & \Big\downarrow {\scriptstyle \pi_{n,\mathcal{F}}} \\
\overline{V}_n & \xrightarrow[\ \boldsymbol{m}\circ(\mathrm{I}\otimes\boldsymbol{S})\circ\boldsymbol{\Delta}\ ]{} & \overline{V}_n
\end{array}
$$

commutes and therefore the diagram

$$\begin{array}{ccc} V_n & \xrightarrow{\eta\circ\epsilon} & V_n \\ \Big\downarrow{\scriptstyle \pi_{n,\mathcal{F}}} & & \Big\downarrow{\scriptstyle \pi_{n,\mathcal{F}}} \\ \overline{V}_n & \xrightarrow[\boldsymbol{m}\circ(\mathrm{I}\otimes\boldsymbol{S})\circ\boldsymbol{\Delta}]{} & \overline{V}_n \end{array}$$

commutes. It follows that $\boldsymbol{S}$ is a left convolution inverse for $\mathrm{I}_{\mathrm{gr}_{\mathcal{F}}(H)}$. Likewise $\boldsymbol{S}$ is a right convolution inverse for $\mathrm{I}_{\mathrm{gr}_{\mathcal{F}}(H)}$. □

The coradical filtration of a bialgebra H over k is a bialgebra filtration for H if and only if H_0 is a sub-bialgebra of H by Lemma 5.6.5.

Lemma 7.9.3. *Suppose H is a Hopf algebra over the field k. Then the coradical filtration of H is a Hopf algebra filtration for H if and only if H_0 is a Hopf subalgebra of H.*

Proof. If the coradical filtration of H is a Hopf algebra filtration then H_0 is a Hopf subalgebra of H by definition. Suppose H_0 is a Hopf subalgebra of H. In light of Lemma 5.6.5 we need only show that $S(H_n) \subseteq H_n$ for all $n \geq 0$, where S is the antipode of H. By assumption $S(H_0) \subseteq H_0$.

Suppose $n > 0$ and $S(H_m) \subseteq H_m$ for all $m < n$. Since $S : H \longrightarrow H^{cop}$ is a coalgebra map, by virtue of Exercises 2.4.7 and 4.1.12 we have

$$S(H_n) = S(H_0 \wedge H_{n-1}) \subseteq S(H_{n-1}) \wedge S(H_0) \subseteq H_{n-1} \wedge H_0 = H_n.$$

Thus $S(H_n) \subseteq H_n$ for all $n \geq 0$ by induction on n. □

Proposition 7.9.4. *Let H be a bialgebra over the field k whose coradical H_0 is a Hopf subalgebra of H. Then $\mathrm{gr}(H)$ is a coradically graded Hopf algebra with the coalgebra structure described at the end of Section 4.4 and the algebra structure described in (5.3).* □

Exercise

Exercise 7.9.1. Show that $H_{n,q} \simeq \mathrm{gr}(H_{n,q})$. See Section 7.3.

7.10 The cofree pointed irreducible Hopf algebra on an algebra

Let $(\pi, T^{co}(A))$ be the cofree bialgebra on an algebra A over the field k defined in Section 5.5. Then the irreducible component $\mathrm{H}_{pi}(A) = T^{co}(A)_{k1}$ of the simple subcoalgebra $k1$ of $T^{co}(A)$ is a sub-bialgebra of $T^{co}(A)$ by part (b) of Proposition 5.1.16. Therefore $\mathrm{H}_{pi}(A)$ is a Hopf algebra by Proposition 7.6.3. Let $\pi_{\mathrm{H}_{pi}(A)} = \pi|\mathrm{H}_{pi}(A)$.

Theorem 7.10.1. *Let A be an algebra over the field k. The pair $(\pi_{\mathrm{H}_{pi}(A)}, \mathrm{H}_{pi}(A))$ satisfies the following:*

(a) *$\mathrm{H}_{pi}(A)$ is a pointed irreducible Hopf algebra over k and the map $\pi_{\mathrm{H}_{pi}(A)} : \mathrm{H}_{pi}(A) \longrightarrow A$ is an algebra map.*
(b) *If B is a pointed irreducible Hopf algebra over k and $f : B \longrightarrow A$ is an algebra map there exists a Hopf algebra map $F : B \longrightarrow \mathrm{H}_{pi}(A)$ determined by $\pi_{\mathrm{H}_{pi}(A)} \circ F = f$.*

Proof. By virtue of our remarks above and Theorem 5.5.1 part (a) follows and, assuming the hypothesis of part (b), there exists a bialgebra map $F : B \longrightarrow T^{co}(A)$ which is determined by $\pi_{\mathrm{H}_{pi}(A)} \circ F = f$. Since $F(1) = 1$, B is pointed irreducible coalgebra, and F is a coalgebra map, $F(B)$ is an irreducible subcoalgebra of $T^{co}(A)$ with simple subcoalgebra $k1$ by Exercise 4.2.4. Thus $F(B) \subseteq T^{co}(A)_{k1} = \mathrm{H}_{pi}(A)$. Since bialgebra maps of Hopf algebras are Hopf algebra maps, our proof is complete. □

Definition 7.10.2. Let A be an algebra over the field k. A *cofree pointed irreducible Hopf algebra on the algebra A over k* is any pair $(\pi_{\mathrm{H}_{pi}(A)}, \mathrm{H}_{pi}(A))$ which satisfies the conclusion of Theorem 7.10.1.

Let $\mathrm{B}_{pi}(A)$ be the sum of all cocommutative subcoalgebras of $\mathrm{H}_{pi}(A)$. Then $\mathrm{B}_{pi}(A)$ is a sub-bialgebra of $\mathrm{H}_{pi}(A)$ by Exercise 5.1.5. Set $\pi_{\mathrm{B}_{pi}(A)} = \pi_{\mathrm{H}_{pi}(A)}|\mathrm{B}_{pi}(A)$. The proof of the following analog of Theorem 7.10.1 for cocommutative Hopf algebras is left to the reader.

Theorem 7.10.3. *Let A be an algebra over the field k. The pair $(\pi_{\mathrm{B}_{pi}(A)}, \mathrm{B}_{pi}(A))$ satisfies the following:*

(a) *$\mathrm{B}_{pi}(A)$ is a cocommutative pointed irreducible Hopf algebra over k and $\pi_{\mathrm{B}_{pi}(A)} : \mathrm{B}_{pi}(A) \longrightarrow A$ is an algebra map.*
(b) *If B is a cocommutative pointed irreducible Hopf algebra over k and $f : B \longrightarrow A$ is also an algebra map then there exists a Hopf algebra*

map $F : B \longrightarrow \mathrm{B}_{pi}(A)$ *determined by* $\pi_{\mathrm{B}_{pi}(A)}\circ F = f$.

□

Definition 7.10.4. Let A be an algebra over the field k. A *cofree cocommutative pointed irreducible Hopf algebra on the algebra A over k* is any pair $(\pi_{\mathrm{B}_{pi}(A)}, \mathrm{B}_{pi}(A))$ which satisfies the conclusion of Theorem 7.10.1.

Exercise

Exercise 7.10.1. Prove Theorem 7.10.3.

7.11 The shuffle algebra

Let A be an algebra over k and let $(\pi_{\mathrm{Sh}(A)}, \mathrm{Sh}(A))$ be the cofree pointed irreducible coalgebra on the vector space A of Section 4.5. We would like to endow $\mathrm{Sh}(A)$ with an algebra structure which together with its coalgebra structure forms a bialgebra (hence a Hopf algebra) structure on $\mathrm{Sh}(A)$. For such an algebra structure $\mathbf{1} = \mathbf{1}_A$ is the multiplicative neutral element and $\pi(\mathbf{1}) = 0$, where $\pi = \pi_{\mathrm{Sh}(A)}$. Thus $\pi : \mathrm{Sh}(A) \longrightarrow A$ is not an algebra map. For this reason we will merely assume that A is a vector space with associative multiplication. Since we are using the general term "algebra over" k in a specific way in this book, for clarity we make the following definition.

Definition 7.11.1. Let A be a vector space over k. An *associative structure on A* is a linear map $m : A\otimes A \longrightarrow A$ which satisfies the associative axiom (2.1).

Apropos of the definition, we write ab for $m(a\otimes b)$, where $a, b \in A$. The reader should find comparison of this section with Section 7.10 interesting.

Suppose A is a vector space over k with an associative structure m. The coalgebra $C = \mathrm{Sh}(A)\otimes\mathrm{Sh}(A)$ is pointed irreducible by part (b) of Exercise 4.1.3 with $\mathrm{G}(C) = \{\mathbf{1}\otimes\mathbf{1}\}$. Let $f_m : C \longrightarrow A$ be the linear map $f_m = m\circ(\pi\otimes\pi) + \epsilon\otimes\pi + \pi\otimes\epsilon$; thus

$$f_m(a\otimes b) = \pi(a)\pi(b) + \epsilon(a)\pi(b) + \pi(a)\epsilon(b)$$

for all $a, b \in \mathrm{Sh}(A)$. Observe that $f_m(\mathbf{1}\otimes\mathbf{1}) = 0$ since $\pi(\mathbf{1}) = 0$.

By Theorem 4.5.1 there is a coalgebra map $\boldsymbol{m} : C \longrightarrow \mathrm{Sh}(A)$ determined by $\pi\circ\boldsymbol{m} = f_m$. Likewise for the zero map $0 : k \longrightarrow A$ there is a coalgebra map $\boldsymbol{\eta} : k \longrightarrow \mathrm{Sh}(A)$ determined by $\pi\circ\boldsymbol{\eta} = 0$.

We claim that $(\mathrm{Sh}(A), \boldsymbol{m}, \boldsymbol{\eta})$ is an algebra and thus $\mathrm{Sh}(A)$ with its coalgebra structure and this algebra structure is a bialgebra over k; hence it is a Hopf algebra over k since $\mathrm{Sh}(A)$ is pointed irreducible. First of all $\boldsymbol{\eta}(1) = \mathbf{1}$ since k and $\mathrm{Sh}(A)$ are pointed irreducible and $\boldsymbol{\eta}$ is a coalgebra map. To show that $\boldsymbol{m}\circ(\mathrm{I}_{\mathrm{Sh}(A)}\otimes\boldsymbol{\eta}) = \mathrm{I}_{\mathrm{Sh}(A)}$ we note $\boldsymbol{m}\circ(\mathrm{I}_{\mathrm{Sh}(A)}\otimes\boldsymbol{\eta}), \mathrm{I}_{\mathrm{Sh}(A)} : C \longrightarrow \mathrm{Sh}(A)$ are coalgebra maps, where $C = \mathrm{Sh}(A)$. Let $f, g : C \longrightarrow A$ be the linear maps defined by $f = \pi\circ(\boldsymbol{m}\circ(\mathrm{I}_{\mathrm{Sh}(A)}\otimes\boldsymbol{\eta}))$ and $g = \pi\circ\mathrm{I}_{\mathrm{Sh}(A)}$. Since $\pi(\mathbf{1}) = 0$ it follows that

$$f(a) = \pi(\boldsymbol{m}(a\otimes\mathbf{1})) = f_m(a\otimes\mathbf{1})$$

$$= \pi(a)\pi(\mathbf{1}) + \epsilon(a)\pi(\mathbf{1}) + \pi(a)\epsilon(\mathbf{1}) = \pi(a) = g(a)$$

for all $a \in \mathrm{Sh}(A)$. Therefore $f = g = \pi$; in particular $f(\mathbf{1}) = g(\mathbf{1}) = \pi(\mathbf{1}) = 0$. Therefore $\boldsymbol{m}\circ(\mathrm{I}_{\mathrm{Sh}(A)}\otimes\boldsymbol{\eta}) = \mathrm{I}_{\mathrm{Sh}(A)}$ by part (a) of Theorem 4.5.1. Likewise $\boldsymbol{m}\circ(\boldsymbol{\eta}\otimes\mathrm{I}_{\mathrm{Sh}(A)}) = \mathrm{I}_{\mathrm{Sh}(A)}$. We have shown that $a\mathbf{1} = a = \mathbf{1}a$ for all $a \in \mathrm{Sh}(A)$.

To show that $\boldsymbol{m}$ satisfies the associative axiom we consider the pointed irreducible coalgebra $C = \mathrm{Sh}(A)\otimes\mathrm{Sh}(A)\otimes\mathrm{Sh}(A)$ whose grouplike element is $\mathbf{1}\otimes\mathbf{1}\otimes\mathbf{1}$ and the coalgebra maps $\boldsymbol{m}\circ(\boldsymbol{m}\otimes\mathrm{I}_{\mathrm{Sh}(A)}), \boldsymbol{m}\circ(\mathrm{I}_{\mathrm{Sh}(A)}\otimes\boldsymbol{m}) :\longrightarrow \mathrm{Sh}(A)$. Let $f, g : C \longrightarrow A$ be defined by

$$f = \pi\circ(\boldsymbol{m}\circ(\boldsymbol{m}\otimes\mathrm{I}_{\mathrm{Sh}(A)})) \text{ and } g = \pi\circ(\boldsymbol{m}\circ(\mathrm{I}_{\mathrm{Sh}(A)}\otimes\boldsymbol{m})).$$

It is easy to see that

$$f - g = (m\circ(m\otimes\mathrm{I}_A) - m\circ(\mathrm{I}_A\otimes m))\circ(\pi\otimes\pi\otimes\pi)$$

and thus $f = g$ since m satisfies the associative axiom. Since $f(\mathbf{1}\otimes\mathbf{1}\otimes\mathbf{1}) = \pi((\mathbf{11})\mathbf{1}) = \pi(\mathbf{1}) = 0$ and $g(\mathbf{1}\otimes\mathbf{1}\otimes\mathbf{1}) = \pi(\mathbf{1}(\mathbf{11})) = \pi(\mathbf{1}) = 0$, by part (b) of Theorem 4.5.1 we conclude that $\boldsymbol{m}\circ(\boldsymbol{m}\otimes\mathrm{I}_{\mathrm{Sh}(A)}) = \boldsymbol{m}\circ(\mathrm{I}_{\mathrm{Sh}(A)}\otimes\boldsymbol{m})$. Thus $(\mathrm{Sh}(A), \boldsymbol{m}, \boldsymbol{\eta})$ is an algebra over k.

Observe that $\pi : \mathrm{Sh}(A) \longrightarrow A$ satisfies

$$\pi(ab) = \pi(a)\pi(b) + \epsilon(a)\pi(b) + \pi(a)\epsilon(b)$$

for all $a, b \in \mathrm{Sh}(A)$. The pair $(\pi, \mathrm{Sh}(A))$ satisfies a universal mapping property.

Theorem 7.11.2. *Let A be a vector space over the field k with an associative structure m. Then the pair $(\pi, \mathrm{Sh}(A))$ satisfies the following:*

(a) $\mathrm{Sh}(A)$ *is a pointed irreducible Hopf algebra over* k *and the linear map* $\pi : \mathrm{Sh}(A) \longrightarrow A$ *satisfies* $\pi(ab) = \pi(a)\pi(b) + \epsilon(a)\pi(b) + \pi(a)\epsilon(b)$ *for all* $a, b \in \mathrm{Sh}(A)$ *and* $\pi(1) = 0$.

(b) *If* B *is a pointed irreducible Hopf algebras over* k *and* $f : B \longrightarrow A$ *is a linear map which satisfies* $f(ab) = f(a)f(b) + \epsilon(a)f(b) + f(a)\epsilon(b)$ *for all* $a, b \in B$ *and* $f(1) = 0$ *then there is a map of Hopf algebras* $F : B \longrightarrow \mathrm{Sh}(A)$ *determined by* $\pi \circ F = f$.

Proof. We need only show part (b). Assume the hypothesis of part (b). By part (b) of Theorem 4.5.1 there is a coalgebra map $F : B \longrightarrow \mathrm{Sh}(A)$ determined by $\pi \circ F = f$. Thus uniqueness follows. We need only show that F is an algebra map. Since B and $\mathrm{Sh}(A)$ are pointed irreducible and F is a coalgebra map, $F(1) = \mathbf{1}$. Let m_B be the multiplication map of B. Then $\boldsymbol{m} \circ (F \otimes F), F \circ m_B : C \longrightarrow \mathrm{Sh}(A)$ are coalgebra maps, where $C = B \otimes B$. Note that C is pointed irreducible and $\mathrm{G}(C) = \{1 \otimes 1\}$. Let $f, g : C \longrightarrow A$ be the maps defined by $f = \pi \circ (\boldsymbol{m} \circ (F \otimes F))$ and $g = \pi \circ (F \circ m_B)$ and let $b, c \in B$. Then the calculations

$$
\begin{aligned}
f(b \otimes c) &= \pi(\boldsymbol{m}(F(b) \otimes F(c))) \\
&= \pi(F(b))\pi(F(c)) + \epsilon(F(b))\pi(F(c)) + \pi(F(b))\epsilon(F(c)) \\
&= f(b)f(c) + \epsilon(b)f(c) + f(b)\epsilon(c) \\
&= f(bc)
\end{aligned}
$$

and

$$
g(b \otimes c) = \pi(F(bc)) = f(bc)
$$

show that $f = g$. Since $f(1 \otimes 1) = g(1 \otimes 1) = g(1 \cdot 1) = g(1) = 0$, by part (b) of Theorem 4.5.1 again $\boldsymbol{m} \circ (F \otimes F) = F \circ m_B$. Thus F is an algebra map. □

Definition 7.11.3. Let A be a vector space over the field k with an associative structure m. A *cofree pointed irreducible Hopf algebra on the vector space* A *with associative structure* m is a pair $(\pi_{\mathrm{Sh}(A)}, \mathrm{Sh}(A))$ which satisfies the conclusion of Theorem 7.11.2. The *shuffle algebra on the vector space* V *over* k is $\mathrm{Sh}(V)$ where $V = A$ and $m = 0$.

Let V be a vector space over the field k. For the remainder of this section we will study the shuffle algebra $\mathrm{Sh}(V)$. We will give an explicit formula for its multiplication which will show that the coalgebra grading on $\mathrm{Sh}(V)$ is a Hopf algebra grading. An important consequence of the preceding theorem is:

Corollary 7.11.4. *Let H be a pointed irreducible strictly graded Hopf algebra over k and let $V = \mathrm{P}(H)$. Then there is a one-one map of Hopf algebras $F : H \longrightarrow \mathrm{Sh}(V)$ such that $F(V) = \mathrm{P}(\mathrm{Sh}(V))$.*

Proof. Write $H = \bigoplus_{n=0}^{\infty} H(n)$ as a strictly graded Hopf algebra. Then $V = \mathrm{P}(H) = H(1)$. Let $\mathrm{pr} : H \longrightarrow V$ be the projection onto V. Then $\mathrm{pr}(ab) = \epsilon(a)\mathrm{pr}(b) + \mathrm{pr}(a)\epsilon(b)$, and therefore

$$\mathrm{pr}(ab) = \mathrm{pr}(a)\mathrm{pr}(b) + \epsilon(a)\mathrm{pr}(b) + \mathrm{pr}(a)\epsilon(b),$$

for all $a, b \in H$. Also $\mathrm{pr}(1) = 0$. By part (b) of Theorem 7.11.2 there is a Hopf algebra map $F : H \longrightarrow \mathrm{Sh}(V)$ determined by $\pi_{\mathrm{Sh}(V)} \circ F = \mathrm{pr}$. Since $f = \mathrm{pr}$ satisfies the requirements of the proof of Proposition 4.5.3, F satisfies the conclusion of the same. □

We now turn our attention to an explicit description of the multiplication of $\mathrm{Sh}(V)$ using the model of the coalgebra $\mathrm{Sh}(V)$ described by (4.11)–(4.15) of Section 4.5. Let $\eta : k \longrightarrow \mathrm{Sh}(V)$ be the coalgebra map defined by $\eta(1) = \mathbf{1}$, where $\mathbf{1} = \mathbf{1}_V$. Suppose $\boldsymbol{m}' : \mathrm{Sh}(V) \otimes \mathrm{Sh}(V) \longrightarrow \mathrm{Sh}(V)$ is a coalgebra map which satisfies $\boldsymbol{m}'(\mathbf{1} \otimes a) = a = \boldsymbol{m}'(a \otimes \mathbf{1})$ for all $a \in \mathrm{Sh}(A)$ and $\boldsymbol{m}'(\mathrm{Sh}(V)(m) \otimes \mathrm{Sh}(V)(n)) \subseteq \mathrm{Sh}(V)(m+n)$ for all $m, n \geq 0$. Then $\pi \circ \boldsymbol{m}' = \pi \otimes \pi + \epsilon \otimes \pi + \pi \otimes \epsilon$ since $V^2 = (0)$. Thus $\boldsymbol{m}' = \boldsymbol{m}$ by part (b) of Theorem 4.5.1. We will find such an $\boldsymbol{m}'$. Our endeavor leads to interesting combinatorics.

First of all suppose that V is finite-dimensional. Then $T(V^*)$ is a bialgebra over k, where $\Delta_{T(V^*)}(v^*) = v^* \otimes 1 + 1 \otimes v^*$ and $\epsilon_{T(V^*)}(v^*) = 0$ for all $v^* \in V^*$. Consider the dual bialgebra $T(V^*)^o$. First of all note that $1_{T(V^*)^o} = \epsilon_{T(V^*)} = \mathbf{1}$. Thus $\mathrm{Sh}(V) = (T(V^*)^o)_{k1_{T(V^*)}}$ which means that the coalgebra $\mathrm{Sh}(V)$ has a bialgebra structure with the product of $T(V^*)^o$. Let $\boldsymbol{m}'$ be the product of $T(V^*)^o$ restricted to $\mathrm{Sh}(V)$. We show that $\boldsymbol{m}' = \boldsymbol{m}$. At this point we follow the notation conventions found after the proof of Proposition 4.5.3.

Let $\{v_1, \ldots, v_\theta\}$ be a basis for V and let $\{v^1, \ldots, v^\theta\}$ be the dual basis for V^*. For $r, n \geq 1$ let $[r] = \{1, \ldots, r\}$ and let $\mathcal{T}_n = [\theta]^n = [\theta] \times \cdots \times [\theta]$ (n-factors) be the set of all n-tuples with components in $[\theta]$. Let $\mathcal{T} = \bigcup_{n=0}^{\infty} \mathcal{T}_n$, where $\mathcal{T}_0 = \{\emptyset\}$. Then $(\mathcal{T}, \bullet)$ is a monoid with neutral element $\emptyset$ and otherwise product given by $(m_1, \ldots, m_r) \bullet (n_1, \ldots, n_s) = (m_1, \ldots, m_r, n_1, \ldots, n_s)$. In particular $\mathcal{T}_m \bullet \mathcal{T}_n = \mathcal{T}_{m+n}$ for all $m, n \geq 0$. For $\mathsf{n} \in \mathcal{T}$ let $|\mathsf{n}|$ be the number of components of n. Thus $\mathsf{n} \in \mathcal{T}_n$ if and only if $|\mathsf{n}| = n$.

Set $v^{\emptyset} = 1_{T(V^*)}$ and $v_{(\emptyset)} = \mathbf{1}$. For $r \geq 1$ and $\mathsf{n} = (n_1, \ldots, n_r) \in \mathcal{T}_r$ let

$$v^{\mathsf{n}} = v^{n_1} \otimes \cdots \otimes v^{n_r} \quad \text{and} \quad v_{(\mathsf{n})} = v_{n_1} \otimes \cdots \otimes v_{n_r}.$$

Then $\{v^{\mathsf{n}} \,|\, \mathsf{n} \in \mathcal{T}\}$ is a basis for $T(V^*)$ and $\{v_{(\mathsf{n})} \,|\, \mathsf{n} \in \mathcal{T}\}$ is a basis for $\mathrm{Sh}(V)$. Note that

$$<v_{(\mathsf{m})}, v^{\mathsf{n}}> = \delta^{\mathsf{n}}_{\mathsf{m}}$$

for all $\mathsf{m}, \mathsf{n} \in \mathcal{T}$. By definition of the product of $T(V^*)$ we have $v^{\mathsf{m}} v^{\mathsf{n}} = v^{\mathsf{m}\bullet\mathsf{n}}$ for all $\mathsf{m}, \mathsf{n} \in \mathcal{T}$ from which the description of the coalgebra structure of $\mathrm{Sh}(V)$ given by

$$\Delta(v_{(\mathsf{n})}) = \sum_{\mathsf{l}\bullet\mathsf{m}=\mathsf{n}} v_{(\mathsf{l})} \otimes v_{(\mathsf{m})}$$

and

$$\epsilon(v_{(\mathsf{n})}) = \delta_{\emptyset,\mathsf{n}}$$

for all $\mathsf{n} \in \mathcal{T}$ follows.

To determine the products $v_{(\mathsf{m})} v_{(\mathsf{n})}$ in the algebra $T(V^*)^o$ we need a description of the coproduct of $T(V^*)$. Let $\mathsf{n} = (n_1, \ldots, n_r) \in \mathcal{T}$ and $S \subseteq [r]$. Suppose S has $u \geq 1$ elements and write $S = \{i_1, \ldots, i_u\}$, where $1 \leq i_1 < \cdots < i_u \leq r$. Set $S(\mathsf{n}) = (n_{i_1}, \ldots, n_{i_u}) \in \mathcal{T}_u$ and set $\emptyset(\mathsf{n}) = \emptyset$. By induction on $|\mathsf{n}|$ the formulas

$$\Delta_{T(V^*)}(v^{\mathsf{n}}) = \sum_{S \subseteq [|\mathsf{n}|]} v^{S(\mathsf{n})} \otimes v^{([|\mathsf{n}|] \setminus S)(\mathsf{n})} \tag{7.12}$$

$$\epsilon_{T(V^*)}(v^{\mathsf{n}}) = \delta^{\mathsf{n}}_{\emptyset} \tag{7.13}$$

hold for all $\mathsf{n} \in \mathcal{T}$. See Exercise 7.11.2.

To compute $v_{(\mathsf{m})} v_{(\mathsf{n})}$ we also need to compute "shuffles" $\mathsf{m}\,S\,\mathsf{n}$ of $\mathsf{m}, \mathsf{n} \in \mathcal{T}$. Let $S \subseteq [|\mathsf{m}| + |\mathsf{n}|]$ have $|\mathsf{m}|$ elements. If m, n, or S is $\emptyset$ then we set $\mathsf{m}\,S\,\mathsf{n} = \mathsf{m}\bullet\mathsf{n}$. Suppose this is not the case and write $\mathsf{m} = (m_1, \ldots, m_r)$, $\mathsf{n} = (n_1, \ldots, n_s)$ and $S = \{i_1, \ldots, i_r\}$, where $1 \leq i_1 < \cdots < i_r \leq r + s$. Then $\mathsf{m}\,S\,\mathsf{n}$ is the $(r+s)$-tuple with i_j^{th} component m_j for all $1 \leq j \leq r$ and the remaining s components are filled by $n_1, \ldots, n_s$ in that order. For $r, s \geq 0$ let $\boldsymbol{S}(r, s)$ be the set of subsets of $[r+s]$ with r elements. Then

$$v_{(\mathsf{m})} v_{(\mathsf{n})} = \sum_{S \in \boldsymbol{S}(|\mathsf{m}|, |\mathsf{n}|)} v_{(\mathsf{m}\,S\,\mathsf{n})} \tag{7.14}$$

for all $\mathsf{m}, \mathsf{n} \in \mathcal{T}$. To verify the equation, let $\mathsf{r} \in \mathcal{T}$. By virtue of (7.12) we have

$$<v_{(\mathsf{m})} v_{(\mathsf{n})}, v^{\mathsf{r}}> = \sum_{S \subseteq [|\mathsf{r}|]} <v_{(\mathsf{m})}, v^{S(\mathsf{r})}><v_{(\mathsf{n})}, v^{([|\mathsf{r}|] \setminus S)(\mathsf{r})}> = |\mathcal{L}| \cdot 1_k,$$

where

$$\mathcal{L} = \{S \subseteq [|\mathsf{r}|] \mid \mathsf{m} = S(\mathsf{r}) \text{ and } \mathsf{n} = ([|\mathsf{r}|]\backslash S)(\mathsf{r})\}.$$

On the other hand the right-hand side of (7.14) applied to v^{r} is $|\mathcal{R}|{\cdot}1_k$

$$\mathcal{R} = \{S \in \boldsymbol{S}(|\mathsf{m}|, |\mathsf{n}|) \mid \mathsf{r} = \mathsf{m}\, S\, \mathsf{n}\}.$$

Observe that $\mathcal{L} = \emptyset = \mathcal{R}$ unless $|\mathsf{m}| + |\mathsf{n}| = |\mathsf{r}|$. An easy exercise shows that $\mathcal{L} = \mathcal{R}$ in this case. We have established (7.14).

We describe the product of (7.14) as a linear map. Recall S_n denotes the group of permutations of $[n]$ under function composition for $n \geq 1$.

Definition 7.11.5. Let $r, s \geq 1$ and $n = r + s$. An *r-shuffle in* S_n is a permutation $\sigma \in S_n$ which satisfies $\sigma^{-1}(1) < \cdots < \sigma^{-1}(r)$ and $\sigma^{-1}(r{+}1) < \cdots < \sigma^{-1}(n)$. The set of r-shuffles in S_n is denoted by $S_n(r)$.

Let $n > 1$, $1 \leq r < n$, and $u_1, \ldots, u_n \in V$. Then the product of (7.14) can be described as

$$(u_1{\otimes}\cdots{\otimes}u_r)(u_{r+1}{\otimes}\cdots{\otimes}u_n) = \sum_{\sigma \in S_n(r)} u_{\sigma(1)}{\otimes}\cdots{\otimes}u_{\sigma(n)} \tag{7.15}$$

and is therefore an $\boldsymbol{m}'$. Therefore the product $\boldsymbol{m}$ of $\mathrm{Sh}(V)$ is described by (7.15). By Exercise 4.5.4 the product rule of (7.14) holds when V is any vector space over k.

Exercises

Exercise 7.11.1. Complete the proof that (7.14) holds.

Exercise 7.11.2. Let A be a bialgebra over the field k and suppose $v_1, \ldots, v_n \in \mathrm{P}(A)$. Set $v_\emptyset = 1$ and for a non-empty set $S = \{i_1, \ldots, i_r\} \subseteq [n]$ with r-elements, where $1 \leq i_1 < \cdots < i_r \leq n$ let $v_S = v_{i_1}{\otimes}\cdots{\otimes}v_{i_r}$. Show that $\Delta(v_1{\cdots}v_n) = \sum_{S \subseteq [n]} v_S {\otimes} v_{[n]\backslash S}$.

Chapter notes

With the exception of Proposition 7.1.10, and subsequent material, the results of Section 7.1 can be found in Sweedler's book [201] and among those the ones about the antipode are in the paper [72] by Heyneman and Sweedler. Proposition 7.1.10 is attributed to Heyneman as an unpublished result [70]. Lemma 7.1.13 and Theorem 7.1.14 are reformulations of some of the ideas of [162, Section 3]. We point out that part (b) of Theorem 7.1.14 was established earlier in Sweedler's book by very different methods.

The q-binomial symbols, or Gaussian binomial coefficients, play an important part in the description of certain quantum groups. There are many treatments of them; see McDonald [130] for example. Parts (a) and (c) of Proposition 7.2.1 is [99, Proposition 6.5.1].

Section 7.3 details a basic method for constructing bialgebras and Hopf algebras which is by generators and relations. This is our method of choice. The construction results in a quotient. An important aspect of the construction is determining a linear base for the quotient. Enter the Diamond Lemma due to Bergman [21, Theorem 1.2], a very useful tool for algebraists. We explain the significance of the examples $H_{n,q}$ and $U_{n,q}$.

The square of the antipode of a commutative or cocommutative Hopf algebra over k has order one. The Hopf algebras which arise in affine or formal group theory are either commutative or cocommutative. Sweedler constructed a 4-dimensional (unpublished) Hopf algebra with antipode of order 4. Larson constructs infinite-dimensional Hopf algebras which show that the order of the antipode can be $2n$ for all positive integers n or infinite [100]. For $n \geq 1$ the n^2-dimensional Hopf algebra $H_{n,q}$ constructed by Taft [205] has antipode of order $2n$. $H_{2,-1}$ is Sweedler's 4-dimensional example. With the advent of quantum groups, Taft reconsidered the details of his construction in light of q-polynomials [208]. Our treatment of $H_{n,q}$ follows [208]. The well-known Hopf algebra $U_{\mathsf{q}}(sl_2)' = U_{n,q}$ is a quotient of $\mathrm{U}_q(sl_2)$ of Chapter 15, Section 15.6. The latter is the basic quantized enveloping algebra.

The dual Hopf algebra of Section 7.4 is described by Heyneman and Sweedler [72] and also in Sweeldler's book [201]. We continue the theme of the relationship between an object and its dual in Proposition 7.4.3.

The free Hopf algebra on a coalgebra is due to Takeuchi [210] and Section 7.5 closely follows his construction. His paper has many interesting results, one of which is the existence of a Hopf algebra whose antipode is not bijective [210, Theorem 18]. The construction of Theorem 7.5.1 is due to Larson [100].

Section 7.6 is comprised of various results on when a bialgebra has an antipode. That a pointed irreducible bialgebra has an antipode can be traced to Heyneman and Sweedler's paper [72, Corollary 3.5.4]; indeed their argument is that used to prove Lemma 6.2.1 when C is a pointed irreducible coalgebra over k. We focus on when pointed bialgebras have antipodes since so many of the bialgebras discussed in this book are pointed. That pointed bialgebras H over k which are sums of pointed irreducible subcoalgebras are Hopf algebras if and only if $\mathrm{G}(H)$ consists of invertible elements is due to Heyneman and Sweedler [72, 3.5.3 Corollary].

Comatrix identities are seen to be important in connection with the existence of an antipode. That $\mathrm{Det}(\mathcal{A})$ is a grouplike element in the commutative case is a well-known fact from the theory of affine groups and is a key part of the construction of the underlying Hopf algebra of the general linear group. For affine groups see Humphreys' book [77] for example. Corollary 7.6.11 was established by Takeuchi [210, Corollary 69].

The notion of (bialgebra) 2-cocycle of Section 7.7 is due to Doi [40, Equation (6)]. Virtually all of the material of Section 7.7 is found in the paper by Doi and Takeuchi [41, Section 1]; the formula for the antipode in part (b) of Proposition 7.7.3 is that of [40, Theorem 1.6(a5)]. The notion of pairing of bialgebras is due to Majid [114, page 9] and the notion of (right) twist of bialgebra is attributed to Drinfel'd. See the paper by Aljadeff, Etingof, Gelaki, and Nikshych [2] for a discussion of twists and properties of Hopf algebras invariant under twists. Also see the paper by Montgomery [136] where she describes properties of Hopf algebras invariant under left 2-cocycle twisting as well as twists.

The construction of Section 7.10 has a dual counterpart in the connected component $\mathrm{H}(C)_{k1}$, where $(\iota, \mathrm{H}(C))$ is the cofree Hopf algebra on the coalgebra C of Section 7.5. The Hopf algebra $\mathrm{Sh}(A)$ of Section 7.11 is $CH(U)$, where $U = A$, defined by Newman and the author [147]. The shuffle algebra $\mathrm{Sh}(V)$ is described in Sweedler's book [201, Chapter XII]. We follow his treatment of the shuffle algebra but use a slightly different universal mapping property description for $\mathrm{Sh}(V)$.

Chapter 8

Hopf modules and co-Hopf modules

The theories of Hopf modules and of relative Hopf modules of Section 9.2 account for some of the deeper results for Hopf algebras. Hopf modules are vector spaces with both a comodule and module structure which are related in a natural way. Although Hopf modules are defined for bialgebras, their real usefulness is in connection with Hopf algebras. Non-zero Hopf modules over a Hopf algebra are free as modules. Applications of this fundamental result yield important structure theorems.

The rational dual of a Hopf module over a Hopf algebra has a structure which we refer to as a co-Hopf module. The rational dual of a co-Hopf module is a Hopf module; thus the notions of Hopf module and co-Hopf module are dual concepts. A Hopf algebra has both Hopf module and co-Hopf module structures. Both are important. Throughout this chapter A is a bialgebra over the field k and H is a Hopf algebra with antipode S over k.

8.1 Definition of Hopf module and examples

We begin with the definition of left Hopf module.

Definition 8.1.1. Let A be a bialgebra over the field k. A *left A-Hopf module* is a triple (M, μ, ρ), where (M, μ) is a left A-module and (M, ρ) is left A-comodule, such that

$$\rho(a{\cdot}m) = a_{(1)}m_{(-1)} \otimes a_{(2)}{\cdot}m_{(0)}$$

for all $a \in A$ and $m \in M$.

Right A-Hopf modules are the left $A^{op\,cop}$-Hopf modules. Thus:

Definition 8.1.2. Let A be a bialgebra over the field k. A *right A-Hopf module* is a triple (M, μ, ρ), where (M, μ) is a right A-module and (M, ρ) is right A-comodule, such that

$$\rho(m{\cdot}a) = m_{(0)}{\cdot}a_{(1)} \otimes m_{(1)}a_{(2)} \tag{8.1}$$

for all $m \in M$ and $a \in A$.

On a categorical level Hopf modules are modules, or comodules, in a certain setting. By virtue of the following lemma left A-Hopf modules are the left A-comodules in the category of left A-modules or equivalently left A-modules in the category of left A-comodules. The tensor product of modules or comodules is given the tensor product structure.

Lemma 8.1.3. *Suppose that A is a bialgebra over the field k. Let M be a vector space over k with a left A-module structure (M, μ) and a left A-comodule structure (M, ρ). Then the following are equivalent:*

(a) *ρ is a left A-module map.*
(b) *μ is a left A-comodule map.*
(c) *(M, μ, ρ) is a left A-Hopf module.*

□

Definition 8.1.4. Let M be a left A-Hopf module. A *left A-Hopf submodule of M* is a subspace N of M which is simultaneously a left A-submodule and a left A-subcomodule of M.

Let M be a left A-Hopf module. Then M and (0) are left A-Hopf submodules of M. It is easy to see that the sum and intersection of left A-Hopf submodules of M is again a left A-Hopf submodule of M. In particular every subset X of M is contained in a unique minimal left A-Hopf submodule of M, which we refer to as the left A-Hopf submodule generated by X.

Recall that the left A-comodule structure on M can be described in terms of the (rational) right A^*-module structure on M defined by

$$m{\leftharpoonup}a^* = \langle a^*, m_{(-1)}\rangle m_{(0)}$$

for all $m \in M$ and $a^* \in A^*$. There are commutation rules for the two module actions on M, namely:

$$(a{\cdot}m){\leftharpoonup}a^* = a_{(2)}{\cdot}(m{\leftharpoonup}(a^*{\prec}a_{(1)})), \tag{8.2}$$

which holds generally, and

$$a{\cdot}(m{\leftharpoonup}a^*) = (a_{(2)}{\cdot}m){\leftharpoonup}(a^*{\prec}S(a_{(1)})), \tag{8.3}$$

which holds when A is a Hopf algebra with antipode S, where $a \in A$, $m \in M$, and $a^* \in A^*$. Recall the right A-module action on A^* is given by $<a^* \prec a, b> = <a^*, ab>$ for all $a^* \in A^*$ and $a, b \in A$.

Proposition 8.1.5. *Suppose that A is a bialgebra over the field k and let M be a left A-Hopf module.*

(a) *Let N be an A-subcomodule of M. Then the A-submodule generated by N is an A-Hopf submodule of M.*
(b) *If M is a finitely generated A-Hopf module then M is a finitely generated A-module.*
(c) *Suppose that A is a Hopf algebra and N is an A-submodule of M. Then the A-subcomodule of M generated by N is an A-Hopf submodule of M.*

Proof. Part (a) follows by (8.2) and part (c) follows by (8.3). To show part (b) let X be a finite subset of M which generates M as an A-Hopf module. Then the subcomodule N of M which X generates is finite-dimensional by part (a) of Theorem 3.2.8 applied to C^{cop}. By part (a) the submodule which N generates is a Hopf submodule L of A. Thus $L = M$ is a finitely generated left A-module. □

Definition 8.1.6. Let A be a bialgebra over the field k. A *map of left A-Hopf modules M and N* is a linear map $f : M \longrightarrow N$ which is a map of left A-modules and of left A-comodules.

Notice that $\mathrm{Ker}(f)$ is a left A-Hopf submodule of M whenever $f : M \longrightarrow N$ is a left Hopf module map. If N is a Hopf submodule of M then the quotient space M/N has a unique left A-Hopf module structure such that the projection $\pi : M \longrightarrow M/N$ is a map of left A-Hopf modules.

We now consider some basic examples of Hopf modules and constructions which produce others.

Example 8.1.7. A with its product m and coproduct Δ has the structure of a left A-Hopf module (A, m, Δ).

The next example, as we shall see in Section 8.2, theoretically accounts for all left A-Hopf modules when A is a Hopf algebra.

Example 8.1.8. Let V be a vector space over the field k. Then $M = A \otimes V$ has the structure of a left A-Hopf module where

$$a \cdot (b \otimes v) = ab \otimes v \quad \text{and} \quad \rho(b \otimes v) = b_{(1)} \otimes (b_{(2)} \otimes v)$$

for all $a, b \in A$ and $v \in V$.

Example 8.1.9. Let $\{M_i\}_{i\in I}$ be an indexed family of left A-Hopf modules. Then the direct sum of vector spaces $\bigoplus_{i\in I} M_i$ with the usual direct sum left A-module and left A-comodule structures is a left A-Hopf module.

Suppose that $V \neq (0)$ is a vector space over k. Let $\{v_i\}_{i\in I}$ be a linear basis for V. Then the Hopf module $M = A\otimes V$ of Example 8.1.8 is a free left A-module with basis $\{1\otimes v_i \,|\, i \in I\}$. Note that $A_i = A{\cdot}(1\otimes v_i)$ is a left A-Hopf submodule of M, and that $M \simeq \bigoplus_{i\in I} A_i$ and $A \simeq A_i$ as left A-Hopf modules.

Example 8.1.10. Let $\{M_i\}_{i\in I}$ be an indexed family of left A-Hopf modules. Regard each M_i as a right (rational) A^*-module. Endow the direct product of vector spaces $M = \prod_{i\in I} M_i$ with the usual direct product left A-module and right A^*-module structures. Then M_r is a left A-submodule of M, and with the submodule and underlying subcomodule structures M_r is a left A-Hopf module.

The details of the last example are left to the reader in Exercise 8.1.4.

The tensor product of two left A-Hopf modules is a left A-Hopf module with *either* the tensor product module structure *or* the tensor product comodule structure, but usually not with both. See Exercise 8.1.5. More generally:

Example 8.1.11. Let M be a left A-Hopf module and N be a left A-module. Then $M\otimes N$ is a left A-Hopf module with

$$a{\cdot}(m\otimes n) = a_{(1)}{\cdot}m\otimes a_{(2)}{\cdot}n \quad \text{and} \quad \rho(m\otimes n) = m_{(-1)}\otimes(m_{(0)}\otimes n)$$

for all $a \in A, m \in M$ and $n \in N$.

Example 8.1.12. Let M be a left A-Hopf module and N be left A-comodule. Then $M\otimes N$ is a left A-Hopf module with

$$a{\cdot}(m\otimes n) = a{\cdot}m\otimes n \quad \text{and} \quad \rho(m\otimes n) = m_{(-1)}n_{(-1)}\otimes(m_{(0)}\otimes n_{(0)})$$

for all $a \in A, m \in M$ and $n \in N$.

We leave the reader with the exercise of defining A-Hopf submodule of a right A-Hopf module and map of right A-Hopf modules.

Exercises

Throughout these exercises A is a bialgebra over the field k and H is a Hopf algebra with antipode S over k.

Exercise 8.1.1. State and prove analogs of the module isomorphism theorems for Hopf modules.

Exercise 8.1.2. Show that the triple (M, μ, ρ), where (M, μ) is a left A-module and (M, ρ) is a left A-comodule, is a left A-Hopf module if and only if (8.2) is satisfied for all $a \in A$, $m \in M$, and $a^* \in A^*$.

Exercise 8.1.3. Suppose that M is a vector space over k which has a left A-module structure (M, μ) and a right A^*-module structure (M, ν) such that (8.2) holds, where we write $\nu(m \otimes a^*) = m \leftharpoonup a^*$. Show that:

(a) M_r is a left A-submodule of M.

(b) (M_r, μ, ρ) is a left A-Hopf module, where ρ is the underlying left A-comodule structure of (M_r, ν).

Exercise 8.1.4. Let $\{M_i\}_{i \in I}$ be an indexed family of left A-Hopf modules. Regard M_i as a rational right A^*-module for each $i \in I$ and let $M = \prod_{i \in I} M_i$ be the vector space direct product of the M_i's with the direct product left A-module structure (M, μ) and right A^*-module structure (M, ν). Use Exercise 8.1.3 to show that M_r is a left A-Hopf module as claimed in Example 8.1.10.

Exercise 8.1.5. Let $H = k[G]$ be the group algebra of a group $G \neq (e)$ over the field k. Endow $M = N = H$ with the left H-Hopf module structure of Example 8.1.7. Show that $M \otimes N$ with the tensor product left H-module and left H-comodule structures is not a left H-Hopf module.

Exercise 8.1.6. Show that the Hopf module of Example 8.1.8 has the tensor product of A-modules and tensor product of A-comodules structures.

Exercise 8.1.7. Let B be a sub-bialgebra of A and suppose that M is a left A-Hopf module. Let N be a left B-submodule of M and a left A-subcomodule of M. Let $(M/N, \rho)$ denote the quotient A-comodule structure and set $L = \rho^{-1}(B \otimes (M/N))$. Show that:

(a) $\rho(L) \subseteq B \otimes L$ and thus $(L, \rho|L)$ is a left B-comodule.

(b) L is a B-submodule of M/N.

(c) L with its left B-module and left B-comodule structures is a left B-Hopf module.

8.2 The structure of Hopf modules

Let (M, μ, ρ) be a left H-Hopf module. The main result of this section is that the underlying left H-module structure (M, μ) of M is free. We will begin to explore some of the implications of this fact.

Definition 8.2.1. Suppose that A is a bialgebra over k and (M, ρ) is a left (respectively right) A-comodule. Then $M_{co\,inv}$ consists of those $m \in M$ which satisfy $\rho(m) = 1 \otimes m$ (respectively $\rho(m) = m \otimes 1$). $M_{co\,inv}$ is the *set of co-invariants of* M and $M^{co\,inv} = (M^r)_{co\,inv}$.

For a left A-Hopf module M observe that

$$\lambda \in M^{co\,inv} \quad \text{if and only if} \quad a^* {\cdot} \lambda = \langle a^*, 1 \rangle \lambda \tag{8.4}$$

for all $a^* \in A^*$.

Let M be a left H-Hopf module. Define a linear map $P : M \longrightarrow M$ by

$$P(m) = S(m_{(-1)}){\cdot}m_{(0)} \tag{8.5}$$

for all $m \in M$.

Lemma 8.2.2. *Suppose that H is a Hopf algebra with antipode S over a field k. Let M be a left H-Hopf module and $P : M \longrightarrow M$ be defined by (8.5). Then:*

(a) $\mathrm{Im}(P) = M_{co\,inv}$ *and* $P^2 = P$.
(b) $P(h{\cdot}m) = \epsilon(h)P(m)$ *for all* $h \in H$ *and* $m \in M$.
(c) $m_{(-1)}{\cdot}P(m_{(0)}) = m$ *for all* $m \in M$.

Proof. We first show part (a). Suppose that $m \in M$. Then the calculation

$$\begin{aligned}
\rho(P(m)) &= \rho(S(m_{(-1)}){\cdot}m_{(0)}) \\
&= S(m_{(-1)})_{(1)}m_{(0)(-1)} \otimes S(m_{(-1)})_{(2)}{\cdot}m_{(0)(0)} \\
&= S(m_{(-1)(1)})_{(1)}m_{(-1)(2)} \otimes S(m_{(-1)(1)})_{(2)}){\cdot}m_{(0)} \\
&= S(m_{(-1)(1)(2)})m_{(-1)(2)} \otimes S(m_{(-1)(1)(1)}){\cdot}m_{(0)} \\
&= S(m_{(-1)(2)(1)})m_{(-1)(2)(2)} \otimes S(m_{(-1)(1)}){\cdot}m_{(0)} \\
&= \epsilon(m_{(-1)(2)})1 \otimes S(m_{(-1)(1)}){\cdot}m_{(0)} \\
&= 1 \otimes S(m_{(-1)}){\cdot}m_{(0)} \\
&= 1 \otimes P(m)
\end{aligned}$$

for all $m \in M$ shows that $\mathrm{Im}(P) \subseteq M_{co\,inv}$. If $m \in M_{co\,inv}$ then $P(m) = S(1){\cdot}m = 1{\cdot}m = m$ which concludes the proof of part (a).

To show part (b) we note for $h \in H$ and $m \in M$ that

$$\begin{aligned}
P(h{\cdot}m) &= S((h{\cdot}m)_{(-1)}){\cdot}(h{\cdot}m)_{(0)} \\
&= S(h_{(1)}m_{(-1)}){\cdot}(h_{(2)}{\cdot}m_{(0)}) \\
&= (S(m_{(-1)})S(h_{(1)})h_{(2)}){\cdot}m_{(0)} \\
&= (S(m_{(-1)})\epsilon(h)1){\cdot}m_{(0)} \\
&= \epsilon(h)P(m).
\end{aligned}$$

Since

$$\begin{aligned} m_{(-1)}\cdot P(m_{(0)}) &= m_{(-1)}\cdot(S(m_{(0)(-1)})\cdot m_{(0)(0)}) \\ &= (m_{(-1)(1)}S(m_{(-1)(2)}))\cdot m_{(0)} \\ &= (\epsilon(m_{(-1)})1)\cdot m_{(0)} \\ &= 1\cdot m \\ &= m \end{aligned}$$

for all $m \in M$ part (c) follows. Our proof is complete. □

By parts (a) and (b) of the preceding lemma $P : M \longrightarrow M_{co\,inv}$ is a linear projection and P is also a map of left H-modules, where $M_{co\,inv}$ is given the trivial left H-module structure.

Theorem 8.2.3. *Suppose that H is a Hopf algebra over the field k and M is a left H-Hopf module. Regard $H\otimes M_{co\,inv}$ as a left H-Hopf module as in Example 8.1.8. Then*

$$H\otimes M_{co\,inv} \longrightarrow M \qquad (h\otimes m \mapsto h\cdot m)$$

is an isomorphism of left Hopf modules.

Proof. We leave the reader with the straightforward exercise of showing that the map $\imath : H\otimes M_{co\,inv} \longrightarrow M$ defined by $\imath(h\otimes m) = h\cdot m$ and the map $\jmath : M \longrightarrow H\otimes M_{co\,inv}$ defined by $\jmath(m) = m_{(-1)}\otimes P(m_{(0)})$, which is well-defined by part (a) of Lemma 8.2.2, are indeed maps of left H-Hopf modules. We will show that $\imath$ and $\jmath$ are inverses.

Let $m \in M$. Then $\imath(\jmath(m)) = \imath(m_{(-1)}\otimes P(m_{(0)})) = m_{(-1)}\cdot P(m_{(0)}) = m$ by part (c) of Lemma 8.2.2. Now let $m \in M_{co\,inv}$. Then $\rho(m) = 1\otimes m$ by definition and $P(m) = m$ by part (a) of the same. For $h \in H$ we use part (b) of Lemma 8.2.2 to calculate

$$\begin{aligned} \jmath(\imath(h\otimes m)) &= \jmath(h\cdot m) \\ &= (h\cdot m)_{(-1)}\otimes P((h\cdot m)_{(0)}) \\ &= h_{(1)}m_{(-1)}\otimes P(h_{(2)}\cdot m_{(0)}) \\ &= h_{(1)}m_{(-1)}\otimes\epsilon(h_{(2)})P(m_{(0)}) \\ &= h1\otimes P(m) \\ &= h\otimes m. \end{aligned}$$

Therefore $\imath$ and $\jmath$ are linear inverses. □

Theorem 8.2.4. *Suppose that H is a Hopf algebra over the field k and M is a non-zero left H-Hopf module. Then:*

(a) *$M_{co\,inv} \neq (0)$.*
(b) *M is a free left H-module; any linear basis for $M_{co\,inv}$ is a basis for M as a left H-module.*
(c) *M is isomorphic to the direct sum of copies of H as a left H-Hopf module.*

Proof. Parts (a) and (b) are immediate consequences of Theorem 8.2.3. We have noted in our discussion after Example 8.1.9 that the left A-Hopf module $H \otimes V$ is the direct sum of copies of H as Hopf modules where V is any non-zero vector space over k. Thus part (c) follows by Theorem 8.2.3 as well. □

Applying Theorem 8.2.4 to Example 8.1.7 yields a rather striking result.

Proposition 8.2.5. *Suppose that H is a Hopf algebra with antipode S over the field k.*

(a) *The only subspaces of H which are both a left ideal and a left coideal of H are (0) and H.*
(b) *Suppose H contains a non-zero finite-dimensional left or right ideal. Then H is finite-dimensional.*

Proof. Regard H as a left H-Hopf module with the structure (H, m, Δ) of Example 8.1.7 and let N be a non-zero left H-Hopf submodule of H. Let $n \in N_{co\,inv}$. Then $\Delta(n) = 1 \otimes n$ which implies $n = 1\epsilon(n)$. Therefore $N = (0)$ or $N = H$ by part (b) of Theorem 8.2.4.

Part (a) now follows since the Hopf submodules of H are the subspaces which are simultaneously left ideals and left coideals. To show part (b) we suppose L is a non-zero finite-dimensional left or right ideal of H. Since right ideals of H are left ideals of the Hopf algebra $H^{op\,cop}$ we may assume that L is a left ideal of H. Let N be the subcomodule of H which L generates. Since L is finite-dimensional N is finite-dimensional by Proposition 3.2.8. Now N is a Hopf submodule of H by part (c) of Proposition 8.1.5. Consequently $H = N$ and is thus finite-dimensional. □

By Theorem 8.2.4 the tensor products of Examples 8.1.11 and 8.1.12 are free left A-modules. In Example 8.1.11 M is left A-Hopf module and N is a left A-module. The module structure on $M \otimes N$ is the tensor product module structure. When A^{op} has an antipode we can reverse the roles of M and N. For a left A-module N we set $N^{(\emptyset)} = (0)$ and for a non-empty set I we set $N^{(I)} = \bigoplus_{i \in I} N_i$, where $N = N_i$ for all $i \in I$.

Proposition 8.2.6. *Suppose that A is a bialgebra over the field k and suppose M, N are left A-modules. Regard $M{\otimes}N$ as a left A-module with the tensor product module structure.*

(a) *Assume the module structure of M extends to a left A-Hopf module. If A has an antipode then $M{\otimes}N \simeq M^{(\mathcal{B})}$, where $\mathcal{B}$ is a linear basis for N.*
(b) *Assume the module structure of N extends to a left A-Hopf module. If A^{op} has an antipode then $M{\otimes}N \simeq N^{(\mathcal{B})}$, where $\mathcal{B}$ is a linear basis for M.*

Proof. To show part (a) we regard $M{\otimes}N$ as a left A-Hopf module according to Example 8.1.11. Let $\mathcal{B}'$ be a linear basis for $M_{co\,inv}$ and let $\mathcal{B}$ be a linear basis for N. It is easy to see that $\{b'{\otimes}b \,|\, b' \in \mathcal{B}', b \in \mathcal{B}\}$ is a linear basis for $(M{\otimes}N)_{co\,inv}$. Part (a) now follows from part (b) of Theorem 8.2.4.

To show part (b) let ς be the antipode of A^{op} and regard $\mathcal{M} = M$ as a left A-module $(M, \bullet)$, where $a{\bullet}m = \epsilon(a)m$ for all $a \in A$ and $m \in M$. Then $\mathcal{M}{\otimes}N$ is a left A-Hopf module where $a{\cdot}(m{\otimes}n) = a_{(1)}{\bullet}m{\otimes}a_{(2)}{\cdot}n = m{\otimes}a{\cdot}n$ and $\rho(m{\otimes}n) = n_{(-1)}{\otimes}(m{\otimes}n_{(0)})$ for all $a \in A$, $m \in M$, and $n \in N$. Note that $\mathcal{M}{\otimes} \simeq M^{\mathcal{B}}$ where $\mathcal{B}$ is described in part (b).

We define linear maps $F : M{\otimes}N \longrightarrow \mathcal{M}{\otimes}N$ and $G : \mathcal{M}{\otimes}N \longrightarrow M{\otimes}N$ by $F(m{\otimes}n) = \varsigma(n_{(-1)}){\cdot}m{\otimes}n_{(0)}$ and $G(m{\otimes}n) = n_{(-1)}{\cdot}m{\otimes}n_{(0)}$ for all $m \in M$ and $n \in N$. It is easy to see that F and G are inverse functions. The calculation

$$\begin{aligned}
F(a{\cdot}(m{\otimes}n)) &= F(a_{(1)}{\cdot}m{\otimes}a_{(2)}{\cdot}n) \\
&= \varsigma((a_{(2)}{\cdot}n)_{(-1)}){\cdot}(a_{(1)}{\cdot}m){\otimes}(a_{(2)}{\cdot}n)_{(0)} \\
&= \varsigma(a_{(2)(1)}n_{(-1)}){\cdot}(a_{(1)}{\cdot}m){\otimes}a_{(2)(2)}{\cdot}n_{(0)} \\
&= (\varsigma(n_{(-1)})\varsigma(a_{(2)(1)})a_{(1)}){\cdot}m){\otimes}a_{(2)(2)}{\cdot}n_{(0)} \\
&= (\varsigma(n_{(-1)})\epsilon(a_{(1)})1){\cdot}m){\otimes}a_{(2)}{\cdot}n_{(0)} \\
&= \varsigma(n_{(-1)}){\cdot}m){\otimes}a{\cdot}n_{(0)} \\
&= a{\cdot}F(m{\otimes}n)
\end{aligned}$$

for all $a \in A$, $m \in M$ and $n \in N$ shows that F is a map of left A-modules. □

Exercises

Throughout these exercises A is a bialgebra over the field k.

Exercise 8.2.1. Let A be a commutative bialgebra over the field k. Show that A is a Hopf algebra if all left A-Hopf modules are zero or free as A-modules. [Hint: Let $g \in G(A)$ and consider $M = Ag$. See Corollary 7.6.11.]

Exercise 8.2.2. Let B be a Hopf subalgebra of A and $x \in A$ satisfy $\Delta(x) = x \otimes a + 1 \otimes x$ for some $a \in G(B)$. Show that $M = B + Bx$ is a free left B-module with basis $\{1\}$ or $\{1, x\}$. [Hint: Show that M is a left A-Hopf module under multiplication and the coproduct and that the quotient left A-Hopf module M/B is in fact a left B-Hopf module.]

8.3 Co-Hopf modules

Let M be a left H-Hopf module. Recall that the left H-module structure on M gives the linear dual M^* a right H-module structure under the right transpose action defined by

$$<\alpha \prec h, m> = <\alpha, h \cdot m>$$

for all $\alpha \in M^*$, $h \in H$, and $m \in M$. The left H-comodule structure on M accounts for a rational right H^*-module action on M. The transpose of the right rational action gives M^* a left H^*-module structure defined by

$$<h^* \succ \alpha, m> = <\alpha, m \leftharpoonup h^*> = <h^*, m_{(-1)}> m_{(0)}$$

for $h^* \in H^*$, $\alpha \in M^*$ and $m \in M$. Let M^r be the (unique) maximal left rational H^*-submodule of M^*, and let ρ^T denote the underlying right H-comodule structure. We will show that M^r is a right H-submodule of M^* and determine a formula for computing $\rho^T(\alpha \prec h)$ for all $\alpha \in M^r$ and $h \in H$.

Let $\alpha \in M^r$ and $m \in M$. Then $<\alpha_{(0)}, m>\alpha_{(1)} = m_{(-1)}<\alpha, m_{(0)}>$, which is the counterpart of (3.13) for left H^*-modules. For $h \in H$ and

$h^* \in H^*$ we calculate

$$\begin{aligned}
<h^*\succ(\alpha\prec h), m> &= <\alpha\prec h, m\leftharpoonup h^*> \\
&= <h^*, m_{(-1)}><\alpha\prec h, m_{(0)}> \\
&= <h^*, m_{(-1)}><\alpha, h\cdot m_{(0)}> \\
&= <h^*, \epsilon(h_{(1)})1m_{(-1)}><\alpha, h_{(2)}\cdot m_{(0)}> \\
&= <h^*, S(h_{(1)(1)})h_{(1)(2)}m_{(-1)}><\alpha, h_{(2)}\cdot m_{(0)}> \\
&= <h^*, S(h_{(1)})h_{(2)(1)}m_{(-1)}><\alpha, h_{(2)(2)}\cdot m_{(0)}> \\
&= <h^*\prec S(h_{(1)}), (h_{(2)}\cdot m)_{(-1)}><\alpha, (h_{(2)}\cdot m)_{(0)}> \\
&= <\alpha_{(0)}, h_{(2)}\cdot m><h^*\prec S(h_{(1)}), \alpha_{(1)}> \\
&= <\alpha_{(0)}\prec h_{(2)}, m><h^*, S(h_{(1)})\alpha_{(1)}>.
\end{aligned}$$

We have shown that $\alpha\prec h \in M^r$ and that

$$\rho^T(\alpha\prec h) = \alpha_{(0)}\prec h_{(2)}\otimes S(h_{(1)})\alpha_{(1)}.$$

Definition 8.3.1. Suppose that H is a Hopf algebra with antipode S over k. A *right H-co-Hopf module* is a triple (M, μ, ρ), where (M, μ) is a right H-module and (M, ρ) is a right H-comodule, such that

$$\rho(m\cdot h) = m_{(0)}\cdot h_{(2)}\otimes S(h_{(1)})m_{(1)} \tag{8.6}$$

for all $m \in M$ and $h \in H$.

The following proposition summarizes our calculations.

Proposition 8.3.2. *Let H be a Hopf algebra over the field k, suppose that (M, μ, ρ) is a left H-Hopf module, and let (M^*, μ^T) denote the right transpose action on M^* induced by (M, μ). Then:*

(a) *M^r is a right H-submodule of (M^*, μ^T).*
(b) *(M^r, μ^T, ρ^T) is a right H-co-Hopf module.*

□

Conversely:

Proposition 8.3.3. *Let H be a Hopf algebra over the field k, suppose that (M, μ, ρ) is a right H-co-Hopf module, and let (M^*, μ^T) denote the left transpose action on M^* induced by (M, μ). Then:*

(a) *M^r is a left H-submodule of (M, μ^T).*
(b) *(M^r, μ^T, ρ^T) is a left H-Hopf module.*

Proof. We will sketch the proof using the formal notation for comodule action and the coproduct in a less detailed way. The reader should compare the following proof to the proof of Proposition 8.3.2.

Let $h \in H$ and $\alpha \in M^r$. Then for all $h^* \in H^*$ and $m \in M$ we have by virtue of (3.13) that

$$\begin{aligned} <(h\succ\alpha)\prec h^*, m> &= <h\succ\alpha, h^*\rightharpoonup m> \\ &= <\alpha, m_{(0)}\prec h><h^*, m_{(1)}> \\ &= <\alpha, m_{(0)}\cdot h_{(3)}><h^*, h_{(1)}S(h_{(2)})m_{(1)}> \\ &= <\alpha, m_{(0)}\cdot h_{(3)}><h^*\prec h_{(1)}, S(h_{(2)})m_{(1)}> \\ &= <\alpha, (m\cdot h_{(2)})_{(0)}><h^*\prec h_{(1)}, (m\cdot h_{(2)})_{(1)}> \\ &= <h^*\prec h_{(1)}, \alpha_{(-1)}><\alpha_{(0)}, m\cdot h_{(2)}> \\ &= <h^*, h_{(1)}\alpha_{(-1)}><h_{(2)}\succ\alpha_{(0)}, m> \end{aligned}$$

which shows that $h\succ\alpha \in M^r$ and $\rho^T(h\succ\alpha) = h_{(1)}\alpha_{(-1)}\otimes h_{(2)}\succ\alpha_{(0)}$. This completes our proof. □

8.4 A basic co-Hopf module and its dual

We have seen that the left H-Hopf module structure (H, m, Δ) on H of Example 8.1.7 is theoretically important. There is an important H-co-Hopf module structure on H also.

Example 8.4.1. (H, μ, Δ) is a right H-co-Hopf module, where $m\cdot h = S(h)m$ for all $m, h \in H$.

To see this note that

$$\Delta(m\cdot h) = \Delta(S(h)m) = S(h_{(2)})m_{(1)}\otimes S(h_{(1)})m_{(2)} = m_{(0)}\cdot h_{(2)}\otimes S(h_{(1)})m_{(1)}$$

for all $a, m \in H$.

By Proposition 8.3.3 it follows that (H^r, μ^T, Δ^T) is a left H-Hopf module. Note that the rational right H^*-module action induced from Δ^T is right multiplication in H^*. Observe that $\mu^T(h\otimes h^*) = h^*\prec S(h)$ for all $h^* \in H^*$ and $h \in H$. By part (c) of Theorem 8.2.4 there is an isomorphism of left H-Hopf modules $F : H\otimes H^{co\,inv} \longrightarrow H^r$ which is defined by $F(h\otimes\lambda) = \lambda\prec S(h)$ for all $h \in H$ and $\lambda \in H^{co\,inv}$. Notice that $H^{co\,inv} \neq (0)$ implies S is one-one. Observe that

$$H^{co\,inv} = \{\lambda \in H^* \mid \lambda h^* = <h^*, 1>\lambda \text{ for all } h^* \in H^*\}. \qquad (8.7)$$

Now suppose that $\lambda \in H^{co\,inv}$ is not zero and define $f : H \longrightarrow H^r$ by $f(h) = F(h \otimes \lambda)$ for all $h \in H$. Then f is a Hopf module map since it is the composition of such. In terms of module structures

$$f(hk) = f(k) \prec S(h) \quad \text{and} \quad f(h \leftharpoonup h^*) = f(h)h^* \tag{8.8}$$

for all $h, k \in H$ and $h^* \in H^*$. Notice that f is one-one since F is one-one.

Now suppose that H is finite-dimensional. Then $H^r = H^*$ by Theorem 3.1.4. In this case $\mathrm{Dim}\, H^{co\,inv} = 1$ and we conclude:

Theorem 8.4.2. *Let H be a finite-dimensional Hopf algebra with antipode S over the field k. Then there is a linear isomorphism $\boldsymbol{f} : H \longrightarrow H^*$ such that*

$$\boldsymbol{f}(hk) = \boldsymbol{f}(k) \prec S(h) \qquad \text{and} \qquad \boldsymbol{f}(h \leftharpoonup h^*) = \boldsymbol{f}(h)h^*$$

for $h, k \in H$ and $h^* \in H^*$. □

The fact that H and H^* are isomorphic as left Hopf modules, which Theorem 8.4.2 expresses in the language of modules, has profound implications for the structure of finite-dimensional Hopf algebras. We will examine them in great detail in Section 10.2. For openers the theorem implies that S is bijective. This is the essence of part (b) of Theorem 7.1.14, the argument for which was completely different.

The exercises below will give the reader a head start on understanding implications of the preceding theorem. We want to formally record a very important consequence of it which will be used in the next chapter. Recall that a finite-dimensional algebra B over k is Frobenius if there is a non-degenerate bilinear form $\beta : B \otimes B \longrightarrow k$ which is associative; that is $\beta(ab, c) = \beta(a, bc)$ for all $a, b, c \in B$.

Corollary 8.4.3. *Finite-dimensional Hopf algebras over the field k are Frobenius.*

Proof. Continuing the notation of the preceding theorem, $\beta : H \otimes H \longrightarrow k$ defined by $\beta(h, k) = \boldsymbol{f}(h) \prec S(k)$ for all $h, k \in H$ is an associative bilinear form which is non-degenerate since S is bijective and $\boldsymbol{f} : H \longrightarrow H^*$ is a linear isomorphism. □

Exercises

Throughout the following exercises H is a *finite-dimensional* Hopf algebra over the field k.

Exercise 8.4.1. The results of this section show that there exists a non-zero $\lambda \in H^*$, unique up to scalar multiple, such that $\lambda h^* = <h^*, 1>\lambda$ for all $h^* \in H^*$. Use this fact to show that:

(a) There exists a non-zero $\lambda' \in H^*$, unique up to scalar multiple, such that $h^*\lambda' = <h^*, 1>\lambda'$ for all $h^* \in H^*$. [Hint: Show that λ for the Hopf algebra $H^{op\,cop}$ is λ' for H.]

(b) There exist non-zero $\Lambda \in H$ and $\Lambda' \in H$, each unique up to scalar multiple, such that $h\Lambda = \epsilon(h)\Lambda$ and $\Lambda' h = \epsilon(h)\Lambda'$ for all $h \in H$. [Hint: Note that $H = (H^*)^*$ as Hopf algebras.]

Exercise 8.4.2. Show that (H, m, ρ) is also a right H-co-Hopf module structure for H, where $\rho(n) = n_{(2)} \otimes S(n_{(1)})$ for all $n \in H$. Describe the resulting isomorphism of left H-Hopf modules $H \simeq H^*$ in terms of module structures.

Exercise 8.4.3. Show that any linear isomorphism $f : H \longrightarrow H^*$ which satisfies (8.8) is a scalar multiple of $\boldsymbol{f}$.

Exercise 8.4.4. Show that H^* is a free left (respectively right) H-module under the transpose of right (respectively left) multiplication in H.

Apropos of the preceding exercise:

Exercise 8.4.5. Suppose that B is a finite-dimensional algebra over k. Show that the following are equivalent:

(a) B^* is a free left B-module under the transpose of right multiplication in H.

(b) B^* is a free right B-module under the transpose of left multiplication in H.

(c) B is a Frobenius algebra.

Chapter notes

Sections 8.1 and 8.2 are based on the notions and results of the paper by Larson and Sweedler [105] and the paper by Sweedler [203]. Theorem 8.2.3, the Fundamental Theorem of Hopf Modules, is essentially that of [105, Proposition 1]. Part (a) of Proposition 8.1.5 is the left version of [203, Proposition 2.6] and part (b) is Corollary 2.7 of the same.

The left Hopf module structure on H^* described in Section 8.4 is one of the most important ones in the theory of finite-dimensional Hopf algebras. Section 8.3 describes a theoretical underpinning for this structure in terms of a structure on H. Theorem 8.4.2 is found in Section 3 of [105].

Chapter 9

Hopf algebras as modules over Hopf subalgebras

Let H be a Hopf algebra over the field k and suppose that K is a Hopf subalgebra of H. Then H is a left K-module under the multiplication of H. A natural question to ask is what is the nature of H as a left K-module.

A fundamental result in the theory of Hopf algebras is that H is a free K-module when H is finite-dimensional. The purpose of the chapter is to prove this result and to establish other conditions under which H is a free K-module.

We begin with the useful special case $H_0 \subseteq K$ in Section 9.1. Generally consideration of H as a K-module involves relative Hopf modules. In Section 9.2 we treat relative Hopf modules with an eye towards setting the stage for a discussion of the freeness question in Section 9.3. In that section we show that pointed Hopf algebras, and finite-dimensional Hopf algebras, are free as modules over their Hopf subalgebras.

Generally Hopf algebras are not free over all of their Hopf subalgebras. An example is constructed in Section 9.4. A question to ask is whether or not they are projective or faithfully flat over their Hopf subalgebras. This will be addressed in the Chapter Notes. Throughout this chapter A is a bialgebra and H is a Hopf algebra with antipode S over k.

9.1 Filtrations whose base term is a Hopf subalgebra

The ideas of this section are based on a very simple result about certain coalgebra filtrations of Hopf algebras.

Proposition 9.1.1. *Let H be a Hopf algebra over the field k and suppose that $\{V_n\}_{n=0}^{\infty}$ is a coalgebra filtration of H such that V_0 is a Hopf subalgebra of H and V_n is a left V_0-submodule of H under multiplication for all $n \geq 0$.*

Then:

(a) *The quotient V_{n+1}/V_n is zero or a free left V_0-module for all $n \geq 0$.*
(b) *V_n is zero or a free left V_0-module for all $n \geq 0$.*
(c) *H is a free left V_0-module.*

Proof. We need only show part (a). Regard H as the left H-comodule (H, Δ). Then subcoalgebras of H, in particular V_n for all $n \geq 0$, are subcomodules of H. Fix $n \geq 0$ and let $(V_{n+1}/V_n, \rho)$ be the quotient H-comodule structure. Now the projection $\pi : V_{n+1} \longrightarrow V_{n+1}/V_n$ is a map of left H-comodules and $\rho \circ \pi = (\mathrm{I}_H \otimes \pi) \circ \Delta$. Since $\Delta(V_{n+1}) \subseteq \sum_{\ell=0}^{n+1} V_{n+1-\ell} \otimes V_\ell$ it follows that $\rho(V_{n+1}/V_n) \subseteq V_0 \otimes (V_{n+1}/V_n)$. Let $(V_{n+1}/V_n, \mu)$ be the quotient left V_0-module structure on V_{n+1}/V_n. Then $(V_{n+1}/V_n, \mu, \rho)$ is easily seen to be a left V_0-Hopf module. At this point we can apply Theorem 8.2.4 to deduce that $V_{n+1}/V_n = (0)$ or V_{n+1}/V_n is a free left V_0-module. □

We go beyond the proof of the preceding proposition, gleaning a bit more from Theorem 8.2.4. Let

$$V_{n+1}^{\natural} = \{v \in V_{n+1} \,|\, \Delta(v) - 1 \otimes v \in V_{n+1} \otimes V_n\}.$$

Then $\pi(V_{n+1}^{\natural}) = (V_{n+1}/V_n, \rho)_{co\,inv}$. As a consequence any linear basis $\{\pi(v_1), \ldots, \pi(v_m)\}$ for $\pi(V_{n+1}^{\natural})$, where $v_1, \ldots, v_m \in V_{n+1}^{\natural}$, is a basis for V_{n+1}/V_n as a left V_0-module.

There is an important application of Proposition 9.1.1 which we wish to pursue at this point. It is an entry point into the theory of finite-dimensional pointed Hopf algebras and anticipates the notion of relative Hopf module. These structures arise in a variety of contexts.

Let A be a bialgebra over k and suppose that B is a sub-bialgebra of A with an antipode. Suppose that $x \in A$ satisfies

$$\Delta(x) = 1 \otimes x + x \otimes a + u,$$

where $a \in \mathrm{G}(B)$, $u \in B \otimes B$, and $xB \subseteq Bx$. Then the sub-bialgebra of A which $B \cup \{x\}$ generates is $B[x] = B + Bx + Bx^2 + \cdots$. Let $V_n = B + Bx + \cdots + Bx^n$ for all $n \geq 0$. Then $\{V_n\}_{n=0}^{\infty}$ is easily seen to be a coalgebra filtration $B[x]$. Furthermore $B = V_0$ and $V_m V_n \subseteq V_{m+n}$ for all $m, n \geq 0$. Now $B[x]_0 \subseteq B$ by Proposition 4.1.2. Therefore the sub-bialgebras V_0 and $B[x]$ of A are Hopf algebras by Lemma 7.6.2. Thus Proposition 9.1.1 applies and we conclude that the quotients V_{n+1}/V_n are zero or free for all $n \geq 0$.

Note that $x^{n+1} + V_n \in (V_{n+1}/V_n)_{co\,inv}$ and generates the quotient as a left B-module. Therefore $V_n = V_{n+1}$ or V_{n+1}/V_n is a free left B-module with generator $x^{n+1} + V_n$. There are two natural cases to consider.

Case 1: $V_{m-1} \neq V_m$ *for all* $m \geq 1$. Here $B[x]$ is a free left B-module with basis $\{1, x, x^2, \ldots\}$.

Case 2: $V_{m-1} = V_m$ *for some* $m \geq 1$. Let n be the least such non-negative integer m. Then $B[x]$ is a free rank n left B-module with basis $\{1, x, x^2, \cdots, x^{n-1}\}$.

Since $x^n \in V_n = V_{n-1}$ it follows that

$$b_0 + b_1 x + \cdots + b_{n-1} x^{n-1} + x^n = 0 \tag{9.1}$$

for some $b_0, \ldots, b_{n-1} \in B$. *Assume further that* $u = 0$; that is

$$\Delta(x) = x \otimes a + 1 \otimes x.$$

It is easy to see that

$$\Delta(x^m) = x^m \otimes a^m + 1 \otimes x^m + u_m$$

for all $m \geq 1$, where $u_m \in \sum_{\ell=1}^{m-1} Bx^{m-\ell} \otimes Bx^\ell$. (We use the convention that a sum is zero if the initial summation index exceeds the final one.) Applying the coproduct to both sides of (9.1) we derive

$$\Delta(b_0) + \sum_{\ell=1}^{n-1} \Delta(b_\ell)(x^\ell \otimes a^\ell + 1 \otimes x^\ell + u_\ell)$$
$$= \left(\sum_{\ell=0}^{n-1} b_\ell x^\ell \right) \otimes a^n + 1 \otimes \left(\sum_{\ell=0}^{n-1} b_\ell x^\ell \right) - u_n.$$

Now $B[x] \otimes B[x]$ is a free left $B \otimes B$-module with basis $\{x^i \otimes x^j\}_{0 \leq i,j < n}$. Comparing the coefficients of $x^0 \otimes x^0$, $x^0 \otimes x^\ell$, $x^\ell \otimes x^0$, and $x^\ell \otimes x^m$ for all $0 < \ell, m < n$, we see that the preceding equation is equivalent to the five equations:

$$\Delta(b_0) = b_0 \otimes a^n + 1 \otimes b_0; \tag{9.2}$$

$$\Delta(b_\ell) = 1 \otimes b_\ell, \tag{9.3}$$

$$\Delta(b_\ell) = b_\ell \otimes a^{n-\ell}, \tag{9.4}$$

$$\Delta(b_\ell) u_\ell = 0 \tag{9.5}$$

for all $0 < \ell < n$; and

$$u_n = 0. \tag{9.6}$$

The last equation implies

$$\Delta(x^n) = x^n \otimes a^n + 1 \otimes x^n. \tag{9.7}$$

Using the axioms $(\mathrm{I}_A \otimes \epsilon) \circ \Delta = (\epsilon \otimes \mathrm{I}_A) \circ \Delta = \mathrm{I}_A$ we conclude: If $1 \leq \ell < n$ then $b_\ell \neq 0$ implies $b_\ell = \alpha_\ell 1$ for some non-zero $\alpha_\ell \in k$, $a^\ell = a^n$, and $u_\ell = 0$. This conditional statement is equivalent to (9.3)–(9.5). Note that $u_\ell = 0$ means that $\Delta(x^\ell) = x^\ell \otimes a^n + 1 \otimes x^\ell$.

Corollary 9.1.2. *Let A be a finite-dimensional bialgebra over a field k of characteristic zero. Then any primitive element of A is zero.*

Proof. Suppose that $x \in A$ is a non-zero primitive element and set $B = k1$. By our calculations above $B[x]$ is a free B-module with basis $\{1, x, \ldots, x^{n-1}\}$ for some $n \geq 1$ and $\Delta(x^n) = x^n \otimes 1 + 1 \otimes x^n$. Computing $\Delta(x^n) = (x \otimes 1 + 1 \otimes x)^n$ by the binomial theorem we have $u_n = \sum_{\ell=1}^{n-1} \binom{n}{\ell} x^{n-\ell} \otimes x^\ell = 0$. Therefore $n = 1$. But then $x \in k$, which has no non-zero primitives, a contradiction. Thus $x = 0$ to start with. □

We will use the calculations for Case 2 in a preliminary analysis of finite-dimensional pointed Hopf algebras in Chapter 15.

Exercise

Exercise 9.1.1. Suppose that A is a bialgebra over the field k and that the characteristic of k is $p > 0$. Let x be a non-zero primitive element of A. If $\{1, x, x^2, \ldots\}$ spans a finite-dimensional subspace of A, find the possibilities for the minimal polynomial of x over k.

9.2 Relative Hopf modules

First a definition.

Definition 9.2.1. Let A be a bialgebra over the field k and suppose that B is a sub-bialgebra of A. A *left relative* (A, B)*-Hopf module* is a triple (M, μ, ρ), where (M, μ) is a left B-module and (M, ρ) is a left A-comodule, such that $\rho(b{\cdot}m) = b_{(1)} m_{(-1)} \otimes b_{(2)}{\cdot}m_{(0)}$ for all $b \in B$ and $m \in M$.

Example 9.2.2. Let B be a sub-bialgebra of a bialgebra A over k and let M be a left B-module. Then $(A \otimes M, \mu, \rho)$ is a left relative (A, B)-Hopf module where $b{\cdot}(a \otimes m) = b_{(1)} a \otimes b_{(2)}{\cdot}m$ for all $b \in B$, $a \in A$, and $m \in M$, and $\rho(a \otimes m) = a_{(1)} \otimes (a_{(2)} \otimes m)$ for all $a \in A$ and $m \in M$.

When $B = A$ note that Example 9.2.2 is Example 8.1.11 with the left A-Hopf module structure of Example 8.1.7. The notion of relative Hopf module has a natural refinement.

Definition 9.2.3. Let A be a bialgebra over the field k, let B be a subalgebra of A, and let C be a subcoalgebra of A. A *left* (C, B)-*module* is a triple (M, μ, ρ), where (M, μ) is a left B-module and (M, ρ) is a left C-comodule, such that $BC \subseteq C$ and $\rho(b{\cdot}m) = b_{(1)}m_{(-1)}\otimes b_{(2)}{\cdot}m_{(0)}$ for all $b \in B$ and $m \in M$.

Building on the definition, we let ${}^{C}_{B}\mathcal{M}$ be the category whose objects are left (C, B)-modules and whose morphisms $f : M \longrightarrow M'$ are maps of underlying left B-modules and left C-comodules under function composition.

Definition 9.2.4. Let A be a bialgebra over the field k, let B be a subalgebra of A, let C be a subcoalgebra of A, and let M be a left (C, B)-module. A *left* (C, B)-*submodule of* M is a left (C, B)-module N which is a left B-submodule and a left C-subcomodule of M.

Let M be a left (C, B)-module. If N is a left (C, B)-submodule of M then M/N with its quotient module and comodule structures is a left (C, B)-module and the projection $\pi : M \longrightarrow M/N$ is a morphism. Suppose that V is a left C-subcomodule of M. Then the left B-module $B{\cdot}V$ generated by V is a left (C, B)-submodule of M. Assume that $M \neq (0)$. Then M contains a simple left C-subcomodule V which must be finite-dimensional by Theorem 3.2.8 applied to C^{cop}. In particular $B{\cdot}V$ is a non-zero left (C, B)-submodule of M which is finitely generated as a left B-module.

Let B' be a subalgebra of A and C' be a subcoalgebra of A. If $B' \subseteq B$, $C \subseteq C'$, and $B'C' \subseteq C'$ then any left (C, B)-module M is a left (C', B')-module, where M is made a left B'-module by restriction and M is made a left C'-comodule by extension. Thus when B is a sub-bialgebra of A all (C, B)-modules are relative (A, B)-Hopf modules.

Whether or not all non-zero left (C, B)-modules are free as left B-modules depends on whether or not certain types are free.

Lemma 9.2.5. *Let A be a bialgebra over the field k. Suppose that B is a subalgebra of A and C is a subcoalgebra of A such that $BC \subseteq C$. Then the following are equivalent:*

(a) *All non-zero left (C, B)-modules are free as left B-modules.*
(b) *All non-zero finite-dimensional left (C, B)-modules are free as left B-modules.*

(c) *All left (C, B)-modules of the form $B{\cdot}V$, where V is a simple left C-comodule, are free as left B-modules.*

Proof. Part (a) implies part (b). Since simple left C-comodules are finite-dimensional, part (b) implies part (c). To complete the proof we need only show that part (c) implies part (a).

Assume the hypothesis of part (c) and let M be a non-zero left (C, B)-module. We have noted that M contains a (C, B)-submodule of the form $B{\cdot}V$, where V is a simple left C-subcomodule of M. Thus M contains a non-zero left (C, B)-submodule which is free left B-submodule.

Call a non-empty subset L of M a partial basis if $B{\cdot}L$ is a left (C, B)-submodule of M which is a free left B-submodule with basis L. Let $\mathcal{L}$ be the set of partial bases of M. Then $\mathcal{L} \neq \emptyset$. Order the elements of $\mathcal{L}$ by inclusion. It is easy to see that $\mathcal{L}$ has a maximal element L by Zorn's Lemma.

We claim that $B{\cdot}L = M$. Suppose not. Then the non-zero left (C, B)-module $M/B{\cdot}L$ contains a non-zero left (C, B)-module M' which is a free left B-module. Let L'' be a basis for M' and choose a subset L' of M such that $\pi|_{L'} : L' \longrightarrow L''$ is bijective. Then $L' \neq \emptyset$, $L{\cup}L'$ is disjoint, and $\pi^{-1}(M') = B{\cdot}(L{\cup}L')$ is a left (C, B)-submodule of M which is a free left B-module with basis $L{\cup}L'$. This contradicts the maximality of L. Therefore $B{\cdot}L = M$ after all. □

We are interested in the case when B is a sub-bialgebra of A with an antipode.

Lemma 9.2.6. *Let A be a bialgebra over the field k. Suppose that B is a sub-bialgebra of A with an antipode and C is a subcoalgebra of A such that $BC \subseteq C$. Let G be the group of invertible elements of $\mathrm{G}(A)$. If $C_0 \subseteq \sum_{g \in G} Bg$ then all non-zero left (C, B)-modules are free left B-modules.*

Proof. Let $(M, \cdot, \rho)$ be a left (C, B)-module of the form $M = B{\cdot}V$, where V is a simple left C-subcomodule of M. Since V is simple $\rho(V) \subseteq D{\otimes}V$ for some simple subcoalgebra D of C by part (d) of Theorem 3.2.11 applied to C^{cop}. Since Bg is a subcoalgebra of A for all $g \in \mathrm{G}(A)$, and $D \subseteq \sum_{g \in G} Bg$, it follows by part (a) of Proposition 3.4.3 that $D \subseteq Bg$ for some $g \in G$. Now g is invertible, thus $Dg^{-1} \subseteq B$. Note that (M, ρ') is a left A-comodule structure on M, where $\rho'(m) = m_{(-1)}g^{-1}{\otimes}m_{(0)}$ for all $m \in M$, and indeed $(M, \cdot, \rho')$ is a left B-Hopf module. Therefore M is a free left B-module by part (b) of Theorem 8.2.4. Our result now follows by Lemma 9.2.5. □

As a consequence of the preceding lemma:

Proposition 9.2.7. *Let A be a bialgebra over the field k. Suppose that B is a sub-bialgebra of A with an antipode and C is a subcoalgebra of A such that $BC \subseteq C$. If $C_0 \subseteq B$ or C_0 is spanned by invertible elements of $\mathrm{G}(A)$, then all non-zero left (C, B)-modules are free as B-modules.* □

We end this section with a result on relative Hopf modules which we use in the following section. We give the tensor product of modules over a bialgebra the tensor product module structure.

Proposition 9.2.8. *Let A be a bialgebra and suppose that B is a sub-bialgebra of A. Suppose further that A^{op} has an antipode, let $(M, \cdot)$ be a left A-module, and let N be a left relative (A, B)-module. Give $M \otimes N$ the tensor product left B-module structure. Then $M \otimes N \simeq N^{(\mathcal{B})}$ as a left B-module, where $\mathcal{B}$ is a linear basis for M.*

Proof. The proof of part (b) of Proposition 8.2.6 can be trivially adapted to apply here. □

9.3 When Hopf algebras free over their Hopf subalgebras

As an application of Proposition 9.2.7:

Theorem 9.3.1. *Let H be a Hopf algebra over the field k and suppose that K is a Hopf subalgebra of H. Suppose that $H_0 \subseteq K$ or that H is pointed. Then all relative (H, K)-Hopf modules are free as K-modules. In particular H is a free left K-module under multiplication.* □

The fact that H is a free left K-module when $H_0 \subseteq K$ is also a consequence of part (c) of Proposition 9.1.1. See Exercise 5.1.21.

Generally a Hopf algebra is not free over its Hopf subalgebras, even when the Hopf algebra is commutative, finitely generated, and the field k is algebraically closed of characteristic zero. See Section 9.4.

For the remainder of this section H is a finite-dimensional Hopf algebra over k and K is any Hopf subalgebra of H. We will show that all left relative (H, K)-modules are free K-modules. As a consequence H is a free left K-module.

To prove this freeness result we use very basic representation theory of finite-dimensional algebras over a field. What we need is found for the most part in [34, Section 59].

Let B be a finite-dimensional algebra over k and suppose that M is a non-zero finite-dimensional left B-module. Then M is said to be indecomposable if whenever M is the direct sum of two submodules one of the two summands must be (0). By the Krull-Schmidt Theorem M is the direct sum of indecomposable submodules and if $M = M_1 \oplus \cdots \oplus M_r = M'_1 \oplus \cdots \oplus M'_{r'}$ are two such decompositions, then $r = r'$ and the summands of the right-hand expression can be rearranged if necessary so that $M_i \simeq M'_i$ for all $1 \leq i \leq r$. A reference to this theorem and a proof is found on page 538 of [186].

For $r \geq 1$ let $M^{(r)} = M \oplus \cdots \oplus M$ be the direct sum of r copies of M. Suppose that N is also a left B-module. An easy consequence of the Krull-Schmidt Theorem is that if $M^{(r)} \simeq N^{(r)}$ for some $r \geq 1$ then $M \simeq N$.

An augmented algebra over k is a pair (A, ν) where A is an algebra over k and $\nu : A \longrightarrow k$ is an algebra map called an augmentation. The following will be a useful test for freeness.

Lemma 9.3.2. *Let B be a finite-dimensional augmented algebra over the field k and let M be a finite-dimensional left B-module such that $M^{(r)}$ is free for some $r \geq 1$. Then M is a free left B-module.*

Proof. Since $M^{(r)}$ is finite-dimensional and free it follows that $M^{(r)} \simeq B^{(n)}$ for some $n \geq 0$. Regard k as a right B-module via pullback along an augmentation. The calculation

$$r\mathrm{Dim}(k \otimes_B M) = \mathrm{Dim}(k \otimes_B M^{(r)}) = \mathrm{Dim}(k \otimes_B B^{(n)}) = n\mathrm{Dim}(k \otimes_B B) = n$$

shows that $rt = n$, where $t = \mathrm{Dim}(k \otimes_B M)$. Therefore $M^{(r)} \simeq B^{(n)} = (B^{(t)})^{(r)}$ which implies that $M \simeq B^{(t)}$. $\square$

The left B-module M is said to be faithful if $b{\cdot}M = (0)$ implies that $b = 0$. Regard B as a left B-module under multiplication and let $P_1, \ldots, P_s$ represent the distinct isomorphism types of indecomposables in a realization of B as a direct sum of indecomposable left B-modules. Assume further that B is Frobenius. Then M is faithful if and only if $M \simeq M_1 \oplus \cdots \oplus M_{s'}$, where $s \leq s'$, the M_i's are indecomposable, and $M_i = P_i$ for all $1 \leq i \leq s$ by [34, 59.3]. We make the important observation that there is an $r \geq 1$ such that $M^{(r)} = F \oplus E$ is the direct sum of submodules, F which is (0) or free and E which is not faithful.

To show this, we write B (respectively M) as a direct sum of indecomposable submodules and let n_i (respectively m_i) be the number of summands of B (respectively M) which are isomorphic to P_i for all $1 \leq i \leq s$. Then $n_i > 0$ for all $1 \leq i \leq s$. We may assume that $m_i > 0$ for all

$1 \leq i \leq s$; else $M^{(1)} = M$ is not faithful. We wish to find $r, t \geq 1$ such that $M^{(r)} \simeq B^{(t)} \oplus E$, where E is not faithful. This boils down to finding $r, t \geq 1$ such that $rm_i \geq tn_i$ for all $1 \leq i \leq s$ with equality holding in at least one instance. Let $r = n_{i_0}$ and $t = m_{i_0}$, where n_{i_0}/m_{i_0} is the maximum of $n_1/m_1, \ldots, n_s/m_s$.

Theorem 9.3.3. *Let H be a finite-dimensional Hopf algebra over the field k and suppose that K is a Hopf subalgebra of H. Then all non-zero left relative (H, K)-modules are free as K-modules; thus H is a free left K-module.*

Proof. All Hopf algebras over k are augmented. All finite-dimensional Hopf algebras H over k are Frobenius by Corollary 8.4.3 and H^{op} has an antipode by part (b) of Proposition 7.1.14.

Let M be a non-zero left relative (H, K)-module. To show that M is free we may assume that M is finite-dimensional by Lemma 9.2.5. There is an $r \geq 1$ such that the left K-module $M^{(r)} = F \oplus E$, where F is (0) or free and E is not faithful. Now M is a free K-module if $M^{(r)}$ is by Lemma 9.3.2. Thus, since the direct sum of left relative (H, K)-modules is a left relative (H, K)-module, we may assume that $M = F \oplus E$. When L is a left K-module we regard $H \otimes L$ as a relative (H, K)-module with the structure of Example 9.2.2.

Now $H \otimes M = (H \otimes F) \oplus (H \otimes E)$. By Proposition 9.2.8 we have that $H \otimes M \simeq M^{(\mathrm{Dim}(H))} = F^{(\mathrm{Dim}(H))} \oplus E^{(\mathrm{Dim}(H))}$ and $H \otimes F \simeq F^{(\mathrm{Dim}(H))}$. Thus by the Krull-Schmidt Theorem $H \otimes E \simeq E^{(\mathrm{Dim}(H))}$. But the K-submodule $K \otimes E$ of $H \otimes E$ is (0) or free by part (a) of Proposition 8.2.6. In the latter case $E^{(\mathrm{Dim}(H))}$, and hence E, is faithful, a contradiction. Therefore $K \otimes E = (0)$ which means that $E = (0)$. We have shown that $M = F$ is a free left K-module. □

There are many applications of the preceding theorem. We will record two at this point and will consider others in subsequent chapters.

Corollary 9.3.4. *Suppose that H is a finite-dimensional Hopf algebra over the field k and K is a Hopf subalgebra of H. Then* $\mathrm{Dim}(K)$ *divides* $\mathrm{Dim}(H)$. □

Let G' be a subgroup of a finite group G. Lagrange's Theorem is that $|G'|$ divides $|G|$, or equivalently $\mathrm{Dim}(k[G'])$ divides $\mathrm{Dim}(k[G])$. Thus Corollary 9.3.4 is a generalization of Lagrange's Theorem.

Since $k[\mathrm{G}(H)]$ is a Hopf subalgebra of H of dimension $|\mathrm{G}(H)|$:

Corollary 9.3.5. *Suppose that H is a finite-dimensional Hopf algebra over the field k. Then $|\mathrm{G}(H)|$ divides* $\mathrm{Dim}(H)$. □

Exercises

In the following exercises H is a Hopf algebra over the field k.

Exercise 9.3.1. Suppose k is algebraically closed and has characteristic 0. Determine all H over k of dimension at most 4. [Hint: Show that H is pointed and thus $\mathrm{Dim}(H_0)$ divides $\mathrm{Dim}(H)$.]

Definition 9.3.6. H is a *simple Hopf algebra over* k if it is a Hopf algebra over k with exactly two Hopf ideals, namely (0) and $\mathrm{Ker}(\epsilon)$.

Since Hopf ideals are bi-ideals, a Hopf algebra over k which is a simple bialgebra is a simple Hopf algebra.

Exercise 9.3.2. Suppose H is a finite-dimensional simple Hopf algebra. Show that:

(a) H is a simple bialgebra.

(b) H^* has exactly two sub-bialgebras and these are Hopf subalgebras.

Exercise 9.3.3. Suppose $\mathrm{Dim}(H) = p$, where p is a prime integer. Show that:

(a) H and H^* are simple Hopf algebras.

(b) H^* has exactly two Hopf subalgebras.

(c) $H \simeq k[\boldsymbol{Z}_p]$ or $\mathrm{G}(H) = \{1\}$. (Thus $H^* \simeq k[\boldsymbol{Z}_p]$ or $\mathrm{G}(H^*) = \{\epsilon\}$.)

Exercise 9.3.4. Suppose k has characteristic $p > 0$ and $\mathrm{Dim}(H) = p$. Then the conclusions of part (c) of Exercise 9.3.3 hold. Determine which of (a)–(c) below is possible by constructing proofs or finding examples:

(a) $H \simeq H^* \simeq k[\boldsymbol{Z}_p]$.

(b) $H \simeq k[\boldsymbol{Z}_p]$ and $\mathrm{G}(H^*) = \{\epsilon\}$.

(c) $\mathrm{G}(H) = \{1\}$ and $\mathrm{G}(H^*) = \{\epsilon\}$.

9.4 An example of a Hopf algebra which is not free over some Hopf subalgebra

Let k be any field. $\mathrm{Rank}_B(M)$ denotes the rank of a free left module M over a commutative algebra B over k. We will construct a Hopf algebra over k which is not free as a left module over one of its Hopf subalgebras. Our construction is based on the following:

Lemma 9.4.1. *Let (A, ν) be an augmented algebra over the field k. Suppose $A = (V)$ is commutative where V is a subspace of A such that $\nu(V) \neq (0)$. Set $B = k + V^2 + V^4 + \cdots$ and $M = V + V^3 + V^5 + \cdots$. Then:*

(a) *B is a subalgebra of A and $BM = M$. (Thus B is a subalgebra of A and A, M are left B-modules under multiplication.)*
(b) *Suppose $A = B \oplus M$ and is a free left B-module. Then $\mathrm{Rank}_B(A) = 2$ and M is a free left B-module with $\mathrm{Rank}_B(M) = 1$.*

Proof. Part (a) is straightforward. Assume the hypothesis of part (b) and let $\{m_i\}_{i \in I}$ be a basis for A as a left B-module. Note $J = B \cap \mathrm{Ker}(\nu)$ is a codimension one ideal of B. By virtue of the calculation

$$A/JA \simeq \bigoplus_{i \in I} Bm_i/Jm_i \simeq \bigoplus_{i \in I} B/J \simeq \bigoplus_{i \in I} k$$

we conclude $\mathrm{Dim}(A/JA) = \mathrm{Rank}_B(A)$. On the other hand $A/JA \simeq B/J \oplus M/JM \simeq k \oplus M/JM$ which means $\mathrm{Rank}_B(A) = 1 + \mathrm{Dim}(M/JM)$.

We show that $\mathrm{Rank}_B(A) = 2$ by establishing $\mathrm{Dim}(M/JM) = 1$. The left B-module structure on M induces a left B/J-module structure on M/JM. Now $V + JM$ generates M/JM as a left B/J-module. Since $\nu(V) \neq (0)$ there is a $v \in V$ satisfying $\nu(v) = 1$. Observe that $1 - v^2 \in J$ as $1 - v^2 \in B$ and $\nu(1 - v^2) = 1 - \nu(v)^2 = 0$.

Let $u \in V$ satisfy $\nu(u) = 0$. Then $u = (1 - v^2)u + (vu)v \in JM$ since $\nu(vu) = \nu(v)\nu(u) = 0$. We have shown $V + JM = kv + JM$ which implies M/JM is a cyclic B/J-module. Thus $\mathrm{Dim}(M/JM) \leq 1$ since $B/J \simeq k$. Now $(0) \neq \nu(V) \subseteq \nu(M)$ and $\nu(JM) = \nu(J)\nu(M) = (0)$ imply $M/JM \neq (0)$. Therefore $\mathrm{Dim}(M/JM) = 1$.

Let $\{a_1, a_2\}$ be a basis for A as a B-module. To complete the proof we need only show that A has a basis of the form $\{1, m\}$ over B, where $m \in M$. Write $a_1 = b_1 \oplus m_1$ and $a_2 = b_2 \oplus m_2$ where $b_1, b_2 \in B$ and $m_1, m_2 \in M$. Now $1 = x_1a_1 + x_2a_2 = (x_1b_1 + x_2b_2) \oplus (x_1m_1 + x_2m_2)$ for some $x_1, x_2 \in B$ which is equivalent to $1 = x_1b_1 + x_2b_2$ and $x_1m_1 + x_2m_2 = 0$. Let $a = -b_2a_1 + b_1a_2$. Since $\begin{pmatrix} x_1 & x_2 \\ -b_2 & b_1 \end{pmatrix} \begin{pmatrix} a_1 \\ a_2 \end{pmatrix} = \begin{pmatrix} 1 \\ a \end{pmatrix}$ and the 2×2 matrix is invertible, it follows that $\{1, a\}$ is a basis for A over B as well. Write $a = b \oplus m$ where $b \in B$ and $m \in M$. A desired basis is $\{1, m\}$. □

Let $n \geq 2$ be an even integer and $\mathcal{A} = k[X]$ be the polynomial algebra over k on the set of indeterminates $X = \{x_{i,j}\}_{1 \leq i,j \leq n}$. Then $\mathcal{A}$ is a bialgebra over k where $\Delta(x_{i,j}) = \sum_{\ell=1}^{n} x_{i,\ell} \otimes x_{\ell,j}$ and $\epsilon(x_{i,j}) = \delta_{i,j}$ for all $1 \leq i, j \leq n$. Observe that $\mathcal{A} = (\mathcal{V}) = \mathcal{B} \oplus \mathcal{M}$, where $\mathcal{V}$ is the span of X, $\mathcal{B} = k + \mathcal{V}^2 + \mathcal{V}^4 + \cdots$, and $\mathcal{M} = \mathcal{V} + \mathcal{V}^3 + \mathcal{V}^5 + \cdots$.

The determinant $d = \begin{vmatrix} x_{1,1} & \cdots & x_{1,n} \\ \vdots & & \vdots \\ x_{n,1} & \cdots & x_{n,n} \end{vmatrix} \in \mathrm{G}(\mathcal{A})$ by Lemma 7.6.10. Since n is even and $d \in \mathcal{V}^n$ it follows that $d - 1 \in \mathcal{B}$. Thus the bi-ideal $\mathcal{I} = (d-1)\mathcal{A}$ has the direct sum decomposition $\mathcal{I} = (d-1)\mathcal{B} \oplus (d-1)\mathcal{M} = (\mathcal{I} \cap \mathcal{B}) \oplus (\mathcal{I} \cap \mathcal{M})$. Let $A = \mathcal{A}/\mathcal{I}$ be the quotient bialgebra and $\Pi : \mathcal{A} \longrightarrow A$ be the projection. Then $A = (V) = B \oplus M$ where $V = \Pi(\mathcal{V})$, $B = k + V^2 + V^4 + \cdots = \Pi(\mathcal{B})$, and $M = V + V^3 + V^5 + \cdots = \Pi(\mathcal{M})$.

Consider the algebra homomorphism $\pi_0 : k[X] \longrightarrow k[x, y, z]$ defined by

$$\pi_0(x_{i,j}) = \begin{cases} x : i = 1, j = 2; \\ y : i = 2, j = 1; \\ z : i = j; \\ 0 : \text{otherwise}. \end{cases}$$

Then $\pi_0(d) = (z^2 - xy)z^{n-2} = z^n - xyz^{n-2}$. Thus π_0 lifts to an onto algebra homomorphism $\pi : A \longrightarrow k[x, y, z]/(z^n - xyz^{n-2} - 1)$. This quotient is the commutative algebra $\mathbf{A} = k[x, y][z]$ over the polynomial ring $k[x, y]$ generated by z subject to the relation $z^n - xyz^{n-2} - 1$. By the Division Algorithm for commutative rings $\mathbf{A}$ is a free $k[x, y]$-module with basis $\{1, z, \ldots, z^{n-1}\}$. Observe that $\mathbf{A} = (\mathbf{V}) = \mathbf{B} \oplus \mathbf{M}$, where $\mathbf{V} = \pi(V)$, $\mathbf{B} = k + \mathbf{V}^2 + \mathbf{V}^4 + \cdots$, and $\mathbf{M} = \mathbf{V} + \mathbf{V}^3 + \mathbf{V}^5 + \cdots$.

We will describe $\mathbf{A}$ in terms of even and odd monomials. We denote cosets with representatives. Note that

$$\{x^a y^b z^c \,|\, 0 \leq a, b, c \text{ and } c < n\}$$

is a linear basis for $\mathbf{A}$. We refer to a monomial in the basis as even if $a+b+c$ is even and as odd otherwise. We call an element of $\mathbf{A}$ even (respectively odd) if it is a linear combination of even (respectively odd) monomials in the basis. By virtue of the formulas

$$z^{n+2m} = x^{m+1} y^{m+1} z^{n-2} + \sum_{\ell=0}^{m} x^{m-\ell} y^{m-\ell} z^{2\ell}$$

and

$$z^{n+2m+1} = x^{m+1} y^{m+1} z^{n-1} + \sum_{\ell=0}^{m} x^{m-\ell} y^{m-\ell} z^{2\ell+1}$$

for all $m \geq 0$ it follows that any monomial $x^a y^b z^c$ is even if $a+b+c$ is even and is odd otherwise. Thus $\mathbf{B}$ is the span of the even monomials and $\mathbf{M}$ is the span of the odd monomials of $\mathbf{A}$.

To construct our example we specialize to the case $n = 2$. First of all note that A is a Hopf algebra with antipode S which satisfies $S(V) \subseteq V$. Therefore $S(B) \subseteq B$ which means that B is a Hopf subalgebra of A. See Exercise 9.4.1. We will show that A is not a free B-module.

Suppose that A is a free B-module. Then M is a cyclic B-module by Lemma 9.4.1. Thus $\mathbf{M}$ is a cyclic $\mathbf{B}$-module. We will show the latter is not the case.

Suppose $\mathbf{M}$ is a cyclic $\mathbf{B}$-module with generator m. Write $m = m_o \oplus m_e z$, where $m_o, m_e \in k[x, y]$ are odd, even respectively. Let $u \in M$. Then $u = bm$ for some $b \in \mathbf{B}$ which can be written $b = u_e \oplus u_o z$ for some $u_e, u_o \in k[x, y]$ which are even, odd respectively. We emphasize that u_e, u_o depend on b. Writing $u = bm = (u_e m_o + u_o m_e(xy+1)) \oplus (u_e m_e + u_o m_o) z \in k[x, y] \oplus k[x, y] z$ and specializing u to $x, z \in \mathbf{M}$ results in the system of equations

$$\begin{aligned} x_e m_o + x_o m_e(xy + 1) &= x \\ x_e m_e + x_o m_o &= 0 \\ z_e m_o + z_o m_e(xy + 1) &= 0 \\ z_e m_e + z_o m_o &= 1 \end{aligned}$$

Observe that $m_e \neq 0$; else by the fourth equation $z_o, m_o \in k$ which is not possible since both are odd. We show the system of equations has no solution. To do this we may assume k is algebraically closed. Since $0 \neq m_e = m_e(x, y)$ there is a non-zero $\alpha \in k$ such that $m_e(x, \alpha) \neq 0$. Substituting α for y in the system of equations yields the four equations

$$\begin{aligned} pq' + p'q(\alpha x + 1) &= x \\ pp' + qq' &= 0 \\ pq'' + p''q(\alpha x + 1) &= 0 \\ pp'' + qq'' &= 1 \end{aligned}$$

for some $p, p', p'', q, q', q'' \in k[x]$, where $q \neq 0$. Observe that $p \neq 0$ by the first equation. By virtue of the last two equations $p | p''(\alpha x + 1)$ and $q | q''$. Now $(\alpha x + 1) \not| p$ by the first equation. Thus $p | p''$ since $p | p''(\alpha x + 1)$.

Write $p'' = pu$ and $q'' = qv$, where $u, v \in k[x]$. From the third equation we deduce $pq(v + u(\alpha x + 1)) = 0$. Thus $v + u(\alpha x + 1) = 0$ since $p, q \neq 0$. From the fourth equation we have $(p^2 - q^2(\alpha x + 1))u = 1$ and hence $p^2 - q^2(\alpha x + 1) \in k$. This is not possible since the preceding polynomial is the difference of one of even degree and one of odd degree. Therefore M is not a cyclic B-module after all.

Proposition 9.4.2. *Let H be the commutative Hopf algebra over the field k generated as an algebra by $X = \{x_{i,j}\}_{1\leq i,j\leq 2}$ subject to the relation*

$$x_{1,1}x_{2,2} - x_{1,2}x_{2,1} = 1$$

and whose coalgebra structure is determined by

$$\Delta(x_{i,j}) = \sum_{\ell=1}^{2} x_{i,\ell}\otimes x_{\ell,j} \quad \text{and} \quad \epsilon(x_{i,j}) = \delta_{i,j}$$

for $1 \leq i, j \leq 2$. Let V be the span of X. Then $B = (V^2)$ is a Hopf subalgebra of H and H is not a free B-module under multiplication. □

Exercise

Exercise 9.4.1. Complete the proof of Proposition 9.4.2. [Hint: Show that there is an algebra homomorphism $S : H \longrightarrow H$ determined by $S(x_{1,1}) = x_{2,2}$, $S(x_{1,2}) = -x_{1,2}$, $S(x_{2,1}) = -x_{2,1}$, and $S(x_{2,2}) = x_{1,1}$. Now use Exercise 7.3.2.]

Chapter notes

Freeness theorems for Hopf algebras are very important in practice, especially for finite-dimensional Hopf algebras. Theorem 9.3.1, the Nichols-Zoeller Theorem [153], was a fundamental breakthrough. Nichols and Zoeller went on to consider the freeness question for infinite dimensional Hopf algebras. In [152] they show that if H is infinite-dimensional and K is a finite-dimensional semisimple Hopf subalgebra of H, then H is a free K-module. Other freeness results, in particular when K is finite-dimensional and H is commutative, are found in [163, 164].

The first example of a Hopf algebra which is not free over one of its Hopf subalgebras was described by Oberst and Schneider in [155]. The example is commutative, cocommutative, and k is not algebraically closed. Another example having the same characteristics was given by Takeuchi in [218]. The example of Proposition 9.4.2 is the one of [166] and is treated in greater generality here. Schneider has shown that the Hopf algebra A of Section 9.4 is not a free B-module for all even $n \geq 2$ in [190].

Relative Hopf modules, which are fundamental in the proof Theorem 9.3.1 by Nichols and Zoeller, were introduced by Takeuchi in [218] who used them to explore when a Hopf algebra is free or projective over a Hopf subalgebra. He shows that commutative Hopf algebras are projective over Hopf subalgebras and are free over Hopf subalgebras under various conditions, in particular if the Hopf subalgebra is locally finite. It was noted in Takeuchi's paper that for relative (H, K)-module to be defined K need only be a right coideal subalgebra of H, that is a subalgebra which satisfies $\Delta(K) \subseteq K \otimes H$, and some freeness or projectivity results were obtained for them.

Building on the results of [218], Masuoka considers right coideal subalgebras K in a finite-dimensional Hopf algebra H over k in [120]. He shows that H is a free left and right K-module if and only if K is a Frobenius algebra.

The question of when H is a faithfully flat K-module is addressed by Takeuchi in [212]. He shows that when H is commutative or cocommutative H is faithfully flat as a left or right K-module.

Section 9.1 is drawn from [159, 164] and Section 9.2 is based on [218]. Lemma 9.2.6 is a slight generalization of [212, Theorem 3.1]. The proof of Theorem 9.3.3 follows the proof of the same in [153].

Chapter 10

Integrals

A good deal of the general theory of finite-dimensional Hopf algebras is found in this chapter. The reason is that many of the deeper results concerning finite-dimensional Hopf algebras are directly linked to integrals. Integrals are certain elements of a bialgebra or certain functionals on a bialgebra. One may think of integrals which are elements of bialgebras as analogs of the element $\Lambda = \sum_{g \in G} g$ of the group algebra $k[G]$ of a finite group G over k. One may think of integrals which are functionals as analogs of Haar measure on a compact group.

Suppose that H is a finite-dimensional Hopf algebra over k. The freeness of Hopf modules over H accounts for the existence of integrals for H in a very natural way. We base our discussion of integrals for H on Theorem 8.4.2, which is a direct consequence of the freeness result. Theorem 8.4.2 is expressed very simply in terms of module structures. Much of this chapter is a careful examination of the implications of this theorem.

We conclude the chapter with a discussion of integrals which are functionals. Throughout this chapter H is a Hopf algebra with antipode S over the field k.

10.1 Definition of integrals for a bialgebra and its dual algebra

In this section A is a bialgebra over k.

Definition 10.1.1. Let A be a bialgebra over k. A *left* (respectively *right*) *integral for* A is an element $\Lambda \in A$ which satisfies $a\Lambda = \epsilon(a)\Lambda$ (respectively $\Lambda a = \Lambda\epsilon(a)$) for all $a \in A$.

We denote the set of left (respectively right) integrals for A by $\int_\ell$ (respectively $\int_r$). Observe that $\int_\ell$ and $\int_r$ are ideals of A. Notice that a non-zero left integral for A generates a one-dimensional left ideal of A and that a non-zero right integral for A generates a one-dimensional right ideal of A.

Recall that when A is finite-dimensional the counit for the bialgebra A^* is given by $\epsilon_{A^*}(a^*) = <a^*, 1>$ for all $a^* \in A^*$. Thus the definitions of left and right integral for A^* are meaningful for A^* in any case.

Definition 10.1.2. Let A be a bialgebra over k. A *left* (respectively *right*) *integral for* A^* is an element $\lambda \in A^*$ such that $a^*\lambda = <a^*, 1>\lambda$ (respectively $\lambda a^* = \lambda<a^*, 1>$) for all $a^* \in A$.

We denote the set of left (respectively right) integrals for A^* by $\int^\ell$ (respectively $\int^r$). Observe that $\int^\ell$ and $\int^r$ are ideals of A^*. Notice that a non-zero left integral for A^* generates a one-dimensional left ideal of A^* and that a non-zero right integral for A^* generates a one-dimensional right ideal of A^*.

Suppose that H is a Hopf algebra over k. Then $\int^r = H^{co\, inv}$ by (8.7), where (H^r, μ^T, Δ^T) is the left H-Hopf module described in Section 8.4. Left or right integrals for H possess useful invariance properties.

Proposition 10.1.3. *Suppose that H is a Hopf algebra with antipode S over the field k.*

(a) *Let $\Lambda \in \int_\ell$. Then $\Lambda_{(1)} \otimes a\Lambda_{(2)} = S(a)\Lambda_{(1)} \otimes \Lambda_{(2)}$ for all $a \in H$.*
(b) *Let $\Lambda \in \int_r$. Then $\Lambda_{(1)}a \otimes \Lambda_{(2)} = \Lambda_{(1)} \otimes \Lambda_{(2)}S(a)$ for all $a \in H$.*

Proof. Part (a) for the Hopf algebra $H^{op\, cop}$ is part (b) for H. Thus we need only show part (a).

Suppose that $\Lambda \in \int_\ell$ and $a \in H$. Then

$$\begin{aligned}
\Lambda_{(1)} \otimes a\Lambda_{(2)} &= \epsilon(a_{(1)})1\Lambda_{(1)} \otimes a_{(2)}\Lambda_{(2)} \\
&= S(a_{(1)(1)})a_{(1)(2)}\Lambda_{(1)} \otimes a_{(2)}\Lambda_{(2)} \\
&= S(a_{(1)})a_{(2)(1)}\Lambda_{(1)} \otimes a_{(2)(2)}\Lambda_{(2)} \\
&= S(a_{(1)})(a_{(2)}\Lambda_{(1)})_{(1)} \otimes (a_{(2)}\Lambda_{(2)})_{(2)} \\
&= S(a_{(1)})\epsilon(a_{(2)})\Lambda_{(1)} \otimes \Lambda_{(2)} \\
&= S(a_{(1)}\epsilon(a_{(2)}))\Lambda_{(1)} \otimes \Lambda_{(2)} \\
&= S(a)\Lambda_{(1)} \otimes \Lambda_{(2)}
\end{aligned}$$

from which part (a) follows. □

Exercises

In the following exercises A is a bialgebra over the field k, unless otherwise specified, and H is a Hopf algebra with antipode S over k.

Exercise 10.1.1. Let $\lambda \in A^*$. Show that:

(a) $\lambda \in \int^\ell$ if and only if $a_{(1)}<\lambda, a_{(2)}> = <\lambda, a>1$ for all $a \in A$.

(b) $\lambda \in \int^r$ if and only if $<\lambda, a_{(1)}>a_{(2)} = <\lambda, a>1$ for all $a \in A$.

Exercise 10.1.2. Suppose that $\lambda \in \int^r$ and $\lambda' \in \int^\ell$. Show that:

(a) $<\lambda, a_{(1)}b>a_{(2)} = <\lambda, ab_{(1)}>S(b_{(2)})$ for all $a, b \in H$. [Hint: Note that

$$\begin{aligned}<\lambda, a_{(1)}b>a_{(2)} &= <\lambda, a_{(1)}b_{(1)}>a_{(2)}b_{(2)}S(b_{(3)}) \\ &= <\lambda, (ab_{(1)})_{(1)}>(ab_{(1)})_{(2)}S(b_{(2)})\end{aligned}$$

for all $a, b \in H$.]

(b) $b_{(1)}<\lambda', ab_{(2)}> = S(a_{(1)})<\lambda', a_{(2)}b>$ for all $a, b \in H$.

Exercise 10.1.3. Suppose that A is an algebra over k and $\lambda \in A^*$. Define a bilinear form $\beta : A \times A \longrightarrow k$ by $\beta(a, b) = <\lambda, ab>$ for all $a, b \in A$.

(a) Show that β is *associative*, that is $\beta(ab, c) = \beta(a, bc)$ for all $a, b, c \in A$.

Suppose further that A is a bialgebra and consider the following statements:

(1) If $\lambda \in \int^\ell$ then $\beta(a \leftharpoonup \alpha_{(1)}, b \leftharpoonup \alpha_{(2)}) = <\alpha, 1>\beta(a, b)$ for all $a, b \in A$ and $\alpha \in A^*$.

(2) If $\lambda \in \int^r$ then $\beta(\alpha_{(1)} \rightharpoonup a, \alpha_{(2)} \rightharpoonup b) = <\alpha, 1>\beta(a, b)$ for all $\alpha \in A^*$ and $a, b \in A$.

Show that:

(b) (1) implies (2). [Hint: Show that (1) for A^{cop} is (2) for A.]

(c) (2) implies (1).

(d) (1) is true. [Hint: Show that

$$<\lambda, (a \leftharpoonup \alpha_{(1)})(b \leftharpoonup \alpha_{(2)})> = <\alpha, (ab)_{(1)}><\lambda, (ab)_{(2)}>$$

for all $\alpha \in A^*$ and $a, b \in A$.]

Exercise 10.1.4. Suppose that k has characteristic 0 and $H = k[X]$ is the polynomial algebra in indeterminant X over k where x is primitive. Show that $\int^\ell = (0) = \int^r$.

Exercise 10.1.5. Suppose that the dual algebra H^* is an integral domain. Show that $\int^\ell \neq (0)$ if and only if $H = k$.

Exercise 10.1.6. Let G be any monoid, finite or infinite, and let $A = k[G]$ be the monoid algebra of G over k. Show that:

(a) $\int^\ell = \int^r$.

(b) $\mathrm{Dim}(\int^\ell) = 1$.

(c) $<\lambda, 1> \neq 0$ for any non-zero $\lambda \in \int^\ell$.

Exercise 10.1.7. Exercise 10.1.6 can be generalized. Suppose that A is cosemisimple as a coalgebra. Show that parts (a)–(c) of Exercise 10.1.6 hold for A.

Exercise 10.1.8. Let G be a finite monoid, let $A = k[G]$ be the monoid algebra of G over k and let $\Lambda = \sum_{g \in G} g$. Show that Λ is a left integral for A if and only if G is a group.

Exercise 10.1.9. In this exercise we consider a uniqueness condition for integrals. Show that:

(a) If Λ and Λ' are non-zero left and right integrals respectively for A then $\epsilon(\Lambda) \neq 0$ if and only if $\epsilon(\Lambda') \neq 0$ in which case $\int_\ell = k\Lambda = \int_r$.

(b) If λ and λ' are non-zero left and right integrals respectively for A^* then $<\lambda, 1> \neq 0$ if and only if $<\lambda', 1> \neq 0$ in which case $\int^\ell = k\lambda = \int^r$.

Exercise 10.1.10. In this exercise we place integrals in a more general context. Recall that non-zero left integrals generate one-dimensional left ideals and non-zero right integrals generate one-dimensional right ideals.

Suppose A is an algebra over k. Recall that $\mathrm{G}(A^o)$ is the set of algebra homomorphisms $\eta : A \longrightarrow k$. For $\eta \in \mathrm{G}(A^o)$ define

$$L_\eta = \{\Lambda \in A \,|\, a\Lambda = \eta(a)\Lambda \text{ for all } a \in A\}$$

and

$$R_\eta = \{\Lambda \in A \,|\, \Lambda a = \Lambda\eta(a) \text{ for all } a \in A\}.$$

Observe that any non-zero $\Lambda \in L_\eta$ generates a one-dimensional left ideal of A and that any non-zero $\Lambda \in R_\eta$ generates a one-dimensional right ideal of A. Show that:

(a) L_η and R_η are ideals of A for all $\eta \in \mathrm{G}(A^o)$.

(b) If $\eta, \eta' \in \mathrm{G}(A^o)$ and $L_\eta = L_{\eta'} \neq (0)$ or $R_\eta = R_{\eta'} \neq (0)$ then $\eta = \eta'$.

(c) If $\Lambda \in A$ generates a one-dimensional left ideal of A then $\Lambda \in L_\eta$ for some $\eta \in \mathrm{G}(A^o)$.

(d) If $\Lambda \in A$ generates a one-dimensional right ideal of A then $\Lambda \in R_\eta$ for some $\eta \in \mathrm{G}(A^o)$.

(e) If $\eta \in \mathrm{G}(A^o)$ and either $\eta(L_\eta) \neq (0)$ or $\eta(R_\eta) \neq (0)$ then $R_\eta L_\eta \neq (0)$.

(f) If $\eta, \xi \in \mathrm{G}(A^o)$ and $R_\xi L_\eta \neq (0)$ then:

(i) $L_\eta = R_\xi$.
(ii) $\mathrm{Dim}(L_\eta) = 1 = \mathrm{Dim}(R_\xi)$.
(iii) L_η is generated by an idempotent.

(g) If $f : A \longrightarrow A$ is an algebra automorphism of A then $f(L_\eta) = L_{\eta \circ f^{-1}}$ and $f(R_\eta) = R_{\eta \circ f^{-1}}$ for all $\eta \in \mathrm{G}(A^o)$.

(h) If $f : A \longrightarrow A^{op}$ is an algebra isomorphism then $f(L_\eta) = R_{\eta \circ f^{-1}}$ and $f(R_\eta) = L_{\eta \circ f^{-1}}$ for all $\eta \in \mathrm{G}(A^o)$.

Exercise 10.1.11. Let $H = H_{2,-1}$. Compute $\int_\ell, \int_r$ and show that $\int_\ell \neq \int_r$.

Exercise 10.1.12. Suppose λ is a right integral for A^*. Show that $A \leftharpoonup \lambda = k1$. [Hint: For $a^* \in A^*$ and $a \in A$ consider $<a^*, a \leftharpoonup \lambda>$.]

Exercise 10.1.13. Suppose that Λ is a left integral for H, λ is a left integral for H^* and $<\lambda, \Lambda> = 1$. Show that $S(a) = (\lambda \prec a) \rightharpoonup \Lambda$ for all $a \in H$. [Hint: Apply $a^* \in H^*$ to both sides of the equation.]

10.2 Existence and uniqueness of integrals for a Hopf algebra

A non-zero left or right integral for H generates a one-dimensional left or right ideal of H. Hopf algebras over k which have a non-zero finite-dimensional left or right ideal are finite-dimensional by Proposition 8.2.5. We have shown:

Proposition 10.2.1. *Suppose that H is a Hopf algebra over the field k which has a non-zero left or right integral. Then H is finite-dimensional.* □

For the rest of this section H is finite-dimensional and $\boldsymbol{f} : H \longrightarrow H^*$ is the isomorphism of Theorem 8.4.2. We begin to discuss some of the implications of the equations of (8.8), namely

$$\boldsymbol{f}(ab) = \boldsymbol{f}(b) \prec S(a) \qquad \text{and} \qquad \boldsymbol{f}(a \leftharpoonup a^*) = \boldsymbol{f}(a)a^*$$

for all $a, b \in H$ and $a^* \in H^*$. In the process we will re-derive some of the results of Section 8.4 from a slightly different perspective. The elements corresponding to the multiplicative identities of H and H^* under the isomorphism $\boldsymbol{f}$ are the basis of our analysis.

Set

$$\lambda = \boldsymbol{f}(1) \quad \text{and} \quad \Lambda = \boldsymbol{f}^{-1}(\epsilon).$$

The calculations

$$\boldsymbol{f}(a) = \boldsymbol{f}(a1) = \boldsymbol{f}(1) \prec S(a) = \lambda \prec S(a)$$

and

$$\boldsymbol{f}(\Lambda \leftharpoonup a^*) = \boldsymbol{f}(\Lambda)a^* = \epsilon a^* = a^*$$

show that

$$\boldsymbol{f}(a) = \lambda \prec S(a) \qquad \text{and} \qquad \boldsymbol{f}^{-1}(a^*) = \Lambda \leftharpoonup a^* \tag{10.1}$$

for $a \in H$ and $a^* \in H^*$. Using (10.1) to rewrite the equations $\boldsymbol{f}^{-1}(\boldsymbol{f}(a)) = a$ and $\boldsymbol{f}(\boldsymbol{f}^{-1}(a^*)) = a^*$ we obtain

$$\Lambda \leftharpoonup (\lambda \prec S(a)) = a \qquad \text{and} \qquad \lambda \prec (S(\Lambda \leftharpoonup a^*)) = a^* \tag{10.2}$$

for $a \in H$ and $a^* \in H^*$. Specializing (10.2) we deduce

$$\Lambda \leftharpoonup \lambda = 1 \qquad \text{and} \qquad \lambda \prec S(\Lambda) = \epsilon. \tag{10.3}$$

Applying ϵ to both sides of the first equation of (10.3) and applying both sides of the second to 1 yields

$$<\lambda, \Lambda> = 1 = <\lambda, S(\Lambda)>. \tag{10.4}$$

Since

$$\lambda a^* = \boldsymbol{f}(1)a^* = \boldsymbol{f}(1 \leftharpoonup a^*) = <a^*, 1>\boldsymbol{f}(1) = <a^*, 1>\lambda$$

we conclude that

$$\lambda \quad \text{is a right integral for} \quad H^*. \tag{10.5}$$

Since $\boldsymbol{f}$ is one-one the calculation

$$\boldsymbol{f}(a\Lambda) = \boldsymbol{f}(\Lambda) \prec S(a) = \epsilon \prec S(a) = \epsilon(a)\epsilon = \boldsymbol{f}(\epsilon(a)\Lambda)$$

for all $a \in H$ shows that

$$\Lambda \quad \text{is a left integral for} \quad H. \tag{10.6}$$

We next show that

$$\mathrm{Dim}(\int^r) \leq 1. \tag{10.7}$$

Suppose that $\lambda' \in \int^r$. Then $\lambda' = \boldsymbol{f}(a)$ for some $a \in H$ since f is onto. Now for all $a^* \in H^*$ we have

$$\lambda' a^* = <a^*, 1>\lambda' = <a^*, 1>\boldsymbol{f}(a) = \boldsymbol{f}(<a^*, 1>a)$$

since λ' is a right integral. On the other hand

$$\lambda' a^* = \boldsymbol{f}(a)a^* = \boldsymbol{f}(a \leftharpoonup a^*).$$

Since $\boldsymbol{f}$ is one-one we conclude that $a \leftharpoonup a^* = <a^*, 1>a$ for all $a^* \in H^*$. Applying ϵ to both sides of this last equation we see that $<a^*, a> = <a^*, 1>\epsilon(a)$ for all $a^* \in H^*$. Therefore $a = \epsilon(a)1$ which means $\lambda' = \epsilon(a)\lambda$.

Theorem 10.2.2. *Suppose that H is a finite-dimensional Hopf algebra with antipode S over the field k. Then:*

(a) *The ideals of left integrals and right integrals for H or for H^* are one-dimensional.*

(b) *Suppose that Λ is a non-zero left or right integral for H and that λ is a non-zero left or right integral for H^*. Then $<\lambda, \Lambda> \neq 0$.*

(c) *Suppose that Λ is a left integral for H and λ is a right integral for H^* such that $<\lambda, \Lambda> = 1$. Then:*

  (i) $\Lambda \leftharpoonup (\lambda \prec S(a)) = a$ *and* $\lambda \prec (S(\Lambda \leftharpoonup a^*)) = a^*$ *for all* $a \in H$ *and* $a^* \in H^*$.

  (ii) $\Lambda \leftharpoonup \lambda = 1$ *and* $\lambda \prec (S(\Lambda)) = \epsilon$.

(d) *Suppose that $\Lambda \in H$ is a non-zero left or right integral for H. Then $(H, \leftharpoonup)$ is a free right H^*-module and $(H, \rightharpoonup)$ is a free left H^*-module with basis $\{\Lambda\}$.*

(e) *Suppose that $\lambda \in H^*$ is a non-zero left or right integral for H^*. Then $(H^*, \prec)$ is a free right H-module and $(H^*, \succ)$ is a left H-module with basis $\{\lambda\}$.*

Proof. To show part (a) we first observe that the ideal of left integrals for H is the ideal of right integrals for H^{op} and that $H \simeq (H^*)^*$. Thus the ideal of left integrals, or right integrals, for a finite-dimensional Hopf algebra can be identified with the ideal of right integrals for the dual of some finite-dimensional Hopf algebra. Thus part (a) follows from (10.5) and (10.7).

To show part (b) we first observe that the left integrals for H^{cop} are the left integrals for H and that the left integrals for H^{op} and $H^{op\,cop}$ are the right integrals for H. Likewise the right integrals for $(H^{op})^*$ are the right integrals for H^* and the left integrals for $(H^{cop})^*$ and $(H^{op\,cop})^*$ are the right integrals for H^*.

Suppose that Λ is a non-zero left or right integral for H and that λ is a non-zero left or right integral for H^*. Then Λ is a left integral for $\mathcal{H}$ and λ is a right integral for $\mathcal{H}^*$, where $\mathcal{H}$ is one of the Hopf algebras H, H^{op}, H^{cop} or $H^{op\,cop}$. Thus $<\lambda, \Lambda> \neq 0$ by part (a), (10.4), (10.5), and (10.6). We have established part (b).

We have shown that part (c) holds for the special case where Λ and λ defined by $\boldsymbol{f}(\Lambda) = \epsilon$ and $\boldsymbol{f}(1) = \lambda$. Now suppose that $\Lambda' \in \int_\ell$ and $\lambda' \in \int^r$ satisfy $<\lambda', \Lambda'> = 1$. Since $\Lambda' = \alpha\Lambda$ and $\lambda' = \beta\lambda$ for some non-zero $\alpha, \beta \in k$ by part (a), it follows that $\alpha\beta = 1$. Therefore part (c) holds for Λ' and λ' as well.

Since H is finite-dimensional, $(H, \leftharpoonup)$ is a free right H^*-module with basis $\{\Lambda\}$ and $(H^*, \prec)$ is a free right H-module with basis $\{\lambda\}$, where Λ

is a non-zero left integral for H and λ is a non-zero right integral for H^* by part (a), (10.2), (10.5) and (10.6). Thus parts (d) and (e) hold in these special cases. The reader is left to modify the proof of part (b) to show that parts (d) and (e) reduce to these special cases. This completes our proof of parts (d) and (e), and hence completes the proof of the theorem. □

Notice that part (c)-(i) of Theorem 10.2.2 implies that the antipode of a finite-dimensional Hopf algebra over a field is bijective. If H is commutative then $\int_\ell = \int_r$. If H is infinite-dimensional then $\int_\ell = (0) = \int_r$ by Proposition 10.2.1. Ordinarily the ideals of left and right integrals for a finite-dimensional Hopf algebra are not the same. See Exercise 10.1.11.

Definition 10.2.3. A *unimodular Hopf algebra over* k is a finite-dimensional Hopf algebra over k such that $\int_\ell = \int_r$.

By part (a) of Theorem 10.2.2 a finite-dimensional Hopf algebra H over k is unimodular if some non-zero $\Lambda \in H$ is both a left and right integral for H. Suppose that $\lambda \in H^*$ is a non-zero left or right integral for H^* and consider the associative bilinear form $\beta : H \times H \longrightarrow k$ defined by $\beta(a, b) = <\lambda, ab>$ for all $a, b \in H$. Since $\beta_\ell(a) = \lambda \prec a$ for all $a \in H$ it follows by part (e) of Theorem 10.2.2 that β is left non-singular. Since H is finite-dimensional β is non-singular. This is a slightly different proof of Corollary 8.4.3 which states that finite-dimensional Hopf algebras over a field are Frobenius algebras.

Exercises

In the following exercises A is an algebra over the field k and H is a Hopf algebra with antipode S over k.

Exercise 10.2.1. Suppose that H is finite-dimensional. Show that:

(a) $S(\int_\ell) = \int_r$ and $S(\int_r) = \int_\ell$.

(b) $S^*(\int^\ell) = \int^r$ and $S^*(\int^r) = \int^\ell$.

[Hint: See Exercise 10.1.10.]

Exercise 10.2.2. Suppose that H is finite-dimensional and $\epsilon(\int_\ell) \neq (0)$ or $\epsilon(\int_r) \neq (0)$. Show that H is unimodular. [Hint: See Exercise 10.1.10.]

Exercise 10.2.3. Suppose that $H = H(0) \oplus \cdots \oplus H(n)$ is a finite-dimensional graded Hopf algebra. Show that $\int_\ell, \int_r \subseteq H(n)$.

Exercise 10.2.4. Suppose that H is finite-dimensional. Show that:

(a) If Λ is a non-zero left or right integral for H and N is a left or right coideal of H then $\Lambda \in N$ implies that $N = H$.

(b) If λ is a non-zero left or right integral for H^* and L is a left or right ideal of H then $<\lambda, L> = (0)$ implies that $L = (0)$.

Exercise 10.2.5. Suppose that H' is also a finite-dimensional Hopf algebra over k. Show that:

(a) If Λ and Λ' are left (respectively right) integrals for H and H' then $\Lambda \otimes \Lambda'$ is a left (respectively right) integral for $H \otimes H'$.

(b) $H \otimes H'$ is unimodular if and only if H and H' are unimodular.

Exercise 10.2.6. Suppose that λ is a non-zero left integral for H^* and that L is a left ideal of H. Show that $<\lambda, L> = (0)$ implies $L = (0)$.

Definition 10.2.4. Let A be an algebra over k. A *completely reducible left A-module* is a left A-module M such that $M = (0)$ or M is the direct sum of simple subcomodules.

Definition 10.2.5. Suppose that A is finite-dimensional algebra over k. Then A is a *semisimple algebra* if all left A-modules are completely reducible.

Exercise 10.2.7. Suppose that A is a finite-dimensional algebra over the field k. There are various ways of saying that A is semisimple. Show that the following are equivalent:

(a) Every left A-module is completely reducible.

(b) Every non-zero left A-module is the sum of simple submodules.

(c) If M is a left A-module and N is a submodule of M, then $M = L \oplus N$ for some submodule L of M.

Exercise 10.2.8. Let G be any finite group and $H = k[G]$ be the group algebra of G over k. Show that:

(a) H is unimodular. [Hint: See Exercise 10.1.8.]

(b) H^* is unimodular.

(c) $\epsilon(\int_\ell) \neq (0)$ if and only if $|G|$ is prime to the characteristic of k.

(d) Maschke's Theorem can be formulated as follows: Suppose that G is a finite group and $H = k[G]$ is the group algebra of G over k. Then all left H-modules are completely reducible if and only if $\epsilon(\int_\ell) \neq (0)$.

10.3 Integrals and semisimplicity

Suppose that H is finite-dimensional. Part (d) of Exercise 10.2.8 suggests a generalization of Maschke's Theorem for H. We formulate and prove such a generalization in this section.

For the proof we find it convenient to give the space of all linear maps $f : M \longrightarrow N$ of left H-modules a left H-module structure in terms of which it is easy to see whether or not f is a module map. Suppose that H is any Hopf algebra with antipode S and M and N are left H-modules. For $a \in H$ and $f \in \mathrm{Hom}(M, N)$ we define $a{\bullet}f \in \mathrm{Hom}(M, N)$ by

$$(a{\bullet}f)(m) = a_{(1)}{\cdot}(f(S(a_{(2)}){\cdot}m))$$

for all $m \in M$. The proof of the following lemma is straightforward and is left to the reader.

Lemma 10.3.1. *Suppose that H is a Hopf algebra with antipode S over the field k. Let M and N be left H-modules. Then:*

(a) *$(\mathrm{Hom}(M, N), \bullet)$ is a left H-module.*
(b) *$a{\cdot}(f(m)) = (a_{(1)}{\bullet}f)(a_{(2)}{\cdot}m)$ for all $a \in H$, $f \in \mathrm{Hom}(M, N)$, and $m \in M$.*
(c) *$f \in \mathrm{Hom}(M, N)$ is a module map if and only if $a{\bullet}f = \epsilon(a)f$ for all $a \in H$.*

□

Theorem 10.3.2. *Suppose that H is a Hopf algebra over the field k. Then the following are equivalent:*

(a) *All left H-modules are completely reducible.*
(b) *H is finite-dimensional and $\epsilon(\int_\ell) \neq (0)$.*
(c) *H is finite-dimensional and $\epsilon(\int_r) \neq (0)$.*
(d) *All right H-modules are completely reducible.*

If any of these conditions hold then H is unimodular.

Proof. Suppose that H is finite-dimensional and $\epsilon(\int_\ell) \neq (0)$ or $\epsilon(\int_r) \neq (0)$. We first show that H is unimodular. Let Λ, Λ' be non-zero left and right integrals respectively for H. The calculation $\epsilon(\Lambda')\Lambda = \Lambda'\Lambda = \Lambda'\epsilon(\Lambda)$ shows that if $\epsilon(\Lambda') \neq 0$ or $\epsilon(\Lambda) \neq 0$ then Λ and Λ' are non-zero scalar multiples of each other. See Exercise 10.1.10. We have also shown that parts (b) and (c) are equivalent.

In light of the equivalence of parts (b) and (c) it suffices to show that parts (a) and (b) are equivalent. For the equivalence of parts (a) and (b) for the Hopf algebra $H^{op\,cop}$ gives the equivalence of parts (d) and (c) for H.

We show part (a) implies part (b). Suppose that H is completely reducible as a left H-module. Then there exists a minimal left ideal L of H such that $L \not\subseteq \mathrm{Ker}(\epsilon)$. Thus $\epsilon(L) \neq (0)$. Now $L \cap \mathrm{Ker}(\epsilon) = (0)$ since L is a minimal left ideal of H. Thus $L \subseteq \int_\ell$. By Proposition 10.2.1 it follows that H is finite-dimensional. Therefore part (a) implies part (b).

We now show part (b) implies part (a) which completes the proof of the theorem. Suppose that H is finite-dimensional and $\epsilon(\int_\ell) \neq (0)$. Let M be a left H-module and N be a submodule of M. By virtue of Exercise 10.2.7 we need only find a submodule L of M such that $M = L \oplus N$. To this end it suffices to find a module map $P : M \longrightarrow N$ which satisfies $P(n) = n$ for all $n \in N$. For then $M = L \oplus N$, where $L = \mathrm{Ker}(P)$.

Let $p : M \longrightarrow N$ be any linear map which satisfies $p(n) = n$ for all $n \in N$. By assumption we may choose $\Lambda \in \int_\ell$ which satisfies $\epsilon(\Lambda) = 1$. Set $P = \Lambda \bullet p$. By part (c) of Lemma 10.3.1 the calculation

$$a \bullet P = a \bullet (\Lambda \bullet p) = (a\Lambda) \bullet p = \epsilon(a)\Lambda \bullet p = \epsilon(a)P$$

shows that $P : M \longrightarrow N$ is a module map. Let $n \in N$. The calculation

$$\begin{aligned} P(n) &= \Lambda_{(1)}\cdot(p(S(\Lambda_{(2)})\cdot n)) \\ &= \Lambda_{(1)}\cdot(S(\Lambda_{(2)})\cdot n) \\ &= (\Lambda_{(1)}S(\Lambda_{(2)}))\cdot n \\ &= \epsilon(\Lambda)1\cdot n \\ &= n \end{aligned}$$

shows that $P(n) = n$. □

When we say that a Hopf algebra H is semisimple we will mean that H is semisimple as an algebra. By part (a) of Theorem 10.2.2 the ideal of left integrals and the ideal of right integrals of a finite-dimensional Hopf algebra are one-dimensional. Recasting Theorem 10.3.2 in this light we have:

Corollary 10.3.3. *Suppose that H is a finite-dimensional Hopf algebra over the field k. Then:*

(a) *If H is semisimple then H is unimodular.*
(b) *H is semisimple if and only if $\epsilon(\Lambda) \neq 0$ for some left or right integral Λ for H (in which case $\epsilon(\Lambda) \neq 0$ for all non-zero left or right integrals for H).*

(c) *H^* is semisimple if and only if $<\lambda, 1> \neq 0$ for some left or right integral λ for H^* (in which case $<\lambda, 1> \neq 0$ for all non-zero left or right integrals for H^*).*

□

The property of semisimplicity is hereditary for Hopf algebras.

Proposition 10.3.4. *Suppose that H is a finite-dimensional semisimple Hopf algebra over the field k. If K is a Hopf subalgebra of H then K is a semisimple.*

Proof. We use the fact that H is a free left K-module under left multiplication which follows by Theorem 9.3.3. Suppose that $\{m_1, \ldots, m_r\}$ is a basis for H as a left K-module. Now there is a left integral Λ for H which satisfies $\epsilon(\Lambda) \neq 0$ by part (b) of Corollary 10.3.3. Write $\Lambda = \Lambda_1 m_1 + \cdots + \Lambda_r m_r$ where $\Lambda_1, \ldots, \Lambda_r \in K$. For $b \in K$ the relation $b\Lambda = \epsilon(b)\Lambda$ implies that $b\Lambda_i = \epsilon(b)\Lambda_i$ for all $1 \leq i \leq r$. Therefore Λ_i is a left integral for K for all $1 \leq i \leq r$. Since $0 \neq \epsilon(\Lambda) = \epsilon(\Lambda_1)\epsilon(m_1) + \cdots + \epsilon(\Lambda_r)\epsilon(m_r)$ it follows that $\epsilon(\Lambda_i) \neq 0$ for some $1 \leq i \leq r$. Thus K is semisimple by part (b) of Corollary 10.3.3 again. □

Exercises

In the following exercises H is a Hopf algebra with antipode S over k.

Exercise 10.3.1. Use integrals to show that quotients of semisimple Hopf algebras are semisimple.

Exercise 10.3.2. Suppose that K is a field extension of k and consider the K-Hopf algebra $K \otimes H$. See Exercise 7.1.8. Show that:

(a) If Λ is a left (respectively right) integral for H then $1 \otimes \Lambda$ is a left (respectively right) integral for $K \otimes H$.

(b) $K \otimes H$ is unimodular if and only if H is unimodular.

(c) $K \otimes H$ is semisimple if and only if H is semisimple.

Exercise 10.3.3. Suppose that G is a finite group and let $H = k[G]$ be the group algebra of G over k. Using integrals show that H^* is always semisimple. (See Exercise 10.1.6.)

Exercise 10.3.4. Using integrals show that $H = H_{2,-1}$ is not semisimple. [Hint: See Exercise 10.1.11.]

Exercise 10.3.5. Suppose that k is a field of characteristic $p > 0$. Let $n \geq 1$ and $\alpha_0, \ldots, \alpha_{n-1} \in k$. Show that:

(a) There is a p^n-dimensional Hopf algebra H over k which is generated as a k-algebra by a primitive element x subject to the relation $x^{p^n} = \alpha_0 x + \alpha_1 x^p + \cdots + \alpha_{n-1} x^{p^{n-1}}$.

(b) H is unimodular (and determine $\int_\ell$).

(c) H is semisimple if and only if $\alpha_0 \neq 0$.

Exercise 10.3.6. Suppose that M and N are left H-modules and $(\mathrm{Hom}(M, N), \bullet)$ is the module structure of Lemma 10.3.1.

(a) Show that $(\mathrm{Hom}(M, N), \bullet)$ is in fact a left H-module.

Let Λ be a non-zero left integral for H and let $\mathrm{Hom}_H(M, N)$ denote the set of module maps $f : M \longrightarrow N$. Show that:

(b) $\Lambda \bullet \mathrm{Hom}(M, N) \subseteq \mathrm{Hom}_H(M, N)$.

(c) If $\epsilon(\Lambda) = 1$ then a linear map $f : M \longrightarrow N$ is a module map if and only if $f = \Lambda \bullet f$.

(d) If H is semisimple then $\mathrm{Hom}_H(M, N) = \Lambda \bullet \mathrm{Hom}(M, N)$.

Exercise 10.3.7. Let Λ be a left integral for H.

(a) Show that $\Lambda_{(1)} \otimes S(\Lambda_{(2)})a = a\Lambda_{(1)} \otimes S(\Lambda_{(2)})$.

(b) Suppose that M and N are left H-modules and $p : M \longrightarrow N$ is linear. Use part (a) to show that $P : M \longrightarrow N$ defined by $P(m) = \Lambda_{(1)} \cdot (p(S(\Lambda_{(2)}) \cdot m))$ for all $m \in M$ is a module map.

Exercise 10.3.8. Suppose that $D : H \longrightarrow H$ is a derivation of H, Λ is a non-zero left integral for H and $v = \Lambda_{(1)} D(S(\Lambda_{(2)}))$.

(a) Show that $av = \epsilon(\Lambda)D(a) + va$ for all $a \in H$.

(b) Suppose that H is semisimple. Use part (a) to show that all derivations of H are inner. (A derivation D of an associative algebra A over k is inner if there is an $a \in A$ such that $D(x) = [a, x] = ax - xa$ for all $x \in A$.)

Exercise 10.3.9. For $a, b \in H$ define $a \triangleright b = a_{(1)} b S(a_{(2)})$. Show that:

(a) $(H, \triangleright)$ is a left H-module.

(b) $ab = (a_{(1)} \triangleright b) a_{(2)}$ for all $a, b \in H$.

(c) $b \in \mathrm{Z}(H)$ if and only if $a \triangleright b = \epsilon(a) b$ for all $a \in H$.

Let Λ be a non-zero left integral for H. Show that:

(d) $\Lambda \triangleright H \subseteq \mathrm{Z}(H)$.

(e) If $\epsilon(\Lambda) = 1$ then $a \in \mathrm{Z}(H)$ if and only if $a = \Lambda \triangleright a$.

(f) If H is semisimple then $\mathrm{Z}(H) = \Lambda \triangleright H$.

Exercise 10.3.10. Suppose H' is also a Hopf algebra over k. Show that $H \otimes H'$ is semisimple if and only if H and H' are semisimple.

10.4 Integrals and the trace function

Throughout this section H is finite-dimensional. We show that the trace function $\mathrm{Tr} : \mathrm{End}(H) \longrightarrow k$ can be expressed in terms of integrals and integrals can be expressed in terms of the trace function. We derive some of the more immediate consequences of this relationship between the trace function and integrals.

Identify $\mathrm{End}(H) = H^* \otimes H$ by setting $<a^* \otimes a, b> = <a^*, b>a$ for all $a^* \in H^*$ and $a, b \in H$. We noted at the end of Section 1.2 that the trace of the linear endomorphism $a^* \otimes a$ of H is given by $\mathrm{Tr}(a^* \otimes a) = <a^*, a>$.

Let $\Lambda \in \int_\ell$ and $\lambda \in \int^r$ satisfy $<\lambda, \Lambda> = 1$. Let $a^* \in H^*$, $a \in H$ and set $f = a^* \otimes a$. Applying a^* to both sides of the first equation of (10.2) we calculate

$$\begin{aligned}
\mathrm{Tr}(f) &= <a^*, a> \\
&= <a^*, \Lambda \leftharpoonup (\lambda \prec S(a))> \\
&= <\lambda \prec S(a), \Lambda_{(1)}><a^*, \Lambda_{(2)}> \\
&= <\lambda, S(a)\Lambda_{(1)}><a^*, \Lambda_{(2)}> \\
&= <\lambda, S((a^* \otimes a)(\Lambda_{(2)}))\Lambda_{(1)}> \\
&= <\lambda, S(f(\Lambda_{(2)}))\Lambda_{(1)}>.
\end{aligned}$$

We have established a special case of:

Theorem 10.4.1. *Suppose that H is a finite-dimensional Hopf algebra over with antipode S over the field k. Let Λ be a left integral for H and λ be a right integral for H^* which satisfy $<\lambda, \Lambda> = 1$. Then*

$$\begin{aligned}
\mathrm{Tr}(f) &= <\lambda, S(\Lambda_{(2)})f(\Lambda_{(1)})> \\
&= <\lambda, (S \circ f)(\Lambda_{(2)})\Lambda_{(1)}> \\
&= <\lambda, (f \circ S)(\Lambda_{(2)})\Lambda_{(1)}>
\end{aligned}$$

for all $f \in \mathrm{End}(H)$.

Proof. By linearity the three formulations of Tr are established once they are shown to hold in the special case $f = a^* \otimes a$. Thus the calculation preceding the statement of the theorem establishes the second. The first is established in a similar manner by applying both sides of the second equation of (10.2) to $a \in H$. Since $\mathrm{Tr}(S^{-1} \circ f \circ S) = \mathrm{Tr}(f)$ the third formulation follows from the second. □

For $a^* \in H^*$ let $\mathrm{L}(a^*), \mathrm{R}(a^*) \in \mathrm{End}(H)$ be defined by (2.15) and for $a \in H$ let $\ell(a), \mathrm{r}(a) \in \mathrm{End}(H)$ be defined by (2.16). Thus

$$\mathrm{L}(a^*)(b) = a^* \rightharpoonup b \quad \text{and} \quad \mathrm{R}(a^*)(b) = b \leftharpoonup a^*$$

and

$$\ell(a)(b) = ab \quad \text{and} \quad \mathrm{r}(a)(b) = ba$$

for all $b \in H$.

Proposition 10.4.2. *Suppose that H is a finite-dimensional Hopf algebra over with antipode S over the field k. Let Λ be a left integral for H and λ be a right integral for H^* which satisfy $<\lambda, \Lambda> = 1$. Then:*

(a) $\mathrm{Tr}(\mathrm{r}(a) \circ S^2 \circ \mathrm{R}(a^*)) = <\lambda, a><a^*, \Lambda>$ *for all $a \in H$ and $a^* \in H^*$.*
(b) *The functional $\lambda_r \in H^*$ defined by $\lambda_r(a) = \mathrm{Tr}(\mathrm{r}(a) \circ S^2)$ for all $a \in H$ is a right integral for H^*.*
(c) $\lambda_r \neq 0$ *if and only if H is semisimple.*
(d) $\mathrm{Tr}(S^2) = <\lambda, 1><\epsilon, \Lambda>$.

Proof. Since $r(1) = \mathrm{I}_H = \mathrm{R}(\epsilon)$ part (d) follows from part (a). Notice part (a) implies $\lambda_r = \epsilon(\Lambda)\lambda$ and thus also implies part (b). Given this formula for λ_r, part (c) follows by part (b) of Corollary 10.3.3. We have reduced the proof of the proposition to showing part (a).

Suppose that $a \in H$ and $a^* \in H^*$. With the observation that $\Delta(c \leftharpoonup a^*) = (c_{(1)} \leftharpoonup a^*) \otimes c_{(2)}$ and $S(c_{(2)})S^2(c_{(1)}) = S(S(c_{(1)})c_{(2)}) = S(\epsilon(c)1) = \epsilon(c)1$ for all $c \in H$, we use the first formulation for Tr in Theorem 10.4.1 to compute

$$\begin{aligned}
\mathrm{Tr}(\mathrm{r}(a) \circ S^2 \circ \mathrm{R}(a^*)) &= <\lambda, S(\Lambda_{(2)})S^2(\Lambda_{(1)} \leftharpoonup a^*)a> \\
&= <\lambda, S((\Lambda \leftharpoonup a^*)_{(2)})S^2((\Lambda \leftharpoonup a^*)_{(1)})a> \\
&= <\lambda, \epsilon(\Lambda \leftharpoonup a^*)1a> \\
&= <\lambda, a><a^*, \Lambda>
\end{aligned}$$

for all $a \in H$ and $a^* \in H^*$. □

As a consequence of part (d) of Proposition 10.4.2 and parts (b) and (c) of Corollary 10.3.3:

Theorem 10.4.3. *Suppose that H is a finite-dimensional Hopf algebra with antipode S over the field k. Then the following are equivalent:*

(a) *H and H^* are semisimple.*
(b) $\mathrm{Tr}(S^2) \neq 0$.

□

Exercises

In the following exercises H is a finite-dimensional Hopf algebra with antipode S over k.

Exercise 10.4.1. Show that the three formulations of $\mathrm{Tr}(f)$ given in Theorem 10.4.1 are manifestations of the same formula by showing that:

(a) $\mathrm{Tr}(f) = <\lambda, S(\Lambda_{(2)})f(\Lambda_{(1)})>$ for H^{op} is $\mathrm{Tr}(f) = <\lambda, (S\circ f)(\Lambda_{(2)})\Lambda_{(1)}>$ when it is expressed in terms of H.

(b) $\mathrm{Tr}(f) = <\lambda, S(\Lambda_{(2)})f(\Lambda_{(1)})>$ for $H^{op\,cop}$ is $\mathrm{Tr}(f) = <\lambda, (f\circ S)(\Lambda_{(2)})\Lambda_{(1)}>$ when expressed in terms of H.

Exercise 10.4.2. Suppose that H is involutory; that is $S^2 = \mathrm{I}_H$. Show that the following are equivalent:

(a) H and H^* are semisimple.

(b) The characteristic of k does not divide $\mathrm{Dim}(H)$.

Exercise 10.4.3. Suppose that Λ is a left or right integral for H, λ is a left or right integral for H^* and $<\lambda, \Lambda> = 1$. Show that $\mathrm{Tr}(S^2) = <\lambda, 1><\epsilon, \Lambda> = \mathrm{Tr}(S^{-2})$.

Exercise 10.4.4. Show $\ell(a^*) = \mathrm{R}(a^*)^*$ and $\mathrm{r}(a^*) = \mathrm{L}(a^*)^*$ for all $a^* \in H^*$.

Exercise 10.4.5. Suppose that A is a bialgebra over k. Verify the commutation relation $\mathrm{R}(a^*)\circ\mathrm{r}(a) = \mathrm{r}(a_{(2)})\circ\mathrm{R}(a_{(1)}\succ a^*)$ for all $a^* \in A^*$ and $a \in A$.

Exercise 10.4.6. Let $a \in \mathrm{G}(H)$ and $\eta \in \mathrm{G}(H^*)$. Show that $\mathrm{Tr}(\ell(a)) = \delta_{1,a}\mathrm{Dim}(H)$ and $\mathrm{Tr}(\mathrm{L}(\eta)) = \delta_{\epsilon,\eta}\mathrm{Dim}(H)$. [Hint: As an initial step establish the first equation for group algebras and recall that H is a free left $k[\mathrm{G}(H)]$-module.]

Exercise 10.4.7. We will study the characters of the regular left and right representations respectively of a finite-dimensional Hopf algebra in Section 10.7. This exercise and the next lay the groundwork for that discussion.

Suppose that A is a finite-dimensional algebra over k. Then $\chi_\ell, \chi_r : A \longrightarrow k$ defined by $<\chi_\ell, a> = \mathrm{Tr}(\ell(a))$ and $<\chi_r, a> = \mathrm{Tr}(\mathrm{r}(a))$ for all $a \in A$ are the characters for the left and right regular representations of A respectively. Show that:

(a) $\chi_\ell, \chi_r \in \mathrm{Cc}(A^*)$ and $<\chi_\ell, 1> = \mathrm{Dim}(A)1 = <\chi_r, 1>$.

(b) If $F : A \longrightarrow A$ is an algebra automorphism then:

(i) $\ell(F(a)) = F \circ \ell(a) \circ F^{-1}$ and $\mathrm{r}(F(a)) = F \circ \mathrm{r}(a) \circ F^{-1}$ for all $a \in A$.
(ii) $\chi_\ell \circ F = \chi_\ell$ and $\chi_r \circ F = \chi_r$.

(c) If $F : A \longrightarrow A^{op}$ is an algebra isomorphism then:

(i) $\ell(F(a)) = F \circ \mathrm{r}(a) \circ F^{-1}$ and $\mathrm{r}(F(a)) = F \circ \ell(a) \circ F^{-1}$ for $a \in A$.
(ii) $\chi_\ell \circ F = \chi_r$ and $\chi_r \circ F = \chi_\ell$.

Exercise 10.4.8. Let Λ be a left integral for H and λ be a right integral for H^* such that $<\lambda, \Lambda> = 1$. Show that:

(a) The equations

$$\begin{aligned} \mathrm{Tr}(\ell(a) \circ r(b)) &= <\lambda, S(\Lambda_{(2)}) a \Lambda_{(1)} b> \\ &= <\lambda, a S(\Lambda_{(2)}) b \Lambda_{(1)}> \\ &= <\lambda, S(b) S(\Lambda_{(2)}) S(a) \Lambda_{(1)}> \end{aligned}$$

hold for all $a, b \in H$.

(b) $\mathrm{Tr}(\ell(a)) = \mathrm{Tr}(\mathrm{r}(a))$ for all $a \in H$; more generally for all $a, b \in H$ the equation $\mathrm{Tr}(\ell(a) \circ r(b)) = \mathrm{Tr}(\ell(b) \circ r(a))$ holds.

(c) If $F : H \longrightarrow H$ is an algebra automorphism or $F : H \longrightarrow H^{op}$ is an algebra isomorphism then $\mathrm{Tr}(\ell(F(a))) = \mathrm{Tr}(\ell(a))$ for all $a \in H$. [Hint: See parts (b) and (c) of Exercise 10.4.7.]

10.5 Integrals and the antipode

Throughout this section H is finite-dimensional. We discuss some of the more important connections between the antipode and integrals for H and H^*.

To begin we recall that $S : H \longrightarrow H^{op\,cop}$ is a bialgebra isomorphism and S^* is the antipode of H^*. Thus $S : H \longrightarrow H^{op}$ and $S^* : H^* \longrightarrow H^{*\,op}$ are algebra isomorphisms. By part (h) of Exercise 10.1.10 it follows that:

Lemma 10.5.1. *Suppose that H is a finite-dimensional Hopf algebra with antipode S over the field k. Then:*

(a) $S(\int_\ell) = \int_r$ *and* $S(\int_r) = \int_\ell$.
(b) $S^*(\int^\ell) = \int^r$ *and* $S^*(\int^r) = \int^\ell$.

□

The antipode and its inverse can be expressed very simply in terms of integrals.

Proposition 10.5.2. *Suppose that H is a finite-dimensional Hopf algebra with antipode S over the field k. Let λ be a non-zero right integral for H^*.*

(a) *Let Λ be a left integral for H such that $<\lambda, \Lambda> = 1$. Then $S^{-1}(a) = \Lambda \leftharpoonup (\lambda \prec a)$ for all $a \in H$.*
(b) *Let Λ be a right integral for H such that $<\lambda, \Lambda> = 1$. Then $S(a) = \Lambda \leftharpoonup (\lambda \succ a)$ for all $a \in H$.*

Proof. Part (a) follows from the first equation of part (c)-(i) of Theorem 10.2.2. Part (a) for the Hopf algebra H^{op} is part (b) for H. □

Let Λ be a non-zero left integral for H. Then $k\Lambda$ is the ideal of left integrals for H by part (a) of Theorem 10.2.2. By part (a) of Exercise 10.1.10 there exists an $\alpha \in \mathrm{G}(H^*) = \mathrm{Alg}(H, k)$ such that

$$\Lambda a = <\alpha, a>\Lambda \tag{10.8}$$

for all $a \in H$. Observe that α is uniquely determined by (10.8). Since any non-zero left integral for H is a scalar multiple of Λ it follows that α does not depend on Λ.

Now let λ be a non-zero right integral for H^*. Then λ is a left integral for $\mathcal{H} = (H^{cop})^* = (H^*)^{op}$. By the argument of the previous paragraph we conclude that there is a $g \in \mathrm{G}(\mathcal{H}^*) = \mathrm{G}(H)$ uniquely determined by

$$a^*\lambda = <a^*, g>\lambda \tag{10.9}$$

for all $a^* \in H^*$. We note that g is determined by (10.9) and that g does not depend on the choice of λ.

The grouplike elements α and g play an important role in the deeper aspects of the relationship between the antipode and integrals.

Definition 10.5.3. The element g of H is the *H-distinguished grouplike element of H* and the element α of H^* is the *H-distinguished grouplike element of H^*.*

The distinguished grouplike elements of H^{op}, H^{cop}, $H^{op\,cop}$ and their dual Hopf algebras are described in terms of the H-distinguished grouplike elements of H and H^* in Exercise 10.5.1.

Let $a \in H$ and $a^* \in H^*$. Observe that $\ell(a)$ and $\mathrm{r}(a)$ are coalgebra automorphisms of H when $a \in \mathrm{G}(H)$ and that $\mathrm{L}(\eta)$ and $\mathrm{R}(\eta)$ are algebra automorphisms of H when $\eta \in \mathrm{G}(H^*)$. Exercise 10.5.2 develops properties of these automorphisms.

Let Λ be a non-zero left integral for H and suppose that α is the H-distinguished grouplike element of H^*. It is easy to see that $\alpha \rightharpoonup \Lambda =$

$L(\alpha)(\Lambda)$ and $\Lambda\leftharpoonup\alpha = R(\alpha)(\Lambda)$ are right integrals for H. By part (a) of Lemma 10.5.1 it follows that $S(\Lambda)$ is a right integral for H as well. Thus $\alpha\rightharpoonup\Lambda$, $\Lambda\leftharpoonup\alpha$, and $S(\Lambda)$ are pairwise linearly dependent. More precise relationships between these right integrals are included in the next result.

Theorem 10.5.4. *Suppose that H is a finite-dimensional Hopf algebra with antipode S over the field k. Let Λ be a left integral for H, let λ be a right integral for H^*, let g be the H-distinguished grouplike element of H, and let α be the H-distinguished grouplike element of H^*. Then:*

(a) $S(\Lambda) = \alpha\rightharpoonup\Lambda$ *and* $S^{-1}(\Lambda) = \Lambda\leftharpoonup\alpha$.
(b) $S^*(\lambda) = \lambda\prec g$ *and* $S^{*\,-1}(\lambda) = g\succ\lambda$.
(c) $S^2(\Lambda) = \alpha\rightharpoonup\Lambda\leftharpoonup\alpha^{-1} = <\alpha, g^{-1}>\Lambda$.
(d) $S^{*\,2}(\lambda) = g^{-1}\succ\lambda\prec g = <\alpha, g^{-1}>\lambda$.
(e) $<\lambda, S^2(a\leftharpoonup\alpha)b> = <\lambda, ba> = <\lambda, aS^2(\alpha^{-1}\rightharpoonup(g^{-1}bg))>$ *for all* $a, b \in H$.
(f) $\Lambda_{(2)}\otimes\Lambda_{(1)} = \Lambda_{(1)}\otimes S^2(\Lambda_{(2)})g$.
(g) $\lambda_{(2)}\otimes\lambda_{(1)} = \alpha(S^{*\,2}(\lambda_{(1)}))\otimes\lambda_{(2)}$.

Proof. We may assume without loss of generality that $<\lambda, \Lambda> = 1$. Thus $<\lambda, S(\Lambda)> = 1$ by part (a) of Exercise 10.5.3.

We first show part (a). We have noted that $S(\Lambda)$ and $\alpha\rightharpoonup\Lambda$ are linearly dependent. The calculation $<\lambda, \alpha\rightharpoonup\Lambda> = <\lambda\alpha, \Lambda> = <\alpha, 1><\lambda, \Lambda> = 1$ shows that $<\lambda, \alpha\rightharpoonup\Lambda> = <\lambda, S(\Lambda)> = 1$. Therefore $S(\Lambda) = \alpha\rightharpoonup\Lambda$. This equation for H^{cop} is the equation $S^{-1}(\Lambda) = \Lambda\leftharpoonup\alpha$ for H. Part (a) for the Hopf algebra $(H^{op\,cop})^*$ is part (b) for H.

To show part (c) we note that Λ and $\alpha\rightharpoonup\Lambda\leftharpoonup\alpha^{-1}$ are linearly dependent since $\alpha\rightharpoonup\Lambda$ and $\Lambda\leftharpoonup\alpha$ are. Since $S^2(\Lambda)$ is a left integral by Lemma 10.5.1, it follows that $S^2(\Lambda)$ and $\alpha\rightharpoonup\Lambda\leftharpoonup\alpha^{-1}$ are linearly dependent. Now

$$<\lambda, \alpha\rightharpoonup\Lambda\leftharpoonup\alpha^{-1}> = <\alpha^{-1}\lambda\alpha, \Lambda> = <\alpha^{-1}, g><\alpha, 1><\lambda, \Lambda> = <\alpha^{-1}, g>.$$

On the other hand, using part (b), we calculate

$$\begin{aligned} <\lambda, S^2(\Lambda)> &= <S^*(\lambda), S(\Lambda)> \\ &= <\lambda\prec g, S(\Lambda)> \\ &= <\lambda, gS(\Lambda)> \\ &= <\lambda, S(\Lambda g^{-1})> \\ &= <\alpha, g^{-1}><\lambda, S(\Lambda)> \\ &= <\alpha^{-1}, g>. \end{aligned}$$

Since $<\alpha^{-1}, g> = <\alpha, g^{-1}>$ we conclude that

$$<\lambda, S^2(\Lambda)> = <\lambda, \alpha\rightharpoonup\Lambda\leftharpoonup\alpha^{-1}> = <\lambda, \alpha(g^{-1})\Lambda> = <\alpha, g^{-1}>$$

and thus $S^2(\Lambda) = \alpha\rightharpoonup\Lambda\leftharpoonup\alpha^{-1} = <\alpha, g^{-1}>\Lambda$. We have shown part (c). Part (d) for $(H^{op\,cop})^*$ is part (c) for H as the $(H^{op\,cop})^*$-distinguished grouplike element for $(H^{op\,cop})^{**} = H^{op\,cop}$ is g.

To show part (e) we may assume $<\lambda, \Lambda> = 1$. Thus $S(a) = \Lambda\leftharpoonup(\lambda\prec S^2(a)) = <\lambda, S^2(a)\Lambda_{(1)}>\Lambda_{(2)}$ for all $a \in H$ by part (a) of Proposition 10.5.2. Now $\Lambda\leftharpoonup\alpha$ is a right integral for H and $\alpha^{-1}\lambda$ is a right integral for H^* such that

$$<\alpha^{-1}\lambda, \Lambda\leftharpoonup\alpha> = <\alpha(\alpha^{-1}\lambda), \Lambda> = <\lambda, \Lambda> = 1.$$

Thus by part (b) of Proposition 10.5.2 we compute, using the fact that $\mathrm{R}(\alpha)$ is an algebra endomorphism of H,

$$\begin{aligned} S(a) &= (\Lambda\leftharpoonup\alpha)\leftharpoonup(a\succ(\alpha^{-1}\lambda)) \\ &= \Lambda\leftharpoonup(\alpha(a\succ(\alpha^{-1}\lambda))) \\ &= <\alpha(a\succ(\alpha^{-1}\lambda)), \Lambda_{(1)}>\Lambda_{(2)} \\ &= <\alpha, \Lambda_{(1)}><a\succ(\alpha^{-1}\lambda), \Lambda_{(2)}>\Lambda_{(3)} \\ &= <\alpha^{-1}\lambda, (\Lambda_{(1)}\leftharpoonup\alpha)a>\Lambda_{(2)} \\ &= <\alpha^{-1}\lambda, (\Lambda_{(1)}(a\leftharpoonup\alpha^{-1}))\leftharpoonup\alpha>\Lambda_{(2)} \\ &= <\alpha\alpha^{-1}\lambda, \Lambda_{(1)}(a\leftharpoonup\alpha^{-1})>\Lambda_{(2)} \\ &= <\lambda, \Lambda_{(1)}(a\leftharpoonup\alpha^{-1})>\Lambda_{(2)} \end{aligned}$$

for all $a \in H$. Thus $<\lambda, S^2(a)\Lambda_{(1)}>\Lambda_{(2)} = <\lambda, \Lambda_{(1)}(a\leftharpoonup\alpha^{-1})>\Lambda_{(2)}$ for all $a \in H$ by part (a) of Proposition 10.5.1. Applying $a^* \in H^*$ to both sides of this equation we obtain $<\lambda, S^2(a)(a^*\rightharpoonup\Lambda)> = <\lambda, (a^*\rightharpoonup\Lambda)(a\leftharpoonup\alpha^{-1})>$ for all $a \in H$. Since $H^*\rightharpoonup H = H$ by part (d) of Theorem 10.2.2 the equation $<\lambda, S^2(a)b> = <\lambda, b(a\leftharpoonup\alpha^{-1})>$ holds for all $a, b \in H$. Thus $<\lambda, S^2(a\leftharpoonup\alpha)b> = <\lambda, ba>$ holds for all $a, b \in H$ which is the first equation of part (e). This first equation for $H^{cop\,op}$ with left integral $S(\Lambda) = \alpha\rightharpoonup\Lambda$ and right integral $(S^*)^{-1}(\lambda) = g\succ\lambda$ for $(H^{cop\,op})^*$ gives the second equation of part (e) for H since α^{-1} is the $H^{cop\,op}$-distinguished grouplike element of $(H^{cop\,op})^*$. We have completed the proof of part (e).

By definition of the coproduct of H^*, the first equation of part (e) is equivalent to the equation of part (g). To show part (f), identify $H = (H^*)^*$ and let $\mathcal{H} = H^*$. Then $S^{*\,-1}(\lambda)$ is a left integral for $\mathcal{H}$ and $S(\Lambda)$ is a right integral for $\mathcal{H}^*$ such that $<S(\Lambda), S^{*\,-1}(\lambda)> = 1$. Using the fact that S is one-one, it is easy to see that part (g) for $\mathcal{H}$ implies part (f) for H. $\square$

An immediate corollary of Theorem 10.5.4 is a way of expressing the H-distinguished grouplike elements of H and H^* in terms of integrals.

Corollary 10.5.5. *Suppose that H is a finite-dimensional Hopf algebra with antipode S over the field k. Let Λ be a left integral for H, and let λ be a right integral for H^* such that $<\lambda, \Lambda> = 1$. Then:*

(a) *$\alpha = \lambda \prec \Lambda$ and $\alpha^{-1} = S(\Lambda) \succ \lambda$, where α is the H-distinguished grouplike element of H^*.*
(b) *$g = \Lambda \leftharpoonup \lambda$ and $g^{-1} = S^*(\lambda) \rightharpoonup \Lambda$, where g is the H-distinguished grouplike element of H.*

Proof. We first show part (a). By the second equation of part (c)-(i) of Theorem 10.2.2 we have $\lambda \prec S(\Lambda \leftharpoonup \alpha) = \alpha$. By part (a) of Theorem 10.5.4 it follows that $S(\Lambda \leftharpoonup \alpha) = S(S^{-1}(\Lambda)) = \Lambda$. Thus $\lambda \prec \Lambda = \alpha$. To complete the proof of part (a) we first remark that $S(\Lambda)$ is a left integral for H^{op} and that α^{-1} is the H^{op}-distinguished grouplike element of $(H^{op})^*$. Using part (a) of Exercise 10.5.3 we see that the preceding equation for H^{op} translates to $\alpha^{-1} = S(\Lambda) \succ \lambda$ for H. Part (a) is established. Part (a) for the Hopf algebra $(H^{op\,cop})^*$ is part (b) for H. $\square$

We continue by assuming the hypothesis of Theorem 10.5.4. Observe that the equations of part (e) of the theorem can be formulated

$$\lambda \circ \ell(a) = \lambda \circ \mathrm{r}(a) \circ S^2 \circ \mathrm{R}(\alpha)$$

and

$$\lambda \circ \mathrm{r}(a) = \lambda \circ \ell(a) \circ S^2 \circ \mathrm{L}(\alpha^{-1}) \circ \ell(g^{-1}) \circ \mathrm{r}(g)$$

for all $a \in H$. By substituting the expression for $\lambda \circ \ell(a)$ into the right-hand side of last equation we calculate

$$\begin{aligned} \lambda \circ \mathrm{r}(a) \circ \mathrm{I}_H &= \lambda \circ \mathrm{r}(a) \\ &= \lambda \circ \mathrm{r}(a) \circ S^2 \circ \mathrm{R}(\alpha) \circ S^2 \circ \mathrm{L}(\alpha^{-1}) \circ \ell(g^{-1}) \circ \mathrm{r}(g) \\ &= \lambda \circ \mathrm{r}(a) \circ S^4 \circ \mathrm{R}(\alpha) \circ \mathrm{L}(\alpha^{-1}) \circ \ell(g^{-1}) \circ \mathrm{r}(g) \end{aligned}$$

since S^2 and $\mathrm{R}(\alpha)$ commute. By Exercise 10.5.4, whenever $f, g \in \mathrm{End}(H)$ and $\lambda \circ \mathrm{r}(a) \circ f = \lambda \circ \mathrm{r}(a) \circ g$ for all $a \in H$ necessarily $f = g$. Thus

$$\mathrm{I}_H = S^4 \circ \mathrm{R}(\alpha) \circ \mathrm{L}(\alpha^{-1}) \circ \ell(g^{-1}) \circ \mathrm{r}(g)$$

or equivalently

$$S^4 = \mathrm{r}(g^{-1}) \circ \ell(g) \circ \mathrm{L}(\alpha) \circ \mathrm{R}(\alpha^{-1}) = \imath_g \circ (\imath_{\alpha^{-1}})^*.$$

For an algebra A over k and an invertible $u \in A$ we denote by $\imath_u$ the algebra automorphism of A defined by $\imath_u(a) = uau^{-1}$ for all $a \in A$. We have shown that S^4 can be described purely in terms of the H-distinguished grouplike elements of H and H^*. By virtue of Exercise 10.5.2 the operators $\imath_g$ and $(\imath_{\alpha^{-1}})^*$ commute.

Theorem 10.5.6. *Suppose that H is a finite-dimensional Hopf algebra with antipode S over the field k. Let g be the H-distinguished grouplike element for H and α be the H-distinguished grouplike element for H^*. Then*

$$S^4 = \imath_g \circ (\imath_{\alpha^{-1}})^*$$

or equivalently

$$S^4(a) = g(\alpha \rightharpoonup a \leftharpoonup \alpha^{-1})g^{-1}$$

for all $a \in H$. In particular $S^{4n} = \mathrm{I}_H$, where n is the least common multiple of the order of g and the order of α. □

The formula $S^4 = \imath_g \circ (\imath_{\alpha^{-1}})^*$ makes connections between S^4 and unimodularity transparent.

Corollary 10.5.7. *Suppose that H is a finite-dimensional Hopf algebra with antipode S over the field k. Then:*

(a) *The order of the linear operator S^4 divides* $\mathrm{Dim}(H)$.
(b) *If H is unimodular then S^4 is inner.*
(c) *If H^* is unimodular then ${S^*}^4$ is inner.*
(d) *If H and H^* are unimodular, in particular if H and H^* are semisimple, then $S^4 = \mathrm{I}_H$.*

Proof. If H is unimodular then $\alpha = \epsilon$ and if H^* is unimodular then $g = 1$. Therefore parts (b)–(d) follow directly from the formula $S^4 = \imath_g \circ (\imath_{\alpha^{-1}})^*$ of part (a) of Theorem 10.5.6.

To show part (a) we recall that finite-dimensional Hopf algebras are free as modules over their Hopf subalgebras by Theorem 9.3.3. As a consequence $\mathrm{Dim}(k[\mathrm{G}(H)])$ divides $\mathrm{Dim}(H)$. Thus the order of any grouplike element of H divides $\mathrm{Dim}(H)$. For all $a \in \mathrm{G}(H)$ this means the order of the bialgebra automorphism $\imath_a$ of H and its transpose divides $\mathrm{Dim}(H)$. In particular the orders of the commuting operators $\imath_g$ and $(\imath_{\alpha^{-1}})^*$ divide $\mathrm{Dim}(H)$; thus part (a) follows. □

For finite-dimensional unimodular Hopf algebras over k whose duals are also unimodular the bound on the order of the antipode given by part (d) of Corollary 10.5.7 generally cannot be improved. See Exercise 10.5.7.

Exercises

In the following exercises H is a finite-dimensional Hopf algebra with antipode S over k.

Exercise 10.5.1. Suppose that g and α are the H-distinguished grouplike elements of H and H^* respectively. Show that the $\mathcal{H}$-distinguished grouplike elements of $\mathcal{H}$ and $\mathcal{H}^*$ are:

(a) g and α^{-1} respectively when $\mathcal{H} = H^{op}$.

(b) g^{-1} and α respectively when $\mathcal{H} = H^{cop}$.

(c) g^{-1} and α^{-1} respectively when $\mathcal{H} = H^{op\,cop}$.

(d) α^{-1} and g^{-1} respectively when $\mathcal{H} = H^*$.

Exercise 10.5.2. Suppose that $\mathcal{G}_c$ (respectively $\mathcal{G}_a$) is the group of coalgebra (respectively algebra) automorphisms of H under composition. Let $a \in \mathrm{G}(H)$ and $\eta \in \mathrm{G}(H^*)$. Show that:

(a) $\ell(a), \mathrm{r}(a) \in \mathcal{G}_c$ and $f : k[\mathrm{G}(H)] \longrightarrow \mathcal{G}_c$ defined by $f(a) = \ell(a)$ is a one-one group map.

(b) $\mathrm{L}(\eta), \mathrm{R}(\eta) \in \mathcal{G}_a$ and $f : k[\mathrm{G}(H)] \longrightarrow \mathcal{G}_a$ defined by $f(\eta) = \mathrm{L}(\eta)$ is a one-one group map.

(c) $<\eta, a>$ is not zero and $<\eta^{-1}, a> = <\eta, a>^{-1} = <\eta, a^{-1}>$.

(d) $\mathrm{R}(\eta)\circ\mathrm{r}(a) = <\eta, a>\mathrm{r}(a)\circ\mathrm{R}(\eta)$. [Hint: See Exercise 10.4.5.]

(e) If f is either $\ell(a)$ or $\mathrm{r}(a)$ and g is either $\mathrm{L}(\eta)$ or $\mathrm{R}(\eta)$ then $= g\circ f = <\eta, a>f\circ g$.

(f) S^2 commutes with $\ell(a), \mathrm{r}(a), \mathrm{L}(\eta)$, and $\mathrm{R}(\eta)$. [Hint: See part (c) of Exercise 10.4.7 and part (c) of Exercise 10.4.7.]

(g) The operators $\imath_a$ and $\imath_\eta$ commute. [Hint: Note that $\imath_a = \ell(a)\circ\mathrm{r}(a^{-1})$. See Exercise 10.4.4.]

Exercise 10.5.3. Suppose that λ is right integral for H^*. Show that:

(a) If Λ is a left integral for H then $<\lambda, \Lambda> = <\lambda, S(\Lambda)>$.

(b) The equation $<\lambda, \Lambda> = <\lambda, S(\Lambda)>$ is not always true when Λ is a right integral for H.

Exercise 10.5.4. Suppose that λ is a non-zero right integral for H^* and $f, g \in \mathrm{End}(H)$. Show that $\lambda\circ\mathrm{r}(a)\circ f = \lambda\circ\mathrm{r}(a)\circ g$ for all $a \in H$ then $f = g$. [Hint: In the special case $g = 0$ we have $0 = <\lambda\circ\mathrm{r}(a)\circ f, b> = <\lambda, f(b)a> = <a\succ\lambda, f(b)>$ for all $a, b \in H$. See part (e) of Theorem 10.2.2.]

Exercise 10.5.5. Suppose that H' is also a finite-dimensional Hopf algebra over k. Let g and α be the H-distinguished grouplike elements of H and H^* respectively and let g' and α' be the H'-distinguished grouplike elements of H'

and H'^* respectively. Show that $g\otimes g'$ and $\alpha\otimes\alpha'$ are the $H\otimes H'$-distinguished grouplike elements of $H\otimes H'$ and $(H\otimes H')^*$ respectively.

Exercise 10.5.6. Let H be the algebra over k generated by symbols a and x subject to the relations

$$a^2 = 1, \quad x^2 = x, \quad \text{and} \quad xa + ax = a - 1.$$

Show that:

(a) Dim(H) = 4.

(b) H is a Hopf algebra where the coalgebra structure is determined by

$$\Delta(a) = a\otimes a \quad \text{and} \quad \Delta(x) = 1\otimes x + x\otimes a.$$

(c) The antipode of H has order 4.

(d) H, H^{op}, H^{cop}, $\mathrm{H}^{op\,cop}$ and H^* are isomorphic as Hopf algebras.

(e) H $\simeq H_{2,-1}$ if the characteristic of k is not 2.

(f) H $\not\simeq H_{2,-1}$ if the characteristic of k is 2.

Exercise 10.5.7. Let H be the algebra over k generated by symbols a, x and y subject to the relations

$$a^2 = 1,\ x^2 = x,\ y^2 = y,\ xa + ax = a - 1 = ya + ay, \quad \text{and} \quad xy + yx = x + y.$$

Show that:

(a) Dim(H) = 8.

(b) H is a Hopf algebra where the coalgebra structure is determined by

$$\Delta(a) = a\otimes a, \quad \Delta(x) = 1\otimes x + x\otimes a \quad \text{and} \quad \Delta(y) = 1\otimes y + y\otimes a.$$

(c) H, H^{op}, H^{cop}, and $\mathrm{H}^{op\,cop}$ are isomorphic as Hopf algebras.

(d) H and H^* are unimodular.

(e) The antipode of H has order 4.

Exercise 10.5.8. Let H be the unimodular eight-dimensional Hopf algebra of Exercise 10.5.7. Show that H has a Hopf subalgebra and a quotient which are not unimodular. (Compare with Proposition 10.3.4 and Exercise 10.3.1.)

Exercise 10.5.9. Suppose that $n = 2m$ and p are positive integers. Assume that k contains a primitive n^{th} root of unity ω. Let H be the algebra over k generated by symbols a and $x_1, \ldots, x_p$ subject to the relations

$$a^n = 1, \quad x_i^2 = 0, \quad x_i a = \omega a x_i$$

for all $1 \leq i \leq p$ and

$$x_i x_j = -x_j x_i$$

for all $1 \leq i \neq j \leq p$. Show that:

(a) $\mathrm{Dim(H)} = n2^p$.

(b) H is a Hopf algebra with antipode S over k where the coalgebra structure is determined by

$$\Delta(a) = a \otimes a \quad \text{and} \quad \Delta(x_i) = 1 \otimes x_i + x_i \otimes a^m$$

for all $1 \leq i \leq p$.

(c) $S^4 = \mathrm{I_H}$.

(d) The H-distinguished grouplike element α of H^* has order $n/(p, n)$.

(e) H^* is unimodular when p is even.

Exercise 10.5.10. Suppose that p and q are positive integers and that k is an algebraically closed field of characteristic 0. Show that there is a finite-dimensional Hopf algebra H with antipode S over k such that:

(a) $S^4 = \mathrm{I}_H$.

(b) The H-distinguished grouplike element g of H has order p.

(c) The H-distinguished grouplike element α of H^* has order q.

[Hint: Exercises 10.5.5 and 10.5.9 may prove useful.]

Exercise 10.5.11. Apropos of Exercise 10.5.10, show that if the characteristic of k is 0 and S has order $n < 4$ then $g = 1$ and $\alpha = \epsilon$.

Exercise 10.5.12. Suppose k has characteristic 0 and H has odd dimension. Show that the following are equivalent:

(a) H and H^* are unimodular.

(b) H and H^* are semisimple.

[Hint: If H and H^* are unimodular then $S^4 = \mathrm{I}_H$. Thus S^2 is diagonalizable with eigenvalues ± 1.]

10.6 Generalized integrals and grouplike elements

Let H be a finite-dimensional. The ideals $\int_\ell$ and $\int_r$ of left and right integrals respectively for H are examples of one-dimensional ideals of H. In this section we consider the set of one-dimensional ideals of H. This set is very closely related to the set of all grouplike elements $\mathrm{G}(H^*)$ of the dual Hopf algebra H^*.

Let A be an algebra over k, let $\eta \in \mathrm{G}(A^o)$, and let

$$L_\eta = \{\Lambda \in A \,|\, a\Lambda = \eta(a)\Lambda \text{ for all } a \in A\}$$

and

$$R_\eta = \{\Lambda \in A \,|\, \Lambda a = \Lambda\eta(a) \text{ for all } a \in A\}$$

be the ideals of A described in Exercise 10.1.10. The reader should note that any non-zero element of L_η (respectively R_η) generates a one-dimensional left (respectively right) ideal of A and that any one-dimensional left (respectively right) ideal of A is contained in L_η (respectively R_η) for some $\eta \in \mathrm{G}(A^o)$.

Generators of one-dimensional left or right ideals are given a special designation.

Definition 10.6.1. Suppose that A is an algebra over the field k. A *generalized left* (respectively *right*) *integral for* A is an element $\Lambda \in A$ such that $\Lambda = 0$ or Λ generates a one-dimensional left (respectively right) ideal of A.

Now suppose that A is a bialgebra over k. For $\eta \in \mathrm{G}(A^o)$ the endomorphisms $\mathrm{L}(\eta), \mathrm{R}(\eta)$ of A of (2.15) which are defined by

$$\mathrm{L}(\eta)(a) = \eta{\rightharpoonup}a \quad \text{and} \quad \mathrm{R}(\eta)(a) - a{\leftharpoonup}\eta$$

for all $a \in A$ are algebra maps. Suppose that $\eta \in \mathrm{G}(A^o)$ is invertible. Then $\mathrm{L}(\eta), \mathrm{R}(\eta)$ are algebra automorphisms of A with inverses $\mathrm{L}(\eta^{-1}), \mathrm{R}(\eta^{-1})$ respectively. By part (g) of Exercise 10.1.10 it follows that:

$$\eta{\rightharpoonup}L_\rho = L_{\rho\eta^{-1}} \quad \text{and} \quad L_\rho{\leftharpoonup}\eta = L_{\eta^{-1}\rho}; \tag{10.10}$$

$$\eta{\rightharpoonup}R_\rho = R_{\rho\eta^{-1}} \quad \text{and} \quad R_\rho{\leftharpoonup}\eta = R_{\eta^{-1}\rho}; \tag{10.11}$$

and

$$\mathrm{Dim}(L_\rho) = \mathrm{Dim}(L_{\rho\eta^{-1}}) \quad \text{and} \quad \mathrm{Dim}(R_\rho) = \mathrm{Dim}(R_{\rho\eta^{-1}}) \tag{10.12}$$

for all $\rho \in \mathrm{G}(A^o)$. Specializing our discussion to finite-dimensional Hopf algebras we note:

Proposition 10.6.2. *Suppose that H is a finite-dimensional Hopf algebra with antipode S over the field k. Then:*

(a) $\eta^{-1}\rightharpoonup\int_\ell = L_\eta = \int_\ell\leftharpoonup\eta^{-1}$ *and also* $\eta^{-1}\rightharpoonup\int_r = R_\eta = \int_r\leftharpoonup\eta^{-1}$ *for all* $\eta \in \mathrm{G}(H^*)$.

(b) $\mathrm{Dim}(L_\eta) = 1 = \mathrm{Dim}(R_\eta)$ *for all* $\eta \in \mathrm{G}(H^*)$.

(c) *One-dimensional left or right ideals of* H *are one-dimensional ideals of* H.

(d) *The associations* $\eta \mapsto L_\eta$ *and* $\eta \mapsto R_\eta$ *describe one-one correspondences between* $\mathrm{G}(H^*)$ *and the set of one-dimensional ideals of* H.

(e) *The number of one-dimensional ideals of* H *divides* $\mathrm{Dim}(H)$.

(f) *Let* α *be the* H*-distinguished grouplike element of* H^*. *Then:*

  (i) $R_\alpha = L_\epsilon = \int_\ell$.

  (ii) $R_{\alpha\eta} = L_\eta = R_{\eta\alpha}$ *for all* $\eta \in \mathrm{G}(H^*)$.

  (iii) α *is in the center of* $\mathrm{G}(H^*)$.

(g) $S(L_\eta) = R_{\eta^{-1}}$ *and* $S(R_\eta) = L_{\eta^{-1}}$ *for all* $\eta \in \mathrm{G}(H^*)$.

Proof. By definition $\int_\ell = L_\epsilon$ and $\int_r = R_\epsilon$. Since $\eta^{-1}\rightharpoonup L_\epsilon = L_\eta = L_\epsilon\leftharpoonup\eta^{-1}$ for all $\eta \in \mathrm{G}(H^*)$ by (10.10) the first set of equations of part (a) follows. The other set follows from (10.11) in the same manner. Since $\int_\ell$ and $\int_r$ are one-dimensional, part (b) follows from (10.12). Part (c) follows from part (b) and our comments preceding the statement of the proposition. We point the reader to part (b) of Exercise 10.1.10 to construct a proof of part (d). By part (d) the number of one-dimensional ideals of H is $|\mathrm{G}(H^*)| = \mathrm{Dim}(k[\mathrm{G}(H^*)])$. This dimension divides $\mathrm{Dim}(H)$ by Corollary 9.3.5. Part (e) is established.

Part (f) follows from part (d), (10.10), and (10.11). Part (g) follows from part (h) of Exercise 10.1.10. This concludes our proof. □

Corollary 10.6.3. *Suppose that* H *is a finite-dimensional Hopf algebra over the field* k. *Then the* H*-distinguished grouplike element* g *of* H *is in the center of* $\mathrm{G}(H)$. □

Corollary 10.6.4. *Suppose that* H *is a finite-dimensional Hopf algebra with antipode* S *over the field* k. *Then the following are equivalent:*

(a) H *is unimodular.*

(b) *Some one-dimensional ideal of* H *is in the center of* H.

(c) *All one-dimensional ideals of* H *are in the center of* H.

Proof. Since H is unimodular if and only if $L_\epsilon = R_\epsilon$, the corollary follows from parts (d) and (f)-(ii) of Proposition 10.6.2. □

Any one-dimensional ideal of an algebra over k is either nilpotent or idempotent. Suppose that H is a finite-dimensional Hopf algebra over k and $\eta \in \mathrm{G}(H^*)$. Then $(L_\eta)^2 = \eta(L_\eta)L_\eta$ by definition. Thus L_η is an idempotent ideal of H if and only if $\eta(L_\eta) \neq (0)$. Now

$$\eta(L_\eta) = \epsilon(\int_\ell)$$

since

$$\eta(L_\eta) = \eta(\eta^{-1} \rightharpoonup \int_\ell) = \eta\eta^{-1}(\int_\ell) = \epsilon(L_\epsilon)$$

which follows by part (a) of Proposition 10.6.2. Therefore L_η is an idempotent ideal of H if and only if $\epsilon(\int_\ell) \neq (0)$.

Recall that H is semisimple if and only if $\epsilon(\int_\ell) \neq (0)$ and that semisimple Hopf algebras are unimodular, both of which follow by Theorem 10.3.2.

Corollary 10.6.5. *Suppose that H is a finite-dimensional Hopf algebra over the field k. Then the following are equivalent:*

(a) *H is semisimple.*
(b) *Some one-dimensional ideal of H is idempotent.*
(c) *All one-dimensional ideals of H are idempotent.*

If any of the conditions holds then all one-dimensional ideals of H are in the center of H. □

Suppose that Λ is a non-zero left integral for H. Then $\{\Lambda\}$ is a basis for $(H, \leftharpoonup)$ as a free left H^*-module and for $(H, \rightharpoonup)$ as a free right H^*-module by part (d) of Theorem 10.2.2. Since $\int_\ell = k\Lambda$ any non-zero generalized left or right integral for H is a scalar multiple of $\eta \rightharpoonup \Lambda$ or $\Lambda \leftharpoonup \eta$ for some $\eta \in \mathrm{G}(H^*)$ by part (a) of Proposition 10.6.2. Since η is invertible we conclude:

Corollary 10.6.6. *Suppose that H is a finite-dimensional Hopf algebra over the field k. Let Λ be a non-zero generalized left or right integral for H. Then $(H, \leftharpoonup)$ is a free right H^*-module and $(H, \rightharpoonup)$ is a free left H^*-module with basis $\{\Lambda\}$.* □

When a generalized integral for H is cocommutative is discussed in Exercise 10.6.7. We now turn our attention to generalized integrals which are functionals. Let A be a bialgebra over k. For $g \in \mathrm{G}(A)$ define

$$L_g = \{\lambda \in A^* \,|\, a^*\lambda = \langle a^*, g\rangle\lambda \text{ for all } a^* \in A^*\}$$

and

$$R_g = \{\lambda \in A^* \mid \lambda a^* = \langle a^*, g\rangle\lambda \text{ for all } a^* \in A^*\}.$$

The reader should observe that the L_g's and R_g's are ideals of A^* and note that any non-zero element of L_g (respectively R_g) generates a one-dimensional left (respectively right) ideal of A^*.

Conversely, suppose that λ generates a one-dimensional left (respectively right) ideal of H^*. By Exercise 10.6.1 it follows that λ belongs to L_g (respectively R_g) for some $g \in \mathrm{G}(A)$.

Observe that $L_g \subseteq A^r$, where A is given the left A-comodule structure (A, Δ). Likewise $R_g \subseteq A^r$, where A is given the right A-comodule structure (A, Δ).

We are interested in the ideals L_g and R_g of H^* when H is a Hopf algebra over k. When H is finite-dimensional the L_g's and the R_g's are the one-dimensional ideals of the Hopf algebra H^* and are therefore treated in this section. Our discussion of generalized left and right integrals for H^* when H is infinite-dimensional takes place in Section 10.9 where we consider existence and uniqueness of left and right integrals for H^*.

Exercises

In the following exercises A is a bialgebra over k and H is a finite-dimensional Hopf algebra with antipode S over k.

Exercise 10.6.1. Suppose that λ generates a one-dimensional left ideal of A^*.

(a) Let $a \in A$ be any element such that $\langle\lambda, a\rangle = 1$ and set $g = \lambda \rightharpoonup a$. Show that $g \in \mathrm{G}(A)$ and that $\lambda \in L_g$. [Hint: There is an algebra map $\eta : A^* \longrightarrow k$ defined by $a^*\lambda = \langle\eta, a^*\rangle\lambda$ for all $a \in A$. Show that $\langle\eta, a^*\rangle = \langle a^*, g\rangle$ for all $a^* \in A^*$.]

(b) Show that $\lambda \rightharpoonup A = kg$. [Hint: See Exercise 10.1.12.]

Notice that the conclusion for A^{cop} in part (a) translates to every one-dimensional right ideal of A^* is contained in R_g for some $g \in \mathrm{G}(A)$.

Exercise 10.6.2. Suppose that g is the H-distinguished grouplike element of H and α is the H-distinguished grouplike element of H^*. Show that $S^2(\Lambda) = \langle\alpha, g^{-1}\rangle\Lambda$ for all generalized left or right integrals for H. (Thus all generalized left or right integrals for H are eigenvectors of S^2 and belong to the same eigenvalue.)

Exercise 10.6.3. Let α be the H-distinguished grouplike element of H^*. Show that:

(a) $S(L_\eta) = L_{(\alpha\eta)^{-1}}$ for all $\eta \in \mathrm{G}(H^*)$.

(b) If $\eta \in \mathrm{G}(H^*)$ then $S(L_\eta) = L_\eta$ if and only if $\eta^2 = \alpha^{-1}$.

(c) The number of one-dimensional ideals of H invariant under S is the number of square roots of α which lie in $\mathrm{G}(H^*)$.

Exercise 10.6.4. Suppose that the characteristic of k is 0 and let p and q be positive integers. Show that there is a finite-dimensional Hopf H over k such that H has p one-dimensional ideals and H^* has q one-dimensional ideals.

Exercise 10.6.5. Suppose that n is a positive integer and k has a primitive n^{th} root of unity ω. Determine the generalized left and right integrals for H and H^*, where $H = H_{n,\omega}$.

Exercise 10.6.6. Suppose that $\eta \in \mathrm{G}(A^*)$ is invertible. Show that:

(a) $\mathrm{Dim}(L_\eta) = \mathrm{Dim}(\int_\ell)$.

(b) $\mathrm{Dim}(L_{\eta\rho}) = \mathrm{Dim}(L_\rho) = \mathrm{Dim}(L_{\rho\eta})$ for all $\rho \in \mathrm{G}(A^*)$.

Exercise 10.6.7. Suppose that $\Lambda \in L_\eta$ is a generalized left integral for H and let g be the H-distinguished grouplike element of H. Show that:

(a) $\eta^{-1}\rightharpoonup\Lambda_{(2)}\otimes\Lambda_{(1)} = \Lambda_{(1)}\otimes S^2(\eta^{-1}\rightharpoonup\Lambda_{(2)})g$. [Hint: $\Lambda = \eta^{-1}\rightharpoonup\Lambda_0$ for some left integral Λ_0 for H. See part (f) of Theorem 10.5.4.]

(b) Λ is cocommutative if and only if H^* is unimodular and $S^2(a) = \eta\rightharpoonup a\leftharpoonup\eta^{-1}$ for all $a \in H$.

10.7 Integrals, the center, and cocommutative elements of the dual

Throughout this section H is finite-dimensional. We use integrals to relate the center of H and the cocommutative elements of H^*. We describe necessary and sufficient conditions for a non-zero right integral for H^* to be cocommutative and for the character χ_H to be a non-zero right integral for H^*. The section ends with a discussion of the counterpart x_H in H for χ_H in H^* and various ways of describing χ_H in terms of integrals.

Let λ be a non-zero left or right integral for H^*. Then $\{\lambda\}$ is a basis for $(H^*, \prec)$ as a right H-module by part (e) of Theorem 10.2.2. Thus every functional $p \in H^*$ has a unique representation $p = \lambda\prec u$ where $u \in H$. It is therefore natural to ask how p and u are related.

Proposition 10.7.1. *Suppose that H is a finite-dimensional Hopf algebra with antipode S over the field k. Let λ be a non-zero right integral for H^**

and let α *be the* H*-distinguished grouplike element of* H^**. Let* $u \in H$ *and set* $p = \lambda \prec u \in H^*$*. Then:*

(a) p *is cocommutative if and only if* $ua = S^2(a \leftharpoonup \alpha)u$ *for all* $a \in H$.
(b) p *is cocommutative and generates* $(H^*, \prec)$ *as a right* H*-module if and only if* u *in invertible,* H *is unimodular, and* $S^2(a) = uau^{-1}$ *for all* $a \in H$.

Proof. Let $p \in H^*$. Then $p \in \mathrm{Cc}(H^*)$ if and only if $<p, ab> = <p, ba>$ for all $a, b \in H$. Thus $p = \lambda \prec u \in \mathrm{Cc}(H^*)$ if and only if $<\lambda, uab> = <\lambda, uba>$ for all $a, b \in H$, which is equivalent to $<\lambda, uab> = <\lambda, S^2(a \leftharpoonup \alpha)ub>$ for all $a, b \in H$ by part (e) of Theorem 10.5.4. As $H \succ \lambda = H^*$ by part (e) of Theorem 10.2.2 part (a) now follows.

To show part (b) we first observe that $p = \lambda \prec u$ generates $(H^*, \prec)$ as a right H-module if and only if u is invertible. Suppose that the equation of part (a) holds. Applying ϵ to both sides of this equation we conclude that $\epsilon(u)\epsilon = \epsilon(u)\alpha$. Thus H is unimodular if $\epsilon(u) \neq 0$, which is the case if u is invertible. Part (b) now easily follows. □

Corollary 10.7.2. *Suppose that* H *is a finite-dimensional unimodular Hopf algebra with antipode* S *over the field* k *and* S^2 *is inner. Suppose* $u \in H$ *is invertible and* $S^2(a) = uau^{-1}$ *for all* $a \in H$*. Then:*

(a) *Let* λ *be a non-zero right integral for* H^**. Then the linear isomorphism* $f : H \longrightarrow H^*$ *defined by* $f(a) = \lambda \prec a$ *for all* $a \in H$ *restricts to an isomorphism* $u\mathrm{Z}(H) \simeq \mathrm{Cc}(H^*)$.
(b) $\mathrm{Dim}(\mathrm{Z}(H)) = \mathrm{Dim}(\mathrm{Cc}(H^*))$.

□

Generally $\mathrm{Dim}(\mathrm{Z}(H)) \neq \mathrm{Dim}(\mathrm{Cc}(H^*))$. See Exercise 10.7.1.

Let λ be a non-zero right integral for H^*. Since $(H^*, \prec)$ is a free right H-module with basis $\{\lambda\}$ by part (e) of Theorem 10.2.2, by part (b) of Proposition 10.7.1 we have:

Corollary 10.7.3. *Suppose that* H *is a finite-dimensional Hopf algebra with antipode* S *over the field* k*. Then a non-zero right integral* λ *for* H^* *is cocommutative if and only if* H *is unimodular and* $S^2 = \mathrm{I}_H$. □

Let $\chi_H : H \longrightarrow k$ be the character of the left regular representation of H. Then

$$<\chi_H, a> = \mathrm{Tr}(\ell(a)) = \mathrm{Tr}(\mathrm{r}(a)) \tag{10.13}$$

for all $a \in H$ which follows by part (b) of Exercise 10.4.8 also since H is a Frobenius algebra.

Proposition 10.7.4. *Suppose that H is a finite-dimensional Hopf algebra with antipode S over the field k. Then χ_H is a non-zero right integral for H^* if and only if H is semisimple and $S^2 = \mathrm{I}_H$.*

Proof. Suppose that $S^2 = \mathrm{I}_H$ and H is semisimple. Then χ_H is a non-zero right integral for H^* by parts (b) and (c) of Proposition 10.4.2. Conversely, suppose that χ_H is a non-zero right integral for H^*. Then $S^2 = \mathrm{I}_H$ by Corollary 10.7.3. Thus $\chi_H = \lambda_r$ and consequently H is semisimple by part (c) of Proposition 10.4.2 again. □

Let $x_H \in H$ be the counterpart of $\chi_H \in H^*$. Then x_H is determined by

$$<a^*, x_H> = \mathrm{Tr}(\ell(a^*))$$

for all $a^* \in H^*$. To study x_H we will use the following general trace formula. Let $\boldsymbol{f} : H \longrightarrow H$ be the linear isomorphism of Theorem 8.4.2.

Lemma 10.7.5. *Suppose that H is a finite-dimensional Hopf algebra with antipode S over the field k. Then $\lambda = \boldsymbol{f}(1)$ is a non-zero right integral for H^* and*

$$\mathrm{Tr}(\ell(\boldsymbol{f}(a))\circ f^*) = <\lambda, S(a_{(2)})f(a_{(1)})>$$

for all $a \in H$ and $f \in \mathrm{End}(H)$.

Proof. That λ is a right integral for H^* follows by (10.5). Furthermore $\boldsymbol{f}(a) = \lambda \prec S(a)$ and $\boldsymbol{f}(a)a^* = \boldsymbol{f}(a \leftharpoonup a^*)$ for all $a \in H$ and $a^* \in H^*$. We make the identification $H = H^{**}$ by defining $<a, a^*> = <a^*, a>$ for all $a \in H$ and $a^* \in H^*$. We make the identification $\mathrm{End}(H^*) = H \otimes H^*$ by defining $<a \otimes a^*, b^*> = <b^*, a>a^*$ for all $a \in H$ and $a^*, b^* \in H^*$. Since

$$\begin{aligned} <\ell(\boldsymbol{f}(a))\circ f^*, a^*> &= <\ell(\boldsymbol{f}(a)), a^*\circ f> \\ &= \boldsymbol{f}(a)(a^*\circ f) \\ &= \boldsymbol{f}(a \leftharpoonup (a^*\circ f)) \\ &= <a^*\circ f, a_{(1)}>\boldsymbol{f}(a_{(2)}) \\ &= <a^*, f(a_{(1)})>\boldsymbol{f}(a_{(2)}) \end{aligned}$$

for all $a^* \in H^*$ we have $\ell(\boldsymbol{f}(a))\circ f^* = f(a_{(1)}) \otimes \boldsymbol{f}(a_{(2)})$. Thus

$$\begin{aligned} \mathrm{Tr}(\ell(\boldsymbol{f}(a))\circ f^*) &= <\boldsymbol{f}(a_{(2)}), f(a_{(1)})> \\ &= <\lambda \prec S(a_{(2)}), f(a_{(1)})> \\ &= <\lambda, S(a_{(2)})f(a_{(1)})> \end{aligned}$$

for all $a \in H$. □

We define a right H-module structure $(H, \blacktriangleleft)$ on H by $a \blacktriangleleft h = S(h_{(2)})ah_{(1)}$ for all $h, a \in H$. Regard H^* as a left H-module under the adjoint action which is given by

$$<a \blacktriangleright \alpha, b> = <\alpha, b \blacktriangleleft a>$$

for all $\alpha \in H^*$ and $a, b \in H$.

Proposition 10.7.6. *Suppose that H is a finite-dimensional Hopf algebra over the field k. Then:*

(a) $\epsilon(x_H) = (\mathrm{Dim}(H))1$.
(b) x_H *is cocommutative.*
(c) $f(x_H) = x_H$ *for any coalgebra automorphism* $f : H \longrightarrow H$ *or for any coalgebra isomorphism* $f : H \longrightarrow H^{cop}$. *In particular* $S(x_H) = x_H$.
(d) *Let* $\boldsymbol{f} : H \longrightarrow H^*$ *be the isomorphism of Theorem 8.4.2 and* $\lambda = \boldsymbol{f}(1)$. *Then* λ *is right integral for* H^* *and* $\boldsymbol{f}(a)(cx_H) = <\lambda, 1 \blacktriangleleft (S^{-1}(c)a)>$ *for all* $a, c \in H$.
(e) $cx_H = \epsilon(c)x_H$ *for all cocommutative elements* $c \in H$.
(f) $x_H^2 = \epsilon(x_H)x_H$.
(g) $<\lambda, x_H> = <\lambda, 1>$ *for any left or right integral* λ *for* H^*.

Proof. Parts (a)–(c) follow by Exercise 10.4.7 and duality. To show part (d) we first observe that

$$<\boldsymbol{f}(a), x_H> = \mathrm{Tr}(\ell(\boldsymbol{f}(a))) = <\lambda, S(a_{(2)})a_{(1)}> = <\lambda, 1 \blacktriangleleft a>$$

for all $a \in H$ by Lemma 10.7.5. Since $<\boldsymbol{f}(a), cx_H> = <\boldsymbol{f}(a) \prec c, x_H> = <\boldsymbol{f}(S^{-1}(c)a), x_H>$ for all $a, c \in H$ part (d) now follows.

To show part (e), suppose that $c \in \mathrm{Cc}(H^*)$. Then $1 \blacktriangleleft c = S(c_{(2)})c_{(1)} = S(c_{(1)})c_{(2)} = \epsilon(c)1$ means that $1 \blacktriangleleft (S^{-1}(c)a) = (1 \blacktriangleleft (S^{-1}(c))) \blacktriangleleft a = \epsilon(c)(1 \blacktriangleleft a)$ since $S^{-1}(c)$ is cocommutative as well and $\epsilon(c) = \epsilon(S^{-1}(c))$. Thus $\boldsymbol{f}(a)(cx_H) = \boldsymbol{f}(a)(\epsilon(c)x_H)$ for all $a \in H$ by part (d). Since $\mathrm{Im}(\boldsymbol{f}) = H^*$ part (e) now follows.

Part (f) follows from parts (b) and (e). To show part (g) first let $\lambda = \boldsymbol{f}(1)$. Then $\mathrm{Tr}(\ell(\lambda)) = <\lambda, 1>$ by Lemma 10.7.5. Thus $<\lambda, x_H> = <\lambda, 1>$. Now $S(x_H) = x_H$ by part (c). Therefore $<S^*(\lambda), x_H> = <S^*(\lambda), 1>$. Thus part (g) follows by part (a) of Theorem 10.2.2 and part (b) of Lemma 10.5.1. □

Proposition 10.7.7. *Suppose that H is a finite-dimensional Hopf algebra with antipode S over the field k. Then*

$$\mathrm{Tr}(S^2) = (\mathrm{Dim}(H))\mathrm{Tr}(S^2|x_H H).$$

Proof. The functional $\lambda_r : H \longrightarrow k$ defined by $<\lambda_r, a> = \mathrm{Tr}(\mathrm{r}(a) \circ S^2)$ for all $a \in H$ is a right integral for H^* by Proposition 10.4.2. This right integral for the Hopf algebra $(H^{op\, cop})^*$ is therefore a left integral λ_ℓ for H^* defined by $<\lambda_\ell, a> = \mathrm{Tr}(\ell(a) \circ S^2)$ for all $a \in H$. By part (g) of Proposition 10.7.6 we have

$$\mathrm{Tr}(S^2) = <\lambda_\ell, 1> = <\lambda_\ell, x_H> = \mathrm{Tr}(\ell(x_H) \circ S^2).$$

Thus to prove the proposition it suffices to establish

$$\mathrm{Tr}(\ell(x_H) \circ S^2) = (\mathrm{Dim}(H))\mathrm{Tr}(S^2 | x_H H). \tag{10.14}$$

To this end we first observe that $\epsilon(x_H) = (\mathrm{Dim}(H))1$ by part (a) of Proposition 10.7.6 and that $\ell(x_H)$ and S^2 commute by part (c) of the same. There are two natural cases to consider. Note also that $x_H^2 = \epsilon(x_H)x_H$ by part (f) of Proposition 10.7.6.

Case 1: $\epsilon(x_H) = 0$.

In this case $x_H^2 = 0$ so $\ell(x_H)$ is a nilpotent operator. Thus $\ell(x_H) \circ S^2$ is nilpotent as well since $\ell(x_H)$ and S^2 commute. Therefore both sides of the equation of (10.14) are 0, so (10.14) holds trivially in Case 1.

Case 2: $\epsilon(x_H) \neq 0$.

In this case $e = (1/\epsilon(x_H))x_H$ is an idempotent. Thus $\ell(e)$ is an idempotent operator which commutes with S^2. Consequently

$$\begin{aligned}\mathrm{Tr}(\ell(x_H) \circ S^2) &= \epsilon(x_H)\mathrm{Tr}(\ell(e) \circ S^2) \\ &= \epsilon(x_H)\mathrm{Tr}(S^2 | \mathrm{Im}(\ell(e))) \\ &= \mathrm{Dim}(H)\mathrm{Tr}(S^2 | x_H H)\end{aligned}$$

by part (a) of Proposition 10.7.6 and thus (10.14) holds in Case 2. □

As a consequence of Theorem 10.4.3 and Proposition 10.7.7:

Theorem 10.7.8. *Suppose that H is a finite-dimensional Hopf algebra over the field k. If H and H^* are semisimple then the characteristic of k does not divide* $\mathrm{Dim}(H)$. □

We conclude this section by relating χ_H to integrals in various ways and by computing $<\chi_H, x_H>$.

Proposition 10.7.9. *Suppose that H is a finite-dimensional Hopf algebra with antipode S over the field k. Let Λ be a left integral for H and λ be a right integral for H^* such that $<\lambda, \Lambda> = 1$. Then:*

(a) $\chi_H = \Lambda \blacktriangleright \lambda = (1 \blacktriangleleft \Lambda) \succ \lambda = \lambda \prec (1 \blacktriangleleft \Lambda)$.
(b) $<\chi_H, x_H> = \mathrm{Dim}(H)\mathrm{Dim}(x_H H)$.

Proof. To show part (a) we need only note for all $a \in H$ that

$$\mathrm{Tr}(\ell(a)) = <\lambda, S(\Lambda_{(2)})a\Lambda_{(1)}> = <\lambda, aS(\Lambda_{(2)})\Lambda_{(1)}>$$

by Theorem 10.4.1 and $\mathrm{Tr}(\ell(a)) = \mathrm{Tr}(\mathrm{r}(a)) = <\lambda, S(\Lambda_{(2)})\Lambda_{(1)}a>$ by part (b) of Exercise 10.4.8 and Theorem 10.4.1 again. As for part (b), we note that $<\chi_H, x_H> = \mathrm{Tr}(\ell(x_H))$ and repeat the argument which establishes (10.14), replacing S^2 with I_H. □

Exercises

In the following exercises H is a finite-dimensional Hopf algebra with antipode S over k.

Exercise 10.7.1. Suppose $H = H_{2,-1}$. Establish that $\mathrm{Dim}(\mathrm{Z}(H)) = 1$ and $\mathrm{Dim}(\mathrm{Cc}(H^*)) = 2$.

Exercise 10.7.2. Suppose that k is algebraically closed and (A, ϵ) is a finite-dimensional augmented algebra over k. The purpose of this exercise is to establish that $A \neq k1$ implies $\mathrm{Dim}(\mathrm{Cc}(A^*)) > 1$, or equivalently that $\mathrm{Dim}(\mathrm{Cc}(A^*)) = 1$ implies that $A = k1$.

Suppose that $\mathrm{Dim}(\mathrm{Cc}(A^*)) = 1$. Show that:

(a) If $\mathcal{M}$ is a maximal ideal of A then $\mathcal{M} = \mathrm{Ker}(\epsilon)$. [Hint: As an algebra $A/\mathcal{M} \simeq \mathrm{M}_n(k)$ for some $n > 0$. The composition $A \longrightarrow \mathrm{M}_n(k) \longrightarrow k$ of the projection followed by the trace function is a cocommutative element of A^*.]

(b) Any functional $p \in A^*$ which vanishes on $\mathcal{M}^2$ belongs to $\mathrm{Cc}(A^*)$.

(c) $A = k1$.

Exercise 10.7.3. Suppose that λ is a non-zero generalized right integral for H^*. Recall from the discussion at the end of Section 10.6 that $\lambda \in R_a$ for some $a \in \mathrm{G}(H)$. Show that:

(a) $\lambda = \lambda_0 \prec a^{-1}$ for some right integral λ_0 for H^*.

(b) λ is cocommutative if and only if H is unimodular and $S^2(x) = a^{-1}xa$ for all $x \in H$.

Exercise 10.7.4. Suppose that $n > 2$ and $\omega \in k$ is a primitive n^{th} root of unity. Compute $\mathrm{Z}(H)$ and $\mathrm{Cc}(H^*)$ for $H = H_{n,\omega}$.

Exercise 10.7.5. Show that:

(a) $<\alpha, ba> = <a_{(1)} \triangleright \alpha, a_{(2)} b>$ for all $\alpha \in H^*$ and $a, b \in H$.

(b) If $\alpha \in H^*$ then $\alpha \in \mathrm{Cc}(H^*)$ if and only if $a \triangleright \alpha = \epsilon(a)\alpha$ for all $a \in H$.

(c) If Λ is a left integral for H then $\Lambda \triangleright H^* \subseteq \mathrm{Cc}(H^*)$.

(d) If H is semisimple and Λ is a non-zero left integral for H then $\Lambda \triangleright H^* = \mathrm{Cc}(H^*)$.

Exercise 10.7.6. For $a \in H$ define $\chi_a : H \longrightarrow k$ by $<\chi_a, b> = \mathrm{Tr}(\ell(b) \circ r(a))$ for all $b \in H$. Observe that $\chi_H = \chi_1$. Show that:

(a) $\chi_a \in \mathrm{Cc}(H^*)$ for all $a \in H$.

(b) $\chi_a(b) = \chi_b(a)$ for all $a, b \in H$. [Hint: See Exercise 10.4.8.]

(c) If Λ is a left integral for H and λ is a right integral for H^* such that $<\lambda, \Lambda> = 1$ then

$$\chi_a = \Lambda \triangleright (\lambda \prec a) = (a \triangleleft \Lambda) \succ \lambda$$

for all $a \in H$. [Hint: See Exercise 10.4.8.]

(d) If Λ is a non-zero left integral for H then $\Lambda \triangleright H^* = \{\chi_a \,|\, a \in H\}$.

(e) If H is semisimple then $\mathrm{Cc}(H^*) = \{\chi_a \,|\, a \in H\}$.

Exercise 10.7.7. Prove the following:

Proposition 10.7.10. *Suppose that H is a finite-dimensional Hopf algebra over the field k. For $a \in H$ let $\chi_a \in H^*$ be defined by $\chi_a(b) = \mathrm{Tr}(\ell(b) \circ r(a))$ for all $b \in H$. Then the following are equivalent:*

(a) *χ_a is a non-zero right integral for H^* for some $a \in H$.*
(b) *H is semisimple and $S^2 = \mathrm{I}_H$.*
(c) *χ_H is a non-zero right integral for H^*.*

Proposition 10.7.10 is a slight improvement on Proposition 10.7.4.

10.8 Integrals and co-semisimplicity

In this section we characterize cosemisimple Hopf algebras in terms of integrals much in the same way as we characterized semisimple Hopf algebras in terms of integrals in Section 10.3. Our discussion here parallels that of Section 10.3. We begin by finding a convenient comodule structure on the space of linear maps between two finite-dimensional right H-comodules.

Lemma 10.8.1. *Suppose that H is a Hopf algebra with antipode S over the field k. Let M and N be finite-dimensional right H-comodules. Then:*

(a) *There is a right H-comodule structure* $(\mathrm{Hom}(M,N),\rho)$ *on* $\mathrm{Hom}(M,N)$ *such that*

$$f_{(0)}(m)\otimes f_{(1)} = f(m_{(0)})_{(0)}\otimes f(m_{(0)})_{(1)}S(m_{(1)})$$

for all $f \in \mathrm{Hom}(M,N)$ *and* $m \in M$.

(b) $f \in \mathrm{Hom}(M,N)$ *is a map of right H-comodules if and only if* $\rho(f) = f\otimes 1$, *or equivalently* $a^*\rightharpoonup f = <a^*,1>f$ *for all* $a^* \in H^*$.

Proof. Since M is a right H-comodule, M^* is a left H-comodule. The two comodule structures are related by (3.13) which we can write

$$\alpha_{(-1)}<\alpha_{(0)},m> = <\alpha, m_{(0)}>m_{(1)}$$

for all $\alpha \in M^*$ and $m \in M$. Since $S : H \longrightarrow H^{cop}$ is a coalgebra map M^* has a right H-comodule structure (M^*,ρ_0) given by $\rho_0(\alpha) = \alpha_{(0)}\otimes S(\alpha_{(-1)})$ for all $\alpha \in M^*$. Regard M^* and N as right H^{op}-comodules and give $M^*\otimes N$ the tensor product H^{op}-comodule structure $(M^*\otimes N,\rho)$. Then $(M^*\otimes N,\rho)$ is a right H-comodule and

$$\rho(\alpha\otimes n) = (\alpha_{(0)}\otimes n_{(0)})\otimes n_{(1)}S(\alpha_{(-1)})$$

for all $\alpha \in M^*$ and $n \in N$.

Now identify $M^*\otimes N$ with $\mathrm{Hom}(M,N)$ by $<\alpha\otimes n, m> = <\alpha,m>n$ for all $\alpha \in M^*, n \in N$ and $m \in M$. To complete the proof of part (a) we need only show for the special case $f = \alpha\otimes n \in \mathrm{Hom}(M,N)$ that

$$\begin{aligned} f_{(0)}(m)\otimes f_{(1)} &= <\alpha_{(0)},m>n_{(0)}\otimes n_{(1)}S(\alpha_{(-1)}) \\ &= n_{(0)}\otimes n_{(1)}S(\alpha_{(-1)}<\alpha_{(0)},m>) \\ &= n_{(0)}\otimes n_{(1)}S(<\alpha,m_{(0)}>m_{(1)}) \\ &= <\alpha,m_{(0)}>n_{(0)}\otimes n_{(1)}S(m_{(1)}) \\ &= f(m_{(0)})_{(0)}\otimes f(m_{(0)})_{(1)}S(m_{(1)}) \end{aligned}$$

for all $m \in M$.

As for part (b), suppose that $f \in \mathrm{Hom}(M,N)$. We let the reader show that if f is a comodule map then $\rho(f) = f\otimes 1$. Conversely, suppose that $\rho(f) = f\otimes 1$. Then

$$\begin{aligned} f(m)_{(0)}\otimes f(m)_{(1)} &= f(m_{(0)})_{(0)}\otimes f(m_{(0)})_{(1)}(\epsilon(m_{(1)})1) \\ &= f(m_{(0)})_{(0)}\otimes f(m_{(0)})_{(1)}S(m_{(1)(1)})m_{(1)(2)} \\ &= f(m_{(0)(0)})_{(0)}\otimes f(m_{(0)(0)})_{(1)}S(m_{(0)(1)})m_{(1)} \\ &= f_{(0)}(m_{(0)})\otimes f_{(1)}m_{(1)} \\ &= f(m_{(0)})\otimes 1m_{(1)} \\ &= f(m_{(0)})\otimes m_{(1)} \end{aligned}$$

for all $m \in M$ shows that $f : M \longrightarrow N$ is a comodule map. □

Theorem 10.8.2. *Suppose that H is a Hopf algebra over the field k. Then the following are equivalent:*

(a) *All left H-comodules are completely reducible.*
(b) $<\lambda, 1> \neq 0$ *for some* $\lambda \in \int^r$.
(c) $H = k1 \oplus C$ *for some subcoalgebra C of H.*
(d) $<\lambda, 1> \neq 0$ *for some* $\lambda \in \int^\ell$.
(e) *All right H-comodules are completely reducible.*

If any of these conditions hold then $\int^r = \int^\ell$.

Proof. First of all suppose that λ, λ' are left and right integrals respectively for H^*. The calculation $<\lambda', 1>\lambda = \lambda'\lambda = \lambda'<\lambda, 1>$ shows that if either $<\lambda', 1> \neq 0$ or $<\lambda, 1> \neq 0$ then λ and λ' are not zero and are scalar multiples of each other. In this case $\int^r$ and $\int^\ell$ are one-dimensional ideals of H^* which are equal.

To complete the proof we need only show that parts (c)–(e) are equivalent. For parts (c)–(e) for the Hopf algebra $H^{op\,cop}$ are collectively parts (a)–(c) for the Hopf algebra H.

We first show that part (c) implies part (d). Suppose that $H = k1 \oplus C$ for some subcoalgebra C of H. Let $\lambda \in H^*$ be the functional which vanishes on C and satisfies $<\lambda, 1> = 1$. Then $a_{(1)}<\lambda, a_{(2)}> = <\lambda, a>1$ for all $a \in H$. Therefore λ is a left integral for H^* by part (a) of Exercise 10.1.1. We have shown part (c) implies part (d).

To show that part (d) implies part (e) we may assume that H^* has a left integral λ which satisfies $<\lambda, 1> = 1$. Let M be a finite-dimensional right H-comodule and suppose that N is a subcomodule of M. To establish part (e) it suffices to find a subcomodule L of M such that $M = N \oplus L$. To this end we need only find a comodule map $P : M \longrightarrow N$ such that $P(n) = n$ for all $n \in N$. For then $M = N \oplus L$, where $L = \mathrm{Ker}(P)$.

Let $p : M \longrightarrow N$ be any linear map which satisfies $p(n) = n$ for all $n \in N$. Regard $\mathrm{Hom}(M, N)$ as a right H-comodule as in Lemma 10.8.1 and set $P = \lambda \rightharpoonup p$. By part (b) of Lemma 10.8.1 the calculation

$$a^* \rightharpoonup P = a^* \rightharpoonup (\lambda \rightharpoonup p) = (a^*\lambda) \rightharpoonup p = <a^*, 1>\lambda \rightharpoonup p = <a^*, 1>P$$

shows that $P : M \longrightarrow N$ is a comodule map. Let $n \in N$. By part (a) of the same we have

$$\begin{aligned} P(n) &= p(n_{(0)})_{(0)}{<}\lambda, p(n_{(0)})_{(1)}S(n_{(1)}){>} \\ &= n_{(0)(0)}{<}\lambda, n_{(0)(1)}S(n_{(1)}){>} \\ &= n_{(0)}{<}\lambda, n_{(1)(1)}S(n_{(1)(2)}{>} \\ &= n_{(0)}{<}\lambda, \epsilon(n_{(1)})1{>} \\ &= n{<}\lambda, 1{>} \\ &= n \end{aligned}$$

which shows that $P(n) = n$. Thus part (d) implies part (e).

Suppose that all right H-comodules are completely reducible. Then $H = H_0$ by Theorem 3.4.10. By part (c) of Proposition 3.4.3 it follows that H_0 is the direct sum of all of its simple subcoalgebras. Thus $H = k1 \oplus C$, where C is the sum of all simple subcoalgebras of H with the exception of $k1$. Thus part (e) implies part (c) and our proof is complete. □

Cosemisimple Hopf algebras have bijective antipodes.

Lemma 10.8.3. *Let H be a Hopf algebra with antipode S over the field k and suppose λ is a right integral for H^*. If C, D are subcoalgebras of H such that $C \cap D = (0)$ then $\lambda(S(C)D) = (0)$.*

Proof. We may assume $\lambda \neq 0$. Our proof is based on (10.17) of the subsequent section which is equivalent to ${<}\lambda, S(a{\leftharpoonup}p)b{>} = {<}\lambda, S(a)(p{\rightharpoonup}b){>}$ for all $a \in H$ and $p \in H^*$. Suppose C, D are subcoalgebras of H which satisfy $C \cap D = (0)$. Then $C + D$ is a direct sum which means there is a $p \in H^*$ such that $p|C = \epsilon|C$ and $p|D = 0$. Therefore $C{\leftharpoonup}p = C$ and $p{\rightharpoonup}D = (0)$ which means $\lambda(S(C)D) = \lambda(S(C{\leftharpoonup}p)D) = \lambda(S(C)(p{\rightharpoonup}D)) = (0)$. □

Corollary 10.8.4. *Let H be a cosemisimple Hopf algebra with antipode S over the field k. Then $S^2(C) = C$ for all simple subcoalgebras C of H. In particular S is bijective.*

Proof. Let C be a simple subcoalgebra of H. Then C is finite-dimensional by Theorem 2.2.3 and $S^2(C)$ is a subcoalgebra of C since S^2 is a coalgebra endomorphism of H. Since C is simple either $S^2(C) \cap C = (0)$ or $S^2(C) \supseteq C$. We establish the latter which is equivalent to $S^2(C) = C$.

Suppose $S^2(C) \cap C = (0)$. By Theorem 10.8.2 there exists a right integral λ for H^* such that $\lambda(1) \neq 0$. By Lemma 10.8.3 it follows that $\epsilon(C)1 \subseteq \lambda(S(C)S^2(C)) = (0)$ which implies $C = (0)$. Thus $S^2(C) \supseteq C$ after all. □

In proving Theorem 10.8.2 we showed that if H is cosemisimple then $\int^r$ and $\int^\ell$ are one-dimensional ideals of H^*. Generally $\mathrm{Dim}(\int^r)$, $\mathrm{Dim}(\int^\ell) \leq 1$ by Theorem 10.9.5 of the next section.

Exercises

In the following exercises H is a Hopf algebra with antipode S over k unless otherwise stated.

Exercise 10.8.1. Show that the conclusion of Lemma 10.8.1 is true if M is finite-dimensional and N is arbitrary.

Exercise 10.8.2. Suppose that H is finite-dimensional and let M and N be finite-dimensional right H-comodules. Regard M and N as rational left H^*-modules. Show that the rational left H^*-module structure on $\mathrm{Hom}(M, N)$ induced by the right H-comodule structure $(\mathrm{Hom}(M, N), \rho)$ of Lemma 10.8.1 is in fact the left H^*-module structure $(\mathrm{Hom}(M, N), \bullet)$ of Lemma 10.3.1.

Exercise 10.8.3. How are Theorems 10.3.2 and 10.8.2 and their proofs related when H is finite-dimensional?

Exercise 10.8.4. Show that the following are equivalent:

(a) The injective hull of $k1$ in H, where $k1$ is regarded as a left coideal of H, is $k1$.

(b) H is cosemisimple.

(c) The injective hull of $k1$ in H, where $k1$ is regarded as a right coideal of H, is $k1$.

Exercise 10.8.5. Suppose that $n > 0$ and that k has a primitive n^{th} root of unity ω. Let $H = H_{n,\omega}$.

(a) Determine $\int^r$ and $\int^\ell$.

(b) Use integrals to show that H is not cosemisimple.

Exercise 10.8.6. Suppose that H^* has a non-zero left integral λ such that $<\lambda, 1> \neq 0$. Let K be a Hopf subalgebra of H. Show that K^* has a non-zero left integral λ' such that $<\lambda', 1> \neq 0$.

Exercise 10.8.7. Find a bialgebra H which is not a cosemisimple coalgebra but can be written $H = k1 \oplus C$ for some subcoalgebra C of H.

Exercise 10.8.8. Let K be a field extension of k and λ be a left integral for H^*. Show that:

(a) $1_K \otimes \lambda$ is a left integral for $(K \otimes H)^* = \mathrm{Hom}_K(K \otimes H, K)$.

(b) H is cosemisimple if and only if $K \otimes H$ is cosemisimple.

10.9 Existence and uniqueness results for integrals of the dual algebra of a Hopf algebra

Let H be a Hopf algebra over the field k. In this section we explore what it means for H^* to have a non-zero right or left integral. The results described here are thus of interest in the infinite-dimensional case.

Suppose that H^* has a non-zero left or right integral. Then H^* has both a non-zero right integral and a non-zero left integral which are unique up to scalar multiple. In this case H satisfies interesting finiteness conditions. The injective hull of $k1$ plays an important role in the theory of integrals for H^*.

We begin with the isomorphism of left H-Hopf modules $F : H \otimes \int^r \longrightarrow H^r$ described in Section 8.4. Recall that $F(a \otimes \lambda) = \lambda \prec S(a)$ for all $a \in H$ and $\lambda \in \int^r$. Again we note that S is one-one whenever $\int^r \neq (0)$.

Lemma 10.9.1. *Suppose that H is a Hopf algebra over the field k. Then:*

(a) *$\int^r \neq (0)$ if and only if there exists a cofinite right coideal N of H such that $N \neq H$.*

(b) *$\int^\ell \neq (0)$ if and only if there exists a cofinite left coideal N of H such that $N \neq H$.*

Proof. By virtue of the isomorphism F we have that $\int^r = (0)$ if and only if $H^r = (0)$. Since H^r consists of all functionals $a^* \in H^*$ which vanish on a cofinite right coideal of H, part (a) follows. Part (a) for the Hopf algebra $H^{op\,cop}$ is part (b) for H. □

Suppose that $\lambda \in \int^r$ and is not zero. Let $\boldsymbol{f} : H \longrightarrow H^r$ be the one-one map of left H-Hopf modules of Section 8.4 described by

$$\boldsymbol{f}(a) = F(a \otimes \lambda) = \lambda \prec S(a) \tag{10.15}$$

for all $a \in H$. Since $\boldsymbol{f}$ a map of left H-comodules it is a map of right H^*-modules. To say that $\boldsymbol{f}$ is a map of right H^*-modules is the same as saying that

$$\boldsymbol{f}(a \leftharpoonup a^*)(b) = \boldsymbol{f}(a)(a^* \rightharpoonup b) \tag{10.16}$$

for all $a, b \in H$. Equation (10.16) has the equivalent formulation

$$a_{(1)} {<} \lambda, S(a_{(2)})b {>} = {<} \lambda, S(a)b_{(1)} {>} b_{(2)} \tag{10.17}$$

for all $a, b \in H$.

Theorem 10.9.2. *Suppose that H is a Hopf algebra over the field k. Then the following are equivalent:*

(a) $\int^r \neq (0)$.
(b) *H is a right co-Frobenius coalgebra over k.*
(c) *The sum of the finite-dimensional left ideals of H^* is a dense subspace of H^*.*
(d) *The injective hull of any finite-dimensional left H-comodule is finite-dimensional.*
(e) *The injective hull of the left coideal $k1$ of H is finite-dimensional.*
(f) *H is a left co-Frobenius coalgebra over k.*
(g) *The sum of the finite-dimensional right ideals of H^* is a dense subspace of H^*.*
(h) *The injective hull of any finite-dimensional right H-comodule is finite-dimensional.*
(i) *The injective hull of the right coideal $k1$ of H is finite-dimensional.*
(j) $\int^\ell \neq (0)$.

Proof. We first show that part (a) implies part (b). Suppose that λ is a non-zero right integral for H^* and $\boldsymbol{f} : H \longrightarrow H^*$ is the map of (10.15) defined for λ. Then $\boldsymbol{f}$ is a one-one map of right H^*-modules. Therefore H is right co-Frobenius by definition. Part (b) implies part (c) by Proposition 3.6.4. Part (c) implies part (d) by Theorem 3.6.5. That part (d) implies part (e) is trivial.

We now show that part (e) implies part (j). Suppose J_ℓ is an injective hull of the left coideal $k1$ of H and is finite-dimensional. We may assume $J_\ell \subseteq H$ by part (c) of Proposition 3.5.10. Since J_ℓ is an injective left coideal of H, $J_\ell \oplus N = H$ for some left coideal of H by Theorem 3.5.8. Since J_ℓ is finite-dimensional N is a cofinite subspace of H. Thus $\int^\ell \neq (0)$ by part (b) of Lemma 10.9.1, and therefore part (e) implies part (j).

To show that part (j) implies part (a), note that we have shown $\int^r \neq (0)$ implies $\int^\ell \neq (0)$. This statement for the Hopf algebra $H^{op\,cop}$ translates to $\int^\ell \neq (0)$ implies $\int^r \neq (0)$ for the Hopf algebra H. Thus part (j) implies part (a).

We have shown that parts (a)–(e) and (j) are equivalent for H, and thus for $H^{op\,cop}$. The equivalence of these parts for $H^{op\,cop}$ translates to the equivalence of parts (f)–(j) and (a) for H. $\square$

Corollary 10.9.3. *Suppose that H is a Hopf algebra over the field k and K is a Hopf subalgebra of H. If H^* has a non-zero left integral then K^* has a non-zero left integral.*

Proof. Assume H^* has a non-zero left integral. Let $\imath : K \longrightarrow H$ be the inclusion and set $f = \imath^*$. Then $f : H^* \longrightarrow K^*$ is an onto algebra map. The sum L of the finite-dimensional left ideals of H^* is a dense subspace of H^* by Theorem 10.9.2. Using part (d) of Theorem 1.3.10 we compute

$$K^* = f(H^*) = f(\overline{L}) = \overline{f(L)}.$$

Since $f(L)$ is a sum of finite-dimensional left ideals of K^*, the sum of all finite-dimensional left ideals of K^* is a dense subspace of K^*. Therefore K^* has a non-zero left integral by Theorem 10.9.2 again. □

The coalgebra H injective as a left or right H-comodule, where $\rho = \Delta$, by part (b) of Proposition 3.5.4. By part (c) of Proposition 3.5.10 any subcomodule of H has an injective hull which is also a subcomodule of H. We will let J_ℓ (respectively J_r) be an injective hull of $k1$ contained in H when H is regarded as a left (respectively right) H-comodule under Δ. We note that the subcomodules of H are the left coideals of H when (H, Δ) is regarded as a left H-comodule and the subcomodules of H are the right coideals of H when (H, Δ) is regarded as a right H-comodule.

Lemma 10.9.4. *Suppose that H is a Hopf algebra with antipode S over the field k and $\int^r \neq (0)$. Then:*

(a) *$S(J_\ell) \simeq J_r$ and $S(J_r) \simeq J_\ell$.*
(b) *$\mathrm{Dim}(J_\ell) = \mathrm{Dim}(J_r)$ and is finite.*

Proof. We have observed that $\int^r \neq (0)$ implies that S is one-one. Thus $S : H \longrightarrow H^{cop}$ is a one-one map of coalgebras. By parts (e) and (i) of Theorem 10.9.2 the vector spaces J_ℓ and J_r are finite-dimensional. Since J_ℓ is an essential extension of $k1$ as a left H-subcomodule it follows that $S(J_\ell)$ is an essential extension of $S(k1) = k1$ as a right H-subcomodule by part (c) of Exercise 3.5.6. Therefore $S(J_\ell) \subseteq J'_r$, where $J'_r \simeq J_r$. This inclusion for the Hopf algebra $H^{op\,cop}$ translates to $S(J'_r) \subseteq J'_\ell$ for H, where $J'_\ell \simeq J_\ell$. Now J_ℓ is finite-dimensional by Theorem 10.9.2. Since $\mathrm{Dim}(J_\ell) \leq \mathrm{Dim}(J_r) \leq \mathrm{Dim}(J_\ell)$ we conclude $\mathrm{Dim}(J_\ell) = \mathrm{Dim}(J_r)$, $S(J_\ell) = J'_\ell$ and $S(J_r) = J'_r$. □

Now we are in a position to prove uniqueness of integrals for H^*.

Theorem 10.9.5. *Suppose that H is a Hopf algebra with antipode S over the field k and $\int^r \neq (0)$ or $\int^\ell \neq (0)$. Then:*

(a) $\mathrm{Dim}(\int^r) = 1 = \mathrm{Dim}(\int^\ell)$.
(b) *The map $f : H \longrightarrow H^r$ of (10.15) defined for any non-zero right integral for H^* is an isomorphism of right H-Hopf modules.*
(c) *Let $\lambda \in \int^r$. Then $<\lambda, J_\ell> = (0)$ implies $\lambda = 0$.*
(d) *Let $\lambda \in \int^\ell$. Then $<\lambda, J_r> = (0)$ implies $\lambda = 0$.*
(e) $S^*(\int^\ell) = \int^r$ *and* $S^*(\int^r) = \int^\ell$.
(f) *S is bijective.*

Proof. Suppose that $\int^r \neq (0)$ or $\int^\ell \neq (0)$. Then $\int^r \neq (0)$ and $\int^\ell \neq (0)$ by Theorem 10.9.2. By part (b) of Proposition 3.5.10 we have $J_r = H \leftharpoonup e$ for some idempotent e of H^*. Since J_r is the injective hull of the subcoalgebra $k1$ of H regarded as a right coideal of H, it follows by Exercise 3.5.7 that $<e, 1> = 1$ and $J'_\ell = e \rightharpoonup H$ is an injective hull of $k1$ regarded as a left coideal of H. Let $e' = \epsilon - e$. Then $<e', 1> = 0$ and $e'e = 0$.

We first show part (a). Let $F : H \otimes \int^r \longrightarrow H^r$ be the isomorphism of left H-Hopf modules of Section 8.4 and let f be the map of (10.15) defined for a fixed non-zero right integral λ for H^*. By (10.16) we have

$$\mathrm{f}(H \leftharpoonup e)(e' \rightharpoonup H) = \mathrm{f}(H)(e \rightharpoonup (e' \rightharpoonup H)) = \mathrm{f}(H)(ee' \rightharpoonup H) = (0).$$

Therefore $F(J_r \otimes \int^r) \subseteq (e' \rightharpoonup H)^\perp$. Since $\mathrm{Dim}(J_r) = \mathrm{Dim}(J'_\ell)$ and is finite by part (b) of Lemma 10.9.4, from

$$H = (e \rightharpoonup H) \oplus (e' \rightharpoonup H) = J'_\ell \oplus (e' \rightharpoonup H)$$

we conclude that $\mathrm{Dim}((e' \rightharpoonup H)^\perp) = \mathrm{Dim}(J'_\ell)$. Thus

$$(\mathrm{Dim}(J_r))(\mathrm{Dim}(\int^r)) = \mathrm{Dim}(F(J_r \otimes \int^r)) \leq \mathrm{Dim}(J_r).$$

Therefore $\mathrm{Dim}(\int^r) \leq 1$ since $J_r \neq (0)$ and is finite-dimensional. Our conclusion $\mathrm{Dim}(\int^r) \leq 1$ for $H^{op\,cop}$ translates to $\mathrm{Dim}(\int^\ell) \leq 1$ for H. We have established part (a). Part (b) is clear in light of part (a).

To show part (d), let $\lambda \in \int^\ell$ and $a^* \in H^*$. Since $<\lambda, a \leftharpoonup a^*> = <a^*\lambda, a> = <a^*, 1><\lambda, a>$ for all $a \in H$ it follows that $<\lambda, H \leftharpoonup e'> = (0)$. Note that $H = J_r \oplus (H \leftharpoonup e')$. Thus part (d) follows. Part (d) for the Hopf algebra $H^{op\,cop}$ is part (c) for H.

To show part (e) we first note that $S^* : H^* \longrightarrow H^{*\,op}$ is an onto algebra map since $S : H \longrightarrow H^{cop}$ is a one-one coalgebra map. Thus $S^*(\int^r) \subseteq \int^\ell$ and $S^*(\int^\ell) \subseteq \int^r$. Let λ be a non-zero right integral for H^*.

Then $<\lambda, J_\ell> \neq (0)$ by part (c) and we may assume $S(J_r) = J_\ell$ by part (a) of Lemma 10.9.4. Therefore $<S^*(\lambda), J_r> = <\lambda, S(J_r)> = <\lambda, J_\ell> \neq (0)$ which means $S^*(\lambda) \neq 0$. Thus $S^*(\int^\ell) = \int^r$ by part (a). This equation for the Hopf algebra $H^{op\,cop}$ translates to $S^*(\int^\ell) = \int^r$ for H.

We finally show part (f). Since S is one-one we need only show that S is onto. Suppose that $S(H) \neq H$. By Theorem 10.9.2 and Theorem 3.6.5 there is a cofinite right coideal P of H such that $S(H) \subseteq P$ and $P \neq H$. In particular $(0) \neq P^\perp \subseteq H^r$. Let λ be a non-zero right integral for H. Then $\lambda \prec S(a) \in P^\perp$ for some non-zero $a \in H$ by part (b). Observe that $(0) = <\lambda \prec S(a), P> = <\lambda, S(a)P>$. Since $S(Ha) = S(a)S(H) \subseteq S(a)P$ it follows that $(0) = <\lambda, S(Ha)> = <S^*(\lambda), Ha> = (0)$. Now $S^*(\lambda)$ is a non-zero left integral for H^* by parts (a) and (e). Therefore $<S^*(\lambda), (Ha) \leftharpoonup H^*> = (0)$. By part (c) of Proposition 8.1.5 we see that left subcomodule $N = (Ha) \leftharpoonup H^*$ of H generated by the left ideal Ha of H is a left ideal of H. Therefore $N = H$ by part (a) of Proposition 8.2.5 which means $S^*(\lambda) = 0$, a contradiction. Consequently $S(H) = H$ and part (f) is established. $\square$

For $\lambda \in \int^r$ there is an interesting connection between $<\lambda, 1>$ and $\mathrm{Dim}(J_\ell)$ in the involutory case.

Lemma 10.9.6. *Suppose that H is an involutory Hopf algebra over the field k. Then*

$$\mathrm{Dim}(J_\ell)<\lambda, v> = \epsilon(v)<\lambda, 1>$$

for all $\lambda \in \int^r$ and $v \in J_\ell$.

Proof. We may as well assume that $\lambda \neq 0$. Note the antipode S of H is bijective and $S = S^{-1}$ since H is involutory. Set $J = J_\ell$.

Let $J_m = J \cap H_m$ for all $m \geq 0$. Then $J_0 = k1$ and $J_0 \subseteq J_1 \subseteq \cdots \subseteq J_r = J$ for some $r \geq 0$ since J is finite-dimensional by Lemma 10.9.4. Choose a basis $\{v_1, \ldots, v_r\}$ for J such that for each $m \geq 0$ the subspace J_m of J has basis $\{v_1, \ldots, v_s\}$ for some $1 \leq s \leq n$.

Now let $v \in J$. Since J is a left coideal of H we may write $\Delta(v) = \sum_{i=1}^n r_i \otimes v_i$ where $r_i \in H$. Fix $1 \leq s \leq n$. Then $v_s \in J_m \backslash J_{m-1}$ for some $m \geq 0$, where $J_{-1} = (0)$. By Exercise 4.1.14 we may write $\Delta(v_s) = v_s \otimes 1 + \nu$, where $\nu \in H_{m-1} \otimes H_m$. With $a = v_s$ and $b = v$ the equation of (10.17) becomes

$$v_s<\lambda, v> + w = \sum_{i=1}^n <\lambda, S(v_s)r_i>v_i \tag{10.18}$$

for some $w \in H_{m-1}$. Clearly $w \in J$; thus $w \in J_{m-1}$. Writing w as a linear combination of $\{v_1, \ldots, v_n\}$ and comparing terms on both sides of (10.18) we conclude that $<\lambda, v> = <\lambda, S(v_s)r_s>$. Therefore

$$\begin{aligned}\mathrm{Dim}(J_\ell)<\lambda, v> &= n<\lambda, v> \\ &= \sum_{i=1}^{n} <\lambda, S(v_i)r_i> \\ &= <\lambda, S(v_{(2)})v_{(1)}> \\ &= <\lambda, S^{-1}(v_{(2)})v_{(1)}> \\ &= <\lambda, \epsilon(v)1> \\ &= \epsilon(v)<\lambda, 1>\end{aligned}$$

and the lemma is proved. □

The hypothesis of the following theorem holds for commutative or cocommutative Hopf algebras over k.

Theorem 10.9.7. *Let H be an involutory Hopf algebra over the field k.*

(a) *Suppose that the characteristic of k is 0. Then H is cosemisimple if and only if $\int^r \neq (0)$.*
(b) *Suppose that the characteristic of k is $p > 0$. Then H is cosemisimple if and only if $\int^r \neq (0)$ and p does not divide* $\mathrm{Dim}(J_\ell)$.

Proof. H is cosemisimple if and only if there exists a right integral λ for H^* such that $<\lambda, 1> \neq 0$ by Theorem 10.8.2. Suppose that H is cosemisimple and let $\lambda \in \int^r$ satisfy $<\lambda, 1> \neq 0$. Then $\lambda \neq 0$. Applying the formula of Lemma 10.9.6 to λ and $v = 1$ we have $(\mathrm{Dim}(J_\ell))1 = 1$. Thus if the characteristic of k is $p > 0$ it follows that p does not divide $\mathrm{Dim}(J_\ell)$.

Conversely, suppose that λ is a non-zero right integral for H^*. By part (c) of Theorem 10.9.5 there is some $v \in J_\ell$ such that $<\lambda, v> \neq 0$. The formula in Lemma 10.9.6 gives $<\lambda, 1> \neq 0$ if the characteristic of k is 0 or if the characteristic of k is $p > 0$ and p does not divide $\mathrm{Dim}(J_\ell)$. This concludes our proof. □

At this point we resume our discussion of one-dimensional left or right ideals of H^*, that is of the non-zero left or right generalized integrals for H^*, which we began at the end of Section 10.6. For $a \in \mathrm{G}(H)$ observe that $\ell_a, r_a : H^* \longrightarrow H^*$ defined by $\ell_a(a^*) = a \succ a^*$ and $r_a(a^*) = a^* \prec a$ for all $a^* \in H^*$ are algebra automorphisms of H^*. Applying ℓ_a and r_a to L_b and R_b we obtain

$$a \succ L_b = L_{ba^{-1}} \quad \text{and} \quad L_b \prec a = L_{a^{-1}b} \tag{10.19}$$

and

$$a \succ R_b = R_{ba^{-1}} \quad \text{and} \quad R_b \prec a = R_{a^{-1}b} \tag{10.20}$$

for all $a, b \in \mathrm{G}(H)$. We leave the proof of the following to the reader.

Proposition 10.9.8. *Suppose that H is a Hopf algebra with antipode S over the field k. Then:*

(a) *L_a and R_a are ideals of H^* for all $a \in \mathrm{G}(H)$, all of the same dimension which is either zero or one.*
(b) *One-dimensional left or right ideals of H^* are one-dimensional ideals of H^*.*

Suppose that $\int^r \neq (0)$.

(c) *The associations $a \mapsto L_a$ and $a \mapsto R_a$ describe bijective correspondences between $\mathrm{G}(H)$ and the set of all one-dimensional ideals of H^*.*
(d) *$L_g = R_1 = \int^r$ for a unique $g \in \mathrm{G}(H)$. Furthermore:*
   (i) *g is in the center of $\mathrm{G}(H)$ and*
   (ii) *$L_{ga} = R_a = L_{ag}$ for all $a \in \mathrm{G}(H)$.*

□

Remark 10.9.9. When H is finite-dimensional the element g of part (d) of Proposition 10.9.8 is the H-distinguished grouplike element of H.

We conclude this section by listing other properties of Hopf algebras H which have a non-zero left or right integral for H^* possess.

Proposition 10.9.10. *Suppose that H is a Hopf algebra over the field k and $\int^r \neq (0)$. Then:*

(a) *If D is a finite-dimensional subcoalgebra of H then $D^{(\infty)}$ is finite-dimensional.*
(b) *If D and E are finite-dimensional subcoalgebras of H then $D \wedge E$ is finite-dimensional.*
(c) *Every left (respectively right) coideal N of H such that $N \neq H$ is contained in a cofinite left (respectively right) coideal P of H such that $P \neq H$.*
(d) *Let J be either J_ℓ or J_r. Then $JH_0 = H = H_0J$.*
(e) *Suppose that H_0 is a Hopf subalgebra of H. Then $H = H_n$ for some $n \geq 0$. More precisely, if J_ℓ or J_r is contained in H_n then $H = H_n$.*

Proof. Since $\int^r \neq (0)$ the sum of the finite-dimensional left ideals of H^* is a dense subspace of H^* by Theorem 10.9.2. Thus parts (a) and (b) follow by parts (a) and (b) respectively of Corollary 3.6.6. Part (e) follows from part (d). For if H is any Hopf algebra and H_0 is a Hopf subalgebra of H then $H_nH_0 = H_n = H_0H_n$ for all $n \geq 0$ by Lemma 7.9.3.

Notice that part (c) for left coideals follows from part (b) of Theorem 3.6.5. Since $\int^\ell \neq (0)$ by Theorem 10.9.2, the hypothesis of the theorem holds for the Hopf algebra $H^{op\,cop}$. Now part (c) for the left coideals of $H^{op\,cop}$ is part (c) for the right coideals of H. Thus part (c) follows for H.

To show part (d) we first show that $H_0J_r = H$. Suppose that $H_0J_r \neq H$ and let $a \in \mathrm{G}(H)$. Then H_0J_ra is a proper right coideal of H. By part (c) there is a cofinite right coideal P of H such that $H_0J_ra \subseteq P$ and $P \neq H$. Therefore $P^\perp \subseteq H^r$. Let λ be a non-zero right integral for H^*. By part (b) of Theorem 10.9.5 we conclude that $P^\perp = \lambda{\prec}S(N)$ for some non-zero left coideal of H, where S is the antipode of H. Observe that $<\lambda, S(N)P> = (0)$. Therefore $<\lambda, S(N)H_0J_ra> = (0)$.

Since $N \neq (0)$ there is some simple left coideal M of H contained in N by part (a) of Proposition 3.2.11. Since $M \neq (0)$ from $(\mathrm{I}_H \otimes \epsilon)\circ\Delta = \mathrm{I}_H$ we deduce $\epsilon(M) \neq (0)$. Therefore there is some $m \in M$ such that $\epsilon(m) = 1$. Now

$$1 = \epsilon(S(m))1 = S(m)_{(1)}S(S(m)_{(2)}) = S(m_{(2)})S^2(m_{(1)}) \in S(M)H_0$$

since $S^2(H_0) \subseteq H_0$. Thus $1 \in S(N)H_0$ which means that $J_ra \subseteq S(N)H_0J_ra$.

We have shown that $<\lambda, J_ra> = (0)$ for all $a \subset \mathrm{G}(H)$. Now $g^{-1}{\succ}L_1 = L_g = R_1$ for some $g \in \mathrm{G}(H)$ by (10.19) and part (d) of Proposition 10.9.8. Thus $g{\succ}\lambda$ is a non-zero left integral for H^*. This means $(0) \neq <g{\succ}\lambda, J_r> = <\lambda, J_rg>$ by part (d) of Theorem 10.9.5, a contradiction. Therefore $H_0J_r = H$ after all.

We have noted that the hypothesis of the theorem holds for the Hopf algebra $H^{op\,cop}$. This last equation for $H^{op\,cop}$ translates to $J_\ell H_0 = H$ for H. Now S is bijective by part (f) of Theorem 10.9.5. Since $\int^\ell \neq (0)$ the hypothesis of the theorem holds for the Hopf algebra H^{cop}. The last two equations for H^{cop} translate to $H_0J_\ell = H$ and $J_rH_0 = H$ respectively for H. This completes the proof of part (d). As for part (e) we note that $J_r \subseteq H_n$ for some $n \geq 0$ since J_r is finite-dimensional and thus $H = JH_0 \subseteq H_nH_0 \subseteq H_n$ when H_0 is a Hopf subalgebra of H. The last inclusion follows by Lemma 7.9.3 again. $\square$

Since H_0 is a Hopf subalgebra of H when H is pointed:

Corollary 10.9.11. *Suppose that H is a pointed Hopf algebra with antipode S over the field k and $\int^r \neq (0)$. Then $H = H_n$ for some $n \geq 0$.* □

Exercises

Throughout the following exercises H is a Hopf algebra over the field k, J_ℓ is an injective hull of $k1$ in H, where (H, Δ) is regarded as a left H-comodule and J_r is an injective hull of $k1$ in H, where (H, Δ) is regarded as a right H-comodule. We refer to these injective hulls by left or right.

Exercise 10.9.1. Let $H = H_{2,-1}$. For $\alpha \in k$ let L_α be the span of 1 and $x + \alpha a$ and let R_α be the span of 1 and $ax + \alpha 1$. Show that:

(a) L_α (respectively R_α) is a left (respectively right) injective hull of $k1$ in H.

(b) $L_\alpha = L_{\alpha'}$ or $R_\alpha = R_{\alpha'}$ if and only if $\alpha = \alpha'$.

(c) The L_α's (respectively the R_α's) account for the left (respectively right) injective hulls of $k1$ in H.

Exercise 10.9.2. Suppose that $t : H \longrightarrow H^{cop}$ is a one-one bialgebra map. Show that $t^*(\int^r) = \int^\ell$ and $t^*(\int^\ell) = \int^r$.

Exercise 10.9.3. Suppose that $H = k[x]$ is the algebra of polynomials in x over k and that x is primitive. Show that H is the left and right injective hull of $k1$.

Exercise 10.9.4. Suppose that k has characteristic 0 and $\int^r \neq (0)$. Show that 0 is the only primitive element of H. [Hint: If $x \in H$ is a non-zero primitive show that the subalgebra $k[x]$ of H is a polynomial algebra. See Exercise 10.9.3 and Corollary 9.1.2.]

Exercise 10.9.5. Show that the following are equivalent:

(a) $J_\ell = H$.

(b) H is pointed irreducible.

(c) $J_r = H$.

Compare this exercise with Exercise 10.8.4.

Exercise 10.9.6. Find a Hopf algebra H over k and a one-one bialgebra map $t : H \longrightarrow H$ such that $t(J_\ell)$ and $t(J_r)$ are not injective hulls.

Exercise 10.9.7. Show that $S^*(L_a) = R_{a^{-1}}$ and $S^*(R_a) = L_{a^{-1}}$ for all $a \in \mathrm{G}(H)$.

Exercise 10.9.8. What is the connection between Propositions 10.6.2 and 10.9.8?

Exercise 10.9.9. Prove the following proposition. Compare with Corollaries 10.6.4 and 10.6.5.

Proposition 10.9.12. *Suppose that H is a Hopf algebra over the field k and $\int^r \neq (0)$. Then:*

(a) *$L_a = R_a$ for all $a \in \mathrm{G}(H)$ if and only if $L_a = R_a$ for some $a \in \mathrm{G}(H)$.*
(b) *H is cosemisimple if and only if $<\lambda, a> \neq 0$ for some $\lambda \in L_a$ or $\lambda \in R_a$, in which case $L_a = R_a$ and $<L_a, a> \neq (0)$ for all $a \in \mathrm{G}(H)$.*

Exercise 10.9.10. Prove the following proposition. [Hint: For part (a), first consider the case where I is a non-zero left ideal of H and λ is a left integral for H^*. Show that λ vanishes on a non-zero left ideal of H which is also a left coideal of H. See part (a) of Proposition 8.2.5.]

Proposition 10.9.13. *Suppose that H is a Hopf algebra over the field k. Then:*

(a) *If λ is a generalized left or right integral for H^* and I is a left or right ideal of H then $<\lambda, I> = (0)$ implies $\lambda = 0$ or $I = (0)$.*
(b) *If λ is a non-zero generalized left or right for H^* and $\lambda \in H^o$ then H is finite-dimensional.*

Exercise 10.9.11. Suppose that $n \geq 1$ and $\omega \in k$ is a primitive n^{th} root of unity. Find the left and right injective hulls for $k1$ in $H = H_{n,\omega}$ and show that they have dimension n.

Exercise 10.9.12. Suppose that S is bijective. Show that $S(J_\ell)$ and $S(J_r)$ are injective hulls.

Exercise 10.9.13. Suppose $a \in \mathrm{G}(H)$. Show that:

(a) The left injective hulls of $k1$ and ka have the same dimension. [Hint: Left or right multiplication by a is a coalgebra automorphism of H.]
(b) The right injective hulls of $k1$ and ka have the same dimension.
(c) If H is pointed and J is either J_ℓ or J_r then

$$H = \bigoplus_{a \in \mathrm{G}(H)} Ja = \bigoplus_{a \in \mathrm{G}(H)} aJ.$$

(d) If H is pointed and finite-dimensional and J is either J_ℓ or J_r then $(\mathrm{Dim}(J))|\mathrm{G}(H)| = \mathrm{Dim}(H)$.

Exercise 10.9.14. Show that $\int^r \neq (0)$ implies $P_{g,h}(H)$ is finite-dimensional for all $g, h \in \mathrm{G}(H)$. More precisely, show that there is a positive integer n such that $\mathrm{Dim}(P_{g,h}(H)) \leq n$ for all $g, h \in \mathrm{G}(H)$. See Exercise 3.6.3.

Exercise 10.9.15. Suppose that $\int^r \neq (0)$ and that the sum of the finite-dimensional left ideals of H^* is the sum of the finite-dimensional right ideals of H^*. Show that:

(a) The sum of the two-sided ideals of H^* is a dense subspace of H^*.

(b) If D is a subcoalgebra of H such that $D \neq H$ there exists a cofinite subcoalgebra E of H such that $D \subseteq E$ and $E \neq H$.

Exercise 10.9.16. Suppose that H is pointed and $\int^r \neq (0)$. Show that $\mathrm{Rad}(H^*)$ is a nilpotent ideal of H^*.

Chapter notes

Left and right integrals for finite-dimensional Hopf algebras H were introduced by Larson and Sweedler [105] where they established their existence and uniqueness in the theorem of Section 2 of that paper. Left and right integrals for the dual algebra H^*, where H is not necessarily finite-dimensional, were introduced by Sweedler [203]. One of the results of this paper is [203, (2.7) Corollary], our Proposition 10.2.1. The remaining results in Section 10.2 are found in [105]. Integrals are analogs of Haar measure which is described in the paper [62] by Haar.

Larson and Sweedler proved the existence and uniqueness of integrals for a finite-dimensional Hopf algebra using Hopf modules. There are proofs by others which do not use Hopf modules: Beattie, Dăscălescu, Grünenfelder, and Năstăsescu [20]; Dăscălescu, Năstăsescu, and Torrecillas [36]; Kauffman and the author [89]; Menini, Torrecillas, and Wisbauer [131]; Stefan [195]; and Van Daele [222].

Section 10.3 from beginning through Corollary 10.3.3 follows the discussion of [105, Section 5] by Larson and Sweedler. Theorem 10.3.3 is [105, Proposition 3] for both left and right H-modules. That semisimplicity is hereditary is a nice application of the Nichols-Zoeller Theorem [152, Theorem 7] which has been noted by several authors.

Let H be a finite-dimensional Hopf algebra over k. The trace function on $\mathrm{End}(H)$, integrals for H, H^{op}, H^{cop}, and related grouplike elements are seen to be closely connected in [174]. Two of these results are presented in Section 10.4: Theorem 10.4.1, which is [174, Theorem 2], and Proposition

10.4.2, which is [174, Propposition 2]. Theorem 10.4.3 is due to Larson and the author and is [102, (3.5) Theorem].

There is a very close connection between integrals for a finite-dimensional Hopf algebra over k and its antipode S as seen in Section 10.5. This was seen early on in the ground breaking paper by Larson and Sweedler [105]. Proposition 10.5.2 is implicit in this paper. A hint of the role of distinguished grouplike elements is seen in Larson's result [101, Theorem 5.5].

Theorem 10.5.4 is a combination of Theorem 3 and Proposition 3 of [174]. The derivation of the formula for S^4 given in Theorem 10.5.6 is modeled on the one found at the end of Section 2 of [174]. The formula is a technical variant of the one for S^4 given in [160, Proposition 6]. That the order of S if finite is due to the author [160, Theorem 1]. Corollary 10.5.7 is due to Larson; see Theorem 5.5 and Corollary 5.7 of [101]. See [174] also.

The results in Section 10.6 are from [160, Section 2] with the exception of part (e) of Proposition 10.6.2, Corollary 10.6.5, and Corollary 10.6.6.

Section 10.7 contains material from [174, Section 5] and the paper [102, Section 5] by Larson and the author. We note that part (a) of Corollary 10.7.2 is implicit in the proof of [174, Corollary 4]. Proposition 10.7.7, which is [103, Theorem 1], is very important. It is at the heart of the proof of Theorem 16.1.2.

The comodule structure of $\mathrm{Hom}(M, N)$ in Section 10.8 is that described by Larson in [101, Section 2] and Lemma 10.8.1 is essentially [101, Proposition 2.3]. Theorem 10.8.2 is basically Sweedler's [203, (3.2) Theorem]. Larson showed that the antipode of a cosemisimple Hopf algebra is bijective and that simple subcoalgebras are invariant under the square of the antipode [101, Theorem 3.3] using character theory for Hopf algebras which he developed in [101]. Our proof of this result, Corollary 10.8.4, is essentially that of [151, Proposition 3.4] and Lemma 10.8.3 is the left version of [151, Proposition 3.1].

Theorem 10.9.2 of Section 10.9 is a combination of several results. The connection between the existence of non-zero integrals for H^* and H^r is due to Sweedler [203]. In particular Lemma 10.9.1 is the combination of the left and right version of [203, (2.14)]. The equivalence of the existence of a non-zero integral for H^* and the finite-dimensionality of the injective hull of $k1$ is due to Sullivan [197, Theorem 1]. We borrow heavily from his paper [197] in Section 10.9.

Lemma 10.9.4 is a special case of [197, Proposition 1]. Part (a) of Theorem 10.9.5 is [197, Theorem 2]. Lemma 10.9.6 is the result of a crucial computation in his paper. Part (a) of Theorem 10.9.7 is [197, Theorem 3]; part (b) results from the proof of Theorem 3. The material from the end of the proof of Theorem 10.9.7 to the end of Section 10.9 captures the essence of [161].

Corollary 10.9.3 is [198, (2.15) Theorem]. We finally note that Takeuchi constructed a proof of uniqueness of integrals for H^* along different lines when H is cocommutative [214, Theorem 1].

Chapter 11

Actions by bialgebras and Hopf algebras

Module actions and comodule actions by bialgebras on algebras and coalgebras and algebras and coalgebras in other categories are all related. Three important constructions involving a bialgebra, the smash product, the smash coproduct, and the biproduct, arise from actions by bialgebras. In this chapter we discuss actions, these constructions, and their categorical contexts. The reader should find this chapter rich in ideas and important concepts.

Categories of fundamental objects and their morphisms we have encountered so far are subcategories of k-Vec, for example: k-Alg, algebras over k and their maps; k-Coalg, coalgebras over k and their maps; ${}_A\mathcal{M}$, left modules over a k-algebra A and their maps; and ${}^C\mathcal{M}$, left comodules over a k-coalgebra and their maps. The category k-Vec has a special object k and there is a means of "multiplying" two vector spaces to form a third by the tensor product operation $\otimes$. Note that k acts as a neutral object for multiplication. k-Vec together with k and $\otimes$ is an example of a monoidal category. We discuss monoidal categories in Section 11.1.

The definitions of algebra and coalgebra over k are made in terms of linear structure maps and commutative diagrams. It is very useful to define algebra and coalgebra in a monoidal category. This is basically done by reciting the definitions in k-Vec, replacing "linear map" by "morphism".

To define bialgebra, a monoidal category needs additional structure so that for algebras A and B in the category $A\otimes B$ is an algebra. Suppose A and B are k-algebras. The tensor product algebra multiplication for $A\otimes B$ is given by $m_{A\otimes B} = m_A\otimes\tau_{A,B}\otimes m_B$, where $\tau_{A,B} : A\otimes B \longrightarrow B\otimes A$ is the twist map which is defined by $\tau_{A,B}(a\otimes b) = b\otimes a$ for all $a \in A$ and $b \in B$. What is needed are morphisms which play the role of twist maps. Braiding structures are made to order for this. We discuss braiding structures in

Section 11.5.

Let H be a bialgebra over k. Then ${}_H\mathcal{M}$ is a monoidal category with k and ordinary tensor product $\otimes$. See Section 5.7 for left H-module structures on k and the tensor product. An algebra in ${}_H\mathcal{M}$ by definition is an H-module algebra. Let A be an H-module algebra. By virtue of the left H-module action on A, a k-algebra structure can be defined on the linear space $A\otimes H$ which is called the smash product and denoted $A\#H$. For example the group algebra over k of the semidirect product of groups is a smash product. Let ${}_H\mathcal{M}$-Alg be the subcategory of ${}_H\mathcal{M}$ consisting of algebras in ${}_H\mathcal{M}$ and their morphisms. Then ${}_H\mathcal{M}$-Alg $\longrightarrow$ k-Alg $(A \mapsto A\#H)$ is functorial. Module algebras and smash products are discussed in Section 11.2.

Now let us consider modules M over A, where A is regarded as an object of ${}_H\mathcal{M}$. These objects M together with module morphisms form a subcategory ${}_A\underline{\mathcal{M}}$ of ${}_H\mathcal{M}$. There is a way of defining a left $A\#H$-module structure on the vector space $M\otimes H$ such that ${}_A\underline{\mathcal{M}} \longrightarrow {}_{A\#H}\mathcal{M}$ $(M \mapsto M\#H)$ which is functorial. See Exercise 11.2.5.

${}^H\mathcal{M}$ is a monoidal category with k and ordinary tensor product $\otimes$. See Section 5.7 for left comodule structures on k and the tensor product. A coalgebra in ${}^H\mathcal{M}$ by definition is an H-comodule coalgebra. Let C be an H-module coalgebra. By virtue of the left H-comodule structure on C, a k-coalgebra structure can be defined on the linear space $C\otimes H$ which is called the smash coproduct and denoted $C\natural H$. The dual algebra of a smash coproduct is a smash product. Let ${}^H\mathcal{M}$-Coalg be the subcategory of ${}^H\mathcal{M}$ whose objects are the coalgebras in ${}^H\mathcal{M}$ and their morphisms. Then ${}^H\mathcal{M}$-Coalg $\longrightarrow$ k-Coalg $(C \mapsto C\natural H)$ is functorial. Comodule coalgebras and smash coproducts are discussed in Section 11.3 and the duality between the smash product and smash coproduct is discussed in Section 11.4.

Consider comodules M over C, where C is regarded as an object of ${}_H\mathcal{M}$. These objects M together with comodule morphisms form a subcategory ${}^C\underline{\mathcal{M}}$ of ${}^H\mathcal{M}$. There is a way of defining a left $C\natural H$-comodule structure on the vector space $M\otimes H$, denoted $M\natural H$, such that ${}^C\underline{\mathcal{M}} \longrightarrow {}^{C\natural H}\mathcal{M}$ $(M \mapsto M\natural H)$ is functorial. See Exercise 11.3.9.

Most important for the theory of Hopf algebras is the category ${}^H_H\mathcal{YD}$, a Yetter-Drinfel'd category, whose objects are triples $(M, \cdot, \rho)$, where $(M, \cdot)$ is an object of ${}_H\mathcal{M}$ and (M, ρ) is an object of ${}^H\mathcal{M}$, which satisfy the compatibility condition

$$h_{(1)}m_{(-1)}\otimes h_{(2)}\cdot m_{(0)} = (h\cdot m)_{(-1)}h_{(2)}\otimes(h\cdot m)_{(0)}$$

for all $h \in H$ and $m \in M$, and whose morphisms are functions which are morphisms of both ${}_H\mathcal{M}$ and ${}^H\mathcal{M}$. Note this compatibility condition is quite different from the compatibility condition

$$h_{(1)}m_{(-1)}\otimes h_{(2)}{\cdot}m_{(0)} = (h{\cdot}m)_{(-1)}\otimes(h{\cdot}m)_{(0)}$$

which defines left H-Hopf modules. When H is a Hopf algebra ${}^H_H\mathcal{YD}$ has a braiding structure.

Suppose A is a bialgebra in ${}^H_H\mathcal{YD}$. Then A has a very complicated structure; in particular it is a left H-module algebra and a left H-comodule coalgebra. The linear space $A\otimes H$ is a k-bialgebra with the smash product algebra structure $A\#H$ and the smash coproduct coalgebra structure $A\natural H$. This bialgebra structure is denoted $A\times H$ and is called a biproduct. The linear maps $A\times H \underset{\pi}{\overset{\jmath}{\leftrightarrows}} H$ defined by $\jmath(h) = 1\otimes h$ and $\pi(a\otimes h) = \epsilon(a)h$ for all $h \in H$ and $a \in A$ are bialgebra maps which satisfy $\pi\circ\jmath = \mathrm{I}_H$. When A is a Hopf algebra in ${}^H_H\mathcal{YD}$ then $A\times H$ is a Hopf algebra over k.

Suppose that H is a Hopf algebra over k, A is a bialgebra over k, and $A \underset{\pi}{\overset{\jmath}{\leftrightarrows}} H$ are bialgebra maps such that $\pi\circ\jmath = \mathrm{I}_H$. Then $A = B\times H$ for some bialgebra B in ${}^H_H\mathcal{YD}$. A very important case arises when K is a Hopf algebra over k such that K_0 is a Hopf subalgebra of K. Here $A = \mathrm{gr}(K)$, $H = K_0$, $\pi : A \longrightarrow H = \mathrm{gr}(K)_0$ is the projection onto the bottom term of the filtration, and $\jmath : H \longrightarrow \mathrm{gr}(K)$ is the inclusion. In this case $\mathrm{gr}(K) = B\times H$ for some Hopf algebra B in ${}^H_H\mathcal{YD}$.

Suppose H is a pointed Hopf algebra over k. Then $H_0 = k[G(H)]$ is a Hopf subalgebra of S. A natural first step for studying H is to pass to $\mathrm{gr}(H) = B\times H_0$, where B is a pointed irreducible Hopf algebra in ${}^{k[G]}_{k[G]}\mathcal{YD}$. The category ${}^{k[G]}_{k[G]}\mathcal{YD}$ plays a very important role in the theory of pointed Hopf algebras.

The connection between Yetter-Drinfel'd categories and biproducts is made in Section 11.6 and a characterization of biproducts is the subject of Section 11.7. Throughout this chapter H is a bialgebra over the field k.

11.1 Monoidal categories

Let $\mathcal{C}$ be a category. We denote the objects of $\mathcal{C}$ by $\mathrm{ob}\,\mathcal{C}$. If $\mathcal{D}$ is also a category then $\mathcal{C}\times\mathcal{D}$ is the category whose objects are ordered pairs (C, D), where $C \in \mathrm{ob}\,\mathcal{C}$ and $D \in \mathrm{ob}\,\mathcal{D}$, and whose morphisms are ordered pairs $(f, g) : (C, D) \longrightarrow (C', D')$, where $f : C \longrightarrow C'$ and $g : D \longrightarrow D'$

are morphisms of $\mathcal{C}$ and $\mathcal{D}$ respectively, with composition defined componentwise. We are interested in certain bifunctors $\mathcal{C}\times\mathcal{C} \longrightarrow \mathcal{C}$ which we denote by $\otimes$. We write $C\otimes D$ for $\otimes((C,D))$ and for a morphism $(f,g):(C,D) \longrightarrow (C',D')$ we write $f\otimes g$ for $\otimes((f,g))$.

Definition 11.1.1. A *monoidal category* is a tuple $(\mathcal{C},\otimes,e,\alpha,\lambda,\rho)$, where $\mathcal{C}$ is a category, $\otimes:\mathcal{C}\times\mathcal{C} \longrightarrow \mathcal{C}$ is a bifunctor, $e\in \mathrm{ob}\,\mathcal{C}$, and

$$\alpha=\{\alpha_{X,Y,Z}\,|\,X,Y,Z\in \mathrm{ob}\,\mathcal{C}\},\quad \lambda=\{\lambda_X\,|\,X\in \mathrm{ob}\,\mathcal{C}\},\quad \rho=\{\rho_X\,|\,X\in \mathrm{ob}\,\mathcal{C}\}$$

are families of natural isomorphisms

$$\alpha_{X,Y,Z}:X\otimes(Y\otimes Z) \longrightarrow (X\otimes Y)\otimes Z,\quad \lambda_X:e\otimes X \longrightarrow X,\quad \rho_X:X\otimes e \longrightarrow X$$

such that

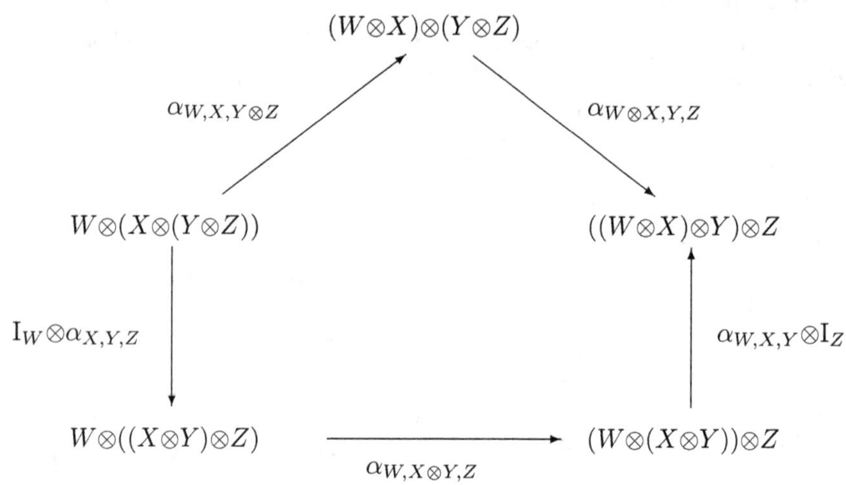

commutes for all $W,X,Y,Z\in \mathrm{ob}\,\mathcal{C}$,

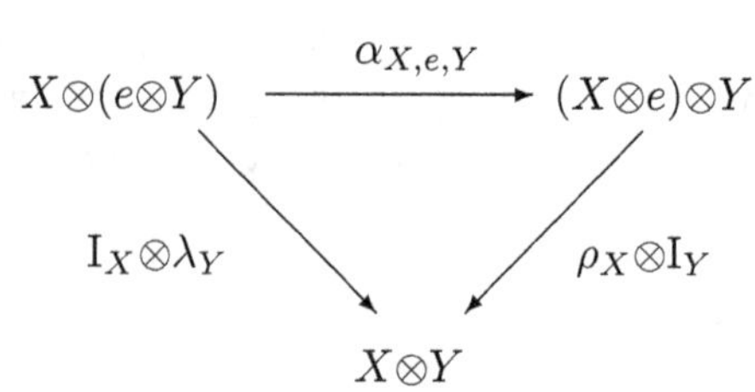

commutes for all $X,Y\in \mathrm{ob}\,\mathcal{C}$, and $\lambda_e=\rho_e$.

Usually we denote a monoidal category by its underlying category $\mathcal{C}$. Our primary example is the category k-Vec with its usual tensor product, with $k = e$, and with the families of natural isomorphisms determined by the usual identifications $U \otimes (V \otimes W) \simeq (U \otimes V) \otimes W$, $k \otimes U \simeq U$, and $U \otimes k \simeq U$ given by $u \otimes (v \otimes w) \mapsto (u \otimes v) \otimes w$, $\alpha \otimes u \mapsto \alpha u$, and $u \otimes \alpha \mapsto \alpha u$ respectively.

Let $\mathcal{C}$ be a monoidal category. Then algebra, module, coalgebra, comodule, bialgebra, Hopf module, and Hopf algebra can be defined in $\mathcal{C}$ essentially by replacing their underlying vector space by an object of $\mathcal{C}$ and replacing "map" by morphism in their respective definitions by commutative diagrams in k-Vec. We will define algebra in $\mathcal{C}$, morphism of algebras in $\mathcal{C}$, and module for an algebra in $\mathcal{C}$. The reader is left with the exercise of making the appropriate definitions for the other structures mentioned above and their morphisms in $\mathcal{C}$.

Definition 11.1.2. Let $\mathcal{C}$ be a monoidal category. An *algebra in* $\mathcal{C}$ is a tuple (A, m, η), where $A \in \operatorname{ob} \mathcal{C}$ and $m : A \otimes A \longrightarrow A$, $\eta : e \longrightarrow A$ are morphisms, such that the diagrams

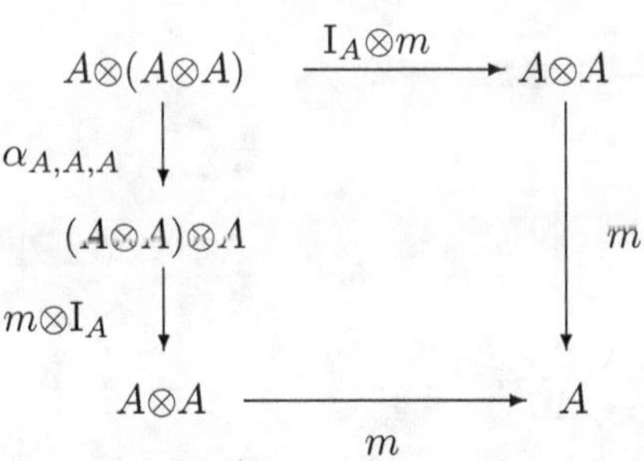

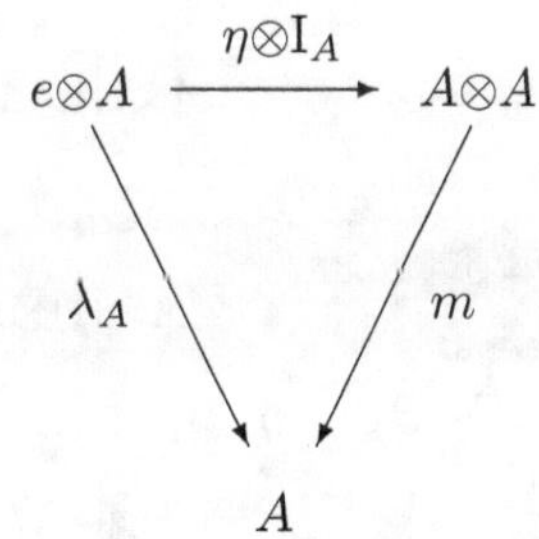

and

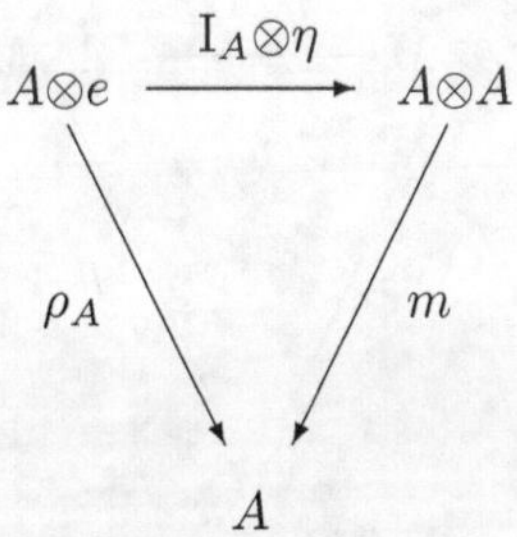

commute.

Definition 11.1.3. Let $\mathcal{C}$ be a monoidal category. A *morphism of algebras* (A, m_A, η_A) *and* (B, m_B, η_B) *in* $\mathcal{C}$ is a morphism $f : A \longrightarrow B$ of $\mathcal{C}$ such that the diagrams

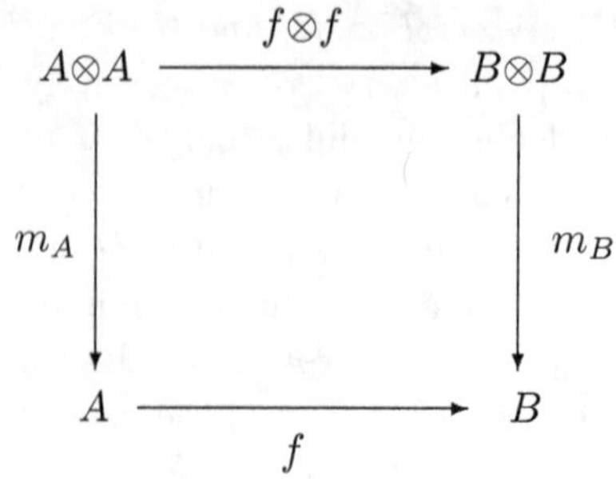

and

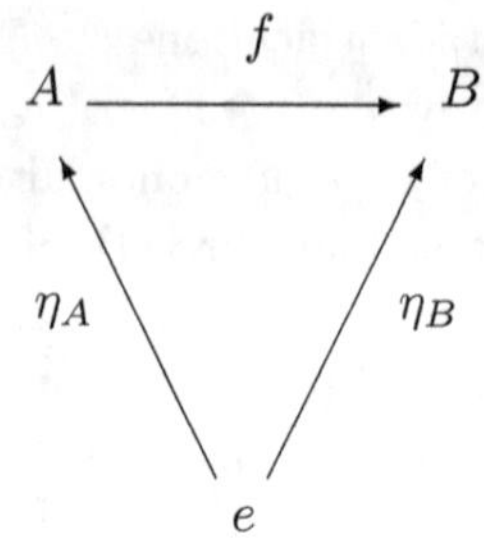

commute.

Definition 11.1.4. Let (A, m, η) be an algebra in a monoidal category $\mathcal{C}$. A *left A-module in $\mathcal{C}$* is a pair (M, μ), where $M \in \operatorname{ob}\mathcal{C}$ and $\mu : A\otimes M \longrightarrow M$ is a morphism, such that the diagrams

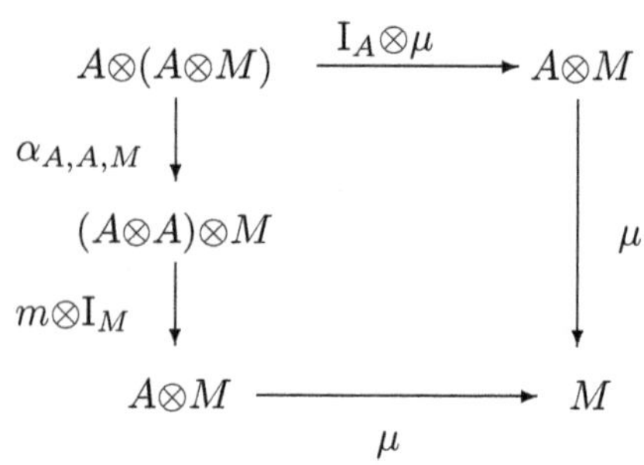

and

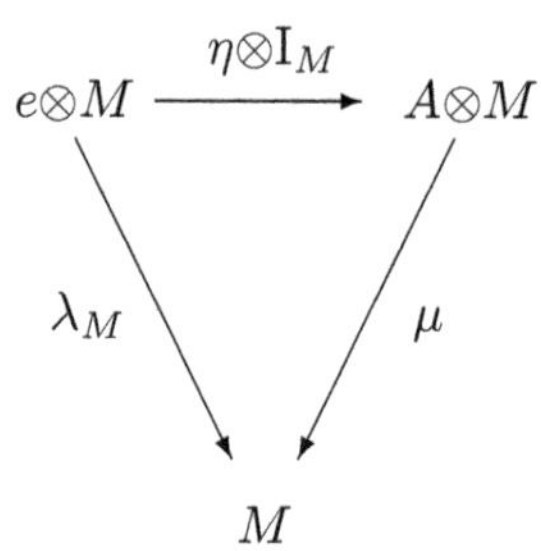

commute.

Ordinarily we denote algebras and the other structures of $\mathcal{C}$ mentioned above by their underlying objects; in particular we denote an algebra (A, m, η) in $\mathcal{C}$ by A.

There are subcategories of k-Vec which inherit a monoidal structure, namely k-Alg, k-Coalg, and hence k-Bialg. See Definitions 2.1.27, 2.1.26, and 5.1.19 respectively. Tensor product structures are given in Definitions 2.3.13, 2.1.23, and 5.1.21 respectively.

We regard k as a coalgebra where $\Delta(1) = 1\otimes 1$ and thus $\epsilon(1) = 1$. We have noted at the end of Section 5.1 that bialgebras over k (that is bialgebras in k-Vec) are the algebras in k-Coalg, or equivalently are the coalgebras in k-Alg.

Let A be a bialgebra over k. Then ${}_A\mathcal{M}$ inherits a monoidal structure from k-Vec also. The tensor product of two left A-modules is the A-module of Definition 5.7.1. We regard k as the left A-module according to $a{\cdot}1 = \epsilon(a)1$ for all $a \in A$. We can think of A as the object of ${}_A\mathcal{M}$ whose left module structure is given by multiplication in A. Likewise ${}^A\mathcal{M}$ inherits a monoidal structure from k-Vec. The tensor product of two left A-comodules is the left A-comodule of Definition 5.7.3. We can think of A as the object of ${}^A\mathcal{M}$ whose left comodule structure is given by the coproduct of A. Now consider a triple (M, μ, ρ), where (M, μ) is a left A-module and (M, ρ) is a left A-comodule. Then (M, μ, ρ) is a left A-Hopf module if and only if ρ is a morphism of ${}_A\mathcal{M}$ which is the case if and only if μ is a morphism of ${}^A\mathcal{M}$ by Lemma 8.1.3.

One final comment about notation. We will usually write $A \in \mathcal{C}$ for $A \in \mathrm{ob}\,\mathcal{C}$.

11.2 Module actions and module algebras, coalgebras

We consider algebras and coalgebras in the monoidal category ${}_H\mathcal{M}$.

Definition 11.2.1. Let H be a bialgebra over the field k. A *left H-module algebra* is an algebra in ${}_H\mathcal{M}$.

Thus a left H-module algebra is an algebra A over k with a left H-module structure which satisfies

$$h{\cdot}(ab) = (h_{(1)}{\cdot}a)(h_{(2)}{\cdot}b) \quad \text{and} \quad h{\cdot}1 = \epsilon(h)1 \tag{11.1}$$

for all $h \in H$ and $a, b \in A$. A very basic example is described as follows.

Example 11.2.2. Let H be a bialgebra over k, let A be an algebra over k, and suppose that $f \in \mathrm{Hom}(H, A)$ is an algebra map with convolution inverse. Then A is a left H-module algebra where $h{\cdot}a = f(h_{(1)})af^{-1}(h_{(2)})$ for all $h \in H$ and $a \in A$.

The fact that the linear map f has a convolution inverse means that (11.1) holds. Now $f : H \longrightarrow A$ is an algebra map $f^{-1} : H \longrightarrow A^{op}$ is an algebra map by part (a) of Proposition 6.1.2 and consequently A is a left H-module under the action. Observe that ideals of A are left H-submodules. For $h \in H$ and $a \in A$ we have the commutation relation

$$f(h)a = (h_{(1)}\cdot a)f(h_{(2)}) \tag{11.2}$$

which is easily seen on evaluating the right-hand side of the equation. Before analyzing this example further we need the notion of invariants of a left H-module.

Definition 11.2.3. Let H be a bialgebra over the field k and suppose that $M \in {}_H\mathcal{M}$. Then $m \in M$ is an *invariant of* M if $h\cdot m = \epsilon(h)m$ for all $h \in H$. The subspace of invariants of M is denoted by M_{inv}.

Returning to Example 11.2.2, let $B = \mathrm{Im}(f)$. Then B is a subalgebra of A. By definition of the module action on A we have $\mathrm{Z}_B(A) \subseteq A_{inv}$. By (11.2) it follows that $A_{inv} \subseteq \mathrm{Z}_B(A)$. Thus $A_{inv} = \mathrm{Z}_B(A)$ and consequently A_{inv} is the center of A when f is onto.

An important special case of Example 11.2.2 is when $H = A$, $f = \mathrm{I}_H$, and H has antipode S. Here $\mathrm{ad}_L : H\otimes H \longrightarrow H$ is the customary notation for the action and the notation $\mathrm{ad}_L h(a)$ is used for $\mathrm{ad}_L(h\otimes a)$. Thus

$$\mathrm{ad}_L h(a) = h_{(1)}aS(h_{(2)}) \tag{11.3}$$

for all $h, a \in H$.

Definition 11.2.4. Let H be a Hopf algebra with antipode S over k. The *left adjoint action of* H *on itself* is the module action $\mathrm{ad}_L : H\otimes H \longrightarrow H$ described above.

Let us consider the special case when H is the group algebra of a group G over k. Let K be a Hopf subalgebra of H. Then $K = k[N]$ is the group algebra of some subgroup N of G by Exercise 7.1.3. Since G is a linearly independent subset of H it follows that K is a left H-submodule under the adjoint action if and only if N is a normal subgroup of G. This example motivates the following definition.

Definition 11.2.5. Let H be a Hopf algebra over k. A *normal Hopf subalgebra of* H is a Hopf subalgebra K of H which is a left H-submodule under the left adjoint action.

Continuing with the group algebra example, suppose that G is the semidirect product of a normal subgroup N and a subgroup L. Then $G = NL$ and $N \cap L = (e)$. Thus every element of $g \in G$ has a unique factorization $g = n\ell$ where $n \in N$ and $\ell \in L$. Observe that

$$(n\ell)(n'\ell') = n\ell n'\ell' = n(\ell n'\ell^{-1})\ell\ell' = n(\mathrm{ad}_L\ell(n'))\ell\ell'$$

for all $n, n' \in N$ and $\ell, \ell' \in L$. The linear isomorphism $f : k[G] \longrightarrow k[N]\otimes k[L]$ defined by $f(g) = n\otimes\ell$, where $g = n\ell$ is the unique factorization of g, transfers the algebra structure of $k[G]$ to $k[N]\otimes k[L]$ and the product on the latter is given by $(n\otimes\ell)(n'\otimes\ell') = n(\ell{\cdot}n')\otimes\ell\ell'$ for all $n, n' \in N$ and $\ell, \ell' \in L$, where $\ell{\cdot}n' = \mathrm{ad}_L\ell(n')$. Observe that $k[N]$ is a left $k[L]$-module algebra under the adjoint action. This example generalizes very nicely.

Proposition 11.2.6. *Let H be a bialgebra over the field k and suppose that A is a left H-module algebra. Then $A\otimes H$ is an algebra over k with unity $1_A\otimes 1_H$ and product defined by $(a\otimes h)(a'\otimes h') = a(h_{(1)}{\cdot}a')\otimes h_{(2)}h'$ for all $a, a' \in A$ and $h, h' \in H$.*

Proof. The reader is left with the exercise of showing that $1_A\otimes 1_H$ is the unity of $A\otimes H$. Associativity follows from the calculations

$$\begin{aligned}
&(a\otimes h)((a'\otimes h')(a''\otimes h''))\\
&= (a\otimes h)(a'(h'_{(1)}{\cdot}a'')\otimes h'_{(2)}h'')\\
&= a(h_{(1)}{\cdot}(a'(h'_{(1)}{\cdot}a'')))\otimes h_{(2)}(h'_{(2)}h'')\\
&= a(h_{(1)(1)}{\cdot}a')(h_{(1)(2)}{\cdot}(h'_{(1)}{\cdot}a''))\otimes h_{(2)}(h'_{(2)}h'')\\
&= a(h_{(1)(1)}{\cdot}a')((h_{(1)(2)}h'_{(1)}){\cdot}a'')\otimes(h_{(2)}h'_{(2)})h''\\
&= a(h_{(1)}{\cdot}a')((h_{(2)}h')_{(1)}{\cdot}a'')\otimes(h_{(2)}h')_{(2)}h''
\end{aligned}$$

and

$$\begin{aligned}
&((a\otimes h)(a'\otimes h'))(a''\otimes h'')\\
&= (a(h_{(1)}{\cdot}a')\otimes h_{(2)}h')(a''\otimes h'')\\
&= a(h_{(1)}{\cdot}a')((h_{(2)}h')_{(1)}{\cdot}a'')\otimes(h_{(2)}h')_{(2)}h''
\end{aligned}$$

for all $a, a', a'' \in A$ and $h, h', h'' \in H$. □

Definition 11.2.7. The algebra of Proposition 11.2.6 is the *smash product of A and H.*

The usual notation for the smash product is $A\#H$ and $a\otimes h$ is written $a\#h$ for all $a \in A$ and $h \in H$. Thus the unity for $A\#H$ is thus denoted $1_A\#1_H$ and the product is written

$$(a\#h)(a'\#h') = a(h_{(1)}{\cdot}a')\#h_{(2)}h' \tag{11.4}$$

for all $a, a' \in A$ and $h, h' \in H$. We note that $(a\#h)(a'\#h') = aa'\#hh'$ whenever $h = 1$ or $a' = 1$. In particular $\imath : A \longrightarrow A\#H$ and $\jmath : H \longrightarrow A\#H$, defined by $\imath(a) = a\#1$ and $\jmath(h) = 1\#h$ for all $a \in A$ and $h \in H$ respectively, are algebra maps.

The tensor product algebra structure on $A \otimes H$ is a smash product. Here $h{\cdot}a = \epsilon(h)a$ for all $h \in H$ and $a \in A$. This left H-module action is determined by $A = A_{inv}$.

Now we turn our attention to coalgebras in ${}_H\mathcal{M}$.

Definition 11.2.8. Let H be a bialgebra over the field k. A *left H-module coalgebra* is a coalgebra in ${}_H\mathcal{M}$.

Thus a left H-module coalgebra is a coalgebra C over k with a left H-module structure such that

$$\Delta(h{\cdot}c) = h_{(1)}{\cdot}c_{(1)} \otimes h_{(2)}{\cdot}c_{(2)} \quad \text{and} \quad \epsilon(h{\cdot}c) = \epsilon(h)\epsilon(c) \tag{11.5}$$

for all $h \in H$ and $c \in C$.

Example 11.2.9. Let C be a bialgebra over k and suppose that H is a sub-bialgebra of C. Regard C as a left H-module under multiplication in C. Then C is a left H-module coalgebra.

We now consider duality between module algebras and module coalgebras. In order to do so we need the notions of right H-module algebra and right H-module coalgebra.

Regard right H-modules as left H^{op}-modules according to $h^{op}{\cdot}m = m{\cdot}h$ for all $h \in H$ and $m \in M$. Then a right H-module algebra is a left H^{op}-module algebra. Thus a right H-module algebra is an algebra A over k with a right H-module structure which satisfies

$$(ab){\cdot}h = (a{\cdot}h_{(1)})(b{\cdot}h_{(2)}) \quad \text{and} \quad 1{\cdot}h = \epsilon(h)1 \tag{11.6}$$

for all $a, b \in A$ and $h \in H$. Likewise a right H-module coalgebra is a coalgebra which is a left H^{op}-module coalgebra. Thus a right H-module coalgebra is a coalgebra C which has a right H-module structure such that

$$\Delta(c{\cdot}h) = c_{(1)}{\cdot}h_{(1)} \otimes c_{(2)}{\cdot}h_{(2)} \quad \text{and} \quad \epsilon(c{\cdot}h) = \epsilon(c)\epsilon(h) \tag{11.7}$$

for all $c \in C$ and $h \in H$.

If M is a left (respectively right) H-module we regard M^* as a right (respectively left) H-module under the transpose action.

Theorem 11.2.10. *Let H be a bialgebra over the field k.*

(a) *If A is a left (respectively right) H-module algebra then A^o is a right (respectively left) H-submodule of A^* and A^o is a right (respectively left) H-module coalgebra.*

(b) *If C is a left (respectively right) H-module coalgebra then C^* is a right (respectively left) H-module algebra.*

Proof. Suppose that A is a left H-module algebra. We show that A^o is a right H-submodule of A^* and that A^o is a right H-module coalgebra. The remainder of the proof is left as an exercise for the reader.

Let $a^o \in A^o$, $h \in H$, and $a, b \in A$. The calculation

$$\begin{aligned} <a^o \prec h, ab> &= <a^o, h{\cdot}(ab)> \\ &= <a^o, (h_{(1)}{\cdot}a)(h_{(2)}{\cdot}b)> \\ &= <{a^o}_{(1)}, h_{(1)}{\cdot}a><{a^o}_{(2)}, h_{(2)}{\cdot}b> \\ &= <{a^o}_{(1)} \prec h_{(1)}, a><{a^o}_{(2)} \prec h_{(2)}, b> \end{aligned}$$

shows that $\Delta_{a^o \prec h}$ exists and $\Delta_{a^o \prec h} = {a^o}_{(1)} \prec h_{(1)} \otimes {a^o}_{(2)} \prec h_{(2)}$. Thus $a^o \prec h \in A^o$ by definition and $\Delta(a^o \prec h) = {a^o}_{(1)} \prec h_{(1)} \otimes {a^o}_{(2)} \prec h_{(2)}$. Since

$$\epsilon_{A^o}(a^o \prec h) = <a^o \prec h, 1> = <a^o, h{\cdot}1> = \epsilon(h)<a^o, 1>$$

we have $\epsilon_{A^o}(a^o \prec h) = \epsilon(h)\epsilon_{A^o}(a^o)$. □

A conclusion we draw from Example 11.2.9 that H is a left H-module coalgebra where the module action is multiplication in H. Thus H^* is a right H-module algebra. Likewise H is a right H-module coalgebra where the right action is again multiplication in H. This means H^* is a left H-module algebra. Observe that H^o is a submodule in any case; thus H^o is a left and right H-module algebra.

Exercises

Throughout these exercises H, H' are bialgebras over the field k.

Exercise 11.2.1. Complete the proof of Theorem 11.2.10.

Exercise 11.2.2. Suppose that H has antipode S, that A is an algebra over k, and $\jmath : H \longrightarrow A$ is an algebra map. Show that $(A, \cdot_\jmath)$ is a left H-module algebra, where $h{\cdot}_\jmath a = \jmath(h_{(1)})a\jmath(S(h_{(2)}))$ for all $h \in H$ and $a \in A$.

Exercise 11.2.3. Let $(M, \cdot)$ be a left H-module and suppose that $\jmath : H' \longrightarrow H$ is a bialgebra map. Let $(M, \cdot_\jmath)$ be the left H'-module structure on A determined by pullback along $\jmath$; thus $h'{\cdot}_\jmath m = \jmath(h'){\cdot}m$ for all $h' \in H'$ and $m \in M$.

(a) If $(M, \cdot)$ is a left H-module algebra show that $(M, \cdot_\jmath)$ is a left H'-module algebra.

(b) If $(M, \cdot)$ is a left H-module coalgebra show that $(M, \cdot_\jmath)$ is a left H'-module coalgebra.

Exercise 11.2.4. Let A, B be algebras in ${}_H\mathcal{M}$ and $f : A \longrightarrow B$ be an algebra morphism. Show that:

(a) $f\#\mathrm{I}_H : A\#H \longrightarrow B\#H$ defined by $(f\#\mathrm{I}_H)(a\#h) = f(a)\#h$ for all $a \in A$ and $h \in H$ is a map of k-algebras.

(b) $A \mapsto A\#H$ and $f \mapsto f\#\mathrm{I}_H$ describes a functor from the category of algebras in ${}_H\mathcal{M}$ and their morphisms to k-Alg.

Exercise 11.2.5. Suppose A is an algebra in ${}_H\mathcal{M}$ and $(M, \bullet)$ is a left A-module in ${}_H\mathcal{M}$. Set $M\#H = M \otimes H$ and $m\#h = m \otimes h$ for all $m \in M$ and $h \in H$. Show that:

(a) $M\#H$ is a left $A\#H$-module, where $(a\#h)\cdot(m\#h') = a\bullet(h_{(1)}\cdot m)\#h_{(2)}h'$ for all $a \in A$, $h, h' \in H$, and $m \in M$.

(b) ${}_A\underline{\mathcal{M}} \longrightarrow {}_{A\#H}\mathcal{M}$ $\quad (M \mapsto M\#H,\ f \mapsto f\otimes\mathrm{I}_H)$ is a functor.

Exercise 11.2.6. Suppose A is an algebra in ${}_H\mathcal{M}$ and $(M, \bullet)$ is a left A-module in ${}_H\mathcal{M}$. Show that:

(a) M is a left $A\#H$-module, where $(a\#h)\cdot m = a\bullet(h\cdot m)$ for all $a \in A$, $h \in H$, and $m \in M$.

(b) $M\#H \longrightarrow M$ defined by $m\#h \mapsto \epsilon(h)m$ is a map of left $A\#H$-modules.

Exercise 11.2.7. Suppose A is a left H-module algebra over k. Show that:

(a) $A\#H$ has the tensor product algebra structure if and only if $h\cdot a = \epsilon(h)a$ for all $h \in H$ and $a \in A$.

(b) $A\#H$ is commutative if and only if A and H are commutative and $h\cdot a = \epsilon(h)a$ for all $h \in H$ and $a \in A$. (Thus $A\#H$ has the tensor product algebra structure when it is commutative.)

Exercise 11.2.8. Let G be a group and $H = k[G]$ be the group algebra of G over k. Suppose M is a left H-module and $\pi : G \longrightarrow \mathrm{End}(M)$ is the resulting representation restricted to G. Show that:

(a) If M is an algebra over k then M is a left H-module algebra if and only if $\pi(g)$ is an algebra endomorphism of M for all $g \in G$.

(b) If M is a coalgebra over k then M is a left H-module coalgebra if and only if $\pi(g)$ is a coalgebra endomorphism of M for all $g \in G$.

Exercise 11.2.9. Let G be a group and $H = k[G]$ be the group algebra of G over k. Suppose M is a left H-module and $M = \bigoplus_{\chi \in \widehat{G}} M_{(\chi)}$, where

$$M_{(\chi)} = \{m \in M \mid g\cdot m = \chi(g)m \ \text{ for all } \ g \in G\}.$$

See Exercise 7.4.6. Show that:

(a) If M is an algebra over k then M is a left H-module algebra if and only if $1 \in M_{(\epsilon)}$ and $M_{(\chi)}M_{(\chi')} \subseteq M_{(\chi\chi')}$ for all $\chi, \chi' \in \widehat{G}$.

(b) If M is a coalgebra over k then M is a left H-module coalgebra if and only if $\epsilon(M_{(\chi)}) = (0)$ when $\chi \neq \epsilon|G$ and $\Delta(M_{(\chi)}) \subseteq \sum_{\chi',\chi'' \in \widehat{G}, \chi'\chi''=\chi} M_{(\chi')} \otimes M_{(\chi'')}$ for all $\chi \in \widehat{G}$.

Compare with Exercise 11.2.8.

11.3 Comodule actions and comodule algebras, coalgebras

We consider algebras and coalgebras in the monoidal category ${}^H\mathcal{M}$.

Definition 11.3.1. Let H be a bialgebra over the field k. A *left H-comodule algebra over k* is an algebra in ${}^H\mathcal{M}$.

Therefore a left H-comodule algebra is an algebra A over k with a left H-comodule structure (A, ρ) such that

$$\rho(1_A) = 1_H \otimes 1_A \quad \text{and} \quad \rho(ab) = a_{(-1)}b_{(-1)} \otimes a_{(0)}b_{(0)} \tag{11.8}$$

for all $a, b \in A$. Note that H is a left H-comodule algebra with left comodule structure (H, Δ).

Definition 11.3.2. Let H be a bialgebra over the field k. A *left H-comodule coalgebra over k* is a coalgebra in ${}^H\mathcal{M}$.

In other words a left H-comodule coalgebra is a coalgebra C over k with a left H-comodule structure which satisfies

$$c_{(1)(-1)}c_{(2)(-1)} \otimes c_{(1)(0)} \otimes c_{(2)(0)} = c_{(-1)} \otimes c_{(0)(1)} \otimes c_{(0)(2)} \tag{11.9}$$

and

$$c_{(-1)}\epsilon(c_{(0)}) = \epsilon(c)1 \tag{11.10}$$

for all $c \in C$. There is an analog of Example 11.2.2.

Example 11.3.3. Let H be a bialgebra over k, let C be a coalgebra over k and suppose that $f \in \mathrm{Hom}(C, H)$ is a coalgebra map which is convolution invertible. Then C with the left H-comodule structure (C, ρ) given by $\rho(c) = f(c_{(1)})f^{-1}(c_{(3)}) \otimes c_{(2)}$ for all $c \in C$ is a left H-comodule coalgebra.

That (11.9) and (11.10) hold for ρ defined in the preceding example follow from the definition of convolution inverse. That (C, ρ) is a left H-comodule follows from the fact that $f : C \longrightarrow H$ and $f^{-1} : C \longrightarrow H^{cop}$ are coalgebra maps; see Proposition 6.1.3. Observe that subcoalgebras are subcomodules in Example 11.3.3. The connection between Example 11.2.2 and Example 11.3.3 is explored in detail in Exercises 11.3.3 and 11.3.4 where a duality is shown.

An important special case of Example 11.3.3 occurs when $H = C$, $f = \mathrm{I}_H$, and H has antipode S. We denote the left H-comodule action by $\mathrm{co\,ad}_L$; thus

$$\mathrm{co\,ad}_L(h) = h_{(1)}S(h_{(3)})\otimes h_{(2)} \tag{11.11}$$

for all $h \in H$.

Definition 11.3.4. Let H be a Hopf algebra with antipode S over the field k. The left H-comodule action $\mathrm{co\,ad}_L$ is the *left coadjoint action on H*.

The dual counterpart of Definition 11.2.5 is:

Definition 11.3.5. Let H be a Hopf algebra with antipode S over k. A *normal ideal of* H is a Hopf ideal I of H which is an H-subcomodule under the left coadjoint action of H on itself.

A duality between normal Hopf subalgebras and normal Hopf ideals is established in Exercise 11.3.6.

Just as an algebra (the smash product) is constructed from a bialgebra H and a left H-module algebra a coalgebra can be constructed from a bialgebra H and a left H-comodule coalgebra.

Proposition 11.3.6. *Let H be a bialgebra over the field k and let C be a coalgebra over k which is a left H-comodule coalgebra. Then $C\otimes H$ has a coalgebra structure given by*

$$\Delta(c\otimes h) = (c_{(1)}\otimes c_{(2)(-1)}h_{(1)})\otimes(c_{(2)(0)}\otimes h_{(2)})$$

and $\epsilon(c\otimes h) = \epsilon(c)\epsilon(h)$ for all $c \in C$ and $h \in H$.

Proof. We will check coassociativity and leave the remaining details to the reader. Let $c \in C$. Then

$$
\begin{aligned}
&((\Delta\otimes\mathrm{I})\circ\Delta)(c\otimes h)\\
&= \Delta(c_{(1)}\otimes c_{(2)(-1)}h_{(1)})\otimes(c_{(2)(0)}\otimes h_{(2)})\\
&= (c_{(1)(1)}\otimes c_{(1)(2)(-1)}(c_{(2)(-1)}h_{(1)})_{(1)})\otimes(c_{(1)(2)(0)}\otimes(c_{(2)(-1)}h_{(1)})_{(2)})\\
&\qquad\otimes(c_{(2)(0)}\otimes h_{(2)})\\
&= (c_{(1)(1)}\otimes c_{(1)(2)(-1)}c_{(2)(-1)(1)}h_{(1)(1)})\otimes(c_{(1)(2)(0)}\otimes c_{(2)(-1)(2)}h_{(1)(2)})\\
&\qquad\otimes(c_{(2)(0)}\otimes h_{(2)})\\
&= (c_{(1)(1)}\otimes c_{(1)(2)(-1)}c_{(2)(-1)}h_{(1)(1)})\otimes(c_{(1)(2)(0)}\otimes c_{(2)(0)(-1)}h_{(1)(2)})\\
&\qquad\otimes(c_{(2)(0)(0)}\otimes h_{(2)})\\
&= (c_{(1)}\otimes c_{(2)(-1)}c_{(3)(-1)}h_{(1)})\otimes(c_{(2)(0)}\otimes c_{(3)(0)(-1)}h_{(2)})\otimes(c_{(3)(0)(0)}\otimes h_{(3)})\\
&= (c_{(1)}\otimes c_{(2)(-1)}h_{(1)})\otimes(c_{(2)(0)(1)}\otimes c_{(2)(0)(2)(-1)}h_{(2)})\otimes(c_{(2)(0)(2)(0)}\otimes h_{(3)})
\end{aligned}
$$

and

$$
\begin{aligned}
&((\mathrm{I}\otimes\Delta)\circ\Delta)(c\otimes h)\\
&= (c_{(1)}\otimes c_{(2)(-1)}h_{(1)})\otimes\Delta(c_{(2)(0)}\otimes h_{(2)})\\
&= (c_{(1)}\otimes c_{(2)(-1)}h_{(1)})\otimes(c_{(2)(0)(1)}\otimes c_{(2)(0)(2)(-1)}h_{(2)(1)})\otimes(c_{(2)(0)(2)(0)}\otimes h_{(2)(2)})\\
&= (c_{(1)}\otimes c_{(2)(-1)}h_{(1)})\otimes(c_{(2)(0)(1)}\otimes c_{(2)(0)(2)(-1)}h_{(2)})\otimes(c_{(2)(0)(2)(0)}\otimes h_{(3)}).
\end{aligned}
$$

□

Definition 11.3.7. The coalgebra of Proposition 11.3.6 is the *smash coproduct of* C *and* H.

Ordinarily the smash coproduct is written $C\natural H$ and $c\natural h$ stands for $c\otimes h$. Thus the coalgebra structure of $C\natural H$ is given by

$$\Delta(c\natural h) = (c_{(1)}\natural c_{(2)(-1)}h_{(1)})\otimes(c_{(2)(0)}\natural h_{(2)}) \text{ and } \epsilon(c\natural h) = \epsilon(c)\epsilon(h) \quad (11.12)$$

for all $c \in C$ and $h \in H$.

We end this section by exploring the duality between H-comodule algebras and H-comodule coalgebras. To do this we need to define right H-comodule algebra and coalgebra. Recall that a right H-comodule (M, ρ) gives rise to a left H^{cop}-comodule (M, ρ^{cop}), where $\rho^{cop}(m) = m_{(1)}\otimes m_{(0)}$ for all $m \in M$. We define a right H-comodule algebra to be a left H^{cop}-comodule coalgebra. Thus a right H-comodule algebra is an algebra A over k with a right H-comodule structure (A, ρ) such that

$$\rho(ab) = a_{(0)}b_{(0)}\otimes a_{(1)}b_{(1)} \text{ and } \rho(1_A) = 1_A\otimes 1_H \quad (11.13)$$

for all $a, b \in A$. Likewise a right H-comodule coalgebra is a coalgebra C with a right H-comodule structure such that

$$c_{(0)(1)} \otimes c_{(0)(2)} \otimes c_{(1)} = c_{(1)(0)} \otimes c_{(2)(0)} \otimes c_{(1)(1)} c_{(2)(1)} \tag{11.14}$$

and

$$\epsilon(c_{(0)}) c_{(1)} = \epsilon(c) 1 \tag{11.15}$$

for all $c \in C$.

Proposition 11.3.8. *Let H be bialgebra over the field k.*

(a) *Suppose that C is a left (respectively right) H-comodule coalgebra. Then C^r is a right (respectively left) H-comodule algebra.*
(b) *Suppose that A is a finite-dimensional left (respectively right) H-comodule algebra. Then A^o is a right (respectively left) H-comodule coalgebra.*

Proof. We sketch a proof of part (a) and leave the remaining details as an exercise. Let C be a left H-comodule coalgebra and suppose that (C, ρ) is its underlying comodule structure. We will show that C^r is a subalgebra of C^*.

Let (C^r, ρ^r) be the comodule structure of the rational dual C^r and suppose $\widehat{\ } : C \longrightarrow C^{**}$ is the evaluation map defined by $\widehat{c}(c^*) = c^*(c)$ for all $c \in C$ and $c^* \in C^*$. Fix $c^* \in C^*$. Then $c^* \in C^r$ if and only if there is a ${\rho^r}_{c^*} \in C^* \otimes H$ which satisfies $(\widehat{c} \otimes \mathrm{I}_H)({\rho^r}_{c^*}) = (\mathrm{I}_H \otimes c^*)(\rho(c))$ for all $c \in C$ in which case ${\rho^r}_{c^*} = \rho^r(c^*)$. See Sections 3.1 and 3.2.

The equation $c_{(-1)}\epsilon(c_{(0)}) = \epsilon(c)1$ for all $c \in C$ shows that ρ^r_ϵ exists and $\rho^r(\epsilon) = \epsilon \otimes 1$. In particular $\epsilon \in C^r$. Let $f, g \in C^r$. Then the calculation

$$\begin{aligned} c_{(-1)} \langle fg, c_{(0)} \rangle &= c_{(-1)} \otimes \langle f, c_{(0)(1)} \rangle \langle g, c_{(0)(2)} \rangle \\ &= c_{(1)(-1)} c_{(2)(-1)} \langle f, c_{(1)(0)} \rangle \langle g, c_{(2)(0)} \rangle \\ &= \langle f_{(0)}, c_{(1)} \rangle \langle g_{(0)}, c_{(2)} \rangle f_{(1)} g_{(1)} \end{aligned}$$

for all $c \in C$ shows that ${\rho^r}_{fg}$ exists and ${\rho^r}_{fg} = f_{(0)} g_{(0)} \otimes f_{(1)} g_{(1)}$. □

In the next section we will establish a duality relationship between smash products and smash coproducts.

Exercises

Throughout these exercises H and H' are bialgebras over the field k.

Exercise 11.3.1. Suppose C is a left H-comodule coalgebra. Show that:

(a) $C\natural H$ has the tensor product coalgebra structure if and only if $\rho(c) = 1{\otimes}c$ for all $c \in C$.

(b) $C\natural H$ is cocommutative if and only if C and H are cocommutative and $\rho(c) = 1{\otimes}c$ for all $c \in C$. (Thus $C\natural H$ has the tensor product coalgebra structure when it is cocommutative.)

Exercise 11.3.2. Let G be a group and $H = k[G]$ be the group algebra of G over k. Let (M, ρ) be a left H-comodule and write $M = \bigoplus_{g \in G} M_g$, where

$$M_g = \{m \in M \mid \rho(m) = g{\otimes}m\}.$$

See Exercise 3.2.12. Show that:

(a) If M is an algebra over k then M is a left H-comodule algebra over k if and only if $1 \in M_1$ and $M_g M_{g'} \subseteq M_{gg'}$ for all $g, g' \in G$.

(b) If M is a coalgebra over k then M is a left H-comodule coalgebra over k if and only if $\epsilon(M_g) = (0)$ unless $g = 1$ and $\Delta(M_g) \subseteq \sum_{h \in G} M_{gh^{-1}}{\otimes}M_h$ for all $g \in G$.

Exercise 11.3.3. Let C be a coalgebra over k. Suppose that $f \in \mathrm{Hom}(C, H)$ is a coalgebra map with an inverse in the convolution algebra.

(a) Show that $F = f^*|H^o : H^o \longrightarrow C^*$ is an algebra map and has an inverse in the convolution algebra $\mathrm{Hom}(H^o, C^*)$.

(b) Let (C, ρ) be the left H-comodule structure of Example 11.3.3. Show that $\rho^*(h^o{\otimes}c^*) = F(h^o_{(1)})c^*F^{-1}(h^o_{(2)})$ for all $h^o \in H^o$. (Thus (C^*, m) is the left H^o-module of Example 11.2.2, where $m = \rho^*|H^o{\otimes}C^*$.)

Exercise 11.3.4. Let A be an algebra over k. Suppose that $f \in \mathrm{Hom}(H, A)$ is an algebra map and has an inverse in the convolution algebra.

(a) Show that $F = f^o : A^o \longrightarrow H^o$ is a coalgebra map and has a convolution inverse in the convolution algebra $\mathrm{Hom}(A^o, H^o)$.

(b) Let (A, m) be the left H-module of Example 11.2.2. Show that $m^*(a^o) = F(a^o_{(1)})F^{-1}(a^o_{(3)}){\otimes}a^o_{(2)}$ for all $a^o \in A^o$. (Thus $m^*(A^o) \subseteq H^o{\otimes}A^o$ and (A^o, ρ) is the left H^o-comodule of Example 11.3.3, where $\rho = m^*|A^o$.)

Exercise 11.3.5. Let (M, ρ) be a left H-comodule, and suppose $\pi : H \longrightarrow H'$ is a bialgebra map. Let (M, ρ_π) be the left H'-module structure on A determined by pushout along π; thus $\rho_\pi(m) = \pi(m_{(-1)}){\otimes}m_{(0)}$ for all $m \in M$. Show that:

(a) If (M, ρ) is a left H-comodule algebra then (M, ρ_π) is a left H'-comodule algebra.

(b) If (M, ρ) is a left H-comodule coalgebra then (M, ρ_π) is a left H'-comodule coalgebra.

Exercise 11.3.6. Let H be a Hopf algebra with antipode S over the field k and suppose K is a Hopf subalgebra of H. Show that:

(a) If K is a normal Hopf subalgebra of H then $K^{\perp}\cap H^{o}$ is a normal ideal of H^{o}.

(b) If H^{o} is a dense subspace of H^{*} and $K^{\perp}\cap H^{o}$ is a normal ideal of H^{o} then K is a normal Hopf subalgebra of H.

Exercise 11.3.7. Let C, D be algebras in ${}^{H}\mathcal{M}$ and $f : C \longrightarrow D$ be a coalgebra morphism. Show that:

(a) $f\natural\mathrm{I}_H : C\natural H \longrightarrow D\natural H$ defined by $(f\natural\mathrm{I}_H)(c\natural h) = f(c)\natural h$ for all $c \in C$ and $h \in H$ is a map of k-coalgebras.

(b) $C \mapsto C\natural H$ and $f \mapsto f\natural\mathrm{I}_H$ describes a functor from the category of coalgebras in ${}^{H}\mathcal{M}$ and their morphisms to k-Coalg.

Exercise 11.3.8. Suppose H has antipode S, that A is a coalgebra over k, and $\pi : A \longrightarrow H$ is a coalgebra map. Show that (A, ρ) is a left H-comodule coalgebra, where $\rho(a) = \pi(a_{(1)})S(\pi(a_{(3)}))\otimes a_{(2)}$ for all $a \in A$.

Exercise 11.3.9. Suppose C is a coalgebra in ${}^{H}\mathcal{M}$ and (M, ϱ) is a left C-comodule in ${}^{C}\mathcal{M}$. Set $M\natural H = M\otimes H$, $\varrho(m) = m_{[-1]}\otimes m_{[0]}$, and $m\natural h = m\otimes h$ for all $m \in M$ and $h \in H$. Show that:

(a) $M\natural H$ is a left $C\natural H$-comodule, where

$$\rho(m\natural h) = (m_{[-1]}\natural m_{[0](-1)}h_{(1)})\otimes(m_{[0](0)}\natural h_{(2)})$$

for all $m \in M$ and $h \in H$.

(b) ${}^{C}\underline{\mathcal{M}} \longrightarrow {}^{C\natural H}\mathcal{M} \quad (M \mapsto M\natural H,\ f \mapsto f\otimes\mathrm{I}_H)$ is a functor.

Exercise 11.3.10. Suppose C is a coalgebra in ${}^{H}\mathcal{M}$ and (M, ϱ) is a left C-comodule in ${}^{C}\mathcal{M}$. Show that:

(a) M is a left $C\natural H$-comodule where $\rho(m) = (m_{[-1]}\natural m_{[0](-1)})\otimes m_{[0](0)}$ for all $m \in M$, where $\varrho(m)$ is written as in Exercise 11.3.9.

(b) $M \longrightarrow M\natural H$ defined by $m \mapsto m\natural 1$ is a map of left $C\natural H$-comodules.

See Exercise 11.2.6.

11.4 Duality between the smash product and smash coproduct

Let C be a coalgebra over k and suppose that (M, ρ) is a left C-comodule. Let (M^{*}, m) be the left C^{*}-module structure on M^{*} given by the transpose action of the rational right C^{*}-module action on M. It is easy to see that $m = \rho^{*}|C^{*}\otimes M^{*}$.

Proposition 11.4.1. *Let H be a bialgebra over the field k and suppose that C is a left H-comodule coalgebra. Regard C^* as a left H^*-module as above and let $(C^*, \cdot)$ be this action restricted to H^o. Then:*

(a) *The algebra C^* with its left H^o-module structure $(C^*, \cdot)$ is a left H^o-module algebra.*

(b) *The one-one map $\imath : C^*\#H^o \longrightarrow (C\natural H)^*$ defined by $\imath(c^*\#h^o) = c^*\otimes h^o$ for all $c^* \in C^*$ and $h^o \in H^o$ is an algebra map.*

Proof. Let $h^o \in H^o$, $c^*, d^* \in C^*$, and $c \in C$. The calculations

$$\begin{aligned}(h^o{\cdot}(c^*d^*))(c) &= h^o(c_{(-1)})(c^*d^*(c_{(0)}))\\ &= h^o(c_{(-1)})c^*(c_{(0)(1)})d^*(c_{(0)(2)})\\ &= h^o(c_{(1)(-1)}c_{(2)(-1)})c^*(c_{(1)(0)})d^*(c_{(2)(0)})\\ &= {h^o}_{(1)}(c_{(1)(-1)}){h^o}_{(2)}(c_{(2)(-1)})c^*(c_{(1)(0)})d^*(c_{(2)(0)})\\ &= ({h^o}_{(1)}{\cdot}c^*)(c_{(1)})({h^o}_{(2)}{\cdot}d^*)(c_{(2)})\\ &= (({h^o}_{(1)}{\cdot}c^*)({h^o}_{(2)}{\cdot}d^*))(c)\end{aligned}$$

and

$$\begin{aligned}(h^o{\cdot}\epsilon)(c) &= h^o(c_{(-1)})\epsilon(c_{(0)})\\ &= h^o(\epsilon(c)1)\\ &= h^o(1)\epsilon(c)\end{aligned}$$

for all $c \in C$ show that $h^o{\cdot}(c^*d^*) = ({h^o}_{(1)}{\cdot}c^*)({h^o}_{(2)}{\cdot}d^*)$ and $h^o{\cdot}\epsilon = \epsilon(h^o)\epsilon$. Thus C^* with its module structure is a left H^o-module algebra. Part (b) is left as an easy exercise in definitions. □

Now suppose that A is a left H-module algebra. We will assume that H has an antipode S. Recall that $\imath : A \longrightarrow A\#H$ and $\jmath : H \longrightarrow A\#H$ defined by $\imath(a) = a\#1$ and $\jmath(h) = 1\#h$ for all $a \in A$ and $h \in H$ respectively are algebra maps. We first characterize $(A\#H)^o$.

Let I be a cofinite ideal of $A\#H$. Then $J = \imath^{-1}(I)$ and $L = \jmath^{-1}(I)$ are cofinite ideals of A and H respectively. For $a \in A$ and $h \in H$ the calculation

$$\begin{aligned}(1\#h_{(1)})(a\#1)(1\#S(h_{(2)})) &= (1\#h_{(1)})(a\#S(h_{(2)}))\\ &= h_{(1)}{\cdot}a\#h_{(2)}S(h_{(3)})\\ &= h{\cdot}a\#1\end{aligned}$$

shows that $\imath(h{\cdot}a) = (1\#h_{(1)})\imath(a)(1\#S(h_{(2)}))$. Therefore the cofinite ideal J of A is also a left H-submodule of A.

Let $A^{\underline{o}}$ be the subspace of A^o consisting of all those functionals on A which vanish on a cofinite ideal I of A which is also a left H-submodule of A. Then $A^{\underline{o}}$ is a subcoalgebra of A^o. Since $I \subseteq \imath(J)(1\#H) + (A\#1)\jmath(L) = J\#H + A\#L$ it follows that the subspace $(A\#H)^o$ of $(A\otimes H)^*$ lies in $A^{\underline{o}}\otimes H^o$.

Proposition 11.4.2. *Let H be a Hopf algebra over the field k and suppose that A is a left H-module algebra. Let (A, μ) be the left H-module structure on A. Then:*

(a) *$\mu^*(A^{\underline{o}}) \subseteq H^o\otimes A^{\underline{o}}$ and the restriction $\rho = \mu^*|A^{\underline{o}}$ gives $A^{\underline{o}}$ a left H^o-comodule structure $(A^{\underline{o}}, \rho)$.*
(b) *With this comodule structure $A^{\underline{o}}$ is a left H^o-comodule coalgebra and the inclusion $(A\#H)^o \hookrightarrow A^{\underline{o}}\natural H^o$ is an isomorphism of coalgebras.*

Proof. We first show part (a). Let $a^o \in A^{\underline{o}}$. Then a^o vanishes on some cofinite ideal I of A which is also a left H-submodule of A. Thus $\mu^*(a^o)$ vanishes on $H\otimes I$. Since I is a cofinite subspace of A it follows that $\mu^*(a^o) \in (H\otimes I)^\perp = H^*\otimes I^\perp \subseteq H^*\otimes A^{\underline{o}}$. See Exercise 11.4.2.

Assume that $\mu^*(a^o) \neq 0$ and write $\mu^*(a^o) = \sum_{i=1}^r f_i\otimes a_i^o$, where r is as small as possible, $f_i \in H^*$, and $a_i^o \in A^o$. Then $\{a_1^o, \ldots, a_r^o\}$ is linearly independent by Lemma 1.2.2. Now let $J' = \cap_{i=1}^r \mathrm{Ker}(f_i)$. Then J' is a cofinite subspace of H. For $h, h' \in H$ and $a \in A$ the calculation

$$\sum_{i=1}^r f_i(hh')a_i^o = a^o(hh'{\cdot}a) = a^o(h{\cdot}(h'{\cdot}a)) = \sum_{i=1}^r f_i(h)a_i^o(h'{\cdot}a)$$

shows that J' is a right ideal of H. Therefore $f_1, \ldots, f_r \in H^o$. We have shown that $\mu^*(A^{\underline{o}}) \subseteq H^o\otimes A^{\underline{o}}$. The reader is left with the exercise of showing that $(A^{\underline{o}}, \rho)$ is a left H^o-comodule, where $\rho = \mu^*|A^{\underline{o}}$.

To establish part (b) we first show that $A^{\underline{o}}$ is a left H^o-comodule coalgebra. Let $a^o \in A^{\underline{o}}$. Then for $h \in H$ and $a, b \in A$ the calculations

$$\begin{aligned}
&a^o{}_{(1)(-1)}a^o{}_{(2)(-1)}(h)a^o{}_{(1)(0)}(a)a^o{}_{(2)(0)}(b)\\
&= a^o{}_{(1)(-1)}(h_{(1)})a^o{}_{(2)(-1)}(h_{(2)})a^o{}_{(1)(0)}(a)a^o{}_{(2)(0)}(b)\\
&= a^o{}_{(1)}(h_{(1)}{\cdot}a)a^o{}_{(2)}(h_{(2)}{\cdot}b)\\
&= a^o((h_{(1)}{\cdot}a)(h_{(2)}{\cdot}b))\\
&= a^o(h{\cdot}(ab))
\end{aligned}$$

and

$$a^o_{(-1)}(h)a^o_{(0)(1)}(a)a^o_{(0)(2)}(b) = a^o_{(-1)}(h)a^o_{(0)}(ab) = a^o(h{\cdot}(ab))$$

show that $a^o{}_{(1)(-1)}a^o{}_{(2)(-1)}\otimes a^o{}_{(1)(0)}\otimes a^o{}_{(2)(0)} = a^o{}_{(-1)}\otimes a^o{}_{(0)(1)}\otimes a^o{}_{(0)(2)}$. The reader is left with the exercise of showing that $a^o_{(-1)}\epsilon_{A^{\underline{o}}}(a^o_{(0)}) = \epsilon_{A^{\underline{o}}}(a^o)1_{H^o}$. This completes our proof that $A^{\underline{o}}$ is a left H^o-module coalgebra.

Suppose further that $h^o \in H^o$. To complete the argument for part (b) we need only show three things: regarding $a^o\otimes h^o \in (A\#H)^*$ that $\Delta_{a^o\otimes h^o}$ exists, that

$$\Delta_{a^o\otimes h^o} = (a^o{}_{(1)}\otimes a^o{}_{(2)(-1)}h^o{}_{(1)})\otimes(a^o{}_{(2)(0)}\otimes h^o{}_{(2)}),$$

and that $\epsilon_{(A\#H)^o} = \epsilon_{A^{\underline{o}}\natural H^o}$. We leave the third as an exercise and establish the first two with the calculation

$$\begin{aligned}
&<a^o\otimes h^o, (a\otimes h)(b\otimes k)> \\
&= (a^o\otimes h^o)(a(h_{(1)}{\cdot}b)\otimes h_{(2)}k) \\
&= a^o{}_{(1)}(a)a^o{}_{(2)}(h_{(1)}{\cdot}b)h^o{}_{(1)}(h_{(2)})h^o(k) \\
&= a^o{}_{(1)}(a)a^o{}_{(2)(-1)}(h_{(1)})a^o{}_{(2)(0)}(b)h^o{}_{(1)}(h_{(2)})h^o{}_{(2)}(k) \\
&= a^o{}_{(1)}(a)a^o{}_{(2)(-1)}h^o{}_{(1)}(h)a^o{}_{(2)(0)}(b)h^o{}_{(2)}(k) \\
&= <a^o{}_{(1)}\otimes a^o{}_{(2)(-1)}h^o{}_{(1)}, a\otimes h><a^o{}_{(2)(0)}\otimes h^o{}_{(2)}, b\otimes k>
\end{aligned}$$

for all $a, b \in A$ and $h, k \in H$. □

Exercises

Exercise 11.4.1. Complete the proof of Proposition 11.4.2.

Exercise 11.4.2. Let U and V be vector spaces over the field k and suppose that $V = V'\oplus V''$ is the direct sum of subspaces, where V'' is a finite-dimensional subspace of V. Show that the canonical inclusion $U^*\otimes V^* \hookrightarrow (U\otimes V)^*$ induces an isomorphism $U^*\otimes V'^{\perp} \simeq (U\otimes V')^{\perp}$. [Hint: $U\otimes V = (U\otimes V')\oplus(U\otimes V'')$ and $(U\otimes V')^* \simeq U^*\otimes V''^*$ since V'' is finite-dimensional.]

11.5 Prebraiding, braiding structures on a monoidal category

Let H be a bialgebra over the field k. In the next section we will consider a monoidal category ${}^H_H\mathcal{YD}$ whose objects have a left H-module and a left H-comodule structure. When H^{op} has an antipode ${}^H_H\mathcal{YD}$ has a braiding structure. This braiding structure is important in connection with the structure of pointed Hopf algebras.

In this section we discuss prebraiding and braiding structures in a monoidal category $\mathcal{C}$ and look at some of the structures which can be made when $\mathcal{C}$ has a braiding.

Definition 11.5.1. Let $(\mathcal{C}, \otimes, e, \alpha, \lambda, \rho)$ be a monoidal category. A *prebraiding for* $\mathcal{C}$ is a collection of morphisms $\gamma = \{\gamma_{X,Y} \mid X, Y \in \operatorname{ob}\mathcal{C}\}$, where

$$\gamma_{X,Y} : X \otimes Y \longrightarrow Y \otimes X$$

are natural in X, Y and such that the diagrams

$$\begin{array}{ccc} X\otimes e & \xrightarrow{\gamma_{X,e}} & e\otimes X \\ & \rho_{X,e} \searrow \quad \swarrow \lambda_{e,X} & \\ & X & \end{array}$$

commute for all $X \in \operatorname{ob}\mathcal{C}$, and

$$\begin{array}{ccc} (X\otimes Y)\otimes Z & \xrightarrow{\gamma_{X\otimes Y,Z}} & Z\otimes(X\otimes Y) \\ \downarrow{\scriptstyle \alpha^{-1}_{X,Y,Z}} & & \downarrow{\scriptstyle \alpha_{Z,X,Y}} \\ X\otimes(Y\otimes Z) & & (Z\otimes X)\otimes Y \\ \downarrow{\scriptstyle \mathrm{I}_X\otimes\gamma_{Y,Z}} & & \uparrow{\scriptstyle \gamma_{X,Z}\otimes\mathrm{I}_Y} \\ X\otimes(Z\otimes Y) & \xrightarrow[\alpha_{X,Z,Y}]{} & (X\otimes Z)\otimes Y \end{array}$$

and

$$\begin{array}{ccc}
X\otimes(Y\otimes Z) & \xrightarrow{\gamma_{X,Y\otimes Z}} & (Y\otimes Z)\otimes X \\
\Big\downarrow{\scriptstyle \alpha_{X,Y,Z}} & & \Big\downarrow{\scriptstyle \alpha^{-1}_{Y,Z,X}} \\
(X\otimes Y)\otimes Z & & Y\otimes(Z\otimes X) \\
\Big\downarrow{\scriptstyle \gamma_{X,Y}\otimes \mathrm{I}_Z} & & \Big\uparrow{\scriptstyle \mathrm{I}_Y\otimes\gamma_{X,Z}} \\
(Y\otimes X)\otimes Z & \xrightarrow[\alpha^{-1}_{Y,X,Z}]{} & Y\otimes(X\otimes Z)
\end{array}$$

commute for all $X, Y, Z \in \mathrm{ob}\,\mathcal{C}$. A *braiding for* $\mathcal{C}$ is a prebraiding for $\mathcal{C}$ consisting of isomorphisms in which case $(\mathcal{C}, \otimes, e, \alpha, \lambda, \rho, \gamma)$ is a *braided monoidal category.* If γ is a braiding for $\mathcal{C}$ and $\gamma^{-1}_{X,Y} = \gamma_{Y,X}$ for all $X, Y \in \mathrm{ob}\,\mathcal{C}$ then the braiding and the braided monoidal category $\mathcal{C}$ are *symmetric.*

For vector spaces U and V over k, let $\tau_{U,V} : U\otimes V \longrightarrow V\otimes U$ be the "twist" map which is defined by $\tau_{U,V}(u\otimes v) = v\otimes u$ for all $u \in U$ and $v \in V$. The family $\{\tau_{U,V} \,|\, U, V \in \mathrm{ob}\,k\text{-Vec}\}$ is a symmetric braiding for k-Vec.

The tensor product, and "opposites", of algebras and coalgebras are defined in monoidal categories with prebraidings. To see how this is done we take our cue from k-Vec.

Let (A, m_A, η_A) and (B, m_B, η_B) be algebras over k. The tensor product algebra structure on $A\otimes B$ may be defined as $(A\otimes B, m_{A\otimes B}, \eta_{A\otimes B})$, where $m_{A\otimes B} = (m_A\otimes m_B)\circ(\mathrm{I}_A\otimes\tau_{A,B}\otimes\mathrm{I}_B)$ and $\eta_{A\otimes B} = \eta_A\otimes\eta_B$. Observe that A^{op} may be described as $(A, m\circ\tau_{A,A}, \eta_A)$.

Let $(\mathcal{C}, \otimes, e, \alpha, \lambda, \rho, \gamma)$ be a monoidal category with a prebraiding γ. For $X, Y \in \mathrm{ob}\,\mathcal{C}$ we define an isomorphism

$$T_{X,Y} : (X\otimes Y)\otimes(X\otimes Y) \longrightarrow (X\otimes X)\otimes(Y\otimes Y)$$

to be the composition

$$
\begin{aligned}
(X\otimes Y)\otimes(X\otimes Y) &\xrightarrow{\alpha_{X\otimes Y,X,Y}} ((X\otimes Y)\otimes X)\otimes Y \\
&\xrightarrow{\alpha^{-1}_{X,Y,X}\otimes \mathrm{I}_Y} (X\otimes(Y\otimes X))\otimes Y \\
&\xrightarrow{(\mathrm{I}_X\otimes\gamma_{Y,X})\otimes \mathrm{I}_Y} (X\otimes(X\otimes Y))\otimes Y \\
&\xrightarrow{\alpha_{X,X,Y}\otimes \mathrm{I}_Y} ((X\otimes X)\otimes Y)\otimes Y \\
&\xrightarrow{\alpha^{-1}_{X\otimes X,Y,Y}} (X\otimes X)\otimes(Y\otimes Y).
\end{aligned}
$$

Suppressing the isomorphisms of the family α we can think of $T_{X,Y}$ as $\mathrm{I}_X\otimes\gamma_{X,Y}\otimes\mathrm{I}_Y$.

For the remainder of this section $\mathcal{C}$ is a monoidal category with prebraiding γ. Let (A, m_A, η_A), (B, m_B, η_B) be algebras in $\mathcal{C}$. Then $(A\otimes B, m_{A\otimes B}, \eta_{A\otimes B})$ is an algebra of $\mathcal{C}$, where $m_{A\otimes B} = (m_A\otimes m_B)\circ T_{A,B}$ and $\eta_{A\otimes B} = (\eta_A\otimes\eta_B)\circ\lambda^{-1}_{e,e}$.

Definition 11.5.2. Let $\mathcal{C}$ be a monoidal category with prebraiding γ. The *tensor product of the algebras* (A, m_A, η_A) *and* (B, m_B, η_B) *in* $\mathcal{C}$ is the algebra $(A\otimes B, m_{A\otimes B}, \eta_{A\otimes B})$ and is denoted by $A\otimes B$.

Note that $(A, m_A\circ\gamma_{A,A}, \eta_A)$ is an algebra in $\mathcal{C}$.

Definition 11.5.3. Let $\mathcal{C}$ be a monoidal category with prebraiding γ. The *opposite algebra of* (A, m, η_A) *in* $\mathcal{C}$ is the algebra $(A, m_A\circ\gamma_{A,A}, \eta_A)$ and is denoted by A^{op}.

When $\mathcal{C} = k$-Vec and γ is the family of "twist" maps then these definitions agree with those made already for the tensor product algebra structure made already for two algebras over k and the "opposite" of an algebra over k.

Let $(C, \Delta_C, \epsilon_C)$ and $(D, \Delta_D, \epsilon_D)$ be coalgebras in $\mathcal{C}$. Then $(C\otimes D, \Delta_{C\otimes D}, \epsilon_{C\otimes D})$ is a coalgebra in $\mathcal{C}$ where $\Delta_{C\otimes D} = T_{C\otimes D}\circ(\Delta_C\otimes\Delta_D)$ and $\epsilon_{C\otimes D} = \lambda_e\circ(\epsilon_C\otimes\epsilon_D)$.

Definition 11.5.4. Let $\mathcal{C}$ be a monoidal category with prebraiding γ. The *tensor product of the coalgebras* $(C, \Delta_C, \epsilon_C)$ *and* $(D, \Delta_D, \epsilon_D)$ *in* $\mathcal{C}$ is the coalgebra $(C\otimes D, \Delta_{C\otimes D}, \epsilon_{C\otimes D})$ and is denoted by $C\otimes D$.

Observe that $(C, \gamma_{C,C} \circ \Delta_C, \epsilon_C)$ is a coalgebra in $\mathcal{C}$.

Definition 11.5.5. Let $\mathcal{C}$ be a monoidal category with prebraiding γ. The *opposite coalgebra of* $(C, \Delta_C, \epsilon_C)$ *in* $\mathcal{C}$ is the coalgebra $(C, T_{C,C} \circ \Delta_C, \epsilon_C)$ and is denoted by C^{cop}.

When $\mathcal{C} = k$-Vec with its braiding by the "twist" maps then these definitions agree with those given for the tensor product of coalgebras and the "opposite" coalgebra.

Exercises

Throughout these exercises $\mathcal{C}$ is a prebraided monoidal category.

Exercise 11.5.1. Suppose that A and B are algebras in $\mathcal{C}$. Show that $A \otimes B$ and A^{op} are algebras in $\mathcal{C}$.

Exercise 11.5.2. Suppose that C and D are coalgebras in $\mathcal{C}$. Show that $C \otimes D$ and C^{cop} are coalgebras in $\mathcal{C}$.

Exercise 11.5.3. Suppose that A and B are bialgebras in $\mathcal{C}$. Show that:

(a) $A \otimes B$ is a bialgebra in $\mathcal{C}$ with the tensor product algebra and coalgebra structures.

(b) A^{op}, A^{cop}, and $A^{op\,cop}$ are bialgebras in $\mathcal{C}$.

Definition 11.5.6. Let $\mathcal{C}$ be a monoidal category with prebraiding γ. Let A and B be bialgebras in $\mathcal{C}$. The *tensor product of bialgebras A and B in* $\mathcal{C}$ is $A \otimes B$ with its tensor product algebra and coalgebra structures in $\mathcal{C}$.

Exercise 11.5.4. Let H be a Hopf algebra with antipode S in $\mathcal{C}$. Which of the results of Section 7.1 have analogs in $\mathcal{C}$?

11.6 Yetter–Drinfel'd modules and biproducts

There is a category ${}^H_H\mathcal{YD}$, whose objects are certain left H-modules and left H-comodules, which is important for the study of Hopf algebras whose coradical is a Hopf subalgebra. Case in point, pointed Hopf algebras. We show that a bialgebra B in the category determines a bialgebra $B \times H$ over k. Its underlying algebra structure is the smash product $B \# H$ and its underlying coalgebra structure is the smash coproduct $B \natural H$.

Further there are bialgebra maps $B\times H \underset{\pi}{\overset{\jmath}{\leftrightarrows}} H$ which satisfy $\pi\circ\jmath = \mathrm{I}_H$.

In the next section we show that a bialgebra A with bialgebra maps $A \underset{\pi}{\overset{\jmath}{\leftrightarrows}} H$ which satisfy $\pi\circ\jmath = \mathrm{I}_H$ is isomorphic to a biproduct when H is a Hopf algebra. Here we discuss elementary aspects of ${}^H_H\mathcal{YD}$ and make several observations about biproducts. The reader is encouraged to do the exercises at the end of this section which develop some very important properties of ${}^H_H\mathcal{YD}$.

The objects of ${}^H_H\mathcal{YD}$ are triples (M,μ,ρ), where (M,μ) is a left H-module and (M,ρ) is a left H-comodule, such that the compatibility condition

$$h_{(1)}m_{(-1)}\otimes h_{(2)}\cdot m_{(0)} = (h_{(1)}\cdot m)_{(-1)}h_{(2)}\otimes(h_{(1)}\cdot m)_{(0)} \tag{11.16}$$

holds for all $h\in H$ and $m\in M$, and the morphisms of ${}^H_H\mathcal{YD}$ are functions $f : M \longrightarrow N$ of underlying vector spaces of objects which are simultaneously maps of left H-modules and left H-comodules.

Example 11.6.1. Let H be a bialgebra over k. A vector space V over k is a left Yetter-Drinfel'd module with the trivial actions (5.6) and (5.7); that is $h\cdot v = \epsilon(h)v$ for all $h\in H$ and $v\in V$ and $\rho(v) = 1\otimes v$ for all $v\in V$.

Definition 11.6.2. Let H be a bialgebra over the field k. Then ${}^H_H\mathcal{YD}$ is a *Yetter–Drinfel'd category*, its objects are *left Yetter–Drinfel'd H-modules*, and its morphisms are *maps of left Yetter–Drinfel'd H modules*.

Example 11.6.3. ${}^k_k\mathcal{YD} = k$-Vec, where for a vector space V over k actions by k are determined by $1_k\cdot v = v$ and $\rho(v) = 1_k\otimes v$ for all $v\in V$.

Subobjects are defined in the obvious way.

Definition 11.6.4. Let H be a bialgebra over the field k and let M be a left Yetter–Drinfel'd H-module. A *Yetter–Drinfel'd submodule of M* is a subspace N of M which is a left H-submodule and a left H-subcomodule of M.

Let M be a left Yetter–Drinfel'd H-module. Then M and (0) are Yetter–Drinfel'd submodules of M. Suppose that V is a subspace of M. Since the intersection of submodules is a submodule and the intersection of subcomodules is a subcomodule, the intersection N of all Yetter–Drinfel'd submodules of M which contain V is again a left Yetter–Drinfel'd submodule of M.

Definition 11.6.5. The Yetter–Drinfel'd submodule N of M just described is the *Yetter–Drinfel'd submodule of M generated by V*. A *finitely generated Yetter–Drinfel'd submodule of M* is a Yetter–Drinfel'd submodule L generated by some finite-dimensional subspace V of M.

The category ${}_H^H\mathcal{YD}$ inherits a monoidal structure from k-Vec. We regard k as an object of ${}_H^H\mathcal{YD}$ by $h{\cdot}1_k = \epsilon(h)$ for all $h \in H$ and $\rho(1_k) = 1_H \otimes 1_k$. Let M, N be objects of ${}_H^H\mathcal{YD}$. Then $M\otimes N$ is an object of ${}_H^H\mathcal{YD}$ with the tensor product module and comodule structures. See Section 5.7.

Now suppose that H is a Hopf algebra with antipode S. Then (11.16) is equivalent to

$$\rho(h{\cdot}m) = h_{(1)}m_{(-1)}S(h_{(3)})\otimes h_{(2)}{\cdot}m_{(0)} \tag{11.17}$$

for all $h \in H$ and $m \in M$. This formulation of (11.16) is very useful in practice. See Exercise 11.6.10 for example.

The monoidal category ${}_H^H\mathcal{YD}$ has a prebraiding structure. Thus algebras and coalgebras are defined in ${}_H^H\mathcal{YD}$. For objects M, N of ${}_H^H\mathcal{YD}$ define

$$\sigma_{M,N} : M\otimes N \longrightarrow N\otimes M$$

by

$$\sigma_{M,N}(m\otimes n) = m_{(-1)}{\cdot}n\otimes m_{(0)}$$

for all $m \in M$ and $n \in N$. The family of $\sigma_{M,N}$'s is a prebraiding structure for ${}_H^H\mathcal{YD}$. This is tantamount to showing that $\sigma_{L,k} = \mathrm{I}_L = \sigma_{k,L}$, $\sigma_{L\otimes M,N} = (\sigma_{L,N}\otimes \mathrm{I}_M)\circ(\mathrm{I}_L\otimes\sigma_{M,N})$, and $\sigma_{L,M\otimes N} = (\mathrm{I}_M\otimes\sigma_{L,N})\circ(\sigma_{L,M}\otimes\mathrm{I}_N)$ for all left Yetter–Drinfel'd H-modules L, M, and N. Here, as is usually done, we suppress the identifications involving α, λ, and ρ in k-Vec. Unless otherwise specified we will regard ${}_H^H\mathcal{YD}$ as a prebraided monoidal category with the prebraiding structure given by the $\sigma_{M,N}$'s.

When the bialgebra H^{op} has antipode ς the $\sigma_{M,N}$'s form a braiding structure for ${}_H^H\mathcal{YD}$. For let $M, N \in {}_H^H\mathcal{YD}$. Then $\sigma_{M,N}$ is a linear isomorphism with inverse given by $\sigma_{M,N}^{-1}(n\otimes m) = m_{(0)}\otimes\varsigma(m_{(-1)}){\cdot}n$ for all $n \in N$ and $m \in M$.

Let (A, m_A, η_A) and (B, m_B, η_B) be algebras in ${}_H^H\mathcal{YD}$, that is A and B are H-module and H-comodule k-algebras. We write $A\underline{\otimes}B$ for the tensor product of algebras in ${}_H^H\mathcal{YD}$ to distinguish it from the tensor product in k-Vec and we write $a\underline{\otimes}b$ for $a\otimes b$. Observe that

$$1_{A\underline{\otimes}B} = 1_A\underline{\otimes}1_B \text{ and } (a\underline{\otimes}b)(a'\underline{\otimes}b') = a(b_{(-1)}{\cdot}a')\underline{\otimes}b_{(0)}b' \tag{11.18}$$

for all $a, a' \in A$ and $b, b' \in B$. Let A^{op} denote the algebra $(A, m_A \circ \sigma_{A,A}, \eta_A)$ in ${}_H^H\mathcal{YD}$ and write $a \underline{\cdot} b$ for the product of $a, b \in A^{\mathsf{op}}$. Then

$$a \underline{\cdot} b = (a_{(-1)} \cdot b) a_{(0)}. \tag{11.19}$$

Thus $A = A^{\mathsf{op}}$ if and only if $ab = (a_{(-1)} \cdot b) a_{(0)}$ for all $a, b \in A$. Now that we have a notion of "opposite product" we are in a position to define a notion of commutator of elements in algebras in ${}_H^H\mathcal{YD}$.

Definition 11.6.6. Let H be a bialgebra over the field k and let A be an algebra in ${}_H^H\mathcal{YD}$. The *braided commutator of* $a, b \in A$ is $[a, b]_{\mathsf{c}} = ab - a \underline{\cdot} b = ab - (a_{(-1)} \cdot b) a_{(0)}$ and $\mathrm{ad}_{\mathsf{c}} a$ is the endomorphism of A defined by $\mathrm{ad}_{\mathsf{c}} a(x) = [a, x]_{\mathsf{c}}$ for all $x \in A$.

Now let $(C, \Delta_C, \epsilon_C)$ and $(D, \Delta_D, \epsilon_D)$ be coalgebras in ${}_H^H\mathcal{YD}$, that is C and D are H-module and H-comodule k-coalgebras. Using the same notation for the tensor product of C and D in ${}_H^H\mathcal{YD}$ as we did for the tensor product of algebras we have

$$\epsilon(c \underline{\otimes} d) = \epsilon(c)\epsilon(d) \text{ and}$$

$$\Delta(c \underline{\otimes} d) = (c_{(1)} \underline{\otimes} c_{(2)(-1)} \cdot d_{(1)}) \otimes (c_{(2)(0)} \underline{\otimes} d_{(2)}) \tag{11.20}$$

for all $c \in C$ and $d \in D$. Let C^{cop} denote the coalgebra $(C, \sigma_{C,C} \circ \Delta_C, \epsilon_C)$ and let Δ^{cop} denote its coproduct. Then

$$\Delta^{\mathsf{cop}}(c) = c_{(1)(-1)} \cdot c_{(2)} \underline{\otimes} c_{(1)(0)}. \tag{11.21}$$

Thus $C = C^{\mathsf{cop}}$ if and only if $c_{(1)} \otimes c_{(2)} = c_{(1)(-1)} \cdot c_{(2)} \underline{\otimes} c_{(1)(0)}$ for all $c \in C$.

A bialgebra in ${}_H^H\mathcal{YD}$ is a tuple $(B, m, \eta, \Delta, \epsilon)$, where (B, m, η) is an algebra in ${}_H^H\mathcal{YD}$ and (B, Δ, ϵ) is a coalgebra in ${}_H^H\mathcal{YD}$, such that the morphisms $\epsilon : B \longrightarrow k$ and $\Delta : B \longrightarrow B \underline{\otimes} B$ are algebra maps. That Δ is an algebra map can be expressed

$$\Delta(1) = 1 \underline{\otimes} 1 \quad \text{and} \quad \Delta(bb') = b_{(1)}(b_{(2)(-1)} \cdot b'_{(1)}) \underline{\otimes} b_{(2)(0)} b'_{(2)} \tag{11.22}$$

for all $b, b' \in B$. A Hopf algebra in ${}_H^H\mathcal{YD}$ is a bialgebra B in ${}_H^H\mathcal{YD}$ with a morphism $S : B \longrightarrow B$ which satisfies (7.1). A Hopf algebra B in ${}_H^H\mathcal{YD}$ has unique such morphism S which is called the antipode of B.

Bialgebras and Hopf algebras in ${}_H^H\mathcal{YD}$ give rise to bialgebras and Hopf algebras in k-Vec, that is bialgebras and Hopf algebras over k. Let $(B, m, \eta, \Delta, \epsilon)$ be a bialgebra in ${}_H^H\mathcal{YD}$. Then (B, m, η) is an algebra in ${}_H\mathcal{M}$ since m and η are module maps. Likewise (B, Δ, ϵ) is a coalgebra in ${}^H\mathcal{M}$ since Δ and ϵ are comodule maps. In particular $B \# H$ and $B \natural H$ are defined. This algebra and coalgebra have the same underlying vector space $B \otimes H$.

Theorem 11.6.7. *Let H be a bialgebra over the field k and suppose that B is a bialgebra in ${}^H_H\mathcal{YD}$. Then:*

(a) $(B{\otimes}H, m_{B\#H}, \eta_{B\#H}, \Delta_{B\natural H}, \epsilon_{B\natural H})$ *is a k-bialgebra.*

(b) *The maps $B{\times}H \overset{\jmath}{\underset{\pi}{\rightleftarrows}} H$ defined by $\pi(b{\otimes}h) = \epsilon(b)h$ and $\jmath(h) = 1{\otimes}h$ respectively for all $b \in B$ and $h \in H$ are k-bialgebra maps which satisfy $\pi{\circ}\jmath = \mathrm{I}_H$.*

Proof. By Lemma 5.1.1, to show part (a) it suffices to show that $\epsilon : B\#H \longrightarrow k$ and $\Delta_{B\#H} : B\#H \longrightarrow (B\#H){\otimes}(B\#H)$ are algebra maps. For $b \in B$ and $h \in H$ we are using the notation $b\natural h = b{\otimes}h = b\#h$. Recall that $\epsilon_{B\natural H}(b\#h) = \epsilon(b)\epsilon(h)$ and $\Delta_{B\natural H}(b\#h) = (b_{(1)}{\otimes}b_{(2)(-1)}h_{(1)}){\otimes}(b_{(2)(0)}{\otimes}h_{(2)})$ for all $b \in B$ and $h \in H$. For $b, b' \in B$ and $h, h' \in H$ recall that $(b\#h)(b'\#h') = b(h_{(1)}{\cdot}b')\#h_{(2)}h'$. It is easy to see that $\epsilon_{B\natural H}$ is an algebra map and that $\Delta_{B\natural H}(1_{B\#H}) = 1_{B\#H}{\otimes}1_{B\#H}$. To complete the proof of part (a), it suffices to show that $\Delta_{B\natural H}$ is multiplicative. Set $\Delta = \Delta_{B\natural H}$. For all $b, c \in B$ and $h, \ell \in H$ we have

$$\begin{aligned}
&\Delta((b\#h)(c\#\ell))\\
&= \Delta(b(h_{(1)}{\cdot}c)\#h_{(2)}\ell)\\
&= \big((b(h_{(1)}{\cdot}c))_{(1)}\#(b(h_{(1)}{\cdot}c))_{(2)(-1)}(h_{(2)}\ell)_{(1)}\big) \otimes \big((b(h_{(1)}{\cdot}c))_{(2)(0)}\#(h_{(2)}\ell)_{(2)}\big)\\
&= \big(b_{(1)}(b_{(2)(-1)}{\cdot}(h_{(1)}{\cdot}c)_{(1)})\#(b_{(2)(0)}(h_{(1)}{\cdot}c)_{(2)})_{(-1)}(h_{(2)}\ell)_{(1)}\big)\\
&\qquad \otimes \big((b_{(2)(0)}(h_{(1)}{\cdot}c)_{(2)})_{(0)}\#(h_{(2)}\ell)_{(2)}\big)\\
&= \big(b_{(1)}(b_{(2)(-1)}{\cdot}(h_{(1)(1)}{\cdot}c_{(1)}))\#b_{(2)(0)(-1)}(h_{(1)(2)}{\cdot}c_{(2)})_{(-1)}(h_{(2)(1)}\ell_{(1)})\big)\\
&\qquad \otimes \big(b_{(2)(0)(0)}(h_{(1)(2)}{\cdot}c_{(2)})_{(0)}\#h_{(2)(2)}\ell_{(2)}\big)\\
&= \big(b_{(1)}((b_{(2)(-1)}h_{(1)}){\cdot}c_{(1)})\#b_{(2)(0)(-1)}(h_{(2)}{\cdot}c_{(2)})_{(-1)}h_{(3)}\ell_{(1)}\big)\\
&\qquad \otimes \big(b_{(2)(0)(0)}(h_{(2)}{\cdot}c_{(2)})_{(0)}\#h_{(4)}\ell_{(2)}\big)\\
&= \big(b_{(1)}((b_{(2)(-1)}h_{(1)}){\cdot}c_{(1)})\#b_{(2)(0)(-1)}(h_{(2)}c_{(2)(-1)})\ell_{(1)}\big)\\
&\qquad \otimes \big(b_{(2)(0)(0)}(h_{(3)}{\cdot}c_{(2)(0)})\#h_{(4)}\ell_{(2)}\big)
\end{aligned}$$

and

$$\begin{aligned}
&\Delta(b\#h)\Delta(c\#\ell)\\
&= \big((b_{(1)}\#b_{(2)(-1)}h_{(1)}){\otimes}(b_{(2)(0)}\#h_{(2)})\big)\big((c_{(1)}\#c_{(2)(-1)}\ell_{(1)}){\otimes}(c_{(2)(0)}\#\ell_{(2)})\big)\\
&= (b_{(1)}\#b_{(2)(-1)}h_{(1)})(c_{(1)}\#c_{(2)(-1)}\ell_{(1)}){\otimes}(b_{(2)(0)}\#h_{(2)})(c_{(2)(0)}\#\ell_{(2)})\\
&= \big(b_{(1)}((b_{(2)(-1)}h_{(1)})_{(1)}{\cdot}c_{(1)})\#(b_{(2)(-1)}h_{(1)})_{(2)}c_{(2)(-1)}\ell_{(1)}\big)\\
&\qquad \otimes \big(b_{(2)(0)}(h_{(2)(1)}{\cdot}c_{(2)(0)})\#h_{(2)(2)}\ell_{(2)}\big)
\end{aligned}$$

$$= \left(b_{(1)}((b_{(2)(-1)(1)}h_{(1)(1)})\cdot c_{(1)})\#b_{(2)(-1)(2)}h_{(1)(2)}c_{(2)(-1)}\ell_{(1)}\right)$$
$$\otimes \left(b_{(2)(0)}(h_{(2)(1)}\cdot c_{(2)(0)})\#h_{(2)(2)}\ell_{(2)}\right)$$
$$= \left(b_{(1)}((b_{(2)(-1)}h_{(1)})\cdot c_{(1)})\#b_{(2)(0)(-1)}h_{(2)}c_{(2)(-1)}\ell_{(1)}\right)$$
$$\otimes \left(b_{(2)(0)(0)}(h_{(3)}\cdot c_{(2)(0)})\#h_{(4)}\ell_{(2)}\right).$$

Part (b) is left as an exercise to the reader. □

Definition 11.6.8. The bialgebra $B\times H$ of Theorem 11.6.7 is the *biproduct of B and H.*

For $b \in B$ and $h \in H$ we let $b\times h$ stand for $b\otimes h = b\#h = b\natural h$.

We continue with the notation of Theorem 11.6.7 and set $A = B\times H$. Suppose that the biproduct $B\times H$ has an antipode S and let S_H be the endomorphism of H defined by $S_H = \pi\circ S\circ\jmath$. Since π is an algebra map and $\jmath$ is a coalgebra map, for $h \in H$ we have

$$\begin{aligned} h_{(1)}S_H(h_{(2)}) &= (\pi\circ\jmath)(h_{(1)})(\pi\circ S\circ\jmath)(h_{(2)}) \\ &= \pi(\jmath(h)_{(1)}S(\jmath(h)_{(2)})) \\ &= \pi(\epsilon(\jmath(h)1)) \\ &= \epsilon(h)1 \end{aligned}$$

and likewise $S_H(h_{(1)})h_{(2)} = \epsilon(h)1$. Thus S_H is an antipode for H.

Let $\pi_B : B\times H \longrightarrow B$ be defined by $\pi_B(b\otimes h) = b\epsilon(h)$ for all $b \in B$ and $h \in H$. Observe that $\pi_B((b\times h)(b'\times h')) = b\pi_B((1\times h)(b'\times h'))$ for all $b, b' \in B$ and $h, h' \in H$. Let S_B be the endomorphism of B defined by $S_B(b) = \pi_B((1\times b_{(-1)})S(b_{(0)}\times 1))$ for all $b \in B$. Applying π_B to first and last expressions of

$$\epsilon(b)1_{B\times H} = (b_{(1)}\times b_{(2)(-1)})S(b_{(2)(0)}\times 1) = (b_{(1)}\times 1)(1\times b_{(2)(-1)})S(b_{(2)(0)}\times 1)$$

yields $\epsilon(b)1_B = b_{(1)}S_B(b_{(2)})$. Thus S_B is a right inverse for I_B.

To continue, endow the subalgebra $B\times 1 = \{b\times 1 \mid b \in B\}$ of A with the coalgebra structure induced by the linear isomorphism $B \simeq B\times 1$ given by $b \mapsto b\times 1$ for all $b \in B$. Then $S_{B\times 1} : B\times 1 \longrightarrow A$ defined by $S_{B\times 1}(b\times 1) = S_B(b)\times 1$ for all $b \in B$ is a right inverse for the inclusion $\imath_{B\times 1} : B\times 1 \longrightarrow A$ in the convolution algebra $\mathrm{Hom}(B\times 1, A)$. To show that S_B is a left inverse for I_B it suffices to find a left inverse $S' : B\times 1 \longrightarrow A$ for $\imath_{B\times 1}$ in the convolution algebra $\mathrm{Hom}(B\times 1, A)$. For then S_B is an inverse for I_B in the convolution algebra $\mathrm{End}(B)$ and must be an antipode for B in light of Exercise 11.6.13.

Let S' be the endomorphism of A defined by $S' = (\jmath{\circ}\pi){*}S$. Then $S'{*}\mathrm{I}_A = \jmath{\circ}\pi$; thus $S' = (\jmath{\circ}\pi){*}S$. Let $b \in B$ and $h \in H$. The calculation

$$\begin{aligned}
S'(b{\times}h) &= (\jmath{\circ}\pi)(b_{(1)}{\times}b_{(2)(-1)}h_{(1)})S(b_{(2)(0)}{\times}h_{(2)}) \\
&= (\epsilon(b_{(1)})1{\times}b_{(2)(-1)}h_{(1)})S(b_{(2)(0)}{\times}h_{(2)}) \\
&= (1{\times}b_{(-1)}h_{(1)})S(b_{(0)}{\times}h_{(2)}) \\
&= (1{\times}b_{(-1)})(1{\times}h_{(1)})S((b_{(0)}{\times}1)(1{\times}h_{(2)})) \\
&= (1{\times}b_{(-1)})(1{\times}h_{(1)})S(1{\times}h_{(2)})S(b_{(0)}{\times}1) \\
&= (1{\times}b_{(-1)})S(b_{(0)}{\times}1)\epsilon(h)
\end{aligned}$$

shows that $S'(b{\times}h) = S'(b{\times}1)\epsilon(h)$. Thus

$$\begin{aligned}
S'(b_{(1)}{\times}1)(b_{(2)}{\times}1) &= S'(b_{(1)}{\times}\epsilon(b_{(2)(-1)})1)(b_{(2)(0)}{\times}1) \\
&= S'(b_{(1)}{\times}b_{(2)(-1)})(b_{(2)(0)}{\times}1) \\
&= (S'{*}\mathrm{I}_A)(b{\times}1) \\
&= (\jmath{\circ}\pi)(b{\times}1) \\
&= \epsilon(b)1{\times}1
\end{aligned}$$

which means that $S'|B{\times}1$ is a left inverse for $\imath_{B\times 1}$ in the convolution algebra $\mathrm{Hom}(B{\times}1, A)$. We have established part (a) of the following:

Theorem 11.6.9. *Let H be a bialgebra over the field k and suppose that B is a bialgebra in ${}^H_H\mathcal{YD}$. Then:*

(a) *If the biproduct $B{\times}H$ has an antipode then H and B have antipodes.*
(b) *Suppose that B and H have antipodes S_B and S_H respectively. Then $B{\times}H$ has an antipode S which is given by $S(b{\times}h) = (1{\times}S_H(b_{(-1)}h))(S_B(b_{(0)}){\times}1)$ for all $b \in B$ and $h \in H$.*

Proof. We need only establish part (b) which is a straightforward exercise left to the reader. □

Assume further that H has antipode S_H. We will describe B in terms of S_H, π, and $\jmath$. Again write $A = B{\times}H$ and let Π be the endomorphisms of A defined by the convolution product $\Pi = \mathrm{I}_A{*}(\jmath{\circ}S_H{\circ}\pi)$. For $b \in B$ and $h \in H$ we have

$$\begin{aligned}
\Pi(b{\times}h) &= (b_{(1)}{\times}b_{(2)(-1)}h_{(1)})(\jmath{\circ}S_H{\circ}\pi)(b_{(2)(0)}{\times}h_{(2)}) \\
&= (b_{(1)}{\times}b_{(2)(-1)}h_{(1)})(1{\times}\epsilon(b_{(2)(0)})S_H(h_{(2)})) \\
&= b{\times}h_{(1)}S_H(h_{(2)}) \\
&= \epsilon(h)(b{\times}1).
\end{aligned}$$

Identifying B with $B\times 1$ we have that $\mathrm{Im}(\Pi) = B$.

Suppose that H is a Hopf algebra with antipode S over k and A is a bialgebra over k such that there are bialgebra maps $A \underset{\pi}{\overset{\jmath}{\leftrightarrows}} H$ which satisfy $\pi\circ\jmath = \mathrm{I}_H$. Let $\Pi = \mathrm{I}_A*(\jmath\circ S\circ\pi)$ and let $B = \mathrm{Im}(\Pi)$. In the next section we characterize A as a biproduct $B\times H$.

Exercises

In the following exercises H is a bialgebra over k.

Exercise 11.6.1. Complete the proof of part (b) of Theorem 11.6.9.

Exercise 11.6.2. Let $\{M_i\}_{i\in I}$ be a family of objects of ${}_H^H\mathcal{YD}$. Show that $M = \bigoplus_{i\in I} M_i$ is an object of ${}_H^H\mathcal{YD}$ with the direct sum module and comodule structures.

Exercise 11.6.3. Let $M_1, \ldots, M_r$ be objects of ${}_H^H\mathcal{YD}$. Show that $M_1\otimes\cdots\otimes M_r$ is an object of ${}_H^H\mathcal{YD}$, where

$$h\cdot(m_1\otimes\cdots\otimes m_r) = (h_{(1)}\cdot m_1)\otimes\cdots\otimes(h_{(r)}\cdot m_r)$$

and

$$\rho(m_1\otimes\cdots\otimes m_r) = m_{1(-1)}\cdots m_{r(-1)}\otimes(m_{1(0)}\otimes\cdots\otimes m_{r(0)})$$

for all $m_i \in M_i$, $1 \leq i \leq r$.

Exercise 11.6.4. Let G be a group and $H = k[G]$ be the group algebra of G over k. Suppose M has a left H-module structure $(M, \cdot)$ and a left H-comodule structure (M, ρ). Write $M = \bigoplus_{g\in G} M_g$, where $M_g = \{m \in M \mid \rho(m) = g\otimes m\}$. See Exercise 3.2.12. Show that $(M, \cdot, \rho)$ is an object of ${}_H^H\mathcal{YD}$ if and only if $h\cdot M_g \subseteq M_{hgh^{-1}}$ for all $h, g \in G$.

Exercise 11.6.5. We expand on Exercises 11.2.5 and 11.3.9. Let A be a bialgebra in ${}_H^H\mathcal{YD}$. Show that:

(a) If $(M, \bullet)$ and $(N, \bullet)$ are objects of ${}_A\underline{\mathcal{M}}$ then $M\otimes N$ is as well, where

$$a\bullet(m\otimes n) = a_{(1)}\bullet(a_{(2)(-1)}\cdot m)\otimes a_{(2)(0)}\bullet n$$

for all $a \in A$, $m \in M$, and $n \in N$.

(b) If M and N are objects of ${}^A\underline{\mathcal{M}}$ then so is $M\otimes N$, where

$$\varrho(m\otimes n) = m_{[-1]}(m_{[0](-1)}\cdot n_{[-1]})\otimes(m_{[0](0)}\otimes n_{[0]})$$

for all $m \in M$ and $n \in N$.

We are in position to define Hopf module for a bialgebra in ${}^H_H\mathcal{YD}$.

Definition 11.6.10. Let H be a bialgebra over k and let A be a bialgebra in ${}^H_H\mathcal{YD}$. A *left A-Hopf module in* ${}^H_H\mathcal{YD}$ is a triple $(M, \bullet, \varrho)$, where $(M, \bullet)$ is an object of ${}_A\underline{\mathcal{M}}$, (M, ϱ) is an object of ${}^A\underline{\mathcal{M}}$, such that $\varrho : M \longrightarrow A{\otimes}M$ is a module morphism, that is

$$\varrho(a{\bullet}m) = a_{(1)}(a_{(2)(-1)}{\cdot}m_{[-1]}){\otimes}a_{(2)(0)}{\bullet}m_{[0]}$$

for all $a \in A$ and $m \in M$.

Exercise 11.6.6. Let A be a bialgebra in ${}^H_H\mathcal{YD}$ and M be an object of ${}^H_H\mathcal{YD}$. Suppose $(M, \bullet)$ is an object of ${}_A\underline{\mathcal{M}}$ and $(M\varrho)$ is an object of ${}^A\underline{\mathcal{M}}$. Show that $(M, \bullet, \varrho)$ is a left A-Hopf module in ${}^H_H\mathcal{YD}$ if and only if $M{\otimes}H$ is a left $A{\times}H$-Hopf module, with left $A\#H$-module structure $M\#H$ and left $A\natural H$-comodule structure $M\natural H$.

Exercise 11.6.7. Let A be a Hopf algebra in ${}^H_H\mathcal{YD}$ and suppose $(M, \bullet, \varrho)$ is a left A-Hopf module in ${}^H_H\mathcal{YD}$. Show that $M_{\mathbf{co\,inv}} = \{m \in M \mid \varrho(m) = 1{\otimes}m\}$ is an object of ${}^H_H\mathcal{YD}$ and that M is a free left A-module with basis any linear basis of $M_{\mathbf{co\,inv}}$. [Hint: Try to mimic the proof of Theorem 8.2.3.]

Exercise 11.6.8. Let G be a finite abelian group, suppose k is algebraically closed and has characteristic 0, and let $H = k[G]$ be the group algebra of G over k. Suppose $(M, \cdot, \rho)$ is an object of ${}^H_H\mathcal{YD}$. Show that $M = \bigoplus_{g\in G, \chi\in\widehat{G}} M_g^\chi$, where

$$M_g^\chi = M_g{\cap}M^\chi, \quad M_g = \{m \in M \mid \rho(m) = g{\otimes}m\}, \quad \text{and}$$

$$M^\chi = \{m \in M \mid g{\cdot}m = \chi(g)m \ \text{ for all } \ g \in G\}.$$

See Exercises 11.6.4 and 7.4.6. Note that M_g^χ is an object of ${}^H_H\mathcal{YD}$.

Exercise 11.6.9. Let G be any abelian group and $H = k[G]$ be the group algebra of G over k. Suppose M has a left H-module structure $(M, \cdot)$ and a left H-comodule structure (M, ρ) such that $M = \bigoplus_{g\in G, \chi\in\widehat{G}} M_g^\chi$, where the summands are defined as in Exercise 11.6.8. Show that $(M, \cdot, \rho)$ is an object of ${}^H_H\mathcal{YD}$.

Exercise 11.6.10. Suppose H has an antipode and $M \in {}^H_H\mathcal{YD}$. Show that:

(a) If N is a subcomodule of M then $H{\cdot}N$ is a Yetter–Drinfel'd submodule of M.

(b) If M is a finitely generated Yetter–Drinfel'd module then M is a finitely generated H-module.

[Hint: See (11.17).]

Exercise 11.6.11. Let A be a Hopf algebra over k. Show that $(A, \mathrm{ad}_L, \Delta)$ and also $(A, m, \mathrm{co\,ad}_L)$ are objects of ${}_A^A\mathcal{YD}$.

Exercise 11.6.12. Let A and B be bialgebras over k and $A \overset{\jmath}{\underset{\pi}{\leftrightarrows}} B$ be bialgebra maps such that $\pi \circ \jmath = \mathrm{I}_B$. Suppose that $(M, \cdot, \rho) \in {}_A^A\mathcal{YD}$. Show that $(M, \cdot_\jmath, \rho_\pi) \in {}_B^B\mathcal{YD}$, where $b \cdot_\jmath m = \jmath(b) \cdot m$ and $\rho_\pi(m) = \pi(m_{(-1)}) \otimes m_{(0)}$ for all $b \in B$ and $m \in M$.

Exercise 11.6.13. Suppose that B is a bialgebra in ${}_H^H\mathcal{YD}$. Assume further that I_B has a convolution inverse S in $\mathrm{End}(B)$.

(a) Show that $S(h \cdot b) = h \cdot S(b)$ for all $h \in H$ and $b \in B$. [Hint: Give $H \otimes B$ the tensor product coalgebra structure over k and consider $\ell, f, r : H \otimes B \longrightarrow B$ defined by

$$\ell(h \otimes b) = h \cdot S(b), \qquad f(h \otimes b) = h \cdot b, \qquad \text{and} \quad r(h \otimes b) = S(h \cdot b)$$

for all $h \in H$ and $b \in B$. Show that ℓ is a left inverse for f and r is a right inverse for f in the convolution algebra $\mathrm{Hom}(H \otimes B, B)$.]

(b) Show that $S(b)_{(-1)} \otimes S(b)_{(0)} = b_{(-1)} \otimes S(b_{(0)})$ for all $b \in B$. [Hint: Give $H \otimes B$ the tensor product algebra structure and consider $L, F, R : B \longrightarrow H \otimes B$ defined by

$$L(b) = b_{(-1)} \otimes S(b_{(0)}), \qquad F(b) = b_{(-1)} \otimes b_{(0)}, \quad \text{and} \quad R(b) = S(b)_{(-1)} \otimes S(b)_{(0)}$$

for all $b \in B$. Show that L is a left inverse for F and R is a right inverse for F in the convolution algebra $\mathrm{Hom}(B, H \otimes B)$.]

By virtue of parts (a) and (b) we conclude that S is an antipode for the bialgebra B in the category ${}_H^H\mathcal{YD}$.

Exercise 11.6.14. Let A be a bialgebra over k, let H be a Hopf algebra with antipode S, and suppose that $A \overset{\jmath}{\underset{\pi}{\leftrightarrows}} H$ are bialgebra maps which satisfy $\pi \circ \jmath = \mathrm{I}_H$.

(a) Show that $(A, \cdot, \rho) \in {}_H^H\mathcal{YD}$, where $h \cdot a = \jmath(h_{(1)}) a \jmath(S(h_{(2)}))$ and $\rho(a) = \pi(a_{(1)}) \otimes a_{(2)}$ for all $h \in H$ and $a \in A$.

(b) Show that $(A, \cdot, \rho) \in {}_H^H\mathcal{YD}$, where $h \cdot a = \jmath(h) a$ and $\rho(a) = \pi(a_{(1)}) S(\pi(a_{(3)})) \otimes a_{(2)}$ for all $h \in H$ and $a \in A$.

Exercise 11.6.15. Let $A \overset{\jmath}{\underset{\pi}{\leftrightarrows}} B$ be maps of bialgebras maps which satisfy $\pi \circ \jmath = \mathrm{I}_B$. Show that:

(a) (A, μ) is a left B-module, where $b \cdot a = \jmath(b) a$ for all $b \in B$ and $a \in A$;

(b) (A, ρ) is a left B-comodule, where $\rho(a) = \pi(a_{(1)}) \otimes a_{(2)}$ for all $a \in A$; and

(c) (A, μ, ρ) is a left B-Hopf module and $A_{co\,inv} = \{a \in A \,|\, \pi(a_{(1)}) \otimes a_{(2)} = 1 \otimes a\}$.

Exercise 11.6.16. Let $\pi : C \longrightarrow D$ be a map of coalgebras, suppose that C, D are also left H-comodules and π is a map of left H-comodules. Show that if C is a left H-comodule coalgebra then so is D.

Exercise 11.6.17. Let A, B be bialgebras in ${}^H_H\mathcal{YD}$ and $f : A \longrightarrow B$ be a morphism of bialgebras. Show that:

(a) $f \times \mathrm{I}_H : A \times H \longrightarrow B \times H$ defined by $(f \times \mathrm{I}_H)(a \times h) = f(a) \times h$ for all $a \in A$ and $h \in H$ is a map of k-bialgebras.

(b) $A \mapsto A \times H$ and $f \mapsto f \times \mathrm{I}_H$ describes a functor from the category of bialgebras in ${}^H_H\mathcal{YD}$ and their morphisms to k-Bialg.

[Hint: See Exercises 11.2.4 and 11.3.7.]

In the next four exercises we consider the pre-braided monoidal category ${}^H_H\mathcal{YD}$ and variations. To do this we introduce two notation conventions.

Let C be a coalgebra over k and suppose (M, ρ) is a left C^{cop}-comodule. Then (M, ρ_r) is a right C-comodule, where $\rho_r = \tau_{C,M} \circ \rho$. This equation can be expressed as $m_{(0)} \otimes m_{(1)} = m_{(0)} \otimes m_{(-1)}$ for all $m \in M$.

Let A be an algebra over k and suppose (M, μ) is a left A^{op}-module. Then (M, μ_r) is a right A-module, where $\mu_r = \mu \circ \tau_{A,M}$. This equation can be expressed by $m \cdot_r a = \mu_r(m \otimes a) = \mu(a \otimes m) = a \cdot m$ for all $a \in A$ and $m \in M$.

Exercise 11.6.18. Recall the ${}^H_H\mathcal{YD}$ is the category described as follows: objects are triples $(M, \cdot, \rho)$, where $(M, \cdot) \in {}_H\mathcal{M}$, $(M, \rho) \in {}^H\mathcal{M}$, and

$$h_{(1)} m_{(-1)} \otimes h_{(2)} \cdot m_{(0)} = (h_{(1)} \cdot m)_{(-1)} h_{(2)} \otimes (h_{(1)} \cdot m)_{(0)} \tag{11.23}$$

for all $h \in H$ and $m, n \in M$, and morphisms $f : (M, \cdot, \rho) \longrightarrow (M', \cdot', \rho')$ are linear maps $f : M \longrightarrow M'$ which are module and comodule maps. Furthermore

(1) $k \in {}^H_H\mathcal{YD}$ with $h \cdot 1_k = \epsilon(h)$ for $h \in H$ and $\rho(1_k) = 1_H \otimes 1_k$

and for $M, N \in {}^H_H\mathcal{YD}$,

(2) the tensor product $M \otimes N \in {}^H_H\mathcal{YD}$, where

$$h \cdot (m \otimes n) = h_{(1)} \cdot m \otimes h_{(2)} \cdot n \text{ and } \rho(m \otimes n) = m_{(-1)} n_{(-1)} \otimes (m_{(0)} \otimes n_{(0)})$$

for all $h \in H$, $m \in M$, and $n \in N$ and

(3) $\sigma_{M,N} : M \otimes N \longrightarrow N \otimes M$, defined by $\sigma_{M,N}(m \otimes n) = m_{(-1)} \cdot n \otimes m_{(0)}$ for all $m \in M$ and $n \in N$, is a morphism.

Show that ${}_H^H\mathcal{YD}$ with the structures defined in (1) and (2) is a monoidal category with prebraiding which is the family of $\sigma_{M,N}$'s described in (3).

Exercise 11.6.19. Let ${}_H\mathcal{YD}^H$ be the category described as follows: objects are triples $(M,\cdot,\rho)$, where $(M,\cdot) \in {}_H\mathcal{M}$, $(M,\rho) \in \mathcal{M}^H$, and

$$h_{(1)}\cdot m_{(0)} \otimes h_{(2)} m_{(1)} = (h_{(2)}\cdot m)_{(0)} \otimes (h_{(2)}\cdot m)_{(1)} h_{(1)} \tag{11.24}$$

for all $h \in H$ and $m, n \in M$, and morphisms $f : (M,\cdot,\rho) \longrightarrow (M',\cdot',\rho')$ are linear maps $f : M \longrightarrow M'$ which are module and comodule maps.

(1) Define a left H-module a right H-comodule structure on k by $h\cdot 1_k = \epsilon(h)$ for $h \in H$ and $\rho(1_k) = 1_k \otimes 1_H$.

For $M, N \in {}_H\mathcal{YD}^H$:

(2) Define a left H-module and a right H-comodule structure on $M\otimes N$ by

$$h\cdot(m\otimes n) = h_{(2)}\cdot m \otimes h_{(1)}\cdot n \quad \text{and} \quad \rho(m\otimes n) = (m_{(0)} \otimes n_{(0)}) \otimes m_{(1)} n_{(1)}$$

for all $h \in H$, $m \in M$, and $n \in N$ and

(3) Define $\sigma_{M,N} : M\otimes N \longrightarrow N \otimes M$ by $\sigma_{M,N}(m\otimes n) = m_{(1)}\cdot n \otimes m_{(0)}$ for all $m \in M$ and $n \in N$.

Show that ${}_H\mathcal{YD}^H$ with the structures defined in (1) and (2) is a monoidal category with prebraiding which is the family of $\sigma_{M,N}$'s described in (3) and show that there is an isomorphism of prebraided monoidal categories ${}_{H^{cop}}^{H^{cop}}\mathcal{YD} \longrightarrow {}_H\mathcal{YD}^H$ given by

$$(M,\cdot,\rho) \mapsto (M,\cdot,\rho_r) \text{ and } f \mapsto f.$$

Exercise 11.6.20. Let ${}^H\mathcal{YD}_H$ be the category described as follows: objects are triples $(M,\cdot,\rho)$, where $(M,\cdot) \in \mathcal{M}_H$, $(M,\rho) \in {}^H\mathcal{M}$, and

$$m_{(-1)} h_{(1)} \otimes m_{(0)}\cdot h_{(2)} = h_{(2)} (m\cdot h_{(1)})_{(-1)} \otimes (m\cdot h_{(1)})_{(0)} \tag{11.25}$$

for all $h \in H$ and $m, n \in M$, and morphisms $f : (M,\cdot,\rho) \longrightarrow (M',\cdot',\rho')$ are linear maps $f : M \longrightarrow M'$ which are module and comodule maps.

(1) Define a right H-module a left H-comodule structure on k by $1_k\cdot h = \epsilon(h)$ for $h \in H$ and $\rho(1_k) = 1_H \otimes 1_k$.

For $M, N \in {}^H\mathcal{YD}_H$:

(2) Define a right H-module and a left H-comodule structure on $M\otimes N$ by

$$(m\otimes n)\cdot h = m\cdot h_{(1)} \otimes n\cdot h_{(2)} \quad \text{and} \quad \rho(m\otimes n) = m_{(-1)} n_{(-1)} \otimes (m_{(0)} \otimes n_{(0)})$$

for all $h \in H$, $m \in M$, and $n \in N$.

(3) Define $\sigma_{M,N} : M\otimes N \longrightarrow N\otimes M$ by $\sigma_{M,N}(m\otimes n) = n\cdot m_{(-1)}\otimes m_{(0)}$ for all $m \in M$ and $n \in N$.

Show that ${}^H\mathcal{YD}_H$ with the structures defined in (1) and (2) is a monoidal category with prebraiding which is the family of $\sigma_{M,N}$'s described in (3) and show that there is an isomorphism of prebraided monoidal categories ${}^{H^{op}}_{H^{op}}\mathcal{YD} \longrightarrow {}^H\mathcal{YD}_H$ given by

$$(M, \cdot, \rho) \mapsto (M, \cdot_r, \rho) \text{ and } f \mapsto f.$$

Exercise 11.6.21. Let $\mathcal{YD}^H_H$ be the category described as follows: objects are triples $(M, \cdot, \rho)$, where $(M, \cdot) \in \mathcal{M}_H$, $(M, \rho) \in \mathcal{M}^H$, and

$$m_{(0)}\cdot h_{(1)}\otimes m_{(1)}h_{(2)} = (m\cdot h_{(2)})_{(0)}\otimes h_{(1)}(m\cdot h_{(2)})_{(1)} \tag{11.26}$$

for all $h \in H$ and $m, n \in M$, and morphisms $f : (M, \cdot, \rho) \longrightarrow (M', \cdot', \rho')$ are linear maps $f : M \longrightarrow M'$ which are module and comodule maps.

(1) Define a right H-module a right H-comodule structure on k by $1_k\cdot h = \epsilon(h)$ for $h \in H$ and $\rho(1_k) = 1_k\otimes 1_H$.

For $M, N \in \mathcal{YD}^H_H$:

(2) Define a right H-module and a left H-comodule structure on $M\otimes N$ by

$$(m\otimes n)\cdot h = m\cdot h_{(2)}\otimes n\cdot h_{(1)} \quad\text{and}\quad \rho(m\otimes n) = (m_{(0)}\otimes n_{(0)})\otimes n_{(1)}m_{(1)}$$

for all $h \in H$, $m \in M$, and $n \in N$.

(3) Define $\sigma_{M,N} : M\otimes N \longrightarrow N\otimes M$ by $\sigma_{M,N}(m\otimes n) = n\cdot m_{(1)}\otimes m_{(0)}$ for all $m \in M$ and $n \in N$.

Show that $\mathcal{YD}^H_H$ with the structures defined in (1) and (2) is a monoidal category with prebraiding which is the family of $\sigma_{M,N}$'s described in (3) and show that there is an isomorphism of prebraided monoidal categories ${}^{H^{op\,cop}}_{H^{op\,cop}}\mathcal{YD} \longrightarrow \mathcal{YD}^H_H$ given by

$$(M, \cdot, \rho) \mapsto (M, \cdot_r, \rho_r) \text{ and } f \mapsto f.$$

We consider implications of H or H^{op} is a Hopf algebra have for ${}^H_H\mathcal{YD}$ and ${}_H\mathcal{YD}^H$. Recall from Proposition 7.1.10 that if H and H^{op} are Hopf algebras then the antipode S for H is bijective and S^{-1} is the antipode for H^{op}.

Exercise 11.6.22. We continue Exercise 11.6.18. Let M, N be objects of ${}^H_H\mathcal{YD}$.

(a) Suppose that H has antipode S. Show that (11.23) is equivalent to

$$\rho(h\cdot m) = h_{(1)}m_{(-1)}S(h_{(3)})\otimes h_{(2)}\cdot m_{(0)} \tag{11.27}$$

for all $h \in H$ and $m \in M$.

(b) Suppose H^{op} has antipode ς. Show that the $\sigma_{M,N}$'s defined for ${}^H_H\mathcal{YD}$ form a braiding for ${}^H_H\mathcal{YD}$ and that $\sigma^{-1}_{M,N}(n\otimes m) = m_{(0)}\otimes\varsigma(m_{(-1)})\cdot n$ for all $n \in N$ and $m \in M$.

Exercise 11.6.23. We continue Exercise 11.6.19. Let M, N be objects of ${}_H\mathcal{YD}^H$.

(a) Suppose that H^{op} has antipode ς. Show that (11.24) is equivalent to

$$\rho(h\cdot m) = h_{(2)}\cdot m_{(0)}\otimes h_{(3)}m_{(1)}\varsigma(h_{(1)}) \tag{11.28}$$

for all $h \in H$ and $m \in M$.

(b) Suppose that H has antipode S. Show that the $\sigma_{M,N}$'s defined for ${}_H\mathcal{YD}^H$ form a braiding for ${}_H\mathcal{YD}^H$ and that $\sigma^{-1}_{M,N}(n\otimes m) = m_{(0)}\otimes S(m_{(1)})\cdot n$ for all $n \in N$ and $m \in M$.

More exercises on Yetter-Drinfel'd modules are found in the exercises of Section 15.5.

11.7 Abstract characterization of biproducts

We pick up where we left off in the previous section. H is a Hopf algebra with antipode S over k, $A = (A, m, \eta, \Delta, \epsilon)$ is a bialgebra over k, and H and A are related by bialgebra maps $A \underset{\pi}{\overset{\jmath}{\leftrightarrows}} H$ which satisfy $\pi\circ\jmath = \mathrm{I}_H$. We first show that $B = \mathrm{Im}(\Pi)$ is a bialgebra in ${}^H_H\mathcal{YD}$ and $A \simeq B\times H$ as bialgebras over k. We then examine B when $A_0\jmath(H) \subseteq A_0$. Our arguments are based on a series of formulas involving Π, $\jmath$, and π. These are easy to establish:

$$\Pi*(\jmath\circ\pi) = \mathrm{I}_A; \tag{11.29}$$

$$\pi\circ\Pi = \eta_H\circ\epsilon_A; \tag{11.30}$$

$$\Pi(\jmath(h)a) = \jmath(h_{(1)})\Pi(a)\jmath(S(h_{(2)})) \tag{11.31}$$

and

$$\Pi(a\jmath(h)) = \epsilon(h)\Pi(a) \tag{11.32}$$

for all $h \in H$ and $a \in A$;

$$\Pi(ab) = a_{(1)}\Pi(b)\jmath(S(\pi(a_{(2)}))) \tag{11.33}$$

for all $a, b \in A$; and

$$\Delta(\Pi(a)) = a_{(1)}\jmath(S(\pi(a_{(3)})))\otimes\Pi(a_{(2)}) \tag{11.34}$$

for all $a \in A$.

Regard A as a right H-comodule by pushout along π; thus $\rho(a) = a_{(1)}\otimes\pi(a_{(2)})$ for all $a \in A$. We show that

$$B = A_{co\,inv} \tag{11.35}$$

and

$$\Pi\circ\Pi = \Pi. \tag{11.36}$$

Let $a \in A$. Then $(\mathrm{I}_A\otimes\pi)(\Delta(\Pi(a))) = \Pi(a)\otimes 1$ by (11.30) and (11.34). Thus $B \subseteq A_{co\,inv}$. Conversely, if $b \in A_{co\,inv}$ then $\Pi(b) = b_{(1)}\jmath(S(\pi(b_{(2)}))) = b\jmath(S(1)) = b$ implies that $b \in B$. We have established (11.35) and in the process (11.36) as well. By virtue of (11.35) it follows that B is a subalgebra of A.

We next define a coalgebra structure on B. Let

$$\Delta_B = (\Pi\otimes\mathrm{I}_A)\circ\Delta|B. \tag{11.37}$$

Since $\Delta(B) \subseteq A\otimes B$ by (11.34) it follows that $\Delta_B(B) \subseteq B\otimes B$. We will regard Δ_B as a linear map $\Delta_B : B \longrightarrow B\otimes B$. Now $(\Pi\otimes\Pi)\circ\Delta = \Delta_B\circ\Pi$ by (11.34) and (11.32). It is easy to see that $(\epsilon|B)\circ\Pi = \epsilon$. Since Π is onto $(B, \Delta_B, \epsilon|B)$ is a coalgebra over k and $\Pi : (A, \Delta, \epsilon) \longrightarrow (B, \Delta_B, \epsilon|B)$ is a coalgebra map.

By part (b) of Exercise 11.6.14 there is a left Yetter-Drinfel'd H-module structure $(A, \cdot_\jmath, \Delta_\pi)$ defined by $h\cdot_\jmath a = \jmath(h)a$ for all $h \in H$ and $a \in A$ and $\Delta_\pi(a) = \pi(a_{(1)})S(\pi(a_{(3)})\otimes a_{(2)}$ for all $a \in A$. By part (a) of the same there is a left Yetter-Drinfel'd H-module structure $(A, \cdot, \rho)$ defined on A by $h\cdot a = \jmath(h_{(1)})a\jmath(S(h_{(2)}))$ for all $h \in H$ and $a \in A$ and $\rho(a) = \pi(a_{(1)})\otimes a_{(2)}$ for all $a \in A$. By (11.31) and (11.32) it follows that B is a Yetter-Drinfel'd submodule of $(A, \cdot, \rho)$ and $\Pi : A \longrightarrow B$ is a map of Yetter-Drinfel'd modules, where A has the structure of the first.

Denoting the structure on B by $(B, \cdot, \rho)$ we have

$$h\cdot b = \jmath(h_{(1)})b\jmath(S(h_{(2)})) \quad \text{and} \quad \rho(b) = \pi(b_{(1)})\otimes b_{(2)} \tag{11.38}$$

for all $h \in H$ and $b \in B$.

We next show that $(B, m|B\otimes B, \eta, \Delta_B, \epsilon|B)$ is a bialgebra in ${}^H_H\mathcal{YD}$. With regard to the Yetter-Drinfel'd H-module structure $(A, \cdot, \rho)$ observe that A is a left H-module algebra with the module structure $(A, \cdot)$ and is a left H-comodule algebra with the comodule structure (A, ρ). See Example 11.2.2 and Exercise 11.3.5. Since B is a Yetter-Drinfel'd submodule of A it follows that B is a left H-module algebra and a left H-comodule algebra. We have noted that $\Pi : (A, \cdot_\jmath, \Delta_\pi) \longrightarrow (A, \cdot, \rho)$ is a map of Yetter-Drinfel'd modules.

Observe that A is a left H-module coalgebra with respect to $(A, \cdot_\jmath)$ and is a left H-comodule coalgebra with respect to (A, Δ_π). See Exercise 11.2.3 and Example 11.3.3. Regarding $\Pi : A \longrightarrow B$ as an onto coalgebra map we conclude that B is a left H-module coalgebra and a left H-comodule coalgebra.

It remains to establish (11.22) to complete our proof that B is a bialgebra of ${}^H_H\mathcal{YD}$. Let $b, b' \in B$. Now B is a subalgebra of A. Using (11.33), (11.29), and (11.34) we calculate

$$\begin{aligned}\Delta_B(bb') &= \Pi(b_{(1)}b'_{(1)})\otimes b_{(2)}b'_{(2)} \\ &= b_{(1)}\Pi(b'_{(1)})\jmath(S(\pi(b_{(2)})))\otimes b_{(3)}b'_{(2)} \\ &= (\Pi*(\jmath\circ\pi))(b_{(1)})\Pi(b'_{(1)})\jmath(S(\pi(b_{(2)})))\otimes b_{(3)}b'_{(2)} \\ &= \Pi(b_{(1)})\jmath(\pi(b_{(2)}))\Pi(b'_{(1)})\jmath(S(\pi(b_{(3)})))\otimes b_{(4)}b'_{(2)} \\ &= \Pi(b_{(1)})(\pi(b_{(2)})\cdot\Pi(b'_{(1)}))\otimes b_{(3)}b'_{(2)}\end{aligned}$$

which establishes (11.22).

The bialgebra A is the biproduct $B\times H$.

Theorem 11.7.1. *Let H be a Hopf algebra with antipode S over the field k, let A be a bialgebra over k, and suppose that $A \underset{\pi}{\overset{\jmath}{\leftrightarrows}} H$ are bialgebra maps which satisfy $\pi\circ\jmath = \mathrm{I}_H$. Let B the subalgebra of A with its structure as a bialgebra in ${}^H_H\mathcal{YD}$ as described above. Then the map $f : B\times H \longrightarrow A$ given by $f(b\times h) = b\jmath(h)$ for all $b \in B$ and $h \in H$ is an isomorphism of bialgebras.*

Proof. Observe that A is a right H-Hopf module under the module action $a\cdot h = a\jmath(h)$ for all $a \in A$ and $h \in H$ and comodule action given by $\rho(a) = a_{(1)}\otimes\pi(a_{(2)})$ for all $a \in A$. Since $A_{co\,inv} = B$ by (11.35), it follows by the counterpart of Theorem 8.2.3 for right Hopf modules that f is a linear isomorphism. The reader is left with the easy exercise of showing that f is also a map of bialgebras. □

We now turn our attention to the coradical filtration of B. Since $\Pi : A \longrightarrow B$ is an onto coalgebra map $\{B_{(n)}\}_{n=0}^{\infty}$ is a filtration of the coalgebra B, where $B_{(n)} = \Pi(A_n)$ for all $n \geq 0$. In particular $B_0 \subseteq B_{(0)}$ by Proposition 4.1.2. Since A_n is a subcoalgebra of A, $B_{(n)}$ is a left H-subcomodule of B for all $n \geq 0$ by (11.34).

From now until the end of the section we assume $A_0\jmath(H) \subseteq A_0$. Since $\jmath(H)$ is a subcoalgebra of A it follows by part (c) of Exercise 5.1.21 that $A_n\jmath(H) \subseteq A_n$ for all $n \geq 0$. The calculation

$$\Pi(A_n) \subseteq A_n\jmath(S(\pi(A_n))) \subseteq A_n\jmath(H) \subseteq A_n$$

shows that

$$\Pi(A_n) \subseteq A_n \tag{11.39}$$

for all $n \geq 0$. Since $B = \Pi(A)$, by (11.36) and (11.39) we have

$$B_{(n)} = A_n \cap B \tag{11.40}$$

for all $n \geq 0$. Next we show

$$B_n \subseteq B_{(n)} \tag{11.41}$$

for all $n \geq 0$ by induction on n. The case $n = 0$ was established above. Suppose $n \geq 1$ and $B_{n-1} \subseteq B_{(n-1)}$. Let $b \in B_n$ and write $\Delta_B(b) = b_{(1)} \otimes b_{(2)}$. Then $\Delta(b) = b_{(1)}\jmath(b_{(2)(-1)}) \otimes b_{(2)(0)}$ by virtue of Theorem 11.7.1. Now $\Delta_B(b) \in B_0 \otimes B + B \otimes B_{n-1} \subseteq A_0 \otimes B + A \otimes (A_{n-1} \cap B)$ by our induction hypothesis. If $a \in A_0$ and $b \in B$ then $a\jmath(b_{(-1)}) \otimes b_{(0)} \in A_0\jmath(H) \otimes A \subseteq A_0 \otimes A$. If $a \in A$ and $b \in A_{n-1} \cap B$ then $a\jmath(b_{(-1)}) \otimes b_{(0)} \in A \otimes A_{n-1}$ since $B_{(n-1)} = A_{n-1} \cap B$ is a subcomodule of B. Thus $\Delta(b) \in A_0 \otimes A + A \otimes A_{n-1}$ which means $b \in B \cap A_n = B_{(n)}$. We have shown $B_n \subseteq B_{(n)}$ which completes the induction argument.

Theorem 11.7.2. *Assume the hypothesis of Theorem 11.7.1 and suppose further that $A_0\jmath(H) \subseteq A_0$. Then:*

(a) *If $B_0 = B_{(0)}$ then the coradical filtration of B is $\{B \cap A_n\}_{n=0}^{\infty}$.*
(b) *Suppose $A_0 = H$, $\jmath$ is the inclusion, A has the structure of a coradically graded bialgebra $A = \bigoplus_{n=0}^{\infty} A(n)$, and $\pi : A \longrightarrow H$ is the projection of A onto $H = A(0)$. Then B is a pointed irreducible coalgebra and has the structure of a coradically graded bialgebra $B = \bigoplus_{n=0}^{\infty} B(n)$ in ${}^H_H\mathcal{YD}$, where $B(n) = B \cap A(n)$ for all $n \geq 0$.*

Proof. We first establish part (a). Assume $B_0 = B_{(0)} = A_0 \cap B$. In light of (11.41) we need only show that $B_{(n)} \subseteq B_n$ for all $n \geq 0$ by induction on n. The case $n = 0$ holds by assumption. Suppose $n \geq 1$ and $B_{(n-1)} \subseteq B_{n-1}$. Let $b \in B_{(n)} = B \cap A_n$. Then $b = \Pi(b)$ since $b \in B$ and $\Delta(b) \in A_0 \otimes A + A \otimes A_{n-1}$ since $b \in A_n$. Since Π is a coalgebra map

$$\begin{aligned}\Delta_B(b) &= \Delta_B(\Pi(b)) \\ &= (\Pi \otimes \Pi)(\Delta(b)) \\ &\in \Pi(A_0) \otimes \Pi(A) + \Pi(A) \otimes \Pi(A_{n-1}) \\ &\subseteq B_0 \otimes B + B \otimes B_{n-1}\end{aligned}$$

which means $b \in B_n$. Therefore $B_{(n)} \subseteq B_n$ and part (a) is established.

Assume the hypothesis for part (b). Then $\Pi(A_0) = k1$ which means $B_{(0)} = k1 = B_0$ by (11.41). By part (a) the coradical filtration of B is $\{B \cap A_n\}_{n=0}^{\infty}$. Our assumption on π means $\Pi(A(n)) \subseteq A(n)$ for all $n \geq 0$. Therefore $\Pi(A(n)) \subseteq A(n) \cap B = B(n)$ for all $n \geq 0$. Since $\sum_{n=0}^{\infty} A(n)$ is direct $\sum_{n=0}^{\infty} B(n)$ is direct. Thus since $\Pi : A \longrightarrow B$ is an onto coalgebra map, $\Pi(A_n) = B_n$ for all $n \geq 0$, and $A = \bigoplus_{n=0}^{\infty} A(n)$ gives A the structure of a coradically graded k-coalgebra, it follows that $B = \bigoplus_{n=0}^{\infty} B(n)$ gives B the structure of a coradically graded k-coalgebra. $1 \in B(0)$ since $1 \in A(0)$. Since $A(m)A(n) \subseteq A(m+n)$ for all $m, n \geq 0$ it follows $B(m)B(n) \subseteq B(m+n)$ holds for all $m, n \geq 0$. Thus $B = \oplus_{n=0}^{\infty} B(n)$ gives B the structure of a graded k-algebra.

We have noted that $B_{(n)} = B \cap A_n$ is a subcomodule of B. The calculation

$$H \cdot B(n) = H \cdot \Pi(A(n)) = \Pi(HA(n)) = \Pi(A(0)A(n)) \subseteq \Pi(A(n)) = B(n)$$

establishes that $B(n)$ is a submodule of B as well. We have shown that the $B(n)$'s are Yetter-Drinfel'd submodules of B which completes our proof. □

Under the hypothesis of part (b) the bialgebra A is a Hopf algebra over k. We shall see in Section 15.5 that B is a Hopf algebra in the category ${}_H^H\mathcal{YD}$. There is a very important situation to which the preceding theorem applies.

Suppose that H is a Hopf algebra over the field k and H_0 is a Hopf subalgebra of H. Important examples, pointed Hopf algebras. Then $\mathrm{gr}(H)$ is a coradically graded Hopf algebra and thus part (b) of Theorem 11.7.2 applies. A first step in determining H is to determine $\mathrm{gr}(H)$ which brings us to the coradically graded pointed irreducible bialgebra B in ${}_H^H\mathcal{YD}$. Much more about this in Section 15.5.

Exercises

Exercise 11.7.1. Complete the proof of Theorem 11.7.1.

Exercise 11.7.2. Suppose $(C, \Delta_C, \epsilon_C)$ is a coalgebra over k and D is a vector space over k. Suppose further that $\Pi : C \longrightarrow D$, $\Delta_D : D \longrightarrow D \otimes D$, and $\epsilon_D : D \longrightarrow k$ are linear maps which satisfy $\Delta_D \circ \Pi = (\Pi \otimes \Pi) \circ \Delta_C$ and $\epsilon_D \circ \Pi = \epsilon_C$. Show that Π onto implies that $(D, \Delta_D, \epsilon_D)$ is a coalgebra over k and Π is a coalgebra map.

Exercise 11.7.3. Show that $\Pi : (A, \cdot_\jmath, \Delta_\pi) \longrightarrow (A, \cdot, \rho)$ as in the proof of Theorem 11.7.1 is a map of Yetter-Drinfel'd modules. This is a key detail in the proof that $(B, m|B \otimes B, \eta, \Delta_B, \epsilon|B)$ is a bialgebra in ${}_H^H\mathcal{D}$.

Chapter notes

Our discussion of monoidal categories in Section 11.1 follows Section 1 of the book by MacLane [129, Chapter VII] with slight notational differences. The notions of H-module algebra, adjoint actions, and smash product of Section 11.2 can be traced to [72, Section 1] by Heyneman and Sweedler. Example 11.2.2 and Definition 11.2.4 are described in [72, Example 1.7.2]. The notion of invariants is dual to the notion of coinvariants, the first manifestation of which is due to Larson and Sweedler [105, Proposition 1]. Part (a) of Exercise 11.2.6 is found in Sommerhäuser's paper [194, Proposition 1].

Section 11.4 is a continuation on the theme of duality relationships. Our discussion of braiding in Section 11.5 follows pages 252–253 of [129].

There is much to be said about Section 11.6. Let H be a bialgebra over k. The category ${}^H_H\mathcal{YD}$ is due to Yetter [229, Section 3] in which ${}^H_H\mathcal{YD}$ is denoted A-cbm, where $A = H$, and its objects are referred to as left crossed bimodules over A. This category, when A is a Hopf algebra over k, is one of many examples given in the paper which account for knot invariants. Variations of ${}^H_H\mathcal{YD}$ are studied in [177] by Towber and the author.

Now suppose H is a Hopf algebra over k and $A = B{\times}H$ is a biproduct which is a Hopf algebra. Majid realized [116, 117] that B is a Hopf algebra in the braided monoidal category ${}^H_H\mathcal{YD}$ and this was the proper theoretical setting for B. The study of Hopf algebras and related structures in ${}^H_H\mathcal{YD}$ is important in connection with biproducts. Biproducts play a fundamental role in the classification of pointed Hopf algebras as we shall see in notes for Chapter 16.

Biproducts were defined, studied, and a universal description was given for them in [168, Section 2]. Theorem 11.6.7 is a consequence of [168, Theorem 1] and Theorem 11.6.9 is [168, Proposition 2].

The results of Section 11.7 from the beginning through the proof of Theorem 11.7.1 are found in [168, Section 3]. Discussion subsequent to the proof follows the paper by [9, Section 2] by Andruskiewitsch and Schneider. Theorem 11.7.2 is essentially a combination of Lemmas 2.1 and 2.4 of [9]. The notion of coradically graded is attributed to Chin and Musson [33, Section 2.2].

Many have contributed to the theory of Hopf algebras in ${}^H_H\mathcal{YD}$. A good many important results for finite-dimensional Hopf algebras over k are shown to hold or have analogs in ${}^H_H\mathcal{YD}$ by arguments intrinsic to ${}^H_H\mathcal{YD}$ or more general categories.

Suppose A is a finite-dimensional Hopf algebra in ${}^H_H\mathcal{YD}$. Results for finite Hopf algebras in a braided tensor category due to Takeuchi [220] have applications to A. The antipode of A is bijective [220, 4.1 Theorem]. He also establishes an analog of Theorem 8.2.3 in [220, 3.4 Theorem] and establishes the existence of integrals for A^* in [220, 4.5 Corollary]. See the paper by Lyubashenko [113] also. At this point the reader is referred to Definition 11.6.10 and Exercise 11.6.6.

In [42] Doi and Takeuchi introduce the notion of bi-Frobenius algebra and derive a formula which specialized to Hopf algebras in ${}_H^H\mathcal{YD}$ is an analog of the formula for S^4 of Theorem 10.5.6. The specialization is that obtained in [23, Theorem 3.6] by Bespalov, Kerler, Lyubashenko, and Turaev who pursue the theory of Hopf algebras in monoidal categories. For a survey of braided Hopf algebras see the article [221] by Takeuchi.

An important result about Hopf algebras in ${}_H^H\mathcal{YD}$ is the following version of Theorem 9.3.3, the Nichols-Zoeller Theorem: If A, B are finite-dimensional Hopf algebras in ${}_H^H\mathcal{YD}$, where B is a Hopf subalgebra of A, then A is a free B-module. This was proved by Scharfschwerdt [188, Theorem 2.2] who attributes the result to Takeuchi as well. Sommerhäuser proves the result [194, Section 1.8] starting with the fact that the k-algebra $A{\times}H$ is a free module over $B{\times}H$, a consequence of Theorem 9.3.3.

Kashina studies the order of the antipode of a semisimple Hopf algebra in ${}_H^H\mathcal{YD}$ in [90]. Sommerhäuser goes in a different direction [194] from the other authors mentioned above and characterizes all Hopf algebras in ${}_H^H\mathcal{YD}$ of a certain important class when H is the group algebra of a cyclic group of prime order under some restrictions on k.

When k has characteristic 0 pointed irreducible Hopf algebras in ${}_H^H\mathcal{YD}$, where H is the group algebra of an abelian group, have an important role to play in the classification of pointed Hopf algebras over k. The structure of these pointed irreducible Hopf algebras is not easy to ascertain. We defer discussion until the notes for Chapter 16.

Chapter 12

Quasitriangular bialgebras and Hopf algebras

Quasitriangular Hopf algebras form an interesting class in their own right and have important connections with topology. A quasitriangular Hopf algebra H has an invertible element $R \in H \otimes H$ which satisfies $\Delta^{cop}(h)R = R\Delta(h)$, or equivalently $\Delta^{cop}(h) = R\Delta(h)R^{-1}$, for all $h \in H$. The element R accounts for solutions to the quantum Yang–Baxter equation. Solutions to this equation are fundamental building blocks for many invariants of knots and links, and in some cases for invariants of 3-manifolds.

As we shall see in this chapter, finite-dimensional quasitriangular Hopf algebras have a rather special structure. There are important special cases, ribbon Hopf algebras and factorizable Hopf algebras. Ribbon Hopf algebras are used to construct 3-manifold invariants. The Drinfel'd double, which is the subject of Chapter 13, is factorizable. Every finite-dimensional Hopf algebra can be embedded in a Drinfel'd double. For this reason alone quasitriangular Hopf algebras are important to study.

We begin with a discussion of the quantum Yang–Baxter and braid equations and relate them to bialgebras. Throughout this chapter H is a bialgebra over the field k.

12.1 The quantum Yang–Baxter and braid equations, Yang-Baxter algebras

Suppose M is a vector space over the field k and $R \in \mathrm{End}(M \otimes M)$. For $1 \leq i < j \leq 3$ define $R_{i,j} \in \mathrm{End}(M \otimes M \otimes M)$ by

$$R_{1,2} = R \otimes \mathrm{I}_M, \quad R_{2,3} = \mathrm{I}_M \otimes R,$$

and

$$R_{1,3} = (\mathrm{I}_M \otimes \tau_{M,M}) \circ (R \otimes \mathrm{I}_M) \circ (\mathrm{I}_M \otimes \tau_{M,M}),$$

where $\tau_{M,M} \in \mathrm{End}(M\otimes M)$ is the twist map given by $\tau_{M,M}(m\otimes n) = n\otimes m$ for all $m, n \in M$. If M' is also a vector space over k recall that the map $\tau_{M,M'} : M\otimes M' \longrightarrow M'\otimes M$ is defined by $\tau_{M,M'}(m\otimes m') = m'\otimes m$ for all $m \in M$ and $m' \in M'$.

Definition 12.1.1. The *quantum Yang–Baxter equation* is

$$R_{1,2}\circ R_{1,3}\circ R_{2,3} = R_{2,3}\circ R_{1,3}\circ R_{1,2}$$

and the *braid equation* is

$$R_{1,2}\circ R_{2,3}\circ R_{1,2} = R_{2,3}\circ R_{1,2}\circ R_{2,3},$$

where M is a vector space over k and $R \in \mathrm{End}(M\otimes M)$.

The braid equation can be expressed very simply as

$$(R\otimes \mathrm{I}_M)\circ(\mathrm{I}_M\otimes R)\circ(R\otimes \mathrm{I}_M) = (\mathrm{I}_M\otimes R)\circ(R\otimes \mathrm{I}_M)\circ(\mathrm{I}_M\otimes R).$$

There is a convenient adaptation of the Heyneman–Sweedler notation for coalgebras which provides a useful way of expressing the quantum Yang–Baxter and braid equations. For $m, n \in M$ write $R(m\otimes n) = m_{[1]}\otimes n_{[2]}$ and let $m = m_{[0]}$ be the result of applying the identity map I_M to m. Then R satisfies the quantum Yang–Baxter equation if and only if

$$\ell_{[0][1][1]}\otimes m_{[1][0][2]}\otimes n_{[2][2][0]} = \ell_{[1][1][0]}\otimes m_{[2][0][1]}\otimes n_{[0][2][2]} \tag{12.1}$$

for all $\ell, m, n \in M$.

Let $B = \tau_{M,M}\circ R$. Then $B(m\otimes n) = n_{[2]}\otimes m_{[1]}$ for all $n, m \in M$ in our notation. Using (12.1) it is easy to see that R satisfies the quantum Yang–Baxter equation if and only if B satisfies the braid equation. In particular there is a one-one correspondence between solutions to the quantum Yang–Baxter equation and solutions to the braid equation.

Suppose M is finite-dimensional. Using the identification $\mathrm{End}(M\otimes M) = \mathrm{End}(M)\otimes \mathrm{End}(M)$ we write $R = \sum_{i=1}^r a_i\otimes b_i$, where $a_i, b_i \in \mathrm{End}(M)$. In this case note that

$$R_{1,2} = \sum_{i=1}^r a_i\otimes b_i\otimes 1, \quad R_{2,3} = \sum_{i=1}^r 1\otimes a_i\otimes b_i, \quad \text{and } R_{1,3} = \textstyle\sum_{i=1}^r a_i\otimes 1\otimes b_i,$$

where $1 = \mathrm{I}_M$.

Let A be any algebra over k and suppose that $R = \sum_{i=1}^r a_i\otimes b_i$, where $a_i, b_i \in A$. For $1 \le i < j \le 3$ we define $R_{i,j} \in A\otimes A\otimes A$ as above and say that R is a solution to the quantum Yang–Baxter equation if the first equation of Definition 12.1.1 is satisfied, where the product in A replaces composition; that is

$$\sum_{i,j,\ell=1}^{r} a_i a_j \otimes b_i a_\ell \otimes b_j b_\ell = \sum_{i,j,\ell=1}^{r} a_j a_i \otimes a_\ell b_i \otimes b_\ell b_j. \tag{12.2}$$

Now let M be any left A-module and define $R_M \in \mathrm{End}(M \otimes M)$ by

$$R_M(m \otimes n) = \sum_{i=1}^{r} a_i \cdot m \otimes b_i \cdot n \tag{12.3}$$

for all $m, n \in M$. Then R satisfies the quantum Yang–Baxter equation if and only if R_M does for all left A-modules M.

We also define $B_M \in \mathrm{End}(M \otimes M)$ by $B_M = \tau_{M,M} \circ R_M$. Thus

$$B_M(m \otimes n) = \sum_{i=1}^{r} b_i \cdot n \otimes a_i \cdot m \tag{12.4}$$

for all $m, n \in M$. Since B_M satisfies the braid equation if and only if R_M satisfies the quantum Yang–Baxter equation, it follows that B_M satisfies the braid equation for all left A-modules M if and only if R satisfies the quantum Yang–Baxter equation.

As this chapter unfolds we will build on the following concept.

Definition 12.1.2. A *quasitriangular algebra over* k is a pair (A, R), where A is an algebra over k and $R \in A \otimes A$ is a solution to the quantum Yang–Baxter equation.

In light of the preceding discussion:

Proposition 12.1.3. *Let A be an algebra over k and let $R \in A \otimes A$. Then the following are equivalent:*

(a) *(A, R) is a quasitriangular algebra over k.*
(b) *R_M is a solution to the quantum Yang–Baxter equation for all left A-modules M.*
(c) *R_A is a solution to the quantum Yang–Baxter equation, where A is regarded as a left A-module under multiplication.*

□

Definition 12.1.4. A *Yang–Baxter algebra over* k is a quasitriangular algebra over (A, R) over k such that R is invertible.

Thus:

Proposition 12.1.5. *Let A be an algebra over the field k and $R \in A\otimes A$. Then (A, R) is a Yang–Baxter algebra over k if and only if R_M is an invertible solution to the quantum Yang–Baxter equation for all left A-modules M.* □

Exercises

Throughout these exercises M and M' are vector spaces over k.

Exercise 12.1.1. Let $R \in \mathrm{End}(M\otimes M)$.

(a) If R is a solution to the quantum Yang–Baxter (respectively braid) equation show that αR is a solution to the quantum Yang–Baxter (respectively braid) equation for all $\alpha \in k$.

(b) If R is an invertible solution to the quantum Yang–Baxter (respectively braid) equation show that R^{-1} is a solution to the quantum Yang–Baxter (respectively braid) equation.

(c) Using (12.1) show that $\tau_{M,M}\circ R$ is a solution to the braid equation if and only if R is a solution to the quantum Yang–Baxter equation.

(d) Using (12.1) show that $R\circ\tau_{M,M}$ is a solution to the braid equation if and only if R is a solution to the quantum Yang–Baxter equation.

(e) Show that $\tau_{M,M}\circ R\circ\tau_{M,M}$ a solution to the quantum Yang–Baxter (respectively braid) equation if and only if R is a solution to the quantum Yang–Baxter (respectively braid) equation.

Exercise 12.1.2. Show that $\alpha \mathrm{I}_{M\otimes M}$, where $\alpha \in k$, and $\tau_{M,M}$ are solutions to the quantum Yang–Baxter and braid equations.

Exercise 12.1.3. Let $R \in \mathrm{End}(M\otimes M)$ and $R' \in \mathrm{End}(M'\otimes M')$. Define $R'' \in \mathrm{End}((M\otimes M')\otimes(M\otimes M'))$ by

$$R'' = (\mathrm{I}_M\otimes\tau_{M,M'}\otimes\mathrm{I}_{M'})\circ(R\otimes R')\circ(\mathrm{I}_M\otimes\tau_{M',M}\otimes\mathrm{I}_{M'}).$$

If R, R' are solutions to the quantum Yang–Baxter (respectively braid) equation, show that R'' is a solution to the quantum Yang–Baxter (respectively braid) equation.

Exercise 12.1.4. Suppose (A, R) is a Yang–Baxter algebra. Show that (A^{op}, R), $(A, R_{2,1})$, and (A, R^{-1}) are as well, where $R_{2,1} = \tau_{A,A}(R)$.

Exercise 12.1.5. Suppose (A, R) and (A', R') are Yang–Baxter algebras. Show that $(A\otimes A', R'')$ is as well, where $R'' = (\mathrm{I}_R\otimes\tau_{A,A'}\otimes\mathrm{I}_{R'})(R\otimes R')$.

Exercise 12.1.6. Apropos of Proposition 12.1.3, suppose A is a finite-dimensional algebra over k, $R = \sum_{i=1}^{r} a_i \otimes b_i \in A \otimes A$, and R_M defined by (12.3) is invertible for all left A-modules M. Show that R is invertible. [Hint: For $a \in A$ let $\ell(a) \in \mathrm{End}(A)$ be as in (2.16). Thus by $\ell(a)(b) = ab$ for all $b \in A$. Let $M = A$ under left multiplication. Observe that $R_M = \sum_{i=1}^{r} \ell(a_i) \otimes \ell(b_i) \in \mathcal{A}$, where the latter is the span of the $\ell(a) \otimes \ell(b)$'s for $a, b \in A$. Since A is finite-dimensional $\mathcal{A}$ is a finite-dimensional subalgebra of $\mathrm{End}(A \otimes A)$ and there is an isomorphism $\mathcal{A} \longrightarrow A \otimes A$ which sends R_M to R.]

12.2 Almost cocommutative Hopf algebras, quasitriangular bialgebras and Hopf algebras

We regard the tensor product of left H-modules M and N as left H-module according to (5.4); that is $h \cdot (m \otimes n) = h_{(1)} \cdot m \otimes h_{(2)} \cdot n$ for all $h \in H$, $m \in M$, and $n \in N$. Let $R \in H \otimes H$. Observe that B_M defined by (12.4) is a map of left H-modules for all $M \in {}_H\mathcal{M}$ if and only if

$$\Delta^{cop}(h) R = R \Delta(h) \tag{12.5}$$

for all $h \in H$. The endomorphisms B_M are invertible if and only if R is invertible. We are led to:

Definition 12.2.1. An *almost cocommutative Hopf algebra over* k is a Hopf algebra H with bijective antipode over k which has an invertible $R \in H \otimes H$ satisfying (12.5) for all $h \in H$.

The square of the antipode of an almost cocommutative Hopf algebra H is an inner algebra automorphism of H.

Proposition 12.2.2. *Let H be an almost cocommutative Hopf algebra with antipode S over the field k and let $R = \sum_{i=1}^{r} a_i \otimes b_i \in H \otimes H$ be as in Definition 12.2.1. Then $u = \sum_{i=1}^{r} S(b_i) a_i$ is invertible and $S^2(h) = uhu^{-1}$ for all $h \in H$. Furthermore $uS(u)$ is in the center of H.*

Proof. We use the Einstein summation convention and write $R = R_i \otimes R^i$, $R^{-1} = r_j \otimes r^j$. In this notation $u = S(R^i) R_i$ and (12.5) is expressed

$$h_{(2)} R_i \otimes h_{(1)} R^i = R_i h_{(1)} \otimes R^i h_{(2)} \tag{12.6}$$

for all $h \in H$. Applying $m \circ \tau_{H,H} \circ (\mathrm{I}_H \otimes S)$ to both sides of (12.6) gives $\epsilon(h)u = S(h_{(2)})uh_{(1)}$ for all $h \in H$. As a consequence

$$\begin{aligned} S^2(h)u &= S^2(h_{(1)})u(\epsilon(h_{(2)})1) \\ &= S^2(h_{(1)})uS(h_{(2)})h_{(3)} \\ &= S(S(h_{(1)})_{(2)})uS(h_{(1)})_{(1)}h_{(2)} \\ &= \epsilon(S(h_{(1)}))uh_{(2)} = uh, \end{aligned}$$

and thus

$$S^2(h)u = uh, \tag{12.7}$$

for all $h \in H$. Applying $m \circ \tau_{H,H} \circ (\mathrm{I}_H \otimes S)$ to both sides of $R_i r_j \otimes R^i r^j = 1 \otimes 1$ yields $S(r^j)ur_j = 1$ from which $S(r^j)S^2(r_j)u = 1$ follows by (12.7). Observe that we have not used the fact that S is bijective at this point. Since S is bijective we can use (12.7) to calculate $1 = S(r^j)ur_j = S^2(S^{-1}(r^j))ur_j = uS^{-1}(r^j)r_j$. We have shown that u has a left inverse and a right inverse. Therefore u is invertible. By (12.7) the equation $S^2(h) = uhu^{-1}$ follows for all $h \in H$.

To see that $uS(u)$ is in the center of H we first note that $S^{-2}(h) = u^{-1}hu$ for all $h \in H$. Applying S to both sides of the preceding equation we have $S^{-2}(S(h)) = S(u)S(h)S(u^{-1})$ for all $h \in H$; thus $S^{-2}(h) = S(u)hS(u)^{-1}$ for all $h \in H$. Since $u^{-1}hu = S^{-2}(h) = S(u)hS(u)^{-1}$ for all $h \in H$ we now conclude that $uS(u)$ is in the center of H. □

We use the Einstein summation convention and write $R = R_i \otimes R^i$. For left H-modules M, N we define $B_{M,N} : M \otimes N \longrightarrow N \otimes M$ by

$$B_{M,N}(m \otimes n) = R^i \cdot n \otimes R_i \cdot m \tag{12.8}$$

for all $m \in M$ and $n \in N$. The same argument which shows that B_M is a left H-module map for all $M \in {}_H\mathcal{M}$ if and only if (12.5) holds shows that $B_{M,N}$ is a map of left H-modules for all $M, N \in {}_H\mathcal{M}$ if and only if the same holds. We are interested in when the $B_{M,N}$'s form a prebraiding for the monoidal category ${}_H\mathcal{M}$. This is the case if and only if (QT.1)–(QT.5) below hold. See Definition 11.5.1 and Exercise 12.2.5.

Definition 12.2.3. A *quasitriangular bialgebra over k* is a pair (H, R), where H be a bialgebra over k and $R \in H \otimes H$, which we write $R = R_i \otimes R^i$, such that:

(QT.1) $\Delta(R_i) \otimes R^i = R_i \otimes R_j \otimes R^i R^j$;
(QT.2) $\epsilon(R_i)R^i = 1$;

(QT.3) $R_i \otimes \Delta(R^i) = R_j R_i \otimes R^i \otimes R^j$;
(QT.4) $R_i \epsilon(R^i) = 1$; and
(QT.5) $\Delta^{cop}(h)R = R\Delta(h)$

for all $h \in H$.

Another formulation of (QT.1) is

$$(\Delta \otimes \mathrm{I}_H)(R) = R_{1,3}R_{2,3} \tag{12.9}$$

and of (QT.3) is

$$(\mathrm{I}_H \otimes \Delta)(R) = R_{1,3}R_{1,2}. \tag{12.10}$$

Let $R \in H \otimes H$.

Remark 12.2.4. The set of $h \in H$ such that (QT.5) holds for R is a subalgebra of H. Thus (QT.5) holds for (H, R) if it holds for algebra generators of H.

Observe that $B_{M,N}$ is invertible for all $M, N \in {}_H\mathcal{M}$ if and only if R is invertible. Suppose that R is invertible. Then the $B_{M,N}$'s give ${}_H\mathcal{M}$ a braiding structure. Note that $B_{M,N}^{-1} = B_{N,M}$ for all $M, N \in {}_H\mathcal{M}$ if and only if $R^{-1} = R_{2,1}$.

Definition 12.2.5. A *triangular bialgebra over* k is a quasitriangular bialgebra (H, R) over k such that R is invertible and $R^{-1} = R_{2,1}$.

Definition 12.2.6. A *quasitriangular* (respectively *triangular*) *Hopf algebra over* k is a quasitriangular (respectively triangular) bialgebra (H, R) over k where H is a Hopf algebra over k.

Definition 12.2.7. A *morphism of quasitriangular bialgebras* (H, R) and (H', R') *over* k is a bialgebra map $f : H \longrightarrow H'$ such that $(f \otimes f)(R) = R'$. A *morphism of quasitriangular Hopf algebras* is a morphism of quasitriangular bialgebras.

Quasitriangular Hopf algebras have rather special properties.

Theorem 12.2.8. *Let (H, R) be a quasitriangular Hopf algebra over k, let S be the antipode of H, and write $R = R_i \otimes R^i$. Then:*

(a) *R is invertible and $R^{-1} = (S \otimes \mathrm{I}_H)(R)$.*
(b) *$R = (S \otimes S)(R)$.*

(c) *R satisfies the quantum Yang-Baxter equation and is a left twist for H.*
(d) *$J = (1\otimes R^i)\otimes(R_i\otimes 1)$ is a left twist for the tensor product Hopf algebra $H\otimes H$.*
(e) *Set $u = S(R^i)R_i$. Then u is invertible, $u^{-1} = R^iS^2(R_i)$, and $S^2(h) = uhu^{-1}$ for all $h \in H$.*
(f) *S is bijective; thus H is almost cocommutative.*

Proof. We first establish part (a). Using (QT.2) and applying $(m\otimes \mathrm{I}_H)\circ(S\otimes \mathrm{I}_H\otimes \mathrm{I}_H)$ and $(m\otimes \mathrm{I}_H)\circ(\mathrm{I}_H\otimes S\otimes \mathrm{I}_H)$ to both sides of the equation of (QT.1) we obtain

$$1\otimes 1 = \epsilon(R_i)1\otimes R^i = S(R_i)R_j\otimes R^iR^j = R_iS(R_j)\otimes R^iR^j.$$

Thus R and $(S\otimes \mathrm{I}_H)(R)$ are inverses. To show part (b) we use (QT.4) and apply $(S\otimes \mathrm{I}_H)\circ(\mathrm{I}_H\otimes m)\circ(\mathrm{I}_H\otimes S\otimes \mathrm{I}_H)$ to both sides of the equation of (QT.3) to obtain

$$1 = S(R_i)\otimes\epsilon(R^i)1 = S(R_i)S(R_j)\otimes S(R^i)R^j.$$

Therefore $(S\otimes S)(R)$ is a left inverse of $(S\otimes \mathrm{I}_H)(R) = R^{-1}$ and thus part (b) follows.

To show part (c) we use (QT.1) to calculate

$$\begin{aligned} R_{1,2}R_{1,3}R_{2,3} &= (R_i\otimes R^i\otimes 1)(R_j\otimes 1\otimes R^j)(1\otimes R_\ell\otimes R^\ell) \\ &= R_iR_j\otimes R^iR_\ell\otimes R^jR^\ell \\ &= R\Delta(R_\ell)\otimes R^\ell \end{aligned}$$

and likewise

$$R_{2,3}R_{1,3}R_{1,2} = R_jR_i\otimes R_\ell R^i\otimes R^\ell R^j = \Delta^{cop}(R_j)R\otimes R^j.$$

Now we use (QT.5) to conclude that $R_{1,2}R_{1,3}R_{2,3} = R_{2,3}R_{1,3}R_{1,2}$. Thus R satisfies the quantum Yang-Baxter equation. We have shown $R_{2,3}R_{1,3}R_{1,2} = R\Delta(R_i)\otimes R^i$. Since $R_{2,3}R_{1,3}R_{1,2} = R_i\otimes R\Delta(R^i)$ as well, R is a left twist for H.

Part (d) follows by an easy calculation using (QT.1) and (QT.3). Part (f) follows from part (e) which we now establish.

The calculations in the proof of Proposition 12.2.2 show that (12.7) holds and $S(r^j)S^2(r_j)u = 1$, where $R^{-1} = r_j\otimes r^j$. Let $v = S(r^j)S^2(r_j)$. Then $vu = 1$ and $v = S(r^j)S^2(r_j) = S(R^i)S^3(R_i) = R^iS^2(R_i)$ by parts (a) and (b). Let $h = S(r_j)r^j$. Then $v = S(h)$. Now $v = S^2(v)$ by part (b). Therefore $v = S^3(h)$. Since

$$S^3(h)uS(u) = uS(h)S(u) = uS(uh) = uS(S^2(h)u) = uS(u)S^3(h)$$

it follows that v and $uS(u)$ commute. Consequently

$$u(S(u)vS(v)) = vuS(u)S(v) = vuS(vu) = 1.$$

We have shown that u has a left inverse v and also has a right inverse. Therefore u in invertible. As (12.7) holds our proof of part (e) is complete. □

Definition 12.2.9. Let (H, R) be a quasitriangular Hopf algebra over k. The *Drinfel'd element of* (H, R) is the element u of Theorem 12.2.8. The *quantum Casimir element of* H is the product $uS(u)$.

By virtue of part (e) of the preceding theorem $S^2(u) = u$ and thus $S(u) = S^3(u) = uS(u)u^{-1}$. In particular u and $S(u)$ commute.

The following reconciles the original definition of quasitriangular Hopf algebra with the one given here.

Corollary 12.2.10. *Let H be a Hopf algebra over k and $R \in H \otimes H$. Then the following are equivalent:*

(a) *(H, R) is quasitriangular.*
(b) *H is almost cocommutative, where R is invertible and satisfies (12.9), (12.10).*

Proof. Part (a) implies part (b) by definition and part (a) of Theorem 12.2.8. Suppose that the hypothesis of part (b) holds. Then (QT.1), (QT.3), and (QT.5) hold. We need only show that (QT.2) and (QT.4) hold. Write $R = R_i \otimes R^i$ and let $e = \epsilon(R_i)R^i$. Applying $\epsilon \otimes \epsilon \otimes \mathrm{I}_H$ to both sides of (QT.1) we have $e = e^2$. Applying $\epsilon \otimes \mathrm{I}_H$ to both sides of the equation $RR^{-1} = 1$ we deduce that e has a right inverse f. Therefore $1 = ef = eef = e$ which gives (QT.2). Starting with (QT.3) in place of (QT.1) one can construct a similar argument which shows that (QT.4) holds. □

Define $f_R, g_R : H^o \longrightarrow H$ by $f_R(p) = (p \otimes \mathrm{I}_H)(R)$ and $g_R(p) = (\mathrm{I}_H \otimes p)(R)$ for all $p \in H^*$. The quasitriangular axioms are reflected in the properties of f_R and g_R.

Proposition 12.2.11. *Let (H, R) be a quasitriangular bialgebra over the field k. Then:*

(a) *$f_R : H^o \longrightarrow H^{cop}$ is a bialgebra map and*
$$(p_{(1)} \rightharpoonup h) f_R(p_{(2)}) = f_R(p_{(1)})(h \leftharpoonup p_{(2)})$$
for all $p \in H^o$ and $h \in H$.

(b) $g_R : H^o \longrightarrow H^{op}$ *is a bialgebra map and*

$$(h{\leftharpoonup}p_{(1)})g_R(p_{(2)}) = g_R(p_{(1)})(p_{(2)}{\rightharpoonup}h)$$

for all $h \in H$ *and* $p \in H^o$.

Proof. We show part (a) and note the reader can mimic our proof to construct an argument for part (b). First of all we extend the definition of f_R to all of H^* and call the resulting linear map f'_R. Then f'_R is multiplicative if and only if (QT.1) holds and $f'_R(\epsilon) = 1$ if and only if (QT.2) holds.

Note that $\epsilon_H \circ f_R = \epsilon_{H^o}$ if (QT.4) holds and (QT.3) implies

$$\begin{aligned}\Delta(f_R(p)) &= p(R_i)\Delta(R^i)\\ &= R_i \otimes R_j p(R^j R^i)\\ &= R_i \otimes R_j p_{(1)}(R^j)p_{(2)}(R^i) = f_R(p_{(2)}) \otimes f_R(p_{(1)})\end{aligned}$$

for all $p \in H^o$. Therefore $f_R : H^o \longrightarrow H^{cop}$ is a bialgebra map.

Let $p \in H^o$ and $h \in H$. Applying $p \otimes \mathrm{I}_H$ to both sides of the equation of (QT.5) we have $p(h_{(2)}R_i)h_{(1)}R^i = p(R_i h_{(1)})R^i h_{(2)}$, or equivalently $p_{(1)}(h_{(2)})p_{(2)}(R_i)h_{(1)}R^i = p_{(1)}(R_i)p_{(2)}(h_{(1)})R^i h_{(2)}$, which can be expressed $(p_{(1)}{\rightharpoonup}h)f_R(p_{(2)}) = f_R(p_{(1)})(h{\leftharpoonup}p_{(2)})$. □

In light of the calculations of the preceding proof:

Corollary 12.2.12. *Let* H *be a bialgebra over the field* k *and* $R \in H \otimes H$. *Suppose that* H^o *is a dense subspace of* H^*. *Then* (H, R) *is a quasitriangular bialgebra over* k *if either part (a) or part (b) of Proposition 12.2.11 holds.* □

Notice that the corollary applies when H is finite-dimensional.

For the remainder of this section (H, R) is a quasitriangular bialgebra over k. Let

$$R_{(\ell)} = \{(\mathrm{I}_H \otimes p)(R) \mid p \in H^*\}, \quad R_{(r)} = \{(p \otimes \mathrm{I}_H)(R) \mid p \in H^*\}. \tag{12.11}$$

Write $R = \sum_{i=1}^r a_i \otimes b_i$, where R is as small as possible. Then $\{a_1, \ldots, a_r\}$ and $\{b_1, \ldots, b_r\}$ are linearly independent sets by Lemma 1.2.2 and are therefore bases for $R_{(\ell)}$ and $R_{(r)}$ respectively. When H^o is a dense subspace of H^* observe that $R_{(\ell)} = \mathrm{Im}(g_R)$ and $R_{(r)} = \mathrm{Im}(f_R)$ and thus $R_{(\ell)}$ and $R_{(r)}$ are sub-bialgebras of H by Proposition 12.2.11. See the paragraph following Definition 1.3.7. It is always the case that the subspaces $R_{(\ell)}$ and $R_{(r)}$ are sub-bialgebras of H. See Exercise 12.2.7.

We note that:

$$\mathrm{Rank}(R) = \mathrm{Dim}(R_{(\ell)}) = \mathrm{Dim}(R_{(r)}). \tag{12.12}$$

Assume further that H has antipode S. Since $R_{(\ell)}$ and $R_{(r)}$ are finite-dimensional sub-bialgebras of H they must be Hopf subalgebras of H by Proposition 7.6.1. Since H is quasitriangular S is bijective by part (f) of Theorem 12.2.8. Write $R = R_i \otimes R^i$. There are equivalent versions of (QT.5) from which we deduce commutation relations for $R_{(\ell)}$ and $R_{(r)}$:

$$aR_i \otimes R^i = R_i a_{(2)} \otimes S(a_{(1)}) R^i a_{(3)}; \tag{12.13}$$

$$R_i a \otimes R^i = a_{(2)} R_i \otimes a_{(1)} R^i S(a_{(3)}); \tag{12.14}$$

$$R_i \otimes aR^i = S^{-1}(a_{(3)}) R_i a_{(1)} \otimes R^i a_{(2)}; \tag{12.15}$$

and

$$R_i \otimes R^i a = a_{(3)} R_i S^{-1}(a_{(1)}) \otimes a_{(2)} R^i \tag{12.16}$$

for all $a \in H$. Recall $R = \sum_{i=1}^{r} a_i \otimes b_i$ where $\{a_1, \ldots, a_r\}$ and $\{b_1, \ldots, b_r\}$ are linearly independent. Fix $1 \leq \ell \leq r$ and let $p \in H^*$ satisfy $p(b_i) = \delta_{\ell,i}$ for all $1 \leq i \leq r$. Setting $h = a$ and applying $\mathrm{I}_H \otimes p$ to both sides of (12.13) and (12.14) we deduce the commutation relations

$$ha_\ell = \sum_{i=1}^{r} a_i p(h,i) \quad \text{and} \quad a_\ell h = \sum_{i=1}^{r} p(i,h) a_i \tag{12.17}$$

for all $h \in H$, where

$$p(h,i) = h_{(2)} p(S(h_{(1)}) b_i h_{(3)}) \quad \text{and} \quad p(i,h) = p(h_{(1)} b_i S(h_{(3)})) h_{(2)}$$

for all $1 \leq i \leq r$ and $h \in H$. Now let $p' \in H^*$ be any functional which satisfies $p'(a_i) = \delta_{i,\ell}$ for all $1 \leq i \leq r$. Applying $p' \otimes \mathrm{I}_H$ to both sides of the equations of (12.15) and (12.16) we obtain the commutations relations

$$hb_\ell = \sum_{i=1}^{r} b_i p'(h,i) \quad \text{and} \quad b_\ell h = \sum_{i=1}^{r} p'(i,h) b_i \tag{12.18}$$

respectively, where $p'(h,i) = h_{(2)} p'(S^{-1}(h_{(3)}) a_i h_{(1)})$ and $p'(i,h) = p'(h_{(3)} a_i S^{-1}(h_{(1)})) h_{(2)}$ for all $h \in H$ and $1 \leq i \leq r$.

Proposition 12.2.13. *Let (H, R) be a quasitriangular Hopf algebra over the field k, and let $R_{(\ell)}, R_{(r)}$ be defined as above. Then:*

(a) *$CR_{(\ell)} = R_{(\ell)} C$ and $CR_{(r)} = R_{(r)} C$ for all subcoalgebras C of H.*

(b) $R_{(\ell)}$, $R_{(r)}$, *and* $H_R = R_{(\ell)}R_{(r)}$ *are finite-dimensional Hopf subalgebras of* H.
(c) $\mathrm{Rank}(R)$ *divides* $\mathrm{Dim}(H_R)$.

Proof. Part (a) follows by the commutation relations of (12.17) and (12.18). We have noted that $R_{(\ell)}$ and $R_{(r)}$ are finite-dimensional sub-bialgebras of H. By part (a) these commute; thus the product H_R is a sub-bialgebra of H. Since finite-dimensional sub-bialgebras of a Hopf algebra are Hopf subalgebras by Proposition 7.6.1 part (b) follows. Part (c) follows by (12.12), part (b), and Corollary 9.3.4. □

Since $R \in H_R \otimes H_R$ the pair (H_R, R) is a quasitriangular. It is minimal in the following sense:

Definition 12.2.14. A *minimal quasitriangular Hopf algebra over* k is a quasitriangular Hopf algebra (H, R) over k such that H is the only Hopf subalgebra K of H which satisfies $R \in K \otimes K$.

We will continue our discussion of H_R in Section 13.3.

Exercises

Throughout these exercises H is a Hopf algebra over k and $R \in H \otimes H$.

Exercise 12.2.1. Complete the proof of Proposition 12.2.11.

Exercise 12.2.2. Suppose (H, R) is a quasitriangular Hopf algebra over k.

(a) Show that $(H^{cop}, R_{2,1})$, $(H^{op}, R_{2,1})$, $(H^{cop\ op}, R)$, and (H^{cop}, R^{-1}) are quasitriangular.
(b) Suppose that $f : H \longrightarrow H'$ is a Hopf algebra map. Show that $(\mathrm{Im}(f), (f \otimes f)(R))$ is quasitriangular and $f : (H, R) \longrightarrow (\mathrm{Im}(f), (f \otimes f)(R))$ is a map of quasitriangular Hopf algebras.

Exercise 12.2.3. Suppose that (H, R) is a quasitriangular, the characteristic of k is zero, and H is pointed irreducible. Show that $R = 1 \otimes 1$ and thus H is cocommutative. [Hint: See Corollary 9.1.2.]

Exercise 12.2.4. Suppose that $(H, \Delta(a))$ is quasitriangular, where $a \in H$. Show that $a = 1$ and thus H is cocommutative.

Exercise 12.2.5. Let A be a bialgebra over k and regard ${}_A\mathcal{M}$ as the monoidal category described at the end of Section 11.1. Let $R \in A \otimes A$ and for $M, N \in {}_A\mathcal{M}$ let $B_{M,N}$ be defined as in (12.8). Show that:

(a) $B_{M,N}$ is a module map for all $M, N \in {}_A\mathcal{M}$ if and only if (QT.5) holds for (A, R).

(b) The $B_{M,N}$'s form a prebraiding for ${}_A\mathcal{M}$ if and only if (QT.1)–(QT.5) hold for (A, R).

Exercise 12.2.6. Let H be a bialgebra over k such that H^o is a dense subspace of H^* and let $R \in H \otimes H$. Show that (H, R) is quasitriangular if

(a) $f_R : H^o \longrightarrow H^{cop}$ is a bialgebra map and (QT.5) holds for (H, R) or

(b) $g_R : H^o \longrightarrow H^{op}$ is a bialgebra map and (QT.5) holds for (H, R).

Exercise 12.2.7. Let (H, R) be a quasitriangular bialgebra over k. Show that $R_{(\ell)}$ and $R_{(r)}$ are sub-bialgebras of H. [Hint: Write $R = \sum_{i=1}^{r} a_i \otimes b_i$, where r is as small as possible. For fixed $1 \leq j \leq r$ choose $p \in H^*$ such that $p(b_i) = \delta_{i,j}$ for all $1 \leq i \leq r$. Note that $\Delta(a_j) = (\mathrm{I}_H \otimes p)((\Delta \otimes \mathrm{I}_H)(R))$.]

Exercise 12.2.8. Suppose H has a bijective antipode and $R \in H \otimes H$ satisfies (QT.5); that is $\Delta^{cop}(h)R = R\Delta(h)$ for all $h \in H$. Then (12.13) – (12.16) hold. Let $\mathrm{f}_R, \mathrm{g}_R : H^* \longrightarrow H$ be defined as in Section 12.2. Thus $\mathrm{f}_R(p) = (p \otimes \mathrm{I}_H)(R)$ and $\mathrm{g}_R(p) = (\mathrm{I}_H \otimes p)(R)$ for all $p \in H^*$. Show that (12.13) – (12.16) can be reformulated as:

$$h\mathrm{g}_R(p) = \mathrm{g}_R(h_{(3)} \succ p \prec S(h_{(1)}))h_{(2)}; \tag{12.19}$$

$$\mathrm{g}_R(p)h = h_{(2)}\mathrm{g}_R(S(h_{(3)}) \succ p \prec h_{(1)}); \tag{12.20}$$

$$h\mathrm{f}_R(p) = \mathrm{f}_R(h_{(1)} \succ p \prec S^{-1}(h_{(3)}))h_{(2)}; \tag{12.21}$$

and

$$\mathrm{f}_R(p)h = h_{(2)}\mathrm{f}_R(S^{-1}(h_{(1)}) \succ p \prec h_{(3)}) \tag{12.22}$$

for all $h \in H$ and $p \in H^*$.

Exercise 12.2.9. Suppose (H, R) is quasitriangular.

(a) Using Exercise 1.2.8 show that there is a unique Hopf subalgebra H' of H which satisfies $R \in H' \otimes H'$ and whenever K is a Hopf subalgebra of H with $R \in K \otimes K$ then $H' \subseteq K$.

(b) Show that (H', R) is quasitriangular and $H' = H_R$.

Exercise 12.2.10. Suppose $n \geq 1$ and k has an nth root of unity q. Let $H = H_{n,q}$ be the Taft algebra described in Section 7.3.

(a) Determine all bialgebra maps $f : H^* \longrightarrow H^{cop}$. [Hint: See Exercise 7.4.3.]

(b) Show that if H admits a quasitriangular structure then $n = 1$ or $n = 2$.

Exercise 12.2.11. We follow up on the preceding exercise. Suppose that the characteristic of k is not 2 and let $H = H_{2,q}$, where $q = -1$. The Taft algebra H

in this case is referred to as Sweedler's Example. For $\alpha \in k$ set

$$R_\alpha = \frac{1}{2}(1\otimes 1 + 1\otimes a + a\otimes 1 - a\otimes a) + \frac{\alpha}{2}(x\otimes x + x\otimes ax + ax\otimes ax - ax\otimes x).$$

Show that:

(a) The quasitriangular structures on H are the (H, R_α)'s.
(b) (H, R_α) is triangular for all $\alpha \in k$.
(c) $\mathrm{Rank}(R_0) = 2$ and (A, R_0) is not minimal quasitriangular.
(d) $\mathrm{Rank}(R_\alpha) = 4$ and (A, R_α) is minimal quasitriangular for all $\alpha \in k\backslash 0$. [Hint: See Lemma 1.2.2.]

12.3 Grouplike and ribbon elements

Throughout this section (H, R) is a quasitriangular Hopf algebra with antipode S over k. For $\eta \in \mathrm{G}(H^o)$ let $g_\eta = (\mathrm{I}_H\otimes\eta)(R)$. Writing $R = R_i\otimes R^i$ we have $g_\eta = R_i\eta(R^i)$. There is a relationship between $\mathrm{G}(H^o)$ and $\mathrm{G}(H)$ via the correspondence $\eta \mapsto g_\eta$.

Proposition 12.3.1. *Let (H, R) be a quasitriangular Hopf algebra over the field k. Then:*

(a) *$g_\eta \in \mathrm{G}(H)$ for all $\eta \in \mathrm{G}(H^o)$.*
(b) *The map $\mathrm{G}(H^o) \longrightarrow \mathrm{G}(H)$ defined by $\eta \mapsto g_\eta$ is a group anti-homomorphism.*
(c) *$(h{\leftharpoonup}\eta)g_\eta = g_\eta(\eta{\rightharpoonup}h)$ for all $\eta \in \mathrm{G}(H^o)$ and $h \in H$.*
(d) *$g_\eta \in \mathrm{Z}(\mathrm{G}(H))$ for all $\eta \in \mathrm{G}(H^o)$.*
(e) *g_η is in the center of H if and only if η is in the center of H^*.*
(f) *For $\eta \in \mathrm{G}(H^o)$ let T_η be the automorphism of H^* defined by $T_\eta(p) = \eta p\eta^{-1}$ for all $p \in H^*$ and let t_η be the automorphism of H defined by $t_\eta(h) = g_\eta h g_\eta^{-1}$ for all $h \in H$. Then $T_\eta = t_\eta^*$.*

Proof. The image of a grouplike element is a grouplike element under a coalgebra map. Thus parts (a)–(c) follow by part (b) of Proposition 12.2.11 since $g_R(\eta) = g_\eta$ for all $\eta \in \mathrm{G}(H^o)$. Part (d) follows from part (c). Part (e) does as well since $p \in H^*$ is in the center of H^* if and only if $h{\leftharpoonup}p = p{\rightharpoonup}h$ for all $h \in H$ and $H = H{\leftharpoonup}\eta$ since η is invertible. Part (f) is an easy consequence of part (c). □

The element $uS(u^{-1})$ of H plays an important role in the theory of quasitriangular Hopf algebras.

Proposition 12.3.2. *Let (H, R) be a quasitriangular Hopf algebra with antipode S over k and let u be the Drinfel'd element of (H, R). Then:*

(a) $\Delta(u) = (u \otimes u)(R_{2,1}R)^{-1} = (R_{2,1}R)^{-1}(u \otimes u)$ *and* $\epsilon(u) = 1$.
(b) $u \in \mathrm{G}(H)$ *if and only if* (H, R) *is triangular.*
(c) $\mathsf{g} = uS(u^{-1}) \in \mathrm{G}(H)$ *and* $S^4(h) = \mathsf{g}h\mathsf{g}^{-1}$ *for all* $h \in H$.

Proof. To show part (a) we write $R = R_i \otimes R^i$. Therefore $u = S(R^i)R_i$. Applying $\mathrm{I}_H \otimes \mathrm{I}_H \otimes \Delta$ to both sides of (QT.1) we obtain

$$\Delta(R_i) \otimes \Delta(R^i) = R_i R_j \otimes R_k R_\ell \otimes R^j R^\ell \otimes R^i R^k.$$

Using (12.7) and part (b) of Theorem 12.2.8 we calculate

$$\begin{aligned} \Delta(u) &= S(R^i_{(2)})R_{i\,(1)} \otimes S(R^i_{(1)})R_{i\,(2)} \\ &= S(R^i R^k)R_i R_j \otimes S(R^j R^\ell)R_k R_\ell \\ &= S(R^k)S(R^i)R_i R_j \otimes S(R^\ell)S(R^j)R_k R_\ell \\ &= S(R^k)S^2(R_j)u \otimes S(R^\ell)S(R^j)R_k R_\ell \\ &= S(R^k)S(R_j)u \otimes S(R^\ell)R^j R_k R_\ell, \end{aligned}$$

and thus

$$\Delta(u) = S(R_j R^k)u \otimes S(R^\ell)R^j R_k R_\ell.$$

Using (12.15) and (QT.5) we derive the commutation relation

$$R_j R^k \otimes h R^j R_k = S^{-1}(h_{(3)})R_j R^k h_{(2)} \otimes R^j R_k h_{(1)} \tag{12.23}$$

for all $h \in H$. Applying $\mathrm{I}_H \otimes \Delta \otimes \mathrm{I}_H$ to both sides of the equation of (QT.3) we obtain $R_\ell \otimes \Delta^2(R^\ell) = R_\ell R_m R_n \otimes R^n \otimes R^m \otimes R^\ell$, hence

$$R_\ell \otimes \Delta^2(S(R^\ell)) = R_\ell R_m R_n \otimes S(R^\ell) \otimes S(R^m) \otimes S(R^n). \tag{12.24}$$

Applying $\mathrm{I}_H \otimes S$ to both sides of $R_k S(R_m) \otimes R^k R^m = 1 \otimes 1$, which follows from part (a) of Theorem 12.2.8, and using part (b) of the same, we obtain

$$R_k R_m \otimes R^m S(R^k) = 1 \otimes 1. \tag{12.25}$$

Using (12.23)–(12.25) and part (b) of Theorem 12.2.8 again, we continue our calculation of

$$\begin{aligned} \Delta(u) &= S(S^{-1}(S(R^n))R_j R^k S(R^m))u \otimes R^j R_k S(R^\ell)R_\ell R_m R_n \\ &= S^2(R^m)S(R^k)S(R_j)S(R^n)u \otimes R^j R_k S^2(R_m)S^2(R_n)u \\ &= S(R_j)R^n u \otimes R^j S(R_n)u \\ &= R^{-1}R_{2,1}^{-1}(u \otimes u) \\ &= (R_{2,1}R)^{-1}(u \otimes u). \end{aligned}$$

Since $R = (S\otimes S)(R)$, and hence $R_{2,1} = (S\otimes S)(R_{2,1})$, both R and $R_{2,1}$ commute with $u\otimes u$ as $S^2(h)u = uh$ for all $h \in H$. We have established part (a).

To see part (b) we deduce from part (a) that $\Delta(S(u)) = (S(u)\otimes S(u))(R_{2,1}R)^{-1}$ and the two factors commute; thus $\Delta(S(u^{-1})) = (S(u^{-1})\otimes S(u^{-1}))(R_{2,1}R)$ and the two factors commute. It now follows that $uS(u^{-1}) \in \mathrm{G}(H)$ by part (a) again. To establish part (c), observe that if $v \in H$ is invertible and $S^2(h) = vhv^{-1}$ for all $h \in H$ then $S^3(h) = S(v^{-1})S(h)S(v)$, or equivalently $S^2(h) = S(v^{-1})hS(v)$ for all $h \in H$. Thus $S^4(h) = vS(v^{-1})h(vS(v^{-1}))^{-1}$ for all $h \in H$. □

Let $R_{(\ell)}$ and $R_{(r)}$ be the subspaces of H defined by (12.11). We have shown that $H_R = R_{(\ell)}R_{(r)}$ is a finite-dimensional Hopf subalgebra of H in part (b) of Proposition 12.2.13. Since $R \in H_R\otimes H_R$ it follows that $u, S(u)$, the grouplike element $uS(u^{-1})$, and the quantum Casimir element $uS(u)$ belong to H_R. Thus to study these elements, we can replace H with any finite-dimensional Hopf subalgebra which contains H_R.

Proposition 12.3.3. *Let (H, R) be a finite-dimensional quasitriangular Hopf algebra with antipode S over the field k, let g and α be the H-distinguished grouplike elements of H and H^* respectively, let u be the Drinfel'd element of (H, R), and let g be the element of part (c) of Proposition 12.3.2. Then:*

(a) $\mathrm{g} = uS(u^{-1}) = gg_{\alpha^{-1}} = g_{\alpha^{-1}}g = (g_\alpha)^{-1}g$.
(b) *If H is unimodular then $g = uS(u^{-1})$ is the H-distinguished grouplike element of H.*
(c) *The quantum Casimir element of H is $uS(u) = g^{-1}g_\alpha u^2 = u^2g^{-1}g_\alpha$.*

Proof. Part (b) is a direct consequence of part (a), as $g_\epsilon = 1$, and part (c) follows from part (a) and the calculation $uS(u) = (uS(u^{-1}))^{-1}u^2$ together with the observation that u commutes with grouplike elements of H. It remains to show part (a).

Let Λ be a non-zero left integral for H and write $R = R_i\otimes R^i$. Then we have $\Lambda_{(2)}R_i\otimes\Lambda_{(1)}R^i = R_i\Lambda_{(1)}\otimes R^i\Lambda_{(2)}$ by (QT.5). Now $\Lambda_{(2)}\otimes\Lambda_{(1)} = \Lambda_{(1)}\otimes S^2(\Lambda_{(2)})g$ by part (f) of Theorem 10.5.4. Therefore $\Lambda_{(1)}R^i\otimes\Lambda_{(2)}R_i = R^i\Lambda_{(1)}\otimes R_iS^2(\Lambda_{(2)})g$. Since Λ is a left ϵ-integral and a right α-integral we deduce that

$$\Lambda_{(1)}\otimes\Lambda_{(2)}S(R^i\leftharpoonup\alpha)R_i = \Lambda_{(1)}\otimes R_iS^2(S^{-1}(R^i)\Lambda_{(2)})g \tag{12.26}$$

from the preceding equation. Since $(H, \leftharpoonup)$ is a free left H-module with basis $\{\Lambda\}$, by part (d) of Theorem 10.2.2, there is a $p \in H^*$ such that

$1 = \Lambda \leftharpoonup p = p(\Lambda_{(1)})\Lambda_{(2)}$. Applying $p \otimes \mathrm{I}_H$ to both sides of the equation of (12.26) we see that $S(R^i \leftharpoonup \alpha)R_i = R_i S(R^i)g$. As $R_i \otimes R^i \leftharpoonup \alpha = R_i g_\alpha \otimes R^i$ by (QT.3), and $S(u) = S(R_i)S^2(R^i) = R_i S(R^i)$, it follows that $ug_\alpha = S(u)g$. Since u and $S(u)$ commute, part (a) follows from this equation and part (d) of Proposition 12.3.1. □

Let $c = uS(u)$ be the quantum Casimir element and let $v \in H$. Consider the following axioms:

(R.1) $v^2 = c$;
(R.2) $S(v) = v$;
(R.3) $\epsilon(v) = 1$; and
(R.4) $\Delta(v) = (v \otimes v)(R_{2,1}R)^{-1}$.

Definition 12.3.4. Let (H, R) be a quasitriangular Hopf algebra with antipode S over k. A *quasi-ribbon element of* (H, R) is a $v \in H$ which satisfies (R.1)–(R.4). A *ribbon element of* (H, R) is a quasi-ribbon element of (H, R) which is in the center of H.

Definition 12.3.5. A *ribbon Hopf algebra over* k is a triple (H, R, v), where (H, R) is a quasitriangular Hopf algebra over k and v is a ribbon element of (H, R).

Quasi-ribbon elements and ribbon elements can be characterized in terms of grouplike elements. Let $\mathrm{QR}(H, R)$ denote the set of quasi-ribbon elements and let $\mathrm{R}(H, R)$ denote the set of ribbon elements of (H, R). Let u be the Drinfel'd element of (H, R) and $\mathrm{g} = uS(u^{-1})$. The set

$$\mathrm{G}(\mathrm{QR}(H, R)) = \{\ell \in \mathrm{G}(H) \mid \ell^2 = \mathrm{g}^{-1}\}$$

corresponds to quasi-ribbon elements and its subset

$$\mathrm{G}(\mathrm{R}(H, R)) = \{\ell \in \mathrm{G}(H) \mid \ell^2 = \mathrm{g}^{-1} \text{ and } S^2(h) = \ell^{-1}h\ell \text{ for all } h \in H\}$$

corresponds to ribbon elements.

Theorem 12.3.6. *Let (H, R) be a quasitriangular Hopf algebra over the field k and let u be the Drinfel'd element of (H, R). Then:*

(a) *The map* $\mathrm{G}(\mathrm{QR}(H, R)) \longrightarrow \mathrm{QR}(H, R)$ *given by* $\ell \mapsto u\ell = \ell u$ *is a set bijection.*
(b) *The map of part (a) restricts to a bijection* $\mathrm{G}(\mathrm{R}(H, R)) \longrightarrow \mathrm{R}(H, R)$.
(c) *Suppose* $v \in \mathrm{R}(H, R)$ *and let* $Z = \{z \in \mathrm{Z}(H) \cap \mathrm{G}(H) \mid z^2 = 1\}$. *Then* Z *is a subgroup of* $\mathrm{G}(H)$ *and there is a bijection* $Z \longrightarrow \mathrm{G}(\mathrm{R}(H, R))$ *given by* $z \mapsto zv$.

(d) $\mathrm{G}(\mathrm{R}(H,R)) = \{\ell \in \mathrm{G}(H) \,|\, \ell u \in \mathrm{Z}(H)$ *and* $S(u) = \ell^2 u\}$.

Proof. Recall that u, and hence $S(u)$, commute with all grouplike elements of H. By part (a) of Proposition 12.3.3 we have $u = \mathrm{g}S(u) = S(u)\mathrm{g}$ and thus $c = uS(u) = \mathrm{g}^{-1}u^2 = u^2\mathrm{g}^{-1}$. Recall that $S^2(h) = uhu^{-1}$ for all $h \in H$ by part (e) of Theorem 12.2.8. We first show part (a).

Let $\ell \in \mathrm{G}(H)$ satisfy $\ell^2 = \mathrm{g}^{-1}$ and set $v = \ell u = u\ell$. Then $v^2 = \ell^2 u^2 = \mathrm{g}^{-1}u^2 = c$, $S(v) = S(u)\ell^{-1} = S(u)\mathrm{g}\ell = u\ell = v$, $\epsilon(v) = \epsilon(\ell)\epsilon(u) = 1$, and $\Delta(v) = (v \otimes v)(R_{2,1}R)^{-1}$; the latter two assertions follow from part (a) of Proposition 12.3.2.

Conversely, suppose that v is a quasi-ribbon element of (H,R). Using (R.2) we see that $v = S^2(v) = uvu^{-1}$ which means that u and v commute. Let $\ell = u^{-1}v$. Thus $\Delta(\ell) = \ell \otimes \ell$ by part (a) of Proposition 12.3.2 and (R.4), and $\epsilon(\ell) = 1$ by (R.3). Using (R.1) we compute $\ell^2 = u^{-2}v^2 = u^{-2}c = u^{-2}u^2\mathrm{g}^{-1} = \mathrm{g}^{-1}$. We have established part (a).

We show part (b). Assume $\ell \in \mathrm{G}(\mathrm{R}(H,R))$. Then ℓu is in the center of H if and only if $\ell uh = h\ell u$, or equivalently $uhu^{-1} = \ell^{-1}h\ell$, for all $h \in H$. As $S^2(h) = uhu^{-1}$ for all $h \in H$, part (b) is established. Part (c) is an easy consequence of parts (a) and (b). In light of the fact that u commutes with all grouplike elements of H, part (d) follows by appealing to the definition of ribbon element and part (a). □

The previous theorem, Lagrange's Theorem for finite groups, and Corollary 9.3.5 imply:

Corollary 12.3.7. *Let (H,R) be a finite-dimensional quasitriangular Hopf algebra over the field k. Then:*

(a) *The number of quasi-ribbon elements of (H,R) is finite.*
(b) *The number of ribbon elements of (H,R) is zero or divides* $\mathrm{Dim}(H)$.

□

By the previous theorem, Corollary 9.3.5, and since every element of a finite group of odd order is the square of exactly one element:

Corollary 12.3.8. *Let (H,R) be a finite-dimensional quasitriangular Hopf algebra over the field k. Suppose that $|\mathrm{G}(H)|$ or* $\mathrm{Dim}(H)$ *is odd. Then (H,R) has a unique quasi-ribbon element.* □

We end this section with a characterization of ribbon elements in terms of cocommutative elements of the dual.

Corollary 12.3.9. *Let (H, R) be a quasitriangular Hopf algebra with antipode S over the field k, let u be the Drinfel'd element of (H, R), and suppose that H is finite-dimensional and unimodular. Let λ be a non-zero right integral for H^* and $\ell \in \mathrm{G}(H)$. Then ℓu is a ribbon element of (H, R) if and only if $\chi = \lambda{\prec}\ell^{-1}$ is cocommutative and $S^*(\chi) = \chi$.*

Proof. χ generates H^* as a free right H-module since λ does by part (e) of Theorem 10.2.2. Thus χ is cocommutative if and only if $S^2(h) = \ell^{-1}h\ell$ for all $h \in H$ by part (a) of Proposition 10.7.1. Now $\lambda(ba) = \lambda(S^2(a)b)$ for all $a, b \in H$ by part (e) of Theorem 10.5.4 and $\lambda{\circ}S = \lambda{\prec}g$ by part (b) of the same, where g is the H-distinguished grouplike element of H. The calculation

$$(\chi{\circ}S)(h) = \lambda(\ell^{-1}S(h)) = \lambda(S(h\ell)) = \lambda(gh\ell) = \lambda(\ell gh)$$

for all $h \in H$ shows that $\chi{\circ}S = \chi$ if and only if $\lambda{\prec}\ell^{-1} = \lambda{\prec}(\ell g)$ which is the case if and only if $\ell^{-1} = \ell g$, or equivalently $\ell^{-2} = g = \mathrm{g}$; the latter follows by part (a) of Proposition 12.3.3 since H unimodular. At this point the corollary follows by part (b) of Theorem 12.3.6. □

12.4 Factorizable Hopf algebras

Throughout this section (H, R) is a quasitriangular Hopf algebra with antipode S over the field k. Let $\mathcal{R} \in H{\otimes}H$ and $f_{\mathcal{R}}, g_{\mathcal{R}} : H^* \longrightarrow H$ be defined by $f_{\mathcal{R}}(p) = (p{\otimes}\mathrm{I}_H)(\mathcal{R})$ and $g_{\mathcal{R}}(p) = (\mathrm{I}_H{\otimes}p)(\mathcal{R})$ for all $p \in H^*$. Let $\beta : H^*{\times}H^* \longrightarrow k$ be the bilinear form given by $\beta(p, q) = (p{\otimes}q)(\mathcal{R})$ for all $p, q \in H^*$. Then $\beta_\ell = f_{\mathcal{R}}$ and $\beta_r = g_{\mathcal{R}}$. Since $\mathrm{Im}(f_{\mathcal{R}})$ and $\mathrm{Im}(g_{\mathcal{R}})$ are both finite-dimensional, $f_{\mathcal{R}}$ and $g_{\mathcal{R}}$ are both isomorphisms or neither are. See Exercise 12.4.1. When H is finite-dimensional $f_{\mathcal{R}}$ is the map of Proposition 12.2.11.

Definition 12.4.1. A *factorizable Hopf algebra over the field k* is a quasitriangular Hopf algebra (H, R) over k such that $f_{R_{2,1}R}$, or equivalently $g_{R_{2,1}R}$, is a linear isomorphism.

Factorizable Hopf algebras are thus finite-dimensional. They are rather special quasitriangular Hopf algebras.

Proposition 12.4.2. *Let (H, R) be a finite-dimensional quasitriangular Hopf algebra over the field k. Then:*

(a) $f_{R_{2,1}R} = f_{R_{2,1}}{*}f_R$ *in the convolution algebra* $\mathrm{Hom}(H^*, H)$.

(b) $\mathrm{Im}(f_{R_{2,1}R}) \subseteq H_R$. *Thus factorizable Hopf algebras are minimal quasitriangular.*

(c) *Suppose that* $x \in H$ *satisfies* $\Delta(x) = (x \otimes x)(R_{2,1}R)$. *Then* $x \leftharpoonup p = x f_{R_{2,1}R}(p \prec x)$ *for all* $p \in H^*$.

(d) *Let* α *be the* H*-distinguished grouplike element of* H^*. *Then* $f_{R_{2,1}R}(\epsilon) = 1 = f_{R_{2,1}R}(\alpha)$. *Thus factorizable Hopf algebras are unimodular.*

Proof. Let $\mathcal{R}, \mathcal{R}' \in H \otimes H$ and write $\mathcal{R} = \mathcal{R}_i \otimes \mathcal{R}^i$, $\mathcal{R}' = \mathcal{R}'_j \otimes \mathcal{R}'^j$. Then

$$f_{\mathcal{R}\mathcal{R}'}(p) = p(\mathcal{R}_i\mathcal{R}'_j)\mathcal{R}^i\mathcal{R}'^j = p_{(1)}(\mathcal{R}_i)p_{(2)}(\mathcal{R}'_j)\mathcal{R}^i\mathcal{R}'^j = f_{\mathcal{R}}(p_{(1)})f_{\mathcal{R}'}(p_{(2)})$$

for all $p \in H^*$ from which part (a) follows. Since $\mathrm{Im}(f_{R_{2,1}}), \mathrm{Im}(f_R) \subseteq H_R$, and the latter is a subalgebra of H, part (b) follows from part (a). As for part (c), for $x \in H$ which satisfies $\Delta(x) = (x \otimes x)(\mathcal{R})$ and for $p \in H^*$ we compute $x \leftharpoonup p = p(x\mathcal{R}_i)x\mathcal{R}^i = x f_{\mathcal{R}}(p \prec x)$. It remains to show part (d).

First of all note that $f_{R_{2,1}R}(\epsilon) = f_{R_{2,1}}(\epsilon)f_R(\epsilon) = 1$ by (QT.2) and (QT.4). Since $\alpha \in \mathrm{G}(H^*)$ we have $f_{R_{2,1}R}(\alpha) = f_{R_{2,1}}(\alpha)f_R(\alpha) = g'_\alpha g_\alpha$, where $g'_\alpha = f_{R_{2,1}}(\alpha) = \alpha(R^i)R_i$ by part (a) again. We will show that $g'_\alpha = (g_\alpha)^{-1}$ which will complete the proof of part (d).

Recall that $(H^{cop}, R_{2,1})$ is a quasitriangular Hopf algebra with antipode S^{-1} over k by Exercise 12.2.2. The Drinfel'd element of $(H^{cop}, R_{2,1})$ is $S^{-1}(R_i)R^i = S^{-1}(S(R^i)R_i) = S^{-1}(u)$. The calculation

$$S^{-1}(u)S^{-1}(S^{-1}(u)) = S^{-1}(S^{-1}(u)u) = S^{-1}(S(u)u) = S^{-1}(uS(u))$$

shows that the quantum Casimir element for $(H^{cop}, R_{2,1})$ is $S^{-1}(uS(u))$. Let g be the H-distinguished grouplike element of H. Then g^{-1} and α are the H^{cop}-distinguished grouplike elements of H^{cop} and $H^{cop\,*}$ respectively by Exercise 10.5.1. By part (c) of Proposition 12.3.3 we have $(g'_\alpha)^{-1}g^{-1} = S^{-1}(uS(u)) = S^{-1}((g_\alpha)^{-1}g) = g^{-1}g_\alpha$ from which we deduce that $g'_\alpha = (g_\alpha)^{-1}$. □

Part (c) of the preceding proposition leads to a description of factorizable Hopf algebras in terms of elements resembling ribbon elements.

Corollary 12.4.3. *Let* (H, R) *be a finite-dimensional quasitriangular Hopf algebra over the field* k. *Suppose that* $x \in H$ *is invertible and* $\Delta(x) = (x \otimes x)(R_{2,1}R)$. *Then the following are equivalent:*

(a) (H, R) *is factorizable.*

(b) $\mathrm{Rank}(\Delta(x)) = \mathrm{Dim}(H)$.

(c) $x \leftharpoonup H^* = H$, *that is* $(H, \leftharpoonup)$ *is a free right* H*-module with basis* $\{x\}$.

□

Under the hypothesis of the corollary $\lambda(x) \neq 0$, where λ is a non-zero right integral for H^* and (H, R) is factorizable. More generally:

Corollary 12.4.4. *Assume the hypothesis of the preceding corollary and suppose that (H, R) is factorizable. Then $\mu(x) \neq 0$ for all non-zero generalized right integrals μ for H^*.*

Proof. There is an $a \in \mathrm{G}(H)$ such that $h{\leftharpoonup}\mu = \mu(h_{(1)})h_{(2)} = \mu(h)a$ for all $h \in H$. Since $0 \neq x{\leftharpoonup}\mu = \mu(x)a$ by Corollary 12.4.3 it follows that $\mu(x) \neq 0$. □

Let

$$\mathrm{C}_r(H) = \{p \in H^* \mid p(hh') = p(S^2(h')h) \text{ for all } h, h' \in H\}.$$

Then C_r is a subalgebra of H^*. Let u be the Drinfel'd element of (H, R). Then $f : \mathrm{Cc}(H^*) \longrightarrow \mathrm{C}_r(H)$ defined by $f(p) = p{\prec}u^{-1}$ is a linear isomorphism of the subspace of cocommutative elements of H^* and $\mathrm{C}_r(H)$. Since $\eta \circ S^2 = \eta$ for all $\eta \in \mathrm{G}(H^*)$ it follows that $\mathrm{G}(H^*) \subseteq \mathrm{Cc}(H^*) \cap \mathrm{C}_r(H)$.

There is an algebra map $F_r : \mathrm{C}_r(H) \longrightarrow \mathrm{Z}(H)$ which is an isomorphism when (H, R) is factorizable. To derive F_r we start with two commutation relations. Writing $R = R_i \otimes R^i = R_j \otimes R^j$, and thus $R_{2,1}R = R^iR_j \otimes R_iR^j$, we have

$$hR^iR_j \otimes R_iR^j = R^iR_jh_{(1)} \otimes S^{-1}(h_{(3)})R_iR^jh_{(2)} \tag{12.27}$$

and

$$R^iR_j \otimes hR_iR^j = S(h_{(1)})R^iR_jh_{(2)} \otimes R_iR^jh_{(3)} \tag{12.28}$$

for all $h \in H$ which follow by (12.15) and (12.13) respectively together with (QT.5).

Proposition 12.4.5. *Let (H, R) be a quasitriangular Hopf algebra over the field k and set $F = f_{R_{2,1}R}$. Then:*

(a) *$F(\mathrm{C}_r(H)) \subseteq \mathrm{Z}(H)$.*
(b) *$F(pq) = F(p)F(q)$ for all $p \in \mathrm{C}_r(H)$ and $q \in H^*$. In particular the restriction $F_r = F|\mathrm{C}_r(H)$ is an algebra map $F_r : \mathrm{C}_r(H) \longrightarrow \mathrm{Z}(H)$.*

Proof. Let $h \in H$ and $p \in \mathrm{C}_r(H)$. Using (12.28) we compute

$$\begin{aligned} hF(p) &= p(R^iR_j)hR_iR^j \\ &= p(S(h_{(1)})R^iR_jh_{(2)})R_iR^jh_{(3)} \\ &= p(S^2(h_{(2)})S(h_{(1)})R^iR_j)R_iR^jh_{(3)} \\ &= p(S(h_{(1)}S(h_{(2)}))R^iR_j)R_iR^jh_{(3)} \\ &= p(R^iR_j)R_iR^jh \\ &= F(p)h \end{aligned}$$

which shows that $F(\mathrm{C}_r(H)) \subseteq \mathrm{Z}(H)$. We have shown part (a) holds.

Now let $p \in \mathrm{C}_r(H)$ and $q \in H^*$. Then

$$\begin{aligned} F(pq) &= pq(R^i R_j)R_i R^j \\ &= p(R^i{}_{(1)}R_{j\,(1)})q(R^i{}_{(2)}R_{j\,(2)})R_i R^j \\ &= p(R^i R_j)q(R^k R_\ell)R_k R_i R^j R^\ell \\ &= q(R^k R_\ell)R_k F(p)R^\ell \\ &= F(p)F(q); \end{aligned}$$

the last equation follows since $F(p) \in \mathrm{Z}(H)$ by part (a). Since $F(\epsilon) = 1$ part (b) is established. □

Continuing, the calculation

$$\begin{aligned} \Delta(F(p)) &= p(R^i R_j)\Delta(R_i R^j) \\ &= p(R^i R_j)R_{i\,(1)}R^j{}_{(1)}\otimes R_{i\,(2)}R^j{}_{(2)} \\ &= p(R^i R^k R_j R_\ell)R_i R^\ell \otimes R_k R^j \\ &= p(S^2(R_\ell)R^i R^k R_j)R_i R^\ell \otimes R_k R^j \\ &= p_{(1)}(S^2(R_\ell)R^i)p_{(2)}(R^k R_j)R_i R^\ell \otimes R_k R^j \\ &= (p_{(1)}\circ S)(S^{-1}(R^i)S(R_\ell))R_i R^\ell \otimes f_{R_{2,1}R}(p_{(2)}) \\ &= (p_{(1)}\circ S)(R^i S(R_\ell))S(R_i)R^\ell \otimes f_{R_{2,1}R}(p_{(2)}) \\ &= f_{R_{2,1}^{-1}R^{-1}}(p_{(1)}\circ S)\otimes f_{R_{2,1}R}(p_{(2)}) \end{aligned}$$

shows that

$$\Delta(F(p)) = f_{R_{2,1}^{-1}R^{-1}}(p_{(1)}\circ S)\otimes f_{R_{2,1}R}(p_{(2)}) \tag{12.29}$$

holds for $p \in \mathrm{C}_r(H)$.

Lemma 12.4.6. *Suppose that (H, R) is a factorizable Hopf algebra over the field k. Then (H^{cop}, R^{-1}) is also.*

Proof. First of all (H^{cop}, R^{-1}) is quasitriangular by part (a) of Exercise 12.2.2. Let λ be a non-zero right integral for H^*. Since (H, R) is factorizable H is unimodular by part (d) of Proposition 12.4.2. Thus $\lambda \in \mathrm{C}_r(H)$ by part (e) of Theorem 10.5.4. For $q \in H^*$ we use part (b) of Proposition 12.4.5 to calculate

$$F(\lambda)F(q) = F(\lambda q) = F(\lambda{<}q, 1{>}) = F(\lambda)\epsilon(F(q)).$$

Since F is onto $\Lambda' = F(\lambda)$ is a right integral for H. Since F is one-one $F(\lambda) \neq 0$. Now let K be any finite-dimensional Hopf algebra over k and suppose that Λ is a non-zero right integral for K.

Since $(K, \leftharpoonup)$ is a free left K^*-module with basis $\{\Lambda\}$ it follows that $\mathrm{Rank}(\Delta(\Lambda)) = \mathrm{Dim}(K)$. Therefore $\mathrm{Rank}(\Delta(\Lambda')) = \mathrm{Dim}(H)$ and we may write $\Delta_{H^*}(\lambda) = \sum_{i=1}^{n} p_i \otimes q_i \in H^* \otimes H^*$, where $n = \mathrm{Dim}(H^*)$. By (12.29) the rank of $\sum_{i=1}^{n} f_{R_{2,1}^{-1}R^{-1}}(p_i \circ S) \otimes f_{R_{2,1}R}(q_i)$ is $\mathrm{Dim}(H)$. Therefore $\{f_{R_{2,1}^{-1}R^{-1}}(p_1 \circ S), \ldots, f_{R_{2,1}^{-1}R^{-1}}(p_n \circ S)\}$ is a basis for H. We have shown that $f_{R_{2,1}^{-1}R^{-1}} : H^* \longrightarrow H$ is a linear isomorphism. □

The previous proposition has interesting consequences for a factorizable Hopf algebra.

Theorem 12.4.7. *Let (H, R) be a factorizable Hopf algebra over the field k and set $F_r = F|\mathrm{C}_r(H)$, where $F = f_{R_{2,1}R}$. Then:*

(a) *$F_r : \mathrm{C}_r(H) \longrightarrow \mathrm{Z}(H)$ is an isomorphism of algebras.*
(b) *$F(\mathrm{G}(H^*)) = \mathrm{G}(H) \cap \mathrm{Z}(H)$. Moreover the restriction*

$$F_r|\mathrm{G}(H^*) : \mathrm{G}(H^*) \longrightarrow \mathrm{G}(H) \cap \mathrm{Z}(H)$$

is an isomorphism of groups. Thus $\mathrm{G}(H^*)$ is isomorphic to a subgroup of $\mathrm{Z}(\mathrm{G}(H))$.
(c) *$\mathrm{G}(H^*)$ is commutative and $|\mathrm{G}(H^*)|$ divides $|\mathrm{G}(H)|$.*

Proof. Since (H, R) is factorizable $F = f_{R_{2,1}R}$ is a linear isomorphism. Therefore $F_r : \mathrm{C}_r(H) \longrightarrow \mathrm{Z}(H)$ is a one-one algebra map by part (b) of Proposition 12.4.5. Now H is unimodular and S^2 is inner by Proposition 12.4.2. Therefore $\mathrm{Dim}(\mathrm{Cc}(H^*)) = \mathrm{Dim}(\mathrm{Z}(H))$ by part (b) of Corollary 10.7.2. We noted in the discussion following Corollary 12.4.4 that $\mathrm{Dim}(\mathrm{Cc}(H^*)) = \mathrm{Dim}(\mathrm{C}_r(H))$. Thus $\mathrm{Dim}(\mathrm{C}_r(H)) = \mathrm{Dim}(\mathrm{Z}(H))$ and part (a) follows.

To show part (b) let $\eta \in \mathrm{G}(H^*)$. Then $\eta \in \mathrm{C}_r(H)$ and $F(\eta) = f_{R_{2,1}R}(\eta) = f_{R_{2,1}}(\eta) f_R(\eta)$ by part (a) of Proposition 12.4.2. Just as $f_{R_{2,1}}(\eta) = g_\eta \in \mathrm{G}(H)$ by (QT.1) and (QT.2) one sees that $f_R(\eta) \in \mathrm{G}(H)$ by (QT.3) and (QT.4). Therefore $F(\eta) \in \mathrm{G}(H)$. We have shown that $F(\mathrm{G}(H^*)) \subseteq \mathrm{G}(H) \cap \mathrm{Z}(H)$.

Conversely, suppose that $a \in \mathrm{G}(H) \cap \mathrm{Z}(H)$. Then $F(\eta) = a$ for some $\eta \in \mathrm{C}_r(H)$. Since

$$0 \neq \Delta(a) = a \otimes a = \Delta(F(\eta)) = f_{R_{2,1}^{-1}R^{-1}}(\eta_{(1)} \circ S) \otimes f_{R_{2,1}R}(\eta_{(2)}),$$

the rank of the latter is one. See (12.29). But the rank of $\Delta(\eta) = \eta_{(1)} \otimes \eta_{(2)}$ is the rank of $(\eta_{(1)} \circ S) \otimes \eta_{(2)}$ since S is onto. Since $f_{R_{2,1}R}$ and $f_{R_{2,1}^{-1}R^{-1}}$ are one-one, the latter is by Lemma 12.4.6, we conclude that the rank of $\Delta(\eta)$ must be one. As $\eta(1) = \epsilon(F(\eta)) = \epsilon(a) = 1$ it follows that $\eta \in \mathrm{G}(H^*)$. We have completed the proof of part (b). Part (c) is an easy consequence of part (b). □

Exercises

Exercise 12.4.1. Let V be a vector space over k, $\mathcal{R} \in V\otimes V$, and let $f_{\mathcal{R}}, g_{\mathcal{R}} : V^* \longrightarrow V$ be defined by $f_{\mathcal{R}}(p) = (p\otimes \mathrm{I}_H)(\mathcal{R})$ and $g_{\mathcal{R}}(p) = (\mathrm{I}_H\otimes p)(\mathcal{R})$ for all $p \in V^*$. Show that both $f_{\mathcal{R}}$ and $g_{\mathcal{R}}$ are isomorphisms or neither are. [Hint: See Exercise 1.3.16.]

Exercise 12.4.2. Let (H, R) be a quasitriangular Hopf algebra over k.

(a) Show that if (H, R) is triangular and factorizable then $H = k$.

(b) Show that the quasitriangular Hopf algebras of Exercise 12.2.11 are not factorizable.

Exercise 12.4.3. Let (H, R) be a finite-dimensional quasitriangular Hopf algebra over k and suppose that λ is a non-zero right integral for H^*. Suppose that H is unimodular. Show that (H, R) is factorizable if and only if $f_{R_{2,1}R}(\lambda)$ is a non-zero right integral for H.

Chapter notes

The quantum Yang-Baxter equation originates in statistical mechanics. See the work of: Yang [226]; Akutsa, Deguchi, and Wadati [223]; Baxter [16]; and Zamolodchikov and Zamolodchikov [230]. The quantum Yang-Baxter equation is equivalent to the braid equation which in turn is very important for the construction of invariants of knots, links, and 3-manifolds. References for invariants are numerous. A very, very small sample: Kauffman [86]; Kauffman and Lins [87]; Reshetikhin [178]; Reshetikihn and Turaev [181]; and Yang and Ge [227, 228].

The quantum Yang-Baxter equation became very well-known with the advent of Drinfel'd's paper [44] in which the term quantum group was coined. For a treatment of quantum groups which includes a strong Hopf algebra flavor the reader is referred to the books by Kassels [94] and Majid [118]. Our adaptation of the Heyneman-Sweedler notation in Section 12.1 in connection with the quantum Yang-Baxter equation is found in the book by Lambe and the author [99, Section 2.2.1].

The notion of almost cocommutative Hopf algebra of Section 12.2 is due to Drinfel'd [45]. Proposition 12.2.2 follows from [45, Proposition 2.2]. Quasitriangular Hopf algebras are defined in Drinfel'd's paper [44, Section 10]. Parts (a)-(c) of Theorem 12.2.8 are [44, Proposition 3.1] and parts (e)-(f) are [169, Theorem 1]. Part (f) shows that if no assumption is made on the antipode S of a quasitriangular Hopf algebra necessarily S is bijective. The material in this section following the statement of Corollary 12.2.12 is found in Section 2 of [171]. The conclusion of Exercise 12.2.10 was established by Gelaki in this Master's Thesis [57].

Not every almost cocommutative Hopf algebra is quasitriangular. Masuoka provided an example in [127].

The notions of ribbon Hopf algebra and ribbon element in Section 12.3 are due to Reshetikhin and Turaev; see (3.3.1) and (3.3.2) of [180]. Proposition 12.3 is based on [171, Lemma 4]. Part (a) of Proposition 12.3.2 and part (c) are found in [45, Section 2]. Part (a) of Proposition 12.3.3 is [45, Proposition 6.1] for H^{op} and part (b) is derived from Section 3 of the same. Proposition 12.3.3 also follows from [169, Theorem 2] which is proved by different methods. Parts (a) and (b) of Theorem 12.3.6 constitute [88, Theorem 1] due to Kauffman and the author. Parts (c) and (d) are [180, (3.6)]. Corollary 12.3.7 is found in [88, Section 3] as is Corollary 12.3.9.

The notion of factorizable Hopf algebra introduced in Section 12.4 is due to Reshetikhin and Semenov-Tian-Shansky [179]. Proposition 12.4.2 is [173, Proposition 3]. Corollary 12.4.3 is [173, Corollary 1] and Corollary 12.4.4 is [173, Corollary 2]. Proposition 12.4.5 is a special case of [45, Proposition 3.3] and part (a) of Theorem 12.4.7 is found in [45].

Chapter 13

The Drinfel'd double of a finite-dimensional Hopf algebra

Some of the most important finite-dimensional Hopf algebras are Drinfel'd, or quantum, doubles. A double is a minimal quasitriangular Hopf algebra $(D(H), \mathcal{R})$ constructed from a finite-dimensional Hopf algebra H over k. Both $H^{*\,cop}$ and H are identified with Hopf subalgebras of $D(H)$ and $D(H)$ is the product of these identifications.

We describe the double, its representations, and study its properties in this chapter. The double is factorizable; therefore it is minimal quasitriangular and $D(H)$ is unimodular. We consider $D(H)$ when H has a quasitriangular or factorizable structure. Of interest is how properties of $H^{*\,cop}$ or H are reflected in $D(H)$ and vice versa. Throughout this chapter H is a finite-dimensional Hopf algebra over k unless otherwise stated. We will find is convenient to use the Einstein summation notation convention and will use it quite often.

13.1 The double and its category of representations

Let H be a bialgebra over k. The context of this section is the formal variation ${}_H\mathcal{YD}^H$ of the prebraided category ${}_H^H\mathcal{YD}$ introduced in Section 11.6. We recount the aspects of Exercise 11.6.19, where ${}_H\mathcal{YD}^H$ is introduced and discussed, needed here.

Objects of ${}_H\mathcal{YD}^H$ are triples $(M, \cdot, \rho)$, where $(M, \cdot)$ is a left H-module and (M, ρ) is a right H-comodule, such that the compatibility condition

$$h_{(1)}\cdot m_{(0)} \otimes h_{(2)} m_{(1)} = (h_{(2)}\cdot m)_{(0)} \otimes (h_{(2)}\cdot m)_{(1)} h_{(1)},$$

which is (11.24), holds for all $h \in H$ and $m \in M$. Morphisms are functions $f : M \longrightarrow N$ of underlying vector spaces which are simultaneously maps of left H-modules and right H-comodules. At this point we introduce some terminology.

Definition 13.1.1. Let H be a bialgebra over k. Then ${}_H\mathcal{YD}^H$ is a *Yetter–Drinfel'd category*, its objects are called *left-right Yetter–Drinfel'd modules*, and its morphisms are called *maps of left-right Yetter–Drinfel'd modules*.

Now suppose that H is a finite-dimensional Hopf algebra over k. We would like to find a Hopf algebra D so that the category ${}_H\mathcal{YD}^H$ can be thought of as the category ${}_D\mathcal{M}$. Let $(M, \cdot, \rho)$ be an object of ${}_H\mathcal{YD}^H$ and suppose $(M, \bullet)$ is a left D-module structure on M. Now M has a left H-module structure and a (rational) left H^*-module structure since it is an object of ${}_H\mathcal{YD}^H$. So that these three structures may be related, we will assume that H and H^* can be identified as subalgebras of D and for all $m \in M$ that $a{\bullet}m = a{\cdot}m$ for all $a \in H$ and $p{\bullet}m = p{\rightharpoonup}m$ for all $p \in H^*$.

Suppose A and B are the subalgebras of D identified with H^* and H respectively. Let us assume $D = AB$; specifically that multiplication $A{\otimes}B \longrightarrow D$ is a linear isomorphism. Use this isomorphism and the identifications of H^* and H with A and B to make an identification $D = H^*{\otimes}H$. Then $\jmath_{H^*} : H^* \longrightarrow D$ defined by $\jmath_{H^*}(p) = p{\otimes}1$ for all $p \in H^*$ and $\jmath_H : H \longrightarrow D$ defined by $\jmath_H(a) = \epsilon{\otimes}a$ for all $a \in H$ are one-one algebra maps. Let $p, q \in H^*$ and $a, b \in H$ and let $m \in M$. Then

$$(p{\otimes}a)(q{\otimes}b) = pq{\otimes}ab \text{ whenever } a = 1 \text{ or } q = \epsilon. \tag{13.1}$$

In particular $p{\otimes}a = (p{\otimes}1)(\epsilon{\otimes}a)$. By virtue of this equation

$$(p{\otimes}a){\bullet}m = p{\rightharpoonup}(a{\cdot}m) \text{ and } ((\epsilon{\otimes}a)(p{\otimes}1)){\bullet}m = a{\cdot}(p{\rightharpoonup}m). \tag{13.2}$$

Using (11.24) we make the calculation

$$\begin{aligned}
a{\cdot}(p{\rightharpoonup}m) &= a{\cdot}m_{(0)}<p, m_{(1)}> \\
&= a_{(1)}{\cdot}m_{(0)}<p, S^{-1}(a_{(3)})a_{(2)}m_{(1)}> \\
&= a_{(1)}{\cdot}m_{(0)}<p{\prec}S^{-1}(a_{(3)}), a_{(2)}m_{(1)}> \\
&= (a_{(2)}{\cdot}m)_{(0)}<p{\prec}S^{-1}(a_{(3)}), (a_{(2)}{\cdot}m)_{(1)}a_{(1)}> \\
&= (a_{(2)}{\cdot}m)_{(0)}<a_{(1)}{\succ}p{\prec}S^{-1}(a_{(3)}), (a_{(2)}{\cdot}m)_{(1)}> \\
&= (a_{(1)}{\succ}p{\prec}S^{-1}(a_{(3)})){\rightharpoonup}(a_{(2)}{\cdot}m)
\end{aligned}$$

from which we deduce the commutation relation

$$((\epsilon{\otimes}a)(p{\otimes}1)){\bullet}m = ((a_{(1)}{\succ}p{\prec}S^{-1}(a_{(3)})){\otimes}a_{(2)}){\bullet}m. \tag{13.3}$$

Suppose

$$(\epsilon{\otimes}a)(p{\otimes}1) = (a_{(1)}{\succ}p{\prec}S^{-1}(a_{(3)})){\otimes}a_{(2)}$$

for all $a \in H$ and $p \in H^*$. Let $p, q \in H^*$ and $a, b \in H$. Since

$$(p \otimes a)(q \otimes b) = (p \otimes 1)(\epsilon \otimes a)(q \otimes 1)(\epsilon \otimes b)$$

the product for D is given by

$$(p \otimes a)(q \otimes b) = p(a_{(1)} \succ q \prec S^{-1}(a_{(3)})) \otimes a_{(2)} b. \tag{13.4}$$

There is a Hopf algebra $D = D(H)$ meeting our requirements. For categorical details the reader is referred to Exercises 13.1.6 and 13.1.7.

Theorem 13.1.2. *Let H be finite-dimensional Hopf algebra with antipode S over the field k. Then:*

(a) *$D(H) = H^{*\,cop} \otimes H$ is a Hopf algebra with algebra structure defined by (13.4) and the tensor product coalgebra structure.*
(b) *$\jmath_H : H \longrightarrow D(H)$ and $\jmath_{H^*} : H^{*\,cop} \longrightarrow D(H)$ defined by $\jmath_H(a) = \epsilon \otimes a$ for all $a \in H$ and $\jmath_{H^*}(p) = p \otimes 1$ for all $p \in H^*$ are one-one Hopf algebra maps.*
(c) *Let $\{h_1, \ldots, h_n\}$ be a basis for H, let $\{h^1, \ldots, h^n\}$ be the corresponding dual basis for H^*, and set $\mathcal{R} = \sum_{i=1}^{n} (\epsilon \otimes h_i) \otimes (h^i \otimes 1)$. Then $(D(H), \mathcal{R})$ is a quasitriangular Hopf algebra over k. Furthermore $\mathcal{R}$ does not depend on the choice of basis.*

Proof. Let $A = H^{*\,cop}$, $B = H$, and $\tau : A \times B \longrightarrow k$ be the evaluation map given by $\tau(p, a) = \langle p, a \rangle$ for all $p \in A$ and $a \in B$. Then $\tau_\ell = \mathrm{I}_A$ and $\tau_r = \mathrm{I}_B$. This means $\tau_\ell : A^{cop} \longrightarrow B^*$ and $\tau_r : B \longrightarrow A^*$ are coalgebra maps. Thus τ is a skew pairing. Since $(S^{-1})^*$ is the antipode of A^*, by Propositions 7.7.3 and 7.7.10 and (7.8) the vector space $A \otimes B = H^{*\,cop} \otimes H$ is a Hopf algebra with the tensor product coalgebra structure and product given by

$$\begin{aligned}
(p \otimes a)(q \otimes b) &= p\tau(q_{(3)}, a_{(1)}) q_{(2)} \otimes a_{(2)} \tau(q_{(1)} \circ S^{-1}, a_{(3)}) b \\
&= p \langle q_{(3)}, a_{(1)} \rangle q_{(2)} \otimes a_{(2)} \langle q_{(1)} \circ S^{-1}, a_{(3)} \rangle b \\
&= p \langle q_{(3)}, a_{(1)} \rangle q_{(2)} \otimes a_{(2)} \langle q_{(1)}, S^{-1}(a_{(3)}) \rangle b \\
&= p(a_{(1)} \succ q \prec S^{-1}(a_{(3)})) \otimes a_{(2)} b;
\end{aligned}$$

the last equation follows since $h \succ q \prec h' = \langle q_{(1)}, h' \rangle q_{(2)} \langle q_{(3)}, h \rangle$ for all $h, h' \in H$ and $q \in H^*$.

Thus the product for $H^{*\,cop} \otimes H$ is that of (13.4). We have established part (a). As for part (b), observe that (13.4) implies (13.1). Hence $\jmath_H$ and $\jmath_{H^*}$ are one-one algebra maps. They are coalgebra maps also since $D(H)$ has the tensor product coalgebra structure.

Assume the hypothesis of part (c). That (QT.1)–(QT.4) hold for $(D(H), \mathcal{R})$ follows by Exercise 13.1.2. That (QT.5) holds for $(D(H), \mathcal{R})$ follows by Exercise 13.1.3. That $\mathcal{R}$ does not depend on the basis follows from Exercise 1.2.9. □

Definition 13.1.3. Let H be a finite-dimensional Hopf algebra over the field k. The *Drinfel'd*, or *quantum, double of* H is the pair $(D(H), \mathcal{R})$ of Theorem 13.1.2.

It will be useful to record formulas for the antipode $S_{D(H)}$ of the double, for $S^2_{D(H)}$, and also compute the Drinfel'd element for $(D(H), \mathcal{R})$.

Corollary 13.1.4. *Let H be a finite-dimensional Hopf algebra with antipode S over the field k. Then:*

(a) $S_{D(H)}(\epsilon \otimes a) = \epsilon \otimes S(a)$ *for all* $a \in H$ *and* $S_{D(H)}(p \otimes 1) = p \circ S^{-1} \otimes 1$ *for all* $p \in H^*$. *Thus* $S_{D(H)}(p \otimes a) = (\epsilon \otimes S(a))(p \circ S^{-1} \otimes 1)$ *for all* $p \in H^*$ *and* $a \in H$.

(b) $S^2_{D(H)}(p \otimes a) = p \circ S^{-2} \otimes S^2(a)$ *for all* $p \in H^*$ *and* $a \in H$.

(c) $u = h^i \circ S^{-1} \otimes (h_i) = h^i \otimes S^{-1}(h_i)$ *is the Drinfel'd element of* $(D(H), \mathcal{R})$, *where* $\{h_1, \ldots, h_n\}$ *is a basis for* H *and* $\{h^1, \ldots, h^n\}$ *is the dual basis for* H^*.

Proof. To calculate $S_{D(H)}$ we may use part (b) of Proposition 7.7.3 or use part (c) of Lemma 7.1.3. We use the latter which implies $\jmath_H \circ S_H = S_{D(H)} \circ \jmath_H$ and $\jmath_{H^*} \circ S_{H^{*\,cop}} = S_{D(H)} \circ \jmath_{H^*}$. Therefore $\epsilon \otimes S_H(a) = S_{D(H)}(\epsilon \otimes a)$ for all $a \in H$ and $S_{H^{*\,cop}}(p) \otimes 1 = S_{D(H)}(p \otimes 1)$ for all $p \in H^*$. Since

$$S_{D(H)}(p \otimes a) = S_{D(H)}((p \otimes 1)(\epsilon \otimes a)) = S_{D(H)}(\epsilon \otimes a) S_{D(H)}(p \otimes 1)$$

and $S_{H^{*\,cop}} = (S^{-1})^*$, part (a) now follows. Part (b) follows directly from part (a).

As for part (c), we write $\mathcal{R} = (\epsilon \otimes h_i) \otimes (h^i \otimes 1)$ as in part (c) of Theorem 13.1.2 and calculate

$$\begin{aligned} u &= S_{D(H)}(\mathcal{R}^i)\mathcal{R}_i \\ &= (S_{D(H)}(h^i \otimes 1))(\epsilon \otimes h_i) \\ &= (h^i \circ S^{-1} \otimes 1)(\epsilon \otimes h_i) \\ &= h^i \circ S^{-1} \otimes h_i = h^i \otimes S^{-1}(h_i); \end{aligned}$$

the last equation follows by part (c) of Exercise 1.2.9. □

Exercises

In the following exercises H, H' are bialgebras over the field k.

Exercise 13.1.1. Suppose that H^{op} is a Hopf algebra with antipode ς. Show that:

(a) End(H) has an algebra structure with unity $\eta\circ\epsilon$ and product given by

$$(T{\cdot}T')(h) = T(h_{(1)})_{(2)}T'(\varsigma(T(h_{(1)})_{(3)})h_{(2)}T(h_{(1)})_{(2)}) \qquad (13.5)$$

for all $T, T' \in \text{End}(H)$ and $h \in H$.

Regard $\mathcal{A} = H^*\otimes H$ as a subspace of End(H) according to $(p\otimes a)(h) = \langle p, h\rangle a$ for all $p \in H^*$ and $a, h \in H$. Show that:

(b) $\mathcal{A}$ is a subalgebra of End(H) and (13.5) is given by

$$(p\otimes a)(q\otimes b) = p(a_{(1)}\succ q\prec\varsigma(a_{(3)}))\otimes a_{(2)}b \qquad (13.6)$$

for all $p, q \in H^*$ and $a, b \in H$.

(c) $(p\otimes a)(q\otimes b) = pq\otimes ab$ whenever $a = 1$ or $q = \epsilon$.

(d) $\imath_H : H \longrightarrow \mathcal{A}$ and $\imath_{H^*} : H^* \longrightarrow \mathcal{A}$ defined by $\imath_H(a) = \epsilon\otimes a$ and $\imath_{H^*}(p) = p\otimes 1$ respectively for all $a \in H$ and $p \in H^*$ are one-one algebra maps and $p\otimes a = \imath_{H^*}(p)\imath_H(a)$ for all $p \in H^*$ and $a \in H$.

Let $\mathcal{D} = H^{o\,cop}\otimes H$. Show that:

(e) $\mathcal{D}$ is a sub-algebra of $\mathcal{A}$ and is a bialgebra with the tensor product coalgebra structure.

(f) $\jmath_H : H \longrightarrow \mathcal{D}$ and $\jmath_{H^o} : H^{o\,cop} \longrightarrow \mathcal{D}$ defined by $\jmath_H(a) = \epsilon\otimes a$ for all $a \in H$ and $\jmath_{H^o}(p) = p\otimes 1$ for all $p \in H^o$ are bialgebra maps.

(g) If H is a Hopf algebra with bijective antipode S (in which case $\varsigma = S^{-1}$) then $\mathcal{D}$ is a Hopf algebra and $\mathcal{D} = D(H)$ when H is finite-dimensional.

In the next two exercises we carefully examine why (QT.1)–(QT.5) hold for the double.

Exercise 13.1.2. Suppose H is finite-dimensional. Let $\{h_1, \cdots, h_n\}$ be a basis for H, let $\{h^1, \cdots, h^n\}$ be the dual basis for H^*, and set $R = (\epsilon\otimes h_i)\otimes(h^i\otimes 1)$. Give $D = H^{*\,cop}\otimes H$ the tensor product coalgebra structure and suppose that D has an algebra structure such that the maps $\jmath_H$ and $\jmath_{H^*}$ of part (b) of Theorem 13.1.2 are algebra maps. For (D, R) show that:

(a) (QT.1) is equivalent to $h_{i\,(1)}\otimes h_{i\,(2)}\otimes h^i = h_i\otimes h_j\otimes h^i h^j$.

(b) (QT.2) is equivalent to $\epsilon(h_i)h^i = \epsilon$.

(c) (QT.3) is equivalent to $h_i\otimes h^i_{(2)}\otimes h^i_{(1)} = h_i h_j\otimes h^j\otimes h^i$.

(d) (QT.4) is equivalent to $<h^i, 1>h_i = 1$.

(e) (QT.1)–(QT.4) hold. [Hint: To establish the equation of part (a), show that

$$h_{i\,(1)} \otimes h_{i\,(2)} <h^i, a> = h_i \otimes h_j <h^i h^j, a>$$

for all $a \in H$. Likewise for part (c) show that

$$<p, h_i> h^i_{(2)} \otimes h^i_{(1)} = <p, h_i h_j> h^j \otimes h^i$$

for all $p \in H^*$. See Exercise 1.2.9.]

Exercise 13.1.3. Assume the hypothesis of Exercise 13.1.2 and set $R = (\epsilon \otimes h_i) \otimes (h^i \otimes 1)$. Here we carefully examine what it means for (QT.5) hold for the double.

(a) Let $a \in H$. Show that $\Delta_D^{cop}(\epsilon \otimes a)R = R\Delta_D(\epsilon \otimes a)$ if and only if

$$a_{(2)} h_i \otimes (\epsilon \otimes a_{(1)})(h^i \otimes 1) = h_i a_{(1)} \otimes h^i \otimes a_{(2)}. \tag{13.7}$$

(b) Let $p \in H^*$. Show that $\Delta_D^{cop}(p \otimes 1)R = R\Delta_D(p \otimes 1)$ if and only if

$$p_{(1)} \otimes h_i \otimes p_{(2)} h^i = (\epsilon \otimes h_i)(p_{(2)} \otimes 1) \otimes h^i p_{(1)}. \tag{13.8}$$

(c) Show that (13.7) holds for all $a \in H$, and (13.8) for all $p \in H^*$, if and only if

$$(a_{(1)} \succ p) \otimes a_{(2)} = (\epsilon \otimes a_{(1)})((p \succ a_{(2)}) \otimes 1) \tag{13.9}$$

for all $a \in H$ and $p \in H^*$.

Suppose further H is a Hopf algebra.

(d) Show that (13.9) is equivalent to $(\epsilon \otimes a)(p \otimes 1) = (a_{(1)} \succ p \prec S^{-1}(a_{(3)})) \otimes a_{(2)}$ for all $a \in H$ and $p \in H^*$.

(e) Show that (QT.5) holds for $(D(H), \mathcal{R})$. See Remark 12.2.4.

Exercise 13.1.4. Suppose H, H' are finite-dimensional Hopf algebras. Show that:

(a) $H \simeq H'$ as Hopf algebras implies $(D(H), \mathcal{R}) \simeq (D(H'), \mathcal{R}')$ as quasitriangular Hopf algebras.

(b) $(D(H), \mathcal{R}) \simeq (D(H'), \mathcal{R}')$ as quasitriangular Hopf algebras implies $\mathcal{R}_{(\ell)} \simeq \mathcal{R}'_{(\ell)}$ as Hopf algebras. See (12.11) and Proposition 12.2.13.

(c) There is a bijection between the isomorphism classes of finite-dimensional Hopf algebras over k and the isomorphism classes of doubles over k given by $[H] \mapsto [(D(H), \mathcal{R})]$.

Exercise 13.1.5. Here we push the details of a solution to the previous exercise.

(a) Suppose H, H' are finite-dimensional Hopf algebras and $f : H \longrightarrow H'$ is an isomorphism of Hopf algebras. Show that $F : (D(H), \mathcal{R}) \longrightarrow (D(H'), \mathcal{R}')$ is an isomorphism of quasitriangular Hopf algebras, where $F = (f^*)^{-1} \otimes f$.

(b) Show that there is an equivalence between the category whose objects are finite-dimensional Hopf algebras over k and whose morphisms are isomorphisms of Hopf algebras and the category whose objects are Drinfel'd doubles of finite-dimensional Hopf algebras and whose morphisms are isomorphisms of quasitriangular Hopf algebras.

Exercise 13.1.6. Suppose H is a finite-dimensional Hopf algebra. Show that

$$ {}_H\mathcal{YD}^H \longrightarrow {}_{D(H)}\mathcal{M} \text{ given by } (M, \cdot, \rho) \mapsto (M, \bullet) \text{ and } f \mapsto f, $$

where $(p \otimes a) \bullet m = p \rightharpoonup (a \cdot m)$ for all $p \in H^*$, $a \in H$, and $m \in M$, is an isomorphism of categories.

Apropos Exercise 13.1.6, the braiding structures of ${}_H\mathcal{YD}^H$ and ${}_{D(H)}\mathcal{M}$ do not correspond under the isomorphism. One remedy is provided by the next exercise.

Exercise 13.1.7. Let ${}_H\mathcal{C}^H = {}_H\mathcal{YD}^H$ as categories. Let M, N be objects of ${}_H\mathcal{C}^H$. We regard $M \otimes N$ as an object of ${}_H\mathcal{C}^H$ by

$$ h \cdot (m \otimes n) = h_{(1)} \cdot m \otimes h_{(2)} \cdot n \text{ and } \rho(m \otimes n) = (m_{(0)} \otimes n_{(0)}) \otimes n_{(1)} m_{(1)} $$

for all $h \in H$, $m \in m$, and $n \in N$; see Proposition 3.8.1 of [99]. We define $\varsigma_{M,N} : M \otimes N \longrightarrow N \otimes M$ by

$$ \varsigma_{M,N}(m \otimes n) = n_{(0)} \otimes n_{(1)} \cdot m $$

for all $m \in M$ and $n \in N$.

(a) Show that ${}_H\mathcal{C}^H$ with the tensor product defined above has the structure of a monoidal category and the $\varsigma_{M,N}$'s form a pre-braiding for it.

(b) Suppose that H has an antipode. Show that ${}_H\mathcal{C}^H$ with the structures of part (a) is a braided monoidal category.

(c) Suppose that H is a finite-dimensional Hopf algebra. Show that the correspondence of Exercise 13.1.6 is an isomorphism ${}_H\mathcal{C}^H \simeq {}_{D(H)}\mathcal{M}$ of braided monoidal categories. Again, the $\varsigma_{M,N}$'s form the braiding on ${}_H\mathcal{C}^H$. The $B_{M,N}$'s form the braiding on ${}_{D(H)}\mathcal{M}$.

Exercise 13.1.8. Let G be a finite group and $H = k[G]$ be its group algebra over k. We set the stage for a structure constant description of $(D(H), \mathcal{R})$.

For $g \in G$ let $e_g \in H^*$ be defined by $e_g(h) = \delta_{g,h}$ for all $h \in G$ and set $b_{(g,h)} = e_g \otimes h$ for all $g, h \in G$. Then $\{e_g \otimes h\}_{(g,h) \in G \times G}$ is a basis for $D(H)$.

(a) Show that $\mathcal{R} = \sum_{g \in G} (\epsilon \otimes g) \otimes (e_g \otimes 1)$.

(b) Find the structure constants for the Hopf algebra $D(H)$ with respect to the basis $\{b_{(g,h)}\}_{(g,h)\in G\times G}$. See Exercise 5.1.16.

Exercise 13.1.9. Let $H = k[G]$ be as in Exercise 13.1.8. Find the structure constants for $D(H^*)$ with respect to the basis $\{c_{(g,h)}\}_{(g,h)\in G\times G}$, where $c_{(g,h)} = g\otimes e_h$.

13.2 Basic properties of the double

We begin with:

Theorem 13.2.1. *Let H be a finite-dimensional Hopf algebra over the field k. Then:*

(a) *$(D(H),\mathcal{R})$ is factorizable;*
(b) *$(D(H),\mathcal{R})$ is minimal quasitriangular; and*
(c) *H is unimodular.*

Proof. In light of Proposition 12.4.2 we need only establish part (a). We continue with the notation of Section 12.4 and the proof of Theorem 13.1.2.

Observe that

$$\begin{aligned}\mathcal{R}_{2,1}\mathcal{R} &= ((h^i\otimes 1)\otimes(\epsilon\otimes h_i))((\epsilon\otimes h_j)\otimes(h^j\otimes 1))\\ &= (h^i\otimes 1)(\epsilon\otimes h_j)\otimes(\epsilon\otimes h_i)(h^j\otimes 1)\\ &= (h^i\otimes h_j)\otimes(\epsilon\otimes h_i)(h^j\otimes 1).\end{aligned}$$

Using the vector space identifications $D(H)^* = H^{**}\otimes H^* = H\otimes H^*$ we calculate

$$f_{\mathcal{R}_{2,1}\mathcal{R}}(h\otimes p) = \langle h^i, h\rangle\langle p, h_j\rangle(\epsilon\otimes h_i)(h^j\otimes 1) = (\epsilon\otimes h)(p\otimes 1).$$

Since

$$\begin{aligned}S_{D(H)}((\epsilon\otimes h)(p\otimes 1)) &= S_{D(H)}(p\otimes 1)S_{D(H)}(\epsilon\otimes h)\\ &= ((S^{-1})^*(p)\otimes 1)(\epsilon\otimes S(h))\\ &= (S^{-1})^*(p)\otimes S(h)\end{aligned}$$

we have the formula $f_{\mathcal{R}_{2,1}\mathcal{R}} = S_{D(H)}^{-1}\circ((S^{-1})^*\otimes S)\circ\tau_{H,H^*}$. In particular $f_{\mathcal{R}_{2,1}\mathcal{R}}$ is the composition of bijective maps and is therefore bijective. □

Even though parts (b) and (c) of the preceding theorem are consequences of part (a), there are more direct justifications for them. Note that $(D(H),\mathcal{R})$ is minimal quasitriangular since the elements $h^i\otimes h_j$, where

$1 \leq i, j \leq n$, form a basis for $D(H)$ and $h^i \otimes h_j = (h^i \otimes 1)(\epsilon \otimes h_j)$. Unimodularity is worked out more directly in Exercise 13.2.2.

Integrals for the double $D(H)$ are expressed very simply in terms of integrals of H and H^*.

Proposition 13.2.2. *Let H be a finite-dimensional Hopf algebra over the field k, let $\lambda \in H^*$ be a left integral, and let $\Lambda \in H$ be a right integral. Then:*

(a) *$\lambda \otimes \Lambda$ is a two-sided integral for $D(H)$.*
(b) *$\Lambda \otimes \lambda$ is a left integral for $D(H)^*$.*
(c) *$\mathrm{G}(D(H)) = \{\beta \otimes a \mid \beta \in \mathrm{G}(H^*),\ a \in \mathrm{G}(H)\}$; moreover the map $\mathrm{G}(H^*) \times \mathrm{G}(H) \longrightarrow \mathrm{G}(D(H))$ given by $(\beta, a) \mapsto \beta \otimes a$ is a group isomorphism.*
(d) *$\alpha \otimes g$ is the $D(H)$-distinguished grouplike element of $D(H)$, where α is the H-distinguished grouplike element of H^* and g is the H-distinguished grouplike element of H.*

Proof. We may assume that $\lambda, \Lambda \neq 0$. Part (b) follows since $D(H)^* \simeq H^{op} \otimes H^*$ as an algebra and $\Lambda \otimes \lambda$ is the tensor product of left integrals for the tensorands. Part (d) is easy to see since $\Lambda' \otimes \lambda'$ is a right integral for $D(H)^*$, where Λ' is a left integral for H and λ' is a right integral for H^*. Part (c) follows by Exercise 2.1.21 and the definition of the product of the double.

It remains to show part (a). Now $D(H)$ is unimodular by part (a) of Theorem 13.2.1. Let $\boldsymbol{\Lambda}$ be a non-zero two-sided integral for $D(H)$ and write $\boldsymbol{\Lambda} = \sum_{i=1}^{n} p_i \otimes a_i$, where n is as small as possible. Then $\{p_1, \ldots, p_n\}$ and $\{a_1, \ldots, a_n\}$ are linearly independent by Lemma 1.2.2. We use part (a) of Theorem 10.2.2, which states that left or right integrals for a finite-dimensional Hopf algebras are unique up to scalar multiple, to complete the proof.

For $p \in H^*$ note that $<p, 1>\boldsymbol{\Lambda} = \sum_{i=1}^{n} pp_i \otimes a_i$. Thus $<p, 1>p_i = pp_i$ for all $1 \leq i \leq n$. Therefore $p_1, \ldots, p_n$ are left integrals for H^*. Consequently $n = 1$. Since $\epsilon(a)\boldsymbol{\Lambda} = \boldsymbol{\Lambda}(\epsilon \otimes a) = p_1 \otimes a_1 a$ we conclude that a_1 is a right integral for H. Thus p_1, a_1 are scalar multiples of λ, Λ respectively. □

Corollary 13.2.3. *Let H be a finite-dimensional Hopf algebra over the field k. Then the following are equivalent:*

(a) *$D(H)$ is semisimple.*
(b) *H and H^* are semisimple.*

(c) *H and H^* are cosemisimple.*
(d) *$D(H)$ is cosemisimple.*

Proof. Let $0 \neq \lambda, \Lambda$ be as in Proposition 13.2.2. Then $\lambda \otimes \Lambda$ and $\Lambda \otimes \lambda$ are non-zero left integrals for $D(H)$ and $D(H)^*$ respectively and $\epsilon_{D(H)}(\lambda \otimes \Lambda) = <\lambda, 1>\epsilon(\Lambda) = \epsilon_{D(H)^*}(\Lambda \otimes \lambda)$. The corollary now follows by Corollary 10.3.3. □

Since H and $H^{*\,cop}$ can be regarded as Hopf subalgebras of $D(H)$ it is not hard to show that:

Proposition 13.2.4. *Let H be a finite-dimensional Hopf algebra over the field k. Then the following are equivalent:*

(a) *$D(H)$ is commutative.*
(b) *H and H^* are commutative.*
(c) *H and H^* are cocommutative.*
(d) *$D(H)$ is cocommutative.*

□

Proposition 13.2.5. *Let H be a finite-dimensional Hopf algebra over the field k. Then the following are equivalent:*

(a) *$D(H)$ is pointed.*
(b) *H and H^* are pointed.*

□

Exercises

In the following exercises H is a finite-dimensional Hopf algebra with antipode S over k.

Exercise 13.2.1. Make the vector space identification $D(H) = H^* \otimes H = \mathrm{End}(H)$; thus $<p, \otimes h, a> = <p, a>h$ for all $p \in H^*$ and $h, a \in H$. Let $u = \mathrm{I}_H$. Show that $g_{\mathcal{R}_{2,1}\mathcal{R}} = {}^u(\ )\circ\tau_{H^*,H}$, where ${}^u(\ )$ is conjugation by u in the convolution algebra $\mathrm{End}(H^{op})$. This is another proof that $(D(H), \mathcal{R})$ is factorizable.

Exercise 13.2.2. Let $\Lambda \in H$ be a right integral and $\lambda \in H^*$ be a left integral. The purpose of this exercise is to show that $\lambda \otimes \Lambda$ is a two-sided integral for $D(H)$ follows from two basic equations involving integrals.

(a) Show that $S^{-1}(\Lambda_{(3)})g^{-1}\Lambda_{(1)} \otimes \Lambda_{(2)} = 1 \otimes \Lambda$, where g is the H-distinguished grouplike element of H. [Hint: See part (f) of Theorem 10.5.4.]

(b) Show that $(\lambda \otimes \Lambda)(p \otimes 1) = <p, 1> \lambda \otimes \Lambda$ for all $p \in H^*$.

(c) Show that $\lambda_{(2)} \otimes \lambda_{(3)} \alpha^{-1} S_{H^*}^{-1}(\lambda_{(1)}) = \lambda \otimes \epsilon$, where α is the H-distinguished grouplike element of H^*. [Hint: See part (g) of Theorem 10.5.4.]

(d) Show that $(\epsilon \otimes a)(\lambda \otimes \Lambda) = \epsilon(a) \lambda \otimes \Lambda$ for all $a \in H$.

(e) Use parts (b) and (d) to show that $\lambda \otimes \Lambda$ is a two-sided integral for $D(H)$.

13.3 Characterizations of the double as a quasitriangular Hopf algebra

$(D(H), \mathcal{R})$ is minimal quasitriangular and $\mathrm{Dim}(D(H)) = (\mathrm{Rank}(\mathcal{R}))^2$. We will show doubles can be characterized in terms of a mapping property and that doubles are those minimal quasitriangular Hopf algebras (A, R) such that $\mathrm{Dim}(A) = (\mathrm{Rank}(R))^2$.

To begin our discussion of the mapping property we let (A, R) be a quasitriangular Hopf algebra over k resume the commentary following Corollary 12.2.12. The map $\mathsf{f}_R : R_{(\ell)}^* \longrightarrow R_{(r)}$ defined by $\mathsf{f}_R(p) = (p \otimes I)(R)$ for all $p \in R_{(\ell)}^*$ is a linear isomorphism by (12.12). For the same reasons the map f_R of part (a) of Proposition 12.2.11 is f_R; thus $\mathsf{f}_R : R_{(\ell)}^* \longrightarrow R_{(r)}^{cop}$ is a bialgebra isomorphism. The commutation relation

$$\mathsf{f}_R(a_{(1)} \succ q \prec S_A^{-1}(a_{(3)}))a_{(2)} = a\mathsf{f}_R(p) \tag{13.10}$$

for all $a \in R_{(\ell)}$ and $p \in R_{(\ell)}^*$ is a key relation in this section. This equation is a minor variation of (12.21) in Exercise 12.2.8. See Exercise 13.3.1.

For the double $(D(H), \mathcal{R})$ there is a relationship between the maps $\jmath_H : H \longrightarrow D(H)$, $\jmath_{H^*} : H^{*\,cop} \longrightarrow D(H)$, and $\mathsf{f}_{\mathcal{R}} : \mathcal{R}_{(\ell)}^* \longrightarrow \mathcal{R}_{(r)}^{cop}$. Define

$$\jmath'_H : H \longrightarrow \mathrm{Im}(\jmath_H) = \epsilon \otimes H = \mathcal{R}_{(\ell)}$$

by $\jmath'_H(h) = \jmath_H(h)$ for all $h \in H$ and

$$\jmath'_{H^*} : H^* \longrightarrow \mathrm{Im}(\jmath_{H^*}) = H^* \otimes 1 = \mathcal{R}_{(r)}$$

by $\jmath'_{H^*}(p) = \jmath_{H^*}(p)$ for all $p \in H^*$. Then $\jmath'_{H^*} \circ (\jmath'_H)^* = \mathsf{f}_{\mathcal{R}}$.

Definition 13.3.1. Let H be a finite-dimensional Hopf algebra over k. An *H-admissible tuple* is a quadruple $(A, R, \imath_R, \pi_R)$, where (A, R) is a finite-dimensional quasitriangular Hopf algebra over k where the maps $\imath_R : H \longrightarrow R_{(\ell)}$ and $\pi_R : H^{*\,cop} \longrightarrow R_{(r)}^*$ are bialgebra isomorphisms such that $\pi_R \circ \imath_R^* = \mathsf{f}_R$.

Thus $(D(H), \mathcal{R}, \jmath'_H, \jmath'_{H^*})$ is an H-admissible tuple. Observe that if $H = R_{(\ell)}$ then $(A, R, \mathrm{I}_H, \mathrm{f}_R)$ is an H-admissible tuple. The main result of this section relates $(D(H), \mathcal{R})$ to all quasitriangular Hopf algebras (A, R) over k such that there is an isomorphism $\imath_R : H \longrightarrow R_{(\ell)}$, where $(A, R, \imath_R, \pi_R)$ is an H-admissible tuple.

Theorem 13.3.2. *Let H be a finite-dimensional Hopf algebra over the field k and let $(D(R), \mathcal{R})$ be the Drinfel'd double of H. Then:*

(a) *$(D(H), \mathcal{R}, \imath_{\mathcal{R}}, \pi_{\mathcal{R}})$ is an H-admissible tuple, where $\imath_{\mathcal{R}} = \jmath'_H$ and $\pi_{\mathcal{R}} = \jmath'_{H^*}$.*
(b) *Suppose (A, R) is a quasitriangular Hopf algebra over k and $(A, R, \imath_R, \pi_R)$ is an H-admissible tuple. Then there is a morphism of quasitriangular Hopf algebras $F : (D(H), \mathcal{R}) \longrightarrow (A, R)$ determined by $F\circ\imath_{\mathcal{R}} = \imath_R$ and $F\circ\pi_{\mathcal{R}} = \pi_R$.*

Proof. We need only establish part (b). Suppose F satisfies the conditions of part (b). Since F is multiplicative,

$$\begin{aligned} F(p\otimes a) &= F((p\otimes 1)(\epsilon\otimes a)) \\ &= F(p\otimes 1)F(\epsilon\otimes a) \\ &= F(\jmath'_{H^*}(p))F(\jmath'_H(a)) \\ &= \pi_R(p)\imath_R(a) \end{aligned}$$

for all $p \in H^*$ and $a \in H$. Uniqueness is established. As far as existence, define $F : D(H) \longrightarrow A$ by $F(p\otimes a) = \pi_R(p)\imath_R(a)$ for all $p \in H^*$ and $a \in H$. Then $F\circ\pi_{\mathcal{R}} = \pi_R$ and $F\circ\imath_{\mathcal{R}} = \imath_R$. Since $\pi_R : H^{*\,cop} \longrightarrow R_{(r)}$ is a coalgebra map $F = m'\circ(\pi_R\otimes\imath_R)$ is a composition of coalgebra maps, where $m' : R_{(r)}\otimes R_{(\ell)} \longrightarrow A$ is the restriction of the multiplication map of A. Therefore F is a coalgebra map.

Let $p, q \in H^*$ and $a, b \in H$. Since π_R and $\imath_R$ are algebra homomorphisms $F(\epsilon\otimes 1) = 1$ and, whenever $a = 1$ or $q = \epsilon$,

$$\begin{aligned} F(p\otimes a)F(q\otimes b) &= \pi_R(p)\imath_R(a)\pi_R(q)\imath_R(b) \\ &= \pi_R(p)\pi_R(q)\imath_R(a)\imath_R(b) \\ &= \pi_R(pq)\imath_R(ab) \\ &= F(pq\otimes ab) \\ &= F((p\otimes a)(q\otimes b)). \end{aligned}$$

Thus to show that F is multiplicative we need only show that $F((\epsilon\otimes a)(q\otimes 1)) = F(\epsilon\otimes a)F(q\otimes 1)$, or equivalently

$$\pi_R(a_{(1)}\succ q\prec S^{-1}(a_{(3)}))\imath_R(a_{(2)}) = \imath_R(a)\pi_R(q). \tag{13.11}$$

Since $\imath_R : H \longrightarrow R_{(\ell)}$ is an isomorphism $\imath_R^* : R_{(\ell)}^* \longrightarrow H^*$ is as well. In particular $q = \imath_R^*(r)$ for some $r \in R_{(\ell)}^*$. As $a{\succ}q{\prec}b = a{\succ}\imath_R^*(r){\prec}b = \imath_R^*(\imath_R(a){\succ}r{\prec}\imath_R(b))$, and $\imath_R : H \longrightarrow R_{(\ell)}$ is a bialgebra map, it follows that (13.11) is equivalent to

$$\mathsf{f}_R(\imath_R(a)_{(1)}{\succ}r{\prec}S_A^{-1}(\imath_R(a)_{(3)}))\imath_R(a)_{(2)} = \imath_R(a)\mathsf{f}_R(r)$$

which is the case by (13.10). We have shown that F is an algebra map and is therefore a bialgebra map.

It remains to show that $(F{\otimes}F)(\mathcal{R}) = R$. Let $\{h_1, \ldots, h_n\}$ be a basis for H and let $\{h^1, \ldots, h^n\}$ be the dual basis for H^*. Write $R = R_\ell{\otimes}R^\ell$. The calculation

$$\begin{aligned}(F{\otimes}F)(\mathcal{R}) &= F(\epsilon{\otimes}h_\jmath){\otimes}F(h^\jmath{\otimes}1)\\ &= \imath_R(h_\jmath){\otimes}\pi_R(h^\jmath)\\ &= \imath_R(h_\jmath){\otimes}\pi_R(\imath_R^*(\imath_R^{*\,-1}(h^\jmath)))\\ &= \imath_R(h_\jmath){\otimes}\mathsf{f}_R((\imath_R^{-1})^*(h^\jmath))\\ &= \imath_R(h_\jmath){\otimes}<(\imath_R^{-1})^*(h^\jmath), R_\ell>R^\ell\\ &= \imath_R(h_\jmath){\otimes}<h^\jmath, \imath_R^{-1}(R_\ell)>R^\ell\\ &= \imath_R(\imath_R^{-1}(R_\ell)){\otimes}R^\ell\\ &= R_\ell{\otimes}R^\ell\end{aligned}$$

shows that $(F{\otimes}F)(\mathcal{R}) = R$. □

The usual categorical argument for uniqueness of universal objects shows that if $(A, R, \imath_R, \pi_R)$ and $(A', R', \imath_{R'}, \pi_{R'})$ satisfy parts (a) and (b) of the preceding theorem then there is an isomorphism $F : (A, R) \longrightarrow (A', R')$ of quasitriangular Hopf algebras determined by $F{\circ}\imath_R = \imath_{R'}$ and $F{\circ}\pi_R = \pi_{R'}$. The preceding theorem provides a mapping characterization of the double.

Corollary 13.3.3. *Let (A, R) be a quasitriangular Hopf algebra over k and set $H = R_{(\ell)}$. Then:*

(a) *There exists a morphism $F : (D(H), \mathcal{R}) \longrightarrow (A, R)$ of quasitriangular Hopf algebras determined by $F{\circ}\jmath_H = \mathrm{I}_H$ and $F{\circ}\jmath_{H^*} = \mathsf{f}_R$.*
(b) $\mathrm{Im}(F) = A_R$.
(c) $\mathrm{Rank}(\mathcal{R}) = \mathrm{Rank}(R)$.

Proof. We apply the previous theorem to the H-admissible tuple $(A, R, \mathrm{I}_H, \mathsf{f}_R)$ and use (12.12). □

The double $(D(H), \mathcal{R})$ is minimal quasitriangular and $\mathrm{Dim}(D(H)) = (\mathrm{Rank}(\mathcal{R}))^2$. Among the minimal quasitriangular Hopf algebras over k doubles are those (A, R) such that $\mathrm{Dim}(A) = (\mathrm{Rank}(R))^2$.

Corollary 13.3.4. *Let (A, R) be a minimal quasitriangular Hopf algebra over k. Then*

(a) $\mathrm{Rank}(R)$ *divides* $\mathrm{Dim}(A)$ *and* $\mathrm{Dim}(A)$ *divides* $(\mathrm{Rank}(R))^2$

and the following are equivalent:

(b) $(A, R) \simeq (D(R_{(\ell)}), \mathcal{R})$.
(c) $(A, R) \simeq (D(H), \mathcal{R})$ *for some finite-dimensional Hopf algebra H over the field k.*
(d) $\mathrm{Dim}(A) = (\mathrm{Rank}(R))^2$.

Proof. That $\mathrm{Rank}(R)$ divides $\mathrm{Dim}(A)$ follows by part (c) of Proposition 12.2.13 since $A = A_R$. That part (b) implies part (c) is clear. Note that part (c) implies part (d) since $\mathrm{Dim}(D(H)) = (\mathrm{Rank}(\mathcal{R}))^2$ and an isomorphism $(A, R) \simeq (A', R')$ of quasitriangular Hopf algebras means that $\mathrm{Rank}(R) = \mathrm{Rank}(R')$ and that $\mathrm{Dim}(A) = \mathrm{Dim}(A')$. It remains to show $\mathrm{Dim}(A)$ divides $(\mathrm{Rank}(R))^2$ and part (d) implies part (b).

Let $H = R_{(\ell)}$. Then by Corollary 13.3.3 there exists a morphism $F : (D(H), \mathcal{R}) \longrightarrow (A, R)$ of quasitriangular Hopf algebras such that $F : D(H) \longrightarrow A$ is onto and $\mathrm{Rank}(\mathcal{R}) = \mathrm{Rank}(R)$. Since F is onto $F^* : A^* \longrightarrow D(H)^*$ is a one-one Hopf algebra map. Thus $\mathrm{Dim}(A^*) = \mathrm{Dim}(\mathrm{Im}(F))$ and the latter divides $\mathrm{Dim}(D(H)^*)$ by Corollary 9.3.4. Thus $\mathrm{Dim}(A)$ divides $\mathrm{Dim}(D(H)) = (\mathrm{Rank}(\mathcal{R}))^2 = (\mathrm{Rank}(R))^2$. If $\mathrm{Dim}(A) = (\mathrm{Rank}(R))^2$ then $\mathrm{Dim}(A) = \mathrm{Dim}(D(H))$ and the onto map F must be an isomorphism of quasitriangular Hopf algebras. □

Apropos of the corollary, assume that the characteristic of the field is not 2 and consider the 4-dimensional triangular Hopf algebras (H, R_α), where $\alpha \in k$, of Exercise 10.1.11. They are not factorizable by Exercise 12.4.2. If $\alpha \neq 0$ then $\mathrm{Rank}(R_\alpha) = \mathrm{Dim}(H)$ and (H, R_α) is minimal quasitriangular. Observe that (H, R_0) is not minimal quasitriangular, $\mathrm{Dim}(H) = (\mathrm{Rank}(R_0))^2$, and (H, R_0) is not a double since it is not factorizable. Thus the equivalence of parts (b)–(d) of Corollary 13.3.4 does not hold for quasitriangular Hopf algebras in general.

Exercise

Exercise 13.3.1. Verify (13.10). Observe that f_R is a bialgebra map by virtue of (QT.1)–(QT.4) and that (13.10) is a reflection of (QT.5).

13.4 The dual of the double

Here we consider when the dual of the double has properties which the double possesses. To understand when the dual is quasitriangular we describe its Hopf algebra structure. We also characterize the grouplike elements of the dual and show that semi-simplicity and commutativity are properties possessed by both the double and its dual or by neither. Throughout this section H is a finite-dimensional Hopf algebra with antipode S over k.

By part (c) of Theorem 13.2.1 the Hopf algebra $D(H)$ is unimodular. By part (d) of Proposition 13.2.2:

Proposition 13.4.1. *Let H be a finite-dimensional Hopf algebra over the field k. Then the following are equivalent:*

(a) *$D(H)^*$ is unimodular.*
(b) *H and H^* are unimodular.*

□

In our next result we use the fact that the maps $\jmath_H : H \longrightarrow D(H)$ and $\jmath_{H^*} : H^{*\,cop} \longrightarrow D(H)$ defined by $\jmath_H(a) = \epsilon \otimes a$ and $\jmath_{H^*}(p) = p \otimes 1$ for all $a \in H$ and $p \in H^*$ are one-one Hopf algebra maps. By part (e) of Theorem 12.2.8 the square of the antipode of $D(H)$ is inner.

Proposition 13.4.2. *Let H be a finite-dimensional Hopf algebra over the field k. Then the following are equivalent:*

(a) *$S^2_{D(H)^*}$ is inner.*
(b) *S^2_H and $S^2_{H^*}$ are inner.*

Proof. Since $\jmath_H : H \longrightarrow D(H)$ and $\jmath_{H^*} : H^{*\,cop} \longrightarrow D(H)$ are one-one Hopf algebra maps the transposes $\jmath_H^* : D(H)^* \longrightarrow H^*$ and $\jmath_{H^*}^* : D(H)^* \longrightarrow H^{op}$ are onto Hopf algebra maps. Suppose $S^2_{D(H)^*}$ is inner. Then $S^2_{H^*}$ and $S^2_{H^{op}} = S_H^{-2}$ are also. Since S_H^{-2} is inner S^2_H is also. We have shown part (a) implies part (b).

Suppose S_H^2 and $S_{H^*}^2$ are inner. Since S_H^2 is inner $S_{H^{op}}^2$ is also. As an algebra $D(H)^* = H^{op}\otimes H^*$ and $S_{D(H)^*}^2 = S_{H^{op}}^2\otimes S_{H^*}^2$; for the latter we use part (b) of Corollary 13.1.4. Therefore $S_{D(H)^*}^2$ is inner. We have shown part (b) implies part (a). □

By part (c) of Theorem 13.1.2 the double is quasitriangular. To determine when $D(H)^*$ is quasitriangular we will describe its Hopf algebra structure. The convenient way to do this is to use the realization of the double as a double crossproduct [114, Example 6.4.4] to describe multiplication in the double. The basis for the multiplication formula is a left module action by H on H^* and a right module action by H^* on H. These are given by

$$h \rightharpoonup\!\!\!\rightharpoonup p = h_{(1)}\succ p \prec S_H^{-1}(h_{(2)}) \tag{13.12}$$

and

$$h \leftharpoonup\!\!\!\leftharpoonup p = S_{H^*}^{-1}(p_{(1)}) \rightharpoonup h \leftharpoonup p_{(2)} \tag{13.13}$$

for all $h \in H$ and $p \in H^*$. The product in $D(H)$ can be expressed as

$$(p\otimes a)(q\otimes b) = p(a_{(1)} \rightharpoonup\!\!\!\rightharpoonup q_{(2)})\otimes(a_{(2)} \leftharpoonup\!\!\!\leftharpoonup q_{(1)})b \tag{13.14}$$

for all $p, q \in H^*$ and $a, b \in H$.

Express the module action of (13.12) as a linear map $H\otimes H^* \longrightarrow H^*$. Its transpose defines a left H^*-comodule structure on $H = H^{**}$ and likewise the action of (13.13) defines a right H-comodule structure on H^*. These are determined by

$$<h_{(-1)}, a> h_{(0)} = S^{-1}(a_{(2)})ha_{(1)} \tag{13.15}$$

and

$$p_{(0)} <q, p_{(1)}> = q_{(2)}pS_{H^*}^{-1}(q_{(1)}) \tag{13.16}$$

for all $h, a \in H$ and $p, q \in H^*$. The first equation holds since the resulting equations are true when all $p \in H^*$ are applied to both sides and likewise the second holds since the resulting equations are true when both sides are applied to all $a \in H$.

Lemma 13.4.3. *Let H be a finite-dimensional Hopf algebra over the field k. Then $D(H)^* = H^{op}\otimes H^*$ as an algebra and its coalgebra structure is determined by $\epsilon_{D(H)^*}(h\otimes p) = \epsilon(h)p(1)$ and*

$$\Delta_{D(H)^*}(h\otimes p) = (h_{(1)}\otimes h_{(2)(-1)}p_{(1)(0)})\otimes(p_{(1)(1)}h_{(2)(0)}\otimes p_{(2)})$$

for all $h \in H$ and $p \in H^*$.

Proof. Let $h \in H$ and $p \in H^*$. Then $\epsilon_{D(H)^*}(a\otimes p) = <a\otimes p, 1_{D(H)}> = \epsilon(a)p(1)$. Establishing the formula for the coproduct can be done by writing the formula for multiplication in $D(H)$ given by (13.14) as a composition of linear maps and computing its transpose. Here we verify the coproduct formula utilizing (13.15) and (13.16).

For all $a, b \in H$ and $q, r \in H^*$ the calculation

$$
\begin{aligned}
&<(h_{(1)}\otimes h_{(2)(-1)}p_{(1)(0)})\otimes(p_{(1)(1)}h_{(2)(0)}\otimes p_{(2)}), (q\otimes a)\otimes(r\otimes b)> \\
&= q(h_{(1)})h_{(2)(-1)}(a_{(1)})p_{(1)(0)}(a_{(2)})r_{(1)}(p_{(1)(1)})r_{(2)}(h_{(2)(0)})p_{(2)}(b) \\
&= q(h_{(1)})r_{(2)}(S_H^{-1}(a_{(1)(2)})h_{(2)}a_{(1)(1)})(r_{(1)(2)}p_{(1)}S_{H^*}^{-1}(r_{(1)(1)}))(a_{(2)})p_{(2)}(b) \\
&= <q(a_{(1)} \rightharpoonup r_{(2)}), h><p, (a_{(2)} \leftharpoonup r_{(1)})b> \\
&= <h\otimes p, (q\otimes a)(r\otimes b)>
\end{aligned}
$$

establishes the formula for $\Delta_{D(H)^*}$. □

Theorem 13.4.4. *Let H be a finite-dimensional Hopf algebra with antipode S over the field k. Then the following are equivalent:*

(a) *$D(H)^*$ has a quasitriangular structure.*
(b) *H and H^* have a quasitriangular structure.*

Let $\{h_1, \ldots, h_n\}$ be a basis for H and $\{h^1, \ldots, h^n\}$ be the dual basis for H^*. Suppose (H, r) and (H^*, R) are quasitriangular. Write $r = r_i\otimes r^i$ and $R = R_j\otimes R^j$. Then $(D(H)^*, \boldsymbol{R})$ is quasitriangular as well, where

$$\boldsymbol{R} = (r^i h_j \otimes R_\ell h^m)\otimes(S^{-1}(h_m)r_i \otimes h^j R^\ell).$$

Furthermore $\boldsymbol{R}$ does not depend on the choice of basis.

Proof. We use the fact that the bialgebra maps $\jmath_H : H \longrightarrow D(H)$ and $\jmath_{H^*} : H^{*\,cop} \longrightarrow D(H)$ are one-one. We have noted their transposes $\jmath_H^* : D(H)^* \longrightarrow H^*$ and $\jmath_{H^*}^* : H^{op} \longrightarrow D(H)^*$ are onto bialgebra maps. Thus if $D(H)^*$ has a quasitriangular structure then H^* and H^{op}, hence H, do also by Exercise 12.2.2. We have shown that part (a) implies part (b).

Suppose that (H, r) and (H^*, R) are quasitriangular. We first establish that $\boldsymbol{R}$ does not depend on the choice of basis. To show part (b) implies part (a) we use $F = f_R$, where $R = \boldsymbol{R}$, to show that $(D(H)^*, \boldsymbol{R})$ is quasitriangular. To establish quasitriangularity it suffices to show that $F : D(H)^{**} \longrightarrow D(H)^{*\,cop}$ is a bialgebra map and that (QT.5) is satisfied for $(\Delta_{D(H)^*}, \boldsymbol{R})$. See Exercise 12.2.6. We make the identifications $D(H)^{**} = D(H)$ and $D(H)^* = H^{op}\otimes H^*$, the latter according to Lemma 13.4.3. Our arguments are based on several identities, a good number of which arise from (13.15) or (13.16). These are (13.17)–(13.26) whose verifications are left as exercises for the reader.

Let $f_{H^*,R} = f_R$ and $g_{H,r} = g_r$. By definition $F(\alpha) = (\alpha \otimes I)(\boldsymbol{R})$ for all $\alpha \in D(H)^{**} = D(H)$. Therefore

$$F(p\otimes a) = S^{-1}(a_{(2)})g_{H,r}(p_{(1)})\otimes p_{(2)}f_{H^*,R}(a_{(1)}) \tag{13.17}$$

for all $a \in H$ and $p \in H^*$. Since F determines $\boldsymbol{R}$ it follows that $\boldsymbol{R}$ does not depend on the choice of basis for H.

Let $p, q \in H^*$ and $a, b \in H$. To show that F is a bialgebra map we will used Proposition 12.2.11 without particular reference. Using (13.17) we see that $F(\epsilon\otimes 1) = 1\otimes\epsilon$ and

$$\begin{aligned}&F(p\otimes a)F(q\otimes b)\\ &= S^{-1}(b_{(2)})g_{H,r}(q_{(1)})S^{-1}(a_{(2)})g_{H,r}(p_{(1)})\otimes p_{(2)}f_{H^*,R}(a_{(1)})q_{(2)}f_{H^*,R}(b_{(1)}).\end{aligned}$$

If $a = 1$ or $q = \epsilon$ observe that

$$F(p\otimes a)F(q\otimes b) = F(pq\otimes ab) = F((p\otimes a)(q\otimes b)).$$

Thus to show that F is multiplicative we need only establish $F(\epsilon\otimes a)F(q\otimes 1) = F((\epsilon\otimes a)(q\otimes 1))$ or

$$\begin{aligned}&g_{H,r}(q_{(1)})S^{-1}(a_{(2)})\otimes f_{H^*,R}(a_{(1)})q_{(2)}\\ &= S^{-1}(a_{(2)(2)})g_{H,r}((a_{(1)}\succ q\prec S^{-1}(a_3))_{(1)})\\ &\qquad \otimes(a_{(1)}\succ q\prec S^{-1}(a_3))_{(2)}f_{H^*,R}(a_{(2)(1)}).\end{aligned}$$

Since $\Delta_{H^*}(h\succ r\prec \ell) = (r_{(1)}\prec \ell)\otimes(h\succ r_{(2)})$ and $r_{(1)}\otimes r_{(2)}\prec h = h\succ r_{(1)}\otimes r_{(2)}$ for all $h, \ell \in H$ and $r \in H^*$, using part (a) of Proposition 12.2.11, we see that the last equation is equivalent to

$$\begin{aligned}&g_{H,r}(q_{(1)})S^{-1}(a_{(2)})\otimes f_{H^*,R}(a_{(1)})q_{(2)}\\ &= S^{-1}(a_{(3)})g_{H,r}(a_{(2)}\succ q_{(1)}\prec S^{-1}(a_{(4)}))\otimes f_{H^*}(a_{(1)})q_{(2)}\end{aligned}$$

which follows from the identity

$$g_{H,r}(q)S^{-1}(a) = S^{-1}(a_{(2)})g_{H,r}(a_{(1)}\succ q\prec S^{-1}(a_{(3)})). \tag{13.18}$$

We have shown that F is an algebra map.

Next we show that F is a coalgebra map. Using (13.17) it follows that $\epsilon_{D(H)^{*\,cop}}\circ F = \epsilon_{D(H)}$. Since F is an algebra map the equation $\Delta_{D(H)^{*\,cop}}\circ F = (F\otimes F)\circ\Delta_{D(H)}$ holds if and only if it holds on algebra generators $p\otimes 1$ and $\epsilon\otimes a$. Now $\Delta_{D(H)^{*\,cop}}(F(p\otimes 1)) = (F\otimes F)(\Delta_{D(H)}(p\otimes 1))$ if and only if

$$\begin{aligned}&(p_{(3)(1)}g_{H,r}(p_{(2)})_{(0)}\otimes p_{(4)})\otimes(g_{H,r}(p_{(1)})\otimes g_{H,r}(p_{(2)})_{(-1)}p_{(3)(0)})\\ &= (g_{H,r}(p_{(3)})\otimes p_{(4)})\otimes(g_{H,r}(p_{(1)})\otimes p_{(2)})\end{aligned}$$

which follows from

$$p_{(2)(1)}g_{H,r}(p_{(1)})_{(0)}\otimes g_{H,r}(p_{(1)})_{(-1)}p_{(2)(0)} = g_{H,r}(p_{(2)})\otimes p_{(1)}. \tag{13.19}$$

Next observe $\Delta_{D(H)^{*\,cop}}(F(\epsilon\otimes a)) = (F\otimes F)(\Delta_{D(H)}(\epsilon\otimes a))$ if and only if

$$\begin{aligned} &f_{H^*,R}(a_{(2)})_{(1)}S^{-1}(a_{(3)})_{(0)}\otimes f_{H^*,R}(a_{(1)}) \\ &\qquad \otimes S^{-1}(a_{(4)})\otimes S^{-1}(a_{(3)})_{(-1)}f_{H^*,R}(a_{(2)})_{(0)} \\ &= S^{-1}(a_{(2)})\otimes f_{H^*,R}(a_{(1)})\otimes S^{-1}(a_{(4)})\otimes f_{H^*,R}(a_{(3)}) \end{aligned}$$

which follows by

$$\begin{aligned} &f_{H^*,R}(a_{(1)})_{(1)}S^{-1}(a_{(2)})_{(0)}\otimes S^{-1}(a_{(2)})_{(-1)}f_{H^*,R}(a_{(1)})_{(0)} \\ &= S^{-1}(a_{(1)})\otimes f_{H^*,R}(a_{(2)}). \end{aligned} \tag{13.20}$$

Thus F is a coalgebra map.

The proof of the theorem will be complete once we show (QT.5) is satisfied for $(D(H)^*, \boldsymbol{R})$. To this end, by Remark 12.2.4 we need only show this equation is satisfied on algebra generators, on $1\otimes p$ and $a\otimes\epsilon$ is sufficient.

Let $p \in H^*$. Since

$$1_{(-1)}\otimes 1_{(0)} = \epsilon\otimes 1, \tag{13.21}$$

and

$$h_i p_{(1)}\otimes p_{(0)}h^i = h_i\otimes h^i p \tag{13.22}$$

we have

$$\Delta_{D(H)^*}(1\otimes p) = (1\otimes p_{(1)(0)})\otimes(p_{(1)(1)}\otimes p_{(2)})$$

and

$$\begin{aligned} &\Delta^{cop}_{D(H)^*}(1\otimes p)\boldsymbol{R} \\ &= ((p_{(1)(1)}\otimes p_{(2)})\otimes(1\otimes p_{(1)(0)}))((r^i h_j\otimes R_\ell h^m)\otimes(S^{-1}(h_m)r_i\otimes h^j R^\ell)) \\ &= (r^i h_j p_{(1)(1)}\otimes p_{(2)}R_\ell h^m)\otimes(S^{-1}(h_m)r_i\otimes p_{(1)(0)}h^j R^\ell) \\ &= (r^i h_j\otimes p_{(2)}R_\ell h^m)\otimes(S^{-1}(h_m)r_i\otimes h^j p_{(1)}R^\ell). \end{aligned}$$

Since

$$h^i p_{(0)}\otimes p_{(1)}S^{-1}(h_i) = ph^i\otimes S^{-1}(h_i) \tag{13.23}$$

we can use (QT.5) for (H^*, R) to conclude

$$\begin{aligned} &\boldsymbol{R}\Delta_{D(H)^*}(1\otimes p) \\ &= ((r^i h_j\otimes R_\ell h^m)\otimes(S^{-1}(h_m)r_i\otimes h^j R^\ell))((1\otimes p_{(1)(0)})\otimes(p_{(1)(1)}\otimes p_{(2)})) \\ &= (r^i h_j\otimes R_\ell h^m p_{(1)(0)})\otimes(p_{(1)(1)}S^{-1}(h_m)r_i\otimes h^j R^\ell p_{(2)}) \\ &= (r^i h_j\otimes R_\ell p_{(1)}h^m)\otimes(S^{-1}(h_m)r_i\otimes h^j R^\ell p_{(2)}) \\ &= (r^i h_j\otimes p_{(2)}R_\ell h^m)\otimes(S^{-1}(h_m)r_i\otimes h^j p_{(1)}R^\ell). \end{aligned}$$

Therefore (QT.5) holds for $(D(H)^*, \boldsymbol{R})$ when $h = 1\otimes p$.

Let $a \in H$. Since

$$\epsilon_{(0)}\otimes\epsilon_{(1)} = \epsilon\otimes 1 \tag{13.24}$$

and

$$h_i a_{(0)}\otimes a_{(-1)}h^i = ah_i\otimes h^i \tag{13.25}$$

we see that

$$\Delta_{D(H)^*}(a\otimes\epsilon) = (a_{(1)}\otimes a_{(2)(-1)})\otimes(a_{(2)(0)}\otimes\epsilon)$$

and

$$\begin{aligned}
&\Delta^{cop}_{D(H)^*}(a\otimes\epsilon)\boldsymbol{R}\\
&= ((a_{(2)(0)}\otimes\epsilon)\otimes(a_{(1)}\otimes a_{(2)(-1)}))((r^ih_j\otimes R_\ell h^m)\otimes(S^{-1}(h_m)r_i\otimes h^jR^\ell))\\
&= (r^ih_ja_{(2)(0)}\otimes R_\ell h^m)\otimes(S^{-1}(h_m)r_ia_{(1)}\otimes a_{(2)(-1)}h^jR^\ell)\\
&= (r^ia_{(2)}h_j\otimes R_\ell h^m)\otimes(S^{-1}(h_m)r_ia_{(1)}\otimes h^jR^\ell).
\end{aligned}$$

Since

$$h^ia_{(-1)}\otimes a_{(0)}S^{-1}(h_i) = h^i\otimes S^{-1}(h_i)a \tag{13.26}$$

we can use (QT.5) for (H, r) to conclude

$$\begin{aligned}
&\boldsymbol{R}\Delta_{D(H)^*}(a\otimes\epsilon)\\
&= ((r^ih_j\otimes R_\ell h^m)\otimes(S^{-1}(h_m)r_i\otimes h^jR^\ell))((a_{(1)}\otimes a_{(2)(-1)})\otimes(a_{(2)(0)}\otimes\epsilon))\\
&= (a_{(1)}r^ih_j\otimes R_\ell h^m a_{(2)(-1)})\otimes(a_{(2)(0)}S^{-1}(h_m)r_i\otimes h^jR^\ell)\\
&= (a_{(1)}r^ih_j\otimes R_\ell h^m)\otimes(S^{-1}(h_m)a_{(2)}r_i\otimes h^jR^\ell)\\
&= (r^ia_{(2)}h_j\otimes R_\ell h^m)\otimes(S^{-1}(h_m)r_ia_{(1)}\otimes h^jR^\ell).
\end{aligned}$$

Therefore (QT.5) holds for $(D(H)^*, \boldsymbol{R})$ when $h = a\otimes\epsilon$ which concludes our argument that (QT.5) holds for $(D(H)^*, \boldsymbol{R})$. This completes our proof of the theorem. □

The grouplike elements of $D(H)$ are characterized in part (c) of Proposition 13.2.2. To characterize the grouplike elements of $D(H)^*$ we use part (b) of Theorem 12.4.7 and Theorem 13.1.2. Recall that $(D(H), \mathcal{R})$ is factorizable by part (a) of Theorem 13.2.1. In our discussion below we follow the notation associated with these results.

Let $\xi \in \mathrm{G}(D(H)^*) = \mathrm{Alg}(D(H), k)$. Since $p\otimes a = (p\otimes 1)(\epsilon\otimes a) = \jmath_{H^*}(p)\jmath_H(a)$ for all $p \in H^*$ and $a \in H$ it follows that $\xi = a\otimes\eta$ for some $a \in \mathrm{G}(H)$ and $\eta \in \mathrm{G}(H^*)$.

Let $a \in \mathrm{G}(H)$ and $\eta \in \mathrm{G}(H^*)$. We compute $F(a \otimes \eta)$, where $F = f_{\mathcal{R}_{21}\mathcal{R}}$ and $\mathcal{R}$ is as in part (c) of Theorem 13.1.2. Since $(\epsilon \otimes a)(\eta \otimes 1) = (a \succ \eta \prec a^{-1}) \otimes a = \eta \otimes a$, the second calculation in the proof of Theorem 13.2.1 shows that $F(a \otimes \eta) = \eta \otimes a$. By part (a) of Theorem 13.2.1 and part (b) of Theorem 12.4.7 we conclude that $a \otimes \eta \in \mathrm{G}(D(H)^*)$ if and only if $\eta \otimes a$ is in the center of $D(H)$. Now $\eta \otimes a$ is in the center if and only if it commutes with generators of $D(H)$, or equivalently if and only if $x(\eta \otimes a) = (\eta \otimes a)x$ where $x = p \otimes 1$ for all $p \in H^*$ and $x = \epsilon \otimes h$ for all $h \in H$. The latter conditions are met if and only if $(h \leftharpoonup \eta)a = a(\eta \rightharpoonup h)$ for all $h \in H$. Collecting results:

Proposition 13.4.5. *Let H be a finite-dimensional Hopf algebra over the field k. Then*

(a) $\mathrm{G}(D(H)^*) \subseteq \{a \otimes \eta \mid a \in \mathrm{G}(H), \eta \in \mathrm{G}(H^*)\}$

and for $a \in \mathrm{G}(H)$ and $\eta \in \mathrm{G}(H^*)$ the following are equivalent:

(b) $a \otimes \eta \in \mathrm{G}(D(H)^*)$.
(c) *$\eta \otimes a$ is in the center of $D(H)$.*
(d) *$(h \leftharpoonup \eta)a = a(\eta \rightharpoonup h)$ for all $h \in H$.*

□

The proof of Corollary 13.2.3 is easily adapted to establish:

Proposition 13.4.6. *Let H be a finite-dimensional Hopf algebra over the field k. Then the following are equivalent:*

(a) *$D(H)^*$ is semisimple.*
(b) *H and H^* are semisimple.*
(c) *H and H^* are cosemisimple.*
(d) *$D(H)^*$ is cosemisimple.*

□

Thus if one of $D(H)$ and $D(H)^*$ is semisimple they both are, and likewise if one is cosemisimple they both are. Compare with Corollary 13.2.3.

Since a finite-dimensional Hopf algebra is commutative (respectively cocommutative) if and only if its dual is cocommutative (respectively commutative), in light of Proposition 13.2.4 we conclude:

Proposition 13.4.7. *Let H be a finite-dimensional Hopf algebra over the field k. Then the following are equivalent:*

(a) *$D(H)^*$ is commutative.*
(b) *H and H^* are commutative.*
(c) *H and H^* are cocommutative.*
(d) *$D(H)^*$ is cocommutative.*

□

Thus if one of $D(H)$ and $D(H)^*$ is commutative they both are, and likewise if one is cocommutative they both are. See Proposition 13.2.4.

Exercises

In the following exercises H is a finite-dimensional Hopf algebra with antipode S over the field k.

Exercise 13.4.1. Establish (13.18). [Hint: Observe that

$$g_{H,r}(q)S^{-1}(a) = g_{H,r}(q_{(1)})(q_{(2)}\rightharpoonup(S_{H^*}(q_{(3)})\rightharpoonup S^{-1}(a)))$$

and that

$$(S_{H^*}(q_{(3)})\rightharpoonup S^{-1}(a)\leftharpoonup q_{(1)})\otimes q_{(2)} = S^{-1}(a_{(2)})\otimes(a_{(1)}\succ q\prec S^{-1}(a_{(3)})).]$$

Exercise 13.4.2. Establish (13.19). [Hint: Show that

$$<p_{(0)}, a>p_{(1)} = S^{-1}(a_{(2)}\leftharpoonup p)a_{(1)}$$

and that

$$S^{-1}(a_{(2)}\leftharpoonup p)\otimes a_{(1)} = S^{-1}(a_{(2)})\otimes(p\rightharpoonup a_{(1)})$$

and then

$$p_{(2)(1)}g_{H,r}(p_{(1)})_{(0)}<g_{H,r}(p_{(1)})_{(-1)}p_{(2)(0)}, a> = g_{H,r}(p_{(2)})<p_{(1)}, a>.]$$

Exercise 13.4.3. Establish (13.20).

Exercise 13.4.4. Establish (13.21)–(13.26). [Hint: See the comments for establishing (13.15) and (13.16).]

Exercise 13.4.5. Suppose that the characteristic of k is not 2 and let H be Sweedler's Example, that is the Taft algebra $H_{2,-1}$ of Section 7.3. We continue the discussion of Example 10.1.11 where its parameterized family $\{(H, R_\alpha)\}_{\alpha\in k}$ of quasitriangular structures is described. There is a Hopf algebra isomorphism

$f : H \longrightarrow H^*$ described in Exercise 7.4.3. For $\alpha, \beta \in k$ let $(D(H)^*, \boldsymbol{R}_{\alpha,\beta})$ be the quasitriangular structure described in Theorem 13.4.4 for $R_\alpha \in H \otimes H$ and $(f \otimes f)(R_\beta) \in H^* \otimes H^*$.

(a) Find $\mathrm{Rank}(\boldsymbol{R}_{\alpha,\beta})$.

(b) Determine when $(D(H)^*, \boldsymbol{R}_{\alpha,\beta})$ is minimal quasitriangular.

Exercise 13.4.6. Here is a small generalization of Proposition 13.4.1. Suppose H, K are finite-dimensional Hopf algebras over k and $\mathcal{H} = H \otimes K$ has a Hopf algebra structure such that the underlying algebra structure is the tensor product algebra structure. Show that $\mathcal{H}$ is unimodular if and only if H and K are.

Exercise 13.4.7. Suppose $f : H \longrightarrow K$ is an onto Hopf algebra map. Show that S_H^2 inner implies that S_K^2 is inner.

13.5 The double of a quasitriangular Hopf algebra

Throughout this section (H, R) is a finite-dimensional quasitriangular Hopf algebra. Let $\pi : D(H) \longrightarrow H$ be defined by

$$\pi(p \otimes h) = f_R(p)h \tag{13.27}$$

for all $p \in H^*$ and $h \in H$. Now $f_R : H^* \longrightarrow H^{cop}$ is a bialgebra map by part (a) of Proposition 12.2.11 and it satisfies (12.21). The calculations $\pi(1_{D(H)}) = \pi(\epsilon \otimes 1) = f_R(\epsilon)1 = 1$ and

$$\begin{aligned}
\pi((p \otimes a)(q \otimes b)) &= \pi(p(a_{(1)} \succ q \prec S^{-1}(a_{(3)})) \otimes a_{(2)}b) \\
&= f_R(p(a_{(1)} \succ q \prec S^{-1}(a_{(3)})))a_{(2)}b \\
&= f_R(p) f_R(a_{(1)} \succ q \prec S^{-1}(a_{(3)}))a_{(2)}b \\
&= f_R(p) a f_R(q) b \\
&= \pi(p \otimes a)\pi(q \otimes b)
\end{aligned}$$

for all $p, q \in H^*$ and $a, b \in H$ show that π is an algebra map. The calculations $\epsilon(\pi(p \otimes a)) = \epsilon(f_R(p))\epsilon(a) = p(1)\epsilon(a) = \epsilon_{D(H)}(p \otimes a)$ and

$$\begin{aligned}
\Delta(\pi(p \otimes a)) &= \Delta(f_R(p)a) \\
&= f_R(p)_{(1)}a_{(1)} \otimes f_R(p)_{(2)}a_{(2)} \\
&= f_R(p_{(2)})a_{(1)} \otimes f_R(p_{(1)})a_{(2)} \\
&= \pi(p_{(2)} \otimes a_{(1)}) \otimes \pi(p_{(1)} \otimes a_{(2)})
\end{aligned}$$

for all $p \in H^*$ and $a \in H$ show that π is a coalgebra map. Therefore π is a map of Hopf algebras. We show that $\pi : (D(H), \mathcal{R}) \longrightarrow (H, R)$ is a map of quasitriangular Hopf algebras. Write $\mathcal{R} = \sum_{i=1}^{n}(\epsilon\otimes h_i)\otimes(h^i\otimes 1)$, where $\{h_1, \ldots, h_n\}$ is a linear basis for H and $\{h^1, \ldots, h^n\}$ is its dual basis for H^*. Then

$$\begin{aligned}(\pi\otimes\pi)(\mathcal{R}) &= \pi(\epsilon\otimes h_i)\otimes\pi(h^i\otimes 1)\\ &= f_R(\epsilon)h_i\otimes f_R(h^i)1\\ &= h_i\otimes <h^i, R_j> R^j\\ &= <h^i, R_j> h_i\otimes R^j\\ &= R_j\otimes R^j\end{aligned}$$

shows that $(\pi\otimes\pi)(\mathcal{R}) = R_j\otimes R^j = R$.

Let $\jmath : H \longrightarrow D(H)$ be the Hopf algebra map given by $\jmath(a) = \epsilon\otimes a$ for all $a \in H$. Then $\pi\circ\jmath = \mathrm{I}_H$. Therefore $D(H) = B\times H$ is a bi-product, where $D(H)_{co\,inv} = B = \mathrm{Im}(\Pi)$; see (11.35) and Theorem 11.7.1. Recall $\Pi = \mathrm{I}_H*(\jmath\circ S\circ\pi)$. Hence

$$\begin{aligned}\Pi(p\otimes a) &= (p_{(2)}\otimes a_{(1)})(\epsilon\otimes S(\pi(p_{(1)}\otimes a_{(2)})))\\ &= p_{(2)}\otimes a_{(1)}S(f_R(p_{(1)})a_{(2)})\\ &= p_{(2)}\otimes a_{(1)}S(a_{(2)})S(f_R(p_{(1)}))\\ &= p_{(2)}\otimes\epsilon(a)S(f_R(p_{(1)}))\end{aligned}$$

which means

$$B = D(H)_{co\,inv} = \{p_{(2)}\otimes S(f_R(p_{(1)}))\,|\,p \in H^*\}. \tag{13.28}$$

To summarize:

Theorem 13.5.1. *Let (H, R) be a finite-dimensional quasitriangular Hopf algebra over the field k. Then:*

(a) $\pi : (D(H), \mathcal{R}) \longrightarrow (H, R)$, *where π is defined by (13.27), is a morphism of quasitriangular Hopf algebras.*
(b) $\pi\circ\jmath = \mathrm{I}_H$, *where $\jmath : H \longrightarrow D(H)$ is the Hopf algebra map given by $\jmath(h) = \epsilon\otimes h$ for all $h \in H$. Thus $D(H) = B\times H$ is a bi-product, where B is given by (13.28).*
(c) *The map $H^* \longrightarrow B$ $(p \mapsto p_{(2)}\otimes S(f_R(p_{(1)})))$ is a linear isomorphism.*

Proof. Since $\mathrm{Dim}(B) = \mathrm{Dim}(H^*)$ as $(\mathrm{Dim}(H))^2 = \mathrm{Dim}(D(H)) = (\mathrm{Dim}(H))(\mathrm{Dim}(B))$ part (c) follows which is all that remains to be proved. $\square$

There is an algebra homomorphism $F : D(H) \longrightarrow H{\otimes}H$ which will be the object of study of the next section. First of all we define $f : D(H) \longrightarrow H$ by

$$f(p{\otimes}h) = S(g_R(p))h \tag{13.29}$$

for all $p \in H^*$ and $h \in H$. The calculation, which uses (12.20),

$$\begin{aligned}
&S(g_R(h_{(1)}{\succ}p{\prec}S^{-1}(h_{(3)})))h_{(2)}\\
&= S(S^{-1}(h_{(2)})g_R(h_{(1)}{\succ}p{\prec}S^{-1}(h_{(3)})))\\
&= S(S^{-1}(h)_{(2)}g_R(S(S^{-1}(h)_{(3)}){\succ}p{\prec}S^{-1}(h)_{(1)}))\\
&= S(g_R(p)S^{-1}(h))\\
&= hS(g_R(p))
\end{aligned}$$

shows that $S{\circ}g_R$ satisfies (12.21). The proof that π is an algebra homomorphism shows that f is as well. Since $S : H \longrightarrow H^{op\,cop}$ is a bialgebra map $S{\circ}g_R : H^o \longrightarrow H^{cop}$ is also by part (b) of Proposition 12.2.11.

We now define $F : D(H) \longrightarrow H{\otimes}H$ by $F = (f{\otimes}\pi){\circ}\Delta_{D(H)}$ which is an algebra map since it is the composite of such. By definition

$$F(x) = f(x_{(1)}){\otimes}\pi(x_{(2)}) \tag{13.30}$$

for all $x \in D(H)$.

It is interesting to see what F does to elements of $D(H)_{co\,inv}$. Let $p \in H^*$. Then

$$\begin{aligned}
F(p_{(2)}{\otimes}S(f_R(p_{(1)}))) &= f(p_{(2)(2)}{\otimes}S(f_R(p_{(1)}))_{(1)}){\otimes}\pi(p_{(2)(1)}{\otimes}S(f_R(p_{(1)}))_{(2)})\\
&= S(g_R(p_{(2)(2)}))S(f_R(p_{(1)(1)})){\otimes}f_R(p_{(2)(1)})S(f_R(p_{(1)(2)}))\\
&= S(g_R(p_{(4)}))S(f_R(p_{(1)})){\otimes}f_R(p_{(3)})S(f_R(p_{(2)}))\\
&= S(g_R(p_{(3)}))S(f_R(p_{(1)})){\otimes}f_R(p_{(2)})_{(1)}S(f_R(p_{(2)})_{(2)})\\
&= S(g_R(p_{(3)}))S(f_R(p_{(1)})){\otimes}\epsilon(f_R(p_{(2)}))1\\
&= S((f_R{*}g_R)(p)){\otimes}1\\
&= S(g_{R_{2,1}R}(p)){\otimes}1.
\end{aligned}$$

Proposition 13.5.2. *Let (H, R) be a finite-dimensional quasitriangular Hopf algebra over the field k and write $R = R_i{\otimes}R^i$. Then:*

(a) $J = (1{\otimes}R^i){\otimes}(R_i{\otimes}1)$ *is a left twist of the tensor product bialgebra $H{\otimes}H$ and $F : D(H) \longrightarrow (H{\otimes}H)^J$ defined by (13.30) is a bialgebra homomorphism.*

(b) $F(p_{(2)}{\otimes}S(f_R(p_{(1)}))) = S(g_{R_{2,1}R}(p)){\otimes}1$ *for all $p \in H^*$.*

Proof. Note that J is a left twist of $H{\otimes}H$ by part (d) of Theorem 12.2.8. Recall that $f,\pi : D(H) \longrightarrow H$ are bialgebra maps. In light of the computations above we need only show that F is a coalgebra map. To do this, for $x \in D(H)$ we compare

$$\begin{aligned}\Delta_{H\otimes H}(F(x)) &= \Delta_{H\otimes H}(f(x_{(1)})\otimes\pi(x_{(2)}))\\ &= (f(x_{(1)})_{(1)}\otimes\pi(x_{(2)})_{(1)})\otimes(f(x_{(1)})_{(2)}\otimes\pi(x_{(2)})_{(2)})\\ &= (f(x_{(1)(1)})\otimes\pi(x_{(2)(1)}))\otimes(f(x_{(1)(2)})\otimes\pi(x_{(2)(2)}))\\ &= (f(x_{(1)})\otimes\pi(x_{(2)(2)}))\otimes(f(x_{(2)(1)})\otimes\pi(x_{(3)}))\end{aligned}$$

and

$$\begin{aligned}(F\otimes F)(\Delta_{D(H)}(F(x))) &= F(x_{(1)})\otimes F(x_{(2)})\\ &= (f(x_{(1)(1)})\otimes\pi(x_{(1)(2)}))\otimes(f(x_{(2)(1)})\otimes\pi(x_{(2)(2)}))\\ &= (f(x_{(1)})\otimes\pi(x_{(2)(1)}))\otimes(f(x_{(2)(2)})\otimes\pi(x_{(3)})).\end{aligned}$$

Once we show $f(x_{(2)})R_i\otimes\pi(x_{(1)})R^i = R_if(x_{(1)})\otimes R^i\pi(x_{(2)})$ our proof will be complete. To this end we may assume $x = p\otimes a$, where $p \in H^*$ and $a \in H$. In this case the preceding equation is

$$S(g_R(p_{(1)}))a_{(2)}R_i\otimes f_R(p_{(2)})a_{(1)}R^i = R_iS(g_R(p_{(2)}))a_{(1)}\otimes R^if_R(p_{(1)})a_{(2)}$$

which, in light of (QT.5) is equivalent to

$$S(g_R(p_{(1)}))R_i\otimes f_R(p_{(2)})R^i = R_iS(g_R(p_{(2)}))\otimes R^if_R(p_{(1)})$$

for all $p \in H^*$. But this equation holds by the following lemma. □

Lemma 13.5.3. *Let (H,R) be a quasitriangular Hopf algebra over the field k. Then $S(g_R(p_{(1)}))R_i\otimes f_R(p_{(2)})R^i = R_iS(g_R(p_{(2)}))\otimes R^if_R(p_{(1)})$ for all $p \in H^*$.*

Proof. For all $p \in H^*$ the left-hand side of the equation is

$$S(R_j{<}p_{(1)}, R^j{>})R_i\otimes({<}p_{(2)}, R_\ell{>}R^\ell)R^i = {<}p, R^jR_\ell{>}S(R_j)R_i\otimes R^\ell R^i$$

and the right-hand side is

$$R_iS(R_j{<}p_{(2)}, R^j{>})\otimes R^i({<}p_{(1)}, R_\ell{>}R^\ell) = {<}p, R_\ell R^j{>}R_iS(R_j)\otimes R^iR^\ell.$$

Thus the equation we need to establish is equivalent to

$$R^jR_\ell\otimes S(R_j)R_i\otimes R^\ell R^i = R_\ell R^j\otimes R_iS(R_j)\otimes R^iR^\ell$$

which is reformulated, on applying $\tau_{H,H}\otimes S^{-1}$ to both sides, as

$$S(R_j)R_i\otimes R^jR_\ell\otimes S^{-1}(R^i)S^{-1}(R^\ell) = R_iS(R_j)\otimes R_\ell R^j\otimes S^{-1}(R^\ell)S^{-1}(R^i).$$

Since $R^{-1} = S(R_i)\otimes R^i = R_i\otimes S^{-1}(R^i)$, by Theorem 12.2.8, the last equation is the quantum Yang-Baxter equation for R^{-1}. Since the quantum Yang-Baxter equation holds for R by part (c) of Theorem 12.2.8, it holds for R^{-1}. See Exercise 12.1.4. □

13.6 The double of a factorizable Hopf algebra

The main purpose of this section is to prove:

Theorem 13.6.1. *Suppose that (H, R) is a finite-dimensional quasitriangular Hopf algebra over the field k and write $R = R_i \otimes R^i$. Then the bialgebra map $F : D(H) \longrightarrow H \otimes H$ of (13.30) is bijective if and only if (H, R) is factorizable in which case $F : D(H) \longrightarrow (H \otimes H)^J$ is a Hopf algebra isomorphism, where $J = (1 \otimes R^i) \otimes (R_i \otimes 1)$.*

Proof. We continue with the notation of the previous section. By Proposition 13.5.2 we need only show F is bijective if and only if (H, R) is factorizable. This we do using the theory of Hopf modules.

Since $D(H) \overset{\jmath}{\underset{\pi}{\leftrightarrows}} H$ are Hopf algebra maps, we can use the algebra map $\jmath$ to give $D(H)$ a right H-module structure by pull-back and we can use the coalgebra map π to give $D(H)$ a right H-comodule $(D(H), \rho_{D(H)})$ structure by pull-back. Using $\pi \circ \jmath = \mathrm{I}_H$ we deduce

$$\begin{aligned} \rho_{D(H)}(x \cdot h) &= \rho_{D(H)}(x \jmath(h)) \\ &= (x \jmath(h))_{(1)} \otimes \pi((x \jmath(h))_{(2)}) \\ &= x_{(1)} \jmath(h_{(1)}) \otimes \pi(x_{(2)} \jmath(h_{(2)})) \\ &= x_{(1)} \cdot h_{(1)} \otimes \pi(x_{(2)}) \pi(\jmath(h_{(2)})) \\ &= x_{(1)} \cdot h_{(1)} \otimes \pi(x_{(2)}) h_{(2)} \end{aligned}$$

which means these structures give $D(H)$ the structure of a right H-Hopf module. Observe that $D(H)_{co\,inv}$ is described by (13.28).

Now regard H as a right H-module under multiplication. Then $H \otimes H$ is a right H-Hopf module with the right tensor product module structure and right H-comodule structure $(H \otimes H, \rho_{H \otimes H})$ described by $\rho_{H \otimes H}(a \otimes b) = (a \otimes b_{(1)}) \otimes b_{(2)}$ for all $a, b \in H$. Observe that $(H \otimes H)_{co\,inv} = H \otimes 1$.

We show that $F : D(H) \longrightarrow H \otimes H$ is a map of right H-modules. Since f, π are algebra maps and $f \circ \jmath = \mathrm{I}_H = \pi \circ \jmath$, the calculation

$$\begin{aligned} F(x \jmath(h)) &= f(x_{(1)} \jmath(h)_{(1)}) \otimes \pi(x_{(2)} \jmath(h)_{(2)}) \\ &= f(x_{(1)}) f(\jmath(h_{(1)})) \otimes \pi(x_{(2)}) \pi(\jmath(h_{(2)})) \\ &= f(x_{(1)}) h_{(1)} \otimes \pi(x_{(2)}) h_{(2)} \end{aligned}$$

shows that F is a module map. Since π is a coalgebra map F is a comodule map.

Since F is a comodule map $F(D(H)_{co\,inv}) \subseteq (H\otimes H)_{co\,inv}$. By Theorem 8.2.3 it follows that F is bijective if and only if the restriction

$$F|D(H)_{co\,inv} : D(H)_{co\,inv} \longrightarrow (H\otimes H)_{co\,inv}$$

is bijective. In light of part (c) of Theorem 13.5.1 and part (b) of Proposition 13.5.2 the latter is the case if and only if $g_{R_{2,1}R} : H^* \longrightarrow H$ is bijective. □

Since the double of a finite-dimensional Hopf algebra over k is factorizable by part (a) of Theorem 13.2.1:

Corollary 13.6.2. *Let H be a finite-dimensional Hopf algebra over the field k, let $\{h_1, \ldots, h_n\}$ be a basis for H and $\{h^1, \ldots, h^n\}$ be the dual basis for H^*. Then $D(D(H)) \simeq (D(H)\otimes D(H))^J$ as Hopf algebras, where*

$$J = ((\epsilon\otimes 1)\otimes(h^i\otimes 1))\otimes((\epsilon\otimes h_i)\otimes(\epsilon\otimes 1)).$$

□

13.7 Quasi-ribbon and ribbon elements of the double

Here we specialize Theorem 12.3.6 to the double $(D(H), \mathcal{R})$. First of all, by part (c) of Proposition 13.2.2 there is a group isomorphism $\mathrm{G}(H^*)\times\mathrm{G}(H) \simeq \mathrm{G}(D(H))$ given by $(\beta, a) \mapsto \beta\otimes a$ and by part (d) of the same $\alpha\otimes g$ is the $D(H)$-distinguished grouplike element of $D(H)$, where α and g are the H-distinguished grouplike elements of H^* and H respectively. Let u be the Drinfel'd element of $(D(H), \mathcal{R})$. Since $D(H)$ is unimodular $\mathsf{g} = uS_{D(H)}(u^{-1}) = \alpha\otimes g$ by parts (a) and (b) of Proposition 12.3.3. We now determine $\mathrm{G}(\mathrm{QR}(D(H), \mathcal{R}))$ and $\mathrm{G}(\mathrm{R}(D(H), \mathcal{R}))$.

By definition $\mathrm{G}(\mathrm{QR}(D(H), \mathcal{R}))$ consists of all $\beta\otimes a \in \mathrm{G}(D(H))$ such that $(\beta\otimes a)^2 = (\alpha\otimes g)^{-1}$, or equivalently $\beta^{-2}\otimes a^{-2} = \alpha\otimes g$. Therefore:

Lemma 13.7.1. *Let H be a finite-dimensional Hopf algebra over the field k, let g and α be the H-distinguished grouplike elements of H and H^* respectively. Then $\mathrm{G}(\mathrm{QR}(D(H), \mathcal{R}))$ consists of the elements $\beta\otimes a$, where*

(a) $\beta \in \mathrm{G}(H^*)$, $a \in \mathrm{G}(H)$ *and*
(b) $\beta^{-2} = \alpha$, $a^{-2} = g$.

□

Now $\mathrm{G}(\mathrm{R}(D(H), \mathcal{R}))$ is the set of $\beta \otimes a \in \mathrm{G}(\mathrm{QR}(D(H), \mathcal{R}))$ such that

$$S^2_{D(H)}(x) = (\beta \otimes a)^{-1} x (\beta \otimes a) \tag{13.31}$$

is satisfied for all $x \in D(H)$. Let $\beta \in \mathrm{G}(H^*)$ and $a \in \mathrm{G}(H)$. We find necessary and sufficient conditions for (13.31) to hold. Since (13.31) holds for all $x \in D(H)$ if and only if it holds for algebra generators, we need only consider the cases $x = p \otimes 1$, where $p \in H^*$ and $x = \epsilon \otimes h$, where $h \in H$. We treat the latter case first.

Suppose $h \in H$. Then $S^2_{D(H)}(\epsilon \otimes h) = \epsilon \otimes S^2_H(h)$ by part (b) of Corollary 13.1.4. Let $\beta \in \mathrm{G}(H^*)$. Since the assignment $h \mapsto \beta^{-1} \rightharpoonup h \leftharpoonup \beta$ defines an algebra automorphism of H, and

$$(h_{(1)} \succ \beta \prec S^{-1}(h_{(3)})) \otimes h_{(2)} = \beta \otimes (\beta^{-1} \rightharpoonup h \leftharpoonup \beta) \tag{13.32}$$

as $\beta \in \mathrm{G}(H^*)$, we calculate

$$\begin{aligned}
&(\beta \otimes a)^{-1}(\epsilon \otimes h)(\beta \otimes a) \\
&= (\beta^{-1} \otimes a^{-1})(\epsilon \otimes h)(\beta \otimes a) \\
&= (\beta^{-1} \otimes a^{-1}h)(\beta \otimes a) \\
&= \beta^{-1}((a^{-1}h)_{(1)} \succ \beta \prec S^{-1}((a^{-1}h)_{(3)})) \otimes (u^{-1}h)_{(2)}a \\
&= \beta^{-1}\beta \otimes \beta^{-1} \rightharpoonup (a^{-1}h) \leftharpoonup \beta \\
&= \epsilon \otimes (\beta^{-1} \rightharpoonup a^{-1} \leftharpoonup \beta)(\beta^{-1} \rightharpoonup h \leftharpoonup \beta)a \\
&= \epsilon \otimes a^{-1}(\beta^{-1} \rightharpoonup h \leftharpoonup \beta)a.
\end{aligned}$$

Therefore (13.31) holds for $\epsilon \otimes h$ if and only if

$$S^2(h) = a^{-1}(\beta^{-1} \rightharpoonup h \leftharpoonup \beta)a. \tag{13.33}$$

Let $p \in H^*$. Then $S^2_{D(H)}(p) = S^{-2}_{H^*}(p) \otimes 1$ by part (b) of Corollary 13.1.4 again. Since $a \in \mathrm{G}(H)$ the assignment $p \mapsto a \succ p \prec a^{-1}$ defines an algebra automorphism of H^*. Thus

$$\begin{aligned}
&(\beta \otimes a)^{-1}(p \otimes 1)(\beta \otimes a) \\
&= (\beta^{-1} \otimes a^{-1})(p\beta \otimes a) \\
&= \beta^{-1}(a^{-1} \succ p\beta \prec S^{-1}(a^{-1})) \otimes a^{-1}a \\
&= \beta^{-1}(a^{-1} \succ p\beta \prec a) \otimes 1 \\
&= \beta^{-1}((a^{-1} \succ p \prec a)(a^{-1} \succ \beta \prec a)) \otimes 1 \\
&= \beta^{-1}((a^{-1} \succ p \prec a)\beta) \otimes 1
\end{aligned}$$

We have shown (13.31) holds for $p{\otimes}1$ if and only if

$$S_{H^*}^{-2}(p) = \beta^{-1}(a^{-1}{\succ}p{\prec}a)\beta. \tag{13.34}$$

Now (13.33) holds for all $h \in H$ if and only if (13.34) holds for all $p \in H^*$; see Exercise 13.7.2. We have established:

Lemma 13.7.2. *Let H be a finite-dimensional Hopf algebra over the field k. Then $\mathrm{G}(\mathrm{R}(D(H), \mathcal{R}))$ consists of the elements $\beta{\otimes}a$, where*

(a) $\beta \in \mathrm{G}(H^*)$, $a \in \mathrm{G}(H)$,
(b) $\beta^{-2} = \alpha$, $a^{-2} = g$, *and*
(c) $S^2(h) = a^{-1}(\beta^{-1}{\rightharpoonup}h{\leftharpoonup}\beta)a$ *for all* $h \in H$.

□

As a consequence of Theorem 12.3.6:

Theorem 13.7.3. *Let H be a finite-dimensional Hopf algebra over the field k, let α and g be the H-distinguished grouplike elements of H^* and H respectively, and let u be the Drinfel'd element of $(D(H), \mathcal{R})$. Then:*

(a) $(\beta, a) \mapsto (\beta^{-1}{\otimes}a^{-1})u$ *describes a bijection between the set of* $(\beta, a) \in \mathrm{G}(H^*){\times}\mathrm{G}(H)$, *where* $\beta^2 = \alpha$ *and* $a^2 = g$, *and the set of quasi-ribbon elements of* $(D(H), \mathcal{R})$.
(b) $(\beta, a) \mapsto (\beta^{-1}{\otimes}a^{-1})u$ *describes a bijection between the set of* $(\beta, a) \in \mathrm{G}(H^*){\times}\mathrm{G}(H)$, *where* $\beta^2 = \alpha$ *and* $a^2 = g$, *and* $S^2(h) = a(\beta{\rightharpoonup}h{\leftharpoonup}\beta^{-1})a^{-1}$ *for all* $h \in H$, *and the set of ribbon elements of* $(D(H), \mathcal{R})$.

□

For $a \in \mathrm{G}(H)$ and $\beta \in \mathrm{G}(H^*)$ let $\imath_a$ and $\jmath_\beta$ be the algebra automorphisms of H defined by $\imath_a(h) = aha^{-1}$ and $\jmath_\beta(h) = \beta{\rightharpoonup}h{\leftharpoonup}\beta^{-1}$ for all $h \in H$. There is a very interesting connection between the formula for S^4 in terms of α and g and the existence of a ribbon element for $(D(H), \mathcal{R})$.

Recall that $S^4(h) = g(\alpha{\rightharpoonup}h{\leftharpoonup}\alpha^{-1})a^{-1}$ for all $h \in H$ by part (b) of Theorem 10.5.6. This formula can be expressed $S^4 = \imath_g{\circ}\jmath_\alpha$. We have shown $(D(H), \mathcal{R})$ has a ribbon element if and only if the formal square root derivation

$$\text{“}S^2 = \sqrt{S^4} = \sqrt{\imath_g{\circ}\jmath_\alpha} = \sqrt{\imath_g}{\circ}\sqrt{\jmath_\alpha} = \imath_{\sqrt{g}}{\circ}\jmath_{\sqrt{\alpha}}\text{”}$$

results in an equation $S^2 = \imath_{\sqrt{g}}{\circ}\jmath_{\sqrt{\alpha}}$, where $\sqrt{g} \in \mathrm{G}(H)$ and $\sqrt{\alpha} \in \mathrm{G}(H^*)$ are elements whose squares are g and α respectively.

Exercises

In the following exercises H is a finite-dimensional Hopf algebra with antipode S over the field k.

Exercise 13.7.1. Verify (13.32).

Exercise 13.7.2. Here we establish the equivalence of (13.33) and (13.34). Let $a \in \mathrm{G}(H)$ and $\beta \in \mathrm{G}(H^*)$.

(a) Show that $\imath_a$ and $\jmath_\beta$ commute.

(b) Show that $\imath_a^* = \jmath_{a^{-1}}$ and $\jmath_\beta^* = \imath_{\beta^{-1}}$.

(c) Use parts (a) and (b) to show that (13.33) and (13.34) are equivalent.

Exercise 13.7.3. Let $n \geq 1$, suppose that k contains a primitive n^{th} root of unity q, and let $H = H_{n,q}$ be the Taft algebra of Section 7.3. Show that $(D(H), \mathcal{R})$ has a ribbon element if and only if n is odd in which case the ribbon element is unique.

13.8 Generalized doubles and their duals

Since the underlying Hopf algebra of the double is explained by a left 2-cocycle, we revisit Section 7.7. Consideration of the dual of the double will lead us to Section 7.8.

Let U and A be Hopf algebras over the field k and suppose $U \times A \xrightarrow{\tau} k$ is a skew pairing of U and A. Then τ is invertible by part (b) or part (c) of Proposition 7.7.9 and $(U \otimes A) \times (U \otimes A) \xrightarrow{\sigma_\tau} k$ defined by

$$\sigma_\tau(u \otimes a, u' \otimes a') = \epsilon(u)\tau(u', a)\epsilon(a')$$

for all $u, u' \in U$ and $a, a' \in A$ is a left 2-cocycle for $U \otimes A$ by Proposition 7.7.10. Thus $D_\tau(U, A) = (U \otimes A)_{\sigma_\tau}$ is a Hopf algebra by Proposition 7.7.3.

Definition 13.8.1. A *generalized double over* k is a Hopf algebra of the form $D_\tau(U, A)$ as described above.

We recall that as a coalgebra $D_\tau(U, A) = U \otimes A$ has the tensor product coalgebra structure and its product is given by (7.8) which we rewrite as

$$(u \otimes a) \cdot (u' \otimes a') = u\tau(u'_{(1)}, a_{(1)})u'_{(2)} \otimes a_{(2)}\tau^{-1}(u'_{(3)}, a_{(3)})a'$$

for all $u, u' \in U$ and $a, a' \in A$. As noted in the proof of Theorem 13.1.2:

Example 13.8.2. Let A be a finite-dimensional Hopf algebra over the field k and $U = A^{*\,cop}$. Then the evaluation map $U \times A \xrightarrow{\tau} k$ which is given by

$\tau(a^*, a) = a^*(a)$ for all $a^* \in U$ and $a \in A$ is a skew pairing of U and A and $D_\tau(U, A) = D(A)$.

We now assume that U, A are finite-dimensional and let $\tau : U \times A \longrightarrow k$ be a bilinear form. Since $(U \otimes A)^* = U^* \otimes A^*$ we may write $\tau_{lin} = R_i \otimes R^i \in U^* \otimes A^*$. Thus $R_i(u)R^i(a) = \tau(u, a)$ for all $u \in U$ and $a \in A$. It is easy to see that the axioms (SP.1)–(SP.4) of Section 7.7 for τ are equivalent to the axioms (QT.1)–(QT.4) respectively of Section 12.2 for τ_{lin}. With σ_τ defined as above note that

$$(\sigma_\tau)_{lin} = 1_{U^*} \otimes R^i \otimes R_i \otimes 1_{A^*} \in U^* \otimes A^* \otimes U^* \otimes A^* = ((U \otimes A) \otimes (U \otimes A))^*.$$

We leave the completion of the proof of the following to the reader:

Proposition 13.8.3. *Let U and A be finite-dimensional Hopf algebras over the field k. Suppose that $U \times A \xrightarrow{\tau} k$ is a skew pairing of U and A and let $\tau_{lin} = R = R_i \otimes R^i \in U^* \otimes A^*$. Then R satisfies (QT.1)–(QT.4), $J = 1_{U^*} \otimes R^i \otimes R_i \otimes 1_{A^*}$ is a left twist for $U^* \otimes A^*$, and $D_\tau(U, A)^* \simeq (U^* \otimes A^*)^J$ as Hopf algebras.* □

Conversely:

Proposition 13.8.4. *Let U, A be Hopf algebras over the field k and suppose $R = R_i \otimes R^i \in U \otimes A$ is invertible and satisfies (QT.1)–(QT.4). Then:*

(a) *$J = 1_U \otimes R^i \otimes R_i \otimes 1_A$ is a left twist for $U \otimes A$;*
(b) *$U^o \times A^o \xrightarrow{\tau} k$ defined by $\tau(u^o, a^o) = u^o(R_i)a^o(R^i)$ for all $u^o \in U^o$ and $a^o \in A^o$ is a skew pairing for U^o and A^o.*
(c) *$((U \otimes A)^J)^o \simeq (U^o \otimes A^o)_{\sigma_\tau}$ as Hopf algebras.*

Proof. Since R is invertible τ is also. Since R satisfies (QT.1)–(QT.4) it follows that τ satisfies (SP.1)–(SP.4). Therefore τ is a skew pairing for U^o and A^o. In light of Proposition 7.8.5 and Exercise 7.4.2 to complete the proof we need only establish part (a). To this end the proof of part (a) of Theorem 12.2.8 carries over verbatim. □

Exercises

Exercise 13.8.1. Complete the proof of Proposition 13.8.3.

Exercise 13.8.2. Complete the proof of Proposition 13.8.4.

Chapter notes

The double was defined by Drinfel'd in various contexts in his ground breaking paper [44]. This chapter is about doubles in one of these contexts, doubles $D(H)$ of Hopf algebras H in the category of finite-dimensional vector spaces over k. Majid showed that $D(H)$ is an example of a double cross product [114, Section 6.4.4]. Here we use the description of $D(H)$ found in [171]. For a survey of products of all types see the paper of Bespalov and Drabant [22].

Majid characterized the left modules of $D(H)$ as the Yetter-Drinfel'd modules in ${}_H^H\mathcal{YD}$ [115]. We use the variant ${}^H_{\mathcal{YD}}$ of ${}_H^H\mathcal{YD}$ to motivate the structure of $D(H)$ in terms of objects of ${}^H_{\mathcal{YD}}$. An identification of ${}_{D(H)}\mathcal{M}$ and ${}_H^H\mathcal{YD}$ is described in [172]. That $D(H) = (H^{*\,cop}\otimes H)_\sigma$, where σ is a 2-cocycle derived from the evaluation map $H^*\otimes H \longrightarrow k$ is due to Doi and Takeuchi [41] and is of the simplest ways to describe multiplication in the double.

As mentioned in the notes for chapter 7, the notion of factorizable Hopf algebra and the fact that $D(H)$ is factorizable is due to Reshetikhin and Semenov-Tian-Shansky [179]. Unimodularity for $D(H)$ was established independently by Hennings [69] and the author [171, Theorem 4]. The remaining material of Section 13.2 and that of Sections 13.3–13.4 is found in [171].

Drinfel'd and Majid showed that there is a Hopf algebra projection from $D(H)$ onto H when H is quasitriangular independently in [45] and [115] respectively. The discussion following the statement of Theorem 13.5.1 through Section 12.4 follows Schneider's paper [191] very closely and the results of Section 12.4 are found there as well. The material of Section 13.7 is included in [88] by Kauffman and the author.

The generalized double of Section 13.8 is merely a formal variant of the Drinfel'd double described in Section 3.2 of Joseph's book [80]. See Section 3.5.1 of the same for an account of the evolution of the concept of Drinfel'd double.

An abstract highest weight theory for the generalized double $D_\tau(U, A)$ when U and A are pointed Hopf algebras over k and the characteristic of k is zero has been developed by Schneider and the author [176]. When A is a finite-dimensional pointed Hopf algebra and A^* is not pointed representations of the double $D(A)$ have been investigated by Krop and the author [96–98]. These are part of the extensive literature on representations of quantum groups.

Chapter 14

Coquasitriangular bialgebras and Hopf algebras

In this chapter we explore dual notions of quasitriangular structures, the coquasitriangular coalgebras, bialgebras, and Hopf algebras. Our method of study will be straightforward; formulate dual versions of basic concepts and results for quasitriangular structures and whenever possible adapt proofs for coquasitriangular structures. Just as quasitriangular structures play an important role in the construction of invariants of knots and links, coquasitriangular structures do as well.

Let (A, R) be a finite-dimensional quasitriangular algebra over k. Then $R \in A \otimes A = (A^* \otimes A^*)^*$ can be regarded as a linear form $R : A^* \otimes A^* \longrightarrow k$ for the dual coalgebra, which we will think of as a bilinear form. A coquasitriangular coalgebra over k is a pair (C, β), where C is a coalgebra over k and $\beta : C \times C \longrightarrow k$ is a bilinear form which satisfies axioms reflecting the fact that R satisfies the quantum Yang–Baxter equation.

An important result for quasitriangular Hopf algebras is the formula for the square of the antipode. We derive the dual version for a coquasitriangular Hopf algebra. Our chapter ends with a universal construction for coquasitriangular bialgebras which is based on the free bialgebra on a coalgebra.

14.1 Coquasitriangular and Yang–Baxter coalgebras

Throughout this section (A, R) is a quasitriangular algebra over k. We define coquasitriangular and Yang–Baxter coalgebras, explore duality between quasitriangular algebras and coquasitriangular coalgebras, and describe an analog of the map R_M defined by (12.3) for coquasitriangular coalgebras.

Consider the bilinear form $\beta : A^o \times A^o \longrightarrow k$ by $\beta(p, q) = (p \otimes q)(R)$ for all $p, q \in A^o$. Observe that β is of finite type, meaning $\mathrm{Dim}(\mathrm{Im}(\beta_\ell))$ is finite. See Exercise 1.3.18. Write $R = \sum_{i=1}^{n} a_i \otimes b_i \in A \otimes A$. Let $c, d, e \in A^o$. Applying $c \otimes d \otimes e$ to both sides of the equation of (12.2) results in

$$\begin{aligned} &\beta(c_{(1)}, d_{(1)})\beta(c_{(2)}, e_{(1)})\beta(d_{(2)}, e_{(2)}) \\ &= \beta(c_{(2)}, d_{(2)})\beta(c_{(1)}, e_{(2)})\beta(d_{(1)}, e_{(1)}). \end{aligned} \tag{14.1}$$

The pair (A^o, β) is a coquasitriangular coalgebra.

Definition 14.1.1. A *coquasitriangular coalgebra over the field* k is a pair (C, β), where C is a coalgebra over k and $\beta : C \times C \longrightarrow k$ is a bilinear form, such that (14.1) holds for all $c, d, e \in C$.

In our definition we do not require β to be of finite type. To summarize the discussion preceding the definition:

Proposition 14.1.2. *Suppose (A, R) is a quasitriangular algebra over the field k and let $\beta : A^o \times A^o \longrightarrow k$ be the bilinear form defined by $\beta(p, q) = (p \otimes q)(R)$ for all $p, q \in A^o$. Then β is of finite type and (A^o, β) is a coquasitriangular coalgebra over k.* □

Let $\beta : U \times V \longrightarrow k$ be a bilinear form. Recall $\beta_{lin} : U \otimes V \longrightarrow k$ is defined by $\beta_{lin}(u \otimes v) = \beta(u, v)$ for all $u \in U$ and $v \in V$. If β is of finite type then $\beta_{lin} \in U^* \otimes V^*$. See Exercise 1.3.18.

The preceding proposition and the next establish a duality between quasitriangular algebras over k and certain coquasitriangular coalgebras over k.

Proposition 14.1.3. *Let (C, β) be a coquasitriangular coalgebra over k and suppose β is of finite type. Then (C^*, β_{lin}) is a quasitriangular algebra over k.*

Proof. Write $\beta_{lin} = \sum_{i=1}^{r} a_i \otimes b_i$, where $a_i, b_i \in C^*$ for all $1 \leq i \leq r$. We need to show that β_{lin} satisfies (12.2).

Now $\beta(c, d) = \beta_{lin}(c \otimes d) = \sum_{i=1}^{r} <a_i, c><b_i, d>$ for all $c, d \in C$. Let $c, d, e \in C$. The left-hand side of the equation of (14.1) is the left-hand side of the equation of (12.2) applied to $c \otimes d \otimes e$ and likewise for the right-hand sides. Thus β_{lin} satisfies (12.2). □

Suppose (C, β) is a coquasitriangular coalgebra over k. Then (C^{cop}, β) and (C, β^{op}) are as well, where $\beta^{op}(c, d) = \beta(d, c)$ for all $c, d \in C$. If β is convolution invertible (C, β^{-1}) is a coquasitriangular coalgebra over k. See Exercise 14.1.1. For a discussion of convolution invertible forms see the discussion at the beginning of Section 7.7. Note that if D is a subcoalgebra of C then $(D, \beta|D\otimes D)$ is a coquasitriangular coalgebra over k.

Definition 14.1.4. A *Yang–Baxter coalgebra over the field k* is a coquasitriangular coalgebra (C, β) over k such that β is a convolution invertible bilinear form.

For two basic results on Yang–Baxter coalgebras the reader is referred to Exercises 14.1.2 and 14.1.3.

Now we turn to the endomorphisms R_M associated with (A, R). Let M be a left A-module, and assume further that M is finite-dimensional or more generally is locally finite. Then there is a right A^o-comodule structure (M, ρ) defined by the equation

$$m_{(0)}{<}m_{(1)}, a{>} = a{\cdot}m$$

for all $m \in M$ and $a \in A$. See Exercises 3.2.14 and 3.2.17. Now let $\beta : A^o{\times}A^o \longrightarrow k$ be the bilinear form described above associated with R and write $R = \sum_{i=1}^r a_i\otimes b_i$. Then

$$\begin{aligned} R_M(m\otimes n) &= \sum_{i=1}^r a_i{\cdot}m\otimes b_i{\cdot}n \\ &= \sum_{i=1}^r m_{(0)}{<}m_{(1)}, a_i{>}\otimes n_{(0)}{<}n_{(1)}, b_i{>} \\ &= m_{(0)}\otimes n_{(0)}\beta(m_{(1)}, n_{(1)}) \end{aligned}$$

for all $m, n \in M$ shows that R_M can be expressed in terms of the A^o-comodule structure on M. For a coalgebra C, bilinear form $\beta : C{\times}C \longrightarrow k$, and a right C-comodule M we define $\beta^M \in \mathrm{End}(M\otimes M)$ by

$$\beta^M(m\otimes n) = m_{(0)}\otimes n_{(0)}\beta(m_{(1)}, n_{(1)}) \tag{14.2}$$

for all $m, n \in M$. Thus $R_M = \beta^M$ for locally finite left A-modules, where β is the bilinear form associated with R.

An analog of Proposition 12.1.3 is:

Proposition 14.1.5. *Let C be a coalgebra over k and $\beta : C{\times}C \longrightarrow k$ be a bilinear form. Then the following are equivalent:*

(a) *(C, β) is a coquasitriangular coalgebra over k.*
(b) *$\beta^M \in \mathrm{End}(M\otimes M)$ is a solution to the quantum Yang–Baxter equation for all right C-comodules M.*
(c) *$\beta^C \in \mathrm{End}(C\otimes C)$ is a solution to the quantum Yang–Baxter equation, where C is regarded as a right C-comodule via Δ.*

Proof. That part (a) implies part (b) is a straightforward exercise, and part (b) clearly implies part (c). To show part (c) implies part (a), suppose that β^C is a solution to the quantum Yang–Baxter equation. Let $c, d, e \in C$. Applying both sides of the quantum Yang–Baxter equation for β^C to $c\otimes d\otimes e$ and then applying $\epsilon\otimes\epsilon\otimes\epsilon$ to both sides of the resulting equation yields (14.1). □

There is a version of the preceding proposition for Yang–Baxter coalgebras. The proof is facilitated by the following:

Lemma 14.1.6. *Let C be a coalgebra over the field k and $\beta : C\times C \longrightarrow k$ be a bilinear form. If $\beta^C \in \mathrm{End}(C\otimes C)$ is an invertible endomorphism then β is convolution invertible.*

Proof. We will reduce to the finite-dimensional case. Suppose the lemma is true when C is finite-dimensional. Let C be any coalgebra over k and suppose β^C is invertible.

Note that if f is a linear automorphism of a vector space V over k and U is a finite-dimensional subspace of V such that $f(U) \subseteq U$ then $f(U) = U = f^{-1}(U)$; thus $f|U : U \longrightarrow U$ has an inverse which is $f^{-1}|U$. Let D be a finite-dimensional subcoalgebra of C. Since $\beta^C(D\otimes D) \subseteq D\otimes D$ we conclude that $(\beta|D\times D)^D = \beta^C|D\otimes D$ is invertible. Thus $\beta|D\times D$ is convolution invertible by assumption. Since C is the union of its finite-dimensional subcoalgebras by Theorem 2.2.3, and any two finite-dimensional subcoalgebras of C are contained in a finite-dimensional subcoalgebra of C, the convolution inverses of the restrictions $\beta|D\times D$'s can be patched to form a convolution inverse for β. We have reduced the lemma to the finite-dimensional case.

Suppose C is finite-dimensional and $\beta^C \in \mathrm{End}(C\otimes C)$ has an inverse. Write $\beta_{lin} = \sum_{i=1}^r a_i\otimes b_i \in C^*\otimes C^*$. For $a \in C^*$ let $\mathrm{L}(a) \in \mathrm{End}(C)$ be defined by (2.15). Thus $\mathrm{L}(a)(c) = a{\rightharpoonup}c$ for all $c \in C$. The calculation

$$\begin{aligned}\beta^C(c\otimes d) &= c_{(1)}\otimes d_{(1)}\beta(c_{(2)}, d_{(2)})\\ &= \sum_{i=1}^r c_{(1)}\otimes d_{(1)}<a_i, c_{(2)}><b_i, d_{(2)}>\\ &= \sum_{i=1}^r (a_i{\rightharpoonup}c)\otimes(b_i{\rightharpoonup}d)\end{aligned}$$

for all $c, d \in C$ shows that $\beta^C = \sum_{i=1}^r \mathrm{L}(a_i)\otimes\mathrm{L}(b_i)$. Let $\mathcal{A}$ be the subalgebra of $\mathrm{End}(C\otimes C)$ generated by the $\mathrm{L}(a)\otimes\mathrm{L}(b)$'s, where $a, b \in C^*$. Then π :

$\mathcal{A} \longrightarrow C^* \otimes C^*$ given by $\pi(\mathrm{L}(a) \otimes \mathrm{L}(b)) = a \otimes b$ for $a, b \in C^*$ is a well-defined linear isomorphism since it is the inverse of the tensor product of such. Note that π is an algebra isomorphism and $\pi(\beta^C) = \beta_{lin}$. Since $\beta^C \in \mathcal{A}$ its inverse must belong to $\mathcal{A}$ as well. Thus β_{lin} has an inverse in $C^* \otimes C^* = (C \otimes C)^*$ which means β is convolution invertible. □

Proposition 14.1.7. *Let C be a coalgebra and $\beta : C \times C \longrightarrow k$ be a bilinear form. Then the following are equivalent:*

(a) *(C, β) is a Yang–Baxter coalgebra.*
(b) *$\beta^M \in \mathrm{End}(M \otimes M)$ is an invertible solution to the quantum Yang–Baxter equation for all right C-comodules M.*
(c) *$\beta^C \in \mathrm{End}(C \otimes C)$ is an invertible solution to the quantum Yang–Baxter equation, where C is regarded as a right C-comodule via Δ.*

Proof. Suppose that (C, β) is a Yang–Baxter coalgebra and let M be a right C-comodule. Since β is convolution invertible β^M and $(\beta^{-1})^M$ are inverses in $\mathrm{End}(M \otimes M)$. In light of Proposition 14.1.5 we need only show that if $\beta^C \in \mathrm{End}(C \otimes C)$ is an invertible then β is convolution invertible. This is the case by Lemma 14.1.6. □

We end with:

Definition 14.1.8. A *map $f : (C, \beta) \longrightarrow (C', \beta')$ of coquasitriangular coalgebras over k* is a map $f : C \longrightarrow C'$ of coalgebras which satisfies $\beta'(f(c), f(d)) = \beta(c, d)$ for all $c, d \in C$; that is $\beta' \circ (f \times f) = \beta$.

Exercises

Exercise 14.1.1. Suppose (C, β) is a coquasitriangular coalgebra over k.

(a) Show that (C^{cop}, β) and (C, β^{op}) are as well, where $\beta^{op}(c, d) = \beta(d, c)$ for all $c, d \in C$.

(b) Suppose that (C, β) is a Yang–Baxter coalgebra over k. Show that (C^{cop}, β), (C, β^{op}), and (C, β^{-1}) are as well. [Hint: To show that (C, β^{-1}) is a Yang–Baxter coalgebra over k show that $\ell, r \in (C \otimes C \otimes C)^*$ defined by

$$\ell(c \otimes d \otimes e) = \beta(c_{(1)}, d_{(1)})\beta(c_{(2)}, e_{(1)})\beta(d_{(2)}, e_{(2)})$$

and

$$r(c \otimes d \otimes e) = \beta^{-1}(c_{(2)}, d_{(2)})\beta^{-1}(c_{(1)}, e_{(2)})\beta^{-1}(d_{(1)}, e_{(1)})$$

for all $c, d, e \in C$ satisfy $\ell * r = 1_{(C \otimes C \otimes C)^*}$.]

Exercise 14.1.2. Prove the following:

Proposition 14.1.9. *Let (A,R) be a Yang–Baxter coalgebra over the field k and let $\beta : A^o{\times}A^o \longrightarrow k$ be the bilinear form defined by $\beta(p,q) = (p{\otimes}q)(R)$ for all $p,q \in A^o$. Then β is of finite type and (A^o,β) is a Yang–Baxter coalgebra over k.* □

Exercise 14.1.3. The dual of a Yang–Baxter coalgebra is a Yang–Baxter algebra in the finite-dimensional case. Prove the following:

Proposition 14.1.10. *Let (C,β) be a Yang–Baxter coalgebra over k and suppose β,β^{-1} are of finite type. Then (C^*,β_{lin}) is a Yang–Baxter algebra over k.*

14.2 Coquasitriangular bialgebras and Hopf algebras

We first define coquasitriangular bialgebra over k and derive some basic properties of these objects. Examples of these bialgebras are duals of quasitriangular bialgebras.

Let (D,R) be a quasitriangular bialgebra over k and consider the dual bialgebra $A = D^o$. Consider the bilinear form $\beta : A{\times}A \longrightarrow k$ defined by $\beta(a,b) = (a{\otimes}b)(R)$ for all $a,b \in A$.

Let $a,b,c \in A$. Applying $a{\otimes}b{\otimes}c$ to both sides of the equations of (QT.1) and (QT.3), doing the same with a in the case of (QT.2) and (QT.4), and likewise with $a{\otimes}b$ in the case of (QT.5), results the following equations:

(CoQT.1) $\beta(ab,c) = \beta(a,c_{(1)})\beta(a,c_{(2)})$;
(CoQT.2) $\beta(1,a) = \epsilon(a)$;
(CoQT.3) $\beta(a,bc) = \beta(a_{(2)},b)\beta(a_{(1)},c)$;
(CoQT.4) $\beta(a,1) = \epsilon(a)$; and
(CoQT.5) $b_{(1)}a_{(1)}\beta(a_{(2)},b_{(2)}) = a_{(2)}b_{(2)}\beta(a_{(1)},b_{(1)})$

for all $a,b,c \in A$.

Definition 14.2.1. A *coquasitriangular bialgebra over k* is a pair (A,β), where A is a bialgebra over k and $\beta : A{\times}A \longrightarrow k$ is a bilinear form satisfying (CoQT.1)–(CoQT.5); that is β is a skew pairing of A with itself which satisfies (CoQT.5).

Let (A,β) be a coquasitriangular bialgebra over k. Since β is a skew pairing of A with itself we note that Proposition 7.7.8 gives formulations of (CoQT.1)–(CoQT.4) in terms of $\beta_\ell, \beta_r : A \longrightarrow A^*$. The preceding discussion about the dual of a quasitriangular bialgebra is summarized by:

Proposition 14.2.2. *Let (A, R) be a quasitriangular bialgebra over k and let $\beta : A^o \times A^o \longrightarrow k$ be the bilinear form defined by $\beta(p, q) = (p \otimes q)(R)$ for all $p, q \in A^o$. Then β is of finite type and (A^o, β) is a coquasitriangular bialgebra over k.* □

Conversely:

Proposition 14.2.3. *Let (A, β) be a coquasitriangular bialgebra over k and suppose β is of finite type. Then $\beta_{lin} \in A^o \otimes A^o$ and (A^o, β_{lin}) is a quasitriangular bialgebra over k.*

Proof. Since β is a skew pairing of A with itself, it follows $\beta_\ell : A \longrightarrow A^*$ and $\beta_r : A \longrightarrow A^{*\,op}$ are algebra maps by Proposition 7.7.8. Write $\beta_{lin} = \sum_{i=1}^n p_i \otimes q_i \in A^* \otimes A^*$ where n is as small as possible. Then $\{p_1, \ldots, p_n\}$ and $\{q_1, \ldots, q_n\}$ are linearly independent by Lemma 1.2.2. Since $\beta_\ell(a) = \sum_{i=1}^n p_i(a) q_i$ and $\beta_r(a) = \sum_{i=1}^n q_i(a) p_i$ for all $a \in A$, the p_i's vanish on $\mathrm{Ker}(\beta_\ell)$ and the q_i's vanish on $\mathrm{Ker}(\beta_r)$. Since β is of finite rank $\mathrm{Im}(\beta_\ell)$ and $\mathrm{Im}(\beta_r)$ are finite-dimensional. Therefore $\mathrm{Ker}(\beta_\ell)$ and $\mathrm{Ker}(\beta_r)$ are cofinite ideals of A. We have shown that $\beta_{lin} \in A^o \otimes A^o$. The fact that $R = \beta_{lin}$ satisfies (QT.1)–(QT.5) is a straightforward exercise which is left to the reader. □

Suppose (A, β) is a coquasitriangular bialgebra over k. Then the pairs $(A^{op\,cop}, \beta)$, (A^{op}, β^{op}), and (A^{cop}, β^{op}) are as well by Exercise 14.2.1. Now β is a skew pairing of A and A and is therefore a pairing of A^{cop} and A. See Definitions 7.7.6 and 7.7.7.

Remark 14.2.4. β is convolution invertible will mean β_{lin} has an inverse in $(A \otimes A)^*$, not necessarily in $(A^{cop} \otimes A)^*$.

If β has an inverse then (A^{cop}, β^{-1}) is a coquasitriangular bialgebra over k; see Exercise 14.2.2.

Lemma 14.2.5. *Suppose that (A, β) is a coquasitriangular bialgebra over k. Then:*

(a) *(A, β) is a coquasitriangular coalgebra over k.*
(b) *Suppose that β is convolution invertible. Then (A, β) is a Yang–Baxter coalgebra over k.*

Proof. Part (b) follows from part (a). To prove part (a), for $c \in A$ either apply $\beta_\ell(c)$ to both sides of the equation of (CoQT.5) and use (CoQT.3) or apply $\beta_r(c)$ to the same and use (CoQT.1). □

Definition 14.2.6. A *coquasitriangular Hopf algebra over* k is a coquasitriangular bialgebra (A, β) over k where A is a Hopf algebra over k.

Our main result is:

Theorem 14.2.7. *Let* (A, β) *be a coquasitriangular Hopf algebra over* k *and let* S *be the antipode of* A. *Then:*

(a) β *is convolution invertible and* $\beta^{-1}(a, b) = \beta(S(a), b)$ *for all* $a, b \in A$.
(b) $\beta(S(a), S(b)) = \beta(a, b)$ *for all* $a, b \in A$.
(c) β *is a left* 2*-cocycle for* A.
(d) $\sigma : (A\otimes A)\times(A\otimes A) \longrightarrow k$ *defined by* $\sigma(a\otimes b, c\otimes d) = \epsilon(a)\beta(c, b)\epsilon(d)$ *for all* $a, b, c, d \in A$ *is a left* 2*-cocycle for* $A\otimes A$.

Proof. We first show part (a). Define $\gamma : A\times A \longrightarrow k$ by $\gamma(a, b) = \beta(S(a), b)$ for all $a, b \in A$. Using (CoQT.1) and (CoQT.2) we calculate

$$\begin{aligned}\beta(a_{(1)}, b_{(1)})\gamma(a_{(2)}, b_{(2)}) &= \beta(a_{(1)}, b_{(1)})\beta(S(a_{(2)}), b_{(2)})\\ &= \beta(a_{(1)}S(a_{(2)}), b)\\ &= \beta(\epsilon(a)1, b)\\ &= \epsilon(a)\epsilon(b),\end{aligned}$$

and thus $\beta(a_{(1)}, b_{(1)})\gamma(a_{(2)}, b_{(2)}) = \epsilon(a)\epsilon(b)$ for all $a, b \in A$. Likewise the equation $\gamma(a_{(1)}, b_{(1)})\beta(a_{(2)}, b_{(2)}) = \epsilon(a)\epsilon(b)$ holds for all $a, b \in A$. Part (a) now follows.

To show part (b) we define $\zeta : A\times A \longrightarrow k$ by $\zeta(a, b) = \beta(S(a), S(b))$ for all $a, b \in A$. The calculation

$$\begin{aligned}\zeta(a_{(1)}, b_{(1)})\gamma(a_{(2)}, b_{(2)}) &= \beta(S(a_{(1)}), S(b_{(1)}))\beta(S(a_{(2)}), b_{(2)})\\ &= \beta(S(a)_{(2)}, S(b_{(1)}))\beta(S(a)_{(1)}, b_{(2)})\\ &= \beta(S(a), S(b_{(1)})b_{(2)})\\ &= \beta(S(a), \epsilon(b)1)\\ &= \epsilon(S(a))\epsilon(b)\\ &= \epsilon(a)\epsilon(b),\end{aligned}$$

which uses (CoQT.3) and (CoQT.4), shows that ζ is a left inverse for γ. Now β is an inverse for γ by part (a). Therefore $\zeta = \beta$ and part (b) is established.

Let $a, b, c \in A$. Since β is convolution invertible by part (a), to show part (c) we need to show $\beta(a_{(1)}, b_{(1)})\beta(a_{(2)}b_{(2)}, c) = \beta(b_{(1)}, c_{(1)})\beta(a, b_{(2)}c_{(2)})$. Applying $\beta_\ell(a)$ to both sides of $c_{(1)}b_{(1)}\beta(b_{(2)}, c_{(2)}) = b_{(2)}c_{(2)}\beta(b_{(1)}, c_{(1)})$ and using (CoQT.3) results in

$$\beta(a_{(2)}, c_{(1)})\beta(a_{(1)}, b_{(1)})\beta(b_{(2)}, c_{(2)}) = \beta(a, b_{(2)}c_{(2)})\beta(b_{(1)}, c_{(1)}).$$

Applying $\beta_r(c)$ to both sides of $b_{(1)}a_{(1)}\beta(a_{(2)}, b_{(2)}) = a_{(2)}b_{(2)}\beta(a_{(1)}, b_{(1)})$ and using (CoQT.1) results in

$$\beta(b_{(1)}, c_{(1)})\beta(a_{(1)}, c_{(2)})\beta(a_{(2)}, b_{(2)}) = \beta(a_{(2)}b_{(2)}, c)\beta(a_{(1)}, b_{(1)}).$$

Since (14.1) holds by Lemma 14.2.5, it now follows that β is a left 2-cocycle. Part (c) is established.

We have noted that β is a skew pairing of A with itself. Since it is convolution invertible by part (a), part (d) follows by Proposition 7.7.10. $\square$

As a consequence of part (a) of Theorem 14.2.7 and Lemma 14.2.5:

Corollary 14.2.8. *A coquasitriangular Hopf algebra over k is a Yang–Baxter coalgebra over k.* $\square$

The reader should compare parts (a)–(d) of Theorem 14.2.7 with parts (a)–(d) of Theorem 12.2.8. Parts (e) and (f) of the latter establish that the antipode S of a quasitriangular Hopf algebra is bijective and describe the nature of S^2. We do the same for the antipode of a coquasitriangular Hopf algebra in the next section.

We end with a characterization of coquasitriangular bialgebras in terms of bialgebra maps and a definition. Suppose (A, β) is a coquasitriangular bialgebra over k. Then β is a skew paring of A with itself. We have noted Proposition 7.7.8 gives formulations of (CoQT.1)–(CoQT.4) in terms of $\beta_\ell, \beta_r : A \longrightarrow A^*$. A consequence of this result is:

Corollary 14.2.9. *Suppose A is a bialgebra over the field k. Then the following are equivalent:*

(a) *A has a coquasitriangular structure.*
(b) *There is a bialgebra map $f : A \longrightarrow A^{o\,op}$ such that*

$$(f(b_{(1)})(a_{(1)}))a_{(2)}b_{(2)} = b_{(1)}a_{(1)}(f(b_{(2)})(a_{(2)}))$$

for all $a, b \in A$, in which case (A, β) is a coquasitriangular bialgebra, where $\beta_r = f$;

(c) *There is a bialgebra map* $f : A \longrightarrow A^{o\,cop}$ *such that*

$$(f(a_{(1)})(b_{(1)}))a_{(2)}b_{(2)} = b_{(1)}a_{(1)}(f(a_{(2)})(b_{(2)}))$$

for all $a, b \in A$, *in which case* (A, β) *is a coquasitriangular bialgebra, where* $\beta_\ell = f$.

□

Definition 14.2.10. A *map of coquasitriangular bialgebras* $f : (A, \beta) \longrightarrow (A', \beta')$ *over* k is a bialgebra map $f : A \longrightarrow A'$ which is also a map of underlying coquasitriangular coalgebras.

Exercises

Exercise 14.2.1. Let (A, β) be a coquasitriangular bialgebra over k. Show that $(A^{op\,cop}, \beta)$, (A^{op}, β^{op}), and (A^{cop}, β^{op}) are as well.

Exercise 14.2.2. Let (A, β) be a coquasitriangular bialgebra over k and suppose β has a convolution inverse. Show that (A^{cop}, β^{-1}) is a coquasitriangular bialgebra over k.

Exercise 14.2.3. Let (A, β) be a coquasitriangular bialgebra over k. For right A-comodules M, N define $\sigma_{M,N} : M \otimes N \longrightarrow N \otimes M$ by

$$\sigma_{M,N}(m \otimes n) = n_{(0)} \otimes m_{(0)} \beta(n_{(1)}, m_{(1)})$$

for all $m \in M$ and $n \in N$. Observe that $\sigma_{M,M} = \beta^M \circ \tau_{M,M}$.

(a) Show that $\mathcal{M}^A$ is a monoidal category with the trivial right comodule structure on k and the tensor product $M \otimes N$ of objects in $\mathcal{M}^A$ is given the tensor product right A^{op}-comodule structure; thus $\rho(m \otimes n) = (m_{(0)} \otimes n_{(0)}) \otimes n_{(1)} m_{(1)}$ for all $m \in M$ and $n \in N$.

(b) Show that the $\sigma_{M,N}$'s give $\mathcal{M}^A$ a prebraiding structure.

(c) Suppose β has an inverse (as is the case when A is a Hopf algebra). Show that the $\sigma_{M,N}$'s give $\mathcal{M}^A$ a braiding structure.

14.3 The square of the antipode of a coquasitriangular Hopf algebra

We examine the square of the antipode of a coquasitriangular Hopf algebra over k, taking our cues from the special case (H^o, β), where (H, R) is a quasitriangular Hopf algebra over k and $\beta(p, q) = (p \otimes q)(R)$ for all $p, q \in$

H^o. Let S be the antipode of H and write $R = R_i \otimes R^i$. By part (e) of Theorem 12.2.8 the Drinfel'd element $u = S(R^i)R_i$ is invertible with inverse $u^{-1} = R^i S^2(R_i)$ and $S^2(h) = uhu^{-1}$ for all $h \in H$.

Applying $p \in H^o$ to the formula for S^2 yields $S^2_{H^o}(p) = \eta' \rightharpoonup p \leftharpoonup \eta$, where $\eta(p) = \beta(p_{(2)}, S_{H^o}(p_{(1)}))$ and $\eta'(p) = \beta(S^2_{H^o}(p_{(2)}), p_{(1)})$. Thinking of u and u^{-1} as elements of $(H^o)^*$ observe that $\eta = u$ and $\eta' = u^{-1}$. The purpose of this section is to prove:

Theorem 14.3.1. *Let (A, β) be a coquasitriangular Hopf algebra over the field k and let S be the antipode of A. Define $\eta \in A^*$ by $\eta(a) = \beta(a_{(2)}, S(a_{(1)}))$ for all $a \in A$. Then:*

(a) *η has an inverse and $\eta^{-1}(a) = \beta(S^2(a_{(2)}), a_{(1)})$ for all $a \in A$.*
(b) *$S^2(a) = \eta^{-1} \rightharpoonup a \leftharpoonup \eta$ for all $a \in A$. Thus S is bijective.*

Proof. The key ingredient for our proof is the equation

$$\beta(a_{(2)}, S(a_{(1)}))1 = \beta(a_{(3)}, S(a_{2}))a_{(4)}S(a_{(1)}) \tag{14.3}$$

for all $a \in A$ which we derive at the end of this section. By virtue of this equation $\eta(a)1 = \eta(a_{(2)})a_{(3)}S(a_{(1)})$, or equivalently $\eta(a)1 = (a_{(2)} \leftharpoonup \eta)S(a_{(1)})$, for all $a \in A$. Consequently

$$a \leftharpoonup \eta = S^2(\eta \rightharpoonup a) \tag{14.4}$$

for all $a \in A$ since

$$a \leftharpoonup \eta = (a_{(3)} \leftharpoonup \eta)S(a_{(2)})S^2(a_{(1)}) = (\eta(a_2)1)S^2(a_{(1)}) = S^2(\eta \rightharpoonup a).$$

Now β has an inverse by part (a) of Theorem 14.2.7. From

$$\begin{aligned}
\beta^{-1}(a_{(3)}, S(a_{(1)}))\eta(a_{(2)}) &= \beta^{-1}(a_{(3)}, S(a_{(1)}))\beta(a_{(2)(2)}, S(a_{(2)(1)})) \\
&= \beta^{-1}(a_{(2)(2)}, S(a_{(1)(1)}))\beta(a_{(2)(1)}, S(a_{(1)(2)})) \\
&= \beta^{-1}(a_{(2)(2)}, S(a_{(1)})_{(2)})\beta(a_{(2)(1)}, S(a_{(1)})_{(1)}) \\
&= \epsilon(a_{(2)})\epsilon(S(a_{(1)})) \\
&= \epsilon(a)
\end{aligned}$$

follows

$$\beta^{-1}(a_{(3)}, S(a_{(1)}))\eta(a_{(2)}) = \epsilon(a) \tag{14.5}$$

for all $a \in A$. Define $\eta' \in A^*$ by $\eta'(a) = \beta^{-1}(S^2(a_{(2)}), S(a_{(1)}))$. By virtue of (14.5) and (14.4)

$$\epsilon(a) = \beta^{-1}(a_{(2)} \leftharpoonup \eta, S(a_{(1)})) = \beta^{-1}(S^2(a_{(2)}), S(a_{(1)}))\eta(a_{(3)}) = \eta' * \eta(a)$$

for all $a \in A$. Therefore η' is a left inverse for η which means it is an inverse for η by Proposition 6.2.3.

Let $a \in A$. Using parts (a) and (b) of Theorem 14.2.7 note that $\eta^{-1}(a) = \beta^{-1}(S^2(a_{(2)}), S(a_{(1)})) = \beta(S^3(a_{(2)}), S(a_{(1)})) = \beta(S^2(a_{(2)}), a_{(1)})$ and that (14.4) implies that $S^2(a) = S^2(\eta \rightharpoonup (\eta^{-1} \rightharpoonup a)) = (\eta^{-1} \rightharpoonup a) \leftharpoonup \eta$. Thus S^2, and therefore S, is bijective. $\square$

It remains to establish (14.3). We will use (CoQT.5), that is the equation

$$b_{(1)}a_{(1)}\beta(a_{(2)}, b_{(2)}) = a_{(2)}b_{(2)}\beta(a_{(1)}, b_{(1)})$$

for all $a, b \in A$. For $a \in A$ we compute

$$\begin{aligned}
&\beta(a_{(3)}, S(a_{(2)}))a_{(4)}S(a_{(1)}) \\
&= \beta(a_{(3)}, S(a_{(2)})_{(1)})a_{(4)}S(a_{(2)})_{(2)}S(S(a_{(2)})_{(3)})S(a_{(1)}) \\
&= \beta(a_{(3)(1)}, S(a_{(2)})_{(1)})a_{(3)(2)}S(a_{(2)})_{(2)}S(S(a_{(2)})_{(3)})S(a_{(1)}) \\
&= \beta(a_{(3)(2)}, S(a_{(2)})_{(2)})S(a_{(2)})_{(1)}a_{(3)(1)}S(S(a_{(2)})_{(3)})S(a_{(1)}) \\
&= \beta(a_{(3)(2)}, S(a_{(2)(2)}))S(a_{(2)(3)})a_{(3)(1)}S(S(a_{(2)(1)}))S(a_{(1)}) \\
&= \beta(a_{(6)}, S(a_{(3)}))S(a_{(4)})a_{(5)}S(S(a_{(2)}))S(a_{(1)}) \\
&= \beta(a_{(4)}, S(a_{(3)}))S(S(a_{(2)}))S(a_{(1)}) \\
&= \beta(a_{(4)}, S(a_{(3)}))S(a_{(1)}S(a_{(2)})) \\
&= \beta(a_{(3)}, S(a_{(2)}))S(\epsilon(a_{(1)})1) \\
&= \beta(a_{(2)}, S(a_{(1)}))1.
\end{aligned}$$

Exercises

Exercise 14.3.1. Equation (14.3) is formally derived from the equation $\epsilon(h)u = S(h_{(2)})uh_{(1)}$ found just after (12.6). Let (H, R) be a quasitriangular Hopf algebra over k with Drinfel'd element u and antipode S. Let (H^o, β) be as at the beginning of this section. Show that (14.3) for H^o is obtained by applying $p \in H^o$ to both sides of the equation $\epsilon(h)u = S(h_{(2)})uh_{(1)}$ for all $h \in H$.

Exercise 14.3.2. Let (A, β) be a coquasitriangular Hopf algebra over k. Suppose β is of finite type and (A^o, β_{lin}) is the quasitriangular Hopf algebra of Proposition 14.2.3. Show that η of Theorem 14.3.1 is the Drinfel'd element of (A^o, β_{lin}) and that the formula in part (b) of the same implies $S^2_{A^o}(p) = \eta p \eta^{-1}$ for all $p \in H^o$.

14.4 The free coquasitriangular bialgebra on a coquasitriangular coalgebra

We construct the free coquasitriangular bialgebra $(\imath, (\mathrm{T}(C,\beta), \boldsymbol{\beta}))$ on a coquasitriangular coalgebra (C, β) over k. We show that (C, β) is a Yang–Baxter coalgebra if and only if $(\mathrm{T}(C,\beta), \boldsymbol{\beta})$ is.

Our construction takes place in a more general context. Let A be a bialgebra over k and suppose that $\beta : A \times A \longrightarrow k$ be a bilinear form which satisfies (CoQT.1)–(CoQT.4). When A is generated by a subcoalgebra C such that $(C, \beta \,|\, C \times C)$ is a quasitriangular coalgebra, we will construct a certain bialgebra quotient of A which has a coquasitriangular structure compatible with β. Since our focus is on (CoQT.5) we are led to consider the difference function $d_\beta : A \times A \longrightarrow A$ defined by

$$d_\beta(a,b) = \beta(a_{(1)}, b_{(1)})a_{(2)}b_{(2)} - b_{(1)}a_{(1)}\beta(a_{(2)}, b_{(2)})$$

for all $a, b \in A$. Note that (CoQT.5) holds for all $a, b \in A$ if and only if $d_\beta = 0$. Likewise for a coalgebra C over k with bilinear form $\beta' : C \times C \longrightarrow k$ we let $\delta_{\beta'} : C \times C \times C \longrightarrow k$ be defined by

$$\begin{aligned}\delta_{\beta'}(c,d,e) = \beta'(c_{(1)}, d_{(1)})\beta'(c_{(2)}, e_{(1)})\beta'(d_{(2)}, e_{(2)}) \\ -\beta'(c_{(2)}, d_{(2)})\beta'(c_{(1)}, e_{(2)})\beta'(d_{(1)}, e_{(1)})\end{aligned}$$

for all $c, d, e \in C$. Note that (C, β') is a coquasitriangular coalgebra over k if and only if $\delta_{\beta'} = 0$.

We will use a number of properties of d_β for our construction. They are listed in the following lemma, the proof of which is left as an exercise. The properties involve four module actions of A on itself.

Recall that $\beta_\ell : A \longrightarrow A^*$ and $\beta_r : A \longrightarrow A^{*\,op}$ are algebra maps by Proposition 7.7.8. Thus A has left A-module structures $(A, \succ_r)$ and $(A, \succ_\ell)$ defined by

$$a \succ_\ell b = \beta_\ell(a) \rightharpoonup b = b_{(1)}\beta(a, b_{(2)})$$

and

$$a \succ_r b = b \leftharpoonup \beta_r(a) = \beta(b_{(1)}, a)b_{(2)}$$

for all $a, b \in A$, and A has right A-module structures $(A, {}_r\!\prec)$ and $(A, {}_\ell\!\prec)$ defined by

$$b\,{}_\ell\!\prec a = b \leftharpoonup \beta_\ell(a) = \beta(a, b_{(1)})b_{(2)}$$

and

$$b\,{}_r\!\prec a = \beta_r(a) \rightharpoonup b = b_{(1)}\beta(b_{(2)}, a)$$

for all $a, b \in A$.

Lemma 14.4.1. *Let A be a bialgebra over the field k and assume that $\beta : A \times A \longrightarrow k$ is a bilinear form which satisfies (CoQT.1)–(CoQT.4). Then:*

(a) $\beta_\ell(a)(d_\beta(b, c)) = -\delta_\beta(c, a, b)$;
(b) $\beta_r(a)(d_\beta(b, c)) = \delta_\beta(b, c, a)$;
(c) $\epsilon(d_\beta(a, b)) = 0$;
(d) $\Delta(d_\beta(a, b)) = d_\beta(a_{(1)}, b_{(1)}) \otimes a_{(2)}b_{(2)} + b_{(1)}a_{(1)} \otimes d_\beta(a_{(2)}, b_{(2)})$;
(e) $d_\beta(1, b) = 0 = d_\beta(a, 1)$;
(f) $d_\beta(ab, c) = (c_{(1)} \succ_r a)d_\beta(b, c_{(2)})d_\beta(a, c_{(1)})(b_r{\prec}c_{(2)})$; *and*
(g) $d_\beta(a, bc) = d_\beta(a_{(2)}, b)(c_\ell{\prec}a_{(1)}) + (a_{(2)} \succ_\ell b)d_\beta(a_{(1)}, c)$

for all $a, b, c \in A$. □

Here is our general construction and context.

Proposition 14.4.2. *Let A be a bialgebra over the field k and suppose $\beta : A{\times}A \longrightarrow k$ is a bilinear form which satisfies (CoQT.1)–(CoQT.4). Suppose that A is generated by a coalgebra C such that $(C, \beta\,|\,C{\times}C)$ is coquasitriangular. Let I be the ideal of A generated by $d_\beta(C, C)$. Then:*

(a) *I is a bi-ideal of A.*
(b) *Let $\mathcal{A} = A/I$ and $\pi : A \longrightarrow \mathcal{A}$ be the projection. There is a coquasitriangular structure $(\mathcal{A}, \boldsymbol{\beta})$ on $\mathcal{A}$ determined by $\boldsymbol{\beta}\circ(\pi{\times}\pi) = \beta$.*

Proof. Let $J = d_\beta(C, C)$. Since C is a subcoalgebra of A it follows that J is a coideal of A by parts (c) and (d) of Lemma 14.4.1. Since $(C, \beta\,|\,C{\times}C)$ is a coquasitriangular coalgebra $\delta_\beta\,|\,C{\times}C{\times}C = 0$. Therefore $\beta_\ell(C)(J) = (0) = \beta_r(C)(J)$ by parts (a) and (b) of Lemma 14.4.1 again.

Now $J^\perp$ is a subalgebra A^* since J is a coideal of A. We have noted that $\beta_\ell : A \longrightarrow A^*$ and $\beta_r : A \longrightarrow A^{*\,op}$ are algebra maps. Therefore $\beta_\ell^{-1}(J^\perp)$ and $\beta_r^{-1}(J^\perp)$ are subalgebras of A. Since both contain C, and C generates A as an algebra, both of these subalgebras are A which means $\beta_\ell(A)(J) = (0) = \beta_r(A)(J)$. Therefore $\beta(A, J) = (0) = \beta(J, A)$.

Proposition 7.7.8 implies $\mathrm{Im}(\beta_\ell), \mathrm{Im}(\beta_r) \subseteq A^o$ and $\beta_\ell : A \longrightarrow A^{o\,cop}$, $\beta_r : A \longrightarrow A^o$ are coalgebra maps. Therefore $\beta_\ell(A)$, $\beta_r(A)$ are subcoalgebras of A^o which means $\beta_\ell(A)^\perp$, $\beta_r(A)^\perp$ are ideals of A. Since both of these sets contain J it follows that they contain I, the ideal which J generates. We have shown $\beta(A, I) = (0) = \beta(I, A)$. Since J is a coideal of A it follows I is as well. Therefore I is a bi-ideal of A. We have shown part (a).

Let $\mathcal{A} = A/I$ be the quotient bialgebra and $\pi : A \longrightarrow \mathcal{A}$ be the projection. Since $\beta(A, I) = (0) = \beta(I, A)$ there is a bilinear form $\boldsymbol{\beta} : \mathcal{A}\times\mathcal{A} \longrightarrow k$ determined by $\boldsymbol{\beta}(\pi(a), \pi(b)) = \beta(a, b)$ for all $a, b \in A$. Since the bialgebra map π is onto $\boldsymbol{\beta}$ satisfies (CoQT.1)–(CoQT.4). In particular Lemma 14.4.1 applies to $\mathcal{A}$ and $\boldsymbol{\beta}$.

To show that (CoQT.5) is satisfied for all $a, b \in \mathcal{A}$ we need only show that $d_{\boldsymbol{\beta}} = 0$. Since π is onto and $d_{\boldsymbol{\beta}}\circ(\pi\times\pi) = \pi\circ d_{\beta}$ it follows that $d_{\boldsymbol{\beta}}(\pi(C), \pi(C)) = (0)$. Since $\pi(C)$ is a subcoalgebra of $\mathcal{A}$, by parts (e) and (f) of Lemma 14.4.1 the set of all $x \in \mathcal{A}$ such that $d_{\boldsymbol{\beta}}(x, \pi(C)) = (0)$ is a subalgebra $\mathcal{B}$ of $\mathcal{A}$. Since $\pi(C)$ generates $\mathcal{A}$ as an algebra and $\pi(C) \subseteq \mathcal{B}$ we conclude that $d_{\boldsymbol{\beta}}(\mathcal{A}, \pi(C)) = (0)$. Using a similar argument and parts (e) and (g) of the lemma we conclude that $d_{\boldsymbol{\beta}}(\mathcal{A}, \mathcal{A}) = (0)$. □

Let (C, β) be a coquasitriangular coalgebra over k and let $(\iota, T(C))$ be the free bialgebra on the coalgebra C. Regarding $T(C)$ as an algebra the pair $(\iota, T(C))$ is the free algebra on the vector space C. See the discussion preceding Theorem 5.3.1. Consider the linear map $\beta_r : C \longrightarrow C^{*\,op}$. By the universal mapping property of the free algebra there is an algebra map $f_o : T(C) \longrightarrow C^{*\,op}$ such that $f_o\circ\iota = \beta_r$. Let $\jmath : C \longrightarrow C^{*\,o}$ be the coalgebra map defined by $\jmath(c)(c^*) = c^*(c)$ for all $c \in C$ and $c^* \in C^*$. Applying the universal mapping property of the tensor bialgebra $(\iota, T(C))$ to the composite of coalgebra maps $f_o^o\circ\jmath : C \longrightarrow T(C)^{o\,cop}$ yields a bialgebra map $F_o : T(C) \longrightarrow T(C)^{o\,cop}$ such that $F_o\circ\iota = f_o^o\circ\jmath$. Let $\beta_o : T(C)\times T(C) \longrightarrow k$ be the bilinear form determined by $\beta_{o\,\ell} = F_o$. By Proposition 7.7.8 the bilinear form β_o satisfies (CoQT.1)–(CoQT.4). We note that

$$\beta_o\circ(\iota\times\iota) = \beta \tag{14.6}$$

since

$$\begin{aligned}\beta_o(\iota(c), \iota(d)) &= F_o(\iota(c))(\iota(d)) \\ &= (f_o^o(\jmath(c)))(\iota(d)) \\ &= \jmath(c)(f_o(\iota(d))) \\ &= \jmath(c)(\beta_r(d)) \\ &= \beta_r(d)(c) \\ &= \beta(c, d)\end{aligned}$$

for all $c, d \in C$. In particular $(\iota(C), \beta_o \,|\, \iota(C)\times\iota(C))$ is a coquasitriangular coalgebra.

Let I be the ideal of $T(C)$ generated by $d_{\beta_o}(\iota(C), \iota(C))$, let $\mathrm{T}(C, \beta) = T(C)/I$, and let $\pi : T(C) \longrightarrow \mathrm{T}(C, \beta)$ be the projection. By Proposition

14.4.2 the ideal I is a bi-ideal and there is a coquasitriangular bialgebra structure $(\mathrm{T}(C,\beta),\boldsymbol{\beta})$ on $\mathrm{T}(C,\beta)$ such that $\boldsymbol{\beta}\circ(\pi\times\pi) = \beta_o$. Let $\imath = \pi\circ\iota$. Then $\imath : C \longrightarrow \mathrm{T}(C,\beta)$ is a coalgebra map since it is the composite of such. Note

$$\boldsymbol{\beta}\circ(\imath\times\imath) = \beta \tag{14.7}$$

since $\boldsymbol{\beta}\circ(\pi\times\pi) = \beta_o$ and $\beta_o\circ(\iota\times\iota) = \beta$; the latter follows by (14.6). Thus $\imath : (C,\beta) \longrightarrow (\mathrm{T}(C,\beta),\boldsymbol{\beta})$ is a map of coquasitriangular coalgebras.

Now suppose that (A,β') is a coquasitriangular bialgebra over k and that $f : (C,\beta) \longrightarrow (A,\beta')$ is a map of coquasitriangular coalgebras. By the universal mapping property of the free bialgebra there is a bialgebra map $F' : T(C) \longrightarrow A$ such that $F'\circ\iota = f$. Since $\beta_o\circ(\iota\times\iota) = \beta = \beta'\circ(f\times f)$ it is easy to see that $F'\circ d_{\beta_o}\circ(\iota\times\iota) = d_{\beta'}\circ(f\times f)$. Since (A,β') is a coquasitriangular bialgebra $d_{\beta'} = 0$. Hence $F'(I) = (0)$ which means that there is a bialgebra map $F : \mathrm{T}(C,\beta) \longrightarrow A$ such that $F\circ\pi = F'$. Therefore $F\circ\imath = f$. We can use (14.7) to calculate

$$\beta'\circ(F\times F)\circ(\imath\times\imath) = \beta'\circ(f\times f) = \beta = \boldsymbol{\beta}\circ(\imath\times\imath).$$

Thus $\beta'(F(c),F(d)) = \boldsymbol{\beta}(c,d)$ for all $c,d \in \imath(C)$. Since $\imath(C)$ generates $\mathrm{T}(C,\beta)$ as an algebra over k it follows that $F : (\mathrm{T}(C,\beta),\boldsymbol{\beta}) \longrightarrow (A,\beta')$ is a map of coquasitriangular bialgebras by Exercise 14.4.2. For the same reason $F\circ\iota = f$ determines F. We have proved:

Theorem 14.4.3. *Suppose that (C,β) is a coquasitriangular coalgebra over the field k. Then the pair $(\imath,(\mathrm{T}(C,\beta),\boldsymbol{\beta}))$ constructed above satisfies the following:*

(a) *$\imath : (C,\beta) \longrightarrow (\mathrm{T}(C,\beta),\boldsymbol{\beta})$ is a map of coquasitriangular coalgebras.*
(b) *If (A,β') is a coquasitriangular bialgebra over the field k and also $f : (C,\beta) \longrightarrow (A,\beta')$ is a map of coquasitriangular coalgebras then there exists a map $F : (\mathrm{T}(C,\beta),\boldsymbol{\beta}) \longrightarrow (A,\beta')$ of coquasitriangular bialgebras determined by $F\circ\imath = f$.*

□

Definition 14.4.4. A *free coquasitriangular bialgebra on a coquasitriangular coalgebra (C,β) over k* is a pair $(\imath,(\mathrm{T}(C,\beta),\boldsymbol{\beta}))$ which satisfies parts (a) and (b) of Theorem 14.4.3.

Corollary 14.4.5. *Suppose that (C,β) is a coquasitriangular coalgebra over the field k and $(\imath,(\mathrm{T}(C,\beta),\boldsymbol{\beta}))$ is a free coquasitriangular bialgebra* *on (C,β). Then β is invertible if and only if $\boldsymbol{\beta}$ is invertible; that is (C,β) is a Yang–Baxter coalgebra over k if and only if $(\mathrm{T}(C,\beta),\boldsymbol{\beta})$ is.*

Proof. If $\boldsymbol{\beta}$ is invertible then β must be since $\imath : (C,\beta) \longrightarrow (\mathrm{T}(C,\beta),\boldsymbol{\beta})$ is a map of coquasitriangular coalgebras. Suppose that β is invertible. To show that $\boldsymbol{\beta}$ is invertible we follow the notation in the proof of Theorem 14.4.3. Now $F_o : T(C) \longrightarrow T(C)^{o\,cop}$ has an inverse in $\mathrm{Hom}(T(C),T(C)^*)$ by Exercise 14.4.4. Note that $\mathrm{Im}(F_o^{-1}) \subseteq T(C)^o$ and $F_o^{-1} : T(C) \longrightarrow T(C)^{o\,op}$ is a bialgebra map by Propositions 6.1.2 and 6.1.3. Let $\eta_o : T(C){\times}T(C) \longrightarrow k$ be the bilinear form defined by $\eta_{o\,\ell} = F_o^{-1}$. Then β_o and η_o are convolution inverses. Since $\eta_{o\,r}^{op} = \eta_{o\,\ell}$, by Proposition 7.7.8 we see that η_o^{op} satisfies (CoQT.1)–(CoQT.4). Now $\eta_{o\,\ell}^{op} = \eta_{o\,r}$. By the same result $\mathrm{Im}(\eta_{o\,r}) \subseteq T(C)^o$ and $\eta_{o\,r} : T(C) \longrightarrow T(C)^{o\,cop}$ is a bialgebra map.

Once we show $\eta_o(I,T(C)) = (0) = \eta_o(T(C),I)$, or equivalently $\eta_{o\,\ell}(I) = (0) = \eta_{o\,r}(I)$, our proof will be complete. For then η_o lifts to a bilinear form on $\mathrm{T}(C,\beta)$ which is a convolution inverse for $\boldsymbol{\beta}$. We establish $\eta_{o\,\ell}(I) = (0)$. The proof that $(0) = \eta_{o\,r}(I)$ is similar and left to the reader.

Since I is generated as an ideal by $J = d_{\beta_o}(\iota(C),\iota(C))$, and $\eta_{o\,\ell}$ is an algebra map, to show $\eta_{o\,\ell}(I) = (0)$ it suffices to show $\eta_{o\,\ell}(J) = (0)$. Since J is a coideal of $T(C)$ and $\eta_{o\,\ell} : T(C) \longrightarrow T(C)^o$ is a coalgebra map, $\eta_{o\,\ell}(J)$ is a coideal of $T(C)^o$ and therefore $\eta_{o\,\ell}(J)^{\perp}$ is a subalgebra of $T(C)$. Since $\iota(C)$ generates $T(C)$ as an algebra, to show that $\eta_{o\,\ell}(J) = (0)$ we need only show that $\eta_{o\,\ell}(J)(\iota(C)) = (0)$.

Let $a,b,c \in \iota(C)$. Using the fact that $(\iota(C),\eta_o\,|\,\iota(C){\times}\iota(C))$ is a coquasitriangular coalgebra, which follows by part (b) of Exercise 14.1.1, we compute

$$\begin{aligned}
&\eta_o(\beta_o(a_{(1)},b_{(1)})a_{(2)}b_{(2)},c)\\
&= \beta_o(a_{(1)},b_{(1)})\eta_o(a_{(2)},c_{(2)})\eta_o(b_{(2)},c_{(1)})\\
&= \beta_o(a_{(1)},b_{(1)})\eta_o(a_{(2)},c_{(2)})\eta_o(b_{(2)},c_{(1)})\eta_o(a_{(3)},b_{(3)})\beta_o(a_{(4)},b_{(4)})\\
&= \beta_o(a_{(1)},b_{(1)})\eta_o(a_{(3)},b_{(3)})\eta_o(a_{(2)},c_{(2)})\eta_o(b_{(2)},c_{(1)})\beta_o(a_{(4)},b_{(4)})\\
&= \beta_o(a_{(1)},b_{(1)})\eta_o(a_{(2)(2)},b_{(2)(2)})\eta_o(a_{(2)(1)},c_{(2)})\eta_o(b_{(2)(1)},c_{(1)})\beta_o(a_{(3)},b_{(3)})\\
&= \beta_o(a_{(1)},b_{(1)})\eta_o(a_{(2)(1)},b_{(2)(1)})\eta_o(a_{(2)(2)},c_{(1)})\eta_o(b_{(2)(2)},c_{(2)})\beta_o(a_{(3)},b_{(3)})\\
&= \beta_o(a_{(1)},b_{(1)})\eta_o(a_{(2)},b_{(2)})\eta_o(a_{(3)},c_{(1)})\eta_o(b_{(3)},c_{(2)})\beta_o(a_{(4)},b_{(4)})\\
&= \eta_o(a_{(1)},c_{(1)})\eta_o(b_{(1)},c_{(2)})\beta_o(a_{(2)},b_{(2)})\\
&= \eta_o(b_{(1)}a_{(1)}\beta_o(a_{(2)},b_{(2)}),c)
\end{aligned}$$

from which $\eta_{o\,\ell}(d_{\beta_o}(a,b))(c) = \eta_o(d_{\beta_o}(a,b),c) = 0$ follows. □

Exercises

Exercise 14.4.1. Prove Lemma 14.4.1.

Exercise 14.4.2. Let (A, β) and (A', β') be coquasitriangular bialgebras over the field k and let $f : A \longrightarrow A'$ be a bialgebra map. Suppose that C, D are subcoalgebras of A which generate A as an algebra. Show that $\beta'(f(c), f(d)) = \beta(c, d)$ for all $c \in C$ and $d \in D$ implies that $\beta'(f(a), f(b)) = \beta(a, b)$ for all $a, b \in A$; that is $f : (A, \beta) \longrightarrow (A', \beta')$ is a map of coquasitriangular bialgebras over k. [Hint: For any subcoalgebras C, D of A show that

$$\mathcal{A} = \{a \in A \,|\, \beta'(f(a), f(d)) = \beta(a, d) \text{ for all } d \in D\}$$

and

$$\mathcal{B} = \{b \in A \,|\, \beta'(f(c), f(b)) = \beta(c, b) \text{ for all } c \in C\}$$

are subalgebras of A.]

Exercise 14.4.3. Let A be a bialgebra over the field k and $\beta : A \times A \longrightarrow k$ be a bilinear form satisfying (CoQT.1)–(CoQT.4). If C is a subcoalgebra of A which generates A as an algebra, show that (A, β) is a coquasitriangular coalgebra if and only if $(C, \beta \,|\, C \times C)$ is a coquasitriangular coalgebra.

Exercise 14.4.4. Show that the bialgebra map $F_o : T(C) \longrightarrow T(C)^{o\,cop}$ in the proof of Corollary 14.4.5 has an inverse in $\mathrm{Hom}(T(C), T(C)^*)$. [Hint: Construct a bialgebra map $G_o : T(C) \longrightarrow T(C)^{o\,op}$ such that $F_o{*}G_o|\iota(C)$ and $G_o{*}F_o|\iota(C)$ are inverses in $\mathrm{Hom}(\iota(C), T(C)^*)$. See Exercise 6.2.1.]

Exercise 14.4.5. Suppose that (C, β) is a coquasitriangular coalgebra over the field k and that $\varsigma : (C, \beta) \longrightarrow (C^{cop}, \beta)$ is a map of coquasitriangular coalgebras. Assume further that $\beta_\ell : C \longrightarrow C^*$ is invertible in the convolution algebra $\mathrm{Hom}(C, C^*)$ and $\beta_\ell^{-1} = \beta_\ell \circ \varsigma$. Show that there exists a pair $(\imath, (\mathrm{H}(C, \varsigma), \boldsymbol{\beta}))$ which satisfies the following property:

(a) $(\imath, (\mathrm{H}(C, \varsigma), \boldsymbol{\beta}))$ is a coquasitriangular Hopf algebra and $\imath : (C, \beta) \longrightarrow (\mathrm{H}(C, \varsigma), \boldsymbol{\beta})$ is a map of coquasitriangular coalgebras, and

(b) if (A, β') is a coquasitriangular Hopf algebra and $f : (C, \beta) \longrightarrow (A, \beta')$ is a map of coquasitriangular coalgebras then there is a map of coquasitriangular Hopf algebras $F : (\mathrm{H}(C, \varsigma), \boldsymbol{\beta}) \longrightarrow (A, \beta')$ determined by $F \circ \imath = f$.

Chapter notes

The notion of Yang–Baxter coalgebra originated with Doi in his paper [40]. Quasitriangular algebras and coquasitriangular coalgebras are described and studied extensively in the book [99] by Lambe and the author. The notion of braided (coquasitriangular) bialgebra is due to Larson and Towber [107] and was treated independently by Hayashi [63]. Significant contributions to the study of these bialgebras were made by Larson and Towber in [107], by Majid in [114, §3.2.3], and by Schauenburg in [189].

Schauenburg derived the formula for the square of the antipode of a coquasitriangular Hopf algebra in [189, §3.3]. His derivation takes place in a rather abstract categorical setting. Our derivation exploits the algebra of the dual of a quasitriangular Hopf algebra in a rather direct and explicit way and is more along the lines of the proof given by Doi [40].

Our treatment of the free coquasitriangular bialgebra in Section 14.4 follows that of [99, §7.4]. There it is considered as one of the many structures associated with solutions to the quantum Yang–Baxter equation. The same bialgebra was constructed by Doi in [40]; however he described it using a different universal mapping property.

Chapter 15

Pointed Hopf algebras

Pointed Hopf algebras include a good number of basic examples of Hopf algebras and they have a very rich theory. In algebraic groups they arise as the underlying Hopf algebras of connected solvable affine groups over algebraically closed fields. In Lie theory they arise as universal enveloping algebras. In quantum group theory they arise as quantized enveloping algebras.

Pointed Hopf algebras are characterized by all rational representations of their dual algebras are one-dimensional. In this chapter we further develop the theoretical framework for studying pointed Hopf algebras and prove some fundamental results about them. Principal players in our discussion are universal enveloping algebras and their counterparts in certain categories of Yetter-Drinfel'd modules.

A pointed Hopf algebra H over k is the crossed product of an indecomposable pointed Hopf algebra U over k and a group algebra $k[G]$. Suppose H is cocommutative and k has characteristic 0. Then the product is a biproduct, where the coalgebra structure is the tensor product coalgebra structure on $U \otimes k[G]$, U is the universal enveloping algebra of $\mathrm{P}(H)$, and $G = \mathrm{G}(H)$. Universal enveloping algebras are pointed irreducible Hopf algebras. A basic technique for studying H, is to pass to the associated graded Hopf algebra.

Let $G = \mathrm{G}(H)$. As we will see in Section 15.5, $\mathrm{gr}(H) = B \times k[G]$, where B is a coradically graded and pointed irreducible Hopf algebra in ${}_{k[G]}^{k[G]}\mathcal{YD}$. Quite often the primitives $B(1)$ of B generate B as an algebra. Such Hopf algebras in ${}_{k[G]}^{k[G]}\mathcal{YD}$ are Nichols algebras and can be thought of as the counterparts of the universal enveloping algebras. Nichols algebras in the category of Yetter-Drinfel'd modules over a bialgebra satisfy a universal mapping property as do the universal enveloping algebras of Lie algebras.

Quantized enveloping algebras are quotients of certain biproducts $B \times k[G]$, where B is a pointed irreducible Hopf algebra in ${}^{k[G]}_{k[G]}\mathcal{YD}$, a Nichols algebra in important cases. To see this we explain the quantized enveloping algebras in terms of the language of ${}^{k[G]}_{k[G]}\mathcal{YD}$ and biproducts.

At this point we make a few remarks about classification of pointed Hopf algebras which takes us well beyond the scope of this book. Suppose k is algebraically closed, has characteristic 0, H is a finite-dimensional Hopf algebra over k, and $\mathrm{G}(H)$ is abelian. Then H is a quotient of a Hopf algebra whose underlying vector space is $B \otimes k[G]$, where B is a Nichols algebra in ${}^{k[G]}_{k[G]}\mathcal{YD}$. There is a very strong connection between the structure of B and Lie theory which suggests even more that Nichols algebras are the counterparts of universal enveloping algebras. More about classification in the chapter notes.

The category of pointed Hopf algebras over k and their morphisms is closed under basic operations. The tensor product of pointed Hopf algebras is pointed by part (c) of Proposition 4.1.7. The property of pointed is hereditary and preserved under quotients.

Proposition 15.0.6. *Hopf subalgebras and homomorphic images of pointed Hopf algebras over the field k are pointed.*

Proof. Let H, K be Hopf algebras over k, where H is pointed. By definition the simple subcoalgebras of H are one-dimensional.

Suppose K is a Hopf subalgebra of H. Since the simple subcoalgebras of K are those of H which are contained in K by part (c) of Proposition 3.4.3, the simple subcoalgebras of K are one-dimensional. Thus K is pointed.

Suppose $f : H \longrightarrow K$ is an onto Hopf algebra map. Then $f(H_0) \supseteq K_0$ by part (a) of Proposition 4.1.7. Therefore K_0 is contained in the span of one-dimensional subcoalgebras of K. By part (a) of Proposition 3.4.3 a simple subcoalgebra of K is contained in one of these summands and is therefore one-dimensional. Thus K is pointed. $\square$

15.1 Crossed products

Crossed products can be viewed as generalizations of smash products and provide interesting examples of algebras in their own right. Pointed Hopf algebras have a crossed product decomposition into an indecomposable Hopf subalgebra and a group algebra. In this section we discuss cross products and establish the decomposition result in the next. We begin with the notion of action on an algebra by a coalgebra.

Definition 15.1.1. Let C be a coalgebra and A be an algebra over the field k. A *measuring of C and A* is a linear map $C\otimes A \longrightarrow A$ $(c\otimes a \mapsto c{\cdot}a)$ such that $c{\cdot}1 = \epsilon(c)1$ and $c{\cdot}(ab) = (c_{(1)}{\cdot}a)(c_{(2)}{\cdot}b)$ for all $c \in C$ and $a, b \in A$. When there is such a map C *measures A to itself.*

Let H be a bialgebra over k. Suppose $H\otimes A \longrightarrow A$ is a measuring and $\sigma : H\times H \longrightarrow A$ is bilinear. Consider the vector space $A\#_\sigma H = A\otimes H$ over k with product defined by

$$(a\otimes h)(b\otimes \ell) = a(h_{(1)}{\cdot}b)\sigma(h_{(2)}, \ell_{(1)})\otimes h_{(3)}\ell_{(2)} \tag{15.1}$$

for all $a, b \in A$ and $h, \ell \in H$.

Remark 15.1.2. Suppose A is a left H-module algebra. The module action is a measuring. If $\sigma(h, \ell) = \epsilon(h)\epsilon(\ell)$ for all $h, \ell \in H$ then (15.1) is the smash product multiplication. If in addition $h{\cdot}a = \epsilon(h)a$ for all $h \in H$ and $a \in A$ then (15.1) is the tensor product of algebras multiplication.

We would like to know when $A\#_\sigma H$ with its product is an associative algebra with unity $1\otimes 1 = 1_A\otimes 1_H$. Suppose this is the case. We find necessary conditions.

Let $a, b, c \in A$ and $h, \ell, m \in H$. From $1\otimes h = (1\otimes h)(1\otimes 1) = \sigma(h_{(1)}, 1)\otimes h_{(2)}$ we deduce

$$\sigma(h, 1) = \epsilon(h)1 \tag{15.2}$$

for all $h \in H$. Since $\sigma(1, 1) = 1$, which follows by (15.2), from $b\otimes \ell = (1\otimes 1)(b\otimes \ell) = (1{\cdot}b)\sigma(1, \ell_{(1)})\otimes \ell_{(2)}$ the equations

$$1{\cdot}b = b \tag{15.3}$$

for all $b \in A$ and

$$\sigma(1, \ell) = \epsilon(\ell)1 \tag{15.4}$$

for all $\ell \in H$ follow. Associativity

$$(a\otimes h)((b\otimes \ell)(c\otimes m)) = ((a\otimes h)(b\otimes \ell))(c\otimes m)$$

translates to

$$\begin{aligned} &a(h_{(1)}{\cdot}(b(\ell_{(1)}{\cdot}c)\sigma(\ell_{(2)}, m_{(1)})))\sigma(h_{(2)}, \ell_{(3)}m_{(2)})\otimes h_{(3)}\ell_{(4)}m_{(3)} \\ &= a(h_{(1)}{\cdot}b)\sigma(h_{(2)}, \ell_{(1)})((h_{(3)}{\cdot}\ell_{(2)}){\cdot}c)\sigma(h_{(4)}\ell_{(3)}, m_{(1)})\otimes h_{(5)}\ell_{(4)}m_{(2)}. \end{aligned}$$

Using the measuring condition $h{\cdot}1 = \epsilon(h)1$ for all $h \in H$, the associativity equation with $a = b = c = 1$ specializes to

$$\begin{aligned}&(h_{(1)}{\cdot}(\sigma(\ell_{(1)}, m_{(1)})))\sigma(h_{(2)}, \ell_{(2)}m_{(2)})\otimes h_{(3)}\ell_{(3)}m_{(3)}\\&= \sigma(h_{(1)}, \ell_{(1)})\sigma(h_{(2)}\ell_{(2)}, m_{(1)})\otimes h_{(3)}\ell_{(2)}m_{(2)}.\end{aligned}$$

Applying $\mathrm{I}_A\otimes\epsilon$ to both sides of the preceding equation yields

$$(h_{(1)}{\cdot}\sigma(\ell_{(1)}, m_{(1)}))\sigma(h_{(2)}, \ell_{(2)}m_{(2)}) = \sigma(h_{(1)}, \ell_{(1)})\sigma(h_{(2)}\ell_{(2)}, m) \qquad (15.5)$$

for all $h, \ell, m \in H$.

Since (15.2) holds the associativity equation with $a = b = 1$ and $m = 1$ specializes to

$$(h_{(1)}{\cdot}(\ell_{(1)}{\cdot}c))\sigma(h_{(2)}, \ell_{(2)})\otimes h_{(3)}\ell_{(3)} = \sigma(h_{(1)}, \ell_{(1)})((h_{(2)}\ell_{(2)}){\cdot}c)\otimes h_{(3)}\ell_{(3)}.$$

Applying $\mathrm{I}_A\otimes\epsilon$ to both sides of the preceding equation results in

$$(h_{(1)}{\cdot}(\ell_{(1)}{\cdot}c))\sigma(h_{(2)}, \ell_{(2)}) = \sigma(h_{(1)}, \ell_{(1)})((h_{(2)}\ell_{(2)}){\cdot}c) \qquad (15.6)$$

for all $h, \ell \in H$ and $c \in A$. The necessary conditions (15.2) – (15.6) are sufficient.

Theorem 15.1.3. *Let H be a bialgebra and A be an algebra over the field k. Suppose $H\otimes A \longrightarrow A$ is a measuring and $\sigma : H{\times}H \longrightarrow A$ is bilinear. Then $A\#_\sigma H$ is an associative algebra with product given by (15.1) and unity $1_A\otimes 1_H$ if and only if:*

(a) $\sigma(h, 1) = \epsilon(h)1 = \sigma(1, h)$;
(b) $1{\cdot}a = a$;
(c) $(h_{(1)}{\cdot}\sigma(\ell_{(1)}, m_{(1)}))\sigma(h_{(2)}, \ell_{(2)}m_{(2)}) = \sigma(h_{(1)}, \ell_{(1)})\sigma(h_{(2)}\ell_{(2)}, m)$; *and*
(d) $(h_{(1)}{\cdot}(\ell_{(1)}{\cdot}a))\sigma(h_{(2)}, \ell_{(2)}) = \sigma(h_{(1)}, \ell_{(1)})((h_{(2)}\ell_{(2)}){\cdot}a)$

for all $h, \ell, m \in H$ and $a \in A$. When σ is convolution invertible (d) is equivalent to

(d') $h{\cdot}(\ell{\cdot}a) = \sigma(h_{(1)}, \ell_{(1)})((h_{(2)}\ell_{(2)}){\cdot}a)\sigma^{-1}(h_{(3)}, \ell_{(3)})$

for all $h, \ell \in H$ and $a \in A$.

Proof. We need only show that conditions (a)–(d) are sufficient for $A\#_\sigma H$ to be an associative algebra as described and that the equations of parts (d) and (d') are equivalent when σ has a convolution inverse.

Observe that $(1_A\otimes 1_H)(a\otimes h) = a\otimes h = (a\otimes h)(1_A\otimes 1_H)$ for all $a \in A$ and $h \in H$ follow by parts (a) and (b). Parts (c) and (d), together with the measuring condition $h{\cdot}(ab) = (h_{(1)}{\cdot}a)(h_{(2)}{\cdot}b)$ for all $h \in H$ and $a, b \in A$, imply

$$
\begin{aligned}
&(a{\otimes}h)((b{\otimes}\ell)(c{\otimes}m))\\
&= a(h_{(1)}{\cdot}(b(\ell_{(1)}{\cdot}c)\sigma(\ell_{(2)}, m_{(1)})))\sigma(h_{(2)}, \ell_{(3)}m_{(2)}){\otimes}h_{(3)}\ell_{(4)}m_{(3)}\\
&= a(h_{(1)}{\cdot}(b(\ell_{(1)}{\cdot}c)))(h_{(2)}{\cdot}\sigma(\ell_{(2)}, m_{(1)}))\sigma(h_{(3)}, \ell_{(3)}m_{(2)}){\otimes}h_{(4)}\ell_{(4)}m_{(3)}\\
&= a(h_{(1)}{\cdot}(b(\ell_{(1)}{\cdot}c)))\sigma(h_{(2)}, \ell_{(2)})\sigma(h_{(3)}\ell_{(3)}, m_{(1)}){\otimes}h_{(4)}\ell_{(4)}m_{(2)}\\
&= a(h_{(1)}{\cdot}b)(h_{(2)}{\cdot}(\ell_{(1)}{\cdot}c))\sigma(h_{(3)}, \ell_{(2)})\sigma(h_{(4)}\ell_{(3)}, m_{(1)}){\otimes}h_{(5)}\ell_{(4)}m_{(2)}\\
&= a(h_{(1)}{\cdot}b)\sigma(h_{(2)}, \ell_{(1)})((h_{(3)}\ell_{(2)}){\cdot}c)\sigma(h_{(4)}\ell_{(3)}, m_{(1)}){\otimes}h_{(5)}\ell_{(4)}m_{(2)}\\
&= ((a{\otimes}h)(b{\otimes}\ell))(c{\otimes}m)
\end{aligned}
$$

for all $a, b, c \in A$ and $h, \ell, m \in H$. Thus the conditions of parts (a) – (d) imply $A\#_\sigma H$ is an associative algebra with the structure described in the theorem.

Suppose that σ has a convolution inverse. Assume part (d) holds. Then

$$
\begin{aligned}
h{\cdot}(\ell{\cdot}a) &= h_{(1)}{\cdot}(\ell_{(1)}{\cdot}a)\sigma(h_{(2)}, \ell_{(2)})\sigma^{-1}(h_{(3)}, \ell_{(3)})\\
&= \sigma(h_{(1)}, \ell_{(1)})((h_{(2)}\ell_{(2)}){\cdot}a)\sigma^{-1}(h_{(3)}, \ell_{(3)})
\end{aligned}
$$

for all $h, \ell \in H$ and $a \in A$ which means (d') holds. Conversely, assume part (d') holds. Then

$$
\begin{aligned}
h_{(1)}{\cdot}(\ell_{(1)}{\cdot}a)\sigma(h_{(2)}, \ell_{(2)}) &= \sigma(h_{(1)}, \ell_{(1)})((h_{(2)}\ell_{(2)}){\cdot}a)\sigma^{-1}(h_{(3)}, \ell_{(3)})\sigma(h_{(4)}, \ell_{(4)})\\
&= \sigma(h_{(1)}, \ell_{(1)})((h_{(2)}\ell_{(2)}){\cdot}a)
\end{aligned}
$$

for all $h, \ell \in H$ and $a \in A$ which means part (d) holds. $\square$

The conditions of parts (a) and (c) of the preceding theorem can be rephrased "σ is a 2-cycle" and parts (b) and (d') can be expressed "A is a twisted left H-module". See [133, §7.1].

We apply Theorem 15.1.3 in Section 15.2 when A is a Hopf algebra and H is a group algebra, hence is cocommutative. In our application $A\#_\sigma H$ is a Hopf algebra with the tensor product coalgebra structure. When A is a bialgebra over k and H is cocommutative there are natural necessary and sufficient conditions for $A\#_\sigma H$ to be a bialgebra with the tensor product coalgebra structure.

Corollary 15.1.4. *Let A, H be bialgebras over the field k where H is cocommutative. Suppose the hypothesis and parts (a)–(d) of Theorem 15.1.3 are satisfied and let $A\#_\sigma H$ be the algebra described therein. Then $A\#_\sigma H$ is a bialgebra with the tensor product coalgebra structure if and only if:*

(a) $\epsilon(h{\cdot}a) = \epsilon(h)\epsilon(a)$ *and* $\epsilon(\sigma(h,\ell)) = \epsilon(h)\epsilon(\ell)$;
(b) $\Delta(h{\cdot}a) = h_{(1)}{\cdot}a_{(1)} \otimes h_{(2)}{\cdot}a_{(2)}$; *and*
(c) $\Delta(\sigma(h,\ell)) = \sigma(h_{(1)},\ell_{(1)}) \otimes \sigma(h_{(2)},\ell_{(2)})$

for all $h,\ell \in H$ *and* $a \in A$.

Proof. Part (a) is necessary and sufficient for $\epsilon_{A\#_\sigma H}$ to be an algebra map. Necessary conditions for $\Delta_{A\#_\sigma H}$ to be an algebra map, which hold when H is any bialgebra, are

$$(h_{(1)}{\cdot}b)_{(1)} \otimes h_{(2)} \otimes (h_{(1)}{\cdot}b)_{(2)} \otimes h_{(3)} = h_{(1)}{\cdot}b_{(1)} \otimes h_{(2)} \otimes h_{(3)}{\cdot}b_{(2)} \otimes h_{(4)}$$

and

$$\begin{aligned}&\sigma(h_{(1)},\ell_{(1)})_{(1)} \otimes h_{(2)}\ell_{(2)} \otimes \sigma(h_{(1)},\ell_{(1)})_{(2)} \otimes h_{(3)}\ell_{(3)}\\ &= \sigma(h_{(1)},\ell_{(1)}) \otimes h_{(2)}\ell_{(2)} \otimes \sigma(h_{(3)},\ell_{(3)}) \otimes h_{(4)}\ell_{(4)}\end{aligned}$$

for all $h,\ell \in H$ and $b \in A$. These conditions are sufficient for $\Delta_{A\#_\sigma H}$ to be an algebra map when H is cocommutative. When H is cocommutative these conditions are equivalent to those of parts (b) and (c) respectively.□

15.2 Pointed Hopf algebras as crossed products

We draw heavily from the material on indecomposable coalgebras in Section 3.7 and on link indecomposable coalgebras in Section 4.8 for this section. Since Section 3.7 and Section 4.8 are very short, we will quote results from them with no particular reference. Other material or results relegated to exercises will be cited. Throughout C is a non-zero coalgebra and H is a pointed Hopf algebra over the field k.

C is indecomposable if whenever $C = D \oplus E$ is the direct sum of subcoalgebras D, E then $D = (0)$ or $E = (0)$. Every indecomposable subcoalgebra of C is contained in a maximal indecomposable subcoalgebra of C; the latter is called an indecomposable component of C. The coalgebra C is the direct sum of its indecomposable components. Indeed, if C is the direct sum of non-zero indecomposable subcoalgebras then the summands must be the indecomposable components. A coalgebra automorphism, or anti-automorphism, permutes the indecomposable components. See Exercise 3.7.7.

Let D, E be subcoalgebras of C. Then $D, E \subseteq D \wedge E$, and hence $D, E \subseteq E \wedge D$. The subcoalgebras D and E are directly linked in C if $D + E$ is a proper subspace of $D \wedge E + E \wedge D$. Two simple subcoalgebras S, S' of C

are said to be linked in C if $S = S'$ or for some $n \geq 1$ there are simple subcoalgebras $S = S_0, \ldots, S_n = S'$ of C such that S_i and S_{i+1} are directly linked in C for all $0 \leq i < n$. If S and S' are linked in C we write $S \sim S'$. Observe that $\sim$ defines an equivalence relation on the set of simple subcoalgebras of C.

The coalgebra C is link indecomposable if its simple subcoalgebras are linked in C. Every link indecomposable subcoalgebra is contained in a maximal link indecomposable subcoalgebra of C. The latter is called a link indecomposable component of C. The link indecomposable components and indecomposable components of C are the same.

Let D be an indecomposable component of C and let S be a simple subcoalgebra of C contained in D. The set of simple subcoalgebras of C contained in D is the equivalence class of S.

We will use the fact that coalgebra automorphisms, or anti-automorphisms, preserve the equivalence relation.

Lemma 15.2.1. *Suppose that C is a coalgebra over the field k and that $f : C \longrightarrow C$ or $f : C \longrightarrow C^{cop}$ is a coalgebra isomorphism. Let S, S' be simple subcoalgebras of C. Then $f(S), f(S')$ are simple subcoalgebras of C and $S \sim S'$ implies $f(S) \sim f(S')$.*

Proof. That $f(S), f(S')$ are simple subcoalgebras of D is easy to see. Let U, V be subspaces of C. Then $f(U \wedge V) \subseteq f(U) \wedge f(V)$ by part (c) of Proposition 2.4.3. Since f^{-1} is a coalgebra isomorphism $f(U) \wedge f(V) = f(f^{-1}(f(U) \wedge f(V))) \subseteq f(f^{-1}(f(U)) \wedge f^{-1}(f(V))) = f(U \wedge V)$. Therefore $f(U \wedge V) = f(U) \wedge f(V)$. Note that $V \wedge_{C^{cop}} U = U \wedge_C V$. Thus if $f : C \longrightarrow D^{cop}$, or equivalently $f : C^{cop} \longrightarrow D$, is a coalgebra isomorphism, $f(U \wedge V) = f(V) \wedge f(U)$. The lemma follows from these general observations. □

When C is pointed its simple subcoalgebras are the kg's, where $g \in \mathrm{G}(C)$.

Lemma 15.2.2. *Let C be a coalgebra over the field k and suppose $g, g' \in \mathrm{G}(C)$. Then kg and kg' are directly linked in C if and only if there is an $x \notin k(g - g')$ such that $\Delta(x) = g \otimes x + x \otimes g'$ or $\Delta(x) = g' \otimes x + x \otimes g$.*

Proof. $kg \wedge kg' = kg \oplus P_{g,g'} = P_{g,g'} \oplus kg'$ by Exercise 2.4.5. □

For $g \in \mathrm{G}(H)$ let $H_{(g)}$ be the indecomposable component containing kg. For $g, g' \in \mathrm{G}(C)$ by slight abuse of notation we write $g \sim g'$ if and only if $kg \sim kg'$. Thus $\sim$ defines an equivalence relation on $\mathrm{G}(H)$.

Proposition 15.2.3. *Let H be a pointed Hopf algebra with antipode S over the field k and $G = \mathrm{G}(H)$. Then:*

(a) $\mathrm{G}(H_{(g)}) = \{g' \in G \,|\, g' \sim g\}$ *for all* $g \in G$.
(b) $N = \mathrm{G}(H_{(1)})$ *is a normal subgroup of* G.
(c) $gH_{(g')} = H_{(gg')} = H_{(g)}g'$ *for all* $g, g' \in G$.
(d) $H_{(g)}H_{(g')} \subseteq H_{(gg')}$ *for all* $g, g' \in G$.
(e) $S(H_{(g)}) = H_{(g^{-1})}$ *for all* $g \in G$.

Proof. We have noted that the set of simple subcoalgebras in an indecomposable component of a coalgebra is the equivalence class of any of the simples in the component. Since $kg \subseteq H_{(g)}$ part (a) follows. To show part (b) we first observe that left or right multiplication by a grouplike element of H is a coalgebra automorphism of H and thus preserves the equivalence relation $\sim$ by Lemma 15.2.1. Let $g, g'', g'' \in G$. As a consequence

$$g \sim g' \text{ implies } g''g \sim g''g' \text{ and } gg'' \sim g'g''. \tag{15.7}$$

Now $S : H \longrightarrow H^{cop}$ is a coalgebra anti-automorphism by part (b) of Proposition 7.1.9 and Corollary 7.6.4. Thus $g \sim g'$ implies $S(g) \sim S(g')$; hence

$$g \sim g' \text{ implies } g^{-1} \sim g'^{-1}. \tag{15.8}$$

Part (b) now follows from part (a), (15.7), and (15.8).

We have noted that the coalgebra automorphisms, in particular left and right multiplication by grouplike elements, and coalgebra anti-automorphisms, of which S is one, permute the indecomposable components of H. As kgg' is a subset of $gH_{(g')}$, $H_{(g)}g'$, and $H_{(gg')}$, these three indecomposable components are the same and part (c) follows. Since kg^{-1} is a subset of the indecomposable component $S(H_{(g)})$, part (e) follows.

The proof of part (d) is a bit more involved. First of all note that the restriction of multiplication $\pi : H_{(g)}\otimes H_{(g')} \longrightarrow H_{(g)}H_{(g')}$ is an onto coalgebra map. Now $H_{(g)}\otimes H_{(g')}$ is pointed and

$$\mathrm{G}(H_{(g)}\otimes H_{(g')}) = \{h\otimes h' \,|\, h \in \mathrm{G}(H_{(g)}),\ h' \in \mathrm{G}(H_{(g')})\}$$

by part (c) of Proposition 4.1.7. By part (a) of the same $\pi((H_{(g)}\otimes H_{(g')})_0) \supseteq (H_{(g)}H_{(g')})_0$. Thus $\mathrm{G}(H_{(g)}H_{(g')}) = \{hh' \,|\, h \in \mathrm{G}(H_{(g)}),\ h' \in \mathrm{G}(H_{(g')})\}$. By part (a) and (15.7) all of the grouplike elements of $H_{(g)}H_{(g')}$ belong to the same equivalence class. Thus $H_{(g)}H_{(g')}$ is link indecomposable which means it is a subspace of an indecomposable component of C. Since $kgg' \in H_{(g)}H_{(g')}$, $H_{(gg')}$ necessarily $H_{(g)}H_{(g')} \subseteq H_{(gg')}$. □

In light of the proposition we are but a few short steps from establishing the decomposition theorem for pointed Hopf algebras. Let $g, g' \in G = \mathrm{G}(H)$. Then $H_{(g)} = H_{(g')}$ if and only if $g \sim g'$. The latter is the case if and only if $g^{-1}g' \sim 1$ by (15.7). Thus

$$\{g' \in G \,|\, H_{(g')} = H_{(g)}\} = gN \tag{15.9}$$

for all $g \in G$ which means the indecomposable components of H are in one-one correspondence with the left cosets of N in G. Now let $\{g_i\}_{i\in I}$ be a complete set of representative for the left cosets N in G. Using (15.9) we conclude $H = \oplus_{i\in I} H_{(g_i)}$. Let $\overline{g} = gN$. Since $H_{(g_i)} = H_{(1)}g_i$ by part (c) of Proposition 15.2.3, there is a linear isomorphism $f : H \longrightarrow H_{(1)}\otimes k[G/N]$ given by

$$f(ag_i) = a\otimes\overline{g_i} \tag{15.10}$$

for all $a \in H_{(1)}$ and $i \in I$. Observe that f is a coalgebra isomorphism, where $H_{(1)}\otimes k[G/N]$ is given the tensor product coalgebra structure. Also $f(1) = 1_{H_{(1)}}\otimes 1_{k[G/N]}$.

By parts (d) and (e) of the preceding proposition $H_{(1)}$ is a Hopf subalgebra of H. By part (c) again $gH_{(1)}g^{-1} = H_{(1)}$. Observe that $k[G/N]$ measures $H_{(1)}$ to itself by

$$\overline{g_i}{\cdot}a = g_i a g_i^{-1} \tag{15.11}$$

for all $i \in I$ and $a \in H_{(1)}$.

We next define a bilinear map $\sigma : k[G/N]\times k[G/N] \longrightarrow H_{(1)}$. Let $i, j \in I$. Then $\overline{g_i}\,\overline{g_j} = \overline{g_\ell}$ for a unique $\ell \in I$. Since $g_ig_jN = g_\ell N = Ng_\ell$ there is an $n_{i\,j} \in N \subseteq H_{(1)}$ such that $g_ig_j = n_{i\,j}g_\ell$. Since ℓ is unique $n_{i\,j}$ is unique. Set

$$\sigma(\overline{g_i}, \overline{g_j}) = n_{i\,j}. \tag{15.12}$$

Consider $H_{(1)}\#_\sigma k[G/N] = H_{(1)}\otimes k[G/N]$ with the product defined by (15.1). For $a, b \in H_{(1)}$ the calculation

$$\begin{aligned}
f((ag_i)(bg_j)) &= f(a(g_ibg_i^{-1})g_ig_j)\\
&= f(a(\overline{g}_i{\cdot}b)n_{i\,j}g_\ell)\\
&= f(a(\overline{g}_i{\cdot}b)\sigma(\overline{g}_i, \overline{g}_j)g_\ell)\\
&= a(\overline{g_i}{\cdot}b)\sigma(\overline{g_i}, \overline{g_j})\otimes\overline{g_\ell}\\
&= a(\overline{g_i}{\cdot}b)\sigma(\overline{g_i}, \overline{g_j})\otimes\overline{g_i}\,\overline{g_j}\\
&= f(ag_i)f(bg_j)
\end{aligned}$$

shows that f is multiplicative. Thus $H_{(1)}\#_\sigma k[G/N]$ is a Hopf algebra. We have shown:

Theorem 15.2.4. *Let H be a pointed Hopf algebra over the field k, let $G = \mathrm{G}(H)$, and let $N = \mathrm{G}(H_{(1)})$. Then:*

(a) *$H_{(1)}$ is a Hopf subalgebra of H and N is a normal subgroup of G.*
(b) *$H_{(1)}\#_\sigma k[G/N]$ is a Hopf algebra, where $k[G/N]$ measures $H_{(1)}$ to itself according to (15.11) and σ is defined by (15.12).*
(c) *The map $f : H \longrightarrow H_{(1)}\#_\sigma k[G/N]$ defined by (15.10) is an isomorphism of Hopf algebras.*

□

We end this section by considering a few examples. There will be more in subsequent sections. A useful tool for analysis is:

Lemma 15.2.5. *Let H be a pointed Hopf algebra over the field k. Then $N = \mathrm{G}(H_{(1)})$ is generated by those $g \in \mathrm{G}(H)$ such that there is an $x \notin k(g-1)$ such that $\Delta(x) = 1{\otimes}x + x{\otimes}g$.*

Proof. $N = \{g \in \mathrm{G}(H) \,|\, g \sim 1\}$ by part (a) of Proposition 15.2.3. Suppose $g \in \mathrm{G}(H)$ satisfies the condition of the lemma. Then kg and $k1$ are directly linked in H. Therefore $g \sim 1$ which means $g \in N$. Thus the subgroup L of $\mathrm{G}(H)$ generated by the $g \in \mathrm{G}(H)$ which satisfy the condition of the lemma is a subgroup of N.

Now suppose $g, g' \in \mathrm{G}(H)$ are directly linked in H. Then $g'g^{-1}$ or gg'^{-1} satisfies the condition of the lemma by Lemma 15.2.2. Therefore $g'g^{-1} \in L$ or $gg'^{-1} \in L$; consequently $g'g^{-1}, gg'^{-1} \in L$ since L is a subgroup. Our conclusion: either $g, g' \in L$ or $g \notin L$, $g' \notin L$.

Let $g \in N$. Then $g = 1$ or for some $n \geq 1$ there are grouplike elements $1 = g_0, \ldots, g_n = g$ such that kg_i and kg_{i+1} are directly linked in H for all $0 \leq i < n$. By induction on $n \geq 1$ it follows that $g_0, \ldots, g_n \in L$. Therefore $N \subseteq L$ which means $N = L$ since $L \subseteq N$. □

The examples of Section 7.3 are indecomposable by the following:

Proposition 15.2.6. *Let H be a pointed Hopf algebra over the field k which is generated as an algebra by $g_1, \ldots, g_n \in \mathrm{G}(H)$, their inverses, and $x_1, \ldots, x_n \in H$ which satisfy $\Delta(x_i) = 1{\otimes}x_i + x_i{\otimes}g_i$ and $x_i \notin k(g_i - 1)$ for all $1 \leq i \leq n$. Then H is indecomposable.*

Proof. Let V_0 be the span of $g_1, \ldots, g_n$ and their inverses and let V_1 be the span of V_0 and $x_1, \ldots, x_n$. Then V_0, V_1 are subcoalgebras of H and $\Delta(V_1) \subseteq V_0 \otimes V_1 + V_1 \otimes V_0$. Therefore V_0, V_1 are the first two terms in a filtration of $C = V_1$. Thus $C_0 \subseteq V_0$ by Proposition 4.1.2 which means $C_0 = V_0$. Since C generates H as an algebra H_0, is a subspace of the subalgebra generated by C_0 by Corollary 5.1.12. Therefore $\mathrm{G}(H)$ is generated by $g_1, \ldots, g_n$ as a group. Now $g_1, \ldots, g_n \in \mathrm{G}(H_{(1)})$ by Lemma 15.2.5. Therefore $\mathrm{G}(H) = \mathrm{G}(H_{(1)}) = N$ which means that H has one indecomposable component by (15.9). □

Let $r, n \geq 1$. We construct an rn^2-dimensional pointed Hopf algebra over k with r indecomposable components, provided k has enough roots of unity.

Let $M = rn$ and suppose that k has a primitive M^{th} root of unity q. Let $H = H_{(r,n,q)}$ be the Hopf algebra over k described as follows. As an algebra H is generated by a and x subject to the relations

$$a^M = 1, \quad x^n = 0, \quad xa = qax$$

and the coalgebra structure of H is given by

$$\Delta(a) = a \otimes a, \quad \Delta(x) = 1 \otimes x + x \otimes a^r.$$

The methods of Section 7.3 can be used to show that $H_{(r,n,q)}$ has basis $\{a^\ell x^m \mid 0 \leq \ell < M, 0 \leq m < n\}$. Thus $\mathrm{Dim}(H_{(r,n,q)}) = rn^2$. The results of Section 7.2 can be used to derive the formula

$$\Delta(a^\ell x^m) = \sum_{u=0}^{m} \binom{m}{u}_Q a^\ell x^{m-u} \otimes a^{\ell + r(m-u)} x^u$$

for all $0 \leq \ell < M$ and $0 \leq m < n$, where $Q = q^r$. This equation and Lemma 15.2.5 imply $N = \mathrm{G}(H_{(1)}) = (a^r)$. Therefore $H_{(r,n,q)}$ has r distinct indecomposable components. Notice when $r = 1$ that $H_{(r,n,q)}$ is a Taft algebra.

Exercises

Exercise 15.2.1. Carry out all of the details concerning the last example and find the indecomposable components of $H_{(r,n,q)}$.

15.3 Cocommutative pointed Hopf algebras; the characteristic 0 case

We assume the notation and results of Section 15.2 and begin with general remarks about cocommutative coalgebras. Let C be a cocommutative coalgebra over the field k. Then C^* is a commutative algebra over k. Suppose S, S' are simple subcoalgebras of C. Then $S^\perp$, $S'^\perp$ are cofinite maximal ideals of C^* by Corollary 2.3.9. Suppose $S \neq S'$. Then $S^\perp \neq S'^\perp$ by the same. Since $S^\perp$ and $S'^\perp$ are comaximal ideals of the commutative algebra C^* we have $S^\perp \cap S'^\perp = S^\perp S'^\perp$. Recall that $V \mapsto V^\perp$ describes a bijective correspondence between the subspaces of C and the closed subspaces of C^* whose inverse is given by $X \mapsto X^\perp$. Using part (a) of Proposition 1.3.8 and part (b) of Proposition 2.4.2 we calculate

$$S + S' = (S + S')^{\perp\perp} = (S^\perp \cap S'^\perp)^\perp = (S^\perp S'^\perp)^\perp = S \wedge S'.$$

Therefore $S \wedge S' = S + S' = S' + S = S' \wedge S$. Our conclusion: $S \sim S'$ if and only if $S = S'$.

We have shown that the indecomposable components of C are irreducible subcoalgebras of C. Since irreducible subcoalgebras are indecomposable in any case:

Proposition 15.3.1. *Let C be a cocommutative coalgebra over the field k. Then the irreducible, link indecomposable, and the indecomposable components of C are one in the same. In particular C is the direct sum of its irreducible components.* □

See Exercise 4.8.5.

Now suppose H is a cocommutative pointed Hopf algebra. We revisit the structures defined for Theorem 15.2.4 in this special case. Let $G = \mathrm{G}(H)$. Note that $N = \mathrm{G}(H_{(1)}) = \{1\}$ by Proposition 15.3.1. Thus $G/N = G$ which means (15.11) becomes $g{\cdot}a = gag^{-1}$ for all $g \in G$ and $a \in H_{(1)}$ and (15.12) becomes $\sigma(g, g') = 1$ for all $g, g' \in G$. Theorem 15.2.4 applied to H, where $F = f^{-1}$, results in:

Theorem 15.3.2. *Let H be a cocommutative pointed Hopf algebra over the field k, let $K = H_{(1)}$, and let $G = \mathrm{G}(H)$. Then:*

(a) *K is a $k[G]$-module algebra, where $g{\cdot}a = gag^{-1}$ for all $g \in G$ and $a \in U$.*

(b) *The smash product $K\#k[G]$ is a Hopf algebra with the tensor product coalgebra structure and there is an isomorphism $F : K\#k[G] \longrightarrow H$ of Hopf algebras determined by $F(a\otimes g) = ag$ for all $a \in K$ and $g \in k[G]$.*

□

The preceding theorem applies to cocommutative Hopf algebras over algebraically close fields since these Hopf algebras are pointed. See Exercise 2.3.35.

By part (d) of Proposition 5.1.15 the space of primitive elements $P(H)$ of a Hopf algebra over any field k is a Lie algebra under associative bracket. There is but one possibility for the Hopf algebra $K = H_{(1)}$ of Theorem 15.3.2 when k has characteristic zero.

Theorem 15.3.3. *Let H be a pointed irreducible Hopf algebra over a field of characteristic zero. Then $H \simeq U(\mathrm{P}(H))$ as Hopf algebras.*

Proof. We do not provide a proof here but instead direct the reader to Montgomery's book [133, Section 5.6] where a short proof is given based on notes by Blattner [24]. □

As a consequence of Theorems 15.3.2 and 15.3.3:

Theorem 15.3.4. *Suppose that k has characteristic zero and H is a cocommutative pointed Hopf algebra over k. Then $H \simeq U(\mathrm{P}(H))\#k[\mathrm{G}(H)]$ as Hopf algebras, where the smash product has the tensor product coalgebra structure.* □

Exercises

Exercise 15.3.1. Show that the conclusion of Theorem 15.3.4 holds when H is a cocommutative Hopf algebra and the field k is algebraically closed of characteristic zero.

15.4 Minimal-pointed Hopf algebras

Examples of pointed Hopf algebras we have constructed so far involve a pair of elements a, x such that

$$\Delta(a) = a\otimes a, \quad \Delta(x) = x\otimes a + 1\otimes x, \quad \text{and} \quad xa = qax \tag{15.13}$$

for some $q \in k\backslash 0$. We begin by analyzing Hopf algebras H over k generated by such a pair. As an algebra H is generated by a, a^{-1} and x. Observe that H is pointed and $\mathrm{G}(H) = (a)$ by Corollary 5.1.14 and that H is indecomposable by Proposition 15.2.3 and (15.9).

We will assume that the field k has characteristic zero for the remainder of this section and will leave the reader to ponder the characteristic $p > 0$ case. The Hopf algebra H is a quotient of the Hopf algebra $F_{(q)}$ over k which as an algebra is generated by symbols a, a^{-1}, and x subject to the relations

$$aa^{-1} = 1, \;\; a^{-1}a = 1, \;\; xa = qax$$

and whose coalgebra structure is determined by

$$\Delta(a) = a \otimes a, \;\; \Delta(x) = x \otimes a + 1 \otimes x.$$

Using the Diamond Lemma (see Section 7.3) one can easily show that $\{a^\ell x^m \mid \ell \in \boldsymbol{Z}, 0 \leq m\}$ is a basis for $F_{(q)}$ as a vector space over k. Using the methods of Section 7.3 one can show that $F_{(q)}$ is a Hopf algebra over the field k with antipode S determined by $S(a) = a^{-1}$ and $S(x) = -xa^{-1}$.

Lemma 15.4.1. *Suppose the field k has characteristic zero and H is a Hopf algebra such that $H \neq H_0$ and is generated by a, x as a Hopf algebra which satisfy (15.13). Then H is isomorphic to the Hopf algebra*

(a) $F_{(q)}$ *or*

q is a primitive n^{th} root of unity for some $n \geq 1$ and H is isomorphic to one of the following Hopf algebras:

(b) $F_{(q,m)} = F_{(q)}/(a^m - 1)$, *where $m \geq 1$ and $n|m$; or*
(c) $F_{(q,n,\alpha)} = F_{(q)}/(x^n - \alpha(a^n - 1))$, *where $n > 1$ and $\alpha \in k$; or*
(d) $F_{(q,n,\alpha,m)} = F_{(q)}/(x^n - \alpha(a^n - 1), a^m - 1)$, *where $n > 1$, $\alpha \in k$, and $n|m$.*

Proof. Note that the antipode S of $F_{(q)}$ satisfies $S(x^\ell) = (-1)^\ell q^{\ell(\ell-1)/2} x^\ell a^{-\ell}$ for all $\ell \geq 1$. We leave the reader with the small exercise of showing that the ideals defining the quotients of parts (b)–(d) are Hopf ideals of $F_{(q)}$. Thus the quotients described in parts (b)–(d) are Hopf algebras over k.

The discussion following the proof of Proposition 9.1.1 is the basis for our proof of the lemma so we start from there. $B = k[\mathrm{G}(H)]$, $\mathrm{G}(H) = (a)$, and $xB \subseteq Bx$ as $xa = qax$. $H = B + Bx + Bx^2 + \cdots$ and $V_n = B +$

$Bx + \cdots + Bx^n$ for $n \geq 0$. Now H is pointed by Corollary 5.1.14 and thus $B = H_0$.

Suppose $V_{m-1} \neq V_m$ for all $m \geq 1$. Then H is a free left B-module with basis $\{1, x, x^2, \ldots\}$. If a has infinite order then $H \simeq F_{(q)}$. Suppose a has finite order m. Now $x = xa^m = q^m a^m x = q^m x$ which implies $q^m = 1$ since $x \neq 0$. Thus q is a primitive n^{th} root of unity for some $n \geq 1$ and $n|m$. In this case $H \simeq F_{(q,m)}$.

Now assume $V_{m-1} = V_m$ for some $m \geq 1$ and let n be the smallest such integer. Then $n > 1$; otherwise $x \in B$ which means $H = B = H_0$. H is a free left B-module with basis $\{1, x, \ldots, x^{n-1}\}$.

Equating the expressions for $\Delta(x^n)$ in (9.7) and Lemma 7.3.1 gives

$$x^n \otimes a^n + 1 \otimes x^n = \Delta(x^n) = \sum_{u=0}^{n} \binom{n}{u}_q x^{n-u} \otimes a^{n-u} x^u$$

and thus

$$\sum_{u=1}^{n-1} \binom{n}{u}_q x^{n-u} \otimes a^{n-u} x^u = 0.$$

Since $\{x^\ell \otimes x^m \,|\, 0 \leq \ell, m < n\}$ is a basis for $H \otimes H$ as a left $B \otimes B$-module, the preceding equation implies $\binom{n}{u}_q = 0$ for all $1 \leq u < n$. By Proposition 7.2.1 this is the case if and only if q is a primitive n^{th} root of unity.

Now $b_0 + b_1 x + \cdots + b_{n-1} x^{n-1} + x^n = 0$ for some $b_0, \ldots, b_{n-1} \in B$ since $V_{n-1} = V_n$. Let $1 \leq \ell < n$. Then (9.3) and (9.4) imply $b_\ell = 0$ or, if not, $b_\ell = \alpha_\ell 1$ for some $\alpha_\ell \in k$ and $a^{n-\ell} = 1$. The latter is not possible since q is a primitive n^{th} root of unity and thus n divides the order of a. Therefore $b_\ell = 0$ for all $1 \leq \ell < n$. We have shown that $x^n = b$ for some $b \in B$ and thus satisfies $\Delta(b) = b \otimes a^n + 1 \otimes b$. Since $\mathrm{G}(H)$ is linearly independent $b = \alpha(a^n - 1)$ for some $\alpha \in k$. If a has finite order m then $n|m$ as argued above. We have shown in the case $V_{n-1} = V_n$ that $H \simeq F_{(q,n,\alpha)}$ or $H \simeq F_{(q,n,\alpha,m)}$. $\square$

Let H be any of the Hopf algebras described in the preceding lemma, identify a and x with their cosets, and let $B = k[\mathrm{G}(H)]$. Using the methods of Section 7.3 one can establish that $H = F_{(q)}$ and $H = F_{(q,m)}$ are free left B-modules with basis $\{1, x, x^2, \ldots\}$ and $H = F_{(q,n,\alpha)}$ and $H = F_{(q,n,\alpha,m)}$ are free with basis $\{1, x, \ldots, x^{n-1}\}$. Furthermore $\mathrm{G}(H) = (a)$ is infinite cyclic when $H = F_{(q)}$ or $H = F_{(q,n,\alpha)}$ and is cyclic or order m when $H = F_{(q,m)}$ or $H = F_{(q,n,\alpha,m)}$. For the record:

Corollary 15.4.2. *The Hopf algebras $F_{(q)}$, $F_{(q,m)}$, and $F_{(q,n,\alpha)}$ of Lemma 15.4.1 are infinite dimensional and* $\mathrm{Dim}(F_{(q,n,\alpha,m)}) = nm$. □

We wish to determine the $z \in H \backslash H_0$ such that $\Delta(z) = z \otimes g + 1 \otimes z$ for some $g \in \mathrm{G}(H)$.

Lemma 15.4.3. *Suppose the field k has characteristic zero, H is a Hopf algebra over k, B is a Hopf subalgebra of $k[\mathrm{G}(H)]$, and $x \in H \backslash B$ satisfies (15.13), where $a \in B$. Assume $r \geq 1$ and $C = B + Bx + \cdots + Bx^r$ is a free left B-module with basis $\{1, x, \ldots, x^r\}$. Suppose $z \in C \backslash B$ satisfies $\Delta(z) = z \otimes g + 1 \otimes z$, where $g \in B$. Then for some $\alpha, \beta \in k$ where $\alpha \neq 0$:*

(a) *$g = a$ and $z = \alpha x + \beta(a - 1)$; or*
(b) *$g = a^n$, q is a primitive n^{th} root of unity, where $1 < n \leq r$, and $z = \alpha x^n + \beta(a^n - 1)$.*

Proof. Let $C(\ell) = Bx^\ell$ for $0 \leq \ell \leq r$ and set $C(\ell) = (0)$ for all $\ell > r$. Then $C = \bigoplus_{\ell=0}^{\infty} C(\ell)$ and is a graded coalgebra by Lemma 7.3.1. Write $z = z(0) \oplus z(1) \oplus \cdots \oplus z(r)$ where $z(\ell) \in C(\ell)$ for all $0 \leq \ell \leq r$. Since $\Delta(z) = z \otimes g + 1 \otimes z$ it follows that $\Delta(z(\ell)) = z(\ell) \otimes g + 1 \otimes z(\ell)$ for all $0 \leq \ell \leq r$ by Exercise 4.4.11.

By assumption $z(\ell) \neq 0$ for some $1 \leq \ell \leq r$. Let ℓ be such an integer. Write $z(\ell) = b_\ell x^\ell$ where $b_\ell \in B$. Then $b_\ell \neq 0$ since $z(\ell) \neq 0$. By Lemma 7.3.1 again

$$b_\ell x^\ell \otimes g + 1 \otimes b_\ell x^\ell = \Delta(z(\ell)) = \sum_{u=0}^{\ell} \binom{\ell}{u}_q \Delta(b_\ell)(x^{\ell-u} \otimes a^{\ell-u} x^u).$$

Since C is a free left B-module with basis $\{1, x, \ldots, x^r\}$ it follows that $C \otimes C$ is a free left $B \otimes B$-module with basis $\{x^u \otimes x^v \mid 0 \leq u, v \leq r\}$. Comparison of coefficients in the expressions for $\Delta(z(\ell))$ above results in the equations

$$1 \otimes b_\ell = \binom{\ell}{\ell}_q \Delta(b_\ell), \tag{15.14}$$

$$b_\ell \otimes g = \binom{\ell}{0}_q \Delta(b_\ell)(1 \otimes a^\ell), \tag{15.15}$$

and

$$0 = \binom{\ell}{u}_q \tag{15.16}$$

for all $1 \leq u < \ell$. From (15.14) we deduce $b_\ell = \epsilon(b_\ell)1$ and thus $a^\ell = g$ follows from (15.15). Since $\mathrm{G}(H)$ is linearly independent $z(0) = \beta(a^\ell - 1)$ for some $\beta \in k$.

Suppose that $\ell > 1$. Then (15.16) holds if and only if q is a primitive ℓ^{th} root of unity by Proposition 7.2.1. This completes our proof. □

At this point we introduce a notion of minimal-pointed Hopf algebra over any field.

Definition 15.4.4. A is *minimal-pointed Hopf algebra over* k is a Hopf algebra H over the field k which is pointed, $H \neq H_0$, and whenever K is a Hopf subalgebra of H such that $K \not\subseteq H_0$ then $K = H$.

Remark 15.4.5. The simple-pointed Hopf algebras of [175] are the minimal-pointed Hopf algebras which are not cocommutative.

Proposition 15.4.6. *Suppose the field k has characteristic zero, H is one of the Hopf algebras of Lemma 15.4.1. Then H is minimal-pointed if and only if:*

(a) $H = F_{(q)}$ *or* $H = F_{(q,m)}$ *and q is not a primitive n^{th} root of unity for all $n > 1$ or*

q is a primitive n^{th} root of unity for some $n > 1$ and for some $\alpha \in k$ and

(b) $H = F_{(q,n,\alpha)}$ *or*
(c) $H = F_{(q,n,\alpha,m)}$, *where $n|m$.*

Furthermore $F_{(q)}$ and $F_{(q,m)}$ contain a minimal-pointed Hopf subalgebra when q is a primitive n^{th} root of unity for some $n > 1$.

Proof. Suppose K is a Hopf subalgebra of H and $K \not\subseteq H_0$. Now K is pointed by Proposition 15.0.6. Since $K \neq K_0$ there is a $z \in K \backslash K_0$ such that $\Delta(z) = z \otimes g + 1 \otimes z$ for some $g \in \mathrm{G}(K)$ by Corollary 4.3.2. Again, we denote cosets of quotients of $F_{(q)}$ by their representatives.

Suppose that if $n > 1$ and q is a primitive n^{th} root of unity. Then $\Delta(x^n) = x^n \otimes a^n + 1 \otimes x^n$ and $xa^n = a^n x$. Thus the subalgebra L of H generated by a^n, a^{-n}, and x^n is a Hopf subalgebra of H. If $H = F_{(q)}$ or $H = F_{(q,m)}$ then $L \not\subseteq k[\mathrm{G}(H)]$ and $L \subseteq B + Bx^n + Bx^{2n} + \cdots \neq H$.

On the other hand, suppose $H = F_{(q)}$ or $H = F_{(q,m)}$ and q is not a primitive n^{th} root of unity for all $n > 1$. Then $g = a$ and $z = \alpha' x + \beta(a - 1)$ for some $\alpha', \beta \in K$, where $\alpha' \neq 0$, by Lemma 15.4.3. Thus $a, x \in K$ which means $K = H$. Likewise if q is a primitive n^{th} root of unity, where $n > 1$, and $H = F_{(q,n,\alpha)}$ or $H = F_{(q,n,\alpha,m)}$, then $g = a$ and z is as just described by Lemma 15.4.3 again. In these two cases $K = H$ also. Hopf subalgebra generated by a^n and x^n is isomorphic to $F_{(1)}$ in the case of $F_{(q)}$ and is isomorphic to $F_{(1,m/n)}$ in the case of $F_{(q,m)}$. $\square$

Let H be a Hopf algebra over k and $g \in \mathrm{G}(H)$. Then $\sigma_g : H \longrightarrow H$ defined by $\sigma_g(h) = ghg^{-1}$ for all $h \in H$ is a Hopf algebra automorphism of H. Suppose $H = \mathcal{U}_q(\mathcal{G})$ is a quantized enveloping algebra. Then σ_{K_i} is diagonalizable and thus $\sigma_{K_i^{-1}} = (\sigma_{K_i})^{-1}$ is also. Corollary 5.1.14 and Exercise 1.2.11 imply that σ_g is diagonalizable for all $g \in \mathrm{G}(H)$. Now suppose H is finite-dimensional and k is algebraically closed. Since $g \in \mathrm{G}(H)$ has finite order σ_g does as well. Therefore σ_g is diagonalizable since k has characteristic zero. The following applies to both the quantized enveloping algebras and finite-dimensional pointed Hopf algebras, the latter when k is algebraically closed.

Theorem 15.4.7. *Suppose the field k has characteristic zero and H is a pointed Hopf algebra over k such that σ_g is diagonalizable for all $g \in \mathrm{G}(H)$. Then:*

(a) *Any Hopf subalgebra K of H such that $K \not\subseteq H_0$ contains a minimal-pointed Hopf subalgebra of H.*
(b) *Suppose K is a minimal-pointed Hopf subalgebra of H. Then K is isomorphic to one of the Hopf algebras described in parts (a)–(c) of Proposition 15.4.6.*

Proof. Suppose K is a Hopf subalgebra of H and $K \not\subseteq H_0$. We have noted that K is pointed by Proposition 15.0.6. Using Corollary 4.3.2 we see that there is a $z \in K\backslash K_0$ such that $\Delta(z) = z \otimes a + 1 \otimes z$ for some $a \in \mathrm{G}(K)$.

Let $V = \{v \in H \,|\, \Delta(v) = v \otimes a + 1 \otimes v\}$. Since $\sigma_a(a) = a$ it follows that $\sigma_a(V) \subseteq V$. Since σ_a is diagonalizable the restriction $\sigma_a|V : V \longrightarrow V$ is diagonalizable and therefore V is the sum of eigenspaces of $\sigma_a|V$. Thus z is the sum of eigenvectors for σ_a. One of these $x \notin H_0$ since $z \notin H_0$. Now $axa^{-1} = \sigma_a(x) = \omega x$ for some $\omega \in k$ which is not zero since σ_a is one-one. Therefore $xa = qax$, where $q = \omega^{-1}$. The Hopf subalgebra L of H generated as an algebra by a, a^{-1}, and x satisfies $L \not\subseteq H_0 \cap L = L_0$ and $L \subseteq K$. Thus L satisfies the hypothesis of Lemma 15.4.1 and Proposition 15.4.6 applies. $\square$

Corollary 15.4.8. *Suppose the field k is algebraically closed has characteristic zero. Then a finite-dimensional Hopf algebra over k is minimal-pointed if and only if $H \simeq F_{(q,n,\alpha,m)}$ where $q \in k$ is a primitive n^{th} root of unity, $n > 1$, $\alpha \in k$, and $n|m$.* □

Exercises

Throughout these exercises H is a finite-dimensional Hopf algebra over k which has characteristic zero.

Exercise 15.4.1. Describe the structure of H^* where H is the Hopf algebra of Corollary 15.4.8.

Exercise 15.4.2. Determine all H of Corollary 15.4.8 such that there is a Hopf algebra projection $H \longrightarrow H_0$.

Exercise 15.4.3. Determine all H such that H and H^* are minimal-pointed.

15.5 Pointed Hopf algebras, biproducts, and Nichols algebras

Suppose A is a bialgebra over k whose coradical is a Hopf subalgebra of A. Pointed Hopf algebras fall into this category. Then A and $\mathrm{gr}(A)$ are Hopf algebras over k by Lemma 7.6.2. The latter is a coradically graded Hopf algebra over k by Proposition 7.9.4. Theorem 11.7.2 applies to $\mathrm{gr}(A) \overset{\jmath}{\underset{\pi}{\leftrightarrows}} H$, where $\jmath$ is the inclusion of $H = A_0$ into $\mathrm{gr}(A)$ and π is the projection onto H. Now by part (b) of the same $\mathrm{gr}(A) \simeq B{\times}H$ and the pointed irreducible Hopf algebra B of ${}_H^H\mathcal{YD}$ has a coradically graded Hopf algebra B structure $B = \bigoplus_{n=0}^{\infty} B(n)$. Such Hopf algebras in ${}_H^H\mathcal{YD}$ play an important role in classification of Hopf algebras over k with coradical H.

Let $V = \mathrm{P}(B) = B(1)$. In this section we study B when it is generated as an algebra by the Yetter-Drinfel'd submodule V of B. Our goal is to show that B is the result of a construction characterized by a universal mapping property. To this end we let H be any bialgebra over k and let V be an object of ${}_H^H\mathcal{YD}$. We begin our construction with the tensor algebra.

Let $(\imath, T(V))$ be the tensor algebra on the vector space V over k, where $T(V) = \bigoplus_{n=0}^{\infty} V^{\otimes n}$ and $\imath$ is the inclusion. Regard $V^{\otimes 0} = k$ as an object of ${}_H^H\mathcal{YD}$ with the trivial actions, that is according to Example 11.6.1, and for $n \geq 1$ regard $V^{\otimes n} = V\otimes\cdots\otimes V$ (n-factors) as an object of ${}_H^H\mathcal{YD}$ as in Exercise 11.6.3 and regard $T(V)$ as an object of ${}_H^H\mathcal{YD}$ with the direct sum structure. See Exercise 11.6.2. With these structures the k-algebra structure maps $m : T(V)\otimes T(V) \longrightarrow T(V)$ and $\eta : k \longrightarrow T(V)$ are morphisms; indeed $(T(V), m, \eta)$ is an H-module algebra and an H-comodule algebra. Therefore $(T(V), m, \eta)$ is an algebra in ${}_H^H\mathcal{YD}$.

Theorem 15.5.1. *Let H be a bialgebra over the field k and let V be an object of ${}_H^H\mathcal{YD}$. Then the pair $(\imath, T(V))$ satisfies the following:*

(a) *$T(V)$ is an algebra in ${}_H^H\mathcal{YD}$ and $\imath : V \longrightarrow T(V)$ is a morphism.*
(b) *If A is an algebra in ${}_H^H\mathcal{YD}$ and $f : V \longrightarrow A$ is a morphism there is a morphism $F : T(V) \longrightarrow A$ of algebras in ${}_H^H\mathcal{YD}$ determined by $F\circ\imath = f$.*

Proof. We need only show that part (b) holds. Suppose A is an algebra in ${}_H^H\mathcal{YD}$ and $f : V \longrightarrow A$ is a morphism. Since $(\imath, T(V))$ is also the tensor algebra in k-Vec there is a map of k-algebras $F : T(V) \longrightarrow A$ determined by $F\circ\imath = f$; see the beginning of Section 5.3. As $F(1) = 1$ and $F(v_1\otimes\cdots\otimes v_n) = f(v_1)\cdots f(v_n)$ for all $n \geq 1$ and $v_1, \ldots, v_n \in V$, an easy calculation shows that F is a map of Yetter-Drinfel'd modules. Since morphisms of algebras in ${}_H^H\mathcal{YD}$ are morphisms of k-algebras, uniqueness follows. □

Definition 15.5.2. Let H be a bialgebra over the field k and let V be an object of ${}_H^H\mathcal{YD}$. A *tensor algebra on the object V in ${}_H^H\mathcal{YD}$* is any pair $(\imath, T(V))$ which satisfies the conclusion of Theorem 15.5.1.

We give $T(V)$ a coalgebra structure which makes it a Hopf algebra in ${}_H^H\mathcal{YD}$. Let $A = T(V)\underline{\otimes}T(V)$. Then $\delta : V \longrightarrow A$ defined by $\delta(v) = \imath(v)\underline{\otimes}1 + 1\underline{\otimes}\imath(v)$ for all $v \in V$ is a morphism. By part (b) of Theorem 15.5.1 there is an algebra morphism $\Delta : T(V) \longrightarrow A$ determined by $\Delta\circ\imath = \delta$. Since $0 : V \longrightarrow k$ is a morphism by the same there is an algebra morphism $\epsilon : T(V) \longrightarrow k$ determined by $\epsilon\circ\imath = 0$. Using Theorem 15.5.1 one can show that $(T(V), \Delta, \epsilon)$ is a coalgebra in ${}_H^H\mathcal{YD}$. Therefore $T(V)$ with its algebra structure and this coalgebra structure is a bialgebra in ${}_H^H\mathcal{YD}$. Observe that $T(V) = \bigoplus_{n=0}^{\infty} V^{\otimes n}$ gives $T(V)$ the structure of a pointed irreducible graded bialgebra by Exercise 15.5.5. By Exercise 15.5.4 it follows that $T(V)$ is a Hopf algebra in ${}_H^H\mathcal{YD}$.

The tensor algebra with this coalgebra structure, which we denote $T_{\text{pi}}(V)$, will be useful to us in the sequel. The pair $(\imath, T_{\text{pi}}(V))$ satisfies a universal mapping property.

Theorem 15.5.3. *Let H be a bialgebra over the field k and let V be an object of ${}_H^H\mathcal{YD}$. Then the pair $(\imath, T_{\rm pi}(V))$ satisfies the following:*

(a) *$T_{\rm pi}(V)$ is a pointed irreducible Hopf algebra in ${}_H^H\mathcal{YD}$, $\imath : V \longrightarrow T_{\rm pi}(V)$ is a morphism such that $\imath(V) \subseteq \mathrm{P}(T_{\rm pi}(V))$.*

(b) *If A is a pointed irreducible Hopf algebra in ${}_H^H\mathcal{YD}$ and $f : V \longrightarrow A$ is a morphism such that $f(V) \subseteq \mathrm{P}(A)$ then there exists a morphism $F : T_{\rm pi}(V) \longrightarrow A$ of Hopf algebras in ${}_H^H\mathcal{YD}$ determined by $F\circ\imath = f$.*

Proof. We need only establish part (b). Suppose A is any Hopf algebra in ${}_H^H\mathcal{YD}$ and that $f : V \longrightarrow A$ is a morphism. By part (b) of Theorem 15.5.1 there is a morphism of algebras $F : T_{\rm pi}(V) \longrightarrow A$ determined by $F\circ\imath = f$. To complete the proof we need only show that F is a coalgebra map.

Consider the morphism $\mathsf{f} : V \longrightarrow A\underline{\otimes}A$ defined by $\mathsf{f} = \Delta_A\circ f$. Now $F\underline{\otimes}F : T_{\rm pi}(V)\underline{\otimes}T_{\rm pi}(V) \longrightarrow A\underline{\otimes}A$ is an algebra morphism by Exercise 15.5.2. By part (b) of Theorem 15.5.1 again the algebra maps $\Delta_A\circ F, (F\underline{\otimes}F)\circ\Delta_{T_{\rm pi}(V)} : T_{\rm pi}(V) \longrightarrow A\underline{\otimes}A$ are the same since $(\Delta_A\circ F)\circ\imath = \mathsf{f} = ((F\underline{\otimes}F)\circ\Delta_{T_{\rm pi}(V)})\circ\imath$. The same argument where f is $0 : V \longrightarrow k$ shows that $\epsilon_A\circ F, \epsilon_{T_{\rm pi}(V)} : T_{\rm pi}(V) \longrightarrow k$ are the same. □

Definition 15.5.4. Let H be a bialgebra over the field k and let V be an object of ${}_H^H\mathcal{YD}$. A *free pointed irreducible Hopf algebra on the object* V is any pair $(\imath, T_{\rm pi}(V))$ which satisfies the conclusion of Theorem 15.5.3.

Let $(\imath, T_{\rm pi}(V))$ be as in the preceding definition. Then $\imath$ is one-one since it is in one case. For convenience we will assume that $\imath$ is the inclusion and thus regard V as a Yetter-Drinfel'd submodule of $T_{\rm pi}(V)$.

We will show that there is a largest quotient of $B(V) = T_{\rm pi}(V)/I$ of $T_{\rm pi}(V)$ which is a bialgebra in ${}_H^H\mathcal{YD}$ and satisfies $\jmath(V) = \mathrm{P}(B(V))$, where $\jmath = \pi\circ\imath$ and $\pi : T_{\rm pi}(V) \longrightarrow T_{\rm pi}(V)/I$ is the projection. Let L be the sum of all graded coideals J of the k-coalgebra $T_{\rm pi}(V)$ such that $J\cap V = (0)$. Then I is a coideal of $T_{\rm pi}(V)$ which satisfies $I\cap V = (0)$ as well. The Yetter-Drinfel'd submodule I' of $T_{\rm pi}(V)$ which I generates is graded and satisfies $I'\cap V = (0)$ by Exercise 15.5.7 and I' is a coideal of $T_{\rm pi}(V)$ by Exercise 15.5.8. Therefore $I = I'$ which means that I is a Yetter-Drinfel'd submodule of $T_{\rm pi}(V)$. The ideal I' which I generates is graded, satisfies $I'\cap V = (0)$, and by Exercise 15.5.6 is a coideal of $T_{\rm pi}(V)$. Therefore $I' = I$ which means I is an ideal of $T_{\rm pi}(V)$ as well. Thus $B(V) = T_{\rm pi}(V)/I$ with its quotient structures is a bialgebra in ${}_H^H\mathcal{YD}$ and the restriction of the projection $\pi : V \longrightarrow T_{\rm pi}(V)/I = B(V)$ is a morphism. Note that $B(V)$

has the structure of a pointed irreducible graded bialgebra in ${}_H^H\mathcal{YD}$, where $B(V)(n) = V^{\otimes n}/(I \cap V^{\otimes n})$ for all $n \geq 0$. Since $B(V)(1) = V/(I \cap V) = V$ we may assume that π is the inclusion.

Observe that V generates $B(V)$ as an algebra; indeed $B(V)(n) = V^n$ for all $n \geq 0$.

We claim that $\mathrm{P}(B(V)) = V$. To see this, let $\mathrm{p} : T_{\mathrm{pi}}(V) \longrightarrow T_{\mathrm{pi}}(V)/I$ be the projection and $b \in \mathrm{P}(B(V))$. By Exercise 4.4.11 all of the components of $\mathrm{p}(x)$ are primitive. Let $n \geq 2$. Since p is an onto map of graded vector spaces, $b(n) = \mathrm{p}(a)$ for some $a \in V^{\otimes n}$. Since p is a coalgebra map and $\mathrm{p}(a)$ is primitive, $(\mathrm{p} \otimes \mathrm{p})(\Delta(a) - (1 \otimes a + a \otimes 1)) = 0$. Thus $\Delta(a) - (1 \otimes a + a \otimes 1) \in \mathrm{Ker}(\mathrm{p} \otimes \mathrm{p}) = I \otimes T_{\mathrm{pi}}(V) + T_{\mathrm{pi}}(V) \otimes I$. Therefore $I' = ka + I$ is a graded coideal of $T_{\mathrm{pi}}(V)$ containing I and $I' \cap V = (0)$. Thus $I' = I$ which means $a \in I$. Therefore $b = b(0) + b(1) = b(1) \in V$ as $b(0) = 0$.

The Yetter-Drinfel'd submodule V has a special role in the structure of $B(V)$.

Definition 15.5.5. Let H be a bialgebra over the field k and let A be an algebra in ${}_H^H\mathcal{YD}$. A Yetter-Drinfel'd submodule V of A *generates a graded subalgebra of* A if $\sum_{n=0}^{\infty} V^n$ is direct.

If V generates a graded subalgebra of A then $B = \sum_{n=0}^{\infty} V^n$ is a k-algebra which is graded. Since the powers V^n are Yetter-Drinfel'd submodules of A it follows that B is a graded subalgebra of ${}_H^H\mathcal{YD}$. See Exercise 15.5.1.

For any bialgebra A in ${}_H^H\mathcal{YD}$ it is always the case that $\mathrm{P}(A)$ is a Yetter-Drinfel'd submodule of A. See Exercise 15.5.10. We have noted that a pointed irreducible bialgebra in ${}_H^H\mathcal{YD}$ is a Hopf algebra in ${}_H^H\mathcal{YD}$.

Theorem 15.5.6. *Let H be a bialgebra over the field k and let V be an object of ${}_H^H\mathcal{YD}$. The pair $(\jmath, B(V))$ satisfies the following:*

(a) *$B(V)$ is a pointed irreducible Hopf algebra in ${}_H^H\mathcal{YD}$, $\jmath : V \longrightarrow B(V)$ is a morphism such that $\jmath(V) \subseteq \mathrm{P}(B(V))$, and $\mathrm{P}(B(V))$ generates a graded subalgebra of $B(V)$.*
(b) *If A is a pointed irreducible Hopf algebra over k, $f : V \longrightarrow A$ is a morphism such that $f(V) \subseteq \mathrm{P}(A)$, and $\mathrm{P}(A)$ generates a graded subalgebra of A there is a morphism of Hopf algebras $F : B(V) \longrightarrow A$ determined by $F \circ \jmath = f$.*

Proof. We need only establish part (b). Let $(\imath, T_{\mathrm{pi}}(V))$ satisfy the conclusion of Theorem 15.5.3. By this theorem there is morphism of Hopf algebras $\mathsf{f} : T_{\rm pi}(V) \longrightarrow A$ in ${}^H_H\mathcal{YD}$ determined by $\mathsf{f}\circ\imath = f$.

Write $B(V) = T_{\rm pi}(V)/I$ as above. Then $I = \bigoplus_{n=2}^{\infty}(I\cap V^{\otimes n})$. Thus

$$\mathsf{f}(I) = \mathsf{f}(\bigoplus_{n=2}^{\infty}(I\cap V^{\otimes n})) = \sum_{n=2}^{\infty}\mathsf{f}(I\cap V^{\otimes n}) \subseteq \sum_{n=2}^{\infty} f(V)^n \subseteq \sum_{n=2}^{\infty}\mathrm{P}(A)^n.$$

Since the latter sum is direct, $\mathsf{f}(I) \subseteq \bigoplus_{n=2}^{\infty}\mathrm{P}(A)^n$. Therefore $\mathsf{f}(I)\cap\mathrm{P}(A) = (0)$. Now $\mathsf{f}(I)$ is a coideal of A since f is a k-coalgebra map and I is a coideal of $T_{\rm pi}(V)$. Therefore $\mathsf{f}(I) = (0)$ by Corollary 4.3.4. See Exercise 4.4.8 also.

Since $\mathsf{f}(I) = (0)$ there is a morphism of bialgebras $F : T_{\rm pi}(V)/I \longrightarrow A$ determined by $F\circ\pi = \mathsf{f}$, where $\pi : T_{\rm pi}(V) \longrightarrow A$ is the projection. Thus $F : B(V) \longrightarrow A$ satisfies $F\circ\jmath = f$. Uniqueness is left as an exercise for the reader. □

Definition 15.5.7. Let H be a bialgebra over the field k and let V be an object of ${}^H_H\mathcal{YD}$. A *Nichols algebra on the Yetter-Drinfel'd module* V is a pair $(\jmath, B(V))$ which satisfies the conclusion of Theorem 15.5.6.

In practice $B(V)$ is called a Nichols algebra and $\jmath$ is assumed to be an inclusion. We have seen in theory it is easy to describe Nichols algebras on objects in ${}^H_H\mathcal{YD}$. In practice it is difficult to find generators and relations for them.

Corollary 15.5.8. *Let H be a bialgebra over the field k, let A be a bialgebra in ${}^H_H\mathcal{YD}$, let $V = \mathrm{P}(A)$ and let $(\jmath, B(V))$ be a Nichols algebra on the Yetter-Drinfel'd submodule V of A. Suppose V generates a graded subalgebra of A and let $B = \bigoplus_{n=0}^{\infty}V^n$. Then there is an isomorphism of bialgebras $F : B(V) \longrightarrow B$ determined by $F\circ\jmath = \mathrm{I}_V$.*

Proof. We apply Theorem 15.5.6 to B and the inclusion $f : V \longrightarrow B$. Note that $F(\jmath(v)) = v$ for all $v \in V = \mathrm{P}(A)$. Since F is an algebra map and $\mathrm{Im}(F)$ generates B as an algebra, F is onto. Now $\jmath(V) = \mathrm{P}(B(V))$; see Exercise 15.5.13. Thus $F|\mathrm{P}(B(V))$ is one-one. Therefore the intersection of the coideal $\mathrm{Ker}(F)$ and $\mathrm{P}(B(V))$ is (0). By Corollary 4.3.4 it follows that F is one-one. □

We end this section by working through two examples. The details should help motivate the generalizations of the quantized enveloping algebras of Section 15.6. For the remainder of this section we assume that the characteristic of k is 0.

The examples will be two finite-dimensional pointed Hopf algebras A over k. We will compute $\mathrm{gr}(A)$ and then the bialgebra B in ${}^H_H\mathcal{YD}$ of the biproduct decomposition $\mathrm{gr}(A) = B{\times}H$, where $H = A_0$. In both cases B will be a Nichols algebra.

Our first example is the Hopf algebra $A = H_{(q,n,\alpha,m)}$ of Lemma 15.4.1. As a k-algebra A is generated by a and x subject to the relations $a^m = 1$, $x^n = \alpha(a^n - 1)$, and $xa = qax$. The coalgebra structure of A is determined by $\Delta(a) = a{\otimes}a$ and $\Delta(x) = x{\otimes}a + 1{\otimes}x$. $\mathrm{Dim}(A) = mn$ and $A_0 = k[(a)]$ is the group algebra of the cyclic group (a) of order m.

Set $H = A_0$ and identify H with $\mathrm{gr}(A)(0)$ via the inclusion map. Since $a \in A_0$ and $x \in A_1$ we may define $\mathsf{a} \in \mathrm{gr}(A)(0)$ and $\mathsf{x} \in \mathrm{gr}(A)(1)$ by $\mathsf{a} = a$ and $\mathsf{x} = x + A_0$. Note that $\mathsf{x} \neq 0$ since $x \in A_1 \backslash A_0$. The relations $\mathsf{a}^m = 1$, $\mathsf{x}^n = 0$, and $\mathsf{xa} = q\mathsf{ax}$ hold. Let $\mathcal{A}$ be the subalgebra of $\mathrm{gr}(A)$ generated by H and x. We deduce from the proof of Lemma 15.4.1 that $\mathcal{A}$ is a free left H-module with basis $\{1, \mathsf{x}, \ldots, \mathsf{x}^{n-1}\}$. Thus $\mathrm{Dim}(\mathcal{A}) = mn = \mathrm{Dim}(A) = \mathrm{Dim}(\mathrm{gr}(A))$ which means $\mathcal{A} = \mathrm{gr}(A)$. In particular $\mathrm{gr}(A) = H_{(q,n,0,m)}$. As $H\mathsf{x}^r \subseteq \mathrm{gr}(A)(r)$ for all $0 \leq r < n$ it follows that $H\mathsf{x}^r = \mathrm{gr}(A)(r)$ for all $0 \leq r < n$. The projection $\pi : \mathrm{gr}(A) \longrightarrow H$ is therefore given by $\pi(\mathsf{a}^\ell \mathsf{x}^r) = \delta_{r,0}\mathsf{a}^\ell$ for all $0 \leq \ell < m$ and $0 \leq r < n$.

Our description of A was very convenient for analysis of $\mathrm{gr}(A)$ by techniques involving left Hopf modules. We now determine the bialgebra B in ${}^H_H\mathcal{YD}$ of part (b) of Theorem 15.5.6. To this end it will be useful to describe A in a slightly different way. Let $c = a^{-1}$ and $v = a^{-1}x$. Observe that $v^r = q^{-(r-1)r/2}a^{-r}x^r$ for all $0 \leq r < n$. As a k-algebra A is generated by c and v subject to the relations

$$c^m = 1, \quad v^n = \alpha'(c^n - 1), \quad \text{and} \quad vc = q^{-1}cv,$$

where $\alpha' = -q^{(n-1)n/2}\alpha$, and its coalgebra structure is determined by

$$\Delta(c) = c{\otimes}c \quad \text{and} \quad \Delta(v) = v{\otimes}1 + c{\otimes}v.$$

Define $\mathsf{c} \in \mathrm{gr}(A)(0)$ and $\mathsf{v} \in \mathrm{gr}(A)(1)$ by $\mathsf{c} = c$ and $\mathsf{v} = v + A_0$. Then $\mathrm{gr}(A)$ as a k-algebra is generated by c and v subject to the relations

$$\mathsf{c}^m = 1, \quad \mathsf{v}^n = 0, \quad \text{and} \quad \mathsf{vc} = q^{-1}\mathsf{cv},$$

and its coalgebra structure is determined by

$$\Delta(\mathsf{c}) = \mathsf{c}{\otimes}\mathsf{c} \quad \text{and} \quad \Delta(\mathsf{v}) = \mathsf{v}{\otimes}1 + \mathsf{c}{\otimes}\mathsf{v}.$$

Its grading is given by $\mathrm{gr}(A)(r) = H\mathsf{v}^r$ for all $0 \leq r < n$ and the projection $\pi : \mathrm{gr}(A) \longrightarrow H$ is given by $\pi(\mathsf{c}^\ell \mathsf{v}^r) = \delta_{r,0}\mathsf{c}^\ell$ for all $0 \leq \ell < m$ and $0 \leq r < n$.

The calculation $(\mathrm{I}_{\mathrm{gr}(A)}\otimes\pi)(\Delta(\mathsf{v})) = \mathsf{v}\otimes\pi(1) + \mathsf{c}\otimes\pi(\mathsf{v}) = \mathsf{v}\otimes 1$ shows that $\mathsf{v} \in B$. Therefore $\mathsf{v} \in \mathrm{gr}(A)(1)\cap B = B(1)$. Since $\mathsf{v}^r \in B(r)$ for all $0 \leq r < n$, $\mathsf{v}^{n-1} \neq 0$, $k1 + k\mathsf{v} + \cdots + k\mathsf{v}^{n-1}$ is direct, and $\mathrm{Dim}(B) = n$, we have $B = B(0)\bigoplus\cdots\bigoplus B(n-1)$ and $B(r) = k\mathsf{v}^r$ for all $0 \leq r < n$. In particular B is a Nichols algebra on $V = B(1) = k\mathsf{v}$. The calculations

$$\rho(\mathsf{v}) = (\pi\otimes\mathrm{I}_{\mathrm{gr}(A)})(\Delta(\mathsf{v})) = \pi(\mathsf{v})\otimes 1 + \pi(\mathsf{c})\otimes\mathsf{v} = \mathsf{c}\otimes\mathsf{v}$$

and

$$\mathsf{c}{\cdot}\mathsf{v} = \mathsf{c}\mathsf{v}\mathsf{c}^{-1} = q\mathsf{v}$$

determine V as an object in ${}^H_H\mathcal{YD}$.

We relate $T_{\mathrm{pi}}(V)$, $B(V)$, $\mathrm{gr}(A) \simeq B(V)\times H$, and A. Let $(\imath, T_{\mathrm{pi}}(V))$ be a free pointed irreducible Hopf algebra on the object V, where $\imath$ is the inclusion. Then $\mathsf{v}^n \in \mathrm{P}(T_{\mathrm{pi}}(V))$ and

$$B(V) = T_{\mathrm{pi}}(V)/(\mathsf{v}^n).$$

See Exercise 15.5.15. The reader is left with the instructive exercise of showing that

$$\mathrm{gr}(A) \simeq (T_{\mathrm{pi}}(V)/(\mathsf{v}^n))\times H \simeq (T_{\mathrm{pi}}(V)\times H)/(\mathsf{v}^n)$$

and

$$A \simeq (T_{\mathrm{pi}}(V)\times H)/(\mathsf{v}^n - \alpha'(\mathsf{c}^n - 1))$$

where in the two quotients of $T_{\mathrm{pi}}(V)\times H$ the elements v and $\mathsf{v}\times 1$ are identified and c and $1\times\mathsf{c}$ are identified. Note that the ideals of $T_{\mathrm{pi}}(V)\times H$ are bi-ideals since $\mathsf{v}^n \in \mathrm{P}_{\mathsf{c}^n,1}(T_{\mathrm{pi}}(V)\times H)$.

Our second example is $A = U_{\mathsf{q}}(sl_2)' = U_{n,q}$ of Section 7.3. As a k-algebra A is generated by symbols a, x, and y subject to the relations $a^n = 1$, $x^n = 0 = y^n$, $xa = qax$, $ya = q^{-1}ay$, and $yx - q^{-1}xy = a^2 - 1$. Its coalgebra structure is determined by $\Delta(a) = a\otimes a$, $\Delta(a) = x\otimes a + 1\otimes x$, and $\Delta(y) = y\otimes a + 1\otimes y$. $\mathrm{Dim}(A) = n^3$, $A_0 = k[(a)]$ is the group algebra generated by the cyclic group (a) of order n, and A has linear basis $\{a^r x^s y^t \,|\, 0 \leq r, s, t < n\}$.

Set $H = A_0$ and identify H with $\mathrm{gr}(A)(0)$ using the inclusion map. Since $a \in H$ and $x, y \in A_1$ we can define $\mathsf{a} \in \mathrm{gr}(A)(0)$ and $\mathsf{x}, \mathsf{y} \in \mathrm{gr}(A)(1)$ by $\mathsf{a} = a$, $\mathsf{x} = x + A_0$, and $\mathsf{y} = y + A_0$. Since $x, y \in A_1\backslash A_0$ it follows that $\mathsf{x}, \mathsf{y} \neq 0$. The relations $\mathsf{a}^n = 1$, $\mathsf{x}^n = 0 = \mathsf{y}^n$, $\mathsf{x}\mathsf{a} = q\mathsf{a}\mathsf{x}$, $\mathsf{y}\mathsf{a} = q^{-1}\mathsf{a}\mathsf{y}$, and $\mathsf{y}\mathsf{x} = q^{-1}\mathsf{x}\mathsf{y}$ hold in $\mathrm{gr}(A)$ as do the equations $\Delta(\mathsf{a}) = \mathsf{a}\otimes\mathsf{a}$, $\Delta(\mathsf{x}) = \mathsf{x}\otimes\mathsf{a} + 1\otimes\mathsf{x}$, and $\Delta(\mathsf{y}) = \mathsf{y}\otimes\mathsf{a} + 1\otimes\mathsf{y}$.

Using the argument to establish Lemma 15.4.1 we see that the Hopf subalgebra $\mathcal{H}$ of $\mathrm{gr}(A)$ generated by H and x is a free left H-module with basis $\{1, \mathsf{x}, \ldots, \mathsf{x}^{n-1}\}$ and likewise the Hopf subalgebra generated by H and y is a free left H-module with basis $\{1, \mathsf{y}, \ldots, \mathsf{y}^{n-1}\}$. Now $\Delta(\mathsf{y}^r) = \sum_{\ell=0}^{r} \binom{r}{\ell}_{q^{-1}} \mathsf{y}^{r-\ell} \otimes \mathsf{a}^{r-\ell}\mathsf{y}^{\ell}$ for all $0 \leq r < n$ by Lemma 7.3.1. Thus $\Delta(\mathsf{y}^r) = \mathsf{y}^r \otimes \mathsf{a}^r + 1 \otimes \mathsf{y}^r$ is not possible for $1 < r < n$ by Proposition 7.2.1. Consequently by the calculations preceding Corollary 9.1.2 the Hopf subalgebra $\mathcal{A}$ of $\mathrm{gr}(A)$ generated by $\mathcal{H}$ and y is a free left $\mathcal{H}$-module with basis $\{1, \mathsf{y}, \ldots, \mathsf{y}^{n-1}\}$. Therefore $\mathrm{Dim}(\mathcal{A}) = n^3 = \mathrm{Dim}(A) = \mathrm{Dim}(\mathrm{gr}(A))$ which means $\mathcal{A} = \mathrm{gr}(A)$.

We have shown that $\mathrm{gr}(A)$ is generated as a k-algebra by a, x, and y. In particular $\{\mathsf{a}^r\mathsf{x}^s\mathsf{y}^t \,|\, 0 \leq r, s, t < n\}$ is a basis for $\mathrm{gr}(A)$ over k. Since $\mathsf{a}^r\mathsf{x}^s\mathsf{y}^t \in \mathrm{gr}(A)(s+t)$ for all $0 \leq r, s, t < n$ it follows that

$$\mathrm{gr}(A) = \bigoplus_{m=0}^{2n-2} \mathrm{gr}(A)(m)$$

and

$$\mathrm{gr}(A)(m) \text{ has basis } \{\mathsf{a}^r\mathsf{x}^s\mathsf{y}^t \,|\, 0 \leq r, s, t < n, s+t = m\} \text{ over } k$$

for all $0 \leq m \leq 2n-2$. Observe that

$$\mathrm{Dim}(\mathrm{gr}(A)(m)) = \begin{cases} (m+1)n & : 0 \leq m < n; \\ (2n-m-1)n : n \leq m \leq 2n-2. \end{cases} \tag{15.17}$$

To compute B we will use a slightly different description of A. Let $c = a^{-1}, u = ya^{-1}$ and $v = xa^{-1}$. Then as a k-algebra A is generated by c, u, and v subject to the relations

$$c^n = 1, \;\; u^n = 0 = v^n, \;\; uc = qcu, \;\; vc = q^{-1}cv, \;\text{ and }\; vu - q^{-1}uv = c^2 - 1.$$

The coalgebra structure of A is determined by

$$\Delta(c) = c \otimes c, \;\; \Delta(u) = u \otimes 1 + c \otimes u, \;\text{ and }\; \Delta(v) = v \otimes 1 + c \otimes v.$$

Let $\mathsf{c} = c$, $\mathsf{u} = u + A_0$, and $\mathsf{v} = v + A_0$. As a k-algebra $\mathrm{gr}(A)$ is generated by $\mathsf{c} \in \mathrm{gr}(A)(0)$, $\mathsf{u}, \mathsf{v} \in \mathrm{gr}(A)(1)$ subject to the relations

$$\mathsf{c}^n = 1, \;\; \mathsf{u}^n = 0 = \mathsf{v}^n, \;\; \mathsf{uc} = q\mathsf{cu}, \;\; \mathsf{vc} = q^{-1}\mathsf{cv}, \;\text{ and }\; \mathsf{vu} - q^{-1}\mathsf{uv} = 0.$$

Furthermore

$$\mathrm{gr}(A)(m) \text{ has basis } \{\mathsf{c}^r\mathsf{u}^s\mathsf{v}^t \,|\, 0 \leq r, s, t < n, s+t = m\} \text{ over } k$$

for all $0 \leq m \leq 2n-2$ and the projection $\pi : \mathrm{gr}(A) \longrightarrow H$ is given by $\pi(\mathsf{c}^r\mathsf{u}^s\mathsf{v}^t) = \delta_{s,0}\delta_{t,0}\mathsf{c}^r$ for all $0 \leq r, s, t < n$.

We next observe that $\mathsf{u}, \mathsf{v} \in B$ since $(\mathrm{I}_{\mathrm{gr}(A)} \otimes \pi)(\Delta(\mathsf{x})) = \mathsf{x} \otimes 1$ for $\mathsf{x} = \mathsf{u}, \mathsf{v}$. For $0 \leq m \leq 2n-2$ let V_m be the span of the monomials $\mathsf{u}^s\mathsf{v}^t$, where $0 \leq s, t < n$ and $s + t = m$. Then $\mathrm{Dim}(V_m) = \mathrm{Dim}(\mathrm{gr}(A)(m))/n$ and $V_m \subseteq \mathrm{gr}(A)(m)$ for all $0 \leq m \leq 2n-2$. Since $\sum_{m=0}^{2n-2} V_m$ is direct,

$$\mathrm{Dim}(\sum_{m=0}^{2n-2} V_m) = \sum_{m=0}^{2n-2} \mathrm{Dim}(\mathrm{gr}(A)(m))/n = \mathrm{Dim}(\mathrm{gr}(A))/n = \mathrm{Dim}(B)$$

which means $\sum_{m=0}^{2n-2} V_m = B$ and $B(m) = V_m$ for all $0 \leq m \leq 2n-2$. Therefore B is a Nichols algebra in $V = V_1$. By virtue of (15.17)

$$\mathrm{Dim}(B(V)(m)) = \begin{cases} m+1 & : 0 \leq m < n; \\ 2n-m-1 & : n \leq m \leq 2n-2. \end{cases}$$

Since V has basis $\{\mathsf{u}, \mathsf{v}\}$ the calculations

$$\rho(\mathsf{u}) = (\pi \otimes \mathrm{I}_{\mathrm{gr}(A)})(\Delta(\mathsf{u})) = \mathsf{c} \otimes \mathsf{u}, \quad \rho(\mathsf{v}) = (\pi \otimes \mathrm{I}_{\mathrm{gr}(A)})(\Delta(\mathsf{v})) = \mathsf{c} \otimes \mathsf{v}$$

and

$$\mathsf{c} \cdot \mathsf{u} = \mathsf{c}\mathsf{u}\mathsf{c}^{-1} = q^{-1}\mathsf{u}, \quad \mathsf{c} \cdot \mathsf{v} = \mathsf{c}\mathsf{v}\mathsf{c}^{-1} = q\mathsf{v}$$

determine V as an object of ${}_H^H\mathcal{YD}$.

An important point to emphasize. The relation $\mathsf{v}\mathsf{u} - q^{-1}\mathsf{u}\mathsf{v} = \mathsf{c}^2 - 1$ can be expressed in terms of the braided commutator

$$\mathrm{ad}_{\mathsf{c}}\mathsf{v}(\mathsf{u}) = \mathsf{v}\mathsf{u} - (\mathsf{v}_{(-1)} \cdot \mathsf{u})\mathsf{v}_{(0)} = \mathsf{v}\mathsf{u} - (\mathsf{c} \cdot \mathsf{u})\mathsf{v} = \mathsf{v}\mathsf{u} - q^{-1}\mathsf{u}\mathsf{v}.$$

Now we put all of the pieces of our construction together. Let $(\imath, T_{\mathrm{pi}}(V))$ be a free pointed irreducible Hopf algebra on the object V, where $\imath$ is the inclusion. Then u^n, v^n, $\mathrm{ad}_{\mathsf{c}}\mathsf{v}(\mathsf{u}) \in \mathrm{P}(T_{\mathrm{pi}}(V))$ and

$$B(V) = T_{\mathrm{pi}}(V)/(\mathsf{u}^n, \mathsf{v}^n, \mathrm{ad}_{\mathsf{c}}\mathsf{v}(\mathsf{u})).$$

Since

$$\mathrm{gr}(A) = B(V) \times H$$

we have

$$\mathrm{gr}(A) = (T_{\mathrm{pi}}(V)/(\mathsf{u}^n, \mathsf{v}^n, \mathrm{ad}_{\mathsf{c}}\mathsf{v}(\mathsf{u}))) \times H = (T_{\mathrm{pi}}(V) \times H)/(\mathsf{u}^n, \mathsf{v}^n, \mathrm{ad}_{\mathsf{c}}\mathsf{v}(\mathsf{u}))$$

and

$$A = (T_{\mathrm{pi}}(V) \times H)/(\mathsf{u}^n, \mathsf{v}^n, \mathrm{ad}_{\mathsf{c}}\mathsf{v}(\mathsf{u}) - (\mathsf{c}^2 - 1)),$$

where in the quotients of $T_{\mathrm{pi}}(V) \times H$ the elements u and v are identified with $\mathsf{u} \times 1$ and $\mathsf{v} \times 1$ respectively and c is identified with $1 \times \mathsf{c}$. Observe that the ideals of $T_{\mathrm{pi}}(V) \times H$ are bi-ideals since $\mathsf{u}^n, \mathsf{v}^n \in \mathrm{P}_{\mathsf{c}^n,1}(T_{\mathrm{pi}}(V) \times H)$ and $\mathrm{ad}_{\mathsf{c}}\mathsf{v}(\mathsf{u}) \in \mathrm{P}_{\mathsf{c}^2,1}(T_{\mathrm{pi}}(V) \times H)$.

Exercises

Throughout these exercises H is a bialgebra over the field k and A is a bialgebra in ${}_H^H\mathcal{YD}$ unless otherwise specified. V is an object of ${}_H^H\mathcal{YD}$ and $(\imath, T_{\rm pi}(V))$ is a free pointed irreducible Hopf algebra on the object V, where $\imath$ is the inclusion.

Exercise 15.5.1. Suppose U, V are Yetter-Drinfel'd submodules of A. Show that UV and $U \wedge V$ are Yetter-Drinfel'd submodules of A.

Exercise 15.5.2. Let A, B be algebras in ${}_H^H\mathcal{YD}$. Show that:

(a) If $f : A \longrightarrow A'$ and $g : B \longrightarrow B'$ are morphisms of algebras in ${}_H^H\mathcal{YD}$ then $f\underline{\otimes}g : A\underline{\otimes}B \longrightarrow A'\underline{\otimes}B'$ defined by $(f\underline{\otimes}g)(a\underline{\otimes}b) = f(a)\underline{\otimes}g(b)$ for all $a \in A$ and $b \in B$ is a morphism of algebras in ${}_H^H\mathcal{YD}$.

(b) $\imath_A : A \longrightarrow A\underline{\otimes}B$ defined by $\imath_A(a) = a\underline{\otimes}1$ and $\jmath_B : B \longrightarrow A\underline{\otimes}B$ defined by $\jmath_B(b) = 1\underline{\otimes}b$ for all $b \in B$ are morphisms of algebras in ${}_H^H\mathcal{YD}$. [Hint: Observe that $(a\underline{\otimes}b)(a'\underline{\otimes}b') = aa'\underline{\otimes}bb'$ for all $a, a' \in A$ and $b, b' \in B$ whenever $b = 1$ or $a' = 1$.]

Exercise 15.5.3. Complete the proof that $(T_{\rm pi}(V), \Delta, \epsilon)$ is a coalgebra in ${}_H^H\mathcal{YD}$.

Exercise 15.5.4. Suppose that the inclusion $\imath : A_0 \longrightarrow A$ has an inverse in the convolution algebra $\mathrm{Hom}(A_0, A)$. Show that A is a Hopf algebra in ${}_H^H\mathcal{YD}$. [Hint: See Exercise 11.6.13.]

Exercise 15.5.5. Let U, U', V, V' be Yetter-Drinfel'd submodules of A. Show that:

(a) $(U\underline{\otimes}V)(U'\underline{\otimes}V')$ is a Yetter-Drinfel'd submodule of $A\underline{\otimes}A$ and

$$(U\underline{\otimes}V)(U'\underline{\otimes}V') \subseteq UU'\underline{\otimes}VV'.$$

(b) If $V \subseteq \mathrm{P}(A)$ then the sequence of Yetter-Drinfel'd submodules $k1 = V^0, V = V^1, V^2 = VV, \ldots$ of A satisfies $\Delta(V^n) \subseteq \sum_{\ell=0}^{n} V^{n-\ell}\underline{\otimes}V^\ell$ for all $n \geq 0$.

(c) If $V \subseteq \mathrm{P}(A)$ then $C = \sum_{\ell=0}^{n} V^\ell$ is a pointed irreducible sub-bialgebra of A in ${}_H^H\mathcal{YD}$.

Exercise 15.5.6. Suppose B, B', and I are Yetter-Drinfel'd submodules of A, B and B' are subcoalgebras of A, and I is a left coideal (respectively right coideal, coideal) of A. Show that BI, IB', and BIB' are left coideals (respectively right coideals, coideals) of the k-algebra A and are Yetter-Drinfel'd submodules of A.

Exercise 15.5.7. Let $M = \bigoplus_{i\in I} M_i$ be the direct sum of objects in ${}_H^H\mathcal{YD}$. Suppose $V = \bigoplus_{i\in I} V_i$, where V_i is a subspace of M_i for all $i \in I$ and let N_i be the Yetter-Drinfel'd submodule of M_i generated by V_i. Show that $N = \bigoplus_{i\in I} N_i$ is the Yetter-Drinfel'd submodule of M generated by V.

Exercise 15.5.8. Let C be a coalgebra in ${}_H^H\mathcal{YD}$. Show that the Yetter-Drinfel'd module generated by a subcoalgebra (respectively left coideal, right coideal, coideal) of the k-coalgebra C is a subcoalgebra (respectively left coideal, right coideal, coideal) of C.

Exercise 15.5.9. Let $C = \bigoplus_{n=0}^{\infty} C(n)$ be a graded coalgebra over the field k. Suppose $g, h \in \mathrm{G}(C)$ and $c = \bigoplus_{n=0}^{\infty} c(n) \in \mathrm{P}_{g,h}(C)$, where $c(n) \in C(n)$ for all $n \geq 0$. Show that $c(n) \in \mathrm{P}_{g,h}(C)$ for all $n \geq 0$.

Exercise 15.5.10. Show that $\mathrm{P}(A)$ is a Yetter-Drinfel'd submodule of A.

Exercise 15.5.11. Let $v \in \mathrm{P}(A)$ and $g \in \mathrm{G}(A)$ satisfy $\rho(v) = a \otimes v$ and $a{\cdot}v = qv$, where $q \neq 0$ and is not a root of unity. Show that $\{1, v, v^2, \ldots\}$ is linearly independent

Exercise 15.5.12. Let $M = km$ be a one-dimensional vector space over k. Here we examine what is means for M, in particular kg where $g \in \mathrm{G}(A)$, to have the structure of an object in ${}_H^H\mathcal{YD}$. Show that:

(a) If $(M, \cdot, \rho)$ is an object of ${}_H^H\mathcal{YD}$ then there are $\ell \in \mathrm{G}(H)$ and $\beta \in \mathrm{G}(H^o)$ such that:

(1) $h{\cdot}m = \beta(h)m$ for all $h \in H$.
(2) $\rho(m) = \ell \otimes m$.
(3) $(\beta \rightharpoonup h)g = g(h \leftharpoonup \beta)$ for all $h \in H$.

(b) If $\ell \in \mathrm{G}(H)$ and $\beta \in \mathrm{G}(H^o)$ satisfy (3) then (1) defines a left H-module structure on M, (2) defines a left H-comodule structure on M, and M is an object of ${}_H^H\mathcal{YD}$ with these structures.

Exercise 15.5.13. Let $(\jmath, B(V))$ be any Nichols algebra on an object V of ${}_H^H\mathcal{YD}$. Show that:

(a) $\jmath(V) = \mathrm{P}(B(V))$.

(b) $B(V) = \bigoplus_{n=0} \jmath(V)^n$ and this direct sum gives $B(V)$ the structure of a graded bialgebra in ${}_H^H\mathcal{YD}$.

(c) Suppose that $B(V) = \bigoplus_{n=0}^{\infty} V_n$ gives $B(V)$ the structure of a graded k-algebra and $\mathrm{P}(B(V)) \subseteq V_1$. Show that $V_n = \jmath(V)^n$ for all $n \geq 0$.

[Hint: Show that $(\jmath, B)$ is a Nichols algebra on V, where $B = \sum_{n=0}^{\infty} \jmath(V)^n$.]

Apropos of the preceding exercise, we set $B(V)(n) = \jmath(V)^n$ for all $n \geq 0$ and will always think of $B(V)$ as a graded bialgebra in ${}_H^H\mathcal{YD}$ with the grading $B(V) = \bigoplus_{n=0}^{\infty} B(V)(n)$.

Exercise 15.5.14. Let $f : U \longrightarrow V$ be a morphism of objects of ${}_H^H\mathcal{YD}$, let $(\jmath_U, B(U))$ and $(\jmath_V, B(V))$ be Nichols algebras on U and V respectively.

(a) Show that there is a morphism of bialgebras $F : B(U) \longrightarrow B(V)$ determined by $F \circ \jmath_U = \jmath_V \circ f$.

(b) Show that $U \mapsto B(U)$ and $f \mapsto F$ is a functor from ${}_H^H\mathcal{YD}$ to the category of pointed irreducible graded bialgebras in ${}_H^H\mathcal{YD}$ and morphisms of graded bialgebras.

Exercise 15.5.15. Suppose $V = kv$ is a one-dimensional object of ${}_H^H\mathcal{YD}$, $\rho(v) = a \otimes v$, where $a \in \mathrm{G}(A)$, and $a \cdot v = qv$ for some non-zero $q \in k$. Show that:

(a) $B(V) = T_{\mathrm{pi}}(V)$ if q is not a primitive n^{th} root of unity for all $n \geq 1$.

(b) $B(V) = T_{\mathrm{pi}}(V)/(v^n)$ if q is a primitive n^{th} root of unity, where $n > 1$.

Exercise 15.5.16. Apropos of Exercise 15.5.15, what is $B(V)$ when $q = 1$?

Exercise 15.5.17. Work out all of the details in the discussion of the two examples in this section. The next exercise is a start.

Exercise 15.5.18. Let A, B be bialgebras in ${}_H^H\mathcal{YD}$.

(a) Suppose $f : A \longrightarrow B$ is a morphism of bialgebras which is an onto map. Show that the map of k-bialgebras $f \times \mathrm{I}_H : A \times H \longrightarrow B \times H$ induces an isomorphism of k-bialgebras $(A \times H)/(J \times H) \simeq B \times H$.

(b) Let J be a bi-ideal of A which is an object of ${}_H^H\mathcal{YD}$. Show that $J \times H$ is a bi-ideal of $A \times H$ and there is an isomorphism of k-bialgebras $(A \times H)/(J \times H) \simeq (A/J) \times H$ determined by $(a \times h) + (J \times H) \mapsto (a + J) \times h$ for all $a \in A$ and $h \in H$.

[Hint: See Exercise 11.6.17.]

The next two exercises are important for Section 15.6.

Exercise 15.5.19. Suppose A is an algebra over k, ϕ is an algebra endomorphism of A, and $v \in A$ satisfies $\phi(v) = qv$ for some $q \in k$. Let D be the linear endomorphism of A defined by $D(x) = vx - \phi(x)v$ for all $x \in A$. Show that

$$D^n(x) = \sum_{m=0}^{n} (-1)^m q^{(m-1)m/2} \binom{n}{m}_q v^{n-m} \phi^m(x) v^m$$

for all $n \geq 0$ and $x \in A$. [Hint: First show that

$$D^n(x) = \sum_{m=0}^{n} (-1)^m \binom{n}{m}_q v^{n-m}\phi^m(x)\phi^{m-1}(v)\cdots\phi^0(v)$$

for all $n \geq 0$ and $x \in A$.]

Exercise 15.5.20. Suppose $v, x \in \mathrm{P}(A)$ satisfy $\rho(v) = a\otimes v$ and $\rho(x) = b\otimes x$, where $a, b \in \mathrm{G}(A)$ are commuting grouplike elements and a is invertible. Assume further $a{\cdot}v = Qv$ for some non-zero $Q \in k$. Let ϕ be the algebra automorphism of A defined by $\phi(u) = a{\cdot}u$ for all $u \in A$ and D be the linear endomorphism of A defined in Exercise 15.5.19 for v and ϕ. Let $n \geq 0$. Show that:

(a)

$$\begin{aligned}&\Delta(v^{n-m}(a^m{\cdot}x)v^m)\\&= \sum_{r=0}^{n-m}\sum_{s=0}^{m} \binom{n-m}{r}_Q \binom{m}{s}_Q v^{n-m-r}((a^{r+m}{\cdot}x)(a^r{\cdot}v^{m-s}))\underline{\otimes} v^{r+s}\\&\quad + \sum_{r=0}^{n-m}\sum_{s=0}^{m} \binom{n-m}{r}_Q \binom{m}{s}_Q v^{n-m-r}((a^r b){\cdot}v^{m-s})\underline{\otimes} v^r(a^m{\cdot}x)v^s\end{aligned}$$

for all $n \geq 0$ and $0 \leq m \leq n$.

(b) $\Delta(D^n(x)) = \mathsf{A} + \mathsf{B}$, where

$$\mathsf{A} = \sum_{t=0}^{n} (-1)^t Q^{(t-1)t/2} \binom{n}{t}_Q \mathsf{A}^{(t)}\underline{\otimes} v^t$$

and $\mathsf{A}^{(t)}$ is

$$\sum_{d=0}^{n-t} \binom{n-t}{d}_Q Q^{(d-1)d/2+td} \left(\sum_{r=0}^{r} (-1)^t Q^{(r-1)r/2} Q^{-tr} \binom{t}{r}_Q\right) v^{n-t-d}(a^{t+d}{\cdot}x)v^d,$$

and

$$\mathsf{B} = \sum_{m=0}^{n}\sum_{r=0}^{n-m}\sum_{s=0}^{m} (-1)^m Q^{(m-1)m/2} \binom{n}{m}_Q \binom{n-m}{r}_Q \binom{m}{s}_Q \mathsf{B}^{(m,r,s)}$$

where

$$\mathsf{B}^{(m,r,s)} = v^{n-m-r}((a^r b){\cdot}v^{m-s})\underline{\otimes} v^r(a^m{\cdot}x)v^s.$$

[Hint: Use part (a) to derive the formula for B and the formula

$$\mathsf{A} = \sum_{m=0}^{n}\sum_{r=0}^{n-m}\sum_{s=0}^{m} (-1)^m Q^{(m-1)m/2} \binom{n}{m}_Q \binom{n-m}{r}_Q \binom{m}{s}_Q \mathsf{A}^{(m,r,s)},$$

where

$$\mathsf{A}^{(m,r,s)} = v^{n-m-r}(a^{r+m}{\cdot}x)(a^r{\cdot}v^{m-s})\underline{\otimes}v^{r+s}.$$

Now use Exercise 7.2.4 and express A in terms of the summation indices $\mathsf{t} = r+s$, $\mathsf{d} = m-s$ and $\mathsf{r} = r$.]

(b) $\mathsf{A} = D^n(x)\underline{\otimes}1$.

Now assume $a{\cdot}x = Rx$ and $b{\cdot}v = Tv$ where $R, T \in k$.

(c) $\mathsf{B} = \sum_{t=0}^{n} \alpha_t v^{n-t}\underline{\otimes}D^t(x)$, where

$$\alpha_t = \binom{n}{t}_Q \left(\sum_{m=0}^{n-t}(-1)^m \binom{n-t}{m}_Q Q^{(m-1)/2}Q^{tm}R^mT^m\right).$$

[Hint: Follow the suggested steps for part (a) where the summation indices are $\mathsf{t} = r+s$, $\mathsf{m} = m-s$, and $\mathsf{s} = s$.]

(d) If $RT = Q^{1-n}$ then $\mathsf{B} = 1\underline{\otimes}D^n(x)$; in this case $D^n(x) \in \mathrm{P}(A)$ and $\rho(D^n(x)) = a^n b\otimes D^n(x)$.

(e) If there are characters $\chi_v, \chi_x \in \widehat{\mathrm{G}(A)}$ such that $g{\cdot}v = \chi_v(g)v$ and $g{\cdot}x = \chi_x(g)x$ for all $g \in \mathrm{G}(A)$ then $v \in A_a^{\chi_v}$, $x \in A_b^{\chi_x}$, and $D^n(x) \in A_{a^nb}^{\chi_v^n\chi_x}$ for all $n \geq 0$. See Exercise 11.6.8 for notation.

See Lemma A.1 of [10].

15.6 Quantized enveloping algebras and their generalizations

Let $\mathcal{G}$ be a semisimple Lie algebra over the field of complex numbers $\boldsymbol{C}$. We give a standard description of the quantized enveloping algebra $\mathrm{U}_q(\mathcal{G})$, where $q \in k$ and satisfies certain restrictions. Quantized enveloping algebras are pointed Hopf algebras. $\mathrm{U}_q(\mathcal{G})$ is the quotient of a biproduct.

The biproduct is determined by a datum $\mathcal{D}$ which, with a family of parameters λ, determines the quotient $\mathrm{U}_q(\mathcal{G})$. Axiomatization of $\mathcal{D}$ and λ gives a more general construction of pointed Hopf algebras and these are important for classification.

Understanding the significance of $\mathrm{U}_q(\mathcal{G})$ requires knowledge of root systems and related structures, particularly of the Cartan matrix of $\mathcal{G}$. Humphreys' book [76] is a well-known reference. The standard description uses just a few facts about the matrix.

Let $\mathcal{A} = (a_{i,j}) \in \mathrm{M}_\vartheta(\boldsymbol{Z})$ be the $\vartheta{\times}\vartheta$ Cartan matrix of $\mathcal{G}$. Its entries satisfy $a_{i,i} = 2$ for all $1 \leq i \leq \vartheta$, $a_{i,j} \leq 0$, and $a_{i,j} = 0$ if and only if $a_{j,i} = 0$

for all $1 \leq i, j \leq \vartheta$. Let $d_i \in \{1, 2, 3\}$ for $1 \leq i \leq \vartheta$ satisfy

$$d_i a_{i,j} = d_j a_{j,i} \tag{15.18}$$

for all $1 \leq i, j \leq \vartheta$.

To define $\mathrm{U}_q(\mathcal{G})$ certain analogs of $(n)_q$ and $\binom{n}{m}_q$ are used. Initially we assume that q is *not* a primitive n^{th} root of unity for all $n \geq 1$.

Let $[n]_q = \dfrac{q^n - q^{-n}}{q - q^{-1}}$ for $n \in \boldsymbol{Z}$. Observe that

$$[n]_q = [n]_{q^{-1}} \quad \text{and} \quad [-n]_q = -[n]_q \tag{15.19}$$

for all $n \in \boldsymbol{Z}$. For $n \geq 0$ let

$$[n]_q! = \begin{cases} 1 & : n = 0 \\ [n]_q [n-1]_q! & : n \geq 1 \end{cases}$$

and let

$$\begin{bmatrix} n \\ m \end{bmatrix}_q = \frac{[n]_q!}{[m]_q! [n-m]_q!}$$

for all $0 \leq m \leq n$. We set $\begin{bmatrix} n \\ m \end{bmatrix}_q = 0$ when $m < 0$. Notice that

$$\begin{bmatrix} n \\ m \end{bmatrix}_q = \begin{bmatrix} n \\ m \end{bmatrix}_{q^{-1}} \tag{15.20}$$

for all $m, n \in \boldsymbol{Z}$ by virtue of (15.19),

$$[n]_q = q^{1-n} (n)_{q^2} \tag{15.21}$$

for all $n \geq 0$, and

$$\begin{bmatrix} n \\ m \end{bmatrix}_q = q^{m(m-n)} \binom{n}{m}_{q^2} \tag{15.22}$$

for all $n, m \in \boldsymbol{Z}$, $n \geq 0$. Now we are in a position to define the Hopf algebra $\mathrm{U}_q(\mathcal{G})$.

As a k-algebra $\mathrm{U}_q(\mathcal{G})$ is generated by symbols

$$K_1, \ldots, K_\vartheta, K_1^{-1}, \ldots, K_\vartheta^{-1}, E_1, \ldots, E_\vartheta, F_1, \ldots F_\vartheta$$

subject to the relations:

(UQ.1) $K_iK_i^{-1} = 1 = K_iK_i^{-1}, \quad 1 \leq i \leq \vartheta.$
(UQ.2) $K_iK_j = K_jK_i, \quad 1 \leq i,j \leq \vartheta.$
(UQ.3) $K_iE_jK_i^{-1} = q^{d_ia_{i,j}}E_j$ and $K_iF_jK_i^{-1} = q^{-d_ia_{i,j}}F_j, \quad 1 \leq i,j \leq \vartheta.$
(UQ.4) $E_iF_j - F_jE_i = \delta_{i,j}\left(\dfrac{K_i - K_i^{-1}}{q^{d_i} - q^{-d_i}}\right), \quad 1 \leq i,j \leq \vartheta.$
(UQ.5)

$$\sum_{r=0}^{1-a_{i,j}} (-1)^r \begin{bmatrix} 1-a_{i,j} \\ r \end{bmatrix}_{q^{d_i}} E_i^{1-a_{i,j}-r}E_jE_i^r = 0, \quad 1 \leq i,j \leq \vartheta \text{ and } i \neq j.$$

(UQ.6)

$$\sum_{r=0}^{1-a_{i,j}} (-1)^r \begin{bmatrix} 1-a_{i,j} \\ r \end{bmatrix}_{q^{d_i}} F_i^{1-a_{i,j}-r}F_jF_i^r = 0, \quad 1 \leq i,j \leq \vartheta \text{ and } i \neq j.$$

The coalgebra structure of $\mathrm{U}_q(\mathcal{G})$ is determined by $\Delta(K_i) = K_i \otimes K_i$,

$$\Delta(E_i) = E_i \otimes 1 + K_i \otimes E_i, \quad \text{and} \quad \Delta(F_i) = F_i \otimes K_i^{-1} + 1 \otimes F_i$$

for all $1 \leq i \leq \vartheta$. That $\Delta(K_i^{-1}) = K_i^{-1} \otimes K_i^{-1}$ for all $1 \leq i \leq \vartheta$ is a consequence of (UQ.1) and the formula for $\Delta(K_i)$.

We will show that $\mathrm{U}_q(\mathcal{G})$ is a bialgebra. This will mean it is a Hopf algebra by Corollary 7.6.7. We encourage the reader to refer to Lusztig's book [112] or Jantzen's subsequent book [78] for motivation of the relations and a detailed treatment of quantized enveloping algebras and their representation theory.

The Hopf algebra $\mathrm{U}_q(\mathcal{G})$ is the quotient of a biproduct. To describe it as such we use a slightly different set of generators and relations. Let

$$k_i = K_i, \quad k_i^{-1} = K_i^{-1}, \quad e_i = E_i, \quad \text{and} \quad f_i = F_iK_i$$

for all $1 \leq i \leq \vartheta$. Then (UQ.1)–(UQ.4) are equivalent to (uq.1)–(uq.f) respectively, where:

(uq.1) $k_ik_i^{-1} = 1 = k_ik_i^{-1}, \quad 1 \leq i \leq \vartheta.$
(uq.2) $k_ik_j = k_jk_i, \quad 1 \leq i,j \leq \vartheta.$
(uq.3) $k_ie_jk_i^{-1} = q^{d_ia_{i,j}}e_j, \ k_if_jk_i^{-1} = q^{-d_ia_{i,j}}f_j, \quad 1 \leq i,j \leq \vartheta.$
(uq.f) $e_if_j - q^{-d_aa_{i,j}}f_je_i = \delta_{i,j}\left(\dfrac{k_i^2 - 1}{q^{d_i} - q^{-d_i}}\right), \quad 1 \leq i,j \leq \vartheta.$

Observe that $\Delta(k_i) = k_i \otimes k_i$,

$$\Delta(e_i) = e_i \otimes 1 + k_i \otimes e_i, \text{ and } \Delta(f_i) = f_i \otimes 1 + k_i \otimes f_i \tag{15.23}$$

for all $1 \leq i \leq \vartheta$.

Let G be the free abelian group with basis $\{k_1, \ldots, k_\vartheta\}$, let $H = k[\mathsf{G}]$, and let V be the vector space with basis $\{e_1, \ldots, e_\vartheta, f_1, \ldots, f_\vartheta\}$. There are characters $\chi_1, \ldots, \chi_\vartheta \in \widehat{\mathsf{G}}$ defined by $\chi_i(k_j) = q^{d_i a_{i,j}}$ for all $1 \leq i, j \leq \vartheta$. Observe that $\chi_i(k_j) = \chi_j(k_i)$ for all $1 \leq i, j \leq \vartheta$. Thus V is a left H-module whose structure is given by

$$k_i{\cdot}e_j = q^{d_i a_{i,j}} e_j = \chi_j(k_i) e_j \quad \text{and} \quad k_i{\cdot}f_j = q^{-d_i a_{i,j}} f_j = \chi_j^{-1}(k_i) f_j$$

for all $1 \leq i, j \leq \vartheta$. Give V the left H-comodule structure determined by

$$\rho(e_i) = k_i \otimes e_i \quad \text{and} \quad \rho(f_i) = k_i \otimes f_i$$

for all $1 \leq i \leq \vartheta$. Then $(V, \cdot, \rho)$ is an object of ${}^H_H\mathcal{YD}$. See Exercise 11.6.9. The notation of Exercise 11.6.8 is very useful for this section. Note $e_i \in V_{k_i}^{\chi_i}$ and $f_i \in V_{k_i}^{\chi_i^{-1}}$ for all $1 \leq i \leq \vartheta$.

Let $B = T_{\mathrm{pi}}(V)$ be the free pointed irreducible Hopf algebra on the object V and $A = B{\times}H$. Let Δ denote the coproduct of the k-Hopf algebra A and let Δ_B denote the coproduct of B. We identify $h \in H$ with $1{\times}h \in A$ and $b \in B$ with $b{\times}1$. By these identifications H is a Hopf subalgebra of A and B is a subalgebra of A. Since $(b{\times}1)(1{\times}h) = b{\times}h$ the product bh is identified with $b{\times}h = b \otimes h$. Hence the elements of A are sums of these products. The product and coproduct of A are therefore determined by

$$hb = (h_{(1)}{\cdot}b)h_{(2)} \tag{15.24}$$

for all $h \in H$, $b \in B$ and

$$\Delta(b) = b_{(1)} b_{(2)(-1)} \otimes b_{(2)(0)} \tag{15.25}$$

for all $b \in B$, where $\Delta_B(b) = b_{(1)} \underline{\otimes} b_{(2)}$. An important implication of (15.24) is

$$gbg^{-1} = g{\cdot}b \tag{15.26}$$

for all $g \in \mathrm{G}(H)$ and $b \in B$. By definition of H the relations (uq.1) and (uq.2) hold in A, and (uq.3) holds in A by virtue of (15.26). Therefore as a k-algebra A is generated by $k_1, \ldots, k_\vartheta$, $k_1^{-1}, \ldots, k_\vartheta^{-1}$, $e_1, \ldots, e_\vartheta$, $f_1, \ldots, f_\vartheta$ subject to the relations (uq.1)–(uq.3).

Observe (15.25) implies

$$\Delta(v) = v \otimes 1 + v_{(-1)} \otimes v_{(0)} \tag{15.27}$$

for $v \in \mathrm{P}(B)$. This means the coproduct of the k_i's, e_i's, and f_i's regarded as elements of A is given by $k_i \in \mathrm{G}(A)$ and (15.23) for all $1 \leq i \leq \vartheta$. Our general remarks about B and A apply to any bialgebra in ${}^H_H\mathcal{YD}$ and its associated biproduct.

Since $\mathrm{ad}_c e_i(b) = e_i b - (k_i{\cdot}b)e_i$ for all $b \in B$, the relation (uq.f) is the same as (uq.4) $\mathrm{ad}_{\mathsf{c}}e_i(f_j) = \delta_{i,j}\left(\dfrac{k_i^2 - 1}{q^{d_i} - q^{-d_i}}\right), \quad 1 \leq i, j \leq \vartheta.$

Therefore (UQ.1)–(UQ.4) are equivalent to (uq.1)–(uq.4). We will show that (UQ.5) is

(uq.5) $(\mathrm{ad}_{\mathsf{c}}e_i)^{1-a_{i,j}}(e_j) = 0, \quad 1 \leq i, j \leq \vartheta$ and $i \neq j$

and (UQ.1)–(UQ.6) are equivalent to (uq.1)–(uq.6), where

(uq.6) $(\mathrm{ad}_{\mathsf{c}}f_i)^{1-a_{i,j}}(f_j) = 0, \quad 1 \leq i, j \leq \vartheta$ and $i \neq j$.

Compare (uq.1)–(uq.6) with Serre's presentation of the universal enveloping algebra of $\mathcal{G}$ over an algebraically closed field of characteristic 0. The reader is directed to [76, §18.1] for example.

Let $n \geq 0$ and $1 \leq i, j \leq \vartheta$. We derive a formula for $(\mathrm{ad}_{\mathsf{c}}e_i)^n(e_j)$. Since $k_i \in \mathrm{G}(H)$ the endomorphism ϕ of B defined by $\phi(b) = k_i{\cdot}b$ for all $b \in B$ is an automorphism of the k-algebra B. Note that $\mathrm{ad}_{\mathsf{c}}e_i = D$, the linear endomorphism of Exercise 15.5.19 defined for $v = e_i$ and ϕ. Let $Q = q^{2d_i}$. Then $k_i{\cdot}v = Qv$. Since $\phi(e_j) = k_i{\cdot}e_j = q^{d_i a_{i,j}}e_j$, we use this exercise to compute

$$\begin{aligned}(\mathrm{ad}_{\mathsf{c}}e_i)^n(e_j) &= \sum_{r=0}^{n}(-1)^r\binom{n}{r}_Q Q^{(r-1)r/2}e_i^{n-r}\phi^r(e_j)e_i^r \\ &= \sum_{r=0}^{n}(-1)^r\binom{n}{r}_Q Q^{(r-1)r/2}q^{d_i a_{i,j} r}e_i^{n-r}e_j e_i^r\end{aligned}$$

from which the formula

$$(\mathrm{ad}_{\mathsf{c}}e_i)^n(e_j) = \sum_{r=0}^{n}(-1)^r\binom{n}{r}_{q^{2d_i}} q^{d_i r(r-(1-a_{i,j}))}e_i^{n-r}e_j e_i^r \tag{15.28}$$

follows.

Suppose $i \neq j$. Then $1 - a_{i,j} \geq 0$. With $n = 1 - a_{i,j}$ the preceding equation is

$$(\mathrm{ad}_{\mathsf{c}}e_i)^{1-a_{i,j}}(e_j) = \sum_{r=0}^{1-a_{i,j}}(-1)^r\begin{bmatrix}1-a_{i,j}\\ r\end{bmatrix}_{q^{d_i}} E_i^{1-a_{i,j}-r}E_jE_i^r$$

by (15.22). Equation (15.28) with e_i, e_j replaced by f_i, f_j respectively is

$$(\mathrm{ad}_{\mathsf{c}}f_i)^n(f_j) = \sum_{r=0}^{n}(-1)^r\binom{n}{r}_{q^{-2d_i}} q^{-d_i r(r-(1-a_{i,j}))}f_i^{n-r}f_jf_i^r.$$

Using the fact that if a, b are elements of an associative algebra, a is invertible, and $aba^{-1} = \omega b$ for some $\omega \in k$, then $(ba)^m = \omega^{(m-1)m/2} b^m a^m$ for all $m \geq 0$ and that $a^\ell b^m = \omega^{\ell m} b^m a^\ell$ for all $\ell, m \geq 0$, one can show that

$$f_i^{n-r} f_j f_i^r = (F_i K_i)^{n-r} (F_j K_j)(F_i K_i)^r = q^{-d_i n(n-(1-a_{i,j}))} F_i^{n-r} F_j F_i^r K_i^n K_j.$$

Therefore

$$(\mathrm{ad}_{\mathsf{c}} f_i)^n (f_j) = \Omega \left(\sum_{r=0}^{n} (-1)^r \binom{n}{r}_{q^{-2d_i}} q^{-d_i r(r-(1-a_{i,j}))} F_i^{n-r} F_j F_i^r \right) K_i^n K_j,$$

where $\Omega = q^{-d_i n(n-(1-a_{i,j}))}$. When $i \neq j$ and $n = 1 - a_{i,j}$, this equation is

$$(\mathrm{ad}_{\mathsf{c}} f_i)^{1-a_{i,j}} (f_j) = \left(\sum_{r=0}^{1-a_{i,j}} (-1)^r \begin{bmatrix} 1 - a_{i,j} \\ r \end{bmatrix}_{q^{-d_i}} F_i^{1-a_{i,j}-r} F_j F_i^r \right) K_i^{1-a_{i,j}} K_j$$

by (15.22). Our proof that (UQ.1)–(UQ.6) are equivalent to (uq.1)–(uq.6) is complete.

By parts (d) and (e) of Exercise 15.5.20 we have:

$$\mathrm{ad}_{\mathsf{c}} e_i (f_j) \in \mathrm{P}(B) \cap B^{\chi_i \chi_j^{-1}}_{k_i k_j} \cap V^2 \tag{15.29}$$

for all $1 \leq i, j \leq \vartheta$.

$$(\mathrm{ad}_{\mathsf{c}} e_i)^{1-a_{i,j}} (e_j) \in \mathrm{P}(B) \cap B^{\chi_i^{1-a_{i,j}} \chi_j}_{k_i^{1-a_{i,j}} k_j} \cap V^{2-a_{i,j}} \tag{15.30}$$

and

$$(\mathrm{ad}_{\mathsf{c}} f_i)^{1-a_{i,j}} (f_j) \in \mathrm{P}(B) \cap B^{\chi_i^{a_{i,j}-1} \chi_j^{-1}}_{k_i^{1-a_{i,j}} k_j} \cap V^{2-a_{i,j}} \tag{15.31}$$

for all $1 \leq i, j \leq \vartheta$, where $i \neq j$. Several conclusions. Each of the elements described in (15.29)–(15.31) spans a one-dimensional coideal which is also a Yetter-Drinfel'd submodule of B. The set of these elements generates a graded ideal J of $B = T_{\mathrm{pi}}(V)$ which satisfies $J \cap V = (0)$, is a coideal, and is a Yetter-Drinfel'd submodule of B. In particular B/J is a pointed irreducible bialgebra in ${}^H_H\mathcal{YD}$ and therefore must be a Hopf algebra in this category.

Using (15.25) and (15.29)–(15.31) we derive:

$$\mathrm{ad}_{\mathsf{c}} e_i (f_j) - \delta_{i,j} \left(\frac{k_i^2 - 1}{q^{d_i} - q^{-d_i}} \right) \in \mathrm{P}_{k_i k_j, 1}(A) \tag{15.32}$$

for all $1 \leq i, j \leq \vartheta$.

$$(\mathrm{ad}_{\mathsf{c}} e_i)^{1-a_{i,j}} (e_j) \in \mathrm{P}_{k_i^{1-a_{i,j}} k_j, 1}(A) \tag{15.33}$$

and

$$(\mathrm{ad}_c f_i)^{1-a_{i,j}}(f_j) \in \mathrm{P}_{k_i^{1-a_{i,j}}k_j,1}(A) \tag{15.34}$$

for all $1 \leq i, j \leq \vartheta$, where $i \neq j$. Since the elements listed in (15.32)–(15.34) are skew primitive, the ideal I of A generated by the set of them is a bi-ideal of A. Thus A/I is a k-bialgebra which must be a k-Hopf algebra since it is pointed irreducible. We have shown

$$\mathrm{U}_q(\mathcal{G}) \simeq A/I = (T_{\mathrm{pi}}(V) \times k[\mathsf{G}])/I. \tag{15.35}$$

At this point we revisit our assumption on q. In light of (UQ.4) the essential requirement is $q^{2d_i} \neq 1$ for all $1 \leq i \leq \vartheta$. This is accomplished if $q^4, q^6 \neq 1$. The symbols $\begin{bmatrix} n \\ m \end{bmatrix}_{q^{d_i}}$ are defined in a way so that (15.22) holds for $0 \leq m \leq n$. See Exercise 15.6.1. The equation of (15.20), which is essential for our calculations, is satisfied. See Exercise 15.6.1 or 7.2.2. When q satisfies the essential requirement, (15.35) holds.

To generalize the construction of $\mathrm{U}_q(\mathcal{G})$ as a quotient of a biproduct we start with the free abelian group G, the Hopf algebra $H = k[\mathsf{G}]$, the elements $k_1, \ldots, k_\vartheta \in \mathsf{G}$, the characters $\chi_1, \ldots, \chi_\vartheta \in \widehat{\mathsf{G}}$, and the vector space V with basis $\{e_1, \ldots, e_\vartheta, f_1, \ldots, f_\vartheta\}$. We want first to associate group elements and characters to the basis for V.

For $1 \leq i \leq \vartheta$ set $v_i = e_i$ and $v_{i+\vartheta} = f_i$, $g_i = g_{i+\vartheta} = k_i$, and $\chi_{i+\vartheta} = \chi_i^{-1}$. Let $\theta = 2\vartheta$. Then V has basis $\{v_1, \ldots, v_\theta\}$ and as an object of ${}^H_H\mathcal{YD}$ is determined by

$$v_i \in V_{g_i}^{\chi_i} \tag{15.36}$$

for all $1 \leq i \leq \theta$.

Let $\mathsf{A} = \begin{pmatrix} \mathcal{A} & 0 \\ 0 & \mathcal{A} \end{pmatrix}$, which is the Cartan matrix for the semisimple Lie algebra $\mathcal{G} \oplus \mathcal{G}$ over $\boldsymbol{C}$. Note that $\mathsf{A} = (\mathsf{a}_{i,j})$ is defined by $\mathsf{a}_{i,j} = \mathsf{a}_{i+\vartheta,j+\vartheta} = a_{i,j}$ and $\mathsf{a}_{i+\vartheta,j} = 0 = \mathsf{a}_{i,j+\vartheta} = 0$ for all $1 \leq i, j \leq \vartheta$. It is easy to see that

$$\chi_i(g_j)\chi_j(g_i) = \chi_i(g_i)^{\mathsf{a}_{i,j}} \tag{15.37}$$

for all $1 \leq i, j \leq \theta$ and

$$\chi_i(g_i) \neq 1 \tag{15.38}$$

for all $1 \leq i \leq \theta$.

To describe the ideal I of $A = T_{\mathrm{pi}}(V) \times H$ generated by the elements of (15.32)–(15.34) in terms of the v_i's and g_i's, we consider the connected components of the graph of A. Its set of vertices is $[\theta] = \{1, \ldots, \theta\}$ and

Definition 15.6.4. The Hopf algebra of part (b) of Theorem 15.6.3 is denoted $\mathrm{U}(\mathcal{D}, \lambda)$.

Apropos of the theorem, by abuse of notation we let 0 denote the linking parameters for $\mathcal{D}$ defined by $\lambda_{i,j} = 0$ for all $1 \leq i, j \leq \theta$. Note that the Hopf algebra of part (a) is $\mathrm{U}(\mathcal{D}, 0) = (T_{\mathrm{pi}}(V)/J)\times H$ and is a biproduct.

Suppose k is algebraically closed. In the generic case, meaning that the $\chi_i(g_i)$'s are not n^{th} roots of unity for all $n \geq 1$, $B(V) = T_{\mathrm{pi}}(V)/J$; thus the quotient is a Nichols algebra. This very important result draws on the work of Lusztig [112] and Rosso [183]. See [176] for a discussion.

Exercises

Exercise 15.6.1. The symbols $\begin{bmatrix} n \\ m \end{bmatrix}_q$ are defined for all $n, m \in \boldsymbol{Z}$ and nonzero $q \in k$ in such a way (15.22) holds for all $0 \leq m \leq n$. We follow [78, Chapter 0].

Let $\boldsymbol{Q}(v)$ be the field of quotients of the algebra of polynomials $\boldsymbol{Q}[v]$ in indeterminate v over the field of rational numbers $\boldsymbol{Q}$. For all $a \in \boldsymbol{Z}$ let

$$[a]_v = \frac{v^a - v^{-a}}{v - v^{-1}}, \quad \begin{bmatrix} a \\ 0 \end{bmatrix}_v = 1, \quad \text{and} \quad \begin{bmatrix} a \\ m \end{bmatrix}_v = \frac{[a]_v[a-1]_v \cdots [a-(m-1)]_v}{[m]_v[m-1]_v \cdots [1]_v}$$

for all $m > 0$. We will use the convention $\begin{bmatrix} a \\ m \end{bmatrix}_v = 0$ whenever $m < 0$. Show that:

(a) $\begin{bmatrix} n \\ n \end{bmatrix}_v = 1$ for all $0 \leq n$.

(b) $\begin{bmatrix} a+1 \\ m \end{bmatrix}_v = v^m \begin{bmatrix} a \\ m \end{bmatrix}_v + v^{m-(a+1)} \begin{bmatrix} a \\ m-1 \end{bmatrix}_v$ for all $a, m \in \boldsymbol{Z}$.

Thus $\begin{bmatrix} a \\ m \end{bmatrix}_v \in \boldsymbol{Z}[v, v^{-1}]$ for all $a, m \in \boldsymbol{Z}$. Let R be any ring with unity, suppose $q \in R$ is invertible, and let $f : \boldsymbol{Z}[v, v^{-1}] \longrightarrow R$ be the homomorphism of rings with unity determined by $f(v) = q$. By definition $\begin{bmatrix} a \\ m \end{bmatrix}_q = f(\begin{bmatrix} a \\ m \end{bmatrix}_v)$ for all $a, m \in \boldsymbol{Z}$. Show that:

(c) $\begin{bmatrix} a \\ m \end{bmatrix}_q = \begin{bmatrix} a \\ m \end{bmatrix}_{q^{-1}}$ for all $a, m \in \boldsymbol{Z}$.

(d) If $q \in k$ is not zero then $\begin{bmatrix} n \\ m \end{bmatrix}_q = q^{m(m-n)} \binom{n}{m}_{q^2}$ for all $0 \leq m \leq n$.

Note that $\begin{bmatrix} n \\ m \end{bmatrix}_q$ for $0 \leq m \leq n$ and $m < 0$ is given by the formula of this section when $0 \neq q \in k$ is not an nth root of unity for all $n \geq 1$.

Exercise 15.6.2. Complete the proof of Theorem 15.6.3.

Exercise 15.6.3. Let H be a Hopf algebra over the field k, let B be a bialgebra in ${}_H^H\mathcal{YD}$, and set $A = B\times H$. We use the identifications in connection with A found in the paragraph preceding (15.25). Let $J \subseteq B$. Show that:

(a) If J is a left H-submodule of B and an ideal of B then JH is an ideal of A.

(b) If J is a left H-subcomodule of B and a coideal of B then JH is a coideal of A.

(c) If J is a Yetter-Drinfel'd submodule of B and a bi-ideal of B then JH is the ideal of A generated by J and is a bi-ideal of A.

(d) If J satisfies the hypothesis of part (c) and $\pi : B \longrightarrow B/J$ is the projection, then $\pi\times \mathrm{I}_H : B\times H \longrightarrow (B/J)\times H$ defined by $(\pi\times\mathrm{I}_H)(b\times h) = \pi(b)\times h$ for $b \in B$ and $h \in H$ is a map of k-bialgebras which induces an isomorphism of k-bialgebras $(B\times H)/JH \longrightarrow (B/J)\times H$.

Exercise 15.6.4. Show that the Hopf algebra $A = H_{(q,n,\alpha,m)}$ of Section 15.5 is a quotient of some $\mathrm{U}(\mathcal{D},\lambda)$ described in Theorem 15.6.3.

Exercise 15.6.5. Show that the Hopf algebra $A = U_{\mathrm{q}}(sl_2)' = U_{n,q}$ of Section 15.5 is a quotient of some $\mathrm{U}(\mathcal{D},\lambda)$ described in Theorem 15.6.3.

Exercise 15.6.6. Here we reconsider the role the Cartan matrix plays in the construction of $\mathrm{U}(\mathcal{D},\lambda)$. Replace $[\theta]$ with a non-empty index set $\mathcal{I}$ (possibly infinite), the Cartan matrix with an array $(a_{i,j})_{i,j\in\mathcal{I}}$ of integers which satisfies $a_{i,i} = 2$ for all $i \in \mathcal{I}$ and $a_{i,j} \leq 0$, $a_{i,j} = 0$ if and only if $a_{j,i} = 0$ for all $i, j \in \mathcal{I}$. Set $i \sim j$ if and only if there is an $r \geq 1$ and $i_0, \ldots, i_r \in \mathcal{I}$ such that $i = i_0, \ldots, i_r = j$ and $a_{i_\ell,i_{\ell+1}} \neq 0$ for all $0 \leq \ell < r$. Can the construction of $\mathrm{U}(\mathcal{D},\lambda)$ be carried out?

Exercise 15.6.7. Let V be an object in ${}_H^H\mathcal{YD}$, let $(\iota, T_{\mathrm{pi}}(V))$ be the free pointed irreducible Hopf algebra in ${}_H^H\mathcal{YD}$ on V, where ι is the inclusion, and let $v \in V$. Suppose $\rho(v) = b\otimes v$.

(a) Show that $\mathrm{ad}_{\mathrm{c}}v(x) = vx - (b{\cdot}x)v$ for all $x \in T_{\mathrm{pi}}(V)$.

Suppose that $u \in V$ and $\rho(u) = a\otimes u$, where $a, b \in \mathrm{G}(H)$ commute and b is invertible. Show that:

(b) $\Delta_{T_{\mathrm{pi}}(V)}(\mathrm{ad}_{\mathrm{c}}v(u)) = \mathrm{ad}_{\mathrm{c}}v(u)\underline{\otimes}1 + 1\underline{\otimes}\mathrm{ad}_{\mathrm{c}}v(u) + (v\underline{\otimes}u - a{\cdot}v\underline{\otimes}b{\cdot}u)$.

(c) $I = ku + kv + kb{\cdot}u + ka{\cdot}v + k\mathrm{ad}_{\mathrm{c}}v(u)$ is a graded coideal of $T_{\mathrm{pi}}(V)$.

(d) If $u, v \neq 0$ then $\mathrm{ad}_{\mathrm{c}}v(u) \in \mathrm{P}(T_{\mathrm{pi}}(V))$ if and only if there are $\alpha, \beta \in k$ such that $a{\cdot}v = \alpha v$, $b{\cdot}u = \beta u$ and $\alpha\beta = 1$.

Let $G = \mathrm{G}(H)$ and assume $b, c \in \mathrm{Z}(G)$. Suppose there are characters $\chi_v, \chi_u \in \widehat{\mathrm{G}(H)}$ such that $g{\cdot}v = \chi_v(g)v$ and $g{\cdot}u = \chi_u(g)u$ for all $g \in G$ and $\chi_v(a)\chi_u(b) = 1$. ($\mathrm{ad}_{\mathrm{c}}v(u) \in \mathrm{P}(T_{\mathrm{pi}}(V))$ if and only if $\chi_v(a)\chi_u(b) = 1$ by part (b).)

(e) Show that $\mathrm{ad}_{\mathrm{c}}v(u) \in \mathrm{P}_{bc,1}(T_{\mathrm{pi}}(V))$.

(f) Suppose that I is a bi-ideal of $T_{\mathrm{pi}}(V){\times}H$, $\lambda \in k$, that $ab \in \mathrm{Z}(H)$, and that $\mathrm{ad}_{\mathrm{c}}v(u) - \lambda(ab - 1) \in I$. Show that $\mathrm{ad}_{\mathrm{c}}v(u) \in I$ if either $ab = 1$ or $\chi_u\chi_v \neq 1$. [Hint: Recall that module action by $g \in \mathrm{G}(T_{\mathrm{pi}}(V))$ on $T_{\mathrm{pi}}(V)$ is conjugation by g with $T_{\mathrm{pi}}(V)$ thought of as a subalgebra of $T_{\mathrm{pi}}(V){\times}H$.]

15.7 Ore extensions and pointed Hopf algebras

Some of the Hopf algebras we have constructed, in particular some quantum groups, are Ore extensions, or iterated Ore extensions, or are quotients of the latter. We describe Ore extensions in this section and indicate how they can be used to construct certain bialgebras with an eye toward applications to Hopf algebras. Throughout this section A and B are algebras over k. We begin with a definition.

Definition 15.7.1. Suppose φ is an automorphism of an algebra A over the field k. A *φ-derivation of A* is an endomorphism δ of A which satisfies

$$\delta(aa') = \delta(a)a' + \varphi(a)\delta(a') \tag{15.44}$$

for all $a, a' \in A$.

For a linear endomorphism φ of A let $[a, b]_\varphi = ab - \varphi(b)a$ for all $a, b \in A$. When φ is an algebra automorphism of A note for a fixed $b \in A$ that $D \in \mathrm{End}(A)$ defined by $D(a) = [a, b]_\varphi$ for all $a \in A$ is an endomorphism described in Exercise 15.5.19.

Example 15.7.2. Suppose A is a subalgebra of B, suppose φ is an algebra automorphism of A, and let $b \in B$. Then $\delta : A \longrightarrow B$ defined by $\delta(a) = [b, a]_\varphi$ for all $a \in A$ satisfies (15.44) for all $a, a' \in A$. Thus δ may be regarded as a φ-derivation of A if $\mathrm{Im}(\delta) \subseteq A$.

In the situation of Example 15.7.2 note that $\mathrm{Im}(\delta) \subseteq A$ if and only if $\delta(S) \subseteq A$, where S generates A as an algebra.

Let δ be a φ-derivation of any algebra A. Then $A[X, \varphi, \delta]$ is an associative algebra over k, where $A[X, \varphi, \delta] = A[X]$ is the k-algebra which is the ring of polynomials in indeterminate X with coefficients in A and whose product is determined by A is a subalgebra of $A[X, \varphi, \delta]$ and

$$Xa = \varphi(a)X + \delta(a) \tag{15.45}$$

for all $a \in A$. See [187, §1.6].

Definition 15.7.3. Let A be an algebra over k. An *Ore extension of* A is algebra $A[X, \varphi, \delta]$ over k as described above, where φ is an algebra automorphism of A and δ is a φ-derivation of A.

Observe that the inclusion map $\imath : A \longrightarrow A[X, \varphi, \delta]$ is a one-one algebra map. The pair $(\imath, A[X, \varphi, \delta])$ satisfies a universal mapping property.

Theorem 15.7.4. *Let A be an algebra over the field k, suppose φ is an algebra automorphism of A, and let δ be a φ-derivation of A. Then the pair $(\imath, A[X, \varphi, \delta])$ satisfies:*

(a) $\imath : A \longrightarrow A[X, \varphi, \delta]$ *is an algebra map and* $X\imath(a) = \imath(\varphi(a))X + \imath(\delta(a))$ *for all* $a \in A$.
(b) *Let B be an algebra over k and $b \in B$. If $f : A \longrightarrow B$ is an algebra map which satisfies $bf(a) = f(\varphi(a))b + f(\delta(a))$ for all $a \in A$ then there is an algebra map $F : A[X, \varphi, \delta] \longrightarrow B$ determined by $F \circ \imath = f$ and $F(X) = b$.*

□

The condition of part (b) of the theorem is satisfied when it is satisfied for generators of the algebra A. See Exercise 15.7.2.

We present a sequence of examples of Ore extensions and iterated Ore extensions culminating in a bialgebra related to many examples of pointed Hopf algebras.

Example 15.7.5. The k-algebra of polynomials $A[X]$ in indeterminate X with coefficients in A is the Ore extension $A[X, \mathrm{I}_A, 0]$ of A.

Let $A = k[c]$ be the algebra of polynomials in indeterminate c over k. For a non-zero scalar $q \in k$ let φ_q be the algebra automorphism of A determined by $\varphi_q(c) = qc$.

Example 15.7.6. Let $q \in k$ be a non-zero scalar. The algebra at the beginning of Section 7.2, which we refer to as $\mathcal{A}_{(q)}$ and regard as generated by symbols c, x subject to the relation $xc = qcx$, is the Ore extension $A[X, \varphi_q, 0]$ of the k-algebra of polynomials $A = k[c]$ in indeterminate c over k, where x is identified with X.

We take this example a step further and use Theorem 15.7.4 to show the existence of a bialgebra structure on $\mathcal{A}_{(q)} = A[X, \varphi_q, 0]$ determined by

$\Delta(c) = c \otimes c$ and $\Delta(x) = c \otimes x + x \otimes 1$. The counit must satisfy $\epsilon(c) = 1$ and $\epsilon(x) = 0$. For the existence of the coproduct let $B = \mathcal{A}_{(q)} \otimes \mathcal{A}_{(q)}$. Let $f : A \longrightarrow B$ be the algebra map determined by $f(c) = c \otimes c$ and set $b = c \otimes X + X \otimes 1$. Then

$$\begin{aligned} bf(c) &= (c \otimes X + X \otimes 1)(c \otimes c) \\ &= c^2 \otimes Xc + Xc \otimes c \\ &= q(c^2 \otimes cX + cX \otimes c) \\ &= q(c \otimes c)(c \otimes X + X \otimes 1) \\ &= f(\varphi_g(c))b + f(\delta(c)) \end{aligned}$$

since $\delta(c) = 0$. We have shown that $bf(c) = f(c)b + f(\delta(c))$; thus $bf(a) = f(a)b + f(\delta(a))$ for all $a \in A$ by Exercise 15.7.2. By Theorem 15.7.4 there is an algebra map $\Delta : A[X, \varphi_q, 0] \longrightarrow \mathcal{A}_{(q)} \otimes \mathcal{A}_{(q)}$ determined by $\Delta \circ \imath = f$, that is $\Delta(c) = c \otimes c$, and $\Delta(X) = b$, that is $\Delta(X) = c \otimes X + X \otimes 1$.

For the existence of a counit we set $B = k$, let $f : A \longrightarrow B$ be the algebra map determined by $f(c) = 1$ and set $b = 0$. Repeating the preceding argument results in the existence of an algebra map $\epsilon : A[X, \varphi_q, 0] \longrightarrow k$ determined by $\epsilon \circ \imath = f$ and $\epsilon(X) = b$. Thus $\epsilon(c) = 1$ and $\epsilon(X) = 0$. The reader is left with the exercise of showing that $(\mathcal{A}_{(q)}, \Delta, \epsilon)$ is a coalgebra.

Since $A[X, \varphi_q, 0] = A[X]$ is a free left A-module with basis $\{1, X, X^2, \ldots\}$ and $A = k[c]$ has basis $\{1, c, c^2, \ldots\}$ over k, the algebra $\mathcal{A}_{(q)}$ has basis $\{c^i x^j\}_{0 \leq i,j}$ over k. The Diamond Lemma can be used to show that the preceding set is a basis for $\mathcal{A}_{(q)}$.

Our next example is an iterated Ore extension $\mathcal{B}_{(q,\mathsf{a})}$, an Ore extension of the Ore extension $\mathcal{A}_{(q^{-1})}$ of $A = k[c]$. Let $q, \mathsf{a} \in k$ where $q \neq 0$.

Example 15.7.7. The bialgebra $\mathcal{B}_{(q,\mathsf{a})}$ over k which, as an algebra, is generated by symbols c, x, y subject to the relations

$$xc = q^{-1}cx, \quad yc = qcy, \quad \text{and} \quad yx = qxy + \mathsf{a}(c^2 - 1)$$

and whose coalgebra structure is determined by

$$\Delta(c) = c \otimes c, \quad \Delta(x) = c \otimes x + x \otimes 1, \quad \text{and} \quad \Delta(y) = c \otimes y + y \otimes 1$$

is an Ore extension of $\mathcal{A}_{(q^{-1})}$ as an algebra and its coalgebra structure is accounted for by Theorem 15.7.4.

By the methods of Section 7.3 one can show that there is a bialgebra $\mathcal{B}_{(q,\mathsf{a})}$ over k satisfying the requirements described in the example. It can be easily shown that $\{c^i x^j y^\ell\}_{0 \leq i,j,\ell}$ is a basis for $\mathcal{B}_{(q,\mathsf{a})}$ with the Diamond

Lemma. The fact that this is a basis is also a result of the realization of $\mathcal{B}_{(q,\mathsf{a})}$ as an Ore extension. Note that $\mathcal{A}_{(q^{-1})}$ is a sub-bialgebra of $\mathcal{B}_{(q,\mathsf{a})}$.

We now justify the claims of the example. Let φ be the algebra automorphism of $\mathcal{A}_{(q^{-1})}$ determined by $\varphi(c) = qc$ and $\varphi(x) = qx$ and let $\delta : \mathcal{A}_{(q^{-1})} \longrightarrow \mathcal{B}_{(q,\mathsf{a})}$ be defined by $\delta(a) = [y,a]_\varphi$ for all $a \in \mathcal{A}_{(q^{-1})}$. Since $\delta(c) = yc - \varphi(c)y = yc - qcy = 0$ and $\delta(x) = yx - \varphi(x)y = yx - qxy = \mathsf{a}(c^2-1) \in \mathcal{A}_{(q^{-1})}$, by Exercise 15.7.1 it follows that $\mathrm{Im}(\delta) \subseteq \mathcal{A}_{(q^{-1})}$. Therefore δ is a φ-derivation of $\mathcal{A}_{(q^{-1})}$. See Example 15.7.5.

Observe that $yc = qcy = \varphi(c)y + \delta(c)$ and $yx = cxy + \mathsf{a}(c^2-1) = \varphi(x)y + \delta(x)$. Thus $ya = \varphi(a)y + \delta(a)$ for all $a \in \mathcal{A}_{(q^{-1})}$ by Exercise 15.7.2. Therefore $\mathcal{B}_{(q,\mathsf{a})} = \mathcal{A}_{(q^{-1})}[X,\varphi,\delta]$ by Theorem 15.7.4, where y is identified with X.

We use Theorem 15.7.4 again to show that the algebra $\mathcal{B}_{(q,\mathsf{a})}$ has the indicated bialgebra structure. For the coproduct let $B = \mathcal{B}_{(q,\mathsf{a})} \otimes \mathcal{B}_{(q,\mathsf{a})}$ and let $f : \mathcal{A}_{(q^{-1})} \longrightarrow B$ be the coproduct of $\mathcal{A}_{(q^{-1})}$ followed by the inclusion and $b = c\otimes y + y\otimes 1$. That $bf(a) = f(\varphi(a))b + f(\delta(a))$ for $a = c, x$ is left as an exercise. Therefore the preceding equation holds for all $a \in \mathcal{A}_{(q^{-1})}$ by Exercise 15.7.2. By Theorem 15.7.4 there is an algebra map $\boldsymbol{\Delta} : \mathcal{A}_{(q^{-1})}[X,\varphi,\delta] \longrightarrow B$ determined by $\boldsymbol{\Delta}\circ\imath = f$ and $\boldsymbol{\Delta}(X) = b$. Thus $\boldsymbol{\Delta}(c) = f(c) = c\otimes c$, $\boldsymbol{\Delta}(x) = f(x) = c\otimes x + x\otimes 1$, and $\boldsymbol{\Delta}(y) = b = c\otimes y + y\otimes 1$. In the last equation y is identified with X. In a similar manner one can show that there is an algebra map $\mathbf{e} : \mathcal{A}_{(q^{-1})}[X,\varphi,\delta] \longrightarrow k$ which accounts for ϵ.

Exercises

Throughout these exercises A, B are algebras over k.

Exercise 15.7.1. Let φ be an algebra automorphism of A, suppose A is a subalgebra of B, and $\delta : A \longrightarrow B$ satisfies (15.44) for all $a \in A$. Show that:

(a) $\mathrm{Ker}(\delta)$ is a subalgebra of A.

(b) If A is a subalgebra of B and is generated by a subset S then $\delta(S) \subseteq A$ implies $\delta(A) \subseteq B$.

Exercise 15.7.2. Let φ be an algebra automorphism of A, let δ be a φ-derivation of A, suppose $f : A \longrightarrow B$ is an algebra map, and let $b \in B$. Show that

$$A_b = \{a \in A \,|\, bf(a) = f(\varphi(a))b + f(\delta(a))\}$$

is a subalgebra of A.

Exercise 15.7.3. Reconsider this section when $k[c]$ is replaced by the group algebra $k[G]$, where $G = (c)$ is infinite and when it is finite. In the latter case, what conditions does q have to satisfy?

Exercise 15.7.4. Let A be a bialgebra over k, let $\eta \in \mathrm{G}(A^o)$, and let $\xi \in A^o$ be a η:ξ-skew primitive. Suppose that η is invertible. Show that φ is an algebra automorphism of A, where $\varphi(a) = \eta \rightharpoonup a$ for all $a \in A$, and that δ is a φ-derivation of A, where $\delta(a) = \xi \rightharpoonup a$ for all $a \in A$.

Exercise 15.7.5. Complete the argument begun in this section that $\mathcal{B}_{(q,\mathsf{a})}$ is a bialgebra over k.

Exercise 15.7.6. We revisit $\mathcal{B}_{(q,\mathsf{a})} = \mathcal{A}_{(q^{-1})}[X, \varphi, \delta]$. Show that

$$\delta(c^m x^n) = q^m \mathsf{a} \left(\left(\frac{1-q^{-n}}{1-q^{-1}} \right) c^{m+2} - \left(\frac{1-q^n}{1-q} \right) c^m \right) x^{n-1}$$

for all $m \geq 0$ and $n \geq 1$.

Chapter notes

The crossed product $A\#_\sigma H$ was defined by Sweedler [200, Section 8] when A is a commutative left H-module algebra. The crossed product $A\#_\sigma H$ was treated in generality by Blattner, Cohen, and Montgomery [25]. The notion of measuring appears in Sweedler's book [201, Section 7]. Our treatment of the crossed product follows that in Montgomery's book [133, Section 7.1].

Section 15.2 is a very slight reworking of the ideas and a slight reformulation of Montgomery's results [134, Section 3] with a few more examples added. The theorems of the brief Section 15.3 are important early results for Hopf algebras. The decomposition of part (b) of Theorem 15.3.2 is attributed to Cartier and Gabriel [39] and Kostant. Theorem 15.3.3 is attributed to Cartier [29] and Kostant (unpublished). See the preface of Sweedler's book [201] for Kostant attributions. Abe's book [1, page 111] and Sweedler's book [201, page 274] are other sources of proofs of Theorem 15.3.3. It should be noted that Theorem 15.3.3 for Hopf algebras arising in algebraic topology was established by Milnor and Moore [132].

Section 15.4 is a slight reformulation of ideas of [175]. Our description of Nichols algebra in Section 15.5 and the mathematics of this section is found in [13, Section 2.1]. The universal mapping property description of the Nichols algebra is implicit in the reference just cited.

We basically follow Jantzen [78, Section 4.1] in our description of the Drinfel'd-Jimbo quantized enveloping algebras [44,81]. The purpose of Section 15.6 is to describe the quantized enveloping algebras the language of Yetter-Drinfel'd categories as was done by Andruskiewitsch and Schneider in the classification program of certain types of pointed Hopf algebras. See their papers [10], where in particular the $(\mathrm{ad}_{\mathsf{c}}(e_i))^{1-a_{i,j}}(e_j)$'s and $(\mathrm{ad}_{\mathsf{c}}(f_i))^{1-a_{i,j}}(f_j)$'s are shown to be primitive in the Appendix, and [12], where the notion of linking parameters is introduced.

Constructing Hopf algebras by Ore extensions is a useful technique. Section 15.7 borrows heavily from the paper [19] by Beattie, Dăscălescu, and Grünenfelder to set up analysis of the example of their paper in Exercise 16.2.1 of Chapter 16, Section 16.2.

The representation theory of quantized enveloping algebras is quite deep and used to study representations of generalized quantized enveloping algebras. For the former see Lusztig's book [112] and the paper by Rosso [182] in particular. For the latter the reader is pointed to the papers by Pei and Rosso [156] and by Andruskiewitsch, Schneider, and the author [7].

Chapter 16

Finite-dimensional Hopf algebras in characteristic 0

We begin by reviewing some of the results discussed for finite-dimensional Hopf algebras H with antipode S over any field. First of all S is bijective and has finite order. Thus S is a diagonalizable operator when k is algebraically closed and has characteristic 0. There exist a left integral $\Lambda \in H$ and a right integral $\lambda \in H^*$ such that $\lambda(\Lambda) = 1$. H is semisimple if and only if $\epsilon(\Lambda) \neq 0$ and H is cosemisimple (or equivalently H^* is semisimple) if and only if $\lambda(1) \neq 0$. By virtue of the trace formula $\mathrm{Tr}(S^2) = \epsilon(\Lambda)\lambda(1)$ it follows that H is semisimple and cosemisimple if and only if $\mathrm{Tr}(S^2) \neq 0$.

The Hopf algebra H is free as a left or right module over any of its Hopf subalgebras. The map $H \longrightarrow H^*$ given by $h \mapsto \lambda{\prec}S(h)$ for all $h \in H$ is an isomorphism of left H-modules. Thus H is a Frobenius algebra.

There are grouplike elements $g \in H$ and $\alpha \in H^*$ which relate Λ, λ, and S. These elements are determined by the equations $\Lambda h = \alpha(h)\Lambda$ for all $h \in H$ and $h^*\lambda = h^*(g)\lambda$ for all $h^* \in H^*$. The relationship between Λ, λ, and S is given by $S^4(h) = g(\alpha \rightharpoonup h \leftharpoonup \alpha^{-1})g^{-1}$ for all $h \in H$.

In this chapter we present a bit more of the extensive theory of two types of finite-dimensional Hopf algebras, semisimple and pointed. The classification of finite-dimensional pointed Hopf algebras whose group of grouplike elements is abelian over an algebraically closed field of characteristic zero is almost complete. At its basis is the theory of Nichols algebras. These are certain pointed irreducible Hopf algebras in a Yetter-Drinfel'd category. A source of semisimple Hopf algebras is biproducts which, as we will see, involve semisimple Hopf algebras in a Yetter-Drinfel'd category.

In Section 16.1 we list various characterizations of semisimple Hopf algebras and briefly consider semisimple Hopf algebras in the Yetter-Drinfel'd category ${}^H_H\mathcal{YD}$. In Section 16.2 we construct a family of n^4-dimensional pointed Hopf algebras which contains infinitely many isomorphism types.

In Section 16.3 we consider some basic classification results, mainly for semisimple Hopf algebras, some of which are presented without proof. This section serves as a modest springboard into the difficult classification problem for semisimple Hopf algebras. Throughout this chapter H is a finite-dimensional Hopf algebra over k.

16.1 Characterizations of semisimple Hopf algebras

Semisimple Hopf algebras are finite-dimensional by part (b) of Proposition 8.2.5. We first characterize finite-dimensional Hopf algebras which are semisimple in the algebraically closed characteristic zero case. To this end we calculate a trace.

Lemma 16.1.1. *Suppose that k is an algebraically closed field, C is a simple coalgebra over k, and T is a diagonalizable coalgebra automorphism of C. Then*

$$\mathrm{Tr}(T) = \left(\sum_{i=1}^{n} \lambda_i\right)\left(\sum_{i=1}^{n} \lambda_i^{-1}\right)$$

where $\lambda_1, \ldots, \lambda_n$ are eigenvalues for T.

Proof. $C \simeq \mathrm{C}_n(k)$ for some $n \geq 1$ by Corollary 2.3.10. Thus we may assume $C = \mathrm{C}_n(k)$. The crux of the proof will be to show that there is a simple left coideal M of C such that $T(M) \subseteq M$. Necessarily $\mathrm{Dim}(M) = n$.

T^* is an algebra automorphism of $C^* = \mathrm{M}_n(k)$. By [186, Corollary 9.122], a consequence of the Skolem-Noether Theorem [186, Corollary 9.121], there is an invertible matrix $u \in \mathrm{M}_n(k)$ such that $T^*(a) = uau^{-1}$ for all $a \in \mathrm{M}_n(k)$. Identify $C^* = \mathrm{M}_n(k)$ with $\mathrm{End}(V)$, where V is an n-dimensional vector space over k. Since k is algebraically closed u has an eigenvalue $\lambda \in k$. Let $v \in V$ be a non-zero vector satisfying $u(v) = \lambda v$. Regard $\mathrm{End}(V)$ and V as left $\mathrm{End}(V)$-modules via function composition and evaluation respectively. Then V is a simple module and the evaluation map $\mathrm{e}_v : \mathrm{End}(V) \longrightarrow V$ given by $\mathrm{e}_v(a) = a(v)$ for all $a \in \mathrm{End}(V)$ is a module map. Therefore $L = \mathrm{Ker}(\mathrm{e}_v) = \{a \in \mathrm{End}(V) \mid a(v) = 0\}$ is a maximal left ideal of $\mathrm{End}(V)$ of codimension $n^2 - n$. Observe that $T^*(L) \subseteq L$. Set $M = L^{\perp}$. Then M is a minimal left coideal of C by Corollary 2.3.8 and $T(M) \subseteq M$ by part (c) of Theorem 1.3.10. Using (1.2) we see that $\mathrm{Dim}(M) = n$.

Since T is diagonalizable and $T(M) \subseteq M$ it follows that the restriction $T|M$ is diagonalizable. Let $\{m_1, \ldots, m_n\}$ be a basis of eigenvectors for

$T|M$ and let $\lambda_1, \ldots, \lambda_n \in k$ satisfy $T(m_i) = \lambda_i m_i$ for all $1 \leq i \leq n$. Then $\lambda_1, \ldots, \lambda_n$ are non-zero scalars since $T|M$ is one-one. For each $1 \leq i \leq n$ write $\Delta(m_i) = \sum_{j=1}^{n} c_{i,j} \otimes m_j$. Then the $c_{i,j}$'s satisfy the comatrix identities and thus span a non-zero subcoalgebra D of C. Since C is simple $D = C$. Since $\mathrm{Dim}(C) = n^2$ necessarily the $c_{i,j}$'s form a basis for C. Applying $T \otimes T$ to both sides of the equation for $\Delta(m_i)$ yields $\sum_{j=1}^{n} \lambda_i c_{i,j} \otimes m_j = \sum_{j=1}^{n} T(c_{i,j}) \otimes \lambda_j m_j$. Therefore $T(c_{i,j}) = \lambda_i \lambda_j^{-1} c_{i,j}$ for all $1 \leq i, j \leq n$. Since $\{c_{i,j}\}_{1 \leq i,j \leq n}$ is a basis for C we calculate

$$\mathrm{Tr}(T) = \sum_{i,j=1}^{n} \lambda_i \lambda_j^{-1} = \left(\sum_{i=1}^{n} \lambda_i\right)\left(\sum_{i=1}^{n} \lambda_i^{-1}\right).$$

□

Theorem 16.1.2. *Let k be a field of characteristic zero and suppose that H is a finite-dimensional Hopf algebra with antipode S over k. Then the following are equivalent:*

(a) *H is cosemisimple.*
(b) $\mathrm{Tr}(S^2) \neq 0$.
(c) *H is semisimple.*
(d) $S^2 = \mathrm{I}_H$.
(e) *$\chi : H \longrightarrow k$ defined by $\chi(a) = \mathrm{Tr}(\mathrm{r}(a))$ for all $a \in H$ is a right integral for H.*

Proof. Part (a) implies part (b). Suppose H is cosemisimple. By Exercise 10.8.8 and Theorem 10.8.2 we may assume that k is algebraically closed. Since H is cosemisimple it is the direct sum of its simple subcoalgebras. Let C be a simple subcoalgebra of H. Then $S^2(C) = C$ by Corollary 10.8.4.

Now S^2 has finite order by Theorem 10.5.6. Since k is algebraically closed of characteristic zero S^2 is diagonalizable. Thus $\mathrm{Tr}(S^2) = \left(\sum_{i=1}^{n} \lambda_i\right)\left(\sum_{i=1}^{n} \lambda_i^{-1}\right)$ where $\lambda_1, \ldots, \lambda_n$ are roots of unity by Lemma 16.1.1. Since the characteristic of k is zero we may assume that $\lambda_1, \ldots, \lambda_n \in \boldsymbol{C}$, the field of complex numbers. See Exercise 16.1.1. Thus

$$\mathrm{Tr}(S^2|C) = \left(\sum_{i=1}^{n} \lambda_i\right)\left(\sum_{i=1}^{n} \lambda_i^{-1}\right) = \left(\sum_{i=1}^{n} \lambda_i\right)\overline{\left(\sum_{i=1}^{n} \lambda_i\right)} = \left|\sum_{i=1}^{n} \lambda_i\right|^2$$

is a non-negative real number. Therefore $\mathrm{Tr}(S^2) = 1 + \sum_C \mathrm{Tr}(S^2|C) \geq 1$, where C runs over the simple subcoalgebras $C \neq k1$ of H. We have shown that $\mathrm{Tr}(S^2) \neq 0$ and thus part (a) implies part (b).

Part (b) implies part (c) by Theorem 10.4.3. To show part (c) implies part (d), assume that H is semisimple. Then H^* is cosemisimple. We have just shown H^* is semisimple; thus H is semisimple and cosemisimple. In particular $\mathrm{Tr}(S^2) \neq 0$.

Now $\mathrm{Tr}(S^2) = (\mathrm{Dim}(H))\mathrm{Tr}(S^2|x_H H)$ by Proposition 10.7.7 and $S^4 = \mathrm{I}_H$ by part (d) of Corollary 10.5.7. Since the characteristic of k is not 2, the last equation implies S^2 is a diagonalizable endomorphism of H with eigenvalues ± 1. Choose a basis of eigenvectors for S^2. Let n_+ be the number of basis vectors belonging to the eigenvalue 1 and let n_- be the number belonging to -1. By the preceding trace formula $n_+ - n_- = (n_+ + n_-)m$ for some integer m which is not zero since $\mathrm{Tr}(S^2) \neq 0$. Squaring both sides of this equation yields

$$-2n_+n_- = (m^2-1)n_+^2 + 2m^2n_+n_- + (m^2-1)n_-^2 \geq 0.$$

Therefore $n_+n_- = 0$. Since $n_+ \neq 0$ necessarily $n_- = 0$. We have shown $S^2 = \mathrm{I}_H$. Thus part (c) implies part (d).

That part (d) implies part (e) follows by part (b) of Proposition 10.4.2. Since $\chi(1) = \mathrm{Dim}(H)1 \neq 0$, part (e) implies part (a) by Theorem 10.8.2.□

We turn our attention to biproducts and in the process broach the rich theory of finite-dimensional Hopf algebras in Yetter-Drinfel'd categories. The context of our discussion is Section 11.6.

Suppose B is a finite-dimensional Hopf algebra in ${}_H^H\mathcal{YD}$. Then $A = B{\times}H$ is a finite-dimensional Hopf algebra over k. Let Λ be a non-zero left integral for A and let $\alpha \in \mathrm{G}(A^*)$ be the A-distinguished grouplike element for A^*. Write $\Lambda = \sum_{i=1}^n b_i{\times}h_i (= \sum_{i=1}^n b_i{\otimes}h_i)$ where n is as small as possible. Then $\{h_1, \ldots, h_n\}$ is linearly independent by Lemma 1.2.2. Recall that $(b{\times}h)(b'{\times}h') = bb'{\times}hh'$ if $h = 1$ or $b' = 1$. In particular $\alpha' \in \mathrm{G}(H^*)$, where $\alpha'(h) = \alpha(1{\times}h)$ for all $h \in H$.

Let $h \in H$. Then

$$\alpha'(h)\Lambda = \alpha(1{\times}h)\Lambda = \Lambda(1{\times}h) = \sum_{i=1}^n b_i{\times}h_ih.$$

Hence $h_1, \ldots, h_n \in R_{\alpha'}$ which is a one-dimensional subspace of H by part (b) of Proposition 10.6.2. Consequently $n = 1$ and $\Lambda = \Lambda_B{\times}\Lambda_H$, where $\Lambda_B = b_1$ and $\Lambda_H = h_1$. Observe that $\epsilon(b)\Lambda = \epsilon(b{\times}1)\Lambda = b\Lambda_B{\times}\Lambda_H$, or equivalently $b\Lambda_B = \epsilon(b)\Lambda_B$, for all $b \in B$. We generalize Definition 10.1.1 from ${}_k^k\mathcal{YD} = k\text{-Vec}$ to ${}_H^H\mathcal{YD}$.

Definition 16.1.3. Let H be a Hopf algebra over the field k and suppose B is a bialgebra in ${}_H^H\mathcal{YD}$. A *left integral* (respectively *right integral*) *for* B is an element $\Lambda_B \in B$ which satisfies $b\Lambda_B = \epsilon(b)\Lambda_B$ (respectively $\Lambda_B b = \epsilon(b)\Lambda_B$) for all $b \in B$.

Thus Λ_B is a left integral for B.

Suppose further that A is semisimple. Then $\alpha = \epsilon$ which means that $\alpha' = \epsilon_H$ and thus Λ_H is a right integral for H. Since $0 \neq \epsilon(\Lambda) = \epsilon_B(\Lambda_B)\epsilon_H(\Lambda_H)$ it follows that $\epsilon_B(\Lambda_B)$, $\epsilon_H(\lambda_H) \neq 0$. In particular H is semisimple and Λ_H is a two-sided integral for H. Let $h \in H$. By virtue of the calculation

$$\begin{aligned}\epsilon_H(h)\Lambda = \epsilon(1\times h)\Lambda = (1\times h)\Lambda &= h_{(1)}\cdot\Lambda_B\times h_{(2)}\Lambda_H \\ &= h_{(1)}\cdot\Lambda_B\times\epsilon(h_{(2)})\Lambda_H = h\cdot\Lambda_B\times\Lambda_H\end{aligned}$$

we see that $h\cdot\Lambda_B = \epsilon_H(h)\Lambda_B$. At this point we have a criterion for the semisimplicity of $A\times H$ in terms of B and H.

Proposition 16.1.4. *Let H be a finite-dimensional Hopf algebra over k and let B be a finite-dimensional Hopf algebra in ${}^H_H\mathcal{YD}$. Then the Hopf algebra $A = B\times H$ is semisimple if and only if*

(a) *H is semisimple and*
(b) *B has a left integral Λ_B which satisfies $\epsilon(\Lambda_B) \neq 0$ and $h\cdot\Lambda_B = \epsilon_H(h)\Lambda_B$ for all $h \in H$.*

Proof. We need only show that the conditions of (a) and (b) imply that $A = B\times H$ is semisimple. Suppose the conditions hold and let Λ_H be a left integral for H such that $\epsilon_H(\Lambda_H) = 1$ and set $\Lambda = \Lambda_B\times\Lambda_H$. Then $\epsilon(\Lambda) = \epsilon_B(\Lambda_B)\epsilon_H(\Lambda_H) \neq 0$. We need only show that $\Lambda = \Lambda_B\times\Lambda_H$ is a left integral for A. Let $b \in B$ and $h \in H$. Since $b\times h = (b\times 1)(1\times h)$ the calculations

$$(b\times 1)\Lambda = b\Lambda_B\times\Lambda_H = \epsilon_B(b)\Lambda_B\times\Lambda_H$$

and

$$\begin{aligned}(1\times h)\Lambda &= h_{(1)}\cdot\Lambda_B\times h_{(2)}\Lambda_H \\ &= h_{(1)}\cdot\Lambda_B\times\epsilon_H(h_{(2)})\Lambda_H \\ &= h\cdot\Lambda_B\times\Lambda_H \\ &= \epsilon_H(h)\Lambda_B\times\Lambda_H\end{aligned}$$

show that Λ is a left integral for A. □

The proposition provides one of many ways of constructing semisimple Hopf algebras.

Exercises

Exercise 16.1.1. Suppose k is a field of characteristic 0, $n \geq 1$ and $q \in k$ is a primitive n^{th} root of unity. Show that there is a one-one map of fields $\boldsymbol{Q}[q] \longrightarrow \boldsymbol{C}$, where $\boldsymbol{Q}$ is the field of rational numbers, the prime field of k. [Hint: Show that q and $e^{2\pi\imath/n}$ have the same irreducible polynomial over $\boldsymbol{Q}$, where $\imath^2 = -1$.]

Exercise 16.1.2. Let G be a group and B be an algebra over k with a vector space decomposition $B = \bigoplus_{g\in G} B_g$ which is a G-graded algebra and coalgebra structure. Suppose $\pi : G \longrightarrow \mathrm{Aut}_{bialg}(B)$ is a homomorphism from the group G to the group of bialgebra automorphisms of B under function composition. Let $H = k[G]$ be the group algebra of G over k.

(a) Regard B as a left H-module via $g{\cdot}b = \pi(g)(b)$ for all $g \in G$ and $b \in B$ and regard B as a left H-comodule by $\rho(b) = g\otimes b$ for all $g \in G$ and $b \in B_g$. Suppose further that $g{\cdot}B_h \subseteq B_{ghg^{-1}}$ for all $g, h \in G$. Show that $(B, \cdot, \rho)$ is an object of ${}_H^H\mathcal{YD}$.

(b) Suppose further that $B_g \neq (0)$ implies $g \in \mathrm{Ker}(\pi)$. Show that the k-bialgebra B is also a bialgebra in ${}_H^H\mathcal{YD}$.

[Hint: See Section 4.4 and Exercises 11.2.8, 11.3.2, and 11.6.4.]

Exercise 16.1.3. Use Exercise 16.1.2 to construct a 12-dimensional semisimple Hopf algebra $A = B{\times}H$ over $\boldsymbol{C}$, where $B = \boldsymbol{C}[\boldsymbol{Z}_3]$ and $H = \boldsymbol{C}[\boldsymbol{Z}_2{\times}\boldsymbol{Z}_2]$, such that neither A nor its dual is a group algebra.

16.2 Isomorphism types of Hopf algebras of the same dimension

There are only finitely many isomorphism classes of semisimple Hopf algebras of the same dimension over an algebraically closed field of characteristic zero; see the chapter notes following Section 16.3. Suppose $n > 2$ and the field k contains a primitive n^{th} root of unity q. There is a family $\{H_q(a)\}_{a\in k}$ of n^4-dimensional Hopf algebras over any field k which accounts for infinitely many isomorphism classes when k is infinite.

We construct the members of this family by generators and relations and use the Diamond Lemma to determine a basis for each of them. Another approach for construction is via Ore extensions which is outlined in the exercises.

Theorem 16.2.1. *Suppose $n > 2$ and the field k contains a primitive n^{th} root of unity q. Let $a \in k$. Then:*

(a) *There exists a Hopf algebra* $H_q(a)$ *over* k *whose structure is determined as follows. As an algebra* $H_q(a)$ *is generated by* c, x, y *subject to the relations*

$$c^{n^2} = 1, \quad x^n = c^n - 1, \quad y^n = c^n - 1,$$

$$xc = q^{-1}cx, \quad yc = qcy, \quad \textit{and} \quad yx = qxy + a(c^2 - 1).$$

As a coalgebra $H_q(a)$ *is determined by*

$$\Delta(c) = c \otimes c \quad \textit{and} \quad \Delta(z) = c \otimes z + z \otimes 1, \quad \textit{where } z = x, y.$$

(b) $H_q(a)$ *is pointed,* $\mathrm{G}(H_q(a)) = (c)$ *and has order* n^2*, and*

$$\mathcal{B} = \{c^i x^j y^\ell \,|\, 0 \leq i < n^2, 0 \leq j, \ell < n\}$$

is a basis for $H_q(a)$ *over* k*. In particular* $\mathrm{Dim}\, H_q(a) = n^4$.

(c) *Suppose* $z \in H_q(a)$ *satisfies* $\Delta(z) = g \otimes z + z \otimes 1$*, where* $g \in \mathrm{G}(H_q(a))$*. Then* $z \in H_q(a)_0$ *or* $g = c$ *and* $z = \alpha(c-1) + \beta x + \gamma y$ *for some* $\alpha, \beta, \gamma \in k$*, where* $\beta \neq 0$ *or* $\gamma \neq 0$.

(d) *Let* $b \in k$*. Then* $H_q(b) \simeq H_q(a)$ *if and only if* $b = ua$ *for some* n^{th} *root of unity* $u \in k$.

Proof. We mimic the construction of the Hopf algebra $H_{n,q}$ of Section 7.3 to establish part (a). Let $\mathcal{C}$ be the coalgebra over k with basis of symbols $\{A, C, X, Y\}$ whose coproduct is given by

$$\Delta(A) = A \otimes A, \quad \Delta(C) = C \otimes C, \quad \text{and} \quad \Delta(Z) = C \otimes Z + Z \otimes A \quad \text{for } Z = X, Y.$$

Necessarily $\epsilon(A) = \epsilon(C) = 1$ and $\epsilon(X) = \epsilon(Y) = 0$. Let $(T(\mathcal{C}), \boldsymbol{\Delta}, \boldsymbol{\epsilon})$ be the coalgebra structure of the free bialgebra on $\mathcal{C}$ and let I be the ideal of $T(C)$ generated by

$$A - 1, \quad C^{n^2} - 1, \quad X^n - (C^n - 1), \quad Y^n - (C^n - 1)$$

$$XC - q^{-1}CX, \quad YC - qCY, \quad \text{and} \quad YX - (qXY + a(C^2 - 1)).$$

Let $\pi : T(\mathcal{C}) \longrightarrow T(\mathcal{C})/I$ be the projection. Note that the relations of part (a) hold for $c = \pi(C)$, $x = \pi(X)$, and $y = \pi(Y)$.

We show that I is a coideal of $T(\mathcal{C})$ and is therefore a bi-ideal of $T(\mathcal{C})$. That $\boldsymbol{\epsilon}(I) = (0)$ is easy to establish since $\boldsymbol{\epsilon}$ is an algebra map which vanishes on generators of I. To show that $\boldsymbol{\Delta}(I) \subseteq I \otimes T(\mathcal{C}) + T(\mathcal{C}) \otimes I$ is a matter of showing that $(\pi \otimes \pi) \circ \boldsymbol{\Delta}$ vanishes on generators of I. That $(\pi \otimes \pi) \circ \boldsymbol{\Delta}$ vanishes on the generators of I listed above is a matter of definition with exception of $X^n - (C^n - 1)$ and $Y^n - (C^n - 1)$. Since $(x\otimes 1)(c\otimes x) = q^{-1}(c\otimes x)(x\otimes 1)$, by Corollary 7.2.2 we can make the calculation

$$\begin{aligned}((\pi\otimes\pi)\circ\boldsymbol{\Delta})(X^n) &= ((\pi\otimes\pi)\circ\boldsymbol{\Delta}(X))^n \\ &= (c\otimes x + x\otimes 1)^n \\ &= c^n\otimes x^n + x^n\otimes 1 \\ &= c^n\otimes(c^n - 1) + (c^n - 1)\otimes 1 \\ &= c^n\otimes c^n - 1\otimes 1 \\ &= ((\pi\otimes\pi)\circ\boldsymbol{\Delta})(C^n - 1)\end{aligned}$$

which shows that $(\pi\otimes\pi)\circ\boldsymbol{\Delta}$ vanishes on $X^n - (C^n - 1)$. Since $(y\otimes 1)(c\otimes y) = q(c\otimes y)(y\otimes 1)$, for the same reasons $(\pi\otimes\pi)\circ\boldsymbol{\Delta}$ vanishes on $Y^n - (C^n - 1)$. Thus I is a bi-ideal of $T(\mathcal{C})$.

Let $H_q(a) = T(\mathcal{C})/I$. Then $H_q(a)$ is a bialgebra over k. We now invoke Corollary 5.1.14 and Proposition 7.6.3 to conclude that $H_q(a)$ is a pointed Hopf algebra with $\mathrm{G}(H_q(a)) = (c)$. In particular we have established part (a).

To complete the proof of part (b) we need only show that $\mathcal{B}$ is a basis for $H_q(a)$. To this end we use the Diamond Lemma and look at the discussion following the proof of Lemma 7.3.1 to formulate an argument.

Order the generators of $T(\mathcal{C}) = k\{A, C, X, Y\}$ by $A < C < X < Y$ and translate the relations for $H_q(a)$ into the substitution rules

$$A \leftarrow 1, \quad C^{n^2} \leftarrow 1, \quad X^n \leftarrow C^n - 1, \quad Y^n \leftarrow C^n - 1,$$

$$XC \leftarrow q^{-1}CX, \quad YC \leftarrow qCY, \quad \text{and} \quad YX \leftarrow qXY + a(C^2 - 1).$$

To establish that $\mathcal{B}$ is a linear basis for $T(\mathcal{C})/I = H_q(a)$ we must resolve the following ten ambiguities, all of which are overlap ambiguities:

$$(C^{\ell}C^{n^2-\ell})C^{\ell} = C^{\ell}(C^{n^2-\ell}C^{\ell}),$$

$$(X^{\ell}X^{n-\ell})X^{\ell} = X^{\ell}(X^{n-\ell}X^{\ell}), \quad (Y^{\ell}Y^{n-\ell})Y^{\ell} = Y^{\ell}(Y^{n-\ell}Y^{\ell}),$$

$$(XC)C^{n^2-1} = XC(C^{n^2-1}), \quad (YC)C^{n^2-1} = Y(CC^{n^2-1}),$$

$$(XX^{n-1})C = X(X^{n-1}C), \quad (YY^{n-1})C = Y(Y^{n-1}C),$$

$$(YX)C = Y(XC),$$

$$(YX)X^{n-1} = Y(XX^{n-1}), \quad \text{and} \quad (Y^{n-1}Y)X = Y^{n-1}(YX).$$

The first eight are easy to resolve. Write "$\Rightarrow$" for "$\overset{S}{\Rightarrow}$". As far as the ninth

$$(YX)X^{n-1} \Rightarrow (qXY + a(C^2-1))X^{n-1} = qXYX^{n-1} + a(C^2-1)X^{n-1}.$$

By induction on m then monomial YX^m reduces to

$$q^m X^m Y + a\left(\left(\sum_{\ell=0}^{m-1} q^{-\ell}\right) C^2 - \left(\sum_{\ell=0}^{m-1} q^{\ell}\right) 1\right) X^{m-1}$$

and thus $(YX)X^{n-1}$ reduces to

$$\begin{aligned} & qX\left(q^{n-1}X^{n-1}Y + a(-q^{-(n-1)}C^2 - (-q^{n-1})1)\right) X^{n-2} \\ & \qquad + a(C^2-1)X^{n-1} \\ & \Rightarrow q^n X^n Y + qa(-q^{-1}C^2 + q^{-1}1)X^{n-1} + a(C^2-1)X^{n-1} \\ & = X^n Y \\ & \Rightarrow (C^n - 1)Y. \end{aligned}$$

Now $Y(XX^{n-1}) \Rightarrow Y(C^n-1) \Rightarrow \cdots \Rightarrow (C^n-1)Y$ since $YC^n \Rightarrow \cdots \Rightarrow q^nC^nY = C^nY$. Therefore the ninth ambiguity is resolved. As for the tenth, observe by induction on m that

$$Y^m X \Rightarrow \cdots \Rightarrow q^m XY^m + a\left(\left(\sum_{\ell=0}^{m-1} q^{2(m-1)-\ell}\right) C^2 - \left(\sum_{\ell=0}^{m-1} q^{\ell}\right) 1\right) Y^{m-1}$$

for all $m \geq 1$ and mimic the resolution of the ninth. We have established that $\mathcal{B}$ is a linear basis for $H_q(a)$ which concludes the proof of part (b).

To show part (c) we let $K = k[\mathrm{G}(H_q(a))] = H_q(a)_0$ and note that $H_q(a)$ is a free left K-module with basis of monomials $\{x^j y^\ell\}_{0\leq j,\ell<n}$. To proceed we derive a coproduct formula for these monomials. Since $(x\otimes 1)(c\otimes x) = q^{-1}(c\otimes x)(x\otimes 1)$ and $(y\otimes 1)(c\otimes y) = q(c\otimes y)(y\otimes 1)$ it follows by (7.2) that

$$\Delta(x^j) = \sum_{u=0}^{j} \binom{j}{u}_{q^{-1}} c^{j-u}x^u \otimes x^{j-u}$$

and

$$\Delta(y^\ell) = \sum_{v=0}^{\ell} \binom{\ell}{v}_{q} c^{\ell-v}y^v \otimes y^{\ell-v}$$

for all $0 \leq j, \ell < n$. Therefore

$$\Delta(x^j y^\ell) = \sum_{u=0}^{j}\sum_{v=0}^{\ell} q^{-(\ell-v)u} \binom{j}{u}_{q^{-1}} \binom{\ell}{v}_{q} c^{j+\ell-(u+v)} x^u y^v \otimes x^{j-u} y^{\ell-v} \quad (16.1)$$

for all $0 \leq j, \ell < n$.

Regard $\mathcal{M} = \boldsymbol{N} \times \boldsymbol{N}$ as a monoid under addition and consider the set $\mathcal{S} = \{(j, \ell) \in \mathcal{M} \,|\, 0 \leq j, \ell < n\}$. For $\mathbf{s} = (j, \ell) \in \mathcal{S}$ let $z^{(\mathbf{s})} = x^j y^\ell$. We have noted that $H_q(a)$ is a free left K-module with basis $\{z^{(\mathbf{s})}\}_{\mathbf{s} \in \mathcal{S}}$. By (16.1) and part (a) of Proposition 7.2.1 we have

$$\Delta(z^{(\mathbf{s})}) = \sum_{\mathbf{u}+\mathbf{v}=\mathbf{s}} u_{(\mathbf{u},\mathbf{v})} z^{(\mathbf{u})} \otimes z^{(\mathbf{v})} \tag{16.2}$$

for all $\mathbf{s} \in \mathcal{S}$, where $u_{(\mathbf{u},\mathbf{v})}$ is a unit in K.

Let $z \in H_q(a)$ and write $z = \sum_{\mathbf{s} \in \mathcal{S}} \alpha_{\mathbf{s}} z^{(\mathbf{s})}$, where $\alpha_{\mathbf{s}} \in K$. Suppose further $g \in \mathrm{G}(H_q(a))$ and $\Delta(z) = g \otimes z + z \otimes 1$. Applying Δ to both sides of the equation for z yields

$$\begin{aligned} &\sum_{\mathbf{s} \in \mathcal{S}} \sum_{\mathbf{u}+\mathbf{v}=\mathbf{s}} \Delta(\alpha_{\mathbf{s}}) u_{(\mathbf{u},\mathbf{v})} (z^{(\mathbf{u})} \otimes z^{(\mathbf{v})}) \\ &\quad = g \otimes \left(\sum_{\mathbf{s} \in \mathcal{S}} \alpha_{\mathbf{s}} z^{(\mathbf{s})} \right) + \left(\sum_{\mathbf{s} \in \mathcal{S}} \alpha_{\mathbf{s}} z^{(\mathbf{s})} \right) \otimes 1 \end{aligned}$$

Now $\{x^{(\mathbf{s})} \otimes x^{(\mathbf{s}')}\}_{\mathbf{s},\mathbf{s}' \in \mathcal{S}}$ is a basis for $H_q(a) \otimes H_q(a)$ as a left $K \otimes K$-module. Comparing coefficients in the preceding equation we conclude $\Delta(\alpha_{\mathbf{s}}) = 0$, and hence $\alpha_{\mathbf{s}} = 0$, unless $\mathbf{s} = (0,0)$, $(1,0)$, or $(0,1)$. Therefore $z = d + \beta x + \gamma y$, where $d \in K$ and $\beta, \gamma \in k$. From

$$\Delta(d) + \beta(c \otimes x + x \otimes 1) + \gamma(c \otimes y + y \otimes 1) = g \otimes (d + \beta x + \gamma y) + (d + \beta x + \gamma y) \otimes 1$$

it follows that $\Delta(d) = g \otimes d + d \otimes 1$, $\beta c = \beta g$, and $\gamma c = \gamma g$. Therefore $\beta = \gamma = 0$ or $g = c$ and $d = \alpha(c - 1)$ for some $\alpha \in k$. We have established part (c).

Suppose $b \in k$ and $f : H_q(b) \longrightarrow H_q(a)$ is an isomorphism of Hopf algebras. Since $x \notin K = H_q(a)_0$ and f is an isomorphism, $f(c) \notin H_q(b)_0$. Since $\Delta(f(x)) = f(c) \otimes f(x) + f(x) \otimes 1$, by part (c) necessarily $f(c) = c$ and $f(x) = \alpha(c-1) + \beta x + \gamma' y$ for some $\beta, \gamma' \in k$ where either $\beta \neq 0$ or $\gamma' \neq 0$. Applying f to both sides of the equation $q^{-1}cx = xc$ yields

$$\begin{aligned} \alpha q^{-1} c(c-1) + \beta q^{-1} cx + \gamma' q^{-1} cy &= \alpha(c-1)c + \beta xc + \gamma' yc \\ &= \alpha c(c-1) + \beta q^{-1} cx + \gamma' qcy. \end{aligned}$$

Since $c \neq 1$ and $q^2 \neq 1$ it follows that $\alpha = 0 = \gamma'$. Therefore $f(x) = \beta x$ and $\beta \neq 0$. Likewise $f(y) = \gamma y$ for some non-zero $\gamma \in k$.

Now apply f to both sides of the equations $x^n = c^n - 1$ and $y^n = c^n - 1$. As $x^n, y^n \neq 0$ we conclude that $\beta^n = 1 = \gamma^n$. Let "$\cdot$" denote the product in $H_q(b)$. Then $y \cdot x = qx \cdot y + b(c^2 - 1)$. Since $f(yx) = f(qxy) + a(c^2 - 1) =$

$q\gamma\beta x{\cdot}y + a(c^2-1)$ and $f(y){\cdot}f(x) = \gamma\beta y{\cdot}x = \gamma\beta(qx{\cdot}y + b(c^2-1))$ it follows that $a = \gamma\beta b$. Therefore $b = ua$ where $u = (\gamma\beta)^{-1}$ and which is an nth root of unity. Conversely, if $b = ua$, where $u \in k$ is an nth root of unity, then there is an isomorphism of Hopf algebras $f : H_q(b) \longrightarrow H_q(a)$ determined by $f(c) = c$, $f(x) = x$, and $f(y) = u^{-1}y$. □

Exercises

Exercise 16.2.1. Section 15.7 almost seems made to order for the Hopf algebra $H_q(a)$. Construct $H_q(a)$ as a quotient of iterated Ore extensions. Note that it is not necessary to use the Diamond Lemma to obtain a basis for $H_q(a)$.

Exercise 16.2.2. Replace the algebra relation $c^{n^2} = 1$ in the description of $H_q(a)$ by $c^N = 1$, where $N \geq 1$. Find necessary and sufficient conditions on N which result in a Hopf algebra of dimension Nn^2. Which values of N result in a family of Hopf algebras of infinitely many isomorphism types as a runs over k?

16.3 Some very basic classification results

Throughout this section k is an algebraically closed field of characteristic zero unless otherwise stated. We are primarily interested in two types of Hopf algebras, pointed and semisimple. The characterization of groups of orders p, p^2, and pq, where p and q are prime integers with $p < q$, are included in the elementary classification results for finite groups. We characterize our two types of Hopf algebras which have dimension p, p^2, or pq.

The pointed case is by far the easier to analyze. Our treatment of the semisimple case involves techniques which we do not fully justify.

Proposition 16.3.1. *Suppose the field k is algebraically closed of characteristic zero. Let H be a pointed Hopf algebra over k of dimension p, p^2, or pq, where p and q are prime integers, $p < q$. Then H is a Taft algebra of dimension p^2 or H is a group algebra.*

Proof. Suppose H is not a group algebra. Then $H \neq H_0$ which means Theorem 15.4.7 applies. Since the dimension of any Hopf subalgebra of H divides the dimension of H by Corollary 9.3.4, it follows that $H = F_{(\omega,n,\alpha,m)}$, where $\omega \in k$ is a primitive n^{th} root of unity, where $n > 1$, and $n|m$. Since $\mathrm{Dim}(H) = mn$ it follows that $m = n = p$ and thus

$H = H_{p,\omega}$.

□

Suppose H is a Hopf algebra of prime dimension p. If $p = 2$ then H is pointed and thus $H = k[\boldsymbol{Z}_2]$ by the previous result. Suppose $p > 2$ and H is not a group algebra. Then H^* is not a group algebra. Therefore H and H^* have unique grouplike elements by Corollary 9.3.5. This means H and H^* are unimodular of odd dimension. By Exercise 10.5.12 it follows that H and H^* are semisimple.

We will show that H is in fact a group algebra. To this end we use a fact from the representation theory of semisimple Hopf algebras. Our discussion begins with some basic ideas from the representation theory of algebras.

Let k be any field, suppose A is an algebra over k, and suppose M is a finite-dimensional left A-module. Then the character of M is the functional $\chi_M : A \longrightarrow \mathrm{End}(M)$ defined by $\chi_M(a) = \mathrm{Tr}(\mathrm{L}_M(a))$ for all $a \in A$, where $\mathrm{L}_M(a)$ is the endomorphism of M given by $\mathrm{L}_M(a)(m) = a{\cdot}m$ for all $m \in M$. Note that $\chi_M \in \mathrm{Cc}(A^o)$. If N is a left A-module also, then it easily follows that $\chi_{M\oplus N} = \chi_M + \chi_N$.

Let $\mathrm{C}_k(A)$ be the span of the χ_M's. Suppose further that A is a bialgebra over k and regard $M{\otimes}N$ as a left A-module according to Definition 5.7.1. Then $\mathrm{L}_{M\otimes N}(a) = \mathrm{L}_M(a_{(1)}){\otimes}\mathrm{L}_N(a_{(2)})$ for all $a \in A$. Hence $\chi_{M\otimes N} = \chi_M{*}\chi_N$, the convolution product of χ_M and χ_N. We have shown $\mathrm{C}_k(A)$ is a subalgebra of A^o when A is a bialgebra over k.

We return to our original context which is that k is algebraically closed and has characteristic 0. Suppose A is a finite-dimensional semisimple algebra over k. We may regard $A = \mathrm{M}_{n_1}(k)\oplus\cdots\oplus\mathrm{M}_{n_r}(k)$ as the direct sum of matrix algebras, where $r, n_1, \ldots, n_r \geq 1$. Therefore $A^o = A^* = \mathrm{C}_{n_1}(k)\oplus\cdots\oplus\mathrm{C}_{n_r}(k)$. The trace functional $\chi_{n_i} \in \mathrm{M}_{n_i}(k)^* = \mathrm{C}_{n_i}(k)$ is a cocommutative element of $\mathrm{C}_{n_i}(k)$. It is easy to see that $\chi_{n_1}, \ldots, \chi_{n_r}$ form a basis for $\mathrm{C}_k(A)$ and that $\mathrm{C}_k(A) = \mathrm{Cc}(A^*)$. When H is a Hopf algebra $\mathrm{C}_k(A)$ is a semisimple algebra over k.

The Class Equation, which is the dimension formula of the next result, has useful applications.

Theorem 16.3.2. *Suppose the field k is algebraically closed and has characteristic 0. Suppose H is a semisimple Hopf algebra over k and suppose $\{e_1, \ldots, e_n\}$ is a complete set of orthogonal idempotents for $\mathrm{C}_k(H)$. Then*

$$\mathrm{Dim}(H) = \sum_{i=1}^{n} \mathrm{Dim}(e_i H^*),$$

each $\mathrm{Dim}(e_iH^*)$ *divides* $\mathrm{Dim}(H)$, *and* $\mathrm{Dim}(e_iH^*) = 1$ *for some* $1 \le i \le n$. $\square$

Apropos of the theorem, the equation $\mathrm{Dim}(e_iH^*) = 1$ is particularly significant. Let λ be a non-zero right integral for H^*. Now $S^2 = \mathrm{I}_H$ by Theorem 16.1.2. The linear map $\boldsymbol{f} : H \longrightarrow H^*$ of Section 10.2 gives rise to a linear isomorphism $f : H \longrightarrow H^*$ defined by $f(a) = \lambda \prec a$ for all $a \in H$. It restricts to an isomorphism $\mathrm{Z}(H) \simeq \mathrm{Cc}(H^*) = \mathrm{C}_k(H)$ by part (a) of Corollary 10.7.2. Suppose $\mathrm{Dim}(e_iH^*) = 1$. Then $e_iH^* = R_{g^{-1}}$ for some $g \in \mathrm{G}(H)$; see Section 10.6. It is a straightforward exercise to show that $R_{g^{-1}} = \lambda \prec kg$. Thus $f(g) \in \mathrm{C}_k(H)$ since $e_iH^* = ke_i \subseteq \mathrm{C}_k(H)$. Since f is one-one and the restriction $f|\mathrm{Z}(H) : \mathrm{Z}(H) \longrightarrow \mathrm{C}_k(H)$ is an isomorphism, $g \in \mathrm{Z}(H)$. Let $\mathcal{I}$ be the set of the one-dimensional ideals of H^* generated by elements of $\mathrm{C}_k(H)$. We have established that

$$\boldsymbol{F} : \mathrm{G}(H) \cap \mathrm{Z}(H) \longrightarrow \mathcal{I} \tag{16.3}$$

given by $\boldsymbol{F}(g) = R_{g^{-1}}$ is a bijection.

Theorem 16.3.3. *Suppose* k *is an algebraically closed field of characteristic* 0 *and* H *is a Hopf algebra over* k *of prime dimension* p. *Then* $H \simeq k[\boldsymbol{Z}_p]$.

Proof. By Theorem 16.3.2 and (16.3) it follows that H contains p central grouplike elements. $\square$

Lemma 16.3.4. *Suppose that* k *is algebraically closed, has characteristic* 0, *and that* H *is a semisimple Hopf algebra of dimension* p^n, *where* $n \ge 1$ *and* p *is a prime integer. Then* H *contains a non-trivial central grouplike element.*

Proof. By Theorem 16.3.2 and (16.3) it follows that H contains at least two central grouplike elements. $\square$

Theorem 16.3.5. *Suppose that* k *is algebraically closed, has characteristic* 0, *and that* H *is a semisimple Hopf algebra of dimension* p^2, *where* p *is a prime integer. Then* H *is a group algebra.*

Proof. We give a partial proof. Since any Hopf subalgebra of H has dimension 1, p, or p^2, by the preceding lemma either H is a group algebra or has a Hopf subalgebra $K \simeq k[\boldsymbol{Z}_p]$ in the center of H. By results we do not go into here the latter forces H to be a group algebra. See [35, §7.6] for a complete proof. $\square$

The case H is semisimple of dimension pq, where p and q are prime integers and $p < q$, is interesting and can be treated by studying the associated Drinfel'd double $D(H)$. The result is:

Theorem 16.3.6. *Suppose that k is algebraically closed, has characteristic 0, and that H is a semisimple Hopf algebra of dimension pq, where p and q are prime integers with $p < q$. Then H or H^* is a group algebra over k.* □

At this point we introduce standard terminology for Hopf algebras.

Definition 16.3.7. A *trivial Hopf algebra* is a finite-dimensional Hopf algebra H over the field k such that H or H^* is isomorphic to a group algebra over k.

The terminology reflects interest in examples of semisimple Hopf algebras which are not group algebras or their duals and by no means is meant to disparage the theory of groups and their representations.

In the study of semisimple Hopf algebras of special dimensions, the case $\mathrm{Dim}(H) = p^3$, where p is a prime integer, would be a logical next step. We end this section with a description of the only 8-dimensional semisimple Hopf algebra over an algebraically closed field of characteristic 0 such that neither it nor its dual is a group algebra. We refer to this Hopf algebra as H_8.

As an algebra H_8 is generated by symbols x, y, and z subject to the relations

$$x^2 = 1, \quad y^2 = 1, \quad yx = xy, \quad zx = yz, \quad zy = xz, \quad \text{and}$$

$$z^2 = \frac{1}{2}(1 + x + y - xy).$$

The coalgebra structure of H_8 is determined by

$$\Delta(x) = x \otimes x, \quad \Delta(y) = y \otimes y, \quad \text{and}$$

$$\Delta(z) = \frac{1}{2}(1 \otimes 1 + 1 \otimes x + y \otimes 1 - y \otimes x)(z \otimes z).$$

Note that $\epsilon(x) = \epsilon(y) = \epsilon(z) = 1$ necessarily follows. The antipode S of H_8 is determined by $S(x) = x$, $S(y) = y$, and $S(z) = z$. Thus $S^2 = \mathrm{I}_H$. However $S \neq \mathrm{I}_H$ since H_8 is neither commutative nor cocommutative. A more direct way to see this is by the calculation $S(xz) = S(z)S(x) = zx = yz$.

Using the Diamond Lemma it is very easy to see that the relations determine an algebra over k with basis $\{x^i y^j z^\ell\}_{1\leq i,j,\ell<2}$. Using the methods of Section 7.3 the reader can show that this 8-dimensional algebra admits a Hopf algebra structure whose coproduct is as described above. Note that x and y generate the Klein 4-group $V \simeq \boldsymbol{Z}_2 \times \boldsymbol{Z}_2$. Since $|\mathrm{G}(H)|$ divides $\mathrm{Dim}(H_8) = 8$ and H_8 is not cocommutative, $\mathrm{G}(H) = V$. Note that 1 and xy are central grouplike elements, and these are the only central grouplike elements. Let $K = k[V]$. Also observe that H_8 is a free left and right K-module with basis $\{1, z\}$.

Let

$$\boldsymbol{\Lambda} = \frac{1}{4}(1 + x + y + xy) = \frac{1}{4}(1 + x)(1 + y)$$

be the two-sided integral for K which satisfies $\epsilon(\boldsymbol{\Lambda}) = 1$. Rewriting the relation

$$z^2 = \frac{1}{2}(1 + x + y - xy) = \frac{1}{2}(1 + x)(1 + y) - xy$$

as

$$z^2 = 2\boldsymbol{\Lambda} - xy$$

it is easy to see that $\Lambda = \dfrac{1}{2}(1 + z)\boldsymbol{\Lambda}$ is a two-sided integral for H_8 which satisfies $\epsilon(\Lambda) = 1$. The preceding equation implies that H_8 is semisimple and cosemisimple.

We analyze the algebra and coalgebra structure of H_8. Generalized integrals for K, that is the one-dimensional ideals of K, play a basic role in the structure analysis. There are four generalized integrals up to scalar multiple:

$$\boldsymbol{\Lambda} = \boldsymbol{\Lambda}_{++} = \frac{1}{4}(1 + x)(1 + y), \quad \boldsymbol{\Lambda}_{+-} = \frac{1}{4}(1 + x)(1 - y),$$

$$\boldsymbol{\Lambda}_{-+} = \frac{1}{4}(1 - x)(1 + y), \quad \text{and} \quad \boldsymbol{\Lambda}_{--} = \frac{1}{4}(1 - x)(1 - y).$$

At this point the reader is encouraged to determine $\mathrm{G}(H_8^*)$ and work out the correspondences of part (d) of Proposition 10.6.2 in detail. We note that $\eta \in \mathrm{G}(H_8^*)$ is determined by $\eta(z)$ as $\eta(z)^2 = \eta(x) = \eta(y) = \pm 1$. In particular $\mathrm{G}(H_8^*) \simeq \boldsymbol{Z}_2 \times \boldsymbol{Z}_2$.

Observe that $\boldsymbol{\Lambda}, \boldsymbol{\Lambda}_{--} \in \mathrm{Z}(H)$. Since $z^2 \in K$ it follows that $L(\boldsymbol{\Lambda}') = k\boldsymbol{\Lambda}' + kz\boldsymbol{\Lambda}'$ is a 2-dimensional left ideal of H_8, where $\boldsymbol{\Lambda}'$ is any one of the four generalized integrals for K, and H_8 is the direct sum of these four left ideals.

Since $|\mathrm{G}(H_8^*)| = 4$ there are four generalized left integrals for H_8 and these are:

$$\Lambda = (1+z)\mathbf{\Lambda}, \quad \Lambda_- = (1-z)\mathbf{\Lambda}, \quad \Lambda_\imath = (1+\imath z)\mathbf{\Lambda}, \quad \text{and} \quad \Lambda_{-\imath} = (1-\imath z)\mathbf{\Lambda},$$

where $\imath^2 = -1$. To see that these are generalized left integrals we use the fact that $z^2\mathbf{\Lambda} = \mathbf{\Lambda}$ and $z^2\mathbf{\Lambda}_{--} = -\mathbf{\Lambda}_{--}$; the latter follows by $z^2 = 2\mathbf{\Lambda} - xy$.

We have shown that $L(\mathbf{\Lambda}) = k\Lambda \oplus k\Lambda_-$ and $L(\mathbf{\Lambda}_{--}) = k\Lambda_\imath \oplus k\Lambda_{-\imath}$ are direct sums of one-dimensional left ideals of H_8 which are necessarily ideals. Since $z\mathbf{\Lambda}_{+-} = \mathbf{\Lambda}_{-+}$ and $z\mathbf{\Lambda}_{-+} = \mathbf{\Lambda}_{+-}$ it is easy to see that $L(\mathbf{\Lambda}_{+-}) \oplus L(\mathbf{\Lambda}_{-+})$ is an ideal of H_8 which must be isomorphic to $\mathrm{M}_2(k)$ as an algebra over k. An algebra isomorphism $\mathrm{M}_2(k) \simeq L(\mathbf{\Lambda}_{+-}) \oplus L(\mathbf{\Lambda}_{-+})$ is determined by the identification

$$\begin{pmatrix} E_{1\,1} & E_{1\,2} \\ E_{2\,1} & E_{2\,2} \end{pmatrix} = \begin{pmatrix} \mathbf{\Lambda}_{+-} & \mathbf{\Lambda}_{+-}z \\ \mathbf{\Lambda}_{-+}z & \mathbf{\Lambda}_{-+} \end{pmatrix}.$$

We now turn to the coalgebra structure of H_8. Since $\Delta(Kz) \subseteq Kz \otimes Kz$, as a coalgebra $H_8 = K \oplus Kz = k1 \oplus kx \oplus ky \oplus kz \oplus Kz$. Necessarily $Kz \simeq \mathrm{C}_2(k)$. Note that the span M of $m_1 = z$ and $m_2 = xz$ is a left coideal of H_8 and must therefore be simple. Let $c_{i,j}$ for $1 \leq i,j \leq 2$ be the elements of $C = Kz$ determined by $\Delta(m_i) = \sum_{j=1}^2 c_{i,j} \otimes m_j$ for $i = 1, 2$. Then the span of the $c_{i,j}$'s is a subcoalgebra of C which must be C since C is simple. In particular the $c_{i,j}$'s form a basis for C. A coalgebra isomorphism $\mathrm{C}_2(k) \simeq C$ is determined by the identifications

$$\begin{pmatrix} e_{1,1} & e_{1,2} \\ e_{2,1} & e_{2,2} \end{pmatrix} = \begin{pmatrix} c_{1,1} & c_{1,2} \\ c_{2,1} & c_{2,2} \end{pmatrix} = \begin{pmatrix} \frac{1}{2}(1+y)z & \frac{1}{2}(1-y)z \\ & \\ \frac{1}{2}(x-xy)z & \frac{1}{2}(x+xy)z \end{pmatrix}.$$

One could construct H_8 by starting with the tensor bialgebra $T(\mathrm{C}_2(k))$ and defining relations among the $e_{i\,j}$'s. These relations would be predicated on the following multiplication table

	$e_{1\,1}$	$e_{1\,2}$	$e_{2\,1}$	$e_{2\,2}$
$e_{1\,1}$	$\mathbf{\Lambda}_{++}$	$\mathbf{\Lambda}_{-+}$	$\mathbf{\Lambda}_{-+}$	$\mathbf{\Lambda}_{++}$
$e_{1\,2}$	$\mathbf{\Lambda}_{+-}$	$-\mathbf{\Lambda}_{--}$	$\mathbf{\Lambda}_{--}$	$-\mathbf{\Lambda}_{+-}$
$e_{2\,1}$	$\mathbf{\Lambda}_{+-}$	$\mathbf{\Lambda}_{--}$	$-\mathbf{\Lambda}_{--}$	$-\mathbf{\Lambda}_{+-}$
$e_{2\,2}$	$\mathbf{\Lambda}_{++}$	$-\mathbf{\Lambda}_{-+}$	$-\mathbf{\Lambda}_{-+}$	$\mathbf{\Lambda}_{++}$

and thus it would be a rather daunting exercise to write down the relations and apply the Diamond Lemma.

We finally note that $C^2 = KzKz = Kz^2 = K$. Thus $H_8 = C \oplus C^2$ as a coalgebra.

In the chapter notes we survey classification results. There are several notions which are important in connection with semisimple Hopf algebras.

Definition 16.3.8. A *Hopf algebra of Frobenius type* is a finite-dimensional Hopf algebra over the field k such that the dimension of a simple H-module divides $\mathrm{Dim}(H)$.

Group algebras of finite groups over algebraically closed fields of characteristic zero are of Frobenius type.

Definition 16.3.9. A *lower semisolvable Hopf algebra over* k is a finite-dimensional Hopf algebra H over k such that for some $s \geq 1$ there is a sequence $H = H_s \supseteq H_{s-1} \supseteq \cdots \supseteq H_0 = k$, where H_i is a normal Hopf subalgebra of H_{i+1} and $H_{i+1}/H_i^+ H_{i+1}$ is trivial for all $1 \leq i < s$. An *upper semisolvable Hopf algebra over* k is finite-dimensional Hopf algebra over k whose Hopf algebra dual is lower semisolvable.

Exercises

Exercise 16.3.1. Show that there are algebra projections $\pi : H_8 \longrightarrow K$ but none are Hopf algebra projections.

Exercise 16.3.2. Construct a Hopf algebra projection $\pi : H_8 \longrightarrow H$ where $H = k[(x)] \simeq k[\boldsymbol{Z}_2]$. (Therefore $H_8 \simeq B \times k[\boldsymbol{Z}_2]$ for some 4-dimensional Hopf algebra B in ${}_H^H\mathcal{YD}$.) Show that π is unique.

Exercise 16.3.3. Describe the Hopf algebra B in ${}_H^H\mathcal{YD}$ of Exercise 16.3.2 explicitly.

Exercise 16.3.4. Let $k = \boldsymbol{C}$. Since H_8 is the only 8-dimensional semisimple Hopf algebra over $\boldsymbol{C}$ which is neither commutative nor cocommutative, then each of H_8^{op}, H_8^{cop}, $H_8^{op\,cop}$, and H_8^* must be isomorphic to H_8. Find isomorphisms.

Exercise 16.3.5. Resolve the following. Let k be any field whose characteristic is not 2. There exists a semisimple cosemisimple 8-dimensional Hopf algebra H over k whose generators and relations and coalgebra structure are the same as those of H_8 and neither H nor H^* is a group algebra.

Chapter notes

Kaplansky in his University of Chicago Lecture Notes made ten conjectures on Hopf algebras [85, Appendix 2]. These have generated a great deal of research over the years. Conjectures five through ten concern finite-dimensional Hopf algebras over k.

His fifth conjecture is that H or H^* semisimple implies that the antipode S of H satisfies $S^2 = \mathrm{I}_H$. This was proved by Larson and the author when k has characteristic 0 [103, Theorem 4] which is part of the more extensive Theorem 16.1.2. In the characteristic $p > 0$ case Etingof and Gelaki showed that H and H^* semisimple implies $S^2 = \mathrm{I}_H$ [46, Theorem 3.1]. They did this by a lifting method which reduced the conjecture in the characteristic $p > 0$ case, when H and H^* are semisimple, to the characteristic 0 case. Their lifting method is quite important and has many applications.

The material of Section 16.1 is drawn from the papers [102] and [103] by Larson and the author with the exception of the results on biproducts. This is found in [168, Section 2].

Kaplansky's tenth conjecture is that for a given positive integer n, which is not divisible by the characteristic of k, that there are only finitely isomorphism classes of Hopf algebras over k of dimension n. This was shown to be the case for semisimple cosemisimple Hopf algebras (and thus semisimple or cosemisimple Hopf algebras in characteristic 0) by Ştefan [196, Theorem 2.2].

Several authors constructed about the same time families of Hopf algebras, with some restrictions on k, which provided counterexamples to the tenth conjecture: Andruskiewitsch and Schneider for $n = p^4$, where p is a prime integer [11, Theorem 0.3]; Beattie, Dăscălescu, and Grünenfelder for $n = p^4$, where $p > 2$ and is a prime integer [18, Theorem 1]; Gelaki for $n = m^2M$, where $2 < m < M$ [58, Corollary 3.3]; and Müller for $n = N^2$, where $N > 2$ [139, Theorem 5.13]. The first three families consist of pointed Hopf algebras. The examples of Andruskwietsch and Schneider follow from classification results established in [11]. The ones in the fourth are not pointed and their duals are not pointed.

Even though the classes contain infinitely many isomorphism types the members are all related in an interesting way. Masuoka showed that the members of each class are 2-cocycle twists of each other; see Section 7.7 [126, Theorem 1]. Etingof and Gelaki constructed an infinite family of Hopf algebras of the same dimension whose members are not isomorphic and no two are related by a twist [49]. We also note that an infinite family of non-isomorphic Hopf algebras of dimension 3 was constructed over certain fields of characteristic zero which are not algebraically closed by Caenepeel, Dăscălescu, and Le Bruyn [27].

The family of Hopf algebras described in Section 16.2 is that of [18] except that p is not required to be a prime integer. In that paper the Hopf algebras $H(a) = H_q(a)$ were constructed using the method of Ore extensions discussed in Section 15.7. We chose to construct them by yet again using the Diamond Lemma. To see how the method of Ore extensions is used to construct other families of pointed Hopf algebras the reader is directed to the papers [19] by Beattie, Dăscălescu, and Grünenfelder and [28] by Caenepeel, Dăscălescu, and Raianu for example.

From this point on the field k is algebraically closed field of characteristic zero. The literature on classification of finite-dimensional Hopf algebras H over k is quite extensive and many have contributed to it. There are basically three types of problems addressed: determine the pointed, semisimple, or all finite-dimensional Hopf algebras of a given dimension n, where n has a certain type of prime factorization (typically $n = p^\ell$, pq, or pq^2 where $\ell \geq 1$ and p, q are distinct primes); determine Hopf algebras of small dimension; and determine all finite-dimensional Hopf algebras of a given type.

We start with semisimple Hopf algebras H over k. There are a good many results at this point. In [231] Zhu establishes the class equation for H and derives Theorem 16.3.3, a result established for the special case of Kac algebras by Kac [82]. Let $n = \mathrm{Dim}(H)$. The case $n = pq$, where p and q are distinct primes, was settled in special cases before in full generality. Masuoka showed that H is trivial when $p = 2$ [120], Gelaki and Westreich showed that H is trivial when $p = 3$ [59, Theorem 3] and obtained partial results for $p = 5, 7$ as well in this paper. The partial results stemmed from [59, Theorem 3.5] which is that when $p = 5, 7$ then H is trivial if and only if H and H^* are of Frobenius type. This is Kaplansky's sixth conjecture [85, Page 45] for the Hopf algebras we are considering. The notion of Frobenius type for Hopf algebras originates the paper [137] by Montgomery and Witherspoon.

Theorem 16.3.6, which is that H is trivial when $p < q$, was obtained by Etingof and Gelaki [47, Theorem 6]. They used the Drinfel'd double $D(H)$ to prove this result. Later Schneider found a shorter and more direct proof based on a result on factorizable Hopf algebras of which the double is a special case [191, Theorem 3.2].

The case $n = p^2$ was settled by Masuoka by [125, Theorem 2] which is our Theorem 16.3.5. Lemma 16.3.4 is [125, Theorem 1]. Masuoka shows that when $n = p^3$, where p is an odd prime, that the non-trivial semisimple Hopf algebras of dimension n over k split into $p+8$ isomorphism classes [121, Theorem 3.1]. The Hopf algebra H_8 of dimension $n = 8 = 2^3$, due to Kac and Paljutkin [83], is shown by Masuoka to be the only 8-dimensional semisimple Hopf algebra over k which is not trivial [123]. Our treatment of H_8 follows Masuoka's description of it.

Kashina shows that when $n = 16 = 2^4$ the nontrivial semisimple Hopf algebras over k split into 16 isomorphism classes [91, Theorem 1.2]. She classifies all non-trivial semisimple Hopf algebras of dimension $n = 2^{n+1}$ with $\mathrm{G}(H) \simeq \boldsymbol{Z}_{2^{n-1}} \times \boldsymbol{Z}_2$. In [93] she describes two families of nontrivial semisimple Hopf algebras of dimension $n = 2^m$ which are biproducts. These families are also found in [194].

Natale classifies in [141] all semisimple Hopf algebras of dimension pq^2, where p and q are distinct primes, which are not simple. We note that semisolvability, a notion for Hopf algebras due to Montgomery and Witherspoon [137], seems to be a property enjoyed by a wide range of examples of semisimple Hopf algebras. Natale shows in [145] that there is one semisimple Hopf algebra of dimension less than 100 which is not lower or upper semisolvable and it has dimension 36.

At this point we refer to the survey papers by Masuoka [128] and Montgomery [136]. The former is more recent and the latter discusses some interesting cases of small dimension. Larson and the author showed that H is trivial when its dimension is odd and less than 20 in their paper [104]; thus we turn to even dimensions in this range. Williams [224] showed the uniqueness of non-trivial semisimple Hopf algebras over k of dimension 8 as did Masuoka later independently [123]. Fuduka found that there are two non-trivial examples of dimension 12 [53], one of which is found in [168]. There are two non-trivial examples of dimension 18 described independently by Masuoka [124] and Sekine [192].

The class of triangular semisimple Hopf algebras has been determined by Etingof and Gelaki [48]. They accomplish this by studying the category of representations of the Hopf algebra as a general object and use Deligne's theorem on Tannakian categories. Studying the category of representations of semisimple Hopf algebras as a general object is an important technique. These categories are examples of fusion categories. See the paper by Etingof, Nikshych, and Ostrik [52].

A semisimple Hopf algebra is called group-theoretical if its category of representations is Morita equivalent to a category of group graded vector spaces. Niksych gives examples of semisimple Hopf algebras which are not group-theoretical in [154].

Now we turn to the classification of all finite-dimensional Hopf algebras H of dimension n over k. General classification results usually build on the same for semisimple and pointed Hopf algebras. Let p be a prime integer. We have noted that $H \simeq k[\boldsymbol{Z}_p]$ when $n = p$ by results of Zhu [231].

Now suppose p is odd. Ng completed the classification in the $n = p^2$ case in [148] and of the $n = 2p$ case in [149]. Classification in the special case $n = 14$ was completed earlier by Beattie and Dăscălescu [17].

Partial results for the $n = p^3$ case are obtained by García [55] and full classification was obtained by García and Vay in [56] when $n = 16 = 2^4$. Higelman and Ng completed classification when $n = 2p^2$ in [74], Cheng and Ng when $n = 4p$, given $p \leq 11$, in [31]. Classification in the special case $n = 12$ was finished earlier by Natale [142].

Now suppose q is an odd prime also and $q \neq p$. Then H is known when $n = pq$ and $2 < p < q < 4p + 11$ by results of Ng [150], an improvement on $2 < p < q < 2p + 1$ in the paper by Etingof and Gelaki [51].

Williams [224] classified all Hopf algebras over k such that $n \leq 11$. Putting the results together gives classification when $n < 32$ with the exception of $n = 24, 27$. At this point we refer the reader to the paper [31] for a discussion of classification in low dimension with references.

One of the more important achievements in the theory of Hopf algebras in the past decade is the classification of certain classes of pointed Hopf algebras by Andruskiewitsch and Schneider [14,15]. Their results are based on theorems in the theory of quantum groups. In particular [14] uses Rosso's work [183].

In [15, Classification Theorem 0.1] they characterize all finite-dimensional pointed Hopf algebras H over an algebraically closed field k of characteristic zero such that $\mathrm{G}(H)$ is a commutative group whose order is not divisible by small primes, namely 2, 3, 5, or 7. $H \simeq \mathrm{U}(\mathcal{D}, \lambda)/I$, where is the Hopf algebra $\mathrm{U}(\mathcal{D}, \lambda)$ of Definition 15.6.4 and I is an ideal of $\mathrm{U}(\mathcal{D}, \lambda)/I$ generated by certain elements of the form $x^n - a$, where $a \in k[\mathrm{G}(H)]$. These elements are determined by a parameter μ which is described, as λ, in terms of $\mathcal{D}$.

Consider the biproduct decomposition $\mathrm{gr}(H) = R \times k[G]$, where $G = \mathrm{G}(H)$ and let $V = \mathrm{P}(R)$. Then $R = \mathrm{B}(V)$, an important fact established in classification. Quantum group theory enters the classification program in the characterization of all objects U in ${}^{k[G]}_{k[G]}\mathcal{YD}$ such that the Nichols algebra $\mathrm{B}(U)$ is finite-dimensional. The work of many comes into play in the proof of the classification theorem of [15]: Andruskiewitsch and Schneider

[9, 10, 12, 13, 15]; De Concini and Kac [37]; De Concini and Procesi [38]; Lusztig [109, 111, 112]; Mueller [138]; Heckenberger [64–66]; Kharchenko [95]; and Rosso [183]. For a cohomological interpretation of classification results see paper by Mastnak and Witherspoon [119].

Analysis of the case $G(H)$ not abelian again involves determining when the Nichols algebra $B(U)$ described above is finite-dimensional. Different techniques are required. Progress is being made by Heckenberger and Schneider [67, 68], by Andruskiewitsch, Heckenberger, and Schneider [5], and by Andruskiewitsch, Fantino, Graña, and Vendramin [3, 4]. We note that all pointed Hopf algebras of dimension $n = pq^2$ were classified by Andruskiewitsch and Natale [6].

A family of finite-dimensional Hopf algebras which has been classified completely is the family of finite-dimensional triangular Hopf algebras over an algebraically closed field of characteristic zero. This was done by Etingof and Gelaki [50].

Bibliography

[1] Abe, E. (1980). *Hopf algebras.* (Translated from the Japanese by Hisae Kinoshita and Hiroko Tanaka.) Cambridge Tracts in Mathematics 74, Cambridge University Press, Cambridge-New York, xii+284 pp.

[2] Aljadeff, E., Etingof, P., Gelaki, S. and Nikshych, D. (2002). On twisting of finite-dimensional Hopf algebras, *J. Algebra* **256**, pp. 484-501.

[3] Andruskiewitsch, N., Fantino, F., Graña, M. and Vendramin, L. (2010). Pointed Hopf algebras over some sporadic simple groups, *C. R. Math. Acad. Sci. Paris* **348**, pp. 605-608.

[4] Andruskiewitsch, N., Fantino, F., Graña, M. and Vendramin, L. (2011). The logbook of Pointed Hopf algebras over the sporadic simple groups, *J. Algebra* **325**, pp. 282-304.

[5] Andruskiewitsch, N., Heckenberger, I. and Schneider, H.-J. (2010). The Nichols algebra of a semisimple Yetter-Drinfeld module, *Amer. J. Math.* **132**, pp. 1493-1547.

[6] Andruskiewitsch, N. and Natale, S. (2001). Counting arguments for Hopf algebras of low dimension, *Tsukuba J. Math.* **25**, pp. 187-201.

[7] Andruskiewitsch, N., Radford, D. and Schneider, H.-J. (2010). Complete reducibility theorems for modules over pointed Hopf algebras, *J. Algebra* **324**, pp. 2932-2970.

[8] Andruskiewitsch, N. and Schneider, H.-J. (1998). Hopf algebras of order p^2 and braided Hopf algebras of order p, *J. Algebra* **199**, pp. 430-454.

[9] Andruskiewitsch N. and Schneider, H.-J. (1998). Lifting of quantum linear spaces and pointed Hopf algebras of order p^3, *J. Algebra* **209**, pp. 658-691.

[10] Andruskiewitsch, N. and Schneider, H.-J. (2000). Finite quantum groups and Cartan matrices, *Adv. Math.* **154**, pp. 1-45.

[11] Andruskiewitsch, N. and Schneider, H.-J. (2000). Lifting of Nichols algebras of type A_2 and pointed Hopf algebras of order p^4. Hopf algebras and quantum groups (Brussels, 1998), in *Lecture Notes in Pure and Appl. Math.* 209, Dekker, New York, pp. 1-14.

[12] Andruskiewitsch, N. and Schneider, H.-J. (2002). Finite quantum groups over abelian groups of prime exponent, *Ann. Sci. École Norm. Sup. (4)* **35**, pp. 1-26.

[13] Andruskiewitsch, N. and Schneider, H.-J. (2002). Pointed Hopf algebras, in *New directions in Hopf algebras*, Math. Sci. Res. Inst. Publ., 43, Cambridge Univ. Press, Cambridge, pp. 1-68.

[14] Andruskiewitsch, N. and Schneider, H.-J. (2004). A characterization of quantum groups, *J. Reine Angew. Math.* **577**, pp. 81-104.

[15] Andruskiewitsch, N. and Schneider, H.-J. (2010). On the classification of finite-dimensional pointed Hopf algebras, *Ann. of Math. (2)* **171**, pp. 375-417.

[16] Baxter, R. J. (1982). *Exactly solved models in statistical mechanics*, Academic Press, Inc. [Harcourt Brace Jovanovich, Publishers], London. xii+486 pp.

[17] Beattie, M. and Dăscălescu, S. (2004). Hopf algebras of dimension 14, *J. London Math. Soc. (2)* **69**, pp. 65-78.

[18] Beattie, M., Dăscălescu, S. and Grünenfelder, L. (1999). On the number of types of finite-dimensional Hopf algebras, *Invent. Math.* **136**, pp. 1-7.

[19] Beattie, M., Dăscălescu, S. and Grünenfelder, L. (2000). Constructing pointed Hopf algebras by Ore extensions, *J. Algebra* **225**, pp. 743-770.

[20] Beattie, M., Dăscălescu, S., Grünenfelder, L. and Năstăsescu, C. (1998). Finiteness conditions, co-Frobenius Hopf algebras, and quantum groups, *J. Algebra* **200**, pp. 312-333.

[21] Bergman, G. M. (1978). The diamond lemma for ring theory, *Adv. in Math.*, **29**, pp. 178–218.

[22] Bespalov, Y. and Drabant, B. (2000). Survey of cross product bialgebras. Hopf algebras and quantum groups (Brussels, 1998), in *Lecture Notes in Pure and Appl. Math.* 209, Dekker, New York, pp. 15-33.

[23] Bespalov, Y., Kerler, T., Lyubashenko, V. and Turaev, V. (2000). Integrals for braided Hopf algebras, *J. Pure Appl. Algebra* **148**, pp. 113-164.

[24] Blattner, R. J. (1983). UCLA lecture notes (unpublished).

[25] Blattner, R. J., Cohen, M. and Montgomery, S. (1986). Crossed products and inner actions of Hopf algebras, *Trans. Amer. Math. Soc.* **298**, pp. 671-711.

[26] Caenepeel, S. and Dăscălescu, S. (1998). Pointed Hopf algebras of dimension p^3, *J. Algebra* **209**, pp. 622-634.

[27] Caenepeel, S. and Dăscălescu, S. and Le Bruyn, L. (1999). Forms of pointed Hopf algebras, *Manuscripta Math.* **100**, pp. 3553.

[28] Caenepeel, S., Dăscălescu, S. and Raianu, S. (2000). Classifying pointed Hopf algebras of dimension 16, *Comm. Algebra* **28**, pp. 541-568.

[29] Cartier, P. (1962). Groupes algébriques et groupes formels. (French) 1962 Colloq. *Théorie des Groupes Algébriques* (Bruxelles, 1962) Librairie Universitaire, Louvain, GauthierVillars, Paris, pp. 87-111.

[30] Chase, S. U. and Sweedler, M. E. (1969). *Hopf algebras and Galois theory.* Lecture Notes in Mathematics, Vol. 97 Springer-Verlag, Berlin-New York ii+133 pp.

[31] Cheng, Y.-L. and Ng, S.-H. (2011). On Hopf algebras of dimension $4p$, *J. Algebra* **328**, pp. 399-419.

[32] Chin, W. and Musson, I. M. (1994). Hopf algebra duality, injective modules and quantum groups, *Comm. Algebra* **22** pp. 4661-4692.

[33] Chin, W. and Musson, I. M. (1996). The coradical filtration for quantized enveloping algebras, *J. London Math. Soc. (2)* **53**, pp. 50-62.

[34] Curtis, C. W. and Reiner, I. (2006). *Representation theory of finite groups and associative algebras.* (Reprint of the 1962 original.) AMS Chelsea Publishing, Providence, RI,. xiv+689 pp.

[35] Dăscălescu, S., Năstăsescu, C. and Raianu, S. (2001). *Hopf algebras, An introduction.* Monographs and Textbooks in Pure and Applied Mathematics, 235. Marcel Dekker, Inc., New York, x+401 pp.

[36] Dăscălescu, S., Năstăsescu, C. and Torrecillas, B. (1999). Co-Frobenius Hopf algebras: integrals, Doi-Koppinen modules and injective objects, *J. Algebra* **220**, pp. 542-560.

[37] De Concini, C. and Kac, V. G. (1990). Representations of quantum groups at roots of 1, in *Operator algebras, unitary representations, enveloping algebras, and invariant theory (Paris, 1989), Progr. Math.*, 92, Birkhuser Boston, Boston, MA, pp. 471-506.

[38] De Concini, C. and Procesi, C. (1993). Quantum groups. *D-modules, representation theory, and quantum groups (Venice, 1992), Lecture Notes in Math.*, 1565, Springer, Berlin, pp. 31-140.

[39] Dieudonné, J. (1973). *Introduction to the theory of formal groups.* Pure and Applied Mathematics, 20. Marcel Dekker, Inc., New York.

[40] Doi, Y. (1993). Braided bialgebras and quadratic bialgebras, *Comm. Algebra* **21**, pp. 1731–1749.

[41] Doi, Y. and Takeuchi, M. (1994). Multiplication alteration by two-cocycles-the quantum version, *Comm. Algebra* **22**, pp. 5715-5732.

[42] Doi, Y. and Takeuchi, M. (2000). Bi-Frobenius algebras, in *New trends in Hopf algebra theory (La Falda, 1999), Contemp. Math.*, 267, Amer. Math. Soc., Providence, RI, pp. 67-97.

[43] Doubilet, P., Rota, G-C. and Stanley, R. (1972). On the foundations of combinatorial theory. VI. The idea of generating function, in *Proceedings of the Sixth Berkeley Symposium on Mathematical Statistics and Probability (Univ. California, Berkeley, Calif., 1970/1971), Vol. II: Probability theory,* Univ. California Press, Berkeley, Calif., pp. 267–318.

[44] Drinfel'd, V. G. (1987). Quantum groups, in *Proceedings of the International Congress of Mathematicians, Vol. 1, 2* (Berkeley, Calif., 1986), Amer. Math. Soc., Providence, RI, pp. 798-820.

[45] Drinfel'd, V. G. (1990). Almost cocommutative Hopf algebras, (Russian) *Algebra i Analiz* **1** (1989), pp. 30–46, translation in *Leningrad Math. J.* **1**, pp. 321-342.

[46] Etingof, P. and Gelaki, S. (1998). On finite-dimensional semisimple and cosemisimple Hopf algebras in positive characteristic, *Internat. Math. Res. Notices* **1998**, pp. 851-864.

[47] Etingof, P. and Gelaki, S. (1998). Semisimple Hopf algebras of dimension pq are trivial, *J. Algebra* **210**, pp. 664-669.

[48] Etingof, P. and Gelaki, S. (2000). The classification of triangular semisimple and cosemisimple Hopf algebras over an algebraically closed field, *Internat. Math. Res. Notices* **2000**, pp. 223-234.

[49] Etingof, P. and Gelaki, S. (2002). On families of triangular Hopf algebras, *Int. Math. Res. Not.* **2002**, pp. 757-768.

[50] Etingof, P. and Gelaki, S. (2003). The classification of finite-dimensional triangular Hopf algebras over an algebraically closed field of characteristic 0, *Mosc. Math. J.* **3**, pp. 37-43, 258.

[51] Etingof, P. and Gelaki, S. (2004). On Hopf algebras of dimension pq, *J. Algebra* **277**, pp. 668-674.

[52] Etingof, P., Nikshych, D. and Ostrik, V. (2005). On fusion categories, *Ann. of Math. (2)* **162**, pp. 581-642.

[53] Fukuda, N. (1997). Semisimple Hopf algebras of dimension 12, *Tsukuba J. Math.* **21**, pp. 43-54.

[54] Galindo, C. and Natale, S. (2007). Simple Hopf algebras and deformations of finite groups, *Math. Res. Lett.* **14**, pp. 943-954.

[55] García, G. A. (2005). On Hopf algebras of dimension p^3, *Tsukuba J. Math.* **29**, pp. 259-284.

[56] García, G. A. and Vay, C. (2010). Hopf algebras of dimension 16, *Algebr. Represent. Theory* **13**, pp. 383-405.

[57] Gelaki, S. (1992). *Topics on quasitriangular Hopf algebras*, M.Sc. thesis, Ben Gurion University of the Negev, Beer Sheva, Israel.

[58] Gelaki, S. (1998). Pointed Hopf algebras and Kaplansky's 10th conjecture, *J. Algebra* **209**, pp. 635-657.

[59] Gelaki, S. and Westreich, S. (2000). On semisimple Hopf algebras of dimension pq, *Proc. Amer. Math. Soc.* **128**, pp. 39-47.

[60] Gillman, L. and Jerison, M. (1960). *Rings of continuous functions.* The University Series in Higher Mathematics, D. Van Nostrand Co., Inc., Princeton, N.J.-Toronto-London-New York ix+300 pp.

[61] Goodearl, K. R. and Warfield, R. B., Jr. (1989) *An introduction to noncommutative Noetherian rings.* London Mathematical Society Student Texts, 16. Cambridge University Press, Cambridge xviii+303 pp.

[62] Haar, A. (1933). Der Massbegriff in der Theorie der kontinuierlichen Gruppen (German), *Ann. of Math.*, **34**, pp. 147-169.

[63] Hayashi, T. (1992). Quantum groups and quantum determinants, *J. Algebra* **152**, pp. 146–165.

[64] Heckenberger, I. (2006). The Weyl groupoid of a Nichols algebra of diagonal type, *Invent. Math.* **164**, pp. 175-188.

[65] Heckenberger, I. (2007). Examples of finite-dimensional rank 2 Nichols algebras of diagonal type, *Compos. Math.* **143**, pp. 165-190.

[66] Heckenberger, I. (2008). Rank 2 Nichols algebras with finite arithmetic root system, *Algebr. Represent. Theory* **11**, pp. 115-132.

[67] Heckenberger, I. and Schneider, H.-J. (2010). Nichols algebras over groups with finite root system of rank two I, *J. Algebra* **324**, pp. 3090-3114.

[68] Heckenberger, I. and Schneider, H.-J. (2010). Root systems and Weyl groupoids for Nichols algebras, *Proc. Lond. Math. Soc. (3)* **101**, pp. 623-654.

[69] Hennings, M. (1996). Invariants of links and 3-manifolds obtained from Hopf algebras, *J. London Math. Soc. (2)* **54**, pp. 594-624.

[70] Heyneman, R. G. (1966). Unpublished result.

[71] Heyneman, R. G. and Radford, D. E. (1974). Reflexivity and coalgebras of finite type, *J. Algebra* **28**, pp. 215–246.

[72] Heyneman, R. G. and Sweedler, M. E. (1969). Affine Hopf algebras. I, *J. Algebra* **13**, pp. 192-241.

[73] Heyneman, R. G. and Sweedler, M. E. (1970). Affine Hopf algebras. II, *J. Algebra* **16**, pp. 271-297.

[74] Hilgemann, M. and Ng, S.-H. (2009). Hopf algebras of dimension p^2, *J. Lond. Math. Soc. (2)* **80**, pp. 295-310.

[75] Hopf, H. (1941). Über die Topologie der Gruppen-Mannigfaltigkeiten und ihre Verallgemeinerungen. (German) *Ann. of Math.* **42**, pp. 22-52.

[76] Humphreys, J. E. (1972). *Introduction to Lie algebras and representation theory*, Graduate Texts in Mathematics Vol. 9. Springer-Verlag, New York-Berlin, xii+169 pp.

[77] Humphreys, J. E. (1975). *Linear algebraic groups*, Graduate Texts in Mathematics No. 21. Springer-Verlag, New York-Heidelberg, xiv+247 pp.

[78] Jantzen, J. C. (1996). *Lectures on quantum groups*, Graduate Studies in Mathematics 6, American Mathematical Society, Providence, RI, viii+266 pp.

[79] Joni, S. A. and Rota, G.-C. (1979). Coalgebras and bialgebras in combinatorics. *Stud. Appl. Math.* **61** pp. 93-139.

[80] Joseph, A. (1995). *Quantum groups and their primitive ideals*, Ergebnisse der Mathematik und ihrer Grenzgebiete (3) [Results in Mathematics and Related Areas (3)], 29. Springer-Verlag, Berlin, x+383 pp.

[81] Jimbo, M. A. (1985). q-difference analogue of $U(\mathrm{g})$ and the Yang-Baxter equation, *Lett. Math. Phys.* **10**, pp. 63-69.

[82] Kac, G. I. (1972). Certain arithmetic properties of ring groups, *Functional Anal. Appl.* **6**, pp. 158-160.

[83] Kac, G. I. and Paljutkin, V. G. (1966). Finite ring groups, (Russian) *Trudy Moskov. Mat. Obšč.* **15**, pp. 224-261.

[84] Kac, V. G. (1990). *Infinite-dimensional Lie algebras*, Third edition. Cambridge University Press, Cambridge, xxii+400 pp.

[85] Kaplansky, I. (1975). *Bialgebras*, Lecture Notes in Mathematics. Department of Mathematics, University of Chicago, Chicago, Ill., iv+57 pp.

[86] Kauffman, L. H. (1991). *Knots and physics*, Series on Knots and Everything, 1. World Scientific Publishing Co., Inc., River Edge, NJ, xii+538 pp.

[87] Kauffman, L. H. and Lins, S. L. (1994). *Temperley-Lieb recoupling theory and invariants of 3-manifolds*, Annals of Mathematics Studies, 134. Princeton University Press, Princeton, NJ, x+296 pp.

[88] Kauffman, L. H. and Radford, D. E. (1993). A necessary and sufficient condition for a finite-dimensional Drinfel'd double to be a ribbon Hopf algebra, *J. Algebra* **159**, pp. 98-114.

[89] Kauffman, L. H. and Radford, D. E. (2000). On two proofs for the existence and uniqueness of integrals for finite-dimensional Hopf algebras, in *New trends in Hopf algebra theory* (La Falda, 1999), Contemp. Math., 267, Amer. Math. Soc., Providence, RI, pp. 177-194.

[90] Kashina, Y. (1999). On the order of the antipode of Hopf algebras in ${}^H_H$YD, *Comm. Algebra* **27**, pp. 1261-1273.
[91] Kashina, Y. (2000). Classification of semisimple Hopf algebras of dimension 16, *J. Algebra* **232**, pp. 617-663.
[92] Kashina, Y. (2003). On semisimple Hopf algebras of dimension 2^m, *Algebr. Represent. Theory* **6**, pp. 393-425.
[93] Kashina, Y. (2003). On two families of Hopf algebras of dimension 2^m, *Comm. Algebra* **31**, pp. 1643-1668.
[94] Kassel, C. (1995). *Quantum groups*, Graduate Texts in Mathematics, 155. Springer-Verlag, New York, xii+531 pp.
[95] Kharchenko, V. K. (1999). A quantum analogue of the Poincaré-Birkhoff-Witt theorem, (Russian) *Algebra Log.* **38**, pp. 476–507, 509, translation in *Algebra and Logic* **38**, pp. 259-276.
[96] Krop, L. and Radford, D. E. (2006). Finite-dimensional Hopf algebras of rank one in characteristic zero. *J. Algebra*, **302**, pp. 214-230.
[97] Krop, L. and Radford, D. E. (2006). Simple modules for the Drinfel'd double of a class of Hopf algebras, in *Groups, rings and algebras*, Contemp. Math., 420, Amer. Math. Soc., Providence, RI, 2006, pp. 229-235.
[98] Krop, L. and Radford, D. E. (2009). Representations of pointed Hopf algebras and their Drinfel'd quantum doubles, *J. Algebra*, **321**, pp. 2567-2603.
[99] Lambe, L. A. and Radford, D. E. (1997). *Introduction to the quantum Yang-Baxter equation and quantum groups: an algebraic approach*, Mathematics and its Applications 423. Kluwer Academic Publishers, Dordrecht,xx+293 pp.
[100] Larson, R. G. (1969). The order of the antipode of a Hopf algebra, *Proc. Amer. Math. Soc.* **21**, pp. 167-170.
[101] Larson, R. G. (1971). Characters of Hopf algebras, *J. Algebra*, **17**, pp. 352-368.
[102] Larson, R. G. and Radford, D. E. (1988). Finite-dimensional cosemisimple Hopf algebras in characteristic 0 are semisimple, *J. Algebra* **117**, 2, pp. 267-289.
[103] Larson, R. G. and Radford, D. E. (1988). Semisimple cosemisimple Hopf algebras, *Amer. J. Math.* **110**, pp. 187-195.
[104] Larson, R. G. and Radford, D. E. (1995). Semisimple Hopf algebras, *J. Algebra* **171**, pp. 5-35.
[105] Larson, R. G. and Sweedler, M. E. (1969). An associative orthogonal bilinear form for Hopf algebras, *Amer. J. Math.* **91**, pp. 75–94.
[106] Larson, R. G. and Taft, E. J. (1990). The algebraic structure of linearly recursive sequences under Hadamard product. Hopf algebras, *Israel J. Math.* **72**, pp. 118–132.
[107] Larson, R. G. and Towber, J. (1991). Two dual classes of bialgebras related to the concepts of "quantum group" and "quantum Lie algebra", *Comm. Algebra* **19**, pp. 3295–3345.
[108] Lin, B. I. (1977). Semiperfect coalgebras, *J. Algebra* **49**, pp. 357-373.
[109] Lusztig, G. (1990). Finite-dimensional Hopf algebras arising from quantized

universal enveloping algebra, *J. Amer. Math. Soc.* **3**, pp. 257-296.

[110] Lusztig, G. (1990). Finite-dimensional Hopf algebras arising from quantized universal enveloping algebra, *J. Amer. Math. Soc.*, **3**, pp. 257-296.

[111] Lusztig, G. (1990). Quantum groups at roots of 1, *Geom. Dedicata* **35**, pp. 89-113.

[112] Lusztig, G. (1993). *Introduction to quantum groups*, Progress in Mathematics 110, Birkhuser Boston, Inc., Boston, MA, xii+341 pp.

[113] Lyubashenko, V. (1995). Modular transformations for tensor categories, *J. Pure Appl. Algebra* **98**, pp. 279-327.

[114] Majid, S. (1990). Quasitriangular Hopf algebras and Yang-Baxter equations, *Internat. J. Modern Phys. A* **5**, pp. 1-91.

[115] Majid, S. (1991). Doubles of quasitriangular Hopf algebras, *Comm. Algebra* **19**, pp. 3061-3073.

[116] Majid, S. (1993). Braided matrix structure of the Sklyanin algebra and of the quantum Lorentz group, *Comm. Math. Phys.* **156**, pp. 607-638.

[117] Majid, S. (1994). Algebras and Hopf algebras in braided categories, in *Advances in Hopf algebras* (Chicago, IL, 1992), Lecture Notes in Pure and Appl. Math. 158, Dekker, New York, pp. 55-105.

[118] Majid, S. (1995). *Foundations of quantum group theory*, Cambridge University Press, Cambridge, x+607 pp.

[119] Mastnak, M. and Witherspoon, S. (2009). Bialgebra cohomology, pointed Hopf algebras, and deformations, *J. Pure Appl. Algebra* **213**, pp. 1399-1417.

[120] Masuoka, A. (1992). Freeness of Hopf algebras over coideal subalgebras, *Comm. Algebra* **20**, pp. 1353–1373.

[121] Masuoka, A. (1995). Self-dual Hopf algebras of dimension p^3 obtained by extension, *J. Algebra* **178**, pp. 791-806.

[122] Masuoka, A. (1995). Semisimple Hopf algebras of dimension $2p$, *Comm. Algebra* **23**, pp. 1931-1940.

[123] Masuoka, A. (1995). Semisimple Hopf algebras of dimension 6, 8, *Israel J. Math.* **92**, pp. 361-373.

[124] Masuoka, A. (1996). Some further classification results on semisimple Hopf algebras, *Comm. Algebra* **24**, pp. 307-329.

[125] Masuoka, A. (1996). The p^n theorem for semisimple Hopf algebras, *Proc. Amer. Math. Soc.* **124**, pp. 735-737.

[126] Masuoka, A. (2001). Defending the negated Kaplansky conjecture, *Proc. Amer. Math. Soc.* **129**, pp. 3185-3192.

[127] Masuoka, A. (2004). Example of almost commutative Hopf algebras which are not coquasitriangular, in *Hopf algebras*, Lecture Notes in Pure and Appl. Math. 237, Dekker, New York, pp. 185191.

[128] Masuoka, A. (2008). Classification of semisimple Hopf algebras, in *Handbook of algebra* 5, Elsevier/North-Holland, Amsterdam, pp. 429-455.

[129] Mac Lane, Saunders (1998). *Categories for the working mathematician. Second edition*, Graduate Texts in Mathematics 5, Springer-Verlag, New York, xii+314 pp.

[130] Macdonald, I. G. (1979). *Symmetric functions and Hall polynomials*, Oxford Mathematical Monographs. The Clarendon Press, Oxford University

Press, New York, viii+180 pp.

[131] Menini, C., Torrecillas, B. and Wisbauer, R. (2001). Strongly rational comodules and semiperfect Hopf algebras over QF rings, *J. Pure Appl. Algebra* **155**, pp. 237-255.

[132] Milnor, J. W. and Moore, J. C. (1965). On the structure of Hopf algebras, *Ann. of Math. (2)* **81**, pp. 211-264.

[133] Montgomery, S. (1993). *Hopf algebras and their actions on rings*, CBMS Regional Conference Series in Mathematics 82. Published for the Conference Board of the Mathematical Sciences, Washington, DC, by the American Mathematical Society, Providence, RI, xiv+238 pp.

[134] Montgomery, S. (1995). Indecomposable coalgebras, simple comodules, and pointed Hopf algebras, *Proc. Amer. Math. Soc.* **123**, pp. 2343-2351.

[135] Montgomery, S. (1998). Classifying finite-dimensional semisimple Hopf algebras, in *Trends in the representation theory of finite-dimensional algebras* (Seattle, WA, 1997), Contemp. Math. 229, Amer. Math. Soc., Providence, RI, pp. 265-279.

[136] Montgomery, S. (2005). Algebra properties invariant under twisting, in *Hopf algebras in noncommutative geometry and physics*, Lecture Notes in Pure and Appl. Math., 239, Dekker, New York, pp. 229-243.

[137] Montgomery, S. and Witherspoon, S. J. (1998). Irreducible representations of crossed products, *J. Pure Appl. Algebra* **129**, pp. 315-326.

[138] Müller, E. (1998). Some topics on Frobenius-Lusztig kernels. I, II, *J. Algebra* **206**, pp. 624-658, 659-681.

[139] Müller, E. (2000). Finite subgroups of the quantum general linear group, *Proc. London Math. Soc. (3)* **81**, pp. 190-210.

[140] Mullin, R. and Rota, G.-C. (1970). On the foundations of combinatorial theory. III. Theory of binomial enumeration, in *Graph Theory and its Applications* (Proc. Advanced Sem., Math. Research Center, Univ. of Wisconsin, Madison, Wis., 1969) (loose errata) Academic Press, New York, pp. 167–213.

[141] Natale, S. (1999). On semisimple Hopf algebras of dimension pq^2, *J. Algebra* **221**, pp. 242-278.

[142] Natale, S. (2002). Hopf algebras of dimension 12, *Algebr. Represent. Theory* **5**, pp. 445-455.

[143] Natale, S. (2002). Quasitriangular Hopf algebras of dimension pq, *Bull. London Math. Soc.* **34**, pp. 301-307.

[144] Natale, S. (2004). On semisimple Hopf algebras of dimension pqr, *Algebr. Represent. Theory* **7**, pp. 173-188.

[145] Natale, S. (2007). *Semisolvability of semisimple Hopf algebras of low dimension*, Mem. Amer. Math. Soc. 186, no. 874, viii+123 pp.

[146] Natale, S. (2010). Semisimple Hopf algebras of dimension 60, *J. Algebra* **324**, pp. 3017-3034.

[147] Newman, K. and Radford, D. E. (1979). The cofree irreducible Hopf algebra on an algebra, *Amer. J. Math.* **101**, pp. 1025-1045.

[148] Ng, S.-H. (2002). Non-semisimple Hopf algebras of dimension p^2, *J. Algebra* **255**, pp. 182-197.

[149] Ng, S.-H. (2005). Hopf algebras of dimension $2p$, *Proc. Amer. Math. Soc.*

133, pp. 2237-2242.

[150] Ng, S.-H. (2008). Hopf algebras of dimension *pq*. II, *J. Algebra* **319**, pp. 2772-2788.

[151] Nichols, W. D. (1994). Cosemisimple Hopf algebras, in *Advances in Hopf algebras* (Chicago, IL, 1992), Lecture Notes in Pure and Appl. Math. 158, Dekker, New York, pp. 135-151.

[152] Nichols, W. D. and Richmond, M. B. (1992). Freeness of infinite-dimensional Hopf algebras, *Comm. Algebra* **20**, pp. 1489–1492.

[153] Nichols, W. D. and Zoeller, M. B. (1989). A Hopf algebra freeness theorem, *Amer. J. Math.* **111**, pp. 381–385.

[154] Nikshych, D. (2008). Non-group-theoretical semisimple Hopf algebras from group actions on fusion categories, *Selecta Math. (N.S.)* **14**, pp. 145-161.

[155] Oberst, U. and Schneider, H.-J. (1974). Untergruppen formeller Gruppen von endlichem Index (German), *J. Algebra* **31**, pp. 10–44.

[156] Pei, Y., Hu, N. and Rosso, M. (2010). Multi-parameter quantum groups and quantum shuffles. I., in *Quantum affine algebras, extended affine Lie algebras, and their applications*, Contemp. Math., 506, Amer. Math. Soc., Providence, RI, pp. 145-171.

[157] Peterson, B. and Taft, E. J. (1980). The Hopf algebra of linearly recursive sequences, *Aequationes Math.* **20**, pp. 1–17.

[158] Radford, D. E. (1973). Coreflexive coalgebras, *J. Algebra* **26** pp. 512–535.

[159] Radford, D. E. (1976). Commutative nearly primitively generated Hopf algebras, *Comm. Algebra* **4**, pp. 823–872.

[160] Radford, D. E. (1976). The order of the antipode of a finite dimensional Hopf algebra is finite, *Amer. J. Math.* **98**, pp. 333-355.

[161] Radford, D. E. (1977). Finiteness conditions for a Hopf algebra with a nonzero integral, *J. Algebra* **46**, pp. 189-195.

[162] Radford, D. E. (1977). Operators on Hopf algebras, *Amer. J. Math.*, **99**, pp. 139-158.

[163] Radford, D. E. (1977/78). Freeness (projectivity) criteria for Hopf algebras over Hopf subalgebras, *J. Pure Appl. Algebra* **11**, pp. 15–28.

[164] Radford, D. E. (1977). Pointed Hopf algebras are free over Hopf subalgebras, *J. Algebra* **45**, pp. 266–273.

[165] Radford, D. E. (1978). On the structure of commutative pointed Hopf algebras, *J. Algebra* **50**, pp. 284-296.

[166] Radford, D. E. (1980). On an analog of Lagrange's theorem for commutative Hopf algebras, *Proc. Amer. Math. Soc.* **79**, pp. 164–166.

[167] Radford, D. E. (1984). On the antipode of a cosemisimple Hopf algebra, *J. Algebra* **88**, pp. 68-88.

[168] Radford, D. E. (1985). The structure of Hopf algebras with a projection, *J. Algebra* **92**, pp. 322-347.

[169] Radford, D. E. (1992). On the antipode of a quasitriangular Hopf algebra, *J. Algebra* **151**, pp. 1-11.

[170] Radford, D. E. (1993). Irreducible representations of $\mathcal{U}_q(g)$ arising from $\mathrm{Mod}^{\bullet}_{C^{1/2}}$, in *Quantum deformations of algebras and their representations* (Ramat-Gan, 1991/1992, Rehovot, 1991/1992), Israel Math. Conf. Proc. 7,

Bar-Ilan Univ., Ramat Gan, pp. 143-170.

[171] Radford, D. E. (1993). Minimal quasitriangular Hopf algebras, *J. Algebra* **157**, pp. 285-315.

[172] Radford, D. E. (1993). Solutions to the quantum Yang-Baxter equation and the Drinfel'd double, *J. Algebra* **161**, pp. 20-32.

[173] Radford, D. E. (1994). On Kauffman's knot invariants arising from finite-dimensional Hopf algebras, in *Advances in Hopf algebras* (Chicago, IL, 1992), Lecture Notes in Pure and Appl. Math. 158, Dekker, New York, pp. 205-266.

[174] Radford, D. E. (1994). The trace function and Hopf algebras, *J. Algebra* **163**, pp. 583-622.

[175] Radford, D. E. (1999). Finite-dimensional simple-pointed Hopf algebras, *J. Algebra* **211**, pp. 686–710.

[176] Radford, D. E. and Schneider, H.-J. (2008). On the simple representations of generalized quantum groups and quantum doubles, *J. Algebra* **319**, pp. 3689-3731.

[177] Radford, D. E. and Towber, J. (1993). Yetter-Drinfel'd categories associated to an arbitrary bialgebra, *J. Pure Appl. Algebra* **87**, pp. 259-279.

[178] Reshetikhin, N. (1991). Invariants of links and 3-manifolds related to quantum groups, in *Proceedings of the International Congress of Mathematicians, Vol. I, II* (Kyoto, 1990), Math. Soc. Japan, Tokyo, pp. 1373-1375.

[179] Reshetikhin, N. and Semenov-Tian-Shansky, M. A. (1988). Quantum RR-matrices and factorization problems, *J. Geom. Phys.* **5**, pp. 533-550.

[180] Reshetikhin, N. and Turaev, V. G. (1990). Ribbon graphs and their invariants derived from quantum groups, *Comm. Math. Phys.* **127**, pp. 1-26.

[181] Reshetikhin, N. and Turaev, V. G. (1991). Invariants of 3-manifolds via link polynomials and quantum groups, *Invent. Math.* **103**, pp. 547-597.

[182] Rosso, M. (1988). Finite-dimensional representations of the quantum analog of the enveloping algebra of a complex simple Lie algebra, *Comm. Math. Phys.* **117** pp. 581-593.

[183] Rosso, M. (1998). Quantum groups and quantum shuffles, *Invent. Math.* **133**, pp. 399-416.

[184] Rota, G.-C. (1964). On the foundations of combinatorial theory. I. Theory of Möbius functions, *Z. Wahrscheinlichkeitstheorie und Verw. Gebiete* **2**, pp. 340–368.

[185] Rota, G-C. (1978) Hopf algebra methods in combinatorics. Problèmes combinatoires et thorie des graphes (Colloq. Internat. CNRS, Univ. Orsay, Orsay, 1976), pp. 363-365, Colloq. Internat. CNRS, 260, CNRS, Paris.

[186] Rotman, J. J. (2002). *Advanced modern algebra*, Prentice Hall, Inc., Upper Saddle River, NJ, xvi+1012+A8+B6+I14 pp.

[187] Rowen, L. H. (1988). *Ring theory. Vol. I*, Pure and Applied Mathematics 127. Academic Press, Inc., Boston, MA, . xxiv+538 pp.

[188] Scharfschwerdt, B. (2001). The Nichols Zoeller theorem for Hopf algebras in the category of Yetter Drinfeld modules, *Comm. Algebra* **29**, pp. 2481-2487.

[189] Schauenburg, P. (1992). On coquasitriangular Hopf algebras and the quantum Yang-Baxter equation, *Algebra Berichte [Algebra Reports]* 67. Verlag Reinhard Fischer, Munich, ii+76 pp.

[190] Schneider, H.-J. (1981). Zerlegbare Untergruppen affiner Gruppen, (German) *Math. Ann.* **255**, pp. 139-158.

[191] Schneider, H.-J. (2001). Some properties of factorizable Hopf algebras, *Proc. Amer. Math. Soc.* **129**, pp. 1891-1898.

[192] Sekine, Y. (1996). An example of finite-dimensional Kac algebras of Kac-Paljutkin type, *Proc. Amer. Math. Soc.* **124**, pp. 1139-1147.

[193] Shudo, T. and Miyamoto, H. (1978). On the decomposition of coalgebras, *Hiroshima Math. J.* **8**, pp. 499-504.

[194] Sommerhäuser, Y. (2002). *Yetter-Drinfel'd Hopf algebras over groups of prime order*, Lecture Notes in Mathematics 1789. Springer-Verlag, Berlin, iv+158 pp.

[195] Ştefan, Ḑ. (1995). The uniqueness of integrals (a homological approach), *Comm. Algebra* **23**, pp. 1657-1662.

[196] Ştefan, Ḑ. (1997). The set of types of n-dimensional semisimple and cosemisimple Hopf algebras is finite, *J. Algebra* **193**, pp. 571-580.

[197] Sullivan, J. B. (1971). The uniqueness of integrals for Hopf algebras and some existence theorems of integrals for commutative Hopf algebras, *J. Algebra* **19**, pp. 426-440.

[198] Sullivan, J. B. (1972). Affine group schemes with integrals, *J. Algebra* **22**, pp. 546-558.

[199] Sullivan, J. B. (1973). A decomposition theorem for pro-affine solvable algebraic groups over algebraically closed fields. *Amer. J. Math.* **95**, pp. 221-228.

[200] Sweedler, M. E. (1968). Cohomology of algebras over Hopf algebras, *Trans. Amer. Math. Soc.* **133**, pp. 205-239.

[201] Sweedler, M. E. (1969) *Hopf algebras*, Mathematics Lecture Note Series W. A. Benjamin, Inc., New York, vii+336 pp.

[202] Sweedler, M. E. (1971). Connected fully reducible affine group schemes in positive characteristic are abelian. *J. Math. Kyoto Univ.* **11**, pp. 51-70.

[203] Sweedler, M. E. (1969). Integrals for Hopf algebras, *Ann. of Math. (2)* **89**, pp. 323–335.

[204] Taft, E. J. (1990). Hurwitz invertibility of linearly recursive sequences, in *Proceedings of the Twentieth Southeastern Conference on Combinatorics, Graph Theory, and Computing* (Boca Raton, FL, 1989). Congr. Numer. 73, pp. 37–40.

[205] Taft, Earl J. (1971). The order of the antipode of finite-dimensional Hopf algebra, *Proc. Nat. Acad. Sci. U.S.A.* **68**, pp. 2631-2633.

[206] Taft, Earl J. (1972). Reflexivity of algebras and coalgebras, *Amer. J. Math.* **94**, pp. 1111–1130.

[207] Taft, Earl J. (1977). Reflexivity of algebras and coalgebras. II, *Comm. Algebra* **5**, pp. 1549–1560.

[208] Taft, E. J. (1982). Noncommutative sequences of divided powers, in *Proceedings of Lie Algebras and Related Topics* (New Brunswick, New Jersey 1981). Lecture Notes in Mathematics, Springer-Verlag, Berlin Heidelberg, New York. 933, 1982, pp. 203–209.

[209] Taft, E. J. and Wilson, R. L. (1974). On antipodes in pointed Hopf algebras, *J. Algebra* **29**, pp. 27–32.

[210] Takeuchi, M. (1971). Free Hopf algebras generated by coalgebras, *J. Math. Soc. Japan* **23**, pp. 561–582.

[211] Takeuchi, M. (1971). There exists a Hopf algebra whose antipode is not injective, *Sci. Papers College Gen. Ed. Univ. Tokyo* **21**, pp. 127–130.

[212] Takeuchi, M. (1972). A correspondence between Hopf ideals and sub-Hopf algebras, *Manuscripta Math.* **7**, pp. 251–270.

[213] Takeuchi, M. (1972). On a semi-direct product decomposition of affine groups over a field of characteristic 0. *Thoku Math. J.* **24**, pp. 453-456.

[214] Takeuchi, M. (1972). On the dimension of the space of integrals of Hopf algebras, *J. Algebra* **21**, pp. 174-177.

[215] Takeuchi, M. (1974). Tangent coalgebras and hyperalgebras. I. *Japan. J. Math.* **42**, pp. 1-143.

[216] Takeuchi, M. (1975). On coverings and hyperalgebras of affine algebraic groups. *Trans. Amer. Math. Soc.* **211**, pp. 249-275.

[217] Takeuchi, M. (1977). Morita theorems for categories of comodules. *J. Fac. Sci. Univ. Tokyo Sect. IA Math.* **24**, pp. 629-644.

[218] Takeuchi, M. (1979). Relative Hopf modules—equivalences and freeness criteria, *J. Algebra* **60**, pp. 452–471.

[219] Takeuchi, M. (1992). Some topics on $\mathrm{GL}_q(n)$, *J. Algebra* **147**, pp. 379-410.

[220] Takeuchi, M. (1999). Finite Hopf algebras in braided tensor categories, *J. Pure Appl. Algebra* **138**, pp. 59-82.

[221] Takeuchi, M. (2000). Survey of braided Hopf algebras, in *New trends in Hopf algebra theory* (La Falda, 1999), Contemp. Math., 267, Amer. Math. Soc., Providence, RI, pp. 301-323.

[222] Van Daele, A. (1997). The Haar measure on finite quantum groups, *Proc. Amer. Math. Soc.* **125**, pp. 3489-3500.

[223] Wadati, M., Deguchi, T. and Akutsu, Y. (1989). Exactly solvable models and knot theory, *Phys. Rep.* **180**, pp. 247-332.

[224] Williams, R. (1988). *Finite-dimensional Hopf algebras*, Ph.D. Thesis, Florida State University, .

[225] Xu, Y. H. and Fong, Y. (1992) *On the decomposition of coalgebras.* Words, languages and combinatorics (Kyoto, 1990), 504522, World Sci. Publ., River Edge, NJ.

[226] Yang, C. N. (1967). Some exact results for the many-body problem in one dimension with repulsive delta-function interaction, *Phys. Rev. Lett.* **19**, pp. 1312-1315.

[227] Yang, C. N. and Ge, M. L., editors. (1989). *Braid group, knot theory and statistical mechanics.* Advanced Series in Mathematical Physics 9. World Scientific Publishing Co., Inc., Teaneck, NJ, x+329 pp.

[228] Yang, C. N. and Ge, M. L., editors. (1994). *Braid group, knot theory and statistical mechanics. II*, Advanced Series in Mathematical Physics 17. World Scientific Publishing Co., Inc., River Edge, NJ, x+467 pp.

[229] Yetter, D. N. (1990). Quantum groups and representations of monoidal categories, *Math. Proc. Cambridge Philos. Soc.* **108**, pp. 261-290.

[230] Zamolodchikov, A. B. and Zamolodchikov, A. B. (1979). Factorized S-matrices in two dimensions as the exact solutions of certain relativistic quan-

tum field theory models, *Ann. Physics* **120**, pp. 253-291.

[231] Zhu, Y. (1994). Hopf algebras of prime dimension, *Internat. Math. Res. Notices* **1994**, pp. 53-59.

Index

SERIES ON KNOTS AND EVERYTHING

Editor-in-charge: Louis H. Kauffman *(Univ. of Illinois, Chicago)*

The Series on Knots and Everything: is a book series polarized around the theory of knots. Volume 1 in the series is Louis H Kauffman's Knots and Physics.

One purpose of this series is to continue the exploration of many of the themes indicated in Volume 1. These themes reach out beyond knot theory into physics, mathematics, logic, linguistics, philosophy, biology and practical experience. All of these outreaches have relations with knot theory when knot theory is regarded as a pivot or meeting place for apparently separate ideas. Knots act as such a pivotal place. We do not fully understand why this is so. The series represents stages in the exploration of this nexus.

Details of the titles in this series to date give a picture of the enterprise.

Published:*

Vol. 1: Knots and Physics (3rd Edition)
by L. H. Kauffman

Vol. 2: How Surfaces Intersect in Space — An Introduction to Topology (2nd Edition)
by J. S. Carter

Vol. 3: Quantum Topology
edited by L. H. Kauffman & R. A. Baadhio

Vol. 4: Gauge Fields, Knots and Gravity
by J. Baez & J. P. Muniain

Vol. 5: Gems, Computers and Attractors for 3-Manifolds
by S. Lins

Vol. 6: Knots and Applications
edited by L. H. Kauffman

Vol. 7: Random Knotting and Linking
edited by K. C. Millett & D. W. Sumners

Vol. 8: Symmetric Bends: How to Join Two Lengths of Cord
by R. E. Miles

Vol. 9: Combinatorial Physics
by T. Bastin & C. W. Kilmister

Vol. 10: Nonstandard Logics and Nonstandard Metrics in Physics
by W. M. Honig

Vol. 11: History and Science of Knots
edited by J. C. Turner & P. van de Griend

*The complete list of the published volumes in the series can also be found at
http://www.worldscibooks.com/series/skae_series.shtml

Vol. 12: Relativistic Reality: A Modern View
edited by J. D. Edmonds, Jr.

Vol. 13: Entropic Spacetime Theory
by J. Armel

Vol. 14: Diamond — A Paradox Logic
by N. S. Hellerstein

Vol. 15: Lectures at KNOTS '96
by S. Suzuki

Vol. 16: Delta — A Paradox Logic
by N. S. Hellerstein

Vol. 17: Hypercomplex Iterations — Distance Estimation and Higher Dimensional Fractals
by Y. Dang, L. H. Kauffman & D. Sandin

Vol. 18: The Self-Evolving Cosmos: A Phenomenological Approach to Nature's Unity-in-Diversity
by S. M. Rosen

Vol. 19: Ideal Knots
by A. Stasiak, V. Katritch & L. H. Kauffman

Vol. 20: The Mystery of Knots — Computer Programming for Knot Tabulation
by C. N. Aneziris

Vol. 21: LINKNOT: Knot Theory by Computer
by S. Jablan & R. Sazdanovic

Vol. 22: The Mathematics of Harmony — From Euclid to Contemporary Mathematics and Computer Science
by A. Stakhov (assisted by S. Olsen)

Vol. 23: Diamond: A Paradox Logic (2nd Edition)
by N. S. Hellerstein

Vol. 24: Knots in HELLAS '98 — Proceedings of the International Conference on Knot Theory and Its Ramifications
edited by C. McA Gordon, V. F. R. Jones, L. Kauffman, S. Lambropoulou & J. H. Przytycki

Vol. 25: Connections — The Geometric Bridge between Art and Science (2nd Edition)
by J. Kappraff

Vol. 26: Functorial Knot Theory — Categories of Tangles, Coherence, Categorical Deformations, and Topological Invariants
by David N. Yetter

Vol. 27: Bit-String Physics: A Finite and Discrete Approach to Natural Philosophy
by H. Pierre Noyes; edited by J. C. van den Berg

Vol. 28: Beyond Measure: A Guided Tour Through Nature, Myth, and Number
by J. Kappraff

Vol. 29: Quantum Invariants — A Study of Knots, 3-Manifolds, and Their Sets
by T. Ohtsuki

Vol. 30: Symmetry, Ornament and Modularity
by S. V. Jablan

Vol. 31: Mindsteps to the Cosmos
by G. S. Hawkins

Vol. 32: Algebraic Invariants of Links
by J. A. Hillman

Vol. 33: Energy of Knots and Conformal Geometry
by J. O'Hara

Vol. 34: Woods Hole Mathematics — Perspectives in Mathematics and Physics
edited by N. Tongring & R. C. Penner

Vol. 35: BIOS — A Study of Creation
by H. Sabelli

Vol. 36: Physical and Numerical Models in Knot Theory
edited by J. A. Calvo et al.

Vol. 37: Geometry, Language, and Strategy
by G. H. Thomas

Vol. 38: Current Developments in Mathematical Biology
edited by K. Mahdavi, R. Culshaw & J. Boucher

Vol. 39: Topological Library
Part 1: Cobordisms and Their Applications
edited by S. P. Novikov & I. A. Taimanov

Vol. 40: Intelligence of Low Dimensional Topology 2006
edited by J. Scott Carter et al.

Vol. 41: Zero to Infinity: The Fountations of Physics
by P. Rowlands

Vol. 42: The Origin of Discrete Particles
by T. Bastin & C. Kilmister

Vol. 43: The Holographic Anthropic Multiverse
by R. L. Amoroso & E. A. Ranscher

Vol. 44: Topological Library
Part 2: Characteristic Classes and Smooth Structures on Manifolds
edited by S. P. Novikov & I. A. Taimanov

Vol. 45: Orbiting the Moons of Pluto
Complex Solutions to the Einstein, Maxwell, Schrödinger and Dirac Equations
by E. A. Rauscher & R. L. Amoroso

Vol. 46: Introductory Lectures on Knot Theory
edited by L. H. Kauffman, S. Lambropoulou, S. Jablan & J. H. Przytycki

Vol. 47: Introduction to the Anisotropic Geometrodynamics
by S. Siparov

Vol. 48: An Excursion in Diagrammatic Algebra: Turning a Sphere from Red to Blue
by J. S. Carter

Vol. 49: Hopf Algebras
by D. E. Radford